Lecture Notes in Electrical Engineering

Volume 743

AF738397

The book series *Lecture Notes in Electrical Engineering* (LNEE) publishes the latest developments in Electrical Engineering - quickly, informally and in high quality. While original research reported in proceedings and monographs has traditionally formed the core of LNEE, we also encourage authors to submit books devoted to supporting student education and professional training in the various fields and applications areas of electrical engineering. The series cover classical and emerging topics concerning:

- Communication Engineering, Information Theory and Networks
- Electronics Engineering and Microelectronics
- Signal, Image and Speech Processing
- Wireless and Mobile Communication
- Circuits and Systems
- Energy Systems, Power Electronics and Electrical Machines
- Electro-optical Engineering
- Instrumentation Engineering
- Avionics Engineering
- Control Systems
- Internet-of-Things and Cybersecurity
- Biomedical Devices, MEMS and NEMS

For general information about this book series, comments or suggestions, please contact leontina.dicecco@springer.com.

To submit a proposal or request further information, please contact the Publishing Editor in your country:

China

Jasmine Dou, Editor (jasmine.dou@springer.com)

India, Japan, Rest of Asia

Swati Meherishi, Editorial Director (Swati.Meherishi@springer.com)

Southeast Asia, Australia, New Zealand

Ramesh Nath Premnath, Editor (ramesh.premnath@springernature.com)

USA, Canada:

Michael Luby, Senior Editor (michael.luby@springer.com)

All other Countries:

Leontina Di Cecco, Senior Editor (leontina.dicecco@springer.com)

**** This series is indexed by EI Compendex and Scopus databases. ****

More information about this series at http://www.springer.com/series/7818

Weijiang Chen · Qingxin Yang · Laili Wang ·
Dingxin Liu · Xiaogang Han · Guodong Meng
Editors

The Proceedings of the 9th Frontier Academic Forum of Electrical Engineering

Volume II

Springer

Editors
Weijiang Chen
State Grid Corporation of China
Beijing, China

Qingxin Yang
Tianjin University of Technology
Tianjin, China

Laili Wang
Xi'an Jiaotong University
Xi'an, Shaanxi, China

Dingxin Liu
Xi'an Jiaotong University
Xi'an, Shaanxi, China

Xiaogang Han
Xi'an Jiaotong University
Xi'an, Shaanxi, China

Guodong Meng
Xi'an Jiaotong University
Xi'an, Shaanxi, China

ISSN 1876-1100 ISSN 1876-1119 (electronic)
Lecture Notes in Electrical Engineering
ISBN 978-981-33-6611-4 ISBN 978-981-33-6609-1 (eBook)
https://doi.org/10.1007/978-981-33-6609-1

This Springer imprint is published by the registered company Springer Nature Singapore Pte Ltd.
The registered company address is: 152 Beach Road, #21-01/04 Gateway East, Singapore 189721, Singapore

Contents

Committees

Honorary Chairman

Chen Weijiang (Academician of the Chinese Academy of Sciences)

Chairman

Prof. Yang Qingxin (Chairman of China Electrotechnical Society, President of Tianjin University of Technology)
Prof. Rong Mingzhe (Vice President of China Electrotechnical Society, Vice President of Xi'an Jiaotong University)

Academic Committees

Chairman
Ma Weiming (Academician of the Chinese Academy of Engineering)

Vice-Chairman
Cheng Yonghong (Xi'an Jiaotong University)
Prof. Yuan Fuxing (Deputy General Manager of Xi'an High Voltage Apparatus Research Institute)

Secretary-General
Prof. Guo Lijun (Director of Editorial Department of Chinese Journal of Electrotechnical Technology)
Prof. Yang Fei (Xi'an Jiaotong University)

Deputy Secretary-General
Prof. Liu Wenfeng (Xi'an Jiaotong University)
Prof. Liu Dingxin (Xi'an Jiaotong University)

Prof. Wang Laili (Xi'an Jiaotong University)
Prof. Xiaogang Han (Xi'an Jiaotong University)
Prof. Li Gengfeng (Xi'an Jiaotong University)
Prof. Meng Guodong (Xi'an Jiaotong University)
Prof. Michael Tai Ching Fang, University of Liverpool, UK
Prof. Anthony B. Murphy, CSIRO Manufacturing, Australia
Prof. Yasunori Tanaka, Kanazawa University, Japan
Prof. Yann Cressault, Université de Toulouse, France
Prof. Jiudun Yan, University of Liverpool, UK

Organizing Committees

Chairman
Prof. Wang Jianhua (Director, State Key Laboratory of Electrical Insulation and Power Equipment, Xi'an Jiaotong University)

Vice-Chairman
Prof. Li Shengtao (Executive Deputy Dean of Xi'an Jiaotong University)
Prof. Wang Jiansheng (General Manager of Xi'an High Voltage Apparatus Research Institute)

Secretary-General
Wang Wenguang (Deputy Director, Editorial Department, China Electrotechnical Society)
Wang Yusheng (Deputy Director, Editorial Department, China Electrotechnical Society)
Jia Yuquan, Shanghai PIBO Information Technology Center, China
Liang Weihong (Xi'an High Voltage Apparatus Research Institute)
Prof. Wang Shihang (Xi'an Jiaotong University)
Prof. Anthony B. Murphy, CSIRO Manufacturing, Australia
Prof. Yann Cressault, Université de Toulouse, France
Prof. Jiudun Yan, University of Liverpool, UK

Research on New SVPWM Control of PMSM Based on Two-Level Inverter

Chunyan Bi and Jianwei Li

Abstract Permanent magnet synchronous motor (PMSM) is widely used in many fields because of its high power density, high efficiency and good control performance. Aiming at the control stability and reduce direct torque control pulsation for PMSM, based on two-level generalized inverter (TGI) technology, a new control strategy of space voltage vector modulation (SVPWM) is proposed. The principle and implementation process of applying TGI technology to PMSM motor are discussed. The switch state function table, the voltage amplitude and voltage vector partition of SVPWM are studied. Finally, the model is built and simulated on MATLAB/Simulink. Through the simulation of the motor control system, the results show that this method can effectively improve the linear modulation region of SVPWM, and prove that the SVPWM control method based on two-level generalized inverter technology has the ability to quickly obtain the system stability, which is more superior than the traditional motor control method.

Keywords PMSM · SVPWM · Two-level inverter

1 Introduction

Permanent magnet synchronous motor (PMSM) is widely used in aerospace, CNC machine tools, ship propulsion and other fields because of its high power density, high efficiency and good control performance [1]. In addition, the torque inertia ratio is large, the loss of stator current and resistance is small, and the rotor parameters can be measured [2]. The control strategies of PMSM mainly include VVVF control, vector control, direct torque control (DTC), decoupling control, as well as adaptive control, neural network control, model predictive control, etc. [3]. Among them,

C. Bi (✉)
School of Electrical and Electronic Information Engineering, Sichuan University Jinjiang College, Jinjiang Avenue 1, Pengshan 620860, China
e-mail: bichunyan19840104@163.com

J. Li
Fuyao Glass Industry Group CO. LTD., Fuyao Industrial Zone II, Fuzhou 350301, China
e-mail: jianwei.li01@fuyaogroup.com

W. Chen et al. (eds.), *The Proceedings of the 9th Frontier Academic Forum of Electrical Engineering*, Lecture Notes in Electrical Engineering 743,
https://doi.org/10.1007/978-981-33-6609-1_1

vector control and direct torque control are widely used. Erjaee [4], Wang [5], Tu [6], respectively adopt vector control mode to realize high-precision, high-speed dynamic response and wide speed range control of the motor, but at the same time, it is pointed out that the control mode relies heavily on the motor parameters, and the change of the motor parameters will have adverse effects on the dynamic characteristics of the control; Chen [7] designed the control regulator by using the multi input and multi output sliding mode variable structure control method, and realized the stator current accurate control by using the current loop tracking pulse width modulation (PWM), but the stator three-phase current has certain distortion and the torque has obvious jump when the load changes.

With the constant progress of power electronics and the increasingly perfecting of modern control theory, the requirements for control performance of PMSM are constantly improved. According to deficiency of traditional SPWM control, how to build PMSM simulation model based on SVPWM [8, 9] has been a hot research topic. On the basis of SVPWM vector control, this paper puts forward a new control method of SVPWM based on SVM technology, and discusses the principle, process and steps of applying SVPWM based on SVM technology to PMSM motor. Finally, the control system of PMSM is simulated by MATLAB/Simulink.

2 Mathematical Model of PMSM

According to the unified theory of motor, the mathematical model of PMSM is established. The voltage balance equation of three-phase winding is [10]:

$$\begin{bmatrix} u_a \\ u_b \\ u_c \end{bmatrix} = \begin{bmatrix} r & 0 & 0 \\ 0 & r & 0 \\ 0 & 0 & r \end{bmatrix} \begin{bmatrix} i_a \\ i_b \\ i_c \end{bmatrix} + p \begin{bmatrix} L & M & M \\ M & L & M \\ M & M & L \end{bmatrix} \begin{bmatrix} i_a \\ i_b \\ i_c \end{bmatrix} + P \begin{bmatrix} \psi \sin\theta \\ \psi \sin(\theta - 2\pi/3) \\ \psi \sin(\theta + 2\pi/3) \end{bmatrix} \quad (1)$$

where, u_a, u_b, u_c are the stator phase winding voltage; r is the winding resistance of per phase; i_a, i_b, i_c are the stator phase winding current; L is winding self-inductance of per phase; M is winding mutual inductance of per phase; p is the differential operator; ψ is motor flux; θ is the rotor position angle; P is the polar logarithm.

3 Principle of Vector Control

Coordinate transformation and space vector pulse width modulation (SVPWM) are two core technologies of vector control system for permanent magnet synchronous motor. The purpose is to decouple the PMSM which is nonlinear, strong coupling and multivariable complex control, so as to facilitate the control. Under the rotating coordinate transformation synchronized with the motor rotary magnetic field, the current is divided into two orthogonal DC components oriented by the excitation current and the torque current, controlled by DC motor and the three-phase current is Decoupling.

According to the different control purposes and occasions, the choice of vector control strategy is also different for PMSM. Generally, there are four control methods as follows: (1) excitation current $i_d = 0$, the given input of the motor only has a given speed, and the stator current is all converted into torque current; (2) power factor $cos\phi = 1$ control, keeping the voltage vector and current vector in the same direction in the whole control process; (3) maximum output power control, control the shaft current i_d to obtain the maximum torque; (4) flux-weakening control, it still needs to speed up when the inverter output voltage is maximum, to maintain voltage balance, needs to decrease torque current, increases excitation current, weaken magnetic field, reduce the magnetic flux. This paper adopts the first control strategy, namely excitation current $i_d = 0$ [11].

3.1 Classical SVPWM DTC

The most important technology of classical SVPWM control is coordinate transformation. Under the basic principle that the magnetomotive force generated in different coordinate systems is completely consistent, the physical model of AC motor is changed equivalently, which is similar to that of DC motor, and then imitating controlled by DC motor. In traditional vector control, the a-b-c coordinate of three-phase is transformed into two-phase coordinate, that is Clark transformation; then the static coordinate is transformed into the changing d-q coordinate, which is called Park transformation. The mathematical model of PMSM in the three-phase static coordinate system is transformed through Clark and Park transformation, the mathematical model of PMSM in coordinate system is obtained. The equations of voltage, flux and torque are [12]:

$$u_d = Ri_d + \frac{d\psi_d}{dt} - \psi_q\omega_e \tag{2}$$

$$u_q = Ri_q + \frac{d\psi_q}{dt} + \psi_d\omega_e \tag{3}$$

$$\psi_d = L_d i_d + \psi_f \tag{4}$$

$$\psi_q = L_q i_q \tag{5}$$

$$Te = \frac{3}{2}p(\psi_d i_d - \psi_q i_d) = \frac{3}{2}p[\psi_f i_q + (L_d - L_q)i_d i_q] \tag{6}$$

Table 1 SVPWM switching state function

SA	SB	SC	VAN	VBN	VCN	VAB	VBC	VCA
0	0	0	0	0	0	0	0	0
1	0	0	$2V_{DC}/3$	$-V_{DC}/3$	$-V_{DC}/3$	V_{DC}	0	$-V_{DC}$
0	1	0	$-V_{DC}/3$	$2V_{DC}/3$	$-V_{DC}/3$	$-V_{DC}$	V_{DC}	0
1	1	0	$V_{DC}/3$	$V_{DC}/3$	$-2V_{DC}/3$	0	V_{DC}	$-V_{DC}$
0	0	1	$-V_{DC}/3$	$-V_{DC}/3$	$2V_{DC}/3$	0	$-V_{DC}$	V_{DC}
1	0	1	$V_{DC}/3$	$-2V_{DC}/3$	$V_{DC}/3$	V_{DC}	$-V_{DC}$	0
0	1	1	$-2V_{DC}/3$	$V_{DC}/3$	$V_{DC}/3$	$-V_{DC}$	0	V_{DC}
1	1	1	0	0	0	0	0	0

where, u_d, u_q are the stator d, q voltage; i_d, i_q are the stator d, q current; L_d, L_q are the stator d, q inductance; ψ_d, ψ_q are the stator d, q flux;R is the stator resistance; ψ_f is rotor permanent magnet flux; T_e is the motor electromagnetism torque; p is the polar logarithm. From the voltage equation and mathematical model of PMSM, it can be seen that the d-axis voltage u_d is not only affected by the d-axis current i_d, but also by the q-axis current i_q, which shows that there is a coupling relationship between the d-axis voltage and the q-axis voltage of PMSM.

The most widely used PMSM motor DTC system adopts three-phase inverter bridge connection, and the switch element adopts IGBT. When the DC-link voltage V_{DC} is known, the different combinations of switch state function S_A,S_B,S_C, and phase voltage and line voltage can be expressed in Table 1.

In the $\alpha-\beta$ coordinate system, the Clark transformation of the three-phase voltage of the motor can obtain as follows:

$$\begin{bmatrix} V_{S\alpha} \\ V_{S\beta} \end{bmatrix} = \frac{2}{3}\begin{bmatrix} 1 & -\frac{1}{2} & -\frac{1}{2} \\ 0 & \frac{\sqrt{3}}{2} & -\frac{\sqrt{3}}{2} \end{bmatrix}\begin{bmatrix} V_{AN} \\ V_{BN} \\ V_{CN} \end{bmatrix} \tag{7}$$

Function expression of composite voltage space vector:

$$U_{out} = U_\alpha + jU_\beta \tag{8}$$

SVPWM controls the on-time of 8 basic space voltage vectors discretely, which makes the space voltage U_{out} generated by U_α and U_β approximate through the synthesis of 8 voltage vectors in the whole cycle space region. Suppose that the voltage vector is in the range of 0°–60° at a certain time, then the four basic voltage space vectors U_0, U_{60}, O_{000} and O_{111} compose U_{out}:

$$\begin{cases} T = T_1 + T_2 + T_0 \\ U_{out} = \frac{T_1}{T}U_0 + \frac{T_2}{T}U_{60} \end{cases} \tag{9}$$

where, T_1,T_2 are the cumulative on time of adjacent switch states in cycle T,T is sampling period, T_0 is the accumulated conduction time of zero state in the cycle, decomposed available:

$$\begin{cases} U_\alpha = \frac{T_1}{T}|U_0| + \frac{T_2}{T}|U_{60}|cos60^\circ \\ U_\beta = \frac{T_2}{T}|U_{60}|sin60^\circ \end{cases} \tag{10}$$

3.2 SVPWM Control System Based on TGI Technology

The results show that although the DTC structure of PMSM is simple, the flux moves along the track, and the speed response is fast, the flux and torque ripple of its output are large. The two-level generalized inverter (TGI) technology is different from the voltage sinusoidal PWM control mode by using the flux sinusoidal SVPWM control mode. The flux sinusoidal SVPWM control mode integrates the switch inverter and the motor, which has the advantages of high voltage utilization, low noise and low torque ripple. Under the same conditions, the output power of motor controlled by SVPWM of TGI technology can reduce the size of power devices, improve the overload capacity of control system, expand the application scope of PMSM in industry, improve the efficiency of motor vector control algorithm, avoid the operation of trigonometric function and anti-trigonometric function with more resource consumption, and realize the maximum use of DSP processor.

4 Establishing of Simulation Model

In this paper, the vector control simulation model of PMSM is built based on the existing modules in Simulink common library and simpowersystem library by using MATLAB 2016a. The SVPWM vector control system based on SVM is mainly composed of several modules, such as current, speed and torque monitoring, Clark, park transformation and its inverse transformation, SVPWM space vector PWM based on SVM, among which the SVPWM vector control module based on SVM mainly includes coordinate transformation module and SVPWM module, and the model building process is shown in Fig. 1.

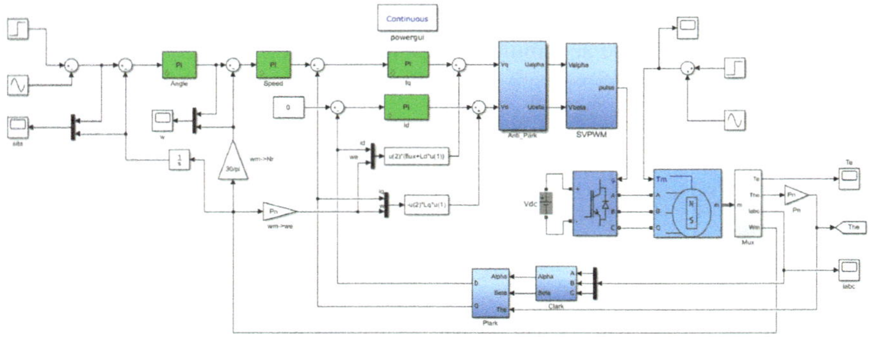

Fig. 1 SVPWM control system based on TGI for PMSM of pure electric vehicle

5 Analysis of Simulation Results

Set up the vector control model of PMSM based on TGI technology in MATLAB/Simulink and carry out simulation analysis. During the simulation, the specific parameters of PMSM need to be specified, as shown in Table 2, the simulation setting time is 2S. The simulation results are shown in Figs. 2, 3 and 4.

Figure 2 shows the current waveform of motor SVPWM control. The motor phase current pulsation is not big and the waveform is reasonable. The phase current impact shows that the system has good robustness.

Figure 3 is a comparison diagram of speed input and motor response. After 0.12 s adjustment, the system input speed and motor response are basically the same, the system response time is short, and the subsequent speed input waveform is consistent with the system response waveform, without obvious fluctuation, speed response speed is fast, and there is no static difference when running in a stable state, which improves the motor speed control accuracy.

Figure 4 is the motor SVPWM torque waveform. From it we can get that at 0.4 s, the speed changes abruptly, the torque fluctuation time is short and the torque recovery is fast, that is, under the condition of motor acceleration, the torque ripple response is rapid and the operation stability is good; at 0.8 s, the load changes suddenly but the torque curve is only slightly floating, that is, under the condition of strong torque

Table 2 Permanent magnet synchronous motor parameter list

Permanent	Value
Polar logarithm Pn	16
Stator resistance R	0.2 Ω
Permanent magnet flux ϕ_f	0.006 H
D axis inductance Ld	0.00015 H
Q axis inductance Lq	0.00009 Wb
Rated speed n	3000 r/min

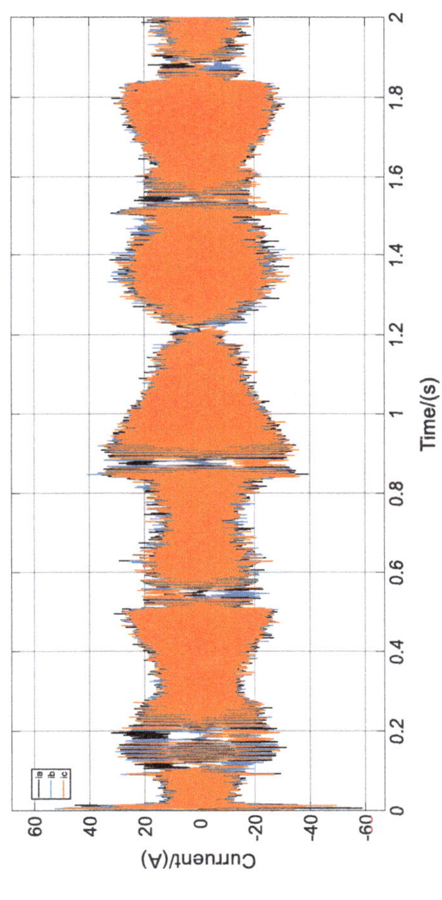

Fig. 2 SVPWM current waveform diagram

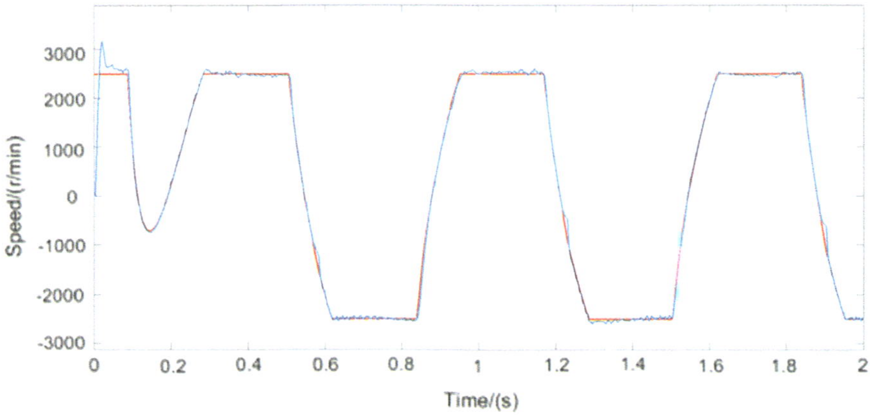

Fig. 3 Comparison diagram of speed input and motor response (red for input and blue for response)

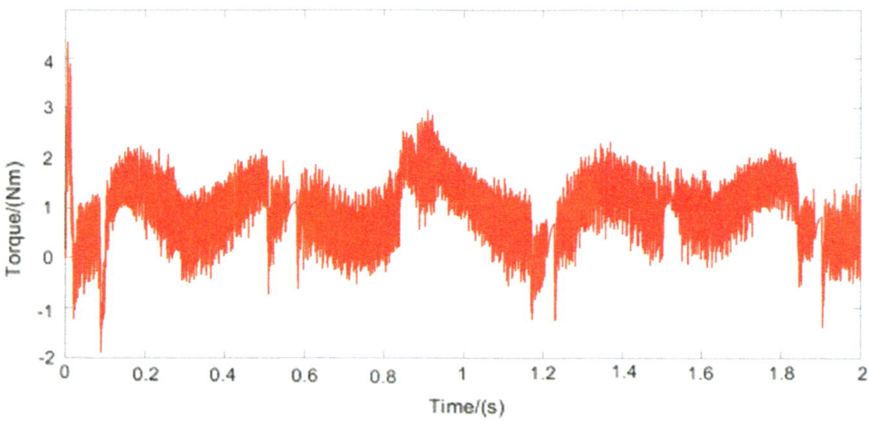

Fig. 4 SVPWM torque waveform diagram

operation, the motor torque ripple is small, the power output is stable and the overall operation stability is good.

6 Conclusion

In this paper, the dynamic control of PMSM is studied systematically, and the following conclusions can be drawn:

(1) A new type of SVPWM control system based on SVM technology is proposed to solve the pulsation problem of direct torque control of motor, through this method, the linear modulation range of SVPWM can be effectively improved;
(2) Through the MATLAB/Simulink simulation of the control system, the results show that the SVPWM control method based on SVM technology can make the system stable quickly;
(3) Motor control is a complex control system. In addition to the realization of motor stability control, it also involves power matching and other technical issues. In the field of control strategy and power matching related to the actual motor, more in-depth research is needed.

Acknowledgements The authors would like to thank China Postdoctoral Science Foundation Grant (2019M662223) for its financial support.

References

1. Qin, Yanzhong, Chen Wei Yan, and Qiang Geng. 2020. Three-vector model predictive current control strategy for permanent magnet synchronous motor drives with parameter error compensation. *Transactions of China Electrotechnical Society* 35 (2): 255–265. (in Chinese).
2. Xu, Yanping, Jibing Wang, Baocheng Zhang, et al. 2018. Three vector model predictive current control of permanent magnet synchronous motor [J]. *Transactions of China Electrotechnical Society* 33 (5): 980–988 (in Chinese).
3. Li, G.H. 2009. Inverse lag synchronization in chaotic system. *Chaos, Solitons & Fractals* 40 (3): 1076–1080.
4. Erjaee, G.H., and S. Momani. 2008. Phase synchronization in fractional differential chaotic system. *Physical Letters A* 372 (14): 2350–2354.
5. Wang, X.Y., and Y.J. He. 2008. Projective synchronization of fractional order chaotic system based on linear separation. *Physical Letters A* 372 (4): 435–441.
6. Tu, Wenyi. 2018. *A thesis submitted in partial fulfillment of the requirements for the Degree of Master of Engineering [D].* Wuhan: Huazhong University of Science & Technology.
7. Chen, X. 2006. Adaptive sliding mode control for discrete time multi-input multi-output systems. *Automatica* 42: 427–435.
8. Zhou, C., G. Yang, and J. Su. 2016. PWM strategy with minimum harmonic distortion for dual three-phase permanent magnet synchronous motor drives operating in the over modulation region. *IEEE Transactions on Power Electronics* 31 (2): 1367–1380.
9. Lu, Siguo, Ouyang Zheng, Meng Honglin, Zhu Chao. 2016. Vector control system of multilevel inverter double star winding shifted by 30 permanent magnet synchronous motor. *Transactions of China Electrotechnical Society* 31 (22): 45–56.
10. Fu, Bo. 2010. *The research on PMSM dynamic decoupling control technique [D].* Harbin Institute of Technology.
11. Wang Xiao, Yougui Guo, Xi Xiao, and Li Chuyun. 2020. Drive scheme of a new brake assist system. *Transactions of China Electrotechnical Society* 35 (supplement 1): 149–154.
12. Rahman, M.F., L. Zhong, and K.W. Lim. 1998. A direct torque controlled interior permanent magnet synchronous motor drive incorptating field weakening. In *IEEE Transactions on IndustrialApplication*, vol. 34, 1246–1253. United States: Institute of Electrical and Electronics Engineers

Overvoltage Protection of Thyristor in Bidirectional Hybrid Circuit Breaker

Chenguang Yang, Xinle Sha, Shun Zou, and Jie Li

Abstract The overvoltage during the reverse recovery of the thyristor is the main cause of device damage, especially for the thyristor used under pulsed high current conditions. For thyristor components in bidirectional hybrid circuit breakers operating at medium voltage levels, this paper first builds a test loop to explore the impact of forward current peak value I_F and turning off di_0/dt_0 on the reverse recovery process of the thyristor. The results show that under the condition of pulsed high current, the carriers stored in the PN junction of the thyristor have a saturation effect, and the peak value of the forward current I_F does not have a significant effect on the reverse recovery process. Turning off di_0/dt_0 is the main factor affecting the reverse recovery characteristics of the thyristor. On this basis, when the protection circuit adopts three different protection schemes: *RC* buffer, varistor and *RC* buffer + varistor, the voltage changes at both ends of the component during the reverse recovery process are calculated sequentially. Finally, a protection scheme using *RC* buffer + varistor can be obtained, which can effectively reduce the voltage rise rate at both ends of the component and achieve effective suppression of the over-voltage peak. *I*t is an effective protection scheme under 10 kV medium voltage level pulse working conditions. Finally, a test circuit was built, and the on–off characteristics of the thyristor components during the short-circuit breaking process were verified to verify the effectiveness of the calculation analysis.

Keywords Thyristor · Hybrid circuit breaker · Reverse recovery · Overvoltage peak · *RC* buffer · Varistor

1 Introduction

Hybrid circuit breakers have both the characteristics of low loss during the on-state process of mechanical switches and the characteristics of no arcing during the off

---

C. Yang (✉) · X. Sha · S. Zou · J. Li
Wuhan Institute of Marine Electric Propulsion, The First Village of Nanhu Automobile School, Wuhan, China
e-mail: 314768335@qq.com

W. Chen et al. (eds.), *The Proceedings of the 9th Frontier Academic Forum of Electrical Engineering*, Lecture Notes in Electrical Engineering 743,
https://doi.org/10.1007/978-981-33-6609-1_2

process of solid-state switches [1]. It is an effective fault protection device in medium voltage DC power systems.

For the short-circuit protection of medium-voltage DC power systems, the research group carried out a research on a natural commutation bidirectional hybrid circuit breaker based on thyristors. The study found that during the natural commutation of short-circuit current, the arc voltage generated after the mechanical switch is turned off is low (only tens of volts), so the number of series-connected devices in the thyristor assembly should be minimized to reduce the on-state voltage drop, to ensure that the short-circuit current can be quickly transferred from the mechanical switch to the thyristor assembly. During the forced shutdown process, when the reverse recovery current in the thyristor component rises to the reverse peak value and enters the falling phase, due to the forced shutdown loop reverse voltage and loop inductance induced electromotive force, a very high overvoltage spike will be generated. Especially in the process of short circuit breaking, the thyristor components work in pulsed high current conditions. The peak forward current exceeds 15kA, and the change rate of turn-off current exceeds 200 A/μs. The resulting overvoltage spikes often reach tens of kilovolts. It is necessary to increase the number of devices connected in series in the thyristor assembly to increase the reverse withstand voltage capability of the assembly. These two requirements contradict each other, and how to use as few serial devices as possible to work reliably during the reverse recovery process has become the key to hybrid circuit breaker design.

The hybrid circuit breakers based on thyristors designed by many scientific research institutes use anti-parallel diodes at both ends of the thyristors for overvoltage protection [2, 3]. However, on the one hand, zero voltage is not conducive to the recovery of the positive blocking characteristics of the thyristor, especially in the power system with a higher voltage level, it is easy to cause the forward breakdown of the thyristor component when the forward voltage is reapplied; On the other hand, under the condition of bidirectional breaking, the anti-parallel diode will increase the on-state voltage drop of the branch, resulting in the failure of the current to be successfully transferred from the mechanical switch to the thyristor assembly. Therefore, it is difficult to apply to medium and high voltage large capacity power systems.

At present, in many projects, *RC* buffers in parallel at both ends of the thyristor are commonly used to suppress the overvoltage peak during the reverse recovery process. Literature [4–6] introduces the selection basis of buffer protection parameters. However, the above studies are all aimed at the protection of thyristor devices under steady-state current. Whether *RC* buffer protection is suitable for overvoltage protection of thyristors under pulsed high current conditions requires further research.

In response to the above problems, this article first introduces the mechanism of reverse recovery overvoltage generation of thyristor components in hybrid circuit breakers; Then a test loop was built to analyze the influencing factors of current changes during the reverse recovery of the thyristor. On this basis, the hyperbolic secant function is used to describe the change trend of the current drop in the thyristor during the reverse recovery process. The changes of the voltage at both ends of the thyristor component are analyzed when using three different protection schemes: *RC* buffer, varistor and *RC* buffer + varistor. Finally, a protection scheme suitable for the reverse recovery process of the thyristor components in the hybrid circuit breaker was proposed, and a test circuit was built to verify the feasibility of the protection scheme.

2 Causes of Reverse Recovery Overvoltage

The main circuit topology of the natural commutation bidirectional hybrid medium voltage DC circuit breaker (hereinafter referred to as the hybrid circuit breaker) is shown in Fig. 1, which mainly includes:

1. The main switch branch including mechanical switch S_1;
2. The bidirectional current transfer branch includes thyristor components T_1 and T_2;
3. The forced commutation branch includes commutation capacitor C_1, commutation inductor L_1 and vacuum trigger switches TVS_1–TVS_4;
4. The voltage-limiting energy consumption branch includes varistor.

Example of breaking the short-circuit current in the direction of Fig. 1. The working process of thyristor component T_1 is as follows: When a short-circuit fault occurs in the system, the short-circuit current *i* rises rapidly, and the mechanical switch S_1 opens after receiving the instruction. At the same time, the thyristor component T_1 is triggered and turned on under the arc voltage of S_1, The short-circuit current *i* begins to transfer to the current transfer branch; Turn on the vacuum trigger switches TVS_2 and TVS_3, the C_1, L_1, and T_1 loops are turned on to generate a reverse pulse current i_c. At this time, the current $i_T = i_c - i$ in the thyristor component T_1 drops rapidly to zero and enters the reverse recovery process. In this process, the current i_T flowing through the thyristor assembly T_1 and the voltage u_T at both ends of the thyristor assembly T_1 are shown in Fig. 2 respectively.

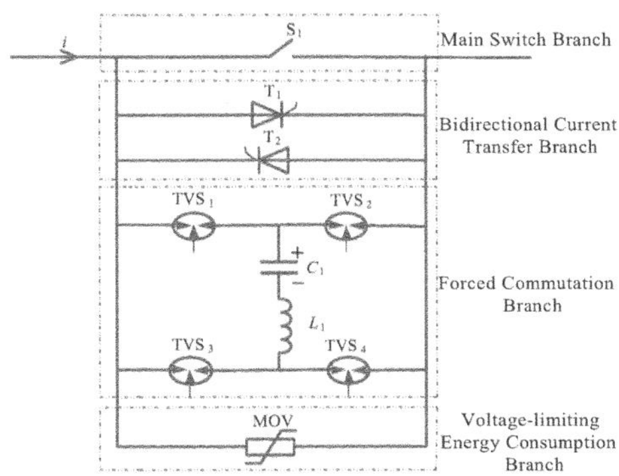

Fig. 1 Main circuit topology of bidirectional hybrid medium voltage DC circuit breaker

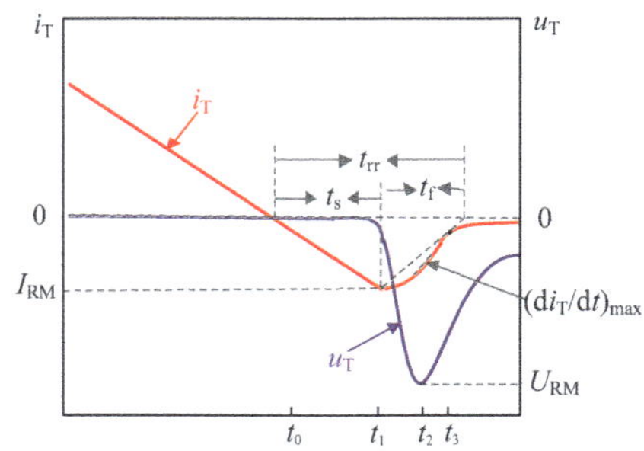

Fig. 2 Current and voltage waveforms during the reverse recovery process

Among them, U_{c1} represents the voltage of the commutation capacitor C_1 at t_0. For fast thyristors, the reverse recovery process time is very short, only a dozen microseconds. In this process, it can be approximated that the voltage of C_1 remains unchanged, $u_{c1} \approx U_{c1}$.

In the forced commutation loop formed by C_1, T_1, and L_1, the voltage on inductor L_1:

$$u_L = L_1 \frac{di_c}{dt} = L_1 \frac{di_T}{dt} + L_1 \frac{di}{dt} \tag{1}$$

In the reverse recovery process, the current change rate in the thyristor assembly is much higher than the system short-circuit current change rate. Therefore, the latter term in formula (1) is much smaller than the former one, and the effect of system short-circuit current i can be omitted in the calculation process. The reverse recovery process is equivalent to the process in which the thyristor component T_1 is forced to turn off under the action of the reverse voltage U_{c1}, and the equivalent circuit is shown in Fig. 3.

Get the equation expression:

$$u_T + L_1 \frac{di_T}{dt} + u_{c1} = 0 \tag{2}$$

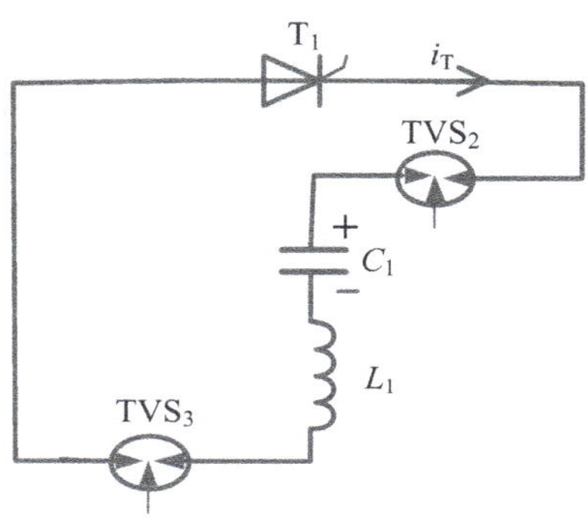

Fig. 3 Equivalent circuit in the forced commutation process

which is:

$$u_T = -\left(u_{c1} + L_1 \frac{di_T}{dt}\right) \tag{3}$$

The turn-off current change rate di_0/dt_0 is approximately equal to:

$$\frac{di_0}{dt_0} = -\frac{U_{C1}}{L_1} \tag{4}$$

In order to better describe the reverse recovery process, the time t_3 corresponding to the intersection of the line of the peak reverse recovery current I_{RM} and nI_{RM} with the x axis is defined as the end of the reverse recovery process, and time $t_f = t_3 - t_2$ represents the current fall time. The time $t_{rr} = t_s + t_f$ represents the reverse recovery time of the thyristor component. During this time period, the integral of the reverse recovery current represents the reverse recovery charge Q_{rr}.

During t_0–t_1, the current in the thyristor assembly drops to the reverse peak value I_{RM} approximately at the rate of change di_0/dt_0. At this stage, the thyristor component does not recover the reverse blocking capability, and the voltage u_T at both ends is equal to the on-state voltage drop, which can be ignored. The Minority carrier stored in the base area during the forward pass are swept out in the form of reverse current. The time $t_s = t_1 - t_0$ represents the carrier storage time.

At t_1, the minority carrier concentration in the base decreases, the device begins to recover its reverse blocking ability, and the current rate of change di_T/dt quickly drops to zero. At this time $u_T(t_1) \approx U_{c1}$。

After that, the space charge area on both sides of the thyristor PN junction gradually widened, and began to extract those minority carriers far away from the PN junction with a lower concentration, the reverse recovery current also began to decrease, and the current rate of change reached the maximum at t_2.

After that, the space charge area on both sides of the thyristor PN junction gradually widened, and the minority carriers that were far from the PN junction with a lower concentration began to be extracted. The rising rate of the reverse recovery current also begins to decrease, and reaches the maximum at t_2. At this time, under the action of the reverse voltage U_{c1} and the induced electromotive force of the commutation inductor L_1, a large reverse overshoot voltage U_{RM} will be generated at both ends of the thyristor component:

$$U_{RM}=u_T(t_2) = -\left[U_{c1} + \left(L_1 \frac{di_T}{dt}\right)_{max}\right] \tag{5}$$

If this value exceeds the reverse breakdown voltage of the thyristor component, the component will be damaged.

3 Analysis of Influence Factors of Current Change

3.1 Experimental Research on Influencing Factors of Current Variation

In order to explore the influencing factors of the current change in the reverse recovery process of the thyristor component under the pulsed high current condition, a test circuit is built as shown in Fig. 4, The detailed parameters of the test are shown in Table 1. During the test, by adjusting the charging voltage of the capacitor C_2, different forward pulse current peak values I_F are generated; By adjusting the charging voltage of the capacitor C_3, different turn-off current change rates of the thyristor component T_1 can be obtained. Among them, the thyristor component T_1 is formed by pressing three fast thyristor devices in series; TVS is used to control the discharge of the converter capacitor C_3.

Adjust the charging voltage of the capacitor C_3 to keep the turn-off current change rate di_0/dt_0 of the thyristor T_1 at 200 A/μs. By controlling the charging voltage of the capacitor C_2, the peak value of the forward current flowing through the thyristor assembly is changed. The current waveform is shown in Fig. 5.

It can be seen from Fig. 5 that if the turn-off current change rate di_0/dt_0 is the same and the forward current peak value I_F is different. The different reverse recovery current curves in the thyristor assembly coincide. Comparing the data in Table 2, it can be seen that the physical quantities t, t_f and Q_{rr} describing the reverse recovery process are also roughly the same. This shows that: different from steady-state current conditions [7], under pulsed high current conditions, the base region of the thyristor has a saturation effect on carrier storage. When the forward current peak I_F reaches a certain level, the number of stored carriers no longer increases with the increase of

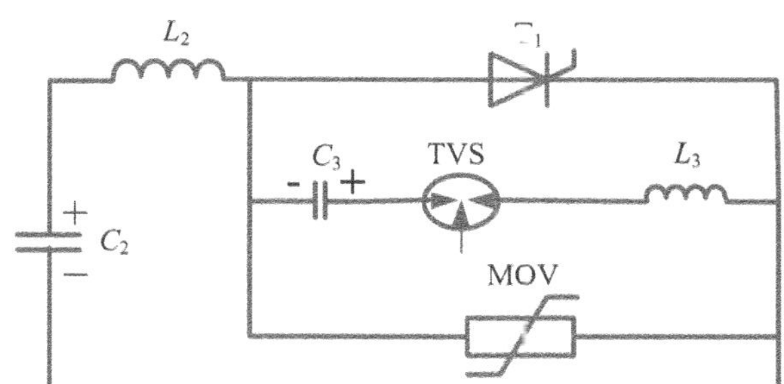

Fig. 4 Experimental circuit topology

Table 1 Test loop parameters

C_2/(mF)	L_2/(μH)	C_3/(μF)	L_3/(μH)
9	160	900	24

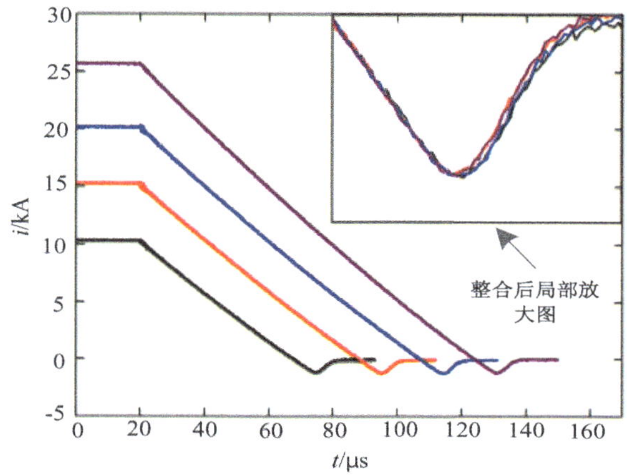

Fig. 5 Current waveform of thyristor reverse recovery process at different I_F with the same di_0/dt_0

Table 2 Changes in thyristor reverse recovery parameters at different I_F with the same di_0/dt_0

I_F/(kA)	I_{RM}/(kA)	t_s/(μs)	t_f/(μs)	Q_{rr}/(μC)
10.48	1.160	6.5962	6.3131	8110
15.53	1.154	6.5388	5.8741	7980
20.78	1.155	6.6501	6.1057	8190
25.74	1.155	6.5229	5.8452	7820

the forward current, so the forward current peak does not have a significant impact on the reverse recovery process.

Adjust the charging voltage of the capacitor C_2 to maintain the peak forward current $I_F = 20$ kA, and adjust the charging voltage of the capacitor C_3 to make the turn-off current change rate di_0/dt_0 different. The current waveform in the reverse recovery process of the thyristor is shown in Fig. 6, and the specific value of the turn-off current change rate di_0/dt_0 is shown in Table 3.

It can be seen from the current waveform in Fig. 6 and the data in Table 3 that when the forward current peak I_F is the same, with the increases of the turn-off current change rate di_0/dt_0: (1) The peak reverse recovery current I_{RM} and reverse recovery charge Q_{rr} increase linearly at the same time; (2) The reverse recovery time t_{rr} gradually decreases.

When the peak forward current is the same, the total amount of carriers stored in the thyristor is the same. As the turn-off current change rate di_0/dt_0 increases, the time for the current to drop to zero decreases, and the number of carriers that are recombined during the current decrease also decreases. Therefore, the proportion of carriers that can be extracted in the reverse recovery process gradually increases, and ultimately leads to a linear increase in the reverse recovery current peak value I_{RM} and the reverse recovery charge Q_{rr}.

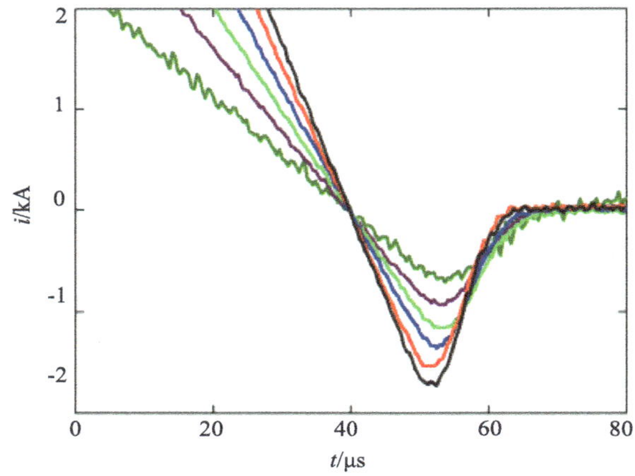

Fig. 6 Current waveform of thyristor reverse recovery process at different di_0/dt_0 with the same I_F

Table 3 Changes in thyristor reverse recovery parameters at different di_0/dt_0 with the same I_F

di_0/dt_0 /(A/μs)	I_{RM} /(kA)	t_s/(μs)	t_f /(μs)	t_{rr} /(μs)	Q_{rr} /(μC)
110	0.696	7.0036	6.7517	13.7553	4830
144	0.829	6.6055	6.3053	12.9108	5400
200	1.155	6.3001	5.8741	12.1742	7820
245	1.355	6.0750	5.3314	11.4064	8270
297	1.533	5.9363	4.8234	10.7597	9470
335	1.713	5.6654	4.5323	10.1977	10,720

3.2 *Fitting Curve of Hyperbolic Secant Function at Current Decreasing Phase*

The peak reverse recovery overvoltage appears in the current drop phase of the reverse recovery process. At this stage, the current change rate gradually increases from zero, then gradually decreases after reaching the maximum value, and finally decreases to zero again when the current drops to reverse leakage current. It is basically consistent with the change trend of the hyperbolic secant function, and the hyperbolic secant function can be used to describe the reverse recovery current in the current falling stage [8, 9]. The mathematical expression is shown in Eq. 6.

$$i_T = -I_{RM} * \operatorname{sech}\left(\frac{t}{\tau}\right) \tag{6}$$

where [10]:

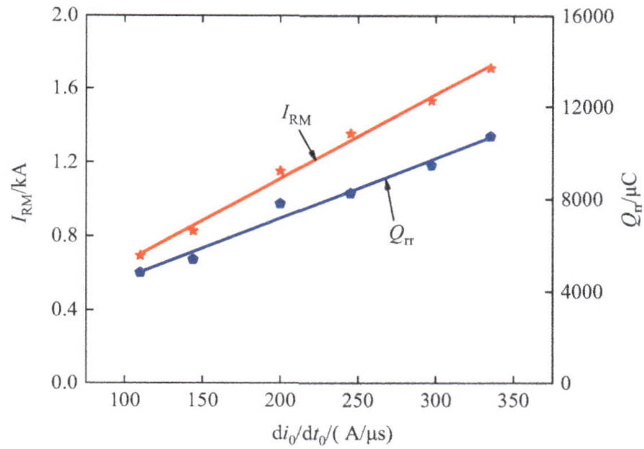

Fig. 7 With the change of di_0/dt_0, the change curve of I_{RM} and Q_{rr}

$$\tau = \frac{(1-n)}{\ln\left(\frac{1}{n} \pm \sqrt{\frac{1}{n^2}-1}\right)} * \left(\frac{2Q_{RR}}{I_{RM}} - \frac{I_{RM}}{di/dt}\right)$$

According to the analysis in the previous section, it can be seen that the turn-off current change rate di_0/dt_0 is the main factor affecting the current change during the reverse recovery process. According to the data in Table 3, the trends of the reverse recovery current peak value I_{RM} and reverse recovery charge Q_{rr} with respect to the turn-off current change rate di_0/dt_0 are shown in Fig. 7, and the corresponding function expressions are fitted according to Eqs. (7) and (8).

$$I_{RM} = f_1(di_0/dt_0) = 0.20509 + 30.00455(di/dt) \tag{7}$$

$$Q_{rr}=f_2(di_0/dt_0)=1989.7084+25.9236*(di/dt) \tag{8}$$

So far, each physical quantity in the reverse recovery process can be expressed as a function of the turn-off current change rate di_0/dt_0. Only need to determine the voltage value U_{c1} of the capacitor C_1 at the time of the current zero crossing, then the current change trend in the thyristor can be accurately described.

4 Overvoltage Protection Scheme

In order to ensure the safe operation of the thyristor components, three protection circuits are proposed, as shown in Fig. 8. Among them, the varistor in (b) and (c) is a kind of voltage sensitive component with nonlinear volt-ampere characteristics. At rated temperature, if the voltage across the varistor is less than its turn-on voltage, the resistance of the device is very large, which is equivalent to an open circuit; If the voltage at both ends exceeds the turn-on voltage of the varistor, the resistance

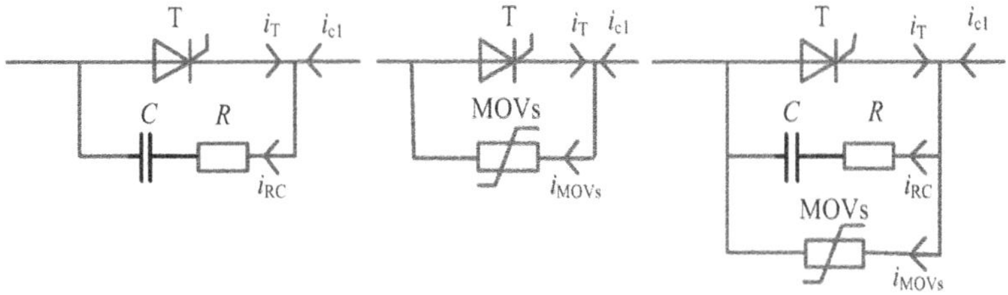

Fig. 8 Different protection schemes of thyristor components

of the device decreases rapidly and a large current flows.Within a certain current range, as long as the residual voltage level of the varistor does not exceed the rated working voltage of the thyristor, the reliable operation of the thyristor device can be guaranteed.

4.1 RC Buffer Protection

When the protection circuit adopts *RC* buffer protection, the equations can be obtained:

$$\begin{cases} u_c + i_{RC}R + L_1 \frac{di_{c1}}{dt} = u_{c1} \\ i_{RC} = i_T + i_{c1} \\ i_{RC} = C\frac{du_c}{dt} \end{cases} \quad (9)$$

And:

$$u_T = -(u_c + i_{RC}R) \quad (10)$$

Simultaneous Eqs. (4), (6), (9), (10), if $u_{c1} = 10$ kV, when $C_l = 500$ μF, $L_1 = 15$ μH, buffer capacitance $C = 1.25$ μF, with the buffer resistance increases, the voltage waveform across the thyristor component during the reverse recovery process is shown in Fig. 9:

For a given buffer capacitor *C* value, there is a unique *R* value to minimize the reverse recovery overvoltage peak. As the value of *R* increases, the peak value of reverse recovery overvoltage increases slowly. When it reaches a certain level, the reverse recovery voltage waveform remains unchanged; As the value of *R* decreases, the peak value of reverse recovery overvoltage shows a substantial increase, and oscillations gradually appear. The reverse recovery voltage rise rate d*u*/d*t* always gradually decreases as the value of *R* decreases.

Under the same conditions, keep the buffer resistance $R = 6\ \Omega$, and gradually increase the value of the buffer capacitor *C*. The voltage waveform during the reverse recovery process is shown in Fig. 10.

Fig. 9 When $C = 1.25$ μf, with the change of buffer resistance R, the voltage waveform during reverse recovery

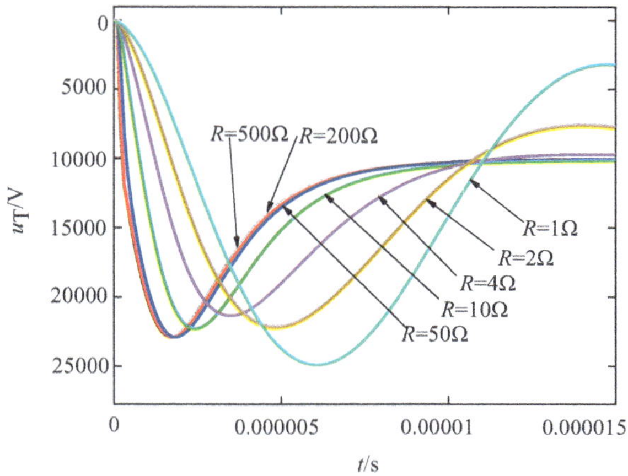

Fig. 10 When $R = 6$ Ω, with the change of buffer capacitance C, the voltage waveform during reverse recovery

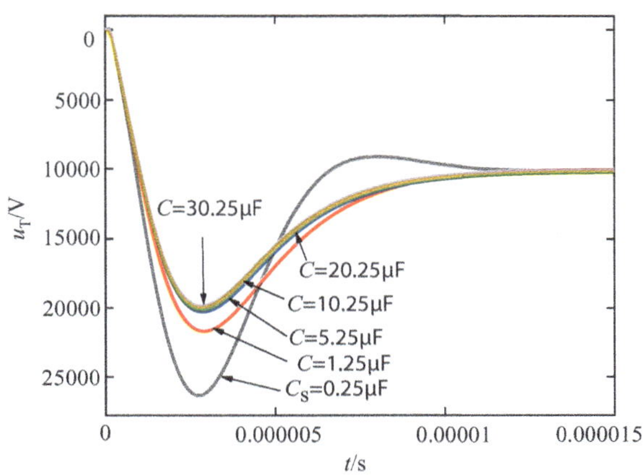

When the value of the buffer resistor R is determined, as the value of the buffer capacitor C increases, the reverse recovery overvoltage peak value and the voltage rise rate du/dt both show a decreasing trend. However, when the capacitance C value increases to a certain extent, the voltage limiting capability tends to be saturated, and the voltage waveform remains basically unchanged.

Considering that only the parameters of the buffer capacitor C and the buffer resistor R can match, the best voltage limiting effect can be achieved. For a given buffer capacitance C value, take $R =$ sqrt (L_1/C) 6, the corresponding buffer parameters are shown in Table 4, and the voltage waveform during the reverse recovery process is shown in Fig. 11.

Realize the parameter matching of buffer capacitance C value and buffer resistance R value, which can greatly reduce the peak value of reverse recovery overvoltage. However, under 10 kV medium voltage pulse high current conditions, only

Table 4 Protection circuit buffer parameters

Serial number	A	B	C	D	E
Buffer capacitor $C(/\mu F)$	0.25	1.25	5.25	15.25	30.25
Buffer resistance $R(/\Omega)$	12	6	3	1	0.70

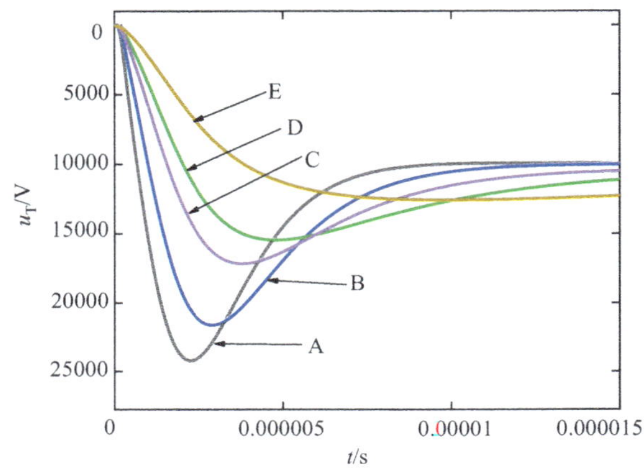

Fig. 11 Voltage waveforms during reverse recovery under different buffer protection parameters

by increasing the value of buffer capacitor C and reducing the value of buffer resistor R can the peak value of reverse recovery overvoltage be limited within a reasonable range. However, on the one hand, the power system will charge the buffer capacitor C in advance. When the mechanical switch S_1 of the hybrid circuit breaker is closed, it will discharge through the C–R–S_1 loop, forming a large pulse current; on the other hand, it will lead to an increase in the volume of the two-way commutating circuit.

Therefore, the RC buffer protection scheme alone is not suitable for the protection of thyristor components under medium-voltage pulse conditions.

4.2 Varistor Protection

If the protection circuit adopts the YH10WL-1.5/3.7 type varistor of Xi'an Shendian Group, the volt-ampere characteristic curve of voltage versus current can be obtained according to the data manual:

$$u_{mov} = f(i_{mov}) \tag{11}$$

Using the protection scheme of YH10WL-1.5/3.7 varistor in parallel at both ends of each device, as shown in Fig. 8b, the equations can be obtained:

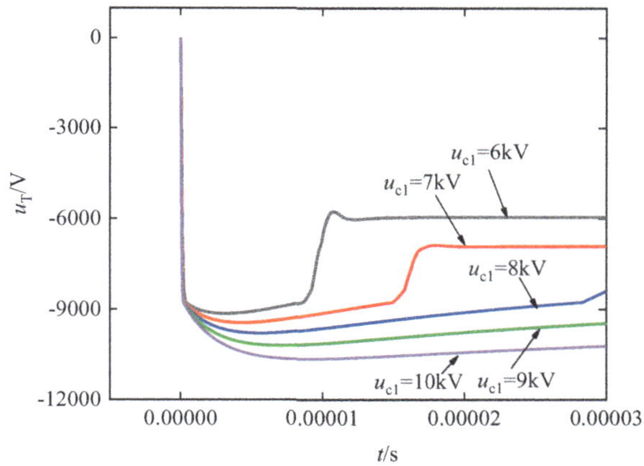

Fig. 12 With the change of u_{c1}, the voltage change curve across the thyristor

$$\begin{cases} u_{mov} + L_1 \dfrac{di_{c1}}{dt} = u_{c1} \\ i_{mov} = i_{c1} + i_T \\ -u_T = u_{mov} \end{cases} \tag{12}$$

Simultaneous Eqs. (4), (6), (11), (12), when $C_1 = 500\,\mu F$, $L_1 = 15\,\mu H$, Obtained when u_{c1} is 6 kV, 7 kV, 8 kV, 9 kV and 10 kV respectively, the voltage change curve at both ends of the thyristor assembly is shown in Fig. 12.

When the voltage across the device in the thyristor assembly reaches the turn-on voltage of the varistor, the voltage rise rate drops rapidly. As u_{c1} increases, the reverse recovery overvoltage peak U_{RM} rises slowly and is clamped within 11 kV. When the voltage across the device in the thyristor assembly is lower than the opening voltage of the varistor, the varistor does not operate, and the reverse voltage is directly applied to both ends of the varistor, which cannot effectively suppress the voltage rise rate, resulting in a very high voltage Abrupt changes occur at both ends of the device in the thyristor assembly.

If the voltage rise rate exceeds the critical rise rate (dv/dt) of the turn-off voltage of the thyristor, it is easy to cause the power breakdown of the device.

Therefore, the protection scheme using the varistor alone can only achieve the suppression of the reverse recovery overvoltage peak after the varistor is turned on.

4.3 RC *Buffer Protection + Varistor Protection*

If the protection circuit adopts the *RC* buffer protection + varistor protection scheme, the equations can be obtained:

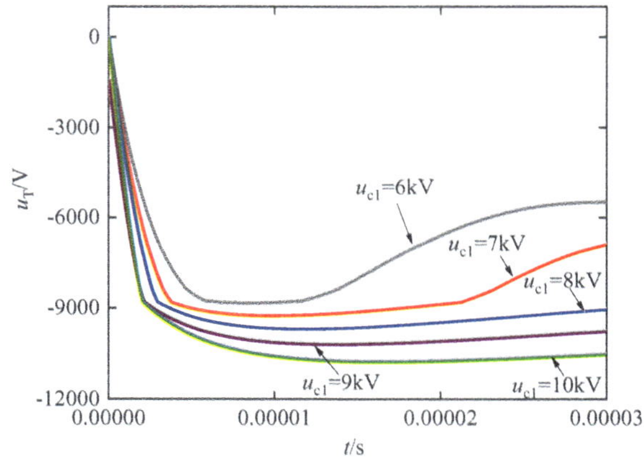

Fig. 13 With the change of u_{c1}, the voltage change curve across the thyristor

$$\begin{cases} u_c + i_{RC}R + L_1\frac{di_{c1}}{dt} = u_{c1} \\ i_{c1} + i_T = i_{RC} + i_{mov} \\ i_{RC} = C\frac{du_c}{dt} \\ u_c + i_{RC}R = f(i_{mov}) \\ -u_T = u_{mov} \end{cases} \tag{13}$$

Simultaneous Eqs. (4), (6), (11), (12), when $C_1 = 500\ \mu F$, $L_1 = 15\ \mu H$. In the protection circuit, the buffer capacitance $C = 1.25\ \mu F$, the buffer resistance $R = 1\ \Omega$, varistor adopts YH10WL-1.5/3.7 type. When u_{c1} is 6 kV, 7 kV, 8 kV, 9 kV and 10 kV respectively, the voltage change curve at both ends of the thyristor assembly is shown in the Fig. 13.

It can be seen from the figure that before the varistor is turned on, the sudden change of the voltage across the thyristor component disappears, and the voltage rise rates are: 1560, 2155, 2573, 2684 V/μs; after the varistor is turned on, The reverse recovery overvoltage peak is clamped within 11 kV.

Therefore, the *RC* buffer protection + varistor protection scheme is an effective technical scheme to realize the protection of thyristor devices under pulsed high current conditions.

5 Verification of on–Off Characteristics of Thyristor Components in DC Breaking Process

After determining the protection scheme of the thyristor component T_1, the test verification of the on–off characteristics for the thyristor component was carried out. In the protection circuit, the buffer capacitance $C = 1.25\ \mu F$, the buffer resistance $R = 1\ \Omega$, varistor adopts YH10WL-1.5/3.7 type. During the on–off process, the

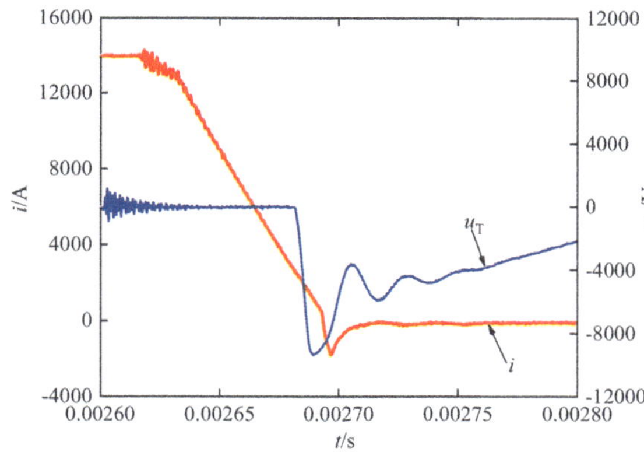

Fig. 14 Voltage and current waveforms of the thyristor assembly during on and off

voltage and current waveforms at both ends of the thyristor component T_1 are shown in Fig. 14 respectively.

The thyristor component T_1 has a flow time of 2 ms, the peak forward current is 15.049 kA, the component begins to decrease at a rate of 299A/μs, the capacitor voltage $U_{c3} = 8.56$ kV at the time of the current zero crossing, and the peak reverse recovery current is 1.59 kA. The peak reverse overvoltage at both ends of the thyristor component is 9.673 kV, and the voltage rise rate is 2125 V/μs. The time from when the current drops to zero to when the forward voltage is applied across the thyristor component is $\Delta t = 179$ μs. The thyristor component was successfully turned off and the forward voltage blocking capability was restored, completing the on–off characteristics verification during the short-circuit breaking process of the medium voltage level system. Therefore, the selected protection scheme can ensure the reliability of the thyristor assembly.

6 Conclusion

1. Under the condition of high pulse current, the minority carriers stored in the base region during the forward pass of the thyristor have become saturated, and the reverse recovery process is basically independent of the forward current peak I_F; Turning off di_0/dt_0 is the main factor affecting the reverse recovery process of the thyristor.
2. In the medium-voltage power system, the *RC* buffer protection can limit the overvoltage peak value of the thyristor component within a safe range only when the buffer capacitance is greatly increased while the buffer resistance is reduced. However, it is not feasible in terms of safety and size.
3. The varistor protection alone can only effectively suppress the reverse recovery overvoltage peak after the varistor is turned on, and cannot suppress the voltage

rise rate at both ends of the thyristor component during the reverse recovery process.

4. The use of RC buffer protection + varistor protection can effectively reduce the rate of rise of the voltage at both ends of the component, and at the same time achieve effective suppression of overvoltage peaks. It is an effective protection scheme for thyristor components under medium-voltage pulse conditions.

References

1. Lü, Wei, Wenjie Wang, Taixun Fang, et al. 2019. Test technology of hybrid HVDC circuit breaker. *High Voltage Engineering* 45 (8): 2425–2433 (in Chinese).
2. Lei, Feng, Ruifeng Gou, Xiaoping Yang, et al. 2018. Hybrid HVDC circuit breaker based on series-connected thyristors with forced zero crossing technique. *High Voltage Engineering* 44 (2): 388–394 (in Chinese).
3. Wandi, Zhou, Xiaoguang Wei, Chong Gao, et al. 2014. Thyristor basedhybrid arc-less high voltage direct current circuit breaker. *Proceedings of the CSEE* 34 (18): 2990–2996 (in Chinese).
4. Xie, Ting, Guangfu Tang, Jianchao Zheng, et al. 2012. Analysis on reverse voltage characteristics of HVDC thyristor valves in the fault state. *Proceedings of the CSEE* 32 (1): 140–146. (in Chinese).
5. Gao, Chong, Jing Zhang, Jianhui Zhou, et al. 2018. Study on the calculation of the commutation overshoot of HVDC valve based on the piecewise fitting of reverse recovery current of thyristor. *Proceedings of the CSEE* 38 (2): 547–554. (in Chinese).
6. Hao, Yong, Yunxia Liu, and Jilun Zeng. 2010. A parameter design of RC buffer for thyristors based on exponential recovery model. *Electrical Engineering* 2010 (2): 20–23. (in Chinese).
7. Ke, Yue, Sun Wei, Liu Longchen, et al. 2017. Correlations between carrier lifetime and reverse recovery characteristics of high power thyristors. *High Voltage Engineering* 43 (12): 3944–3949 (in Chinese).
8. Chokhawala, R.S., E.I. Carroll. 1991. A buffer design tool for PN junction reverse recovery using a more accurate simulation of the reverse recovery waveform. *IEEE Transaction on Industry Applications* 27 (1): 73–84.
9. Frederick, M.M., and J.R. Harold. 1976. The recovered charge characteristics of high power thyristors. *IEEE Transaction on Industry Applications* 12 (3): 305–311.
10. Lee, C.W., and S.B. Park. 1988. Design of a thyristor buffer circuit by considering the reverse recovery process. *IEEE Transactions on Power Electronics* 3 (4): 440–446.

Analysis and Treatment of Common Faults in OTN System Equipment

Can Qi, Hongfei Xu, Ying Wan, Yi Zhang, Hanshuo Duan, and Kunrui Tong

Abstract As a transmission system with large capacity and suitable for many types of services, the dense wavelength division system makes up for the shortcomings of the traditional communication network and is more suitable for the current development needs of the power system communication network. However, after the actual operation, many failures occurred in the OTN system. This paper analyzes the common faults of power communications OTN and puts forward a series of effective preventive measures.

Keywords OTN · Failure Analysis · Troubleshooting

1 Introduction

OTN (Optical Transport Network) is simply a next-generation backbone transport network that works on the optical layer. It is based on wavelength division multiplexing technology. It spans the traditional electrical domain, e.g. digital transmission, and optical domain, e.g. analog transmission, and realizes the standardization of optical domain and electrical domain [1]. Not only that, but it also solves lots of problems, like small business capacity, weak networking ability, and inadequate protection ability of traditional optical transmission networks [2]. As we all know, OTN solves these problems through a series of protocols, so its internal hierarchy is more complicated than traditional optical transmission. At present, the OTN network of Jibei Electric Company Limited elect power Company mainly carries various services such as external information network, dispatching data network two-plane, and data communication network. With the increase in the number of services and the construction of new sites, maintenance personnel will face more challenges [3, 4]. A timely and practical summary of O & M's experience is a vital link to improve O & M capabilities.

C. Qi (✉) · H. Xu · Y. Wan · Y. Zhang · H. Duan · K. Tong
State Grid JiBei Electric Power Co., Ltd, Beijing 100053, China
e-mail: 1164999428@qq.com

W. Chen et al. (eds.), *The Proceedings of the 9th Frontier Academic Forum of Electrical Engineering*, Lecture Notes in Electrical Engineering 743,
https://doi.org/10.1007/978-981-33-6609-1_3

In this paper, we analyze and summarize the past faults and establish a new fault handling process to make the fault handling more lean and standardized. First, we compared the difference between the OTN system and other optical transmission systems. Then, we established a new fault handling model. Finally, through the loss/ODU-LOFLOM alarm processing, the validity of the model is verified.

2 Related Work

2.1 Characteristics of the OTN Network

The OTN network operated and maintained by State Grid Jibei Electric Company Limited has the characteristics of comprehensive coverage, multiple cross-planes, and extensive business volume [5]. At the same time, due to the long operating period of the system, the backward planning in the early stage and the lack of aggregation nodes, the daily maintenance of the network is more burdensome than the traditional SDH network. In the face of such a network, it is worth considering how to overcome difficulties and improve the level of operation and maintenance while maintaining the same number of operators.

2.2 User Experience Work

Although these OTN types of equipment currently in use are of strong networking capabilities, the OTN network hierarchy still is more complicated than the SDH equipment. A failure of particular equipment often generates a large number of alarms throughout the network. For example, the OTU_K LOS alarm will generate LOF, LOM and other alarms and may also affect the following sub-channels [6]. In this case, to solve the unclear resource status, the operation need to use the alarm tracing method to check the fault points one by one, which adds many difficulties to the daily work of the operation and maintenance personnel, prolongs the time for eliminating the fault and threatens the stability and reliability of the OTN network.

Although the OTN network currently supports device performance detection, the internal relationship is still complicated, including nine levels. For example, OPU_k, ODU_k, OTU_k, and OCh [2]. Compared with the traditional optical transmission network, there are considerable differences in equipment management, service configuration, troubleshooting, and network planning. Therefore, solely relying on past experiences, it is difficult for the operation and maintenance personnel to complete the operation and maintenance work.

Based on these above, we can already see that OTN's complex hierarchy makes it difficult to see internal relationships. Not only that, OTN not only cares about traditional optical power, bit error, and other alarms but also needs to pay attention to performance alarms such as spectrum characteristics and signal-to-noise ratio [7]. Correspondingly, a variety of invalid alarms will be generated, which is very likely to overwhelm the actual root alarm [8].

In order to solve the above problems, this article aims to improve the efficiency of operation and maintenance and improve the quality of operation and explores new ideas to solve the various problems faced by the current OTN network operation of State Grid Jibei Information &Telecommunication Company.

3 Findings

In this section, we present our empirical findings. By our research interests, we first outline the generalized data collection practices in State Grid Jibei Information & Telecommunication Company, i.e., what types of operation and maintenance experience that is collected in the different development phases, and by whom. Second, we summed up the past lessons. Finally, we formulated a series of new ideas to improve the efficiency of operation and the quality of maintenance [9], aiming at the network structure and business model of OTN different from the traditional optical transmission system.

3.1 Establish a Complete Operation and Maintenance System

- Network management specification: clarify the authority, responsibilities, and skill requirements of network management personnel in various departments to avoid network security problems.
- Daily operation and maintenance specifications: clarify the alarm inspection cycle and related content to prevent accidents caused by personnel negligence; clarify the network management data maintenance cycle and related details to avoid the impact of network management due to data redundancy.
- Fault handling specifications: clarify fault handling principles and procedures and improve fault handling efficiency.

3.2 Establish a Complete Operation and Maintenance Manual

As an O & M worker, you must be familiar with the common types of failures in OTN networks [10]. Details as follows:

- The optical cable failure. The details include fiber optic cable interruption, excessive bit errors, and high line loss.

- Equipment failure. The details include: OTU, FIU, ODU/OMU BA/LA SCC, and other components are damaged. The regular operation of the board is affected due to the environment and humidity.
- Pigtail failure. The details include broken fiber pigtails, substantial loss of fiber pigtails, and dust on the flange joints or fiber end faces.
- A power failure caused by the equipment room environment. The details include power outage in the equipment room, long-term power outage of the storage battery, and equipment power interface board failure.
- The network management system failure. Specifically, the monitoring interruption caused by the physical connection between the device and network management is interrupted.

By categorizing the types of faults, O & M personnel can break down the faults one by one through theoretical training and on-site practical operations. Once the device has the above problems, the operation and maintenance personnel can query the relevant alarm information and performance in a timely and effective manner, such as querying basic parameters such as alarm, performance overrun, and optical power. At the same time, we can analyze the abnormal conditions, such as alarms and performance data, to make preliminary fault judgments. Effectively avoid prolonging the troubleshooting time due to the ability of O & M personnel.

3.3 Establish Perfect Rules for Routine Maintenance

In addition to being able to master the skills for troubleshooting, the operation and maintenance personnel should do an excellent job in the daily maintenance of the equipment finding hidden dangers in time and preventing them from happening [11].Details as follows:

- Regularly clean the fan to prevent the high temperature from affecting device performance. Clean the air filter at least once during the spring and autumn inspections to ensure that the device dissipates heat properly.
- Regularly test the conventional telephone to ensure that once the network is interrupted, you can also ensure the interconnection of equipment through the conventional telephone.
- Regularly conduct communication safety training for operation and maintenance personnel to avoid damage to the equipment due to human factors. Firstly, for board failures, the board must be replaced to solve the problem. Operation and maintenance personnel must remember to wear an anti-static wrist strap before the operation to avoid static electricity affecting the device. Secondly, for the failure caused by the fiber core problem, the problem must be solved by replacing the fiber core. When inserting or removing the optical fiber, avoid facing the eyes directly to avoid personal injury. At the same time, the unplugged optical fiber should be kept clean with a dust cap. Before re-inserting, if the optical fiber is dusty, please remember to turn off the laser before cleaning it with the fiber box. Fourth, for the

faults that need to be resolved through hardware loopback, the optical attenuator should be appropriately selected according to the threshold range of the received optical power of the board, and then the input and output optical ports of the board should be looped back through the fiber jumper.

3.4 Establish a Smooth Troubleshooting Process

3.4.1 Fault Handling Principles

According to the information and communication operation management method issued by State Grid Corporation, the fault location of OTN equipment should follow the idea of "external first, then internal; site first, then board; line first, then branch."

3.4.2 Troubleshooting Process

(1) Check the network management side to determine the cause of the failure. First, select root alarms based on network management alarms. Second analyze the flow of service signals. Finally troubleshoot points according to the flow of service signals.
(2) On-site inspection to determine the cause of failure. The operation and maintenance personnel at the fault site determine the cause of the fault by using optical power meters, OTDR, and other instruments.

4 Research Validity

4.1 Signal Loss Alarm

4.1.1 Alarm Phenomenon

Channel 192.9 of the port in the LO2 board at site A simultaneously reports "low threshold of input optical power" and "signal loss alarm" (Fig. 1).

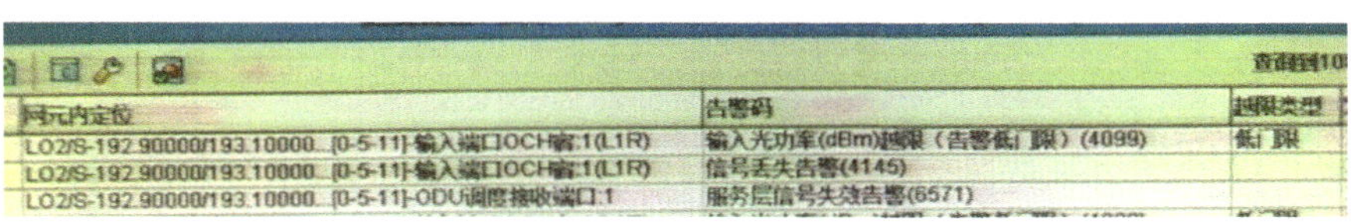

Fig. 1 Alarm content of the site A port 1

4.1.2 Fault Location

The NMS checks the optical power of the port and then finds that the current port is receiving light regularly. However, after looking at the historical performance, it is found that there is no light phenomenon in this channel during each acquisition period (specifically, it is – 60 dB for receiving light), and finally, it can be judged that there is a flashing phenomenon in this channel. The historical performance of the optical port is shown in Fig. 2.

Next, the O & M personnel scanned the light from the upstream Site B multiple times on the local OPM board and found that there was a flash in the 192.9 channel. So the faulty equipment could be locked at the upstream Site B. At the same time, the NMS checks that the corresponding port 1 of the upstream site B is regular. Therefore, the fault point was initially located on the pigtail from port 1 of site B to port 9 of the VUMX board (Fig. 3).

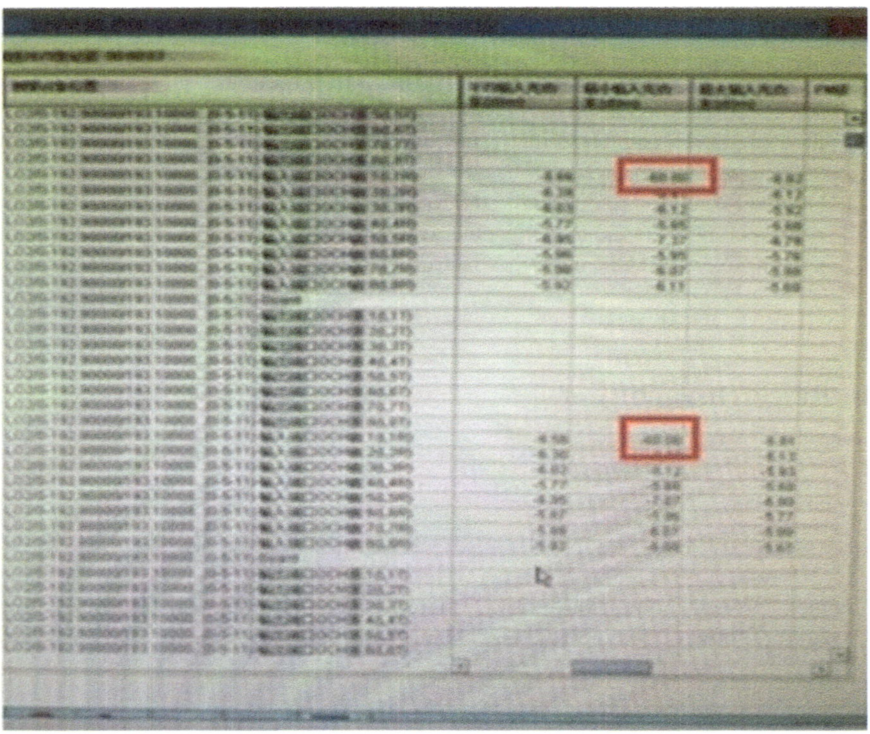

Fig. 2 Historical performance of the site

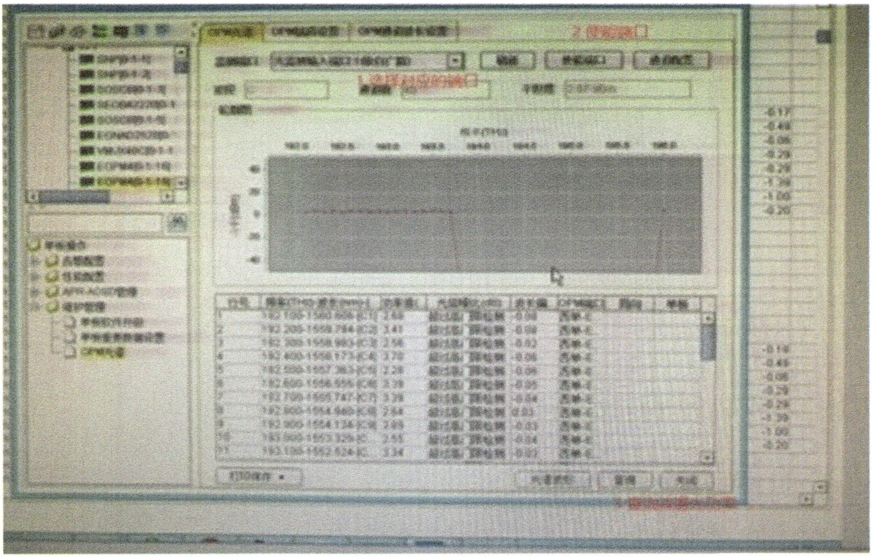

Fig. 3 Results of OPM sweep of port one at site B

4.1.3 On-site Troubleshooting

According to the preliminary conclusions of the troubleshooting, the O & M personnel brought tools to Site B. First, use an optical power meter to measure the presence of light at site A, but the optical power has been jumping from − 4 to − 6 dB. The average light emission of the device should be stable at − 4 dB. Furthermore, the jump is an abnormal phenomenon, so the module is judged to be faulty. However, because there is no flashing, there is a difference from the network management phenomenon, so continue to measure the light reception corresponding to the VUMX board port. The optical power meter shows no light, and finally, it can be judged that the pigtail and the module are faulty at the same time. The fault was eliminated after the O & M personnel replaced the pigtails and modules.

In summary, when the NMS reports a non-optical alarm, it should first determine which type of failure it belongs to. It can be divided into single-channel failure, multi-channel failure, and primary optical failure. If it is a single-channel fault, you should first determine whether the fault occurs at the local end or upstream. If the leading light fails, you should check whether there is any abnormality in the receiving light of the local amplification board. If it is a multi-channel fault, it is most likely to occur between the local demultiplexer board and the circuit board or the signal upstream circuit board to the multiplexer board.

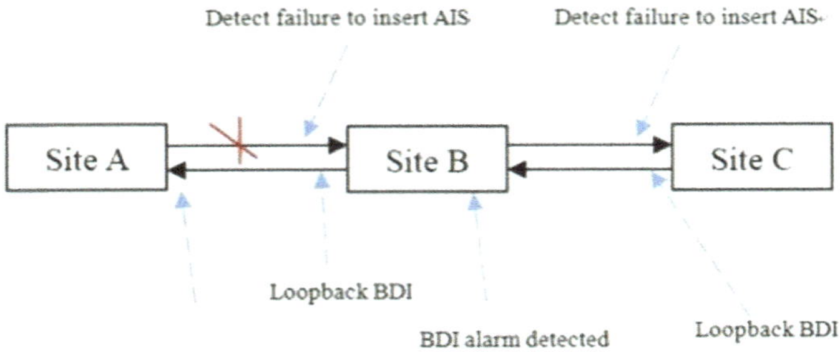

Fig. 4 Positioning method

4.2 ODU-LOFLOM Alarm

4.2.1 Alarm Phenomenon

The ODU-LOFLOM alarm means that the frame positioning byte or multi-frame positioning byte of the ODU is wrong or missing. And it is commonly used in the service board scheduling transmission port.

4.2.2 Fault Location

For ODU-LOFLOM alarms, O & M personnel often have a misunderstanding, that is, the ODU_K dispatch receiving port in the alarm is the direction that the service board sends the signal to the cross-board and the OUDK dispatch sending port is the business board receiving the cross-board signal. However, the sending and receiving mentioned here are contrary to the literal meaning, so the operation and maintenance personnel are extremely confused. Therefore, when locating a fault, you should first locate the faulty site based on the BDI and PM unavailability time, then determine the faulty board through cross loopback and finally locate the cause of the fault. The specific fault location process is shown in Fig. 4.

4.2.3 Summary of Failure Causes

Because ODU-LOFLOM is a time slot type alarm, there are a variety of alarm causes and types. In order to facilitate operation and maintenance personnel to quickly and quickly deal with faults, this article will classify the causes of the alarms and give solutions. ODU-LOFLOM alarm causes and solutions are shown in Table 1.

Table 1 ODU-LOFLOM alarm causes and solutions

Reasons	Fault location basis	Disposal plan
Defect of NE version	Defect of NE version or mismatch with other NE version Software version upgrade	Defect of NE version
The backplane is damaged	If the alarm board has been inserted or removed recently, the backplane may drop, which may cause the service board to fail to fix the frame	Repair the backplane
Single-board FPGA program failure		Frequent flushing of FPGA alarms Single-board IC reset
Board failure	ODU-LOFLOM alarms are reported on all cross-boards	Re-insert the board, if it is invalid, replace the board
Signal degradation	Signal degradation will cause a large number of errors before and after error correction, resulting in frame failure	optimization of service signals
Cross-board failure	The corresponding cross-board failure causes the signal from the backplane to fail	Reset or replace the cross-board
Abnormal cross/black cross	The cross information in the device is inconsistent with the cross information in the NMS/The actual cross of the device is incomplete or the configuration is incorrect	Reset the service board and cross-board after deleting the abnormal cross

5 Conclusion

In this paper, we explore the difference between OTN and traditional optical transmission network, summarizes the problems existing in operation and maintenance process, breaks down one by one, finally explores new ideas to solve the various problems faced by the current OTN network operation and maintenance process of State Grid Jibei Electric Company Limited Elect Power Company. Improve O & M quality and achieve lean O & M.

Acknowledgements This work is supported by the Science and Technology Project of State Grid Jibei Power Company Limited (No. 52018E190075).

References

1. Wei, Gao, Li-Xia Zhang, and Yon Mi. 2011. Research on the application of the DWDM system in the Shanxi power communication network. *Shanxi Electric Power* 4: 61–63 (in Chinese).
2. Xie. Qinghui. 2010. Maintenance analysis of DWDM dense wavelength division optical transmission equipment. *Inner Mongolia Science Technology and Economy* 5: 106–110 (in Chinese).
3. Han, S.I., J.P. Fonseka, and E.M. Dowling. 2014. Constrained turbo block convolutional codes for 100 G and beyond optical transmissions. *IEEE Photonics Technology Letters* 10: 995–998.
4. Khodashenas, P.S. 2016. Benefit evaluation of all-optical sub-channel add–drop-in flexible optical networks. *IEEE Photonics Journal* 8: 1–9.
5. Gorse, S.S. 2018. OTN interface standards for rates beyond 100 Gb/s. *Journal of Lightwave Technology* 36: 19–26.
6. da Silva, J.C. 2019. OTN Network planning over DWDM using computational intelligence. *2019 21st International Conference on Transparent Optical Networks* (8): 1–4.
7. Tang, Xiao Yanbing. 2010. Analysis and treatment of several faults in power communication system maintenance. *Zhong National High-Tech Enterprise* 2X: 126–127.
8. Wang, Xiaowei and Jiay Guo. 2009. Discussion on the maintenance of the SDH optical fiber transmission network. *Heilongjiang Science and Technology Information* 2X: 15.
9. Shi, Jianhua. 2019. Application of OTN technology in-service route optimization of electric power communication network. *Electronic Technology and Software Engineering* 14: 22–23.
10. Zhen, Xuewei, and Qiming Hu. 2019. Some thoughts on critical technologies and the main faults of the 100G OTN system. *Information Communication* (7): 156–157(2019).
11. Xu, Tianwen. 2019. Error analysis in OTN optical transmission system. *China New Communications* (10), 118–121(2019). (in Chinese).

An Automatic Respirator Oil Level Reading Method Based on Instance Segmentation

Rui Han, Li Liu, Feiran Li, Peng Jiang, Shuang Liu, and Guolong Teng

Abstract The promotion of intelligent video surveillance technology has promoted the demand for intelligent image recognition algorithm in the power industry. Power transformer is a very important equipment in substation. In order to realize the automatic reading of the oil level for the transformer respirator, a reading method based on instance segmentation is proposed. Firstly, a Mask Intersection-over-Union (MaskIoU) regression module is added to the standard Mask R-CNN model in order to refine the mask confidence and improve the segmentation accuracy. Then the improved model is used to generate bounding boxes and segmentation mask of the cup and the oil. Finally, morphological operations are applied to the mask and oil level is computed based on pixel ratio of these two kinds of mask. Respirator images collected from substations are used to conduct experiments and the results show that the proposed method is effective in practical application.

Keywords Instance segmentation · Respirator oil seal · Mask R-CNN · MaskIoU regression module

1 Introduction

The transformer respirator is an important component of power transformers. Large power transformers put into use in substations are generally equipped with the respirator. The function of the respirator is to remove impurities and moisture that enter the transformer through air and prevent the dielectric strength decrease of the transformer oil [1]. The respirator is mainly composed of the upper silicone tank and the lower oil seal cup. Before entering the silicone tank, the humid air will pass through the oil seal cup located below, so the oil cover avoids the direct and continuous

R. Han · L. Liu · F. Li · P. Jiang · S. Liu
State Grid Zhejiang Electric Power Co., Ltd. Electric Power Research Institute, Hangzhou 310006, People's Republic of China

G. Teng (✉)
College of Electrical Engineering, Zhejiang University, Hangzhou 310027, People's Republic of China
e-mail: 21910098@zju.edu.cn

W. Chen et al. (eds.), *The Proceedings of the 9th Frontier Academic Forum of Electrical Engineering*, Lecture Notes in Electrical Engineering 743,
https://doi.org/10.1007/978-981-33-6609-1_4

contact of the silicone tank with the humid air, thereby extending the silicone service life. If the oil level in the oil seal cup becomes too low, the stroke of air in the oil will be shorter, and the filtering capacity of the oil seal cup will be greatly reduced [2–4]. In order to ensure that the oil level in the oil seal cup is normal, it is necessary to manually check whether the oil level is normal in the substation. With the rise of video surveillance, inspection robots and other technologies, it is possible to utilize digital image processing technology to complete the automatic identification of the status of power equipment.

Recently, researchers at home and abroad have proposed a lot of power equipment state recognition methods based on image processing technology [5–7]. In terms of respirator state recognition, [8] uses homomorphic filtering to enhance the blue area of the image, and then determines the position of the respirator through threshold segmentation and connected area extraction. [9] uses the deep learning model SSD [10] to detect the respirator, and then uses the HSV feature transformation to judge the status of the silica gel. Most of these methods only detect the discoloration of the silicone of the respirator, but ignore the importance of the oil level of the oil seal cup. In addition, they either detects by hand-designed features, or only achieves bounding-box-level object detection, which is easy to be affected by complicated environmental background.

This paper proposed an automatic reading method of respirator oil level based on instance segmentation. Specifically, an improved Mask R-CNN [11] instance segmentation model is used to simultaneously generate the bounding box and the segmentation mask of objects of interest. Then the mask is post-processed to read the oil level of the oil seal cup. At the end of this paper, comparative experiments are carried out on transformer Respirator images collected in substations, and the results verify the effectiveness of the proposed method.

2 Improved Mask R-CNN

2.1 Architecture

Current instance segmentation methods can be roughly divided into two categories: top-down and bottom-up. The top-down method usually extends some popular object detectors, that is, first obtains the object bounding box, and then predicts the mask in each box region. The bottom-up method first predicts the category label of each pixel of the image, and then groups these pixels to obtain the final instance segmentation result. The top-down method is more robust and can make full use of the results of object detection, which is more advantageous especially in industrial applications where data is relatively scarce.

Mask R-CNN belongs to a top-down image instance segmentation model (see Fig. 1). Its basic framework is Faster R-CNN [12], on top of which an additional branch is added to predict the mask for each region of interest (RoI). It also replaces

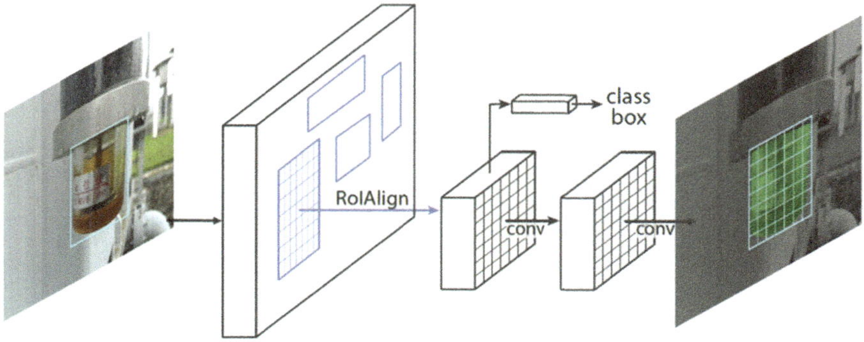

Fig. 1 The standard Mask R-CNN framework for instance segmentation

the RoIPooling layer with RoIAlign layer to further improve the prediction accuracy. The forward propagation process of Mask R-CNN is a cascade of two stages: the first stage is the Region Proposal Network (RPN), whose function is to distinguish foreground from background and generate object candidate regions; the second stage is called R-CNN stage, which uses RoIAlign to extract features with fixed spatial sizes from the region proposals, and then uses these features to predict object categories, bounding boxes, and masks.

2.2 Mask Score Calculation

In the original Mask R-CNN algorithm, the confidence of the predicted instance segmentation mask is equal to the classification confidence of the instance. Assuming that there are K + 1 categories of targets to be detected, the softmax layer of the R-CNN stage outputs a set of confidence values for each candidate region $\{s_1, s_{1,} s_2, \ldots s_{K+1}\}$, where the subscript i represents the ith category and K + 1 represents the background category, so the mask confidence is:

$$s_{mask} = \max_{i=1,2\ldots K,K+1} s_i \quad (1)$$

This mask confidence calculation method has certain defects because the classification confidence only characterizes the possibility that the object in the candidate box belongs to category c, but doesn't reflect the quality and integrity of the segmentation mask. Unreasonable mask confidence will adversely affect subsequent inference. Let MaskIoU be the intersection over union (IoU) of the predicted mask and the true mask. In theory, s_{mask} should reflect both the object category confidence and MaskIoU indicators. However, it is difficult to achieve this goal at the same time with a single objective function. In order to solve this problem, we decouple the above two indicators, so that the model can independently output the predicted value of MaskIoU s_{iou} and the category confidence s_{cls}. The final mask confidence s_{mask} is the product of the two:

$$s_{mask} = s_{cls} \cdot s_{iou} \tag{2}$$

Since s_{cls} has been implemented in the standard Mask R-CNN framework, next we will illustrate the implementation of the novel MaskIoU regression module.

2.3 MaskIoU Regression Module

The improved Mask R-CNN network structure is shown in Fig. 2, in which the blue dashed box represents the designed MaskIoU regression module, whose function is to output the IoU of the true mask and the predicted mask. This module consists of four convolutional layers with a 3 × 3 kernel and three fully connected (FC) layers. The output dimension of the first two FC layers is 1024, and K for the last one. The input of this module consists of two parts: the predicted mask and the feature F extracted by the RoIAlign layer. Due to mismatch in spatial size, the predicted mask is first max-pooled by a factor of 2, and then concatenated with F in the channel dimension.

In the inference stage, we first adopt the standard Faster R-CNN inference process. Specifically, assuming that N predicted bounding boxes are output in the R-CNN stage, followed by the non-maximum suppression (NMS) operation to remove duplicated detections. Then top-k bounding boxes are selected from the remaining

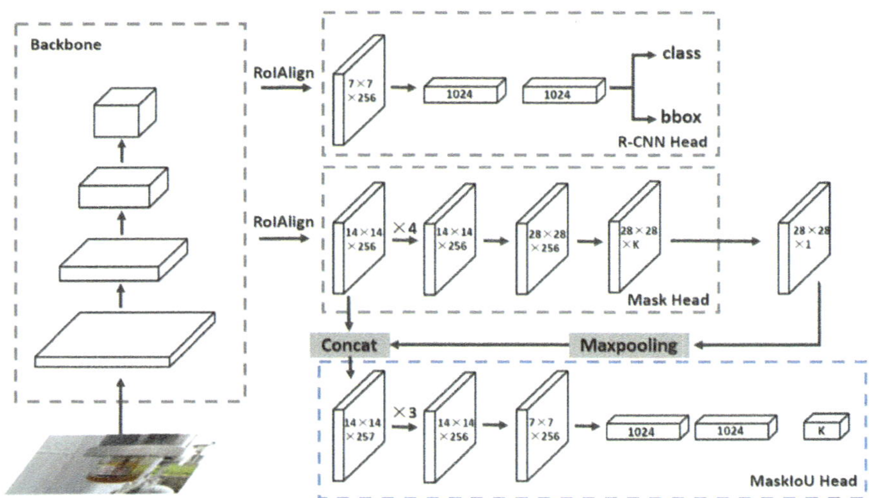

Fig. 2 Network architecture of the improved Mask R-CNN

according to the classification score. After the above standard process, these k RoIs are feed to the mask branch and MaskIoU regression module to predict masks with MaskIoU. Finally, the MaskIoU will be multiplied by the corresponding classification score and becomes the final mask confidence.

2.4 *Objective Function*

In the training phase, the improved Mask R-CNN uses a multi-task objective function. After adding the aforementioned MaskIoU regression module, the objective L of each RoI is formulated as follows:

$$L = \lambda_1 L_{cls} + \lambda_2 L_{box} + \lambda_3 L_{mask} + \lambda_4 L_{iou} \tag{3}$$

where L_{cls}, L_{box}, L_{mask}, L_{iou} denote classification loss, bounding box regression loss, mask and MaskIoU regression loss respectively, and λ_1–λ_4 are the weights of each part.

The classification loss is defined as follows:

$$L_{cls}(p, c) = -\log p_c \tag{4}$$

where p is the probability distribution of K + 1 categories, which is output by the softmax layer of the network, and c is the corresponding label.

For bounding box regression, the regression target and the predicted value are denoted as $t = \left(t_x, t_y, t_w, t_h\right)$ and $v = \left(v_x, v_y, v_w, v_h\right)$ respectively. The bounding box regression loss is defined as the smooth L1 distance of t and v:

$$L_{box}(t, v) = \sum_{i \in \{x,y,w,h\}} smooth_{L_1}(t_i - v_i) \tag{5}$$

smooth L1 distance is defined as follows:

$$smooth_{L_1}(x) = \begin{cases} 0.5x^2, \; |x| < 1 \\ |x| - 0.5, \; |x| \geq 1 \end{cases} \tag{6}$$

The mask branch outputs K binary masks for each input RoI, $\{M_1, M_2, \ldots M_K\}$, each with a resolution of $m \times m$. Assuming that the true category associated with the input RoI is c, then only the cth output mask M_c contributes to the mask loss. The method of calculating the mask is as follows: M_c first goes through a sigmoid activation function, mapping the value of pixels to the range of 0 ~ 1, and the mask loss is obtained by calculating the average binary cross entropy with the true mask M^*.

For MaskIoU regression loss, we adopt L2 distance:

$$L_{iou} = (t_{iou} - s_{iou})^2 \tag{7}$$

where t_{iou} is the true IoU between M^* and M_c, s_{iou} is the IoU predicted by the MaskIoU regression module. Note that the MaskIoU regression module can be embedded in the standard Mask R-CNN framework during training, so that the entire network can be trained end-to-end.

3 Implementation Details

This section introduces the implementation details of the proposed respirator oil level reading system. The improved Mask R-CNN discussed above is used to segment the oil seal cup area and the oil area in the image. Next, we need to read the oil level with the obtained segmentation masks.

3.1 Feature Extraction Network

In the field of computer vision, many deep learning-based algorithms use Residual Network (ResNet) [13] as the basic feature extractor. The emergence of ResNet solves the problem that the accuracy decreases when the number of stacked layers of the deep convolutional network increases to a certain degree. With the residual-learning mechanism, a deeper network obtains stronger feature expression ability. However, a large model may not converge well trained on such small datasets. In this paper, the relatively small ResNet-34 is used as the backbone network, which also saves a lot of computation.

Feature Pyramid Network (FPN) [14] is a network structure that can fuse features of different levels of the backbone network. It fuses high- and low-dimensional features top-down layer by layer, and output the fused features sideways. FPN can further improve the performance of ResNets. In the subsequent experiments, the feature extractor of the model is ResNet-34-FPN by default.

3.2 Post Processing

Due to the problems of complicated background and color distortion in real images, there may be segmentation noises in the object mask directly generated by the model, which has a negative effect on the subsequent procedure. This paper uses traditional algorithms to perform post-processing on segmentation results. The post-processing mainly consists of two steps: the first one is to perform the morphological opening operation of the segmentation mask [15] to remove the noise in the segmentation result while keeping the overall mask basically unchanged; the second one is to use the flooding algorithm to fill the holes. The above-mentioned post-processing can effectively improve the quality of the segmentation mask. Finally, the relative value of the oil level can be easily obtained by calculating the pixel number ratio of the oil and the oil seal cup mask.

4 Experiment

In this section, experiments are designed to verify the effectiveness of the proposed framework. Our dataset contains 2130 images collected from dozens of substations in Zhejiang Province, which is randomly split to 1278 training images, 426 verification images and 426 testing images. There are two kinds of targets to be detected, the respirator oil seal cup and the oil. The annotations of these two categories of targets include bounding box label, category label and mask label. Models are implemented in PyTorch, trained on 4 Titan X GPUs and tested on a single GPU.

4.1 Training Configuration

As a common practice, we use a backbone network which has been pre-trained on ImageNet. To address the problem of data scarcity, we further adopt the strategy of transfer learning: train the instance segmentation model on the MS COCO before finetune it on our own dataset. We train the model 20 K iterations by SGD optimizer, with a mini-batch size of 16, a momentum of 0.9 and a weight decay of 0.0005. The initial learning rate is 0.02, multiplying 0.1 at iteration 12 and 18 K. We use common data augmentations such as color jittering and geometry transforms to prevent over-fitting. The input image is scaled with the principle that the short side is not less than 800, and the long side is not more than 1333.

4.2 Results

The average precision (AP) with different IoU (0.5 and 0.75) thresholds is adopted as the evaluation metric, and we use subscript b and m to distinguish bounding box and mask. In order to verify the effectiveness of the modified Mask R-CNN, we provide some comparative experiments. The baseline model is Faster R-CNN with ResNet-34-FPN as the backbone network, and the mask branch and the MaskIoU regression branch are added in sequence to the baseline model. The evaluation results on testing set are shown in Table 1. It indicates that the mask branch additionally predicts the segmentation mask while improves the accuracy of bounding box detection. Moreover, our novel MaskIoU regression branch can further improve the accuracy of

Table 1 Results of the comparison experiment

Model	AP_b 50	AP_b 75	AP_m 50	AP_m 75
Baseline	0.838	0.576	–	–
+Mask	0.875	0.603	0.816	0.542
+MaskIoU	0.891	0.606	0.824	0.549

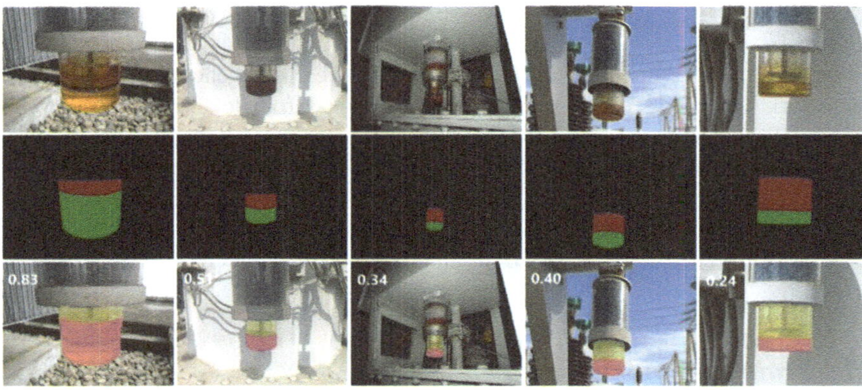

Fig. 3 Results on the testing set. top row: original images, middle row: ground truths, bottom row: output of the proposed framework, the values on top left denote the predicted oil level

mask prediction. We think this performance gain may come from multi-task training. Figure 3 shows the prediction results of some testing images.

5 Conclusion

This paper proposes an automatic respirator oil level reading framework based on instance segmentation and introduce a novel MaskIoU regression module to the standard Mask R-CNN model. The experimental results show that the modified Mask R-CNN improves the accuracy of segmentation with a large margin and can be adopted in practice. In the future research, the accuracy and efficiency of the proposed framework can be further optimized, thus can be applied to other substation equipment status recognition scenarios.

Acknowledgements This paper is supported by State Grid Zhejiang Electric Power Company Science and Technology Project (5211DS18003D).

References

1. Sun, Z., and J. Zhang. 2013. Analysis of transformer oil filling equipment state and configuration of respirator. *Anhui Electric Power* 30 (3): 1–3+17. (in Chinese).
2. Jiang, X., L. Xie, and G. Liu. 2017. Analysis and countermeasures for frequent oil deficiency in oil seal cup of transformer breather. *Guangdong Electric Power* 30 (10): 136–140 (in Chinese).
3. Liang, Y. 2016. Research on replacement and operation maintenance of main transformer breathing apparatus. *China High-Tech Enterprises* 2016 (26): 134–135 (in Chinese).
4. Ruan, S., K. Dong, and Y. Qiu. 2018. Development and application of new breathing apparatus for oil-immersed transformer. *Telecom Power Technology* 35 (08): 47–49 (in Chinese).
5. Wei, Y., L. Tong, and L. Luo. 2019. An exterior defects detecting method of main transformer based on convolutional neural networks. *Zhejiang Electric Power* 38 (04): 61–68 (in Chinese).
6. Shao, J., Y. Yan, and D. Qi. 2016. The substation switch detection and state recognition based on hough forests. *Automation of Electric Power Systems* 40 (11): 115–120 (in Chinese).
7. Yang, S., W. Xiong, S. Zhang, et al. 2019. Research on image detection of power transmission and transformation equipment based on adaptive threshold canny algorithm. *Electronic Design Engineering* 27 (15): 31–36 (in Chinese).
8. Xin, M., X. Xiao, B. Liu, et al. 2018. Automatic detection of transformer respirator based on image processing. *Mathematics in Practice and Theory* 48 (2): 127–133 (in Chinese).
9. Li, R., D. Xu, D. Zhai, et al. 2020. Intelligent defect identification method of transformer dehumidifier based on HSV transformation and object detection. https://doi.org/10.13336/j.1003-6520.hve.20200302010. Last accessed 20 June 2020. (in Chinese).
10. Liu, W., et al. 2016. SSD: Single Shot MultiBox Detector. In *ECCV 2016*, vol. 9905, ed. B. Leibe, J. Matas, N. Sebe, and M. Welling. Lecture Notes in Computer Science, 21–37. Cham: Springer.
11. He, K., G. Gkioxari, P. Dollár, and R. Girshick. 2017. Mask R-CNN. In *International Conference on Computer Vision*, 2980–2988. IEEE, Venice.
12. Ren, S., K. He, R. Girshick, et al. 2017. Faster R-CNN: towards real-time object detection with region proposal networks. *IEEE Transactions on Pattern Analysis and Machine Intelligence* 39 (6): 1137–1149.
13. He, K., X. Zhang, S. Ren, et al. Deep residual learning for image recognition. In *IEEE Conference on Computer Vision and Pattern Recognition*, 770–778. IEEE, Las Vegas.
14. Lin, T.Y., P. Dollár, R. Girshick, et al. Feature pyramid networks for object detection. In *IEEE Conference on Computer Vision and Pattern Recognition*, 936–944. IEEE, Honolulu.
15. Yin, X., and J. Wang. 2008. Research on color image morphology and its application. *Computer Engineering* 2008 (17): 271–273 (in Chinese).

Research on the Topology of Medium Voltage DC Hybrid Current Limiter

Jianying Zhong, Wenkui Liu, Xiao Li, Longlong Wang, Sumin Pang, and Peng Zhao

Abstract With the rapid development of DC technology, the low resistance and strong instantaneous impact of DC fault caused great pressure to the power grid's construction and operation. The suppression and buffering of DC fault energy has positive significance to the safe and economic operation of DC grid. Taking advantage of thyristor's advantages in voltage and current withstanding level, cost, interruption ability and reliability, and the fast interruption ability of high-speed switch, a medium voltage DC hybrid current limiter topology is proposed in this paper. The current limiting principle of hybrid current limiter is analyzed, and the ability of current limiting combined with the model analysis results of the current limiter's key components is simulated and verified. Furthermore, the key influencing factors of the current limiting characteristics are analyzed. Finally, according to the characteristics of DC system, an integrated topology scheme of hybrid DC current limiting and interruption is proposed. The results show that the proposed topology is feasible and effective in field of current limiting capability and cost application.

Keywords DC power grid · Hybrid current limiter · Thyristor

1 Introduction

It will be the rapid development stage of DC grid technology and the initial stage of DC grid construction in next 10–20 years. In order to alleviate the impact on the equipment in the system when the short-circuit fault occurs in the DC system, effectively limit the rising rate and amplitude of the fault current, and greatly reduce the amplitude of the breaking fault current of the DC circuit breaker, it is extremely important to carry out research work on DC current limiting technology [1–3].

J. Zhong · W. Liu · X. Li (✉) · S. Pang
State Grid Pinggao Group Co., LTD., Pingdingshan 467001, China
e-mail: 18742509178@163.com

L. Wang · P. Zhao
State Key Laboratory of Electrical Insulation and Power Equipment, Xi'an Jiaotong University, Xi'an 710049, China

W. Chen et al. (eds.), *The Proceedings of the 9th Frontier Academic Forum of Electrical Engineering*, Lecture Notes in Electrical Engineering 743,
https://doi.org/10.1007/978-981-33-6609-1_5

The ideal current limiter should have the following characteristics: (1) Low-resistance state during normal operation and high-resistance state when fault occurs. (2) It can quickly limit the first peak value of fault current after fault occurs. (3) Low active or reactive loss under rated conditions. (4) It has self-resetting function. (5) The smaller volume and the lower cost. With the continuous improvement of the parameter level of the switch device, the design and topology optimization of the solid-state current limiter and the hybrid current limiter have become the hotspots of current research [4–7]. The most remarkable feature of the solid-state current limiter is the introduction of power electronic devices including IGBT, IGCT, GTO and thyristor on the basis of keeping the traditional inductor and capacitor.

electronic devices including IGBT, IGCT, GTO and thyristor on the basis of keeping the traditional inductor and capacitor. Compared with the solid-state current limiter, the hybrid current limiter adds a main circuit high-speed switch. Contact resistance is only 10^{-6}, which is much smaller than the on-state resistance of the solid-state current limiter, so it is considered to be one of the most likely achieving current limiting devices.

In this paper, taking advantage of thyristor's advantages in voltage and current withstanding level, cost, interruption ability and reliability, a medium voltage DC hybrid current limiter topology is proposed, which combines the ability to fast interruption of high-speed switch. Furthermore, an integrated scheme of hybrid current limiter and circuit breaker is proposed, and its current limiting and interrupting principle is studied [8–10].

2 The Topology Principle of Hybrid Current Limiter

The topology of the DC hybrid current limiter is shown in Fig. 1, where FS is a highspeed switch, C is a pre-charge transfer capacitor, VT is a thyristor, D is diode, and R is a current limiting resistor. The current limiting process is as follows: The high-speed switch FS is in a closed state and the capacitor C pre-charge under normal work conditions. In the case of a short circuit fault, the thyristor is triggered on. At this time, the reverse discharge of capacitor makes the main circuit current drop to

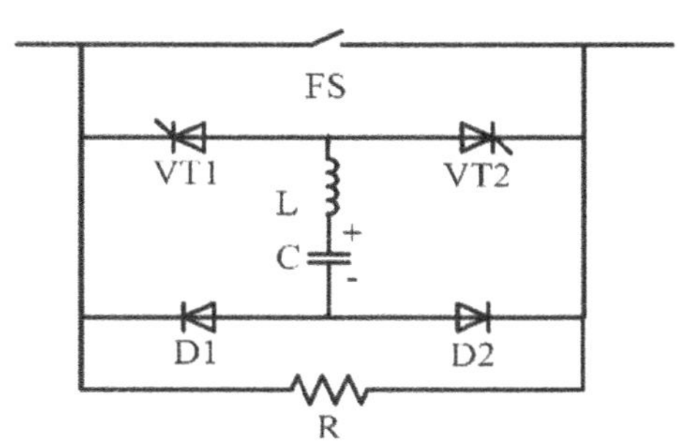

Fig. 1 Hybrid current limiter topology

zero. After that, the high-speed switch FS quickly opens and forms a break, and the voltage at both ends of the capacitor is positive, which makes the fault current transfer to the current limiting resistance branch to complete the current limiting.

3 Research on Hybrid Current Limiter Model

3.1 Research on Component Model of Hybrid Current Limiter Topology

The core components include arc generator, high-speed switch, thyristor and so on for the proposed topology of the DC hybrid current limiter in this paper, which are the key factors affecting the current transfer and over-voltage characteristics during the current limiting process. And it is necessary to exactly describe the characteristics of device in the simulation model to improve the accuracy of the simulation results [11, 12].

The High-Speed Switch Model. The high-speed switch model is very important to accurately describe the relationship between arc voltage and current of switch and evaluate the time and speed of current transfer. The parameters of the switch model mainly include the impedance at closing, the volt ampere characteristics of the arc in the arcing stage and the parameters of the external trigger module. Among them, the volt ampere characteristic of the arc is closely related to the non-linear resistance of the arc, which needs to be determined by the parameters of interrupting current, contact structure, opening speed and insulation medium of the switch. Based on the above analysis, a high-speed switch model is established as shown in Fig. 2.

The Arc Generator Model. Arc generator, whose parameters such as striking arc current, arc volt ampere characteristics, post arc conduction recovery time are very im-portant to accurately simulate fault current transfer characteristics under different working conditions, is the key device of fault current transfer process. Figure 3 shows the arc generator model established in the simulation calculation. There are three parts: (1) The pre arc module. It is used to simulate the state that the current of the

Fig. 2 High-speed switch simulation model

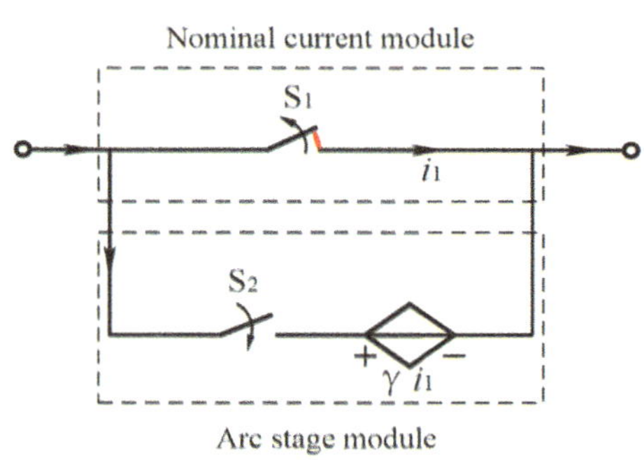

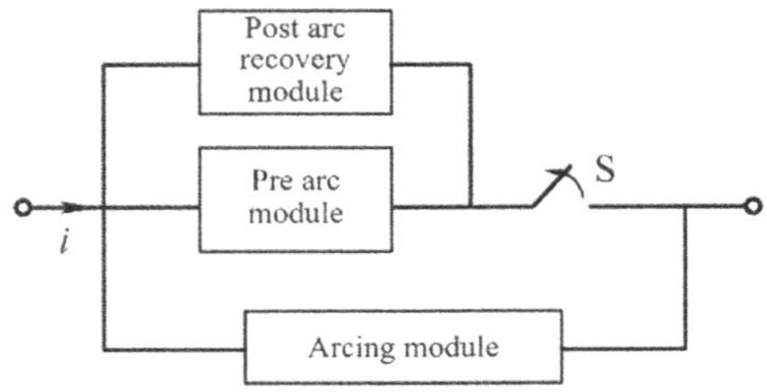

Fig. 3 Simulation model of arc generator

arc generator is less than the critical striking arc current; (2) The arcing module. When the current exceeds the critical striking arc current, the current is transferred from the pre arc module to the controlled voltage source, the arc voltage at both ends of the arc generator is simulated by the voltage of the controlled voltage source. (3) The post arc recovery module. It is used to simulate the post arc recovery conduction stage.

The Thyristor Model. High-power thyristor makes the over-voltage characteristics of hybrid current limiter during the current limiting process closely related to the current cut-off characteristics of semiconductor devices. The research results show that the residual charge in the base area does not disappear immediately with the current passing zero after the thyristor flows through a large instantaneous pulse current, and it gradually decreases through migration, recombination and diffusion until it is completely cut off. Therefore, the establishment of a suitable mathematical thyristor model to describe the cut-off characteristics after the current zero crossing can provide a theoretical basis for the evaluation of the over-voltage and the formulation of the control sequence during the current limiting process. The thyristor model is established and shown in Fig. 4, and the model consists of forward current module and reverse recovery module.

Figure 5 shows the comparison between simulation and experimental results of thyristor cut-off characteristics under the condition of current peak value of 1.1 kA and half wave period of 6 ms. It can be seen that the current characteristics of the thyristor model are in good agreement with the experimental results under the same circuit parameters after current zero crossing. And the error between the current

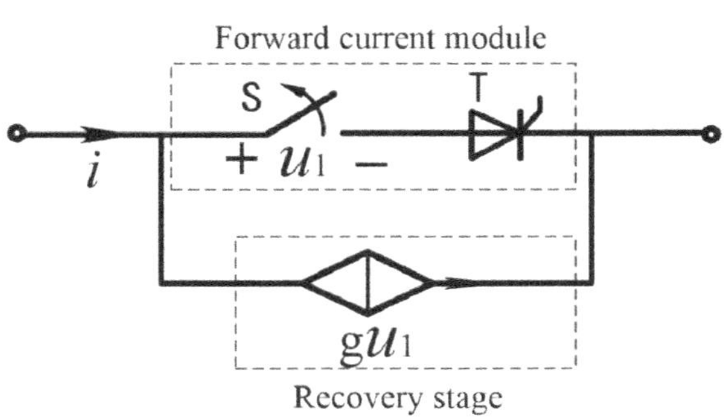

Fig. 4 Thyristor simulation model

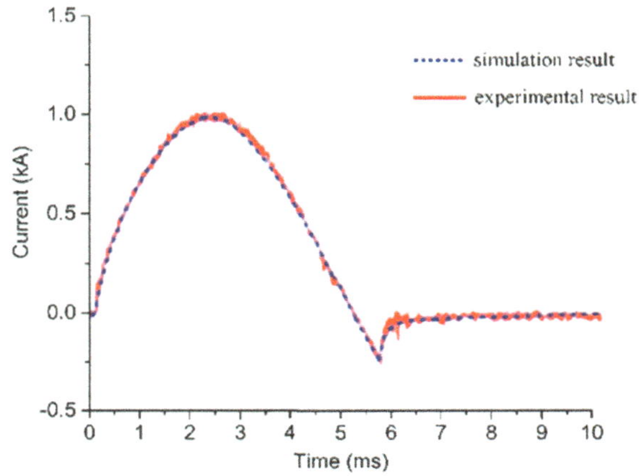

Fig. 5 Thyristor current simulation

peak value and time of reverse cut-off by simulation and the cut-off process in the experiment is very small.

3.2 *Simulation of Hybrid Current Limiter Topology*

According to the established components model of hybrid current limiter, the current limiting characteristics of hybrid current limiter is simulated. The simulation parameters are listed in Table 1. The DC system with rated voltage of 200 kV, is taken as the research object, the simulation results of current and voltage in current limiting process are shown in Fig. 6 according to the simulation parameters in Table 1. It can be seen that the system current is borne by the high-speed switch under normal

Table 1 Simulation parameters of hybrid current limiter topology

Style	Parameter	Value	Remark
Topology component parameter	L	30 μH	
	C	100 μF	Pre-charge voltage 10 kV
	R	10 Ω	
System parameter	Line resistance	2 Ω	
	Line inductance	8 mH	Fault current rise rate: di/dt = 17 A/us
	Load	50 Ω	
Control time sequence parameter	Fault time t_0	1 ms	
	Switch on time t_1	1.16 ms	Transfer current threshold I_0 = 11 kA
	Transfer branch on time t_2	1.3 ms	

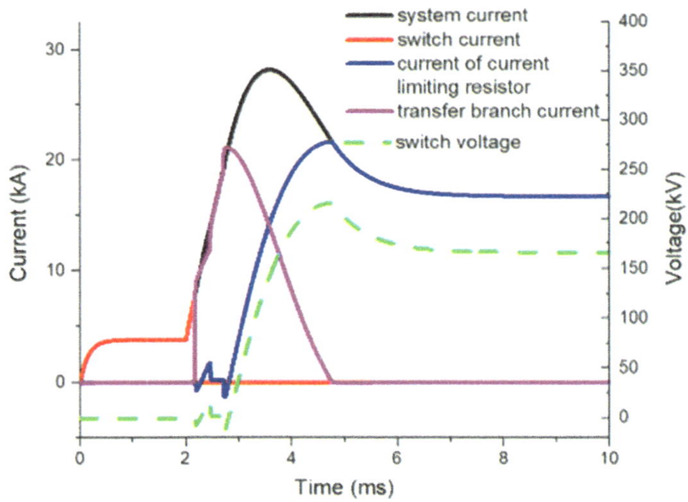

Fig. 6 Simulation of current and voltage in current limiting process

working conditions. The high-speed switch opens when the fault current is detected, the current transfers to the transfer branch. The current is finally fully loaded on the current limiting resistance branch through the transfer of the transfer branch, and then the current limiting action is completed, which is consistent with the theoretical expectation. The simulation result shows that the system current is limited to less than 20 kA for about 4 ms after the short-circuit fault of the current limiter, and the voltage peak value at both ends of the high-speed switch is less than 200 kV.

4 Influencing Factors of Current Limiting Characteristics

The effects of current limiting resistance value, fault current rise rate, capacitance value, and system voltage on the current limiting process of the hybrid current limiter are studied.

4.1 The Effect of Current Limiting Resistance R on the Current Limiting

The effect of the current limiting resistor R on the current limiting characteristics is shown in Fig. 7. It can be seen that the current limiting time is basically the same under different current limiting resistance R, and the larger R is, the stronger the

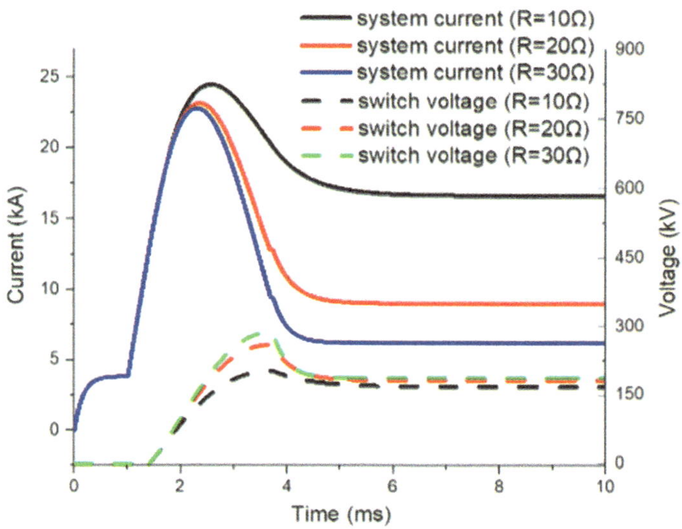

Fig. 7 Effect of current limiting resistance R on current limiting

limiting ability of fault current is. At the same time, the increasing of R can lead to the increasing of transient peak voltage at both ends of the switch.

4.2 *The Effect of Rising Rate of Fault Current on the Current Limiting*

Figure 8 shows the effects of fault current rising rate di/dt on current limiting characteristics. It can be seen that the rising rate of fault current di/dt affects the current limiting time of the switch. The peak value of system fault current and voltage at both ends of the switch increases with the increase of di/dt. However, the current limiting process is fast and the current limiting time is shorter.

4.3 *The Effect of Capacitance C on the Current Limiting*

The effect of the transfer capacitor on the current limiting characteristics are illustrated in Fig. 9. The larger C is, the higher the peak value of fault current is, and the longer the current limiting process takes. At the same time, the voltage rising rate at both ends of the high-speed switch decreases with the increase of C, and the voltage withstand requirement of the switch decreases. Therefore, on the premise of meeting the withstand voltage requirements of high-speed switch, the value of capacitor C

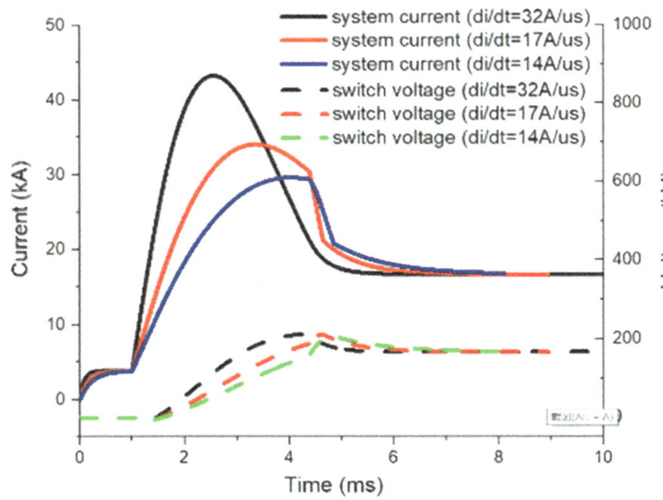

Fig. 8 Effect of fault current rise rate on current limiting

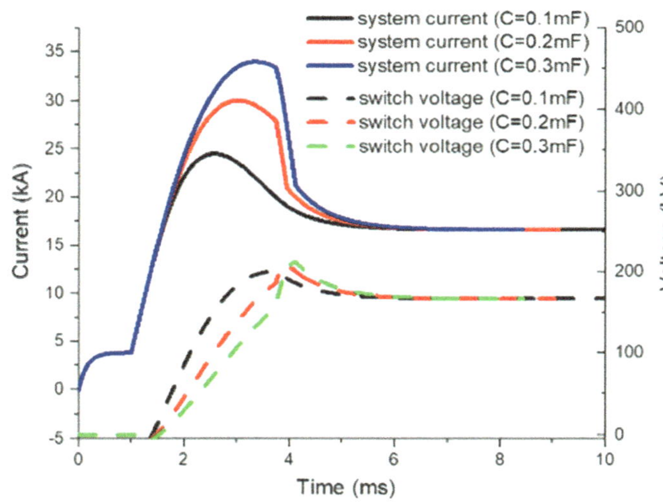

Fig. 9 Effect of transfer capacitance on current limiting

should be reduced as far as possible to decrease the peak value of fault current and shorten the current limiting time.

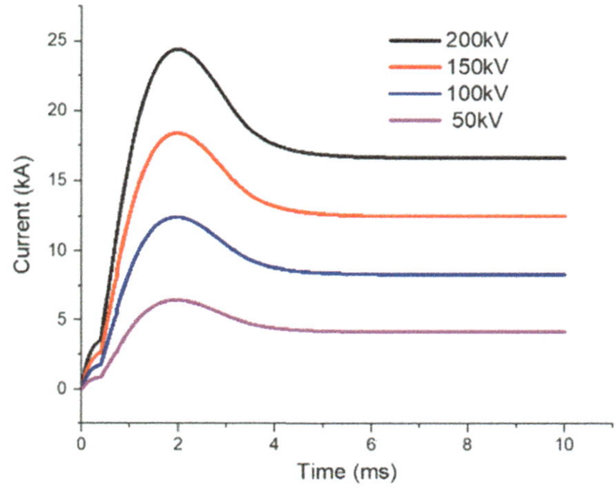

Fig. 10 Effect of system voltage on current limiting process

4.4 The Effect of System Voltage on the Current Limiting

The schematic diagram of system voltage and current limiting characteristics in case of system fault is shown in Fig. 10. It can be seen that the peak value of fault current increases with the increase of system voltage. When the system voltage is 200 kV, the minimum fault current is limited to about 25 kA. Also, fault current and the current limiting response time increases gradually with the increase of system voltage.

5 Integrated Scheme of Hybrid Current Limiter and Circuit Breaker

Although the common hybrid current limiter topology has the characteristics of fast current limiting speed and simple structure, it still needs the cooperation of DC circuit breaker to complete fault current interruption. A new hybrid topology which combines current limiting and interruption is proposed, as shown in Fig. 11. T_1, T_2 and T_3 are thyristors, which are used to control the fault current transfer to the

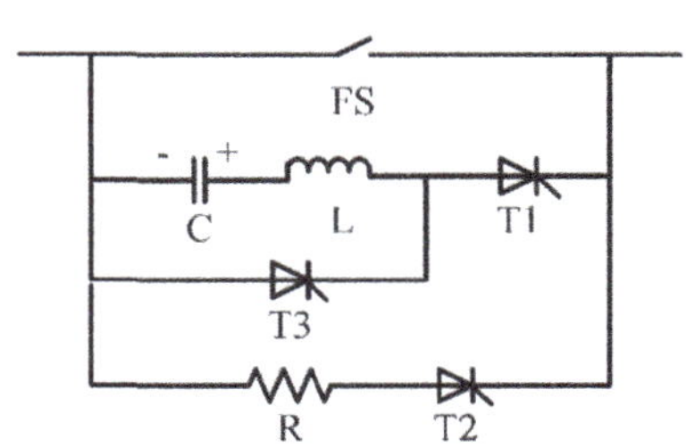

Fig. 11 New hybrid topology combining current limiting and interruption

capacitor in two stages to complete the current limiting and interrupting process [13, 14].

Under normal circumstances, the system current is borne by high-speed switch. Firstly, the T_1 in transfer circuit is controlled to be on after the short-circuit fault occurs, and current zero crossing point is generated in high speed switch by using reverse oscillating current. After the capacitor voltage is reversed, T_2 in current limiting branch is turned on, the fault current is transferred from T_1 to T_2, and the capacitor is charged in reverse. T_3 is controlled to be on after T_1 is cut off, and LC oscillation is used to reverse the capacitor voltage again. After T_3 is cut off, the control thyristor T_1 is turned on again, and the current of the current limiting branch T_2 is quickly transferred to the capacitor. After that, T_2 is cut off and the voltage at both ends of FS increases as the system current charges the capacitor. Finally, the system current drops to zero and the interrupting process is completed.

The main feature of this topology is that after the current is transferred to the current limiting resistor, the fault current is limited to a lower level, so the pre charging capacitor in the LC reverse oscillation circuit can be used to realize the fault current transfer again, and complete the interrupting process. Therefore, this topology can complete the current interruption after realizing the system fault current limiting. It is a new topology combining the current limiting and interrupting functions.

6 Conclusion

In this paper, a medium voltage DC hybrid current limiter topology is proposed. The topology principle of current limiter is introduced, the models including high-speed switch model, arc model and thyristor model, are studied. And the correctness of the theoretical expectation is verified by the current limiter topology simulation results. The influences of current limiting resistance value, fault current rising rate, capacitance value and system voltage on current limiting process are analyzed. At last, an integrated scheme of hybrid current limiter and circuit breaker is put forward, and its principle of current limiting and interruption is introduced.

References

1. Falcome, C.A., J.E. Bechlerm, W.E. Mekolites, et al. 1974. Current limiting device—a utility's need [C]. *IEEE Power Engineering* 93 (6): 1768–1775.
2. Xin, Y. 2015. Review on superconducting fault current limiters [J]. *Southern Power System Technology* 9 (3): 1–9 (in Chinese).
3. Song, Q., B. Zhao, W.H. Liu, and R. Zeng. 2013. An overview of research on smart DC distribution power network. *Proceedings of the CSEE* 33 (25): 9–19 (in Chinese).
4. Tang, G.F., X. Luo, and X.G. Wei. 2013. Multi-terminal HVDC and DC-grid technology. *Proceedings of the CSEE* 33 (10): 8–17 (in Chinese).

5. Xue, Y.L., Z. Xu, and Q.R. Tu. 2012. Modulation and control for a new hybrid cascaded multilevel converter with DC blocking capability. *IEEE Transactions on Power Delivery* 27 (4): 2227–2237.
6. Chang, B., O. Cwikowshi, M. Barnes, et al. 2014. Point-to-point two-level converter system fault analysis [C]. In *7th IET International Conference on Power Electronics, Machines and Drives.*
7. Yang, J., J.E. Fletcher, and J. O'Reilly. 2012. Short-circuit and ground fault analyses and location in VSC-based DC network cables. *IEEE Transactions on Industrial Electronics* 59 (10): 3827–3837.
8. Li, B., and J.W. He. 2016. Research on the DC fault isolating technique in multi-terminal DC system. *Proceedings of the CSEE* 36 (1): 87–95 (in Chinese).
9. Wu, J., L.Z. Yao, Z.B. Wang, et al. 2015. The study of MMC topologies and their DC fault current blocking capacities in DC grid. *Proceedings of the CSEE* 35 (11): 2671–2694 (in Chinese).
10. Zhang, J.P., C.Y. Zhao, H.F. Sun, et al. 2014. Improved topology of modular multilevel converter and application. *Transactions of China Electrotechnical Society* 29 (8): 173–179 (in Chinese).
11. Debnath, S., J. Qin, B. Bahrani, et al. 2015. Operation, control, and applications of the modular multilevel converter: a review. *IEEE Transactions on Power Electronics* 30 (1): 37–53.
12. Nami, A., L.W. Wang, F. Dijkhuizen, 2013. Five level cross connected cell for cascaded converters C]. In *2013 15th European Conference on Power Electronics and Applications.*
13. Li, X.Q., W.H. Liu, Q. Song, et al. 2013. An enhanced MMC topology with DC fault ride-through capability [C]. In *IECON 2013 39th Annual Conference of the IEEE Industrial Electronics Society.*
14. Jiang, D.Z., C. Zhang, H. Zheng, et al. 2014. A scheme of current limiting hybrid DC circuit breaker. *Automation of Electric Power System* 38 (4): 65–71 (in Chinese).

Wind Farm Short-Term Power Prediction Based on Multiple Intelligent Algorithms

Xuan Liu, Lu Zheng, Ke Su, Jun Hu, Jianli Zhao, and Yaodong Hu

Abstract This paper illustrates several intelligent algorithms to build a short-term wind power forecasting model using data of a particular Wind Farm in Inner Mongolia. Advantages and backwards of the four different neural network models have been carefully discussed. Calculation methods and formulas are provided to prove the result. A combination prediction method is proposed in order to make a more accurate distribution power prediction by optimizing the information of multiple single models. Therefore, after analyzing the actual wind power data, the most accurate forecasting model is selected to provide an effective reference for power dispatching, operation and equipment maintenance. In order to integrate theory with practice, a wind farm in Inner Mongolia is chosen to make the short-term power prediction using the different intelligent method discussed in this paper. The most accurate prediction algorithm has been proved by real-time data.

Keywords Wind farm power prediction · BP neural network · Multiple intelligent algorithms

1 Introduction

Short-term wind power prediction is a primary requirement for efficient large-scale integration of wind generation in power systems and electricity markets. The choice of an appropriate prediction model among the numerous available models is not trivial, and has to be based on an objective evaluation of model performance [1]. There are three steps in wind power forecasting: firstly determining wind speed from a model; then calculating the wind power output forecast or prediction; and finally regional forecasting or upscaling or downscaling [2]. Wind speed series usually

X. Liu (✉) · L. Zheng · K. Su · J. Zhao · Y. Hu
Inner Mongolia Electric Power Science & Research Institute, Hohhot, Inner Mongolia, China
e-mail: blesslucine@163.com

J. Hu
Electrical Engineering and Automation Department, Tsinghua University, Haidian District, Beijing, China

W. Chen et al. (eds.), *The Proceedings of the 9th Frontier Academic Forum of Electrical Engineering*, Lecture Notes in Electrical Engineering 743,
https://doi.org/10.1007/978-981-33-6609-1_6

have complex features, such as non-linearity and volatility, which makes the wind energy forecasting highly difficult [3]. Wind power prediction is significant for power resource dispatching, grid operation and equipment maintenance [4]. In this paper, data of power generation in Mingyang Wind Farm is selected as the verification object and the short-term prediction models are established to compared and verified.

Short-term wind power forecasting is based on daily (24 h) units. Since there is a large amount of data provided, the BP neural network has the advantage ability of learning and predicting large-scale parameters. Therefore, the BP neural network is selected as the basic prediction model. However, the existing training algorithms of neural networks have certain limitations in parameter optimization problem. For a single prediction model, large error may appear at some points; optimized models by algorithms reduce the occurrence of large error, and improve the prediction accuracy to a certain degree [5]. In this paper, the conventional BP neural network model is optimized by four algorithms, includes Crisscross Optimization (CSO) algorithm, Particle Swarm Optimization (PSO) algorithm, Whale Optimization Algorithm (WOA) and Entropy algorithm. Compared the prediction results and accuracy of BP neural network model with other integration models, in order to further improve the accuracy of the prediction model, an intelligent fusion wind power short-term prediction model based on the entropy method has been constructed. Finally, according to analysis, the effectiveness of the intelligent fusion prediction model is verified.

2 Prediction Models with Multiple Intelligent Algorithms

2.1 BP Neural Network Model

Back-propagation (BP) neural network is essentially a multilayer feedforward network using a back propagation learning algorithm. Its network structure is shown in Fig. 1, u and y are the input and output vectors of the network. Each neuron is represented by a node. The network consists of nodes in input layer, hidden layer, and output layer. The hidden layer can be one layer or multiple layers, and the nodes from the front to the back are connected by weight. The learning of BP network is a kind of supervised learning. Its algorithms consist of forward propagation and back propagation.

Forward propagation means the input signal transmit from the input layer to the output layer through the hidden layer. If the output layer reaches the ideal output, the learning process is over; otherwise, it turns to back propagation. Back propagation is to substitute the error signal (the difference between the sample output and the network output) back to the original connection path. The gradient descent method is used to adjust the thresholds of the weights of the neurons in each layer to reduce the error signal. The method and calculation steps are as follows:

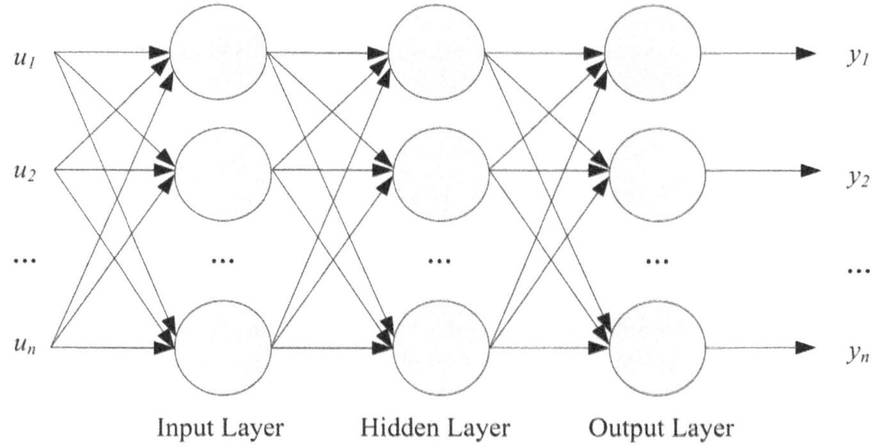

Fig. 1 Structure of BP neural network

Step 1. Set the initial weight of the neural network to a small random non-zero value;
Step 2. Given the input/output sample pairs, calculate the output of the network;
Step 3. Get objective function of network;
Step 4. Discriminant the objective function;
Step 5. According to the objective function, calculate the back propagation of error according to the gradient descent method, adjust the weights layer by layer, and then return to Step 2.

BP network is a kind of global approximation network. It represents the non-linear mapping from the input space to the output space through the composite mapping of several simple non-linear processing units, in order to reach the complex non-linear processing capabilities.

Although the BP algorithm can theoretically approximate any non-linear function and has a strong learning ability, it has two major limitations in the prediction of wind power output. The first limitation is the slow convergence speed. The weight parameter adjustment in the BP algorithm is carried out along the direction of the gradient of the error function. However, due to the serious morbidity of the network error function matrix, the direction of the minimum point of these gradient deviates is facing the error surface. The search path of the value parameter to the minimum point naturally increases the learning time of the BP algorithm, which causes the BP algorithm to converge slowly, easy to fall into local minima. In theory, its training is approximated downward along the slope of the error surface. For a complex network, there are many local minima points. In the process of network training, once caught in the local minima, it is difficult to escape the current algorithm. In fact, the result is usually the best choices within particular area. Therefore, the limitations of conventional BP algorithms have serious influences on the generalization ability of neural networks.

2.2 Optimized Neural Network Model by CSO

As an advanced artificial intelligence algorithm, crisscross optimization (CSO) algorithm has been proved the most efficient way in large-scale parameter optimization problems, which can effectively avoid the local optimization problem of conventional intelligent algorithms. CSO algorithm has the unique advantage of strong global convergence capabilities. This paper uses a crossbar algorithm to optimize the weights and thresholds of the neural network. The key part of the vertical and horizontal crossover algorithm consists of horizontal crossover and vertical crossover. The horizontal crossover updates the particles in a random pairing manner. The vertical crossover is similar to the mutation phase of the genetic algorithm, introducing a dimensional level mutation based on the crossover probability.

Crossover Operation. Before executing the crossover operator, perform random and random combination of all particles within the group, and then perform horizontal crossover on all dimensions of the combined particles in order to updating the particles. Assuming that the d dimension of the parent particles $X(i)$ and $X(j)$ intersects laterally, they produce the next generation according to the following equation.

$$MS_{hc}(i,d) = r_1 \times X(i,d) + (1 - r_1) \times X(j,d) + c_1 \times [X(i,d) - X(j,d)] \quad (1)$$

$$MS_{hc}(j,d) = r_2 \times X(j,d) + (1 - r_2) \times X(i,d) + c_2 \times [X(j,d) - X(i,d)] \quad (2)$$

$i, j \in N(1, M)$; $d \in N(1, D)$; r_1 and r_2 are random values distributed in the interval $[0, 1]$; c_1 and c_2 are random values distributed in the interval $[-1, 1]$; M denotes Particle size; D denotes variable dimension.

As can be seen from the above equation, the horizontal cross is mainly made up of two parts: the first part is a real valued arithmetic which crosses similar to the genetic algorithm; the second part adds an expansion coefficient.

Vertical Cross Operation. As can be seen from the above equation, the horizontal cross is mainly made up of two parts: the first part is a real-valued arithmetic which crosses similar to the genetic algorithm; the second part adds an expansion coefficient.

Many intelligent algorithms have local optimization problems in the later iterations. This precocious problem is often caused by the dimension of some particles falling into the local optimization during the entirety update process. Each dimension of the particles in the group represents a parameter to be optimized by the algorithm. Because the units of all parameters are not uniform, the value ranges of the parameters will also be different. Therefore, before performing crossover, synchronize all dimensions of the particles then perform vertical crossover between different dimensions of all particles according to the crossover probability P_v. The difference from the horizontal intersection is that only one sub-generation is generated for each vertical intersection, the purpose is to make the dimension which trapped in the local optimum jump out of the local optimum without destroying the normal dimension. When the d_1th dimension and the d_2th dimension of particle $X(i)$ intersect longitudinally, the offspring data are generated according to the following equation.

$$MS_{vc}(i, d_1) = r \cdot X(i, d_1) + (1 - r) \cdot X(i, d_2) \quad (3)$$

$i \in N(1, M); d_1, d_2 \in N(1, D); r \in [0, 1]$.

The equations above show the longitudinal intersection generates a mean solution through the intersection of the particle dimension level, which can get rid of a certain dimension trapped in a local optimum, thereby maintaining and increasing population diversity. Since not all dimensions are stagnant or locally optimal, a cross probability is a set for the vertical crossover operator to control the amount of crossover dimensions in the population and effectively ensure the convergence speed of the algorithm.

Optimized Neural Network Parameters by CSO. In the prediction process of BP neural network, the gradient descent method is used to adjust the weights and thresholds. It is easy to fall into a regional optimum result. When the number of influencing factors and learning samples increases, the amount of calculations of the neural network will increase sharply. It makes the convergence speed slower. The CSO algorithm has the characteristics of strong search ability, which provides the possibility to overcome the above disadvantages of neural networks.

Assume that the number of nodes in the input layer, the hidden layer and the output layer of the neural network is *n*, *h*, *m* respectively, and the variable dimension is $D = n \times h + m \times h + h + m$. By using the mean square error equation (fitness function) to measure the fitness value of each particle:

$$fit = \frac{1}{N} \sum_{k=1}^{N} \sum_{t=1}^{m} (p_t - \hat{p}_t)^2 \quad (4)$$

p_t denotes the actual output of the neural network, $\hat{p}_t$ denotes the target output of the neural network, N denotes amount of training samples.

The training process of optimized neural network model by CSO algorithm is shown as Fig. 2.

2.3 *Optimized Neural Network Model by PSO*

Particle swarm optimization (PSO) is an evolutionary computing technology proposed by Dr. Eberhart and Dr. Kennedy in 1995. Compared with other optimization algorithms, PSO has certain advantages in parameter selection and convergence

Fig. 2 Flowchart of optimized neural network model by CSO

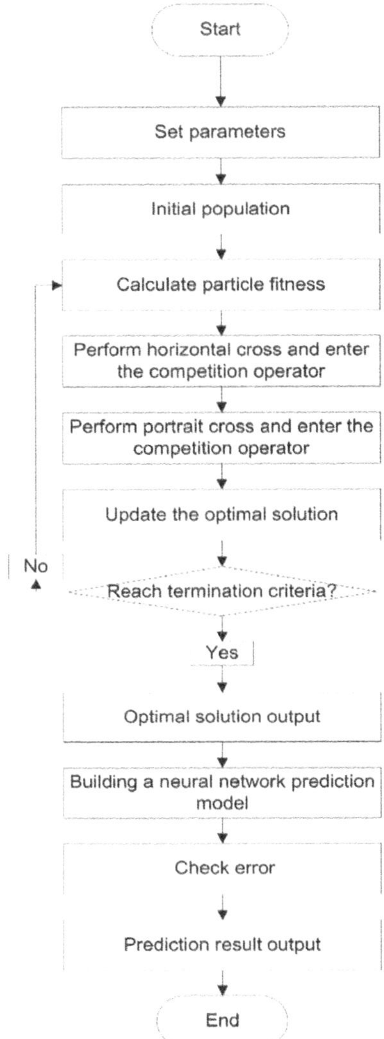

speed. The algorithm is derived from the study of bird predation behavior. The basic idea is to find the optimal solution between the information transfer and information sharing of the group individuals. The algorithm is as follows:

$$v_{id}^{k+1} = w \cdot v_{id}^{k} + c_1 \cdot r_1 \cdot \left(p_{id}^{k} - x_{id}^{k}\right) + c_2 \cdot r_2 \cdot \left(p_{gd}^{k} - x_{id}^{k}\right) \tag{5}$$

$$x_{id}^{k+1} = x_{id}^{k} + v_{id}^{k+1} \tag{6}$$

i denotes the ith particle: M denotes amount of particles in the group; D denotes the dth dimension of the particle, which is the dth parameter optimized by the algorithm; w denotes inertia weight factor, its value is non-negative, the value affects the overall optimization ability.

In order to avoid premature convergence and ensure its convergence to the entire optimum matrix, the weight w is generally linear between the maximum $W_{\max}$ and the minimum $W_{\min}$. In order to get the decreasing change, usually choose $W_{\max} = 0.9$ and $W_{min} = 0.4$; k denotes the algebra optimized at this time; v_{id}^{k} denotes the space velocity of particle i in the dth dimension at k time; c_1 and c_2 are acceleration factors, usually choose $c_1 = c_2 = 2$; r_1 and r_2 are random values distributed in the interval [0, 1];p_{id}^{k} denotes the dth dimension value of the historically optimal solution of particle i, which is the historically optimal solution of a single particle i in the optimized dth parameter.p_{gd}^{k} denotes the minimum value of p_{id}^{k}; x_{id}^{k} denotes the position of particle i at k time.

During each iteration, each particle needs to calculate its adaptive value according to the objective function, which can be mean square error, variance or standard deviation etc. Then, the optimal position of the current particle p_{id}^{k} and the optimal position of the group p_{gd}^{k} are determined according to the adaptive value, and the speed and position of each particle are adjusted according to Eqs. (5) and (6). The end condition is that the number of iterations reaches the set value K_{max} or the optimal position searched by the group meets a preset minimum adaptive value. This paper selects $K_{\max} = 300$. The process is shown in Fig. 3.

2.4 Optimized Neural Network Model by WOA

Whale optimization algorithm (WOA) is a heuristic optimization algorithm proposed by Mirjalili of Griffith University in Australia. It has the advantages such as easy to operate, using less adjustment parameters. The algorithm mainly simulates three steps of humpback whale surrounding preys, hunting preys and searching preys. Whale Optimization Algorithm assumes that the current best candidate solution is the location of the target prey or close to the target prey. When the best search parameter is defined, other search parameter will update their positions to the best search parameter. Its search parameter location equation is updated as follows:

$$D = |X(t) - CX_{best}(t)| \tag{7}$$

$$X(t+1) = X_{best}(t) - AD \tag{8}$$

t denotes current iteration value; A and C are coefficients; D denotes distance between the humpback whale and its prey.

This process simulates the spiral movement of a real humpback whale. In other words, the whale position $X(T_{\max})$ that can be updated is the maximum number of

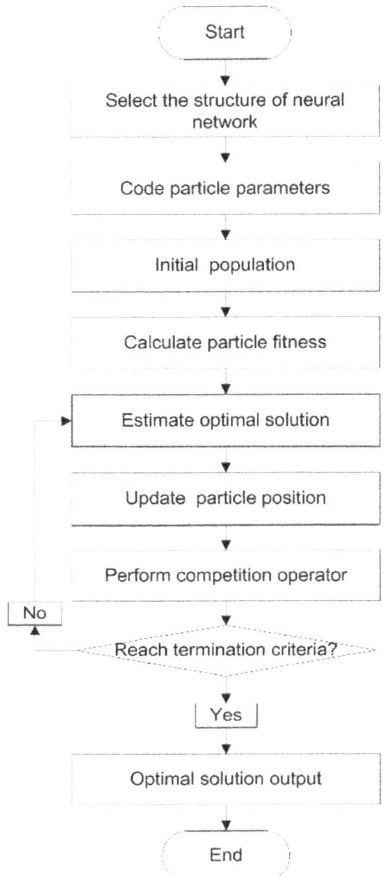

Fig. 3 Flowchart of optimized neural network model by PSO

iterations; r_1 and r_2 are random values distributed in the interval [0, 1], it decreases linearly in the iteration. During each iteration, the position of X_{best} is updated when there is a better solution.

Use the spiral equation between the humpback whale and its prey to update the position of the humpback whale. When the value of A is out of the interval $[-1, 1]$, the whale algorithm forces the new search value to substitute the original humpback whale parameters, the random choice parameter is referred as the new best choice by updating its position. So the whale algorithm has an all domain search capability.

2.5 Entropy Method

The combination prediction method was first proposed by Bates and Granger. The basic idea is to calculate the weighted average value of several prediction results to increase the accuracy of the prediction. The combination prediction method is based on using the maximum information. It optimized the information of multiple single models. Therefore, in most cases, the purpose of increasing the accuracy of prediction results can be achieved through the combination prediction. The basic process is as follows:

Suppose there are Nth prediction models, each model is based on its own information set, the information set of the Jth model at time is I_{jn}: (I_{on}, I_{jn}), where I_{on} is the information set shared by all models, and I_{jn} is the jth prediction. The information set is unique to the model, two of any information sets are assumed to be independent from each other. Therefore, all the information in the entire domain can be represented by the set U_n. If U_n is known, the prediction value y_n and its prediction value in next period y_{n+1} can be estimated by the method of least squares, and its expression is:

$$E_u = E\{y_{n+1}|U_n\} = a(B)z_n + \sum_j \beta_j(B)x_{jn} \tag{9}$$

Among them, $a(B)$ and $\beta_j(B)$ are lag operator polynomials, z_n is a specific expression of the shared information set, and x_{jn} is a specific expression of the information set unique to each model. In fact, we actually cannot get all the information according to the equation above.

It can be seen that the Jth prediction model is merely the application of the information set I_{jn}: (I_{on}, I_{jn}). Therefore, it is a natural idea to effectively combine known information sets to improve the accuracy of prediction. If the information set is known, the combined prediction can be expressed as:

$$E = E\left\{y_{n+1}\middle|I_{on}, I_{jn}, j = 1, 2, \ldots, k\right\} = \sum_j w_j(a(B)z_n + \sum_j \beta_j(B)x_{jn}) \tag{10}$$

w_j denotes the weight of the Jth prediction model.

The determination of weight is the key factor of combination prediction. At present, there are two methods of determining weights: optimal combination method and non-optimal combination method. The optimal combination method constructs the objective function according to a certain criterion and solves the objective function under certain constraints to obtain the weighting coefficient of the combination. The non-optimal combination method determines the weight according to the degree of variation caused by errors of individual prediction models. Larger the variation is, smaller its weight should be. In the same way, smaller the variation is, larger its weight should be. Although these two methods have been widely used, there are still some problems. For example, if the objective function solution is complicated and a strong constraint of non-negative weight is required, weight of the solution is not optimal. More importantly, two methods mentioned cannot reflect the basic idea of combined prediction, which is gathering all the existing information to achieve the maximum use of information. In addition, whether the weight coefficient should be normalized is still controversial. Therefore, the application of the entropy method is a good solution to determine weight of the combination prediction.

Entropy is a thermodynamic concept which refers to thermal energy that cannot be used for generate power. It was later introduced by Shannon into information theory. In information theory, entropy refers to the level of system uncertainty, the degree of disorder and the direction of information volume. On the contrary, larger the entropy value is, the less information remains in the system and the level of system uncertainty gets higher. The calculation equation for the entropy value is:

$$H = -k \sum_{i=1}^{n} p_i \ln p_i \tag{11}$$

H denotes entropy value of the system, p_i denotes probability of each state of the system, k is a constant.

According to the principle of entropy, the magnitude of the relative error entropy predicted by each single model in the combined prediction can indicates the model stability. The larger the relative error entropy is, the more stable can the prediction be. Thus, weight of the combination model should be larger. On the other hand, if the prediction relative error entropy is smaller, the amount of effective prediction information provided by this model is smaller and the variation level of this prediction error is larger. If this prediction model is more unstable, its weight in the combined model can be smaller. Overall, the weight of the combined prediction can be determined by the entropy method. The general steps are as follows:

Step 1. Calculate the proportion p_i of the relative prediction error of the jth prediction model at time t.

$$p_{it} = \frac{e_{jt}}{\sum_{t=1}^{n} e_{jt}} \tag{12}$$

e_{jt} denotes relative error of load prediction, $j = 1, 2, \ldots, k$; k denotes amount of prediction models.

Step 2. Calculate entropy value of the relative prediction error of the jth prediction model.

$$H_j = -\frac{1}{\ln n} \sum_{t=1}^{n} p_{jt} \ln p_{jt} \tag{13}$$

Step 3. Calculate the coefficient D_j of variation of the prediction relative error of the jth prediction model.

$$D_j = 1 - H_j \tag{14}$$

Step 4. Calculate the weight coefficient W_j of the jth prediction model.

$$W_j = \frac{1}{k-1}\left(1 - \frac{D_j}{\sum_{j=1}^{k} D_j}\right) \tag{15}$$

k denotes amount of prediction models.

3 Results and Analysis

For the evaluation criteria of prediction results, the average absolute percentage error ($MAPE$) is used to analyze the accuracy. The specific calculation method is:

$$MAPE = \frac{1}{n}\sum_{i=1}^{n}\left|\frac{(y_i - y_i')}{y_i}\right| \times 100\% \tag{16}$$

y_i denotes the actual power output; y_i' denotes predicted output power; n denotes amount of sequence data.

The statistical results of each model's average absolute error index $MAPE$ are shown Fig. 4, and the database is from Mingyang Wind Farm on Jan. 20, 2019.

Average values and average accuracy rates of the five forecasting models are shown in Table 1. Although the forecasting results of the BP neural network is better at some moments, but its prediction accuracy is difficult to be satisfied from the perspective of the entire results. The conventional BP neural network is a large-scale parameter optimization method; it relies on the gradient descent to adjust the weights and thresholds. Sometimes, the local optimal situations may exits and the optimal network parameters cannot be obtained. Therefore, the prediction model of the BP neural network has a relatively low accuracy. The PSO-BP and the WOA-BP model has different advantages in forecasting, in this case, even if both models share almost the same changing tendency, the WOA model predicts a slightly more accurate result then the PSO model. However, the WOA model still suffers from the problem of poor exploration and local optima stagnation [6]. Despite of the entropy model, the CSO algorithm has proved itself the most efficient way in large-scale parameter optimization problems. It can effectively avoid the local optimization problems of conventional intelligent algorithms comparing with the conventional BP, PSO and

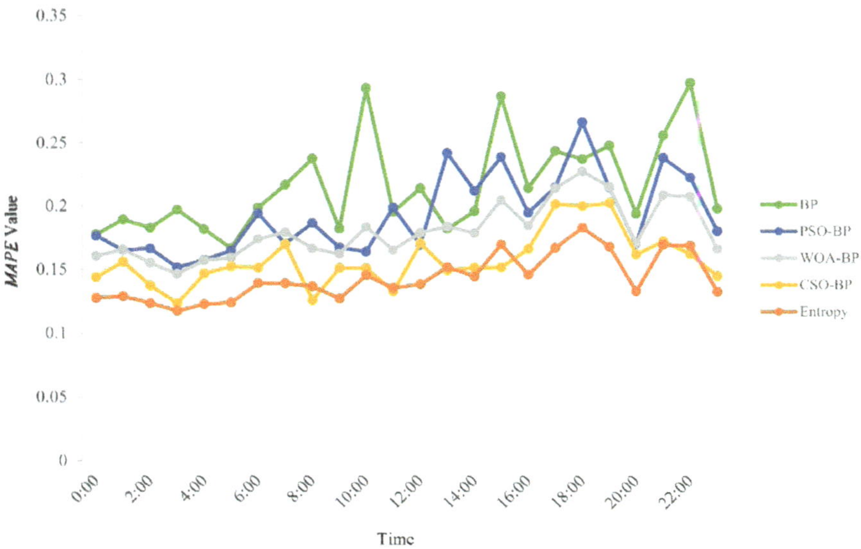

Fig. 4 *MAPE* values of each prediction model

Table 1 Accuracy rate of each prediction model

Index/model type	BP	PSO-BP	WOA-BP	CSO-BP	Entropy
Average value of *MAPE*	0.2155	0.1922	0.1792	0.1569	0.1429
Average accuracy rate	78.45%	80.78%	82.08%	84.31%	85.71%

WOA models. So, in order to meet the short-term wind power prediction accuracy requirements, the solution proposed in this paper is mainly calculated using the combined entropy method to optimize the prediction model. By analyzing maximum information and optimizing information of multiple single models, it acquires the most accurate rate and the steadiest changing tendency than the other prediction models.

4 Conclusion

According to the analysis result, the BP neural network model has the lowest prediction accuracy rate, which average accuracy rate is only 78.45%. The PSO-BP model and the WOA-BP prediction model are better than the BP neural network model, the rate of the WOA-BP prediction model is slightly higher than the PSO-BP model, respectively rated as 80.78% and 82.08%. In the short-term prediction, the predicted values of CSO-BP model are relatively close to the actual values at each prediction

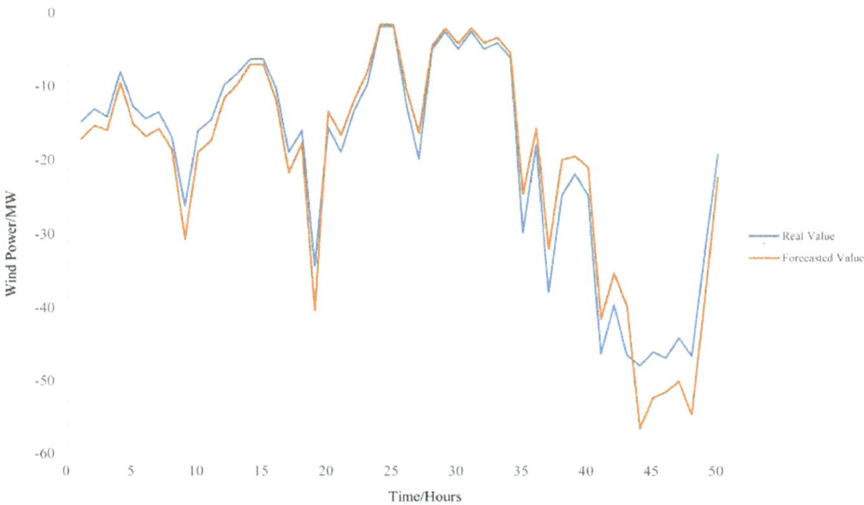

Fig. 5 Comparison of prediction results and actual wind power value

point with the average accuracy rate at 84.31%. Furthermore, it can be seen from Table 1, the forecasting result of the entropy method is the closest to the actual values with accuracy rate at 85.71%, which is the most accurate prediction method among all the other contrast algorithms. Besides, the short-term wind power prediction model can be applied in practice.

The wind power prediction results by entropy algorithm and actual wind power data of Mingyang Wind Farm are compared from 0:00 on February 1, 2019 to 1:00 on February 3, 2019. The comparison curves are shown in Fig. 5.

It can be seen from Fig. 5, testing time form February 1, 2019 to February 3, 2019 has been divided into 50 parts (hours), tendency curve of prediction result remains almost the same with the real testing value, which means the theory of the prediction model established by this paper can effectively predict the actual wind power. The probabilities for the occurrence of high prediction errors depending on the weather stability [7], when the real-time power changes rapidly, the prediction model can self-study the changing tendency and then change its prediction result in time. While the wind farm output power turns stable, the prediction result can be maintained under relatively high-precision results. Overall, the Wind Farm's wind power prediction accuracy calculated under the entropy model can reach 85.1%, ranking the highest among all the other prediction method. By predicting the wind power, the request for electricity can be cautiously controlled, and their precision has a direct effect on consistency and productivity [8]. This prediction accuracy rate meets the current renewable energy standard requirements of short-term wind power prediction in China.

References

1. Madsen, H., P., Pinson, G. Kariniotakis, H.A. Nielsen, T.S. 2005. Nielsen. Standardizing the performance evaluation of short-term wind power prediction models. *Wind engineering* 29 (6): 475–489.
2. Foley, A.M., P.G. Leahy, A. Marvuglia, and E.J. McKeogh. 2012. Current methods and advances in forecasting of wind power generation. *Renewable Energy* 37 (1): 1–8.
3. Qu, Z., W. Mao, K. Zhang, W. Zhang, and Z. Li. 2019. Multi-step wind speed forecasting based on a hybrid decomposition technique and an improved back-propagation neural network. *Renewable Energy* 133: 919–929.
4. Sun, W., and Y. Wang. 2008. Short-term wind speed forecasting based on fast ensemble empirical mode decomposition, phase space reconstruction, sample entropy and improved back-propagation neural network. *Energy conversion and Management* 157, 1–12.
5. Wang, J., K. Fang, W. Pang, and J. Sun. 2017. Wind power interval prediction based on improved PSO and BP neural network. *Journal of Electrical Engineering & Technology* 12 (3): 989–995.
6. Salgotra, R., U. Singh, and S. Saha. 2019. On some improved versions of whale optimization algorithm. *Arabian Journal for Science and Engineering* 44 (11): 9653–9691.
7. Samadianfard, S., S. Hashemi, K. Kargar, M. Izadyar, A. Mostafaeipour, A. Mosavi, N. Narjes, and S. Shamshirband. 2020. Wind speed prediction using a hybrid model of the multi-layer perceptron and whale optimization algorithm. *Energy Reports* 6: 1147–1159.
8. Kariniotakis, G. N., and P. Pinson, Uncertainty of short-term wind power forecasts a methodology for on-line assessment. In: *2004 International Conference on Probabilistic Methods Applied to Power Systems*, 729–736. IEEE, Ames, IA, USA.

Algorithm and Simulation of Guided Electric Vehicle Routing Problem

Siyuan Fan, Weiyao Mei, Xuemei Diao, Lijun Diao, Daomeng Cai, and Chunhui Miao

Abstract Based on the study of the vehicle routing problem and ant colony algorithm, this thesis stablish the guided electric vehicle routing model with battery capacity, cargo loading, soft time window constraints. This model manages to formulate the operation rule using fast, low energy consumption and punctuality as indicators. Based on the mathematical model constructed, the model is solved by ant colony algorithm, and the improved artificial potential field method is used to avoid obstacles, so as to arrange a reasonable driving route for the vehicle. The simulation is performed to prove the feasibility of the model and locate the recharging station. The fuel logistic vehicle model and the guided electric vehicle model are compared and analyzed. Through the calculation, the cost and mileage of the later are close to that of the former; meanwhile the emission of the latter is less than that of the former, indicating the advantages of the guided electric vehicle model.

Keywords Guided electric vehicle · Vehicle routing problem · Ant Colony algorithm · Artificial potential field algorithm

1 Introduction

Logistics mainly covers the storage, transportation, and distribution of goods. Traditional distribution methods have low efficiency and high costs. Route optimization of electric logistics vehicles can improve resource utilization and mitigate pollution, thereby bringing economic benefits [1]. The study targets guided electric logistics vehicles in warehouses or ports. This operating environment is characterized by known environmental variables such as obstacle and customer coordinates.

The logistics vehicle routing problem can be included in the vehicle routing problem, so we first need to understand the mechanism of the vehicle routing problem.

S. Fan · W. Mei · X. Diao · L. Diao (✉)
School of Electrical Engineering, Beijing Jiaotong University, Beijing 100044, China
e-mail: ljdiao@bjtu.edu.cn

D. Cai · C. Miao
CSSC System Engineering Research Institute, Beijing 100094, China

W. Chen et al. (eds.), *The Proceedings of the 9th Frontier Academic Forum of Electrical Engineering*, Lecture Notes in Electrical Engineering 743,
https://doi.org/10.1007/978-981-33-6609-1_7

The Vehicle Routing Problem (VRP) can be described as arranging a reasonable route to let the fleet start from the distribution center, pass through all customers in sequence, and return to the distribution center. The design of this route should satisfy some constraints and achieve certain goals [2]. Among them, the VRP that only meets the load constraints (Capacitated Vehicle Routing Problem, CVRP) is the most basic VRP. Vehicle Routing Problem (VRP) was first proposed by Dantzig in 1959 [2]. Since 1990, research has focused on the modern heuristic algorithms, including tabu search algorithm proposed by Glover in 1986 [3], and ant colony algorithm proposed by Dorigo in 1991 [4]. Their models are easy to build and can handle complex problems.

In the context of advocating green energy, scholars began to focus on the vehicle routing problem of pure electric logistics vehicles. Liu Huaxu [5] analyzed the operation mode of co-distribution, considered the model of electric vehicles with a driving range and charging time constraints, but lacked a comparison with fuel logistics vehicles. Afroditi [6] considered the influence of battery residual capacity constraints and time windows on the VRP, but could not meet the need for guided logistics vehicles to avoid obstacles. This paper improves on the guided electric vehicles routing problem (GEVRP) model proposed in the literature [6] by combining it with the obstacle avoidance algorithm, then uses ant colony algorithm to solve it.

2 Logistic Vehicle Routing Problem Based on Ant Colony Algorithm

Methods for solving VRP problems are exact algorithms and heuristic algorithms. Among the modern heuristic algorithms, tabu search algorithm has climbing ability, but it is complex [3]; genetic algorithm converges quickly, but the ability of dealing problem with multiple constraints is poor. Ant colony algorithm has high efficiency, strong search capability, positive feedback, and robustness, but it is easy to fall into the local optimum [7]. The problem to be studied is a multi-constrained problem that requires combinations with other algorithms, so ant colony algorithm is selected. When ants look for food, they will leave pheromones on the path they pass, and the strength of pheromone can affect the choices of other ants. Pheromones can also evaporate over time; therefore, the accumulation of pheromones on a shorter path is faster than volatilization, so ants are more likely to select it. After several iterations, the final algorithm can converge to an optimal solution [7].

Each ant chooses its next step from among the cities it has not yet visited based on the heuristic information and the pheromone concentration of the path from city i to city j. Pseudo random proportional action choice code is applied to improve ant colony algorithm [8]. At moment t, the transition probability of ant k is as follows [9]:

$$j = \begin{cases} \arg\max\left\{\left[\tau_{ij}(t)\right]^{\alpha}\left[\eta_{ij}(t)\right]^{\beta}\right\} & q<q_0 \\ P_{ij}^{k}(t) & q \geq q_0 \end{cases} \tag{1}$$

$$P_{ij}^{k}(t) = \begin{cases} \frac{\tau_{ij}(t)^{\alpha}\eta_{ij}(t)^{\beta}}{\sum\limits_{k\in tovisit_k}\tau_{ik}(t)^{\alpha}\eta_{ik}(t)^{\beta}} & j \in tovisit_k \\ 0 & j \notin tovisit_k \end{cases} \tag{2}$$

q_0 is a random number, $q \in (0, 1)$. If $q<q_0$, the city with the highest probability is chosen, otherwise it is chosen according to formula (2). $tovisit_k$ is a set that has not been visited by ant k; α is the pheromone coefficient; β is the heuristic coefficient.

To imitate the nature of pheromones that evaporate over time, after a group of ants complete the task, this paper uses the Ant-cycle model [9] to update pheromones:

$$\tau_{ij}(t+n) = (1-\rho)\tau_{ij}(t) + \rho\Delta\tau_{ij} \quad \Delta\tau_{ij}(t+n) = \sum_{k=1}^{M}\Delta\tau_{ij}^{k} \tag{3}$$

In the formula, ρ is the pheromone volatilization coefficient, it must be less than 1; $\Delta\tau_{ij}$ is the total amount of pheromone left on the path.

3 The Establishment of Guided Electric Logistics Vehicle Routing Problem

Compared with the traditional CVRP model, the Guided Electric Vehicle Routing Problem (GEVRP) model adds a charging station. The electric logistics vehicle will go to the charging station when it runs out of power.

The mathematical description of the GEVRP model is as follows: Suppose that $\mathbf{G} = \mathbf{N} \cup \mathbf{F}$, where the vertex set $\mathbf{N}$ is the set of the customer set $\mathbf{N_0} = \{1, 2, \ldots i, \ldots j, \ldots, \boldsymbol{n}\}$ and the distribution center {O}; $\mathbf{F}$ is a set of charging stations. $\mathbf{A} = \{(i, j), \ \forall i, j \in \mathbf{N}, i \neq j\}$ includes all arcs connecting vertices in $\mathbf{N}$. $\mathbf{V} = \{1, 2, \ldots, V\}$ is a fleet of vehicles. Each arc (i, j) is associated with travel time t_{ij}, speed v_{ij}, and distance d_{ij}. The mathematical model is developed as follows:s

$$\min Z = \sum_{i\in F, j\in G}\sum_{k\in V} C_e \tau x_{ijk} + \sum_{(i,j)\in A}\sum_{k\in V} C_t\left(t_{ij} + \mathrm{r} + \sum_{i\in G}\delta_i \Big/ s\right)x_{ijk} + \sum_{i\in N_0}\sum_{k\in V} Z_p \tag{4}$$

$$\sum_{j\in N} x_{0jk} = 1 \quad \forall k \in \mathbf{V} \quad \sum_{j\in G}\sum_{k\in V} x_{ijk} = 1 \quad \forall i \in \mathbf{N_0} \tag{5}$$

$$\sum_{j\in N_0}\delta_j \sum_{i\in G} x_{ijk} \leq \mathrm{M} \quad \forall k \in \mathbf{V} \tag{6}$$

$$y_{jk} \leq y_{jk} - \tau d_{ij} x_{ijk} + Q(1 - x_{ijk}) \quad \forall i \in \mathbf{G}, \forall j \in \mathbf{N_0}, \forall k \in \mathbf{V} \tag{7}$$

$$y_{jk} \geq \min\{\tau d_{i0}, \tau(d_{ij} + d_{j0})\} \tag{8}$$

$$y_{jk} < Q \quad \forall i \in \mathbf{F} \tag{9}$$

$$t_{ij} = d_{ij}/v_{ij} \quad \forall i, j \in \mathbf{G} \tag{10}$$

In these formulas, δ_i is the customer's demand (tons), $i \in \mathbf{N_0}$; x_{ijk} represents the binary variable of vehicle flow on arcs i, j, $\forall i, j \in \mathbf{G}, \forall k \in \mathbf{V}$. If the vehicle k go to j from i, x_{ijk}=1, otherwise, x_{ijk}=0; y_{jk} is the remaining electric quantity of the vehicle when leaving i, at the initial moment y_{jk} = Q.

Formula (5) is the objective function, which requires the sum of the total time cost, the penalty cost for violating the time window and the power loss to be as small as possible. Suppose the time window specified by the customer i is $[ET_i, LT_i]$, the soft time window constraint allows vehicles to arrive earlier or later than the specified time, but at a penalty cost $Z_p = a \times \max\{ET_i - t_i, 0\} + b \times \max\{t_i - LT_i, 0\}$, a and b are the artificial penalty coefficients. Formula (6) ensures that each customer is served by one vehicle; formula (7) represents the on-board capacity constraint; formula (8) (9) (10) describes the rules of power consumption and charging; formula (11) represents the calculation method of travel time.

4 Guided Logistics Vehicles Use Improved Artificial Potential Field Method to Avoid Obstacles

If the obstacle cannot be avoided when planning the route for unmanned logistics vehicle, it will cause a collision. So this paper introduces Artificial Potential Field algorithm (APF). It is characterized by small computational complexity, a smooth generated path, which is suitable for tasks with high real-time performance. The principle of APF algorithm is to construct a virtual force field in the operating environment of the vehicle, in which the vehicle is subjected to the resultant effect of the virtual forces. The target point exerts an attractive force on the object and guides the object move towards it. The obstacle generates a repulsive force on the object to avoid collision with it [10].

As the distance between the logistics vehicle and the target decreases, the attractive force will converge to zero. The attractive field function is as follows:

$$F_{at}(X) = -\mathrm{grad}[U_{at}(X)] = k\rho \tag{11}$$

In the formula, k is the attractive force coefficient; ρ is the relative distance between the vehicle and the target.

A disadvantage of APF algorithm is the problem of goal non reachable with obstacles nearby (GNRON). Its performance is that when there are obstacles near the target, the repulsion and attraction will increase at the same time when the vehicle approaches the target. When the direction of the resultant force changes, it is difficult for the vehicle to reach the target. This drives us to add an attractive force related to the distance between the vehicle and the target when defining the repulsive field function, to ensure that the target is always the minimum point of the entire potential field [10]. The improved repulsion function is as follows:

$$F_{re}(X) = \begin{cases} F_{re1} + F_{re2} & \rho \leq \rho_0 \\ 0 & \rho > \rho_0 \end{cases} \tag{12}$$

$$F_{re1}=\eta\left(\frac{1}{\rho}-\frac{1}{\rho_0}\right)\frac{1}{\rho^2}(X - X_{goal})^n \tag{13}$$

$$F_{re2}=\frac{\mathrm{n}}{2}\eta\left(\frac{1}{\rho}-\frac{1}{\rho_0}\right)^2\frac{\partial\rho}{\partial X}(X - X_{goal})^{n-1} \tag{14}$$

In these formulas, η is the repulsion gain coefficient; ρ_0 is the obstacle influence distance. When the distance between the obstacle and the vehicle is greater than ρ_0, the repulsion force is 0, so ρ_0 can be modified to define the size of the obstacle. n is a random number greater than zero. Vector F_{re1} is a repulsive force, vector F_{re2} is an attractive force. As the logistics vehicle gets closer to the target, F_{re1} gradually approaches zero.

A route-planning model is built on MATLAB to verify the effectiveness of the improved APF algorithm. The simulation result is shown as a figure below.

In Fig. 1, points are marked every 0.2 m, and the curve formed by all points is the simulation path of the logistics vehicle. This path is a smooth and beautiful curve. Among the four obstacle points, (9, 9) is very close to the target (10, 10), but the logistics vehicle can bypass the obstacle instead of stagnating, indicating that the improved APF algorithm can solve the GNRON problem.

5 Simulation and Analysis

This article builds a simulation system in MATLAB R2018b software, and uses a set of fictitious factory data to perform simulation to verify the model proposed above, and compare the GEVRP model with the fuel CVRP model.

15 customers and one obstacle point are located in the warehouse. The oil price is 0.75 CNY/km. The goods are delivered from the distribution center to the customer. The maximum capacity of the battery is 60 Ah, the discharge rate is 1C, the charging current is 80 A, the charging efficiency is 83%, the operating voltage is 24 V, and

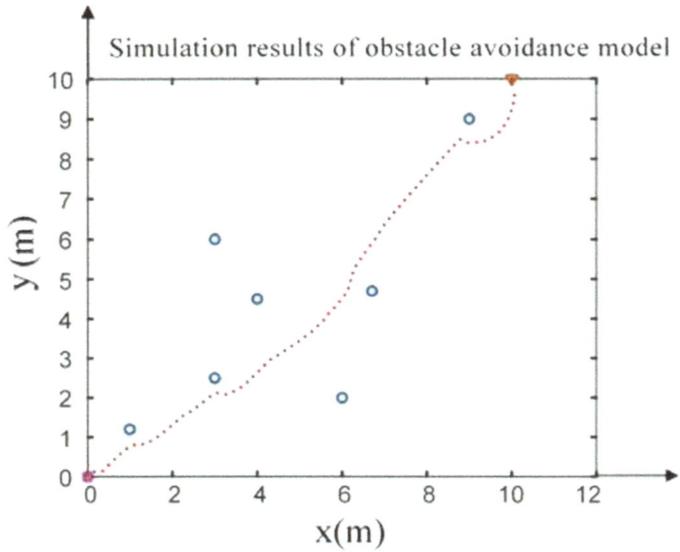

Fig. 1 Simulation results of obstacle avoidance model

the power is 2000 W. It can be calculated that the charging time is 52.65 min and the operating current is 83.3A.

5.1 Comparison Between GEVRP Model and CVRP Model

When the logistics vehicle is a fuel vehicle, that is, the entire VRP problem has no charging behavior, the CVRP model can be applied to find the route of fuel vehicles.

The total cost is 129.60 CNY, the corresponding time is 150.54 min, and the average distance traveled by each vehicle is 1473.07 m. Time costs and penalty costs account for a large proportion of the total cost, so vehicles should try to avoid violating the time window to reduce the penalty cost. The following Fig. 2(a) is the road map when two vehicles cooperate.

Since vehicle charging changes the time it takes logistics vehicles to reach the customer, and the penalty cost and time cost also change, thus choosing an appropriate charging station location plays an important role in minimizing route costs. An optimal charging station location (150, 250) is selected among multiple candidate sites after multiple simulation experiments. Figure 2b shows the simulation path when the charging station is at (150, 250).

Enlarge the figure to see that the route avoids the preset obstacles, proving that APF algorithm is effective. However, the route of the logistics vehicle is not as smooth as imagined. This is because when the vehicle, obstacle and target are in the

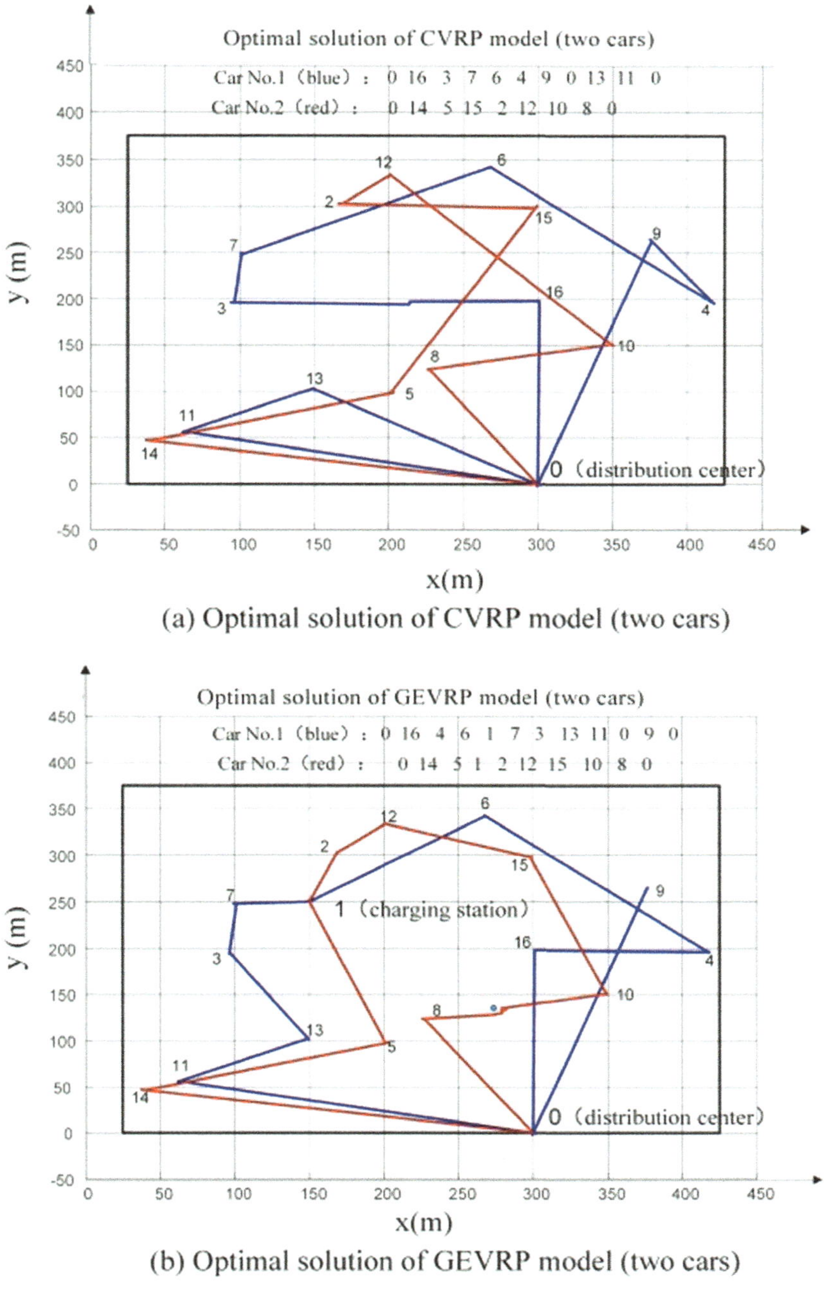

(a) Optimal solution of CVRP model (two cars)

(b) Optimal solution of GEVRP model (two cars)

Fig. 2 Optimal solutions of CVRP and GEVRP model (two cars)

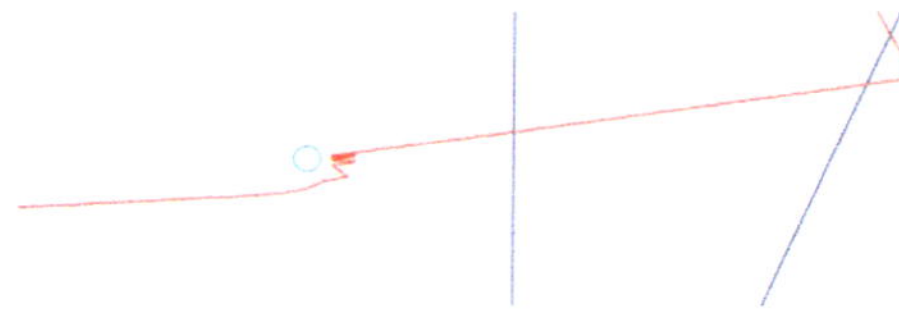

Fig. 3 Obstacle avoidance of GEVRP model (two cars)

same line, the vehicle may be subjected to zero resultant force at a certain point, thus falling into a local minimum point and unable to move forward [11] (Fig. 3).

Assuming that the electricity price is 0.5 CNY/kWh, the total cost is 123.51 CNY, the time corresponding to the minimum total cost is 204.06 min, and the average distance traveled by each vehicle is 1496.53 m. Compared with the solution results of the CVRP model, the cost of the GEVRP model is reduced by 4.7%, the average mileage of each vehicle is increased by 1.59%, and the time is increased by 35.56%. The main reason for the large increase in time is the existence of charging time and the travel time to the charging station. So companies need to use multiple vehicles to make up for the time loss caused by charging.

5.2 Calculation of Carbon Dioxide Emissions

From the perspective of economy, the cost of electricity is cheaper than gasoline, and it is less affected by the rise in oil prices. From the perspective of low-carbon, there is a lot of carbon dioxide and other greenhouse gases in the exhaust of fuel vehicles, which are the main substances that form PM2.5. Electric vehicles mostly use clean energy, which causes less pollution.

This paper calculates the carbon emission model based on the fuel consumption and travel mileage method recommended in the IPCC National Greenhouse Gas Inventory Guidelines report. The formula for calculating the carbon dioxide emissions of gasoline vehicles and electric vehicles is as follows [2]:

$$E=\sum F_k(\mu+\omega) \tag{15}$$

$$E'=\alpha\sum F_{ke}\cdot\lambda \tag{16}$$

In these formulas, $F_k=f\cdot d_{ij}$, f is the fuel consumption per kilometer, d_{ij} is the distance between i and j, μ is fuel conversion coefficient in the fuel processing stage; ω is the exhaust emission coefficient. α is the ratio of the electric power consumption of the vehicle to the thermal power generation, according to the National Bureau of Statistics data, α is 73.32%; $F_{ke}=e\cdot d_{ij}$, e is the power consumption per kilometer; λ is the power conversion coefficient of the thermal power plant.

Select the Jiangling lightweight truck with fuel consumption of 0.08 L/km and the Geely electric truck with an energy consumption of 0.29 kWh/km. The simulation experiments above provide the average mileage of each vehicle in two models. This paper refers to the fuel emission coefficients and electric energy conversion coefficients compiled on literature [1]. μ is 2.73, ω is 1.38 and λ is 0.6808. It is calculated that an electric vehicle emits 0.2313 kg of carbon dioxide, fuel logistics vehicle can emits 0.4922 kg. The former can save 53% of emissions. Therefore, whether it is from the perspective of energy saving or pollution reduction, electric logistics vehicles are an attractive alternative.

6 Conclusion

This paper combines ant colony algorithm and APF algorithm to establish the GEVRP model and formulates the overall utilization of the optimal logistics vehicle operation rules. Use C language to write GEVRP and CVRP programs to verify the effectiveness of the model, and use the improved APF algorithm to avoid obstacles.

Through the research and analysis above, we can see that in terms of algorithms, the frequency of search stagnation in the improved ant colony algorithm is reduced, and it is capable of simulating CVRP and GEVRP models. In terms of obstacle avoidance, the improved APF algorithm no longer has the problem of GNRON. In terms of the solution results, although the result of GEVRP model takes more time, the cost is reduced by 4.7%, it also can bring great emission reduction benefits.

References

1. Liu, Y.X. 2018. *Research on optimization of distribution path of pure electric logistics vehicle from the perspective of low carbon transportation*. China: Liaoning (in Chinese).
2. Dantzig, G.B., and J.H. Ramser. 1959. The truck dispatching problem. *Management Science* 6 (1): 80–91.
3. Glover, F. 1986. Future paths for integer programming and links to artificial intelligence. *Pergamon* 13 (5): 533–549.
4. Dorigo, M., V. Maniezzo, and A. Colorni. 1996. Ant system: optimization by a colony of cooperating agents. *IEEE Transactions on Systems, Man, and Cybernetics. Part B, Cybernetics: a Publication of the IEEE Systems, Man, and Cybernetics Society* 26 (1): 29–41.
5. Liu, H.X. 2012. *Research on scheduling optimization of common delivery based on electric vehicle technical feature*. China: Beijing (in Chinese).
6. Afroditi, A., M. Boile, and S. Theofanis. 2014. Electric vehicle routing problem with industry constraints: trends and insights for future research. *Transportation Research Procedia* (3): 452–459.
7. Wen, W.B., and W. Du. 2002. An abstract on the ant colony algorithms. *Automation in Petro-Chemical Industry* 1: 19–22 (in Chinese).
8. Dorigo, M., and L.M. Gambardella. 1997. Ant colony system: A cooperative learning approach to the traveling salesman problem. *IEEE Transactions on Evolutionary Computation* 1 (1): 53–66.

9. Yang, J.F. 2007. *Research and application of ant colony algorithm*. China: Zhejiang (in Chinese).
10. Wang, H.L., W.P. Fu, and Z.D. Fang. 2002. A path plan method for mobile robot based on improved potential field function. *Machine Tool & Hydraulics* (6): 67–68 + 71. (in Chinese)
11. Luo, Q.Y., H. Zhang, and H. Wang. 2011. Application of improved artificial potential field approach in local path planning for mobile robot. *Computer Engineering and Design* 32(4): 1411–1413 + 1418 (in Chinese).

Estimating Contrast of State of Health for Lithium-Ion Battery Based on Accumulated Residual Energy

Bingxiang Sun, Yuzhe Chen, Shichang Ma, Zhengtao Cui, and Zhanguo Wang

Abstract With the widespread use of electric vehicles, the deficiencies of battery management systems have gradually emerged, among which the short driving range as well as the short service life were prominent issues in electric vehicles. Taking 2.75 A 18,650 NCM material lithium-ion battery as the research object, cycle test based on the full range of state of charge (SOC). It mainly proposes to use the available energy parameters of the battery to characterize the health of the battery, as the number of battery cycles increases, the available energy of the battery decreases. A new definition method to characterize SOH_E with the available energy of the battery and SOH_{ER} with the accumulated residual energy is proposed. The SOH_{ER} is estimated and analyzed by analyzing the relationship between the accumulated residual energy of the battery and the number of use cycles. Under the premise of knowing only part of the data, a new improved similar triangle method is designed through comparative analysis which is used to effectively control the error within 0.13 to get the relationship between SOH_{ER} and energy.

Keywords Lithium-Ion battery · State of health · Available energy · Accumulated residual energy · Comparative analysis

1 Introduction

The life of lithium-ion batteries has been divided into calendar life and cycle life so far [1], the battery cycle life refers to the number of charge and discharge cycles that the battery can perform at a standard current rate. The battery life test mentioned in this article refers to the battery's cycle life. The state of health (SOH) of the power battery, also known as the state of life of the battery, is a characteristic parameter of the battery

B. Sun (✉) · Y. Chen · S. Ma · Z. Cui · Z. Wang
National Active Distribution Network Technology Research Center (NANTEC), Beijing Jiaotong University, Beijing 100044, China
e-mail: bxsun@bjtu.edu.cn

Collaborative Innovation Center of Electric Vehicles in Beijing, Beijing Jiaotong University, Beijing 100044, China

W. Chen et al. (eds.), *The Proceedings of the 9th Frontier Academic Forum of Electrical Engineering*, Lecture Notes in Electrical Engineering 743,
https://doi.org/10.1007/978-981-33-6609-1_8

health state [2], which reflects the ability of the battery to store and output charge in the current state. At present, the commonly used estimation methods for battery SOH in domestic and foreign literatures include definition methods, electrochemical impedance spectroscopy analysis methods, intelligent algorithms, and model-based methods. Among them, the electrochemical impedance spectroscopy analysis method analyzes the battery AC impedance spectroscopy at different frequencies. Understand the state of the chemical substances in the battery, and then evaluate the external characteristics of the battery [3], but the collection of battery internal parameters requires special equipment, and the parameter analysis process is complicated [4].Intelligent algorithms mainly include Kalman filtering method, particle filtering method, support vector machine and artificial neural network method, etc. [5], which requires high precision for the model, requires a large amount of data for algorithm analysis, and is redundant and tedious. Among them, the electrochemical model, based on the chemical reaction mechanism inside the battery, can accurately describe the different chemical reaction stages of the battery and the characteristics of the substance changes [6–8], with high accuracy, but the model research is more complicated and contains a large number of parameter differential equations. Therefore, the method of defining SOH with different characteristic parameters of the battery in the study is worthy of in-depth study.

So far, the prediction of the decay trajectory of SOH is still the difficulty of current research [9, 10]. There is currently no strict definition of the definition of SOH, and there are many ways to define the battery SOH in the literature. For example: capacity angle, internal resistance angle, power angle, etc. According to the basic definition of battery SOH, by collecting battery performance parameters, the current stage of SOH is obtained [11]. Generally, when estimating the SOH of the battery, the SOH is generally defined based on the capacity [12, 13]. However, when these characteristics are defined for SOH, they cannot directly reflect the driving range of electric vehicles. It is well known that the battery will inevitably aging and declining during use. The obvious thing is that the available energy will decrease with the increase of the number of cycles. Consequently, this paper proposes a new method to define SOH based on energy.

2 SOH Definition

2.1 Define SOH_E Based on Energy

This paper proposes a new definition method for SOH from the perspective of energy, and defines SOH from the available energy of battery cells, as shown in the following formula (1)

$$SOH_E = \frac{E_{NK}}{E_{N0}} \tag{1}$$

Table 1 Basic parameters of battery

Project description	Performance parameter	Project description	Performance parameter
Rated capacity	2750 mAh@1C	Typical capacity	2900mAh@0.2C
Rated voltage	3.6 V	Energy density	206 W/kg(1C)
Charging cut-off voltage	4.20 ± 0.05 V	Discharging cut-off voltage	2.50 ± 0.05 V
Maximum charging current	1C (2750 mA)2°C	Maximum discharging current	3C (8250 mA) 25 °C

2.2 *Define SOH_{ER} Based on Energy*

The available energy of the battery decreases with the increase of the number of cycles, the ratio of the accumulated remaining energy to the accumulated energy of the entire life cycle defines the battery health state SOH, as shown in the following formula (2):

$$SOH_{\mathrm{ER}} = \frac{\sum E_R}{\sum E_{Total}} \tag{2}$$

According to the proposed definition method, this paper designs a corresponding charge and discharge cycle test experiment for the battery. When the capacity declines to 70% of the initial capacity in the experiment, it is considered that its life is over.

3 Experimental Design

3.1 *Battery Platform Introduction*

The experimental battery used in this paper is the 2.75 A, 18,650 ternary material battery, the basic parameters are shown in Table 1.

The battery test platform is shown in Fig. 1.

3.2 *Experimental Test*

After the battery is activated, it must first be calibrated for its capacity. The specific cycle test procedures are shown in Table 2. The capacity and energy released by the cell in the initial state and the capacity and energy released after 1100 cycles are shown in Table 3.

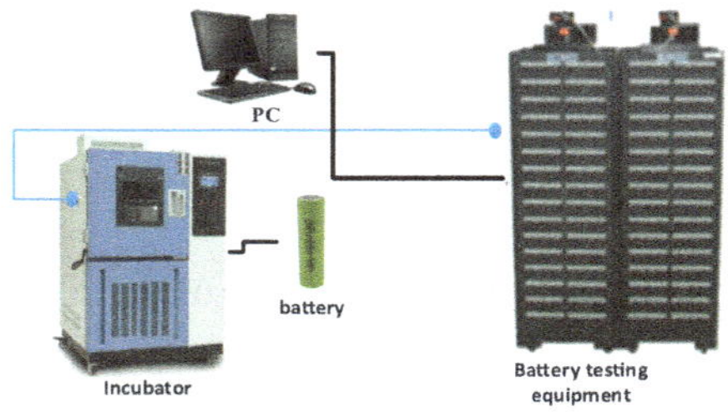

Fig. 1 Lithium battery test platform and data collection

Table 2 Cycle life test protocol

Cycle life	Test procedure
Cycle Step	1. At 25 °C, charge at a constant current of 1/2C to 4.2 V, and then charge at a constant voltage until the current is less than or equal to 1/25C, and let it stand for 30 min 2. Discharge with a constant current of 1C to the cut-off voltage of 2.5 V, let stand for 15 min, and record the capacity and energy value of each release 3. Repeat steps 1 and 2 100 times 4. Perform battery performance test every 100 times 5. The battery capacity declined to 70%, and the experiment ended

Table 3 Main available capacity and energy in the initial and ending states

	Initial	Ending
Available capacity (Ah)	2.83	1.77
Available energy (Wh)	10.10	5.73

The relationship between the energy released by the cell each time is shown in Fig. 2, and the relationship between the cumulative residual energy and the number of cycles is shown in Fig. 3.

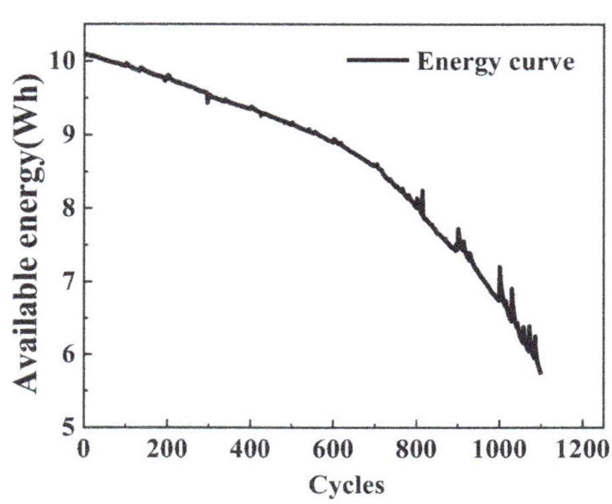

Fig. 2 Single available energy curve with the number of cycles

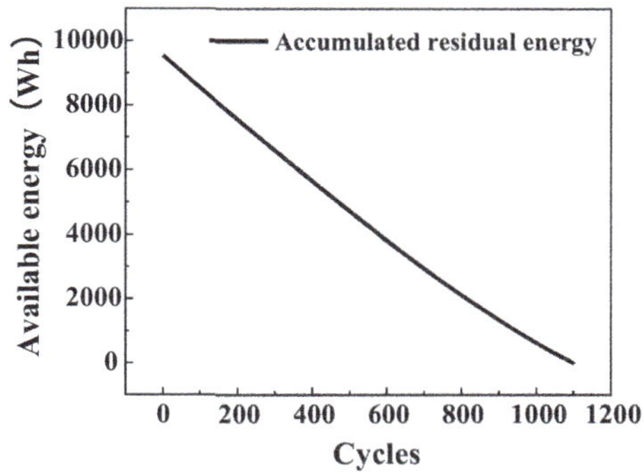

Fig. 3 Cumulative available energy curve with the number of cycles

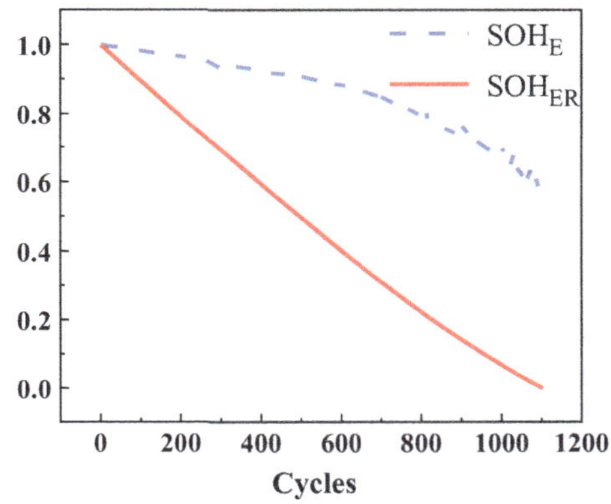

Fig. 4 SOH_{ER} and SOH_E

The single available residual energy and the cumulative available residual energy are calculated according to Eq. (1), and their corresponding SOH are respectively expressed by SOH_E and SOH_{ER}, as shown in Fig. 4.

4 SOH_{ER} Estimation Analysis

4.1 Selection of Reference Value

This paper characterizes the battery SOH by the ratio of the battery's accumulated available energy to the total energy. The change process of the available energy of the available energy of the battery cell with the number of cycles is shown in Fig. 5, the area X1 represents the cumulative available remaining energy, X represents the total available energy, then the expression of SOH_{ER} is as follows (3). In the figure, E_{N0} represents the initial available energy of the battery, E_{NK} represents the available energy after K cycles of the battery, and E_{Nend} represents the available energy of the

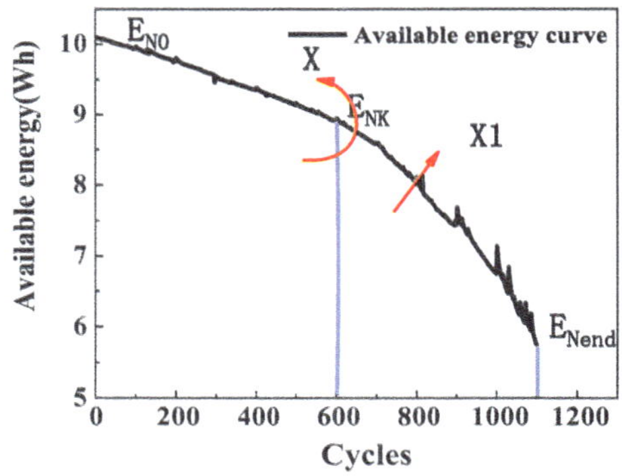

Fig. 5 Available energy curve

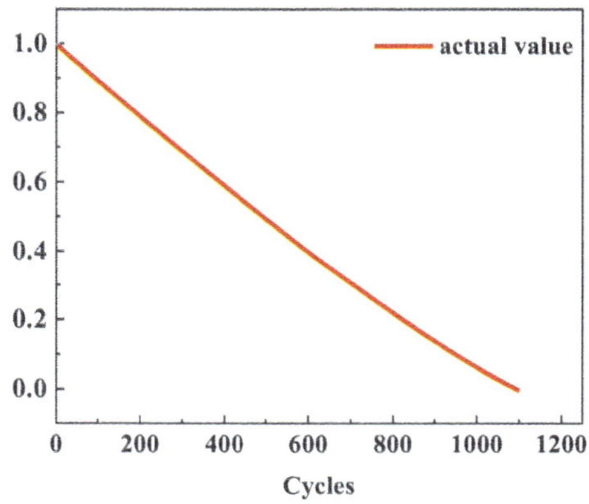

Fig. 6 Real value relationship curve

battery when the cutoff condition is reached.

$$SOH_{ER} = \frac{X1}{X} \tag{3}$$

Accumulate the real data of the battery measured by the cycle experiment, calculate its true value SOH_{ER} according to formula (3), and get its relationship curve with the number of cycles, as shown in Fig. 6, and use it as the reference value.

4.2 Estimated SOH with Known Full Life Cycle Data

When the battery's full life cycle data is known, it is fitted according to the actual measured data curve, and the relationship curve of $\sum E_R = f(SOH_E)$ is obtained, as shown in Fig. 7, and the relative error is shown in Fig. 8, where the maximum value

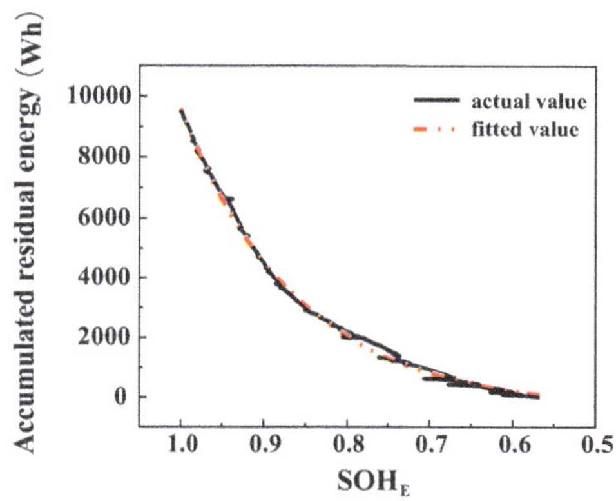

Fig. 7 Accumulated residual energy fitting curve

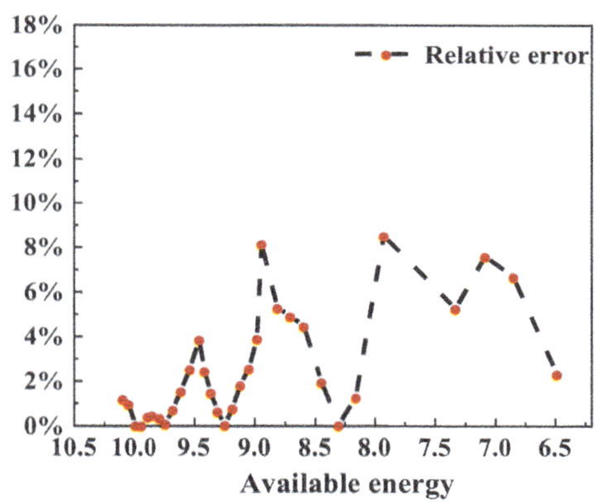

Fig. 8 Relative error curve

does not exceed 9%, and the accuracy is high, which can accurately reflect the true situation.

However, it is difficult to realize the full life cycle data of the battery in the actual process. Therefore, it is necessary to further study the estimated SOH for the known part of the battery data. The following paper gives specific research methods.

4.3 Estimated SOH from Known Partial Period Data

4.3.1 Similar Triangles

It is known from Fig. 5 above that the relationship between the available energy of the battery and the number of cycles changes. The paper uses the idea of mathematical similar triangle, as shown in Fig. 9. The relationship of SOH_{ER} obtained from the

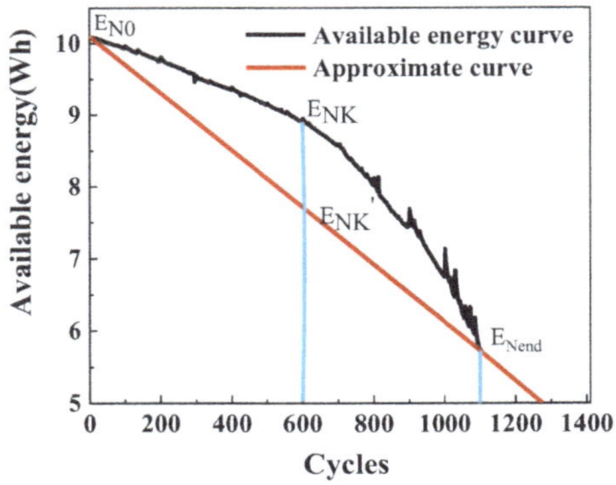

Fig. 9 Real curve and approximate curve

real available energy change curve is shown in the following formula (4).

$$SOH_{ER} = \frac{\sum_{i=N_K}^{N_{end}} E_i}{\sum_{i=N_0}^{N_{end}} E_i} = \frac{\sum E_R}{\sum E_{Total}} \tag{4}$$

Figure 10 is to convert the approximate image of the energy curve into a mathematically similar triangle image, and the solution of the triangle method is used to find the correspondence between SOH_{ER} and the energy. As an auxiliary triangle X1, X1, X2, and X3 in the figure respectively represent their corresponding areas.

As can be seen from Eq. (4), the expressions corresponding to X2 and X3 in the figure are as follows:

$$\text{X2} = \sum_{i=N_K}^{N_{end}} E_i = \Sigma E_R$$

$$\text{X3} = \sum_{i=N_0}^{N_K} E_i \tag{5}$$

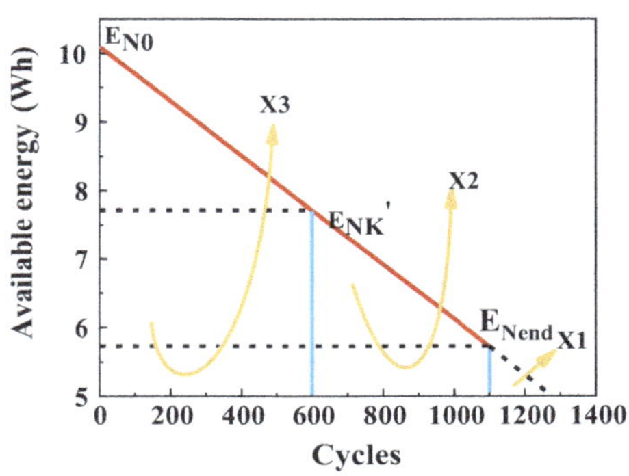

Fig. 10 Approximate available energy curve

$$X2 + X3 = \sum_{i=N_0}^{N_{end}} E_i = \Sigma E_{Total}$$

Combining formula (4) and formula (5), the following relationship can be obtained:

$$SOH_{ER} = \frac{X2}{X_2 + X_3} \tag{6}$$

From the principle of similar triangles:

$$\frac{X1}{X1 + X2} = \left(\frac{E_{N_{end}}}{E_{N_K}}\right)^2 \tag{7}$$

$$\frac{X1 + X2}{X1 + X2 + X3} = \left(\frac{E_{N_K}}{E_{N_0}}\right)^2 \tag{8}$$

$$\frac{X1}{X1 + X2 + X3} = \left(\frac{E_{N_{end}}}{E_{N_0}}\right)^2 \tag{9}$$

Sort out the expressions of SOH_{ER} and energy, as shown in the following formula (10),

$$SOH_{ER} = \frac{E_{Nk}^2 - E_{Nend}^2}{E_{N0}^2 - E_{Nend}^2} \tag{10}$$

According to the above formula, the relationship curve of SOH_{ER} is obtained, and it is compared with the reference value as shown in Fig. 11, and the absolute error is shown in Fig. 12, it could be seen that the maximum value is about 0.28.

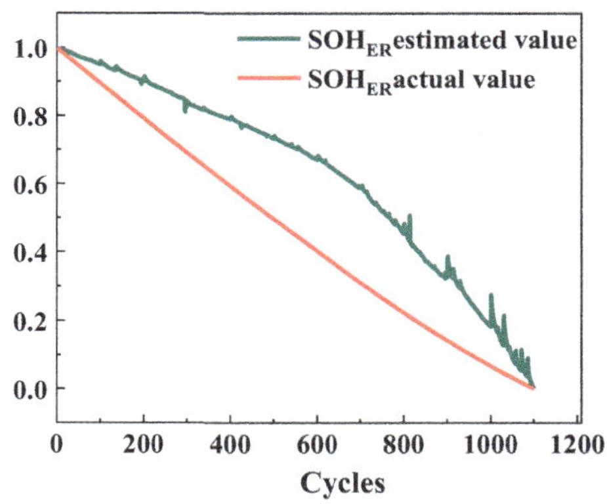

Fig. 11 Fitting SOH_{ER} relationship curve

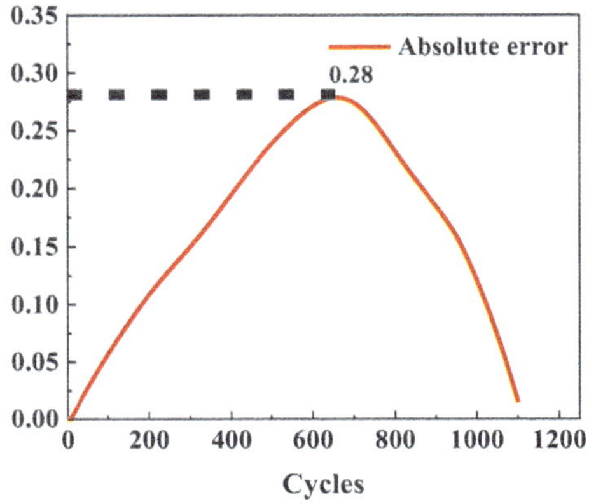

Fig. 12 Absolute error

4.3.2 Improve Similar Triangles Similar Triangles

Figure 9 shows that the actual energy change curve is an approximately linear convex function, and the above-mentioned similar triangle method is to obtain the relational expression based on the direct substitution curve. There is a certain deviation from the original curve, which makes the overall estimated value too large, so in order to reduce the estimate Error, need to add compensation function on the basis of approximate estimation curve. As shown in Fig. 13, the sum of the area S1 and S3 enclosed by the curve in the figure is as close to S2 as possible, that is, the error of (E–E') ΔE reaches a local minimum, and it is desired to obtain the functional relationship of the true curve which is more difficult. In order to be able to determine the position of the curve in the figure, under the premise of knowing part of the entire hundred data, this article proposes an idea to determine the number of cycles N corresponding to different available energies through interpolation, and determine the original truth by N For the available energy E_{NK} of the curve, shift the approximate curve in Fig. 9 upward $(E_{NK}-E_{NK'})/2$ to obtain a new curve, and then use the adjusted

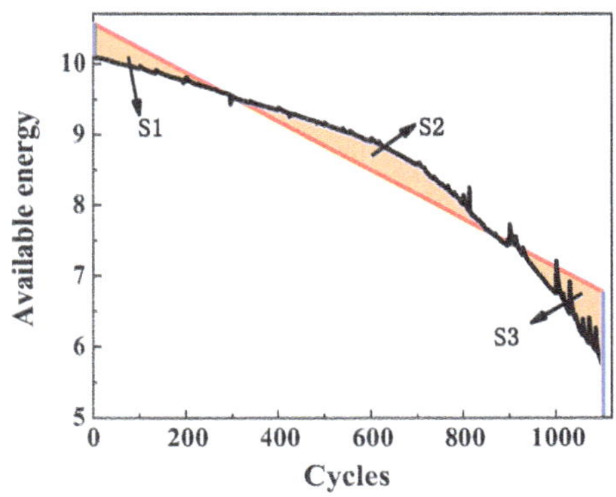

Fig. 13 Available energy curve and linear fitting curve

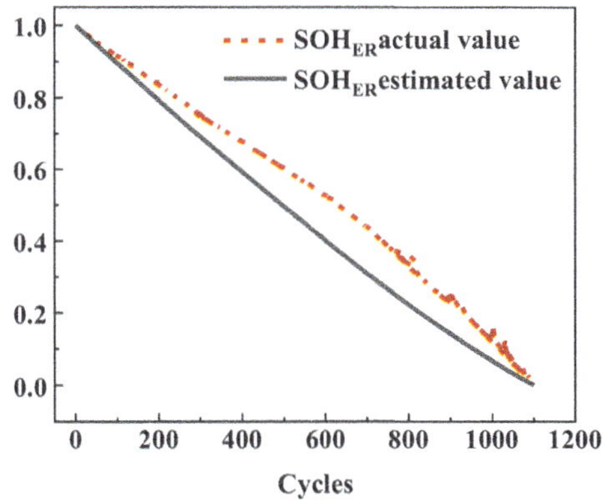

Fig. 14 The true and estimated value of SOH_{ER}

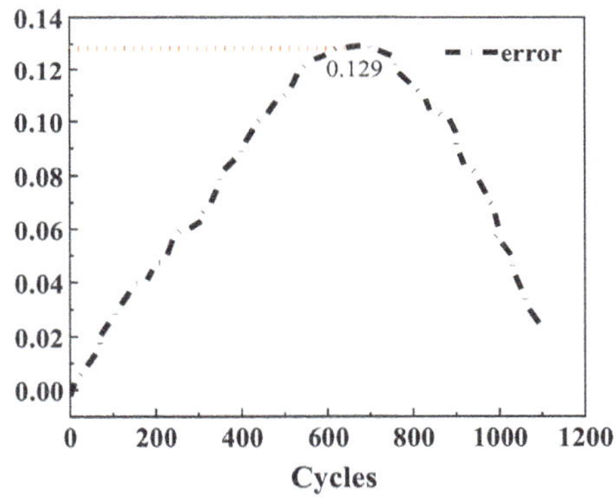

Fig. 15 Absolute error

curve to obtain the SOH_{ER} estimated value by formula (10). The specific results are shown in Fig. 14. The absolute error is shown in Fig. 15. It can be seen that the error accuracy is effectively improved after improvement, and the maximum error is 0.13.

4.3.3 Interval Map

When the remaining accumulated energy is not measurable, the specific formula is shown in formula (11), and the SOH with the actual value range of X–100% is mapped to the SOH_{ER} with the actual range of 0–100%, as shown in Fig. 16, where X% refers to the cut-off condition of battery life, this paper selects 70%. The absolute error is shown in Fig. 17, and the maximum error reaches 0.34. It could be seen that the error of only the mapping method is relatively large, and the estimation of SOH_{ER} is meaningful.

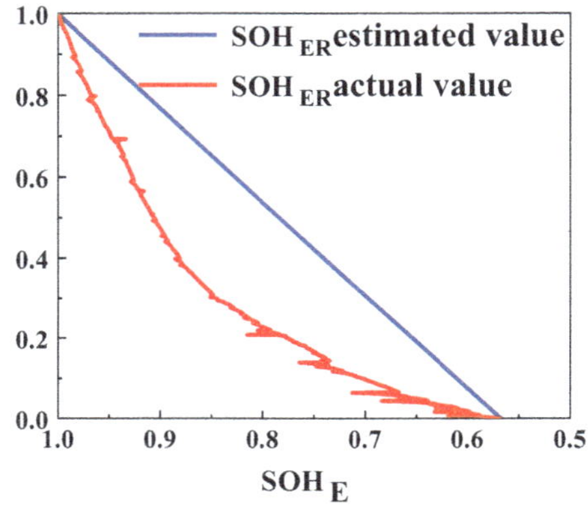

Fig. 16 The true and estimated value of SOH_{ER}

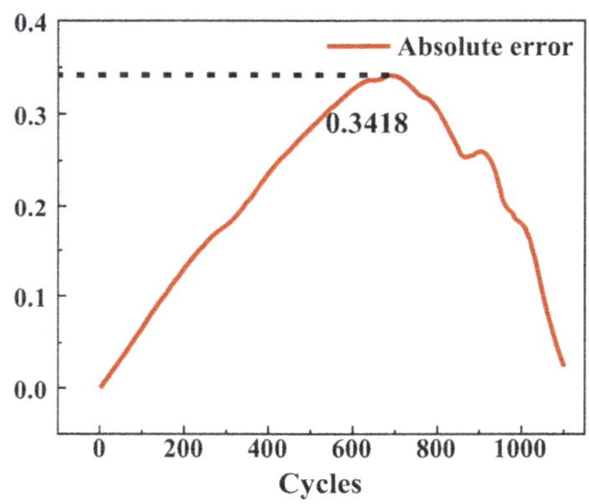

Fig. 17 Absolute error

$$SOH_{ER} = \frac{\frac{E_{NK}}{E_{N0}} - X\%}{1 - X\%} \tag{11}$$

4.4 Comparison of Estimation Methods

Comparing the above four methods, as shown in Fig. 18, The full-range data direct fitting method has the highest precision, but it needs to be established under the condition that the full life cycle data can be measured; although the calculation method of the interval map is simple, the error is large; the similar triangle method only needs to know the initial and the ending time The available energy of Relatively greatly simplifies data processing, but there are still certain deviations, so the improved method proposed on this basis effectively improves the accuracy of error, making it controlled within 0.13.

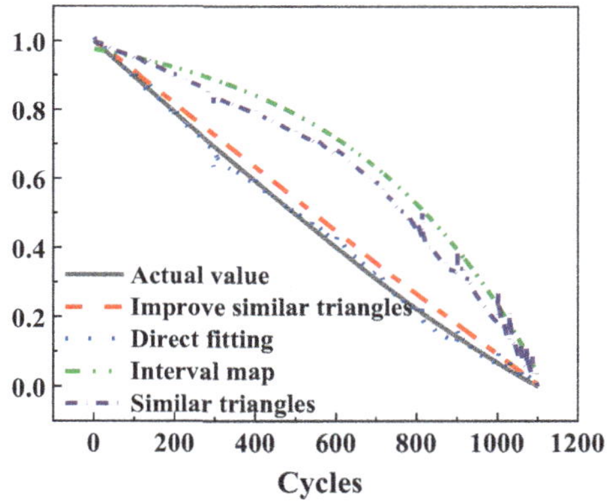

Fig. 18 Relationship curve of SOH_{ER} with cycle times

5 Conclusion

This paper proposes a new definition method for battery SOH based on energy. Experimental data is obtained by conducting cycle life test experiments on 18,650 and 2.75 A batteries. In the method of studying the estimated battery SOH_{ER}, it is divided into two categories to explain separately. When the full life cycle data of the battery is known, it is fitted according to the real data to obtain $\sum E_R = f(SOH_E)$ relation curve, the maximum error value does not exceed 9%; When only part of the cycle data of the battery is known, based on the analysis and comparison of the above three methods, the improved similar triangle method is used to verify that the error can be stabilized within 0.13, which effectively improves the accuracy of the error, and the feasibility and effectiveness of the method are confirmed by experiments. which provides a basis for the future management and control of battery systems and battery secondary utilization strategies.

Acknowledgements I sincerely thank my tutor, Bingxiang Sun, for giving me a lot of care and help, both in scientific research and in life. During laboratory work and paper writing, I also got help from senior students. I would like to express my heartfelt thanks to them again.

This work is supported by the 'National Key R&D Program of China' (Grant NO.: 2018YFB0104400) and the National Natural Science Foundation of China (Grant No. 51907005).

References

1. Rongrong Liang, Liwei Qin, Jiuzhi Zhao. 2018. Research on SOH estimation of battery for industrialized electric vehicle. *Automobile Applied Technology* (15): 17 + 36 (in Chinese).
2. Li, Guangdi, Haohua Lv, Jun Yuan, and Bo. Li. 2016. Review of life research on electric vehicle Li-ion cell. *Chinese Journal of Power Sources* 40 (6): 1312–1314 (in Chinese).
3. Han, Xiaoli. 2011. *Research on electrochemical impedance spectroscopy measurement method of lithium ion battery [D]*. Institute of Technology: Harbin (in Chinese).

4. Xu, Xinmin, Lian Wang, Huiling Shi. 2015. Research on battery aging life based on electrochemical impedance spectroscopy. *Chinese Journal of Power Sources* 39 (12): 2579–2583 (in Chinese).
5. Zhang, Shaohong. 2019. Research on SOH estimation method of Li-ion battery for electric vehicle [D]. Taiyuan University of Technology (in Chinese).
6. Li, Junfu, Lixin Wang, Chao Lyu, Han Wang, Xuan Liu. 2016. New method for parameter estimation of an electrochemical-thermal coupling model for $LiCoO_2$ battery. *Journal of Power Sources* 307.
7. Xin, Zhou, Jeffrey L. Stein, Tulga Ersal. 2017. Battery state of health monitoring by estimation of the number of cyclable Li-ions. *Control Engineering Practice* 66.
8. Energy; New Energy Study Results from Missouri University of Science and Technology Described [A single particle model with chemical/mechanical degradation physics for lithium ion battery State of health (Soh) Estimation]. *Energy Weekly News* 2019.
9. Farmann, A., W. Waag, A. Marongiu, et al. 2015. Critical review of on-board capacity estimation techniques for lithium-ion batteries in electric and hybrid electric vehicles. *Journal of Power Sources* 281: 114–130.
10. Berecibar, M., I. Gandiaga, I. Villarreal, et al. 2016. Critical review of state of health estimation methods of Li-ion batteries for real applications. *Renewable & Sustainable Energy Reviews* 56 (3): 572–587.
11. Tao, Deng, and Weixing Luo. 2018. Research on estimation methods of battery SOH for electric vehicle. *Modern Manufacturing Engineering* (5): 43–49 (in Chinese).
12. Widodo, A., M.C. Shim, W. Caesarendra, et al. 2011. Intelligent prognostics for battery health monitoring based on sample entropy. *Expert Systems with Applications* 38 (9): 11763–11769.
13. Andre, D., A. Nuhic, T. Soczka-Guth, et al. 2013. Comparative study of a structured neural network and an extended Kalman filter for state of health determination of lithium-ion batteries in hybrid electric vehicles. *Engineering Applications of Artificial Intelligence* 26 (3): 951–961.

Peak Power Estimation Considering Initial Polarization

Bingxiang Sun, Guoli Yin, Xitian He, Zhenlin Zhu, and Jian Wu

Abstract It is very important to obtain the peak power reference value with high precision for parameter test and evaluation. In the industry, the estimation of peak power is usually called state of charge (SOP) estimation. At present, the research on the peak power is mainly based on the steady state condition, but the vehicle has been in the dynamic condition in the actual driving process, so the establishment of the peak power reference value acquisition method suitable for the dynamic condition is helpful to promote the development of the industry. In this paper, based on the consideration of initial polarization, the second-order RC model, the multiple linear regression model and the support vector machine model are used to estimate the peak power of the battery.

Keywords Peak power estimation · Initial polarization voltage · Second order RC model

1 Introduction

The core component of new energy vehicle is lithium-ion power battery [1].The state of the battery directly affects the performance of the car. Among many parameters of lithium-ion battery, the peak power SOP is one of the most important parameters [2].SOP directly determines the power capacity of battery and is an important reference index for acceleration, deceleration and uphill of vehicle. The peak power capacity of battery is affected by many factors, such as temperature, voltage, current,

B. Sun (✉) · X. He · Z. Zhu · J. Wu
National Active Distribution Network Technology Research Center (NANTEC), Beijing Jiaotong University, Beijing 100044, China
e-mail: bxsun@bjtu.edu.cn

Collaborative Innovation Center of Electric Vehicles in Beijing, Beijing Jiaotong University, , Beijing 100044, China

G. Yin
Cangzhou Power Supply Branch of State Grid, Hefbei Electric Power Co., Ltd, , Cangzhou, China
e-mail: 904041186@qq.com

W. Chen et al. (eds.), *The Proceedings of the 9th Frontier Academic Forum of Electrical Engineering*, Lecture Notes in Electrical Engineering 743,
https://doi.org/10.1007/978-981-33-6609-1_9

etc. [3].At present, the discussion of peak power does not consider the influence of initial polarization voltage on the peak power capability of the battery. In this paper, the influence of initial polarization voltage is taken into account, and the peak power capacity of the battery is estimated. The polarization of Li-ion battery mainly includes concentration polarization, electrochemical polarization and ohmic polarization. Ohmic polarization, as the name implies, refers to the polarization caused by the battery resistance, which occurs instantaneously; because the speed of electron motion is greater than the speed of electrochemical reaction on the electrode, electrochemical polarization is generated; concentration polarization is generated because the diffusion speed of lithium ion in the electrolyte is less than the speed of electrochemical reaction [4].

2 Peak Power Estimation Model

2.1 Second Order RC Model

The equivalent circuit model of battery mainly includes Thevenin model, multi-level RC model and Rint model. The structure of Thevenin model is simple, and the parallel RC part can simulate the polarization process of the battery very well. Multi order RC model is to add parallel module of RC on the basis of Thevenin model. The higher the order is, the higher the accuracy of the model is, but the calculation becomes more complicated [5].In this paper, the second-order RC model is chosen as the estimation model (Fig. 1).

The transient formula of the second-order RC model is as follows.

$$\begin{cases} U_{P1}(t) = U_{P1}(0)e^{-\frac{t}{\tau 1}} + R_{P1}I\left(1 - e^{-\frac{t}{\tau 1}}\right) \\ U_{P2}(t) = U_{P2}(0)e^{-\frac{t}{\tau 2}} + R_{P2}I(1 - e^{-\frac{t}{\tau 2}}) \\ \tau 1 = R_{P1}C_{P1} \\ \tau 2 = R_{P2}C_{P2} \\ U_0(t) = U_{OCV} - IR_0 - U_{P1}(t) - U_{P2}(t) \end{cases} \tag{1}$$

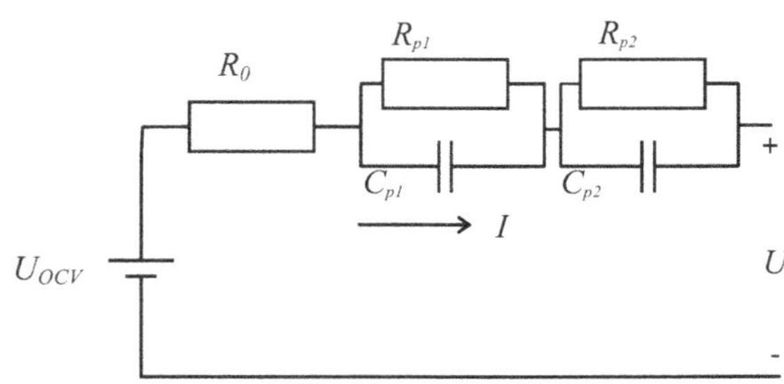

Fig. 1 Second order RC model

In order to realize the online estimation of peak power, the first step is to realize the online acquisition of model parameters. By fitting the experimental data, the relationship between ohmic internal resistance, polarization internal resistance and polarization capacitance of the second-order RC model and discharge multiplier, polarization voltage and SOC can be obtained. After the parameters of the battery are identified, the relationship between the parameters of the model and discharge rate, initial polarization voltage, SOC is obtained:

$$\begin{cases} R(I,S,U_{P0}) = -1.155\times 10^{-4}\times I - 3.49\times 10^{-5}\times S + 0.03067 - 0.0124\times U_{P0} \\ R_{P1}(S) = 0.007318\times e^{-0.02438\times S} + 0.00316 \\ R_{P2}(I,S) = 0.115\times \mathrm{e}^{-1.07I} + 0.04009\times e^{-0.0908S} + 0.00674 \\ C_{P1}(S) = 172.7\times e^{0.01452S} + 1.079\times 10^{-7}e^{0.2395S} \\ C_{P2}(I) = 1927\times e^{0.0206I} - 1622 \end{cases} \tag{2}$$

2.2 Multiple Linear Regression

Multiple linear regression estimation is to obtain the relationship between one dependent variable and multiple independent variables through a large number of experiments and data analysis, and then predict the variables. Multivariate linear regression estimation method has the advantages of simple structure, few parameters and fast operation speed. However, due to the serious collinearity among SOC, internal resistance and temperature, if SOC, internal resistance and temperature are selected as the input of prediction model, it is not suitable to use multivariate linear regression method. Let the random variable be y, γ_0, γ_1, γ_2, ... γn is the regression coefficient, The general variable is x_1, x_2, x_3, ... xn. Then the linear regression model can be expressed as:

$$y = \gamma_0 + \gamma_1 x_1 + \gamma_2 x_2 + \cdots + \gamma_n x_n \tag{3}$$

In this paper, SOC, polarization voltage, discharge time, discharge rate and ohmic internal resistance are selected as input variables. 26 pairs of training data and 10 pairs of verification data.

2.3 Support Vector Machine

Support vector machine (SVM) is based on VC dimension theory and structural risk minimization [6]. It finds the maximum margin hyperplane by classifying the samples, transforms the sample input to the high-dimensional feature space by kernel function transformation, and solves the nonlinear problem of the sample space by linear learning of the feature space. Therefore, the support vector machine algorithm

Fig. 2 The nonlinear regression principle of SVM

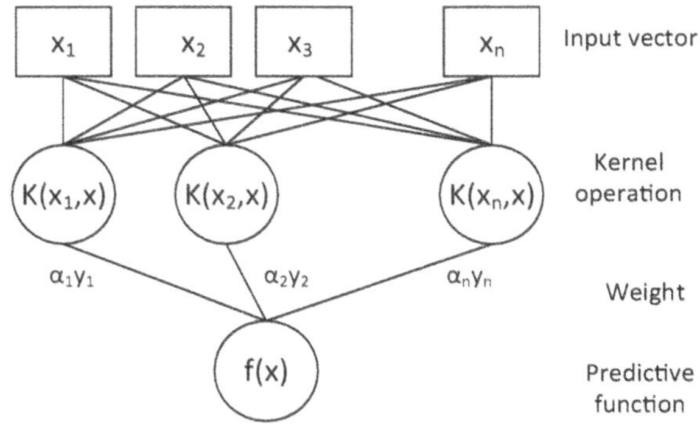

can be regarded as convex quadratic optimization [7]. Support vector machine model has the advantages of excellent learning and generalization ability, short training time, strong extrapolation ability, and no excessive dependence on large samples and historical data [8]. Therefore, SVM is used to solve the nonlinear problem between peak power and parameters. The nonlinear regression principle of SVM is as follows (Fig. 2).

3 Estimation Results

In this paper, the peak power is characterized by the peak current value, and the peak power of 26,650 battery is measured and estimated. The real value of peak power is obtained by the method of constant power method, different initial polarization states are made by using different current discharge for 5 s, and then the high rate discharge is directly carried out on the battery without standing, so as to obtain the peak power value under polarization state.

3.1 Second Order RC Model Estimation Results

Use Matlab to build the model, and estimate the peak power of the battery at 80%, 50%, 40%, 30%, 20%, 10% SOC in polarized state and non polarized state. The estimation results and errors are shown in Figs. 3, 4 and 5.The estimation error of the model is within ±8%, and the estimation result is generally less than the real value, which may be due to the internal temperature rise in the process of high rate discharge of the battery.

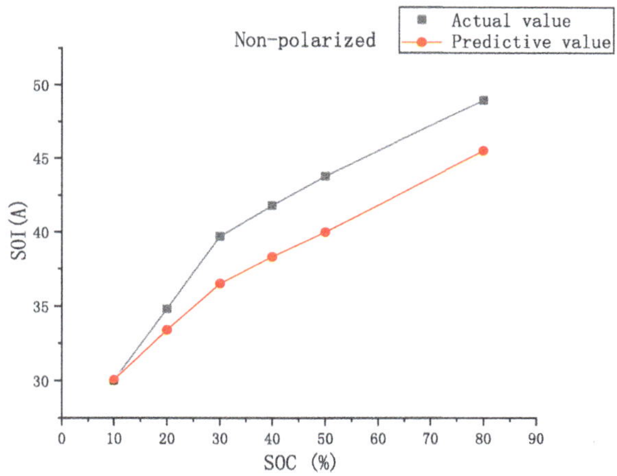

Fig. 3 Comparison between estimated results and real values in unpolarized state

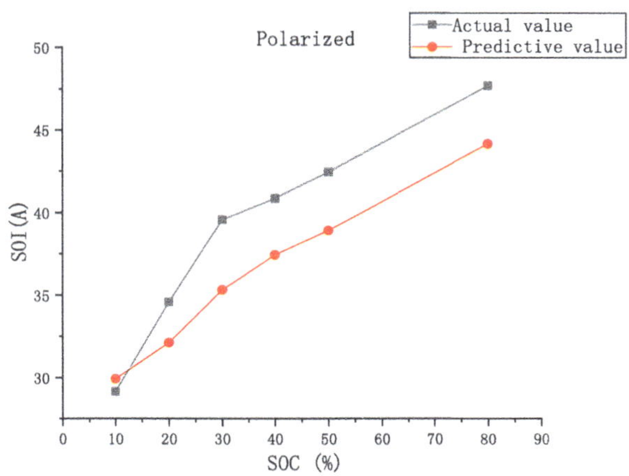

Fig. 4 Comparison between estimated results and real values under polarization

3.2 Multiple Linear Regression Estimation Results

According to the training data, the regression coefficient is obtained. The regression equation is used to estimate the validation set data, and the peak current of lithium-ion battery estimated by the validation set is obtained. The peak current data estimated by the fitting regression equation is compared with the measured data. The estimation results are shown in Figs. 6 and 7. It can be found that the estimated value of the model is very close to the real value, and the relative error is mainly within $\pm 2\%$. It can be seen that the fitting degree of multiple linear regression model is very high, the root mean square error is 0.541A, so it can be considered that multiple linear regression model can estimate the peak current of lithium-ion battery.

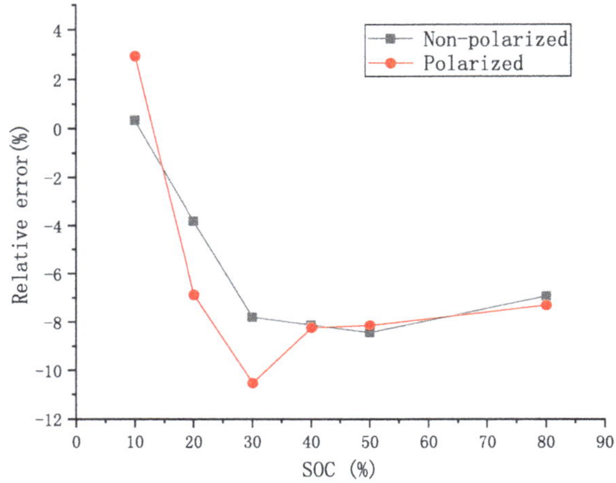

Fig. 5 Relative error between estimated result and real value

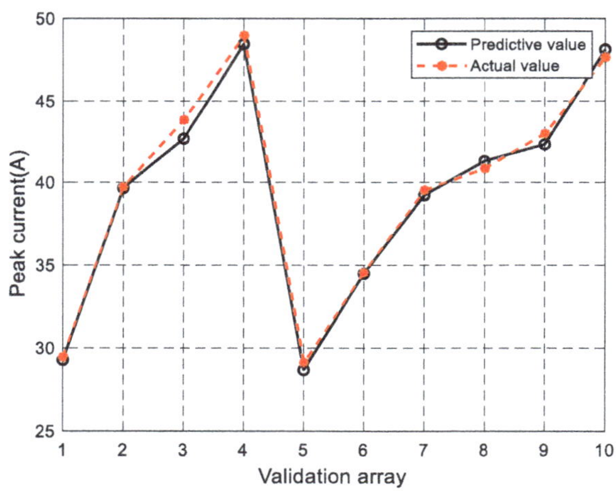

Fig. 6 Comparison between the current value estimated by the model and the real value

3.3 Support Vector Machine Estimation Results

The predicted results are shown in Figs. 8 and 9. $R^2 = 0.9913$, RMSE = 0.573a, the relative error of the model is mainly within ±0.8%.It can be seen that SVM model can better predict the peak power of the battery.

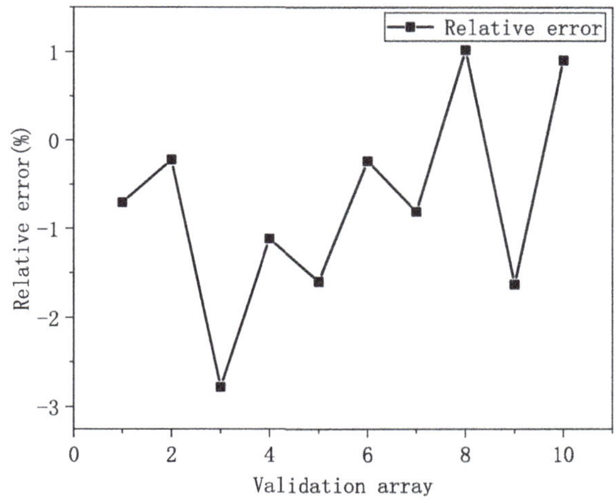

Fig. 7 Relative error between estimated value and real value

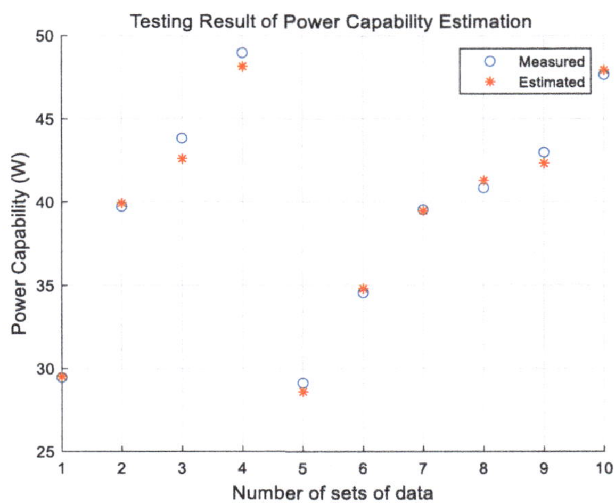

Fig. 8 SVM prediction results

4 Comparison of Different Estimation Methods

In this paper, the multiple linear regression method and support vector machine are used to estimate the peak power of the battery under the same working condition, which is used to compare the second-order RC model horizontally. In this paper, the mean absolute deviation (MAPE) and root mean square error (RMSE) [9] are introduced to make quantitative analysis of three estimation methods.

$$MAPE = \frac{1}{n}\sum\nolimits_{k=1}^{n}\left|\frac{f_k - y_k}{y_k}\right| \times 100\% \tag{4}$$

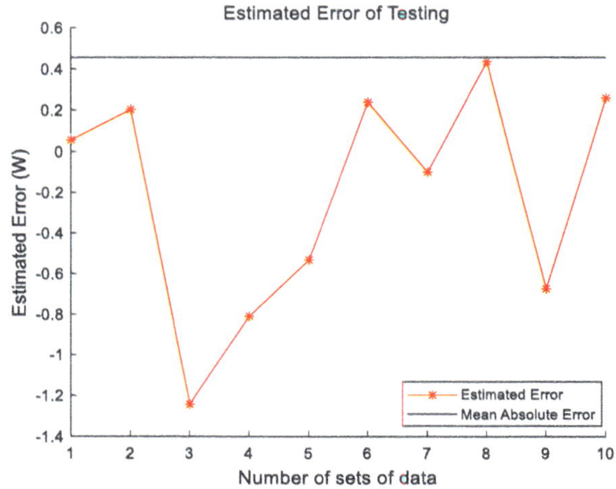

Fig. 9 Error between SVM prediction result and real value

Table 1 Comparison Of three estimation methods

Statistical indicators	Second order RC model	Multivariate linear estimation	Support vector machine
MAPE	5.67	1.1	0.4807
RMSE	2.26	0.541	0.573
Maximum absolute error	3.8	1.22	1.27

$$RMSE = \sqrt{\frac{\sum_{1}^{n} (f_k - y_k)^2}{n}} \tag{5}$$

where y_k represents the true value of peak current and f_k is the estimated value of the model. The comparison results are shown in Table 1.

5 Conclusion

The prediction accuracy based on SVM model is the highest, the average absolute percentage error is 0.48%, followed by multiple linear regression model, the average absolute percentage error is 1.1%. The accuracy of the second-order RC model is the lowest, and the average absolute percentage error is 5.67%, which is within the acceptable range. The predicted results of the second-order RC model are generally lower than the real values, which may be due to the temperature rise in the battery during the high rate discharge process, which improves the material activity and reduces the internal resistance of the battery. The second-order RC model can further improve the test accuracy by combining with the thermal model, and there is a lot of room for improvement.

Acknowledgements I sincerely thank my mentor Sun Bingxiang for her guidance and help. At the same time, I would like to thank all the teachers and students in the power battery Laboratory for their care of me.

This work is supported by the 'National Key R&D Program of China'(Grant NO.: 2018YFB0104400) and the National Natural Science Foundation of China (Grant No. 51907005).

References

1. Fotouhi, A., et al., A review on electric vehicle battery modelling: From Lithium-ion toward Lithium–Sulphur. Renewable and Sustainable Energy Reviews, 2016. 56.
2. Yurkovich, B.J., et al. 2009. A technique for dynamic battery model identification in automotive applications using linear parameter varying structures. *Control Engineering Practice* 17 (10): 1190–1201.
3. Zheng, FangDan. 2017. Research on multi time scale lithium ion battery condition assessment technology based on data driven, 2017, Beijing Jiaotong University (in Chinese).
4. Lindbergh, G., et al. 2010. Analysis of the Polarization in a Li-Ion Battery Cell by Numerical Simulations. *Journal of the Electrochemical Society* 157 (11): A1236–A1246.
5. Guo, Feng, et al, Modeling and Simulation of Ni MH power battery power prediction based on MATLAB / Simulink. Power technology, 2018. 42(05): 699–700+704 (in Chinese).
6. V, C., The nature of statistical learning theory. IEEE Transactions on Neural Networks, 1997. 8(6).
7. Burges, C.J.C. 1998. A Tutorial on support vector machines for pattern recognition. Data Mining and Knowledge Discovery 2 (2).
8. Singh, Himali Singh, Rajesh Kumar Tripathy, and Ram Bilas Pachori. 2020. Detection of sleep apnea from heart beat interval and ECG derived respiration signals using sliding mode singular spectrum analysis. Digital Signal Processing 104.
9. Deiss, L., et al., Tuning support vector machines regression models improves prediction accuracy of soil properties in MIR spectroscopy. Geoderma 365 2020.

A Single Phase Boost Inverter with Reduced Leakage Current for Photovoltaic Energy System

Xuefeng Hu, Shunde Jiang, Hao Shen, Zikang Long, He Cheng, and Lusheng Ge

Abstract In this paper, an integrated transformer-less single phase boost inverter (SPBI) with high reliability for photovoltaic energy system is proposed. The proposed SPBI requires only three active switches and has the inherent features of shoot-through immunity and reduced leakage current, which improves reliability and cost of power conversion. Moreover, the proposed inverter can output ac voltage higher than the input dc voltage. The unified SPWM technique for the proposed inverter is also presented. The steady-state analysis, operating principles, and parameter selection guidelines are presented in detail for the proposed inverter. A laboratory prototype is built and tested to verify the performance of the proposed inverter.

Keywords Single phase · Boost inverter · High reliability · PV

1 Introduction

Recently, renewable energy sources such as solar energy, wind energy, fuel cell power generation technologies have been widely valued and employed [1–3]. Photovoltaic (PV) energy is considered as an excellent candidate to replace the conventional fossil fuel due to no CO_2 emissions. Generally, a line frequency transformer is often adopted in PV applications for providing galvanic isolation and voltage ratio transformations. However, the line frequency transformers increase the size, weight and cost of the PV system and reduce the efficiency and power density. Therefore, transformerless inverters with decreased number of components, low cost, and high efficiency are preferred choices for these applications. However, additional care must be taken to avoid safety hazards such as ground fault currents and leakage currents, e.g., via the parasitic capacitance between the PV panel and/or its frame and ground. The traditional full bridge inverter (FBI) with four active switches (H4) is often used

X. Hu · S. Jiang (✉) · H. Shen · Z. Long · H. Cheng · L. Ge
Anhui University of Technology, Ma'anshan, Anhui Province 243032, China
e-mail: 1443370484@qq.com

W. Chen et al. (eds.), *The Proceedings of the 9th Frontier Academic Forum of Electrical Engineering*, Lecture Notes in Electrical Engineering 743,
https://doi.org/10.1007/978-981-33-6609-1_10

for PV power generation systems because it has good tradeoff between efficiency, complexity and cost. However, the output peak ac voltage of FBI does not exceed the input dc voltage, so many PV modules are often connected in series as input sources.

In order to solve above problems, many single-stage inverters are proposed [4–7], many switching devices are used in these inverter and they have other problems. A novel single-phase current source H5(CH5) inverter has been presented in which a switch is added between the input and the bridge arms [5]. Bases on the H5 topology, a switch branch is added between the input and the midpoints of bridge arms [6]. A H6-type transformerless inverter for grid-tried PV system has been proposed for suppressing the leakage current [7], and it has the ability to inject reactive power into utility grid with low harmonic distortion. However, H6-type converter need more switches. The main drawback is the shoot-through problem between the two switches in a leg phase, which is a common reason reducing the reliability of the conventional inverters. Thus, the dual-buck inverter topologies without shoot through problem are discussed in [8, 9], which can realize higher AC output voltage than the DC input voltage. Nevertheless, there is still the problem of leakage current to ground. Following this trend, an integrated single phase boost inverter (SPBI) based on the dual-buck structure is proposed.

2 Operating Principle and Analysis of the Proposed Topology

2.1 The Topology of Proposed Inverter

The circuit topology of the proposed single phase boost inverter is shown in Fig. 1. The inverter can be viewed as an organic combination of a conventional boost converter and a dual-buck half-bridge inverter.

Fig. 1 The proposed inverter topology

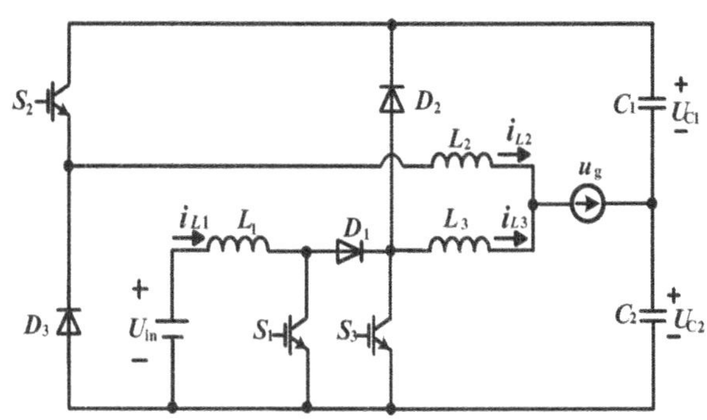

Fig. 2 PWM control strategy

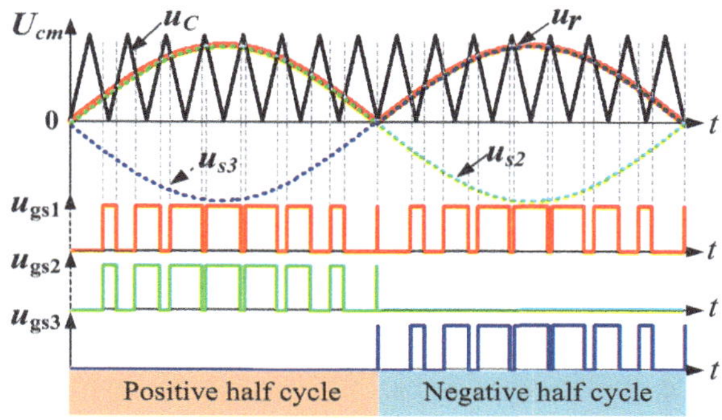

In this section, the sinusoidal pulse modulation (SPWM) for the proposed single phase boost inverter is analyzed, which is called half-period modulation mode. There are six main working modes of the inverter. The corresponding equivalent circuits are presented in Fig. 2.

(1) Mode I [t_0–t_1 Fig. 3a]: At $t = t_0$, the input inductor L_1 are charged by the DC-side power supply U_{in} through the switch S_1, so the inductor current i_{L1} increases linearly. Meanwhile, the bus capacitor C_1 transfers energy to the load passes by the switch S_2 and inductor L_2. So the inductor current i_{L2} increases linearly. The mode ends when S_1 and S_2 are turned off at t_1.

(2) Mode II [t_1–t_2 Fig. 3 (b)]: At $t = t_1$, the input DC power supply Uin and the voltage of inductor L_1 are in series, charging the capacitors C_1 and C_2. The inductor current i_{L2} can not change abruptly, so i_{L2} supplies power to the output through diode D_3, and i_{L2} decreases linearly. At $t = t_2$, when the inductor current i_{L1} decreases to zero, D_1 is turned off naturally and the mode terminates.

(3) Mode III [t_2–t_3 Fig. 3c]: At $t = t_2$, the current of boost inductor L_1 remains zero during this stage. The inductor current i_{L2} supply power to the load through bus capacitor C_2 and diode D_3, and i_{L2} decreases linearly. At $t = t_3$, the switches S_1, S_2 are turned on and the mode III is terminated. Then, the next switching period begins in the positive half cycle.

(4) Mode IV [t_4–t_5 Fig. 3(d)]: At $t = t_4$, the DC power supply U_{in} charges the inductor L_1 through S_1, the boost inductor current i_{L1} increases linearly. Capacitance C_2 supplies power to the load through S_3 and L_3, and i_{L3} increases linearly. The mode ends when S_1 and S_3 are turned off at t_5.

(5) Mode V [t_5–t_6 Fig. 3e]: In this mode, the input DC power supply U_{in} and the voltage of inductor $L1$ are connected in series to charge capacitors C_1 and C_2. The filter inductor current i_{L3} supplies power to the load through D_2, and i_{L3} decreases linearly. At $t = t_6$, the boost inductor current i_{L1} drops to zero, and the diode D_1 naturally turns off.

(6) Mode VI [t_5-t_6 Fig. 3(f)]: In this mode, the current of inductor L_1 is equal to zero. The current i_{L3} flows through the load through C_1 and D_2, and it decreases

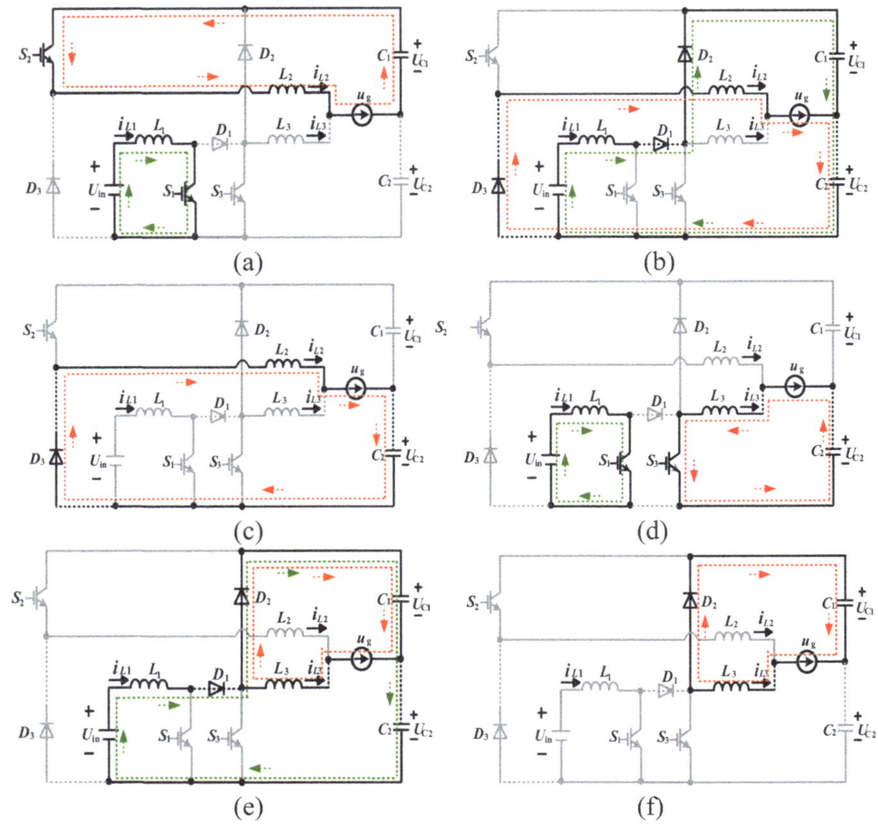

Fig. 3 Equivalent circuit of each mode. Positive half cycle: **a** Mode I. **b** Mode II. **c** Mode III. Negative half cycle: **d** Mode IV. **e** Mode V. **f** Mode VI

linearly. At $t = t_7$, the mode ends when the switches S_1 and S_3 are turned on. The next mode begins in the negative half cycle. Voltage Gain and Parameters of Inverter

2.2 *Analysis of Voltage Gain*

The current waveform of the inductor L_1 working in DCM is shown in Fig. 4. According to the regular symmetry sampling rule, the conducting time of switch S_1 can be expressed as:

$$t_{on} = mT_S|\sin(\omega t)| \tag{1}$$

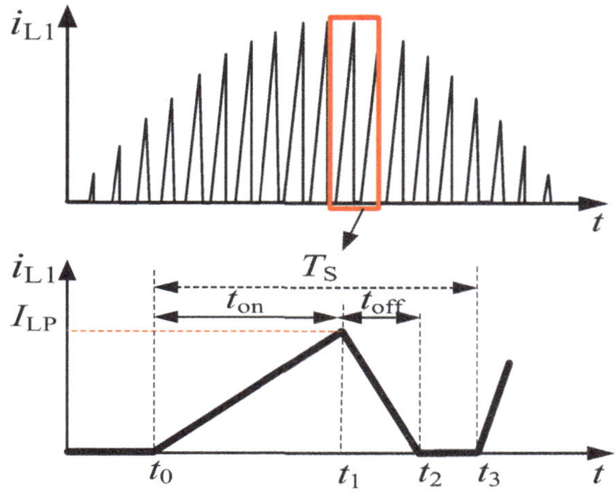

Fig. 4 The current waveform of input Inductance at DCM

Since the inductor L_1 of the inverter is operated in DCM, for the convenience of analysis, it is considered that the voltage of capacitors C_1 and C_2 is equal, that is $U_{C1} = U_{C2} = U_C$. According to the flux (volt-second) balance condition on the inductor L_1, the following formula can be expressed as:

$$U_{in}t_{on} = (U_{C1} + U_{C2} - U_{in})t_{off} = (2U_C - U_{in})t_{off} \tag{2}$$

If all device losses of the inverter are neglected, the input power of the inverter is equal to the output power, so that the following formula can be obtained:

$$I_{in} = \frac{P_o}{U_{in}} = \frac{\left(U_o/\sqrt{2}\right)^2}{U_{in}R_o} \tag{3}$$

U_o is amplitude of the output voltage of the inverter. It can be expressed as:

$$U_O = mU_{C1} \tag{4}$$

The average value of the inductor current is:

$$I_{L1} = \frac{(t_{on} + t_{off})\Delta i}{2T} \tag{5}$$

Δi is the variation of inductance current, and it is expressed as:

$$\Delta i = \frac{U_{in}}{L_1} \times t_{on} \tag{6}$$

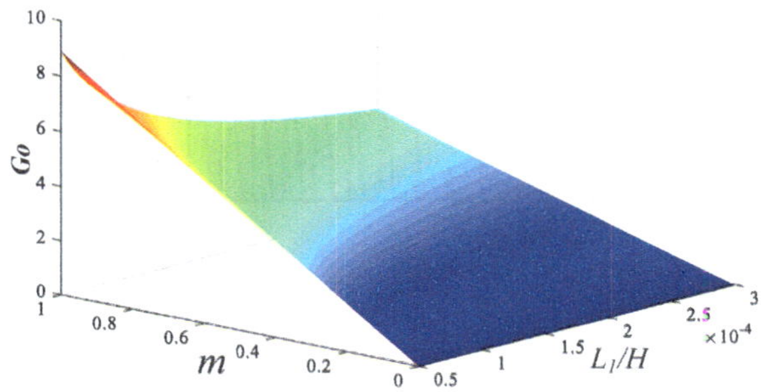

Fig. 5 The relationship between voltage gain G_O and inductance and modulation ratio

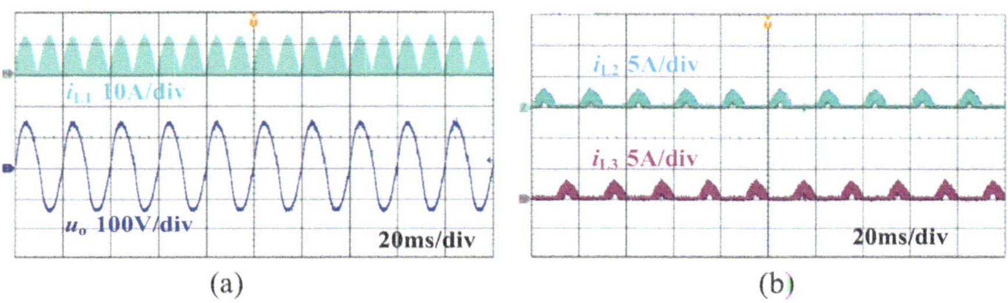

Fig. 6 Experimental waveforms. **a** The currents i_{L1} and the voltage u_o. **b** The currents i_{L2} and i_{L3}

According to formulas (1)–(6), and the gain G_O between the input voltage and the output voltage is mG_C, then the gain of the inverter can be expressed as:

$$G_O = \frac{U_o}{U_{in}} = \frac{m}{4}\left(1 + \sqrt{1 + \frac{8R_O}{L_1 f_s}}\right) \tag{7}$$

Figure 5 shows the three-dimensional relationships of voltage gain G_o versus inductance L_1 and modulation ratio m when $f_S = 20$ kHz, $R_O = 50\ \Omega$ are given. As can be seen from Fig. 6, the G_o increases when the m increases or the L_1 decreases within a certain range.

3 Experimental Results

In order to prove the correctness of theoretical analysis of the proposed inverter, an experimental prototype is built and verified by experimental results, $V_{in} = 45$ V, $L_1 = 110\mu$ H, $L_2 = L_3 = 3$ mH, $C_1 = C_2 = 220\ \mu$F.

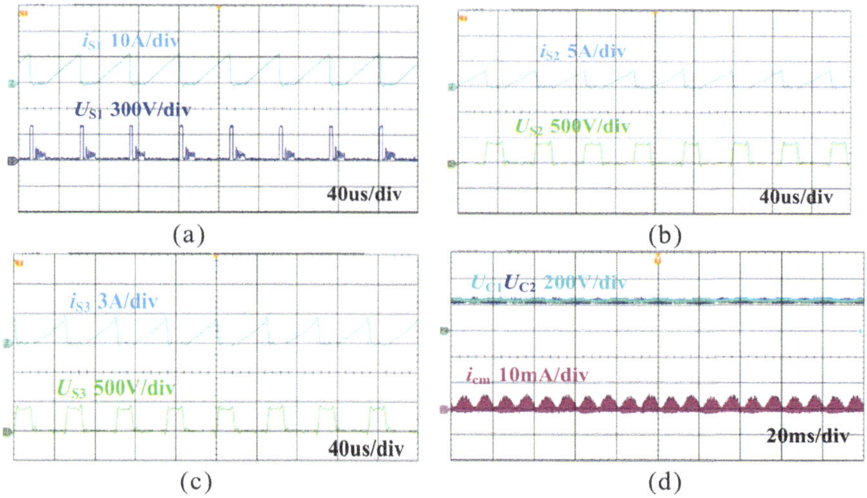

Fig. 7 Experimental waveforms. **a** The currents i_{S1} and the voltage U_{S1}. **b** The currents i_{S2} and the voltage U_{S2}. **c** The currents i_{S3} and the voltage U_{S3}. **d** The currents i_{Cm} and the voltage U_{C1}, U_{C2}

Figure 6a is the experimental waveform of the inductor current i_{L1} and the output voltage u_o. It can be seen from the current waveform of the inductor L_1 that the inductance current i_{L1} is operated in discontinuous mode (DCM). Figure 6b shows experimental waveforms of two filter inductor currents i_{L2} and i_{L3}. It can be seen that the filter inductor current i_{L2} flows only through the positive half cycle, and the inductor current i_{L3} only passes through the negative half cycle.

Figure 7a–c shows the experimental current and voltage stress waveforms of the switches S_1, S_2 and S_3. The maximum voltage stress of each active switch is the sum of the bus capacitor voltages U_{C1} and U_{C2}, liking conventional half-bridge inverter. The maximum current stress of the switch S_1 is equal to the amplitude of the input inductor current i_{L1}, which is approximately 12A. The maximum current stress of the switch S_2, S_3 is equal to the peak value of inductance current i_{L2}, i_{L3}, respectively. Figure 7d shows the voltage U_{C2}, U_{C3} and the common-mode current i_{cm}. There is a good automatic balance between the voltages of U_{C2} andU_{C3}. The leakage current from the presented SPBI, which may easily be seen from Fig. 7d, is very low, and it is about 5 mA. Thus the experimental results are basically consistent with the theoretical analysis.

4 Conclusion

This paper proposes an integrated transformer-less single phase boost inverter (SPBI) with high reliability, in which only three switches and three diodes are used, reducing the number of switches and cost of power generation system. Moreover, there is no shoot-through between the switches in a leg phase and no required dead-time in the SPWM signals, which not only greatly improves the reliability of power system but also reduces the output ac voltage distortion. Fourthly, the SPBI can suppress the leakage current effectively due to its inherent nature, enhancing the safety of the power conversion. The simple unified SPWM strategy for the SPBI is also presented. The effectiveness of the proposed inverter configurations is demonstrated by the experimental results of a 250 W prototype in the laboratory.

Acknowledgements The authors gratefully acknowledge the National Natural Science Foundation (51577002),the Natural Science Foundation of Anhui Province of China (1408085ME80), and the Top-notch Personnel Foundation of the Anhui Higher Education Institutions of China (gxbjZD13) for their financial support.

References

1. Xue, Y., L. Chang, S.B. Kjaer, et al. 2004. Topologies of single-phase inverters for small distributed power generators. *IEEE Transactions on Power Electronics* 19 (5): 1305–1314.
2. Azri, M., N.A. Rahim, and M.F.M. Elias. 2014. Transformerless DC/AC converter for grid-connected PV power generation system. *Arabian Journal for Science & Engineering* 39 (11): 7945–7956.
3. Ciobotaru, M., and R. Teodorescu. 2005. Control of single-stage single-phase PV inverter. Epe Journal, 16 (3), 20–26.
4. Victor, M., F. Greizer, et al. Method of converting a direct current voltage from a source of direct current voltage, a direct current voltage from a source of direct current voltage, into alternating current voltage. In: U.S. Patent 7 411 802 B2 (2005).
5. Guo, X. 2017. A novel CH5 inverter for single-phase transformerless photovoltaic system applications. *IEEE Transactions on Circuits & Systems II Express Briefs* 64 (10): 1197–1201.
6. Xiao, H., S. Xie, Y. Chen, et al. 2011. An optimized transformerless photovoltaic grid-connected inverter. *IEEE Trans. Ind. Electron.* 58 (5): 1887–1895.
7. Khan, M. 2015. H6-type transformerless single-phase inverter for grid-tied photovoltaic system. *Power Electronics Iet* 8 (4), 636–644.
8. Khan, A.A., and H. Cha. 2017. Dual-buck structured high reliability and high efficiency single-stage buck-boost inverters. *IEEE Transactions on Industrial Electronics* 99: 1–1.
9. Yao, Z., L. Xiao, and Y. Yan. 2009. Dual-buck full-bridge inverter with hysteresis current control. *IEEE Transactions on Power Electronics* 56 (8), 3153–3160.

Study on Fractional Order Modeling and Equivalent Stress of AC–DC Superposition Condition for Lithium-Ion Batteries

Bingxiang Sun, Jia Liu, Haijun Ruan, Weige Zhang, Hao Li, and Jiaju Wang

Abstract Lithium-ion battery is an electrochemical system with strong nonlinear characteristics. It shows different response states under different working conditions and doesn't conform to the superposition theorem. In this paper, 2.5Ah ternary lithium-ion battery for electric vehicles is taken as the research object. Fractional order equivalent circuit model is established based on electrochemical impedance spectrum. Its impedance fitting error reaches below 10E−5, which can well simulate the impedance response characteristics inside the battery. Based on the Grunwald–Letnikov fractional differential definition of the output expression of the model is derived, through the sinusoidal excitation at different frequencies for simulation verification, the model relative open circuit voltage error rate is less than 0.2%, with higher accuracy. In addition, the equivalent impact stress under AC–DC superposition condition is analyzed for the battery. It is concluded that the equivalent impact stress for the lithium-ion battery is closer to the superimposed peak condition under the AC–DC superimposed condition of the low and medium frequency.

Keywords Lithium-ion battery · Electrochemical impedance spectroscopy · Fractional order model · AC–DC superposition condition · Equivalent stress

B. Sun (✉) · W. Zhang · H. Li · J. Wang
School of Electrical Engineering, Beijing Jiaotong University, National Active Distribution Network Technology Research Center, Collaborative Innovation Center of Electric Vehicles in Beijing, , Beijing, China
e-mail: bxsun@bjtu.edu.cn

J. Liu
State Grid Beijing Electric Power Company, Qianmen West Street, Xicheng District, Beijing, China

H. Ruan
Faculty of Engineering and the Environment, University of Southampton, Southampton SO14 1BJ, UK

W. Chen et al. (eds.), *The Proceedings of the 9th Frontier Academic Forum of Electrical Engineering*, Lecture Notes in Electrical Engineering 743,
https://doi.org/10.1007/978-981-33-6609-1_11

1 Introduction

As the core component of the "three-electric" system of electric vehicles, the power battery system has become the most important focus of current research [1]. The development of power battery mainly experienced lead-acid battery, nickel-metal hydride battery, lithium-ion battery and fuel cell [2, 3]. Lithium-ion battery has become the main choice of vehicle energy storage system due to its advantages of long life, high safety, high energy density and low self-discharge rate. However, the dynamic characteristics of lithium-ion battery under complex working conditions are subject to great uncertainty. The ability to accurately simulate the external characteristics of the battery under different working conditions is an important link in parameter identification, thermal performance analysis and dynamic system simulation, so an accurate battery model needs to be established. The equivalent circuit model is widely used in battery management system, which describes the internal dynamic response of the battery through simple combination of electrical components such as resistance and capacitance. The main types include Rint model, Thevenin model, Second-order RC model, PNGV model and GNL model [4]. These models are usually based on the response characteristics of the battery under time domain excitation. Compared with the electrochemical models, they do not pay much attention to the reaction process inside the battery but mainly focus on the output accuracy of the model. However, in the monitoring and analysis of various battery states, it is hoped that more characteristic parameters can reflect the changes of battery state while keeping the model simple. Therefore, the fractional equivalent circuit model with strong physical significance and able to explain the internal state to a certain extent is established in this paper, starting from the electrochemical impedance spectroscopy test which reflects more information inside the battery.

The key to determine the accuracy of the model is to understand the variation rule of battery parameters in different states. Jiang et al. [5] analyzed the relationship between battery internal resistance and temperature based on electrochemical impedance spectroscopy tests at different temperatures and verified it through Arrhenis formula. Sun et al. [6] divided different SOC intervals and different ambient temperatures, tested the stress of lithium-ion batteries in cycling conditions and discussed the relationship between battery capacity and internal resistance as well as the correlation between cycling results of different SOC intervals. Huang et al. [7] applied DC superposition AC current frequency sweep excitation to the lithium-ion battery to obtain the battery impedance spectrum data and analyzed the difference in impedance characteristics under charging and discharging directions. But he did not study the response characteristics of the battery after the AC–DC superposition. The above literatures mostly analyzed the response characteristics between battery parameters and temperature or SOC, but seldom studied the parameter characteristics under different current stresses and the equivalent impact stress of AC–DC superposition on the battery.

Therefore, a fractional order equivalent circuit model which can accurately reflect the characteristics of the battery and has certain physical significance is established

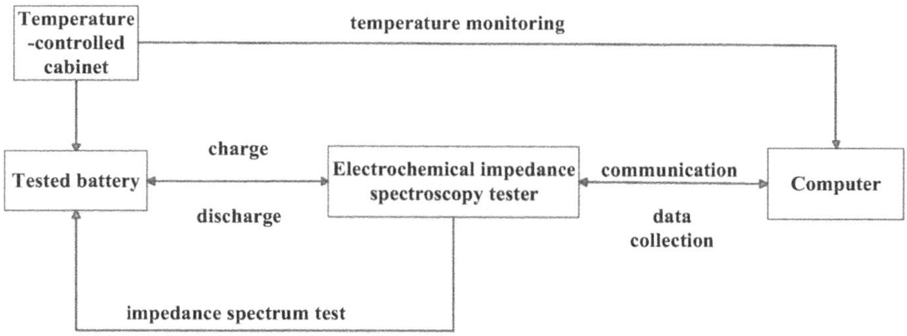

Fig. 1 Device connection diagram

in this paper. Then, the real time state of the battery under the AC–DC superposition condition is taken as the research target, and the impedance data under different states are obtained by pre-test, and then the external characteristics under the AC–DC superposition condition are simulated to determine the equivalent impact stress of the battery under the superposition condition with which amplitude is closer. It has certain significance for parameter estimation of lithium-ion battery model, power system simulation and battery health state estimation.

2 Experiment

2.1 Experimental Platform and Battery Test

The experimental equipment mainly includes the tested battery, the VMP-300 electrochemical workstation, the Su Rui RGD-500 temperature-controlled cabinet and the computer. The specific equipment is connected as shown in the Fig. 1.

The main test object is 2.5 A ternary battery, which is a 18,650 cylindrical battery with a diameter of 18 mm and a height of 65 mm. It has good multiplier characteristics. The specific parameter characteristics are shown in Table 1.

2.2 Experimental Procedure

The test process includes frequency domain impedance spectrum test, constant frequency sinusoidal test and AC–DC superposition test. According to the characteristics of lithium-ion battery, when the temperature is low, the response characteristics of the battery under excitation of different amplitude conditions are more different. So the fixed ambient temperature is 5 °C and the state of charge (SOC) for the battery

Table 1 Parameters of Battery

Performance	Parameters	Unit
Nominal capacity	2500@1C	mAh
Nominal voltage	3.7	V
Charging cut-off voltage	4.20 ± 0.05	V
Discharge cut-off voltage	2.75 ± 0.05	V
maximum charging current	2C (5000)	mA
maximum charging current	5C (12,500)	mA
Working temperate	0–45	°C

is 50% for analysis. As for the AC–DC superposition condition, the design of the superposition condition includes 1C DC superposition 0.5C/1 s periodic square wave pulse (1C/DC + 0.5C/1 s/AC) and 1C DC superposition 0.5C/0.2 s periodic square wave pulse (1C/DC + 0.5c/0.2 s/AC). The rate is the physical quantity representing the charging and discharging speed, whose unit is "C". If a battery with a capacity of 2.5 A is charged with 1.25A current, the charging rate is 1.25/2.5 = 0.5C. This paper refers to the battery test manual and FreedomCAR, USABC and other test standards. The specific test procedures are in Table 2.

Table 2 Test procedure

Test procedure	
Electrochemical impedance spectroscopy test with different amplitudes	(I) Charge the battery in the standard way, standing for 1 h (II)1/3C constant current discharge to 50%SOC, standing for 1 h (III) The sweep frequency range is set to 50 kHz-1 Hz, and the sweep frequency amplitude is adjusted to 0.5C for testing, and standing for 20 min (IV) Change the sine amplitude in step III to 1C and 1.5C respectively for the impedance spectrum sweep frequency test at different current rate
Fixed frequency test	(I) Maintain the SOC at 50% (II) Apply 20 Hz sinusoidal excitation to the battery through the electrochemical workstation, and collect the excitation and response data of the battery (III) Change the frequency in II to 50 Hz and 100 Hz, and collect data
AC–DC superposition condition test	(I) Maintain the SOC at 50% (II) Apply 1C/DC + 0.5C/1 s/AC excitation condition, collect data and stand for 1 h (III) Repeat step II, change the pulse period of AC square wave to 0.2 s, and standing for 1 h

Electrochemical impedance spectroscopy needs to meet the principle of quasi-steady state. When the sine frequency sweep amplitude is large, it needs to ensure that the battery can quickly recover to the original state, so the frequency sweep time needs to be short enough. The sweep frequency range is set above 1 Hz. According to literature [8], the concentration polarization process inside the battery has not started yet. At this time, the parameter values of the battery under the amplitude of different working conditions within the frequency range are identified for stress analysis. In fact, concentration polarization will occur when DC with certain amplitude is superimposed. In order to simulate the actual state, the concentration polarization parameters of the battery were identified in advance through small value test and set as fixed during stress analysis. At the same time, in order to reduce the influence of concentration polarization process, the model simulation test is controlled in a short time range. The reaction process within the cell of this frequency range will be analyzed in detail below.

3 Result Analysis

3.1 Data Processing and Model Building

First, according to the electrochemical impedance spectrum obtained by the test, the appropriate electrical components are selected to describe the impedance characteristics, and the impedance spectrum fitting software Zsimpwin is used to build the model. Figure 2 shows the impedance spectrum fitting results, *iter* is the number of iterations, *Chsq* is the fit value. The smaller the two, the better. *Chsq* reaches 7.58E–5. It can be seen that the model built in this paper can well simulate the battery

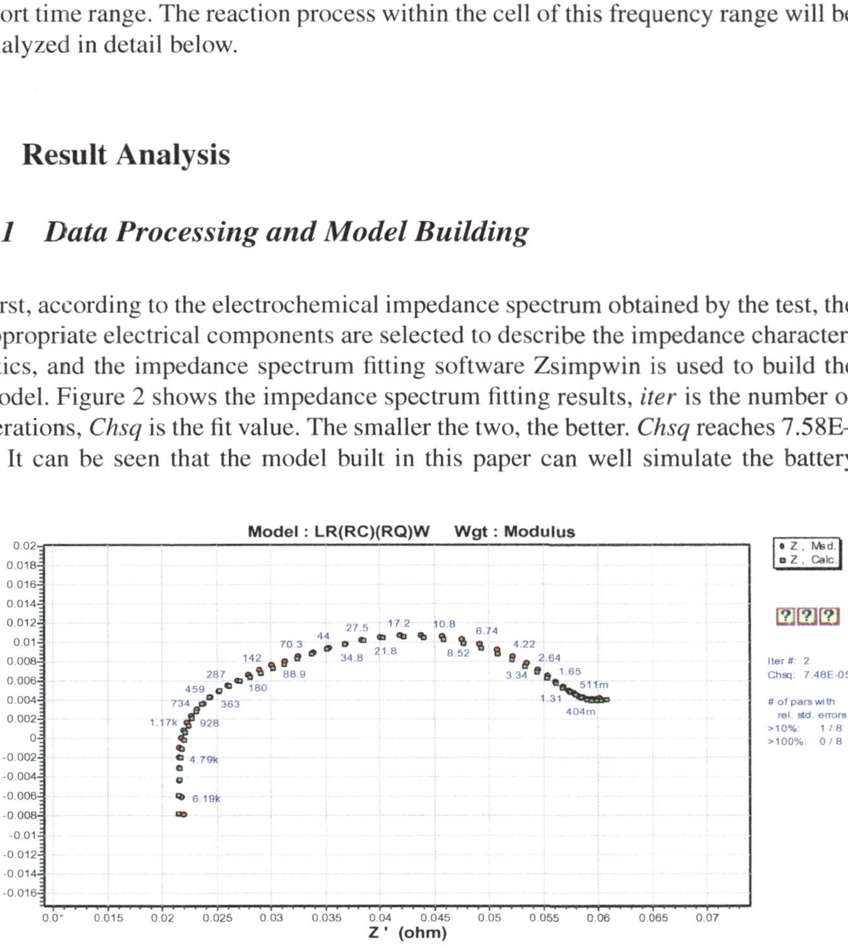

Fig. 2 The fitting result of impedance spectrum

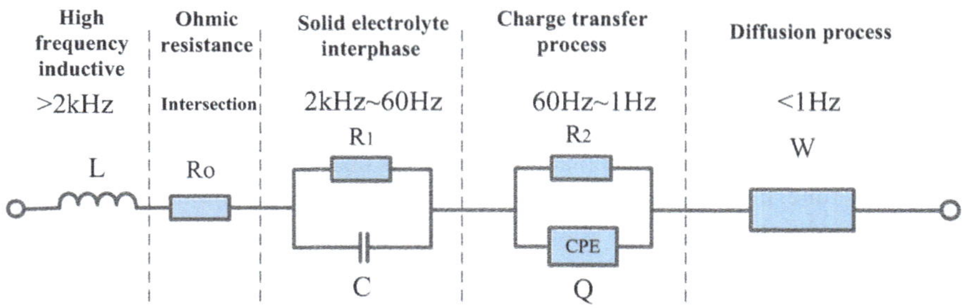

Fig. 3 Equivalent impedance structure

impedance characteristics. The model is shown in Fig. 3. The reaction process of the battery in different frequency range corresponds to different impedance behavior. It contains a constant phase element Q, which can simulate the capacitive dispersion effect. In the process of modeling and derivation of the battery, the impedance expression of the constant phase element will have fractional differential operator, so it is also called fractional order element. The corresponding model is the fractional order model.

In the fractional order model shown in Fig. 3, the Warburg element W, which represents the concentration diffusivity polarization process, keeps a fixed value during the study. It is also a constant phase element, and the fractional order operator is usually set as 0.5 [9]. Summarize the shorter time scales more than about 1 Hz electrochemical behavior in the impedance characteristics of different amplitude excitation condition. Then the external characteristic of AC–DC superposition is simulated to analyze its equivalent stress to the battery. Therefore, it is necessary to test the impedance characteristics of the battery at different amplitudes. The possible equivalent excitation amplitude stresses in the AC–DC superposition conditions studied in this paper include 0.5C, 1C, 1.5C. The corresponding electrochemical impedance spectrum is shown in Fig. 4.

3.2 *Fractional Order Equivalent Circuit Modeling*

According to the impedance spectrum characteristics of the research object, a fractional order equivalent circuit model that can describe the reaction process of the battery is established, as shown in Fig. 5. The output expression of the battery external voltage is derived below. The Laplace transform of fractional order components is as follows (1):

$$Z_Q = 1/Q \cdot s^{\alpha} \tag{1}$$

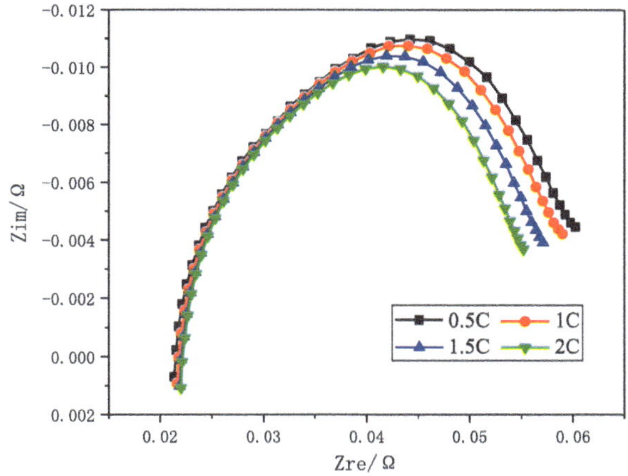

Fig. 4 Impedance spectrum characteristics of batteries at different current-rate

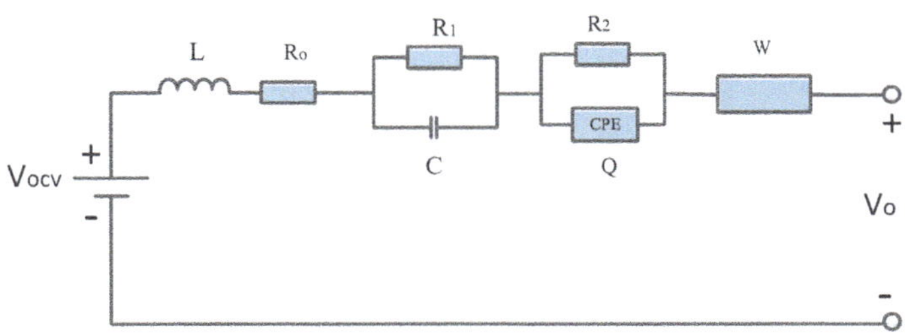

Fig. 5 Fractional order equivalent circuit model

Q is the value of the constant phase element, α is the fractional differential operator, According to Kirchhoff's voltage law and Laplace transform, the input and output transfer function of the battery system is expressed as Eq. (2):

$$\frac{V_{ocv}(s) - V_o(s)}{I(s)} = Z(s) = s \cdot L + R_0 + \frac{R_1}{1 + s \cdot R_1 C_1} + \frac{R_2}{1 + s^{\alpha_2} \cdot R_2 Q_2} + \frac{1}{s^{0.5} \cdot W} \tag{2}$$

Equation (2) is divided into Eq. (3):

$$\begin{aligned}\frac{V_{ocv}(s) - V_o(s)}{I(s)} &= \frac{LR_1R_2CQW \cdot s^{\alpha+2.5} + LR_1CW \cdot s^{2.5} + R_1R_2CQ \cdot s^{\alpha+1}}{R_1R_2CQW \cdot s^{\alpha+1.5} + R_1CW \cdot s^{1.5} + R_2QW \cdot s^{\alpha+0.5} + W \cdot s^{0.5}} \\ &+ \frac{(LR_2QW + R_oR_1R_2CQW) \cdot s^{\alpha+1.5} + (LW + R_oR_1CW + R_1R_2CW) \cdot s^{1.5}}{R_1R_2CQW \cdot s^{\alpha+1.5} + R_1CW \cdot s^{1.5} + R_2QW \cdot s^{\alpha+0.5} + W \cdot s^{0.5}} \\ &+ \frac{(R_oR_2QW + R_1R_2QW) \cdot s^{\alpha+0.5} + R_2Q \cdot s^{\alpha} + (R_oW + R_1W + R_2W) \cdot s^{0.5} + R_1C \cdot s + 1}{R_1R_2CQW \cdot s^{\alpha+1.5} + R_1CW \cdot s^{1.5} + R_2QW \cdot s^{\alpha+0.5} + W \cdot s^{0.5}}\end{aligned} \tag{3}$$

The time domain discrete expression of output voltage involves fractional differential calculation. There are three main ways of defining fractional calculus: GrÜnwald-Letnikov (G-L) definition, Riemann–Liouville (R-L) definition and Caputo definition. G-L definition has a discrete form, which is simple and easy to understand and convenient for direct mathematical implementation [10, 11]. Therefore, the method defined by G-L is mainly used to deal with the fractional order problems involved in this paper. The specific definition is introduced below.

The GrÜnwald-Letniko of the α order derivative of given function $f(t)$ is defined as:

$$ {}_{t_0}^{GL}\mathfrak{D}_t^{\alpha} f(t) = \lim_{h \to 0} \frac{1}{h^{\alpha}} \sum_{j=0}^{[(t-t_0)/h]} (-1)^j \binom{\alpha}{j} f(t - jh) \tag{4} $$

$[\cdot]$ means take the nearest integer.

So the fractional order definition of G-L is actually a generalization of the multi-order formula. Its expression is characterized by conciseness and understandability.

The deformation derivation of Eq. (3) is shown in Eq. (5):

$$ \begin{aligned} &\left(R_1 R_2 C Q W \cdot \mathfrak{D}^{\alpha+1.5} + R_1 C W \cdot \mathfrak{D}^{1.5} + R_2 Q W \cdot \mathfrak{D}^{\alpha+0.5} + W \cdot \mathfrak{D}^{0.5}\right) \\ &[V_{ocv}(t) - V_o(t)] = \left(L R_1 R_2 C Q W \cdot \mathfrak{D}^{\alpha+2.5} + L R_1 C W \cdot \mathfrak{D}^{2.5}\right. \\ &\left.+ R_1 R_2 C Q \cdot \mathfrak{D}^{\alpha+1}\right) \cdot I(t) + ((L R_2 Q W + R_o R_1 R_2 C Q W) \\ &+ \cdot \mathfrak{D}^{\alpha+1.5} + (L W + R_o R_1 C W + R_1 R_2 C W) \cdot \mathfrak{D}^{1.5}\big) \cdot I(t) \\ &+ \big((R_o R_2 Q W + R_1 R_2 Q W) \cdot \mathfrak{D}^{\alpha+0.5} + R_2 Q \cdot \mathfrak{D}^{\alpha} \\ &+ \big(R_o W + R_1 W + R_2 W \cdot \mathfrak{D}^{0.5} + R_1 C \cdot \mathfrak{D}^1 + 1\big) \cdot I(t)\big) \end{aligned} \tag{5} $$

According to the above equation, it can be seen that there is a non-integer order differential ($\mathfrak{D}_t^{\alpha}$) of voltage and current. Therefore, G-L fractional order definition is adopted to solve the problem. First define four dimensional variables: α, a, β, b. α, a represent the degree vector and the coefficient vector of the denominator of the transfer function. β, b represent the degree vectors and coefficient vectors of the molecular polynomial of the transfer function. As shown in Eq. (6):

$$ \begin{aligned} &\alpha = [\alpha + 1.5, 1.5, \alpha + 0.5, 0.5] \\ &a = [R_1 R_2 C Q W, R_1 C W, R_2 Q W, W] \\ &\beta = [\alpha + 2.5, 2.5, \alpha + 1, \alpha + 1.5, 1.5, \alpha + 0.5, \alpha, 0.5, 1, 0] \\ &b = \begin{bmatrix} L R_1 R_2 C Q W, L R_1 C W, R_1 R_2 C Q, (L R_2 Q W + R_o R_1 R_2 C Q W), \\ (L W + R_o R_1 C W + R_1 R_2 C W), (R_o R_2 Q W + R_1 R_2 Q W), R_2 Q, \\ (R_o W + R_1 W + R_2 W), R_1 C, 1 \end{bmatrix} \end{aligned} \tag{6} $$

According to the battery characteristics, rewrite $[V_{ocv}(t) - V_o(t)]$ as polarization voltage $V_p(t)$. Then, according to Eq. (6) and the fractional order definition of G-L, the fractional derivative of both sides of the equation can be obtained:

$$\begin{aligned} &\text{Left: } a_i\mathfrak{D}^{\alpha_i} V_p(t) = \frac{a_i}{h^{\alpha_i}} \sum_{j=0}^{N} (-1)^j \binom{\alpha_i}{j} V_p(t - jh) \\ &\text{Right: } b_i\mathfrak{D}^{\beta_i} I(t) = \frac{b_i}{h^{\beta_i}} \sum_{j=0}^{N} (-1)^j \binom{\beta_i}{j} I(t - jh) \end{aligned} \tag{7}$$

h is sampling interval. V_p is polarization voltage (the open circuit voltage of the battery minus the terminal voltage). I is charge and discharge current. i is the ith fractional order operation. j is the jth data point. N is the data length indicating that the current state is related to the historical data of the number of N. Adding the weight operator can change the influence degree of historical data to the current moment.

In formula (6), i in the numerator and denominator of multidimensional variables can be taken as 4 and 10 respectively in this model. Therefore, the whole system time-domain discrete formula of the model can be further obtained according to G-L definition:

$$\sum_{i=1}^{4}\sum_{j=0}^{N} \frac{a_i}{h^{\alpha_i}} (-1)^j \binom{\alpha_i}{j} V_p(t - jh) = \sum_{i=1}^{10}\sum_{j=0}^{N} \frac{b_i}{h^{\beta_i}} (-1)^j \binom{\beta_i}{j} I(t - jh) \tag{8}$$

By separating the left side $j = 0$ in Eq. (8), the numerical solution form of $V_p(t)$ can be obtained:

$$V(t) = \frac{1}{\sum_{i=1}^{4} \frac{a_i}{h^{\alpha_i}}} \left\{ \sum_{i=1}^{10}\sum_{j=0}^{N} \frac{b_i}{h^{\beta_i}} (-1)^j \binom{\beta_i}{j} I(t - jh) - \sum_{i=1}^{4}\sum_{j=1}^{N} \frac{a_i}{h^{\alpha_i}} (-1)^j \binom{\alpha_i}{j} V_p(t - jh) \right\} \tag{9}$$

Therefore, the discretization form of fractional order model output is shown in Eq. (10):

$$V(k) = \frac{1}{\sum_{i=1}^{4} \frac{a_i}{h^{\alpha_i}}} \left\{ \sum_{i=1}^{10} \sum_{j=0}^{N} \frac{b_i}{h^{\beta_i}} (-1)^j \binom{\beta_i}{j} I(k-j) - \sum_{i=1}^{4} \sum_{j=1}^{N} \frac{a_i}{h^{\alpha_i}} (-1)^j \binom{\alpha_i}{j} V_p(k-j) \right\} \tag{10}$$

The output expression of the fractional order model is used for model simulation output to judge the accuracy of parameters and model construction.

4 Model Validation and Stress Analysis

4.1 *Fractional Order Model Simulation Verification*

In order to verify the accuracy of the model, it is verified under sinusoidal excitation at three frequencies of 100 Hz, 50 Hz and 20 Hz, respectively. As shown in Fig. 6, the simulation results of battery external characteristics under the 4A amplitude are presented. Meanwhile, the simulation root mean square error value (RMSE) and the relative open circuit voltage error rate are summarized in Table 3.

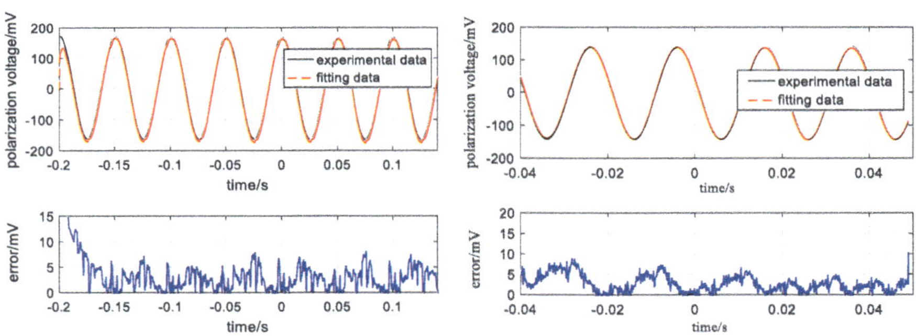

Fig. 6 Simulation results of fractional order model for 20 Hz and 50 Hz

Table 3 Error analysis

Frequency/Hz	RMSE/mV	Error rate/%
20	3.370	0.091
50	3.541	0.096
100	5.422	0.147

It can be seen that the fractional-order model built under different frequency excitation has a RMSE within 6 mV and a relative open circuit voltage error rate of less than 0.2%, which is of high accuracy.

4.2 Analysis of the Equivalent Stress Under the AC–DC Superposition Condition

The working condition is shown in Fig. 7, the 1C DC superimposed 0.5C/1 s square pulse apply to the battery, then the equivalent stress may be the peak, the average or valley value after superimposed.

The equivalent stress of the battery may be 1.5C/1C/0.5C, so the model parameters under the three excitation amplitudes are identified by electrochemical impedance spectroscopy. The results of parameter identification of 0.5C, 1C and 1.5C are respectively brought into the model for the simulation calculation of the external voltage for the battery and the comparative analysis of the actual collected data are as follows (Fig. 8).

According to the simulation error comparison of different equivalent model parameters for the above AC–DC superposition conditions, it can be found that the equivalent stress of the 1C DC with the 0.5C/1 s square wave is closer to the peak value of the superposition condition.

Change the square wave period to 0.2 s, and continue to analyze 1C DC superimposed 0.5C/0.2 s square wave pulse applying to the battery (Fig. 9).

It can be seen that the simulation error under the peak value is still the smallest after the period is changed to 0.2 s, indicating that the equivalent impact stress on the battery under this period is also closer to the peak value.

Fig. 7 Testing condition

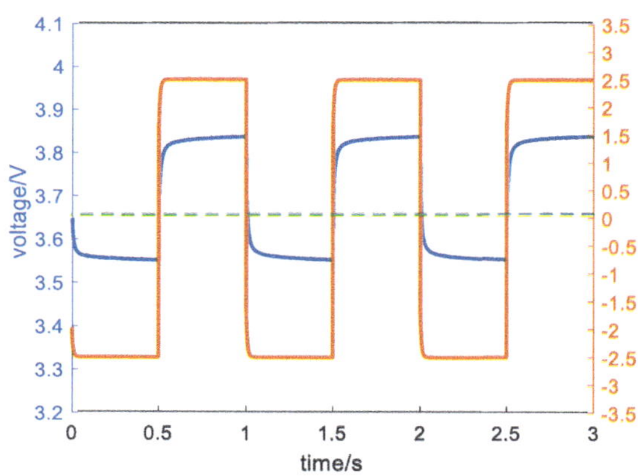

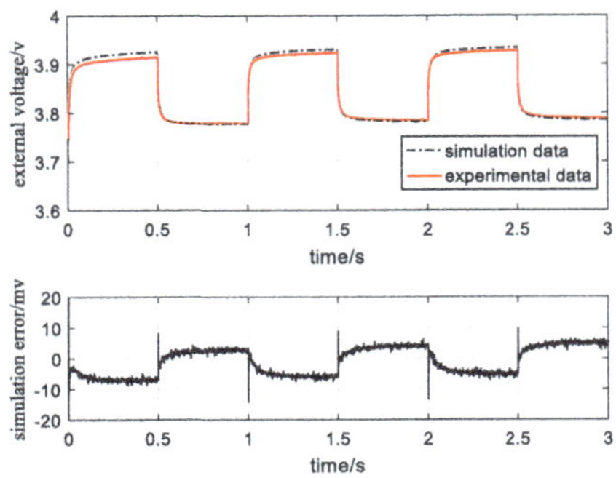

(a) Simulation result of peak parameter

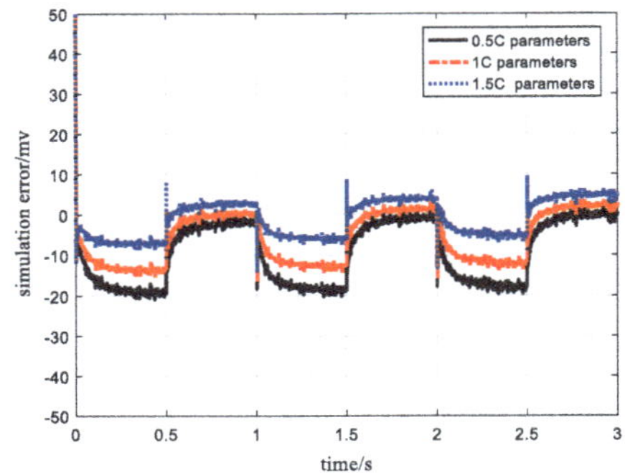

(b) Error comparison

Fig. 8 Simulation results of different stress parameters

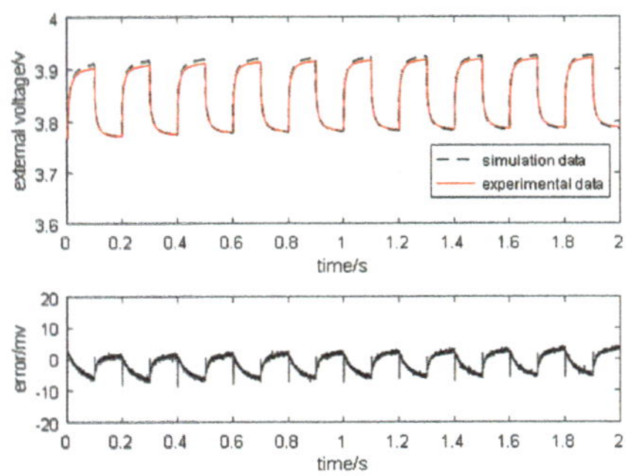

(a) Simulation result of peak parameter

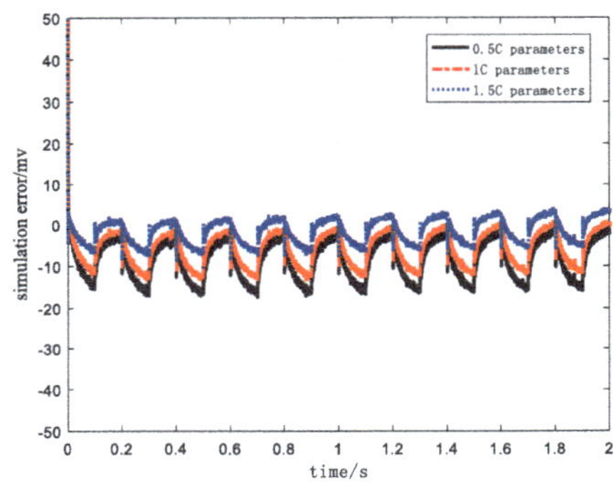

(b) Error comparison

Fig. 9 Simulation results of different stress parameters

5 Conclusion

In this paper, 2.5 A ternary lithium-ion battery is taken as the research object. Based on the electrochemical impedance spectrum test method which reflects a lot of interface information, the fractional order equivalent circuit model is established. The impedance fitting error of the model is below 10e−5, which can well simulate the internal impedance characteristics of the battery. According to the G-L definition method of fractional order, the output voltage formula of the model is derived, and the sinusoidal excitation at different frequencies is simulated. The error of the model is within 6 mV, and the error rate is less than 0.2%, which is of high precision. On this basis, the equivalent impact stress of the battery under AC–DC superposition condition is analyzed, and the conclusion is drawn that when the low-frequency AC–DC superposition condition is applied to the battery system, the stress for the battery is closer to the equivalent impact effect of the peak value after superposition, which is of certain significance for model parameter estimation and battery health state estimation.

Acknowledgements This work is supported by the "National Key R&D Program of China" (Grant NO.: 2017YFB1201005) and the National Natural Science Foundation of China (Grant No. 51907005).

References

1. Che, Tulin, Weishi Liu, and Bin Wang. 2019. Development status and trend of new energy commercial vehicle power battery. *Heavy-duty Car* 5, 37–39.
2. Yao, Lejing. 2019. Research progress and Prospect of new energy vehicle power battery. *Contemporary Chemical Research* 10: 5–7.
3. Cheng, Hao. 2019. A brief analysis of the development trend of new energy vehicles. *Practical Automobile Technology* 22: 1–2.
4. Yang, Jie, Ting Wang, and Du. Chunyu. 2019. A review of lithium-ion battery model research. *Energy Storage Science and Technology* 8: 58–64.
5. Jiang, Jiuchun, Wei Shi, and Yanru Zhang. 2014. Impedance spectrum parameter analysis of lithium iron phosphate power battery. *Journal of Beijing Institute of Technology* 34, 470–474 (2014).
6. Sun, Bingxiang, Jia Liu, and Zhiqiang Han. 2020. Analysis of performance correlation and temperature applicability of li-ion batteries under different fading paths. *Transactions of China Electrotechnical Society* 1–11 (2020).
7. Huang, Jun, Jianbo Zhang, and Zhe Li. 2014. Exploring differences between charge and discharge of $LiMn_2O_4$/Li half-cell with dynamic electrochemical impedance spectroscopy. *Electrochimica Acta* 131: 228–235.
8. Alavi, S.M.M., C.R. Birkl, and D.A. Howey. 2015. Time-domain fitting of battery electrochemical impedance models. *Journal of Power Sources* 288: 345–352.
9. Minghui, Hu., Yunxiao Li, and Shuxian Li. 2018. Lithium-ion battery modeling and parameter identification based on fractional theory. *Energy* 165: 153–163.
10. Bian, Jingji. 2019. *Modeling of Lithium-ion power battery based on fractional order theory [D]*. Beijing: Beijing Jiaotong University.
11. Bryan, H., E.O. Mark, T. Bernard, V. Vincent, F. Isabelle, and M. Marco. 2009. Determination of effective capacitance and film thickness from constant-phase-element parameters. *Electrochimica Acta* 55: 6218–6227.

Analysis Method of Power System Communication Fault Based on Dynamic Fuzzy Petir

Zilan Zhao, Ran Yu, Meng Yu, Jiaojiao Zhang, Ying Wan, and Hongfei Xu

Abstract The characteristics of electric power communication field faults show many characteristics such as multi-level, multi Association, and uncertainty, which brings difficulties to the accurate diagnosis of field faults. This paper, taking the fuzzy Petri net theory as the core, introduces the coloring rules of repository, token and change, puts forward a modeling and reasoning method to improve the fuzzy Petri net, and applies it to the fault diagnosis in the field of electric power communication, to describe the running condition of the fault more intuitively and accurately. Through the experiment simulation, the linear correlation analysis is carried out to improve the reasoning result of the fuzzy Petri net and the normalized data of historical fault statistics, and the correctness, validity, and feasibility of the algorithm are verified. It has good feasibility and practicability and can provide auxiliary analysis support for the fault diagnosis of the electric power communication field.

Keywords Electric power communication network · Fuzzy petri · Fault diagnosis

1 Introduction

In recent years, the scale of electric power communication networks has grown rapidly, effectively improved the safety level of power grid dispatching and control, enterprise production and operation management, and ensured the reliability of various types of information transmission in power systems in harsh climate environments. But at the same time, the degrees of coupling among communication networks are getting closer and closer, and the complexity of the network topology is also rapidly rising, which has an adverse effect on on-site operation and maintenance and troubleshooting. With the rapid development of computer technology and automation technology, professional network management systems, smart devices, sensors and other equipment have been popularized and applied in communication networks, providing the required multi-source data for on-site fault diagnosis. If

Z. Zhao (✉) · R. Yu · M. Yu · J. Zhang · Y. Wan · H. Xu
State Grid JiBei Electric Power Co., Ltd, Beijing 100053, China
e-mail: zhaozilan0609@163.com

W. Chen et al. (eds.), *The Proceedings of the 9th Frontier Academic Forum of Electrical Engineering*, Lecture Notes in Electrical Engineering 743,
https://doi.org/10.1007/978-981-33-6609-1_12

the multi-source data that characterizes the operating status of the communication network can be effectively used and identified, it can provide technical support for fault diagnosis at the communication site. In order to reduce and avoid the cascading reaction that may be induced by a single point of failure, it is necessary to be able to quickly and timely detect potential hidden dangers in the communication network, accurately diagnose and eliminate the risk of failure, and improve the safety and reliability of the communication system [1, 2]. Therefore, it is imperative to research and establish a set of efficient and reasonable communication field fault diagnosis methods. In the electric power communication networks, the relationship between the communication equipment and the optical cable is a mesh topology. The Petri network model is a combined model of graphical representations, which can make qualitative and quantitative analysis of various activities and failures of the communication network, similar to the occurrence and processing of communication network failures [3–5]. With further research, experts and scholars have proposed Fuzzy Petri Net (FPN), which is more in line with human normal thinking habits and cognitive methods, and can solve network diagnosis problems by intuitive expressions and matrix operation methods.

Based on the theory of fuzzy Petri net, this paper proposes a method for on-site fault diagnosis of dynamic electric power communication network, which effectively avoids the error caused by human subjective factors in the process of fault diagnosis of fuzzy Petri net and improves the accuracy of the reasoning process. The simulation results show that the method has good adaptability and versatility, and provides a new idea for communication field fault diagnosis.

2 Background

FPN is proposed based on the combination of fuzzy mathematical ideas and traditional Petri nets. It can accurately and clearly describe and analyze the uncertain fuzzy state information of the elements in the simulated system. The main improvement of FPN is that each library is given a real number between [0, 1] as its confidence value, and each transition is represented by a CF (confidence Factor) to indicate the probability of its occurrence. The inference process is represented by the trigger of changes in inference Petri nets [6–10].

A typical fuzzy Petri net is a seven-tuple:

$$SFPN = (P, T, I, O, \alpha, \beta, \tau) \tag{1}$$

where: $\mathrm{P} = (\mathrm{p1}, \mathrm{p2}, \ldots, (\mathrm{pn}))$ is finite set of nodes of FPN Place; $\mathrm{T} = (\mathrm{t1}, \mathrm{t2}, \ldots, \mathrm{tm})$ is a finite set of nodes of FPN Transition; $\mathrm{I} : \mathrm{P} \rightarrow \mathrm{T}$ is the input matrix, which refers to the mapping from FPN Place to FPN transition, directed arc set to the transition; $\mathrm{O} : \mathrm{T} \rightarrow \mathrm{P}$ is the output matrix, indicating the mapping from FPN transition to FPN place, which is the directed arc set to the place; $\alpha : \mathrm{P} \rightarrow [0, 1]$ indicating the corresponding confidence level for the place P, $\beta : \mathrm{T} \rightarrow [0, 1]$ which indicates the confidence level for the transition T; τ is the ignition threshold for the transition T, which is a function with the value [0,1] in T.

3 An Improved Algorithm Based on Fuzzy Petri Net

The value of α, τ in a typical fuzzy Petri net directly affects the expression of the dynamic characteristics of the fuzzy Petri net, because it is directly related to the ignition capability of the fuzzy transition and the state change of the fuzzy place. The faults in the power communication site are complex and diverse, and there are many correlations among them. The fuzzy Petri net model built by it will be complicated and huge, which directly leads to the failure of the fault information and the expression of the propagation path to be unclear. In view of the multi-level, multi-variable, non-linear and multi-association fault characteristics presented in the power communication site, this paper introduces the coloring rules of the place, token and transition on the basis of the fuzzy Petri net theory, and proposes a kind of coloring rule suitable for the communication site Improved dynamic fuzzy Petri net algorithm for fault diagnosis [11–15].

The Dynamic Fuzzy Petri Net (DFPN) algorithm is defined as a 12-tuple,

$$SDFPN = (U, V, I, O, K, Y, M, w, P, \alpha, 2) \tag{2}$$

Among them,

(1) $U = (u1, u2, \ldots, un)$, $n > 0$ is a collection of finite non-empty warehouses, indicating a collection of communication field failures.
(2) $U = (v1, v2, \ldots, vm)$, $m > 0$ is a finite set of non-empty transitions, which represents the status change or behavior of the communication site fault, reflecting the replacement of the fault transmission phase. $U \cup V \neq \emptyset$ and $U \cap V \neq \emptyset$, it means that there is at least one place and transition, and they are two different types of elements.
(3) $I : U \times V \neq \emptyset$ is the input matrix, indicating a directed arc from the place u i to the transition v j, which is the input arc of the transition v j, and u i is the input place of the transition v j. In the formula,$i = 1, 2, \ldots, n, j = 1, 2, \ldots, m$.
(4) $O : U \times V$ is the input matrix, which represents a directed arc from transition vjgc toui, it is the output arc of transition vj, and ui is the output place of transition vj. In the formula, $i = 1, 2, \ldots, n, j = 1, 2, \ldots, m$.
(5) $K = (k1, k2, \ldots, kn)$, representing a limited set of Token coloring.
(6) Y is a collection of colors, for $\forall u \in U$, Y(u) is a collection of all possible colors of the place U, and different colors represent different types of places; (m) is all the places on the place U Token color set, and different colors represent different confidence levels; Y(v) is the change v all possible color sets, and whether the change color changes color indicates whether ignition occurs.

(7) M = (m1, m2, ..., mn), T is the distribution vector of the sign of the place, mi represents the number and color of the tokens corresponding to the place ui, i = 1,2,......n. The number of tokens indicates the realization of the local state represented by the library. When there is a token, it indicates that a fault has occurred. The number of tokens indicates the number of fault paths and the severity of the fault. Different colors indicate different levels of confidence. M0 is the initial identification vector, indicating the initial state of the system.

(8) w = (w1, w2, ..., wn) T is the weight vector of the failure event of the storehouse, which reflects the degree of influence of the input storehouse on the transition rule v, where ∀u ∈ V, uknk = 1 = 1.

(9) P = (p1, p2, ..., pn) is the failure probability set of the model tree of the storehouse event, pi represents the probability of the occurrence of the fault event represented by the storehouse ui.

(10) $\alpha = (\alpha 1, \alpha 2, \ldots, \alpha n)$ T is the fault event confidence vector, αi represents the fault event ui confidence of the degree of truth, with a certain degree of ambiguity,α i ∈ [0, 1], i, 2,..., n. The initial value is represented by $\alpha 0$, and is based on historical data, expert knowledge, etc.

4 Simulation Implementation and Analysis of the Algorithm

4.1 Modeling Rules of Dynamic Fuzzy Petri Nets

The modeling process of dynamic fuzzy Petri nets mainly includes analyzing the topological structure of the power communication networks, and judging the logical association of faults according to the fuzzy rules of the faults. At the same time, in order to more clearly and accurately express the fault propagation characteristics and fault fuzzy information, the dynamic fuzzy Petri net is modeled through the definition of the place coloring rules and the setting of the fault confidence level according to the Token coloring rules. As shown in Fig. 1.

(1) Coloring rules of the place

If × → = 1, × → = 0, then set the place as the initial place, which is the root cause of the failure, and use "" to indicate the state of the place. If it contains Token coloring, it means that the root cause of the fault is found. If × → = 1, × → = 1 then the location is set as the intermediate location, which is the failure mode, which is indicated by "". If × → = 0, × → = 1, then set the location as the final location, which is the system failure, which is indicated by "". As shown in the example in Fig. 2, u1, u2, u3 are the initial locations, u4 is the intermediate locations, and u5 is the final locations.

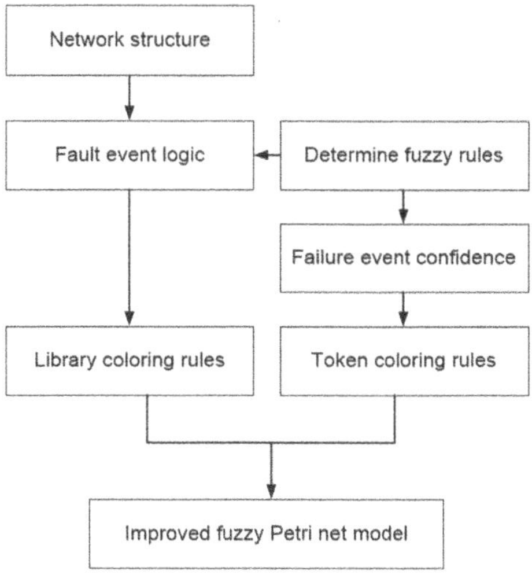

Fig. 1 Modeling flow chart of dynamic fuzzy Petri net

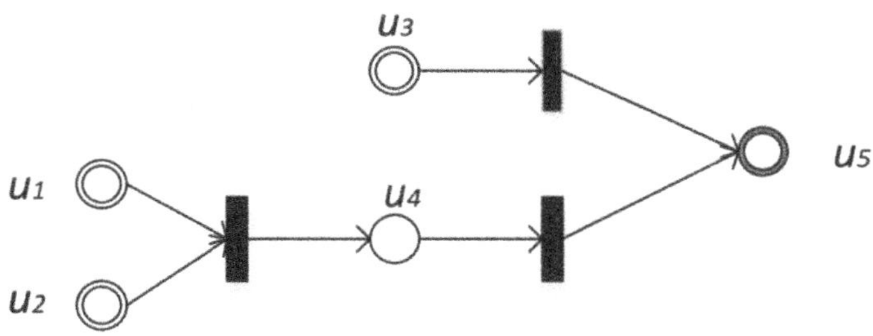

Fig. 2 Example of coloring rules in the library

(2) Token coloring rules

In the on-site fault diagnosis research of electric power communication, the range of the fault confidence is set to (0–1). To visualize the fault information, this paper introduces the Token coloring to represent the confidence level of the fault event (Table 1).

4.2 *Dynamic Fuzzy Petri Net Fault Diagnosis Method*

The dynamic fuzzy Petri fault diagnosis method mainly has the following steps, see Fig. 3.

Table 1 Token coloring rules

Confidence level	Ranges	Color graphics	Color	Meaning
Level 1	[0.91–1.00]	●	Red	Serious failure
Level 2	[0.81–0.90]	⬟	Pink	Important fault
Level 3	[0.71–0.80]	✦	Yellow	General failure
Level 4	[0.31–0.70]	▲	Blue	potential hazards
Level 5	[0.00–0.30]	●	Green	No effect

(1) Determine the weight of each input location, the confidence and threshold of the transition, the confidence of initial place and fault probability based on history operating data, expert experience knowledge and actual monitoring data, etc. Use the confidence matrix method to infer confidence level of the failure events of each place, to provide necessary conditions for the positive and negative intelligent reasoning of the failure.
(2) Based on the confidence and fuzzy transition rules of each place, the potential enablement transition is judged, and the potential enablement transition sequence is obtained, which in turn provides a evidence for the positive and negative intelligent inference ignition judgment.
(3) When no fault occurs in the system, the online monitoring equipment predicts the possible fault symptoms in the network, and the forward intelligent reasoning is used to evaluate the fault status to achieve the evaluation of the system fault severity and the description of the propagation path.
(4) When the system fails, confirm the phenomenon that has occurred, perform reverse intelligent reasoning, trace the most likely source of the failure, and provide guidance for the emergency repair of the communication equipment.

4.3 Experimental Verification

In this paper, taking the transmission equipment failure at the power communication site as an example, the expert experience knowledges are used to map the fuzzy rules of the fuzzy rule set to the initial state of the corresponding dynamic fuzzy Petri net failure model, as shown in Fig. 4.

Figure 4 Dynamic fuzzy Petri net fault model diagram extracts a total of 792 fault records related to the transmission equipment from the historical operation data of the power communication site. Among them, the fault diagnosis inference data and actual fault records about "Poor positioning accuracy of P11 workbench". There are 724 data in total, as shown in Table 2.

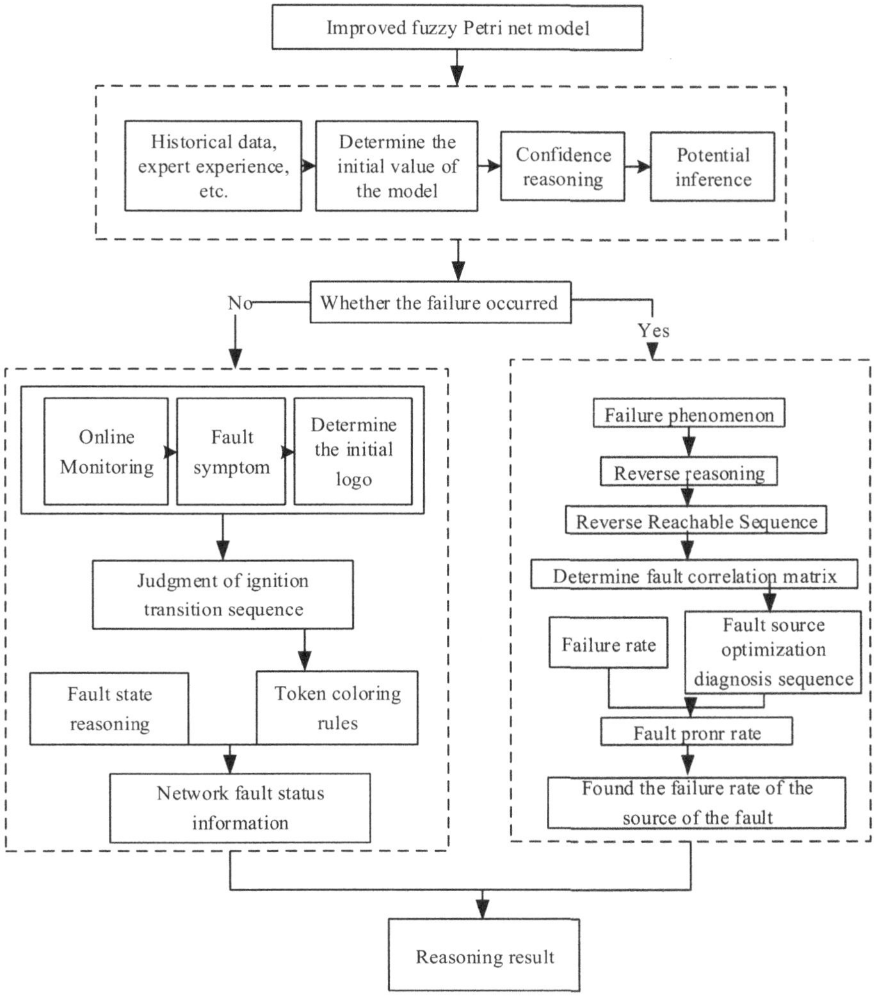

Fig. 3 Fault diagnosis inference graph of dynamic fuzzy Petri net

It can be known from the above experimental verification that this paper uses dynamic fuzzy Petri net model to conduct fault reasoning diagnosis on field faults in power communication. The reasoning results have a strong linear correlation with the actual historical faults on the spot, and have good practical value.

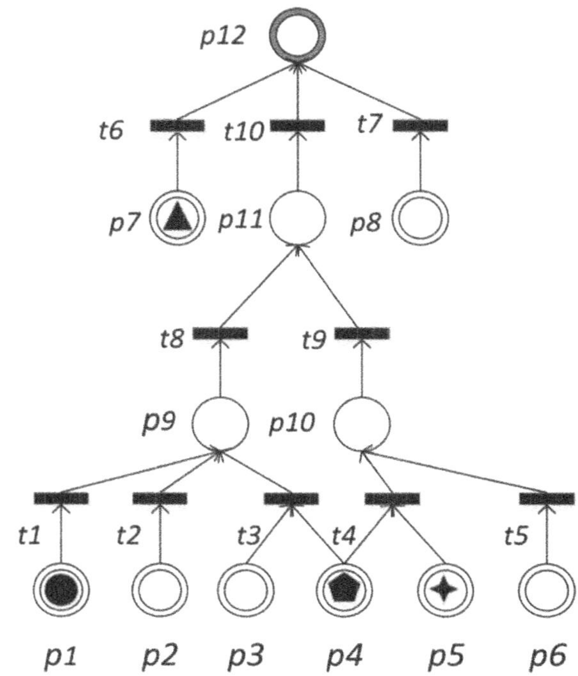

Fig. 4 Dynamic fuzzy Petri net failure model diagram

Table 2 Statistics results

Serial number	Fault phenomenon	Cause of issue	Reasoning result	Historical fault statistics
1	p_{11}	$\{p_1\}$	5.0	322
2	p_{11}	$\{p_2\}$	5.3	305
3	p_{11}	$\{p_3, p_4\}$	0.31	43
4	p_{11}	$\{p_4, p_5\}$	0.4	19
5	p_{11}	Others	0	35

5 Conclusion

Aiming to the problems that on-site fault diagnosis of electric power communication are difficult to solve because of its multi-level, multi-variable, non-linear and multi-correlation characteristics, this paper takes fuzzy theory and Petri net theory as the core, comprehensively uses fault Petri net technology and fuzzy Petri net technology to conduct in-depth research on fault status evaluation and diagnosis of electric power communication site, and proposes a fault model of dynamic fuzzy fault Petri net and modeling rules, to apply to fault reasoning in electric power communication sites.

The algorithm simulation is used to verify the improved fuzzy Petri net model algorithm. By comparing the reasoning results with the normalization of historical fault statistics, the correctness, effectiveness and feasibility of the algorithm are verified, which has good feasibility and practicability. It can provide auxiliary analysis support for the fault diagnosis of the electric power communication site. But at the same time, there is a certain deviation in the diagnosis of the causes of the faults at 2 and 5. The cause of this difference will be studied in the future, and the model algorithm will be further improved, so that the fault diagnosis of the electric power communication site can achieve a more ideal effect.

Acknowledgements This work is supported by the Science and Technology Project of State Grid Jibei Power Company Limited (No. 52018E190075).

References

1. Guo, Chuangxin, Chuanbo Zhu, Yiyi Cao, et al. 2006. Research status and development trend of power system fault diagnosis. *Power System Automation* 30 (8): 98–103. (in Chinese).
2. Bi, Tianshu, Yixin Ni, and Qixun Yang. 2000. Review of the application of artificial intelligence technology in fault diagnosis of transmission network. *Automation of Electric Power Systems* 24 (2): 11–16. (in Chinese).
3. Shi Jian, Weiwei Miao, Haiyang Wu. 2016. Research on LS-SVM-based performance degradation evaluation and prediction model for power communication networks. *Computer and Digital Engineering, 44*: 610–614 (in Chinese).
4. Yang Chunfa. 2004. Research on power grid fault diagnosis method based on Petri net. North China Electric Power University (Beijing) North China Electric Power University (in Chinese).
5. Shi J, Wang SW. 2014. Research on reliability evaluation of power communication network. In *International Conference on Power System Technology*, 1743–1748. IEEE.
6. Zhang Yang, Zheng-you He, Liang Guo. 2010. A power system fault diagnosis method based on mutual-Information Network. In *Asia-Pacific Power and Energy Engineering Conference, APPEEC2010*, Chengdu, China, 1–4. (in Chinese).
7. Xu, Biao, Xianggen Yin, Yang Wang, et al. 2018. Power grid fault diagnosis method based on fuzzy time Petri net. *Power System Automation* 2: 15–19. (in Chinese).
8. Liu, Xinrui, Yiwei Gao, and Zhiliang Wang. 2016. Fault diagnosis method of distribution network based on improved time fuzzy Petri network. *Journal of Northeastern University (Natural Science Edition)* 37 (11): 1526–1529. (in Chinese).
9. Xie, Min, Yaxiong Wu, Yuanyuan Yan, et al. 2015. Power grid fault diagnosis based on improved dynamic adaptive fuzzy Petri net and BP algorithm. *Proceedings of the CSEE* 35 (12): 3008–3017. (in Chinese).
10. Luo, X., and M. Kezunovic. 2008. Implementing fuzzy reasoning Petri Nets for fault section estimation. *IEEE Transactions on Power Delivery* 23 (2): 676–685.
11. He, Z.Y., J.W. Yang, Q.F. Zeng, et al. 2014. Fault Section Estimation for Power Systems Based on Adaptive Fuzzy Petri Nets. *International Journal of Computational Intelligence Systems* 7 (4): 605–614.
12. Sun, J., S.Y. Qin, and Y.H. Song. 2004. Fault diagnosis of electric power systems based on fuzzy Petri nets. *IEEE Transactions on Power Systems* 19 (4): 2053–2059.
13. Zhang, Y., Y. Zhang, F. Wen, et al. 2016. A fuzzy Petri net based approach for fault diagnosis in power systems considering temporal constraints. *International Journal of Electrical Power & Energy Systems* 78 (6): 215–224.

14. Tong, X., H. Xie, and M. Sun. 2013. Power system fault diagnosis model based on layered fuzzy Petri Net considering temporal constraint checking. *Automation of Electric Power Systems* 37 (6): 63–68.
15. Huang YC, CM Huang, CC Liao et al. 2000. A new intelligent fast Petri-net model for fault section estimation of distribution systems. In *International Conference on Power System Technology, 2000. Proceedings. Power*, vol. 1, 217–222. IEEE.

Partial Discharge Characteristics and Development Process of GIS Insulator with Diverse Defects

Yulun Chen, Ziru Zha, Quanhao Li, Guanjun Zhang, and Dingge Yang

Abstract Gas insulated switchgear (GIS) is a widely used apparatus in both middle voltage system and high voltage system. As for the defects diagnosis and condition monitoring of GIS, ultra-high frequency (UHF) detection is generally preferred. However, the conventional research on the diagnosis of partial discharge (PD) evolution is not precise enough. Thus the aim of this paper is to study the development process of the PD in various ways. An experimental platform is designed to complete both UHF and pulse current detection simultaneously, and typical artificial defects are prepared. Analysis of PD signals using the algorithm of Convolutional Neural Networks (CNN) has been carried out. An intensified Charge Coupled Device (ICCD) camera is also utilized in the whole system to capture the process of the discharge. The results show that it is effective in identifying the PDs of different types. Another thing observably is that whatever type of defects, the main frequency of the UHF signals is a constant which equals to 250 MHz, but the intrinsic mechanism of it needs to be further studied.

Keywords GIS · Partial discharge · Pulse current method · Ultra-high frequency method · Insulator

1 Introduction

Nowadays GIS is widely utilized in many fields from middle voltage system even to ultra-high voltage system [1, 2]. And when monitoring or diagnosing the GIS, features like in-situation, easy to operate, non-intrusive, accuracy, etc., get attentions [3, 4]. So far researchers have already proposed various methods to diagnose the equipment, including acoustic method [5], optical method [6], decomposition

Y. Chen (✉) · Z. Zha · Q. Li · G. Zhang
Xi'an Jiaotong University, Xi'an 710049, China
e-mail: yulunchen@foxmail.com

D. Yang
State Grid Shaanxi Electric Power Research Institute, Shaanxi Power Grid Co. Ltd, Xi'an 710199, China

W. Chen et al. (eds.), *The Proceedings of the 9th Frontier Academic Forum of Electrical Engineering*, Lecture Notes in Electrical Engineering 743,
https://doi.org/10.1007/978-981-33-6609-1_13

compound analysis [7], pulse current detection [8], residual charge detection [9], UHF detection and so on.

Plenty of scholars domestic and abroad have been carrying out research on PD in GIS. Si et al. [10] investigated the characteristic of acoustic signals generated by different artificial defects. Song et al. [11] designed a new optical sensor array positioning the site of PD. Ren et al. [12] investigated the basic characteristics of light emission during PDs in SF_6 gas. Ding et al. [13] diagnosed the PD of the GIS by analyzing the decomposition compound of SF_6. Iwabuchi et al. [14] measured the residual charge on a GIS disc spacer with an electrostatic probe. With all these things above, the most vital thing is that the accuracy of the discrimination of the defect types using only one method is not high enough and cannot fulfill the growing demand.

Confronted with this situation, the main aim of this paper is to further arise the precision of the discrimination of different defects. The method of UHF as well as pulse current detection are chosen to achieve the goal of accurate identification. Besides an ICCD camera is set to capture the process of discharge.

2 Experimental Setups

2.1 Experimental Apparatus

In order to facilitate the study, an experimental platform as shown in the Fig. 1 is built. The measurement system consists of a UHF measurement system, pulse current detection device, ICCD camera and air pressure monitoring system. The PD instrument is the PDcheck, which converts the coupling current signal of the branch

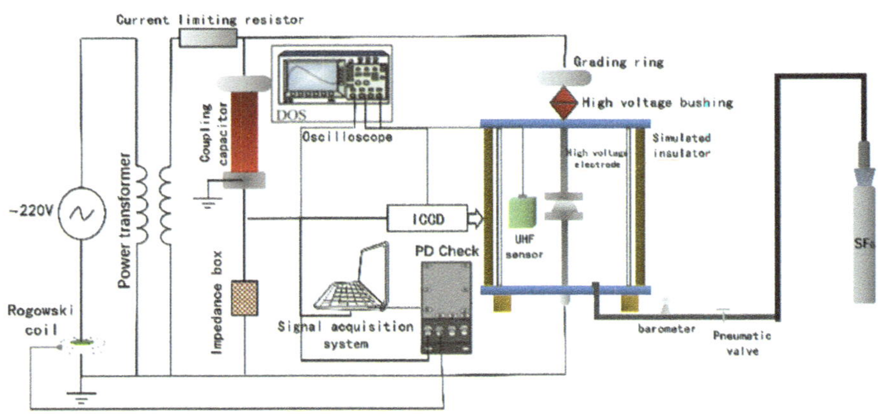

Fig. 1 Schematic figure of PD measurement circuit

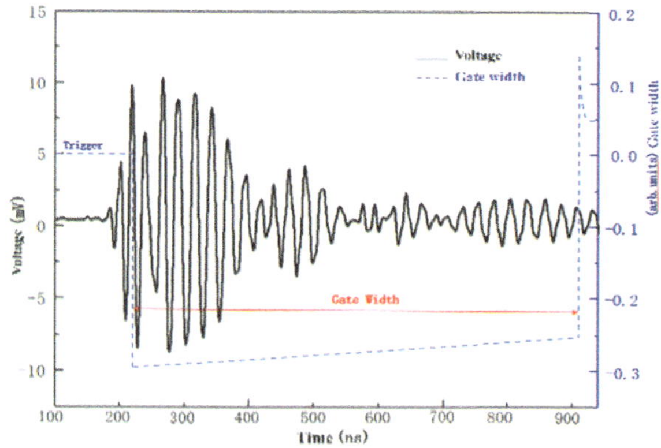

Fig. 2 Sequence diagram of ICCD camera

into a voltage signal and obtains the phase information of the signal through the induction of the Rogowski coil. The UHF sensor is placed on the outer wall of the chamber close to the insulator. The sensor port is connected to the oscilloscope via a high-pass filter. During the experiment, the sealed chamber is filled with SF_6 gas, and the ICCD camera is set up at the position facing the test unit. The ICCD camera trigger terminal is connected to the external trigger port of the oscilloscope, and the camera is triggered by the rising edge of the UHF pulse measured by the oscilloscope as shown in Fig. 2.

2.2 *Preparation of Defects*

The circular truncated GIS insulator model used in the experiment is shown in the Fig. 3. The upper and lower bottom radius of the insulator are 5 and 10 mm, and the height is 6 mm. The composition of insulating material is shown in Table 1.

The internal air void defect model of the insulator is shown in the Fig. 4a. The

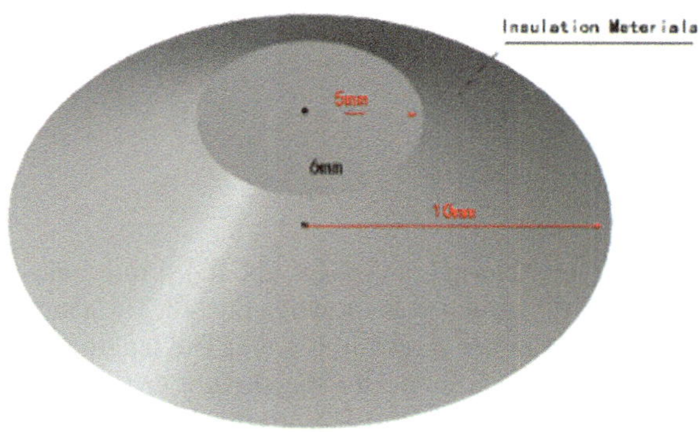

Fig. 3 Circular truncated GIS insulator model

Table 1 The component of the insulating material

Component	Quality score (%)
Epoxy resin	22
Alumina	70
Hardener	8

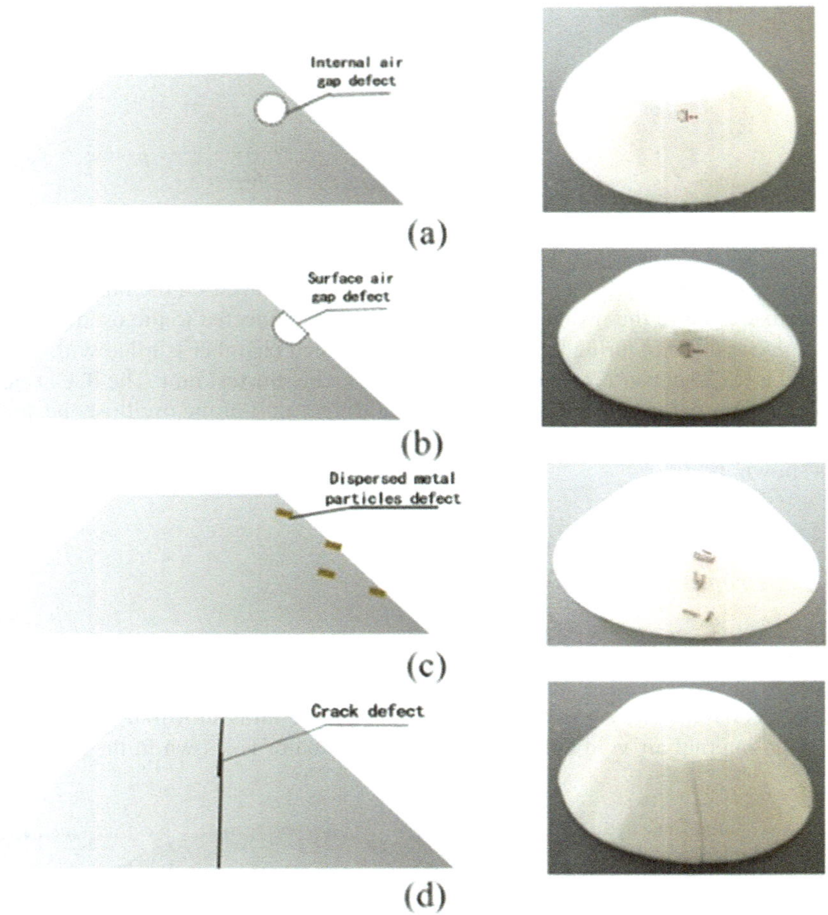

Fig. 4 4 types of defects

radius of the internal air void is about 1.5 mm, placed about 1 mm deep below the surface of the insulator as shown in Fig. 4b. The e metal particle defects on the surface of the insulator are shown in Fig. 4c. The length of the dispersed metal particles are about 1 mm. The model of the surface crack defect of the insulator is shown in Fig. 4d. So as to distinguish from the scratches on the surface of the insulator, this experiment adopts the method of applying stress to the insulator to crack from the inside and then bond to prepare the crack defect.

3 Results

The UHF signals and the pulse current signals are measured simultaneously by corresponding detectors and meanwhile the Lichtenberg figures are also taken by ICCD camera. During measurement, at least 10 sets of signals are recorded per voltage level. Tables 2, 3, 4 and 5 give the corresponding results: typical phase resolved partial discharge (PRPD) patterns, equivalent time length- equivalent frequency (T-F) map, i.e. Classification Fingerprints Pattern (CFP) as well as typical UHF waveforms.

As shown in Table 2. In the spectrum at PDIV which is 2 kV, the positive and negative half cycles start to discharge and a peak appears in the positive half cycle. In the spectrum at 4.5 kV, rabbit-ear-shape pattern appears in the positive half cycle spectrum, indicating that discharge gaps are formed among metal particles. When the applied voltage reaches 5.7 kV, the discharge gap characteristics are weakened, and the discharge is concentrated at the bottom of the spectrum, indicating that the discharge gap among metal particles has reached breakdown.

The PDIV of air void defects in insulators is 2.68 kV as shown in Table 3 and the discharge in negative half period is stronger than that in positive half period. With the applied voltage rising, a rabbit-ear-shape PRPD pattern appeared in the negative half period, indicating that the air gap defect has begun to discharge continuously. From

Table 2 PRPD pattern and CFP as well as UHF signals of dispersive metal particles defect

Defect type	Pulse current method results			UHF results
	Voltage (kV)	PRPD pattern	T-F pattern (CFP)	PD and flashover waveform
Metal particles	2.00	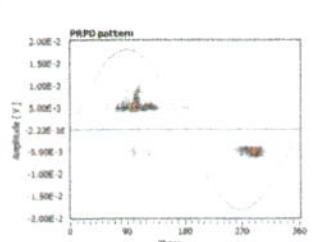	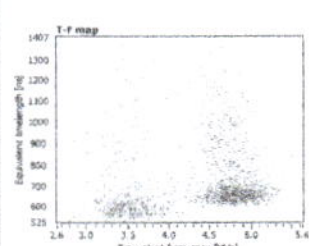	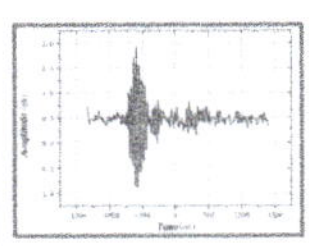
	4.50	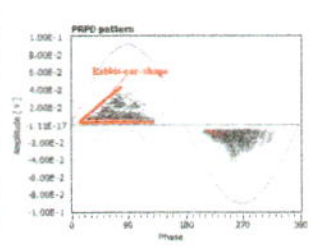	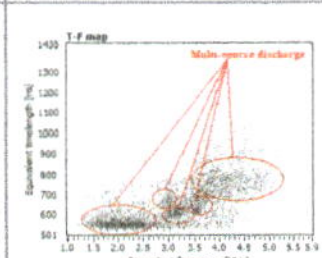	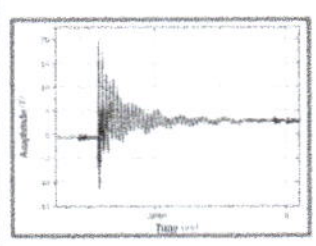
	5.70	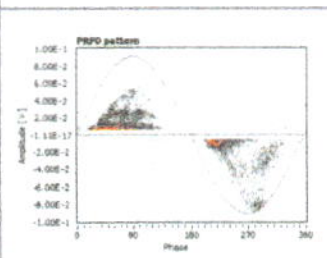	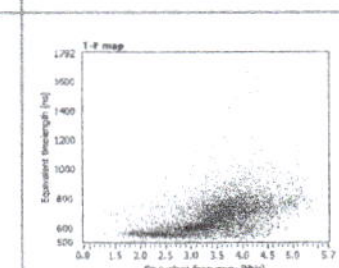	

Table 3 PRPD pattern and CFP as well as UHF signals of internal air void defect

Defect type	Pulse current method results			UHF results
	Voltage (kV)	PRPD pattern	T-F pattern (CFP)	PD and flashover waveform
Internal air void	2.68	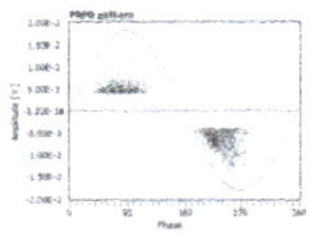	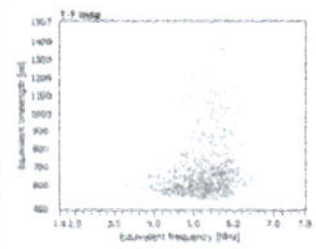	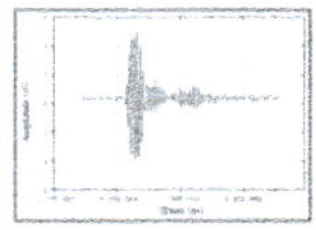
	3.34	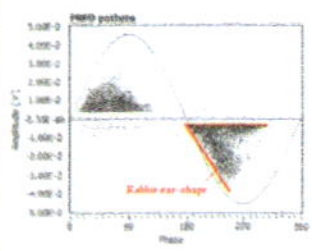	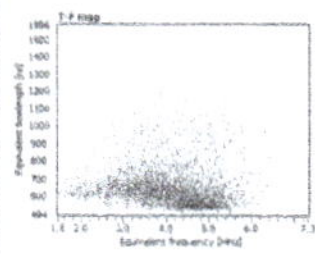	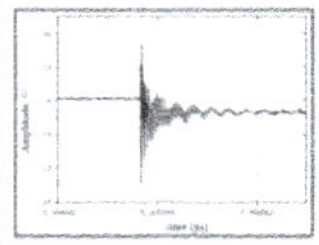
	5.07	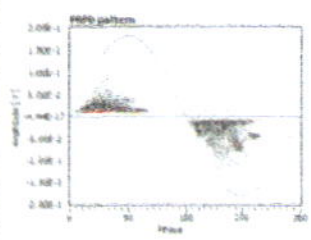	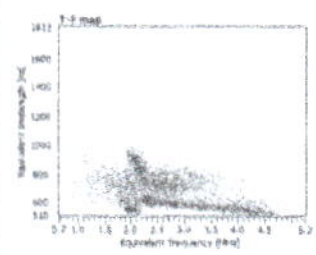	

Table 4 PRPD pattern and CFP as well as UHF signals of surface air void defect

Defect type	Pulse current method results			UHF results
	Voltage (kV)	PRPD pattern	T-F pattern (CFP)	PD and flashover waveform
Surface air void	2.43	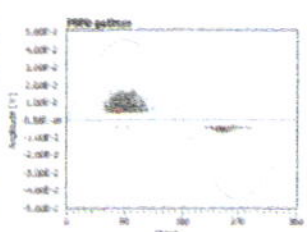	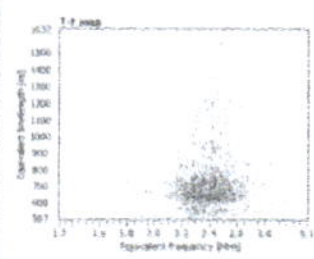	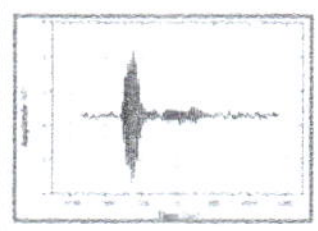
	6.80	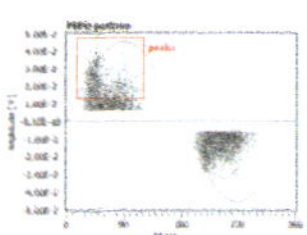	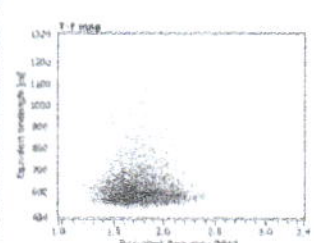	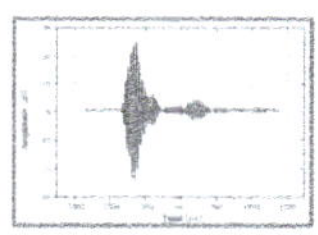
	10.20	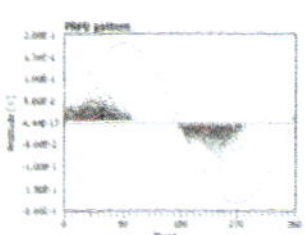	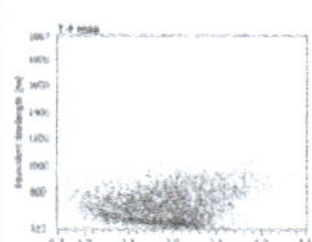	

the typical UHF waveforms, the amplitude of flashover signal is obviously higher than that of PD signal.

The PDIV of surface air void defects in insulator is 2.43 kV as shown in Table 4. On the contrary to the internal air void, obvious discharge occurs in positive half

Table 5 PRPD pattern and CFP as well as UHF signals of crack defect

Defect type	Pulse current method results			UHF results
	Voltage (kV)	PRPD pattern	T-F pattern (CFP)	PD and Flashover waveform
Crack defect	4.78	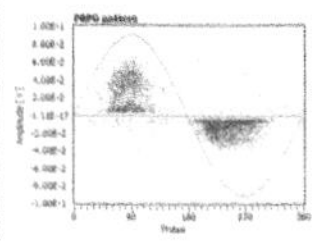	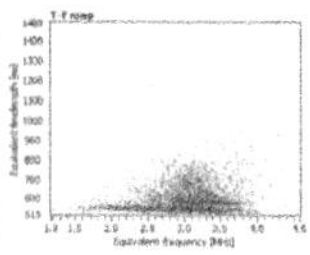	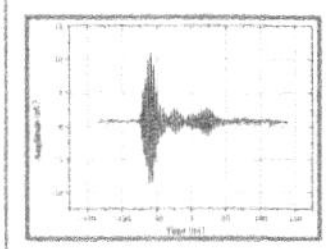
	7.98	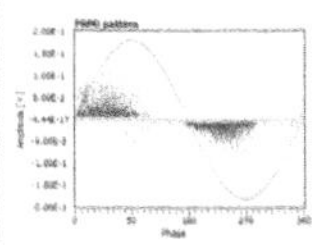	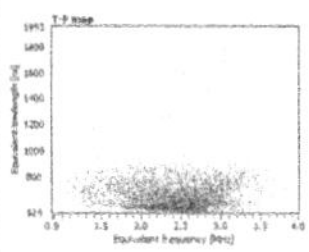	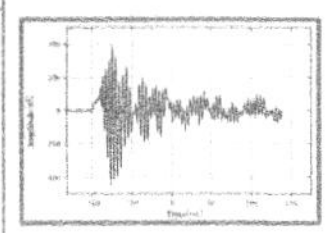
	10.30	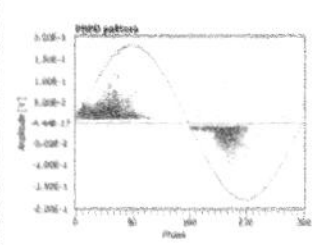	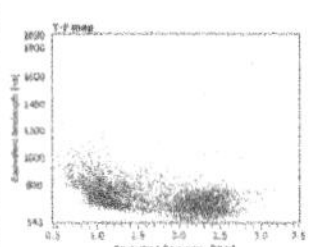	

period and weak discharge occurs in negative half period. With the applied voltage increasing, discharge intensity in both the positive and negative half period are obviously enhanced. Superposition of two peaks in the PRPD spectrum at 6.8 kV is observed, indicating that the surface air gap has begun to discharge continuously.

The PDIV of crack defects is 4.78 kV as shown in Table 5. With the applied voltage increasing, discharge intensity is enhanced. In facts, we found that crack has almost no effect on the PD, which mainly happens in the narrow air gap between the insulator and cathode electrode, thus no specific characteristics are observed.

Figure 5 illustrates ICCD photos of four different defects showing how the PD develops. The first and second column show the fluorescence image of internal void and surface air void respectively. No matter what kind of air void, PD evades from the air void, thus it has lower luminance than other areas. From the image of metal particles, the discharge creeping from cathode to anode can be observed distinctly while the area of luminescence is exactly where the metal particles are distributed. However, no PD associated with crack defect can be observed.

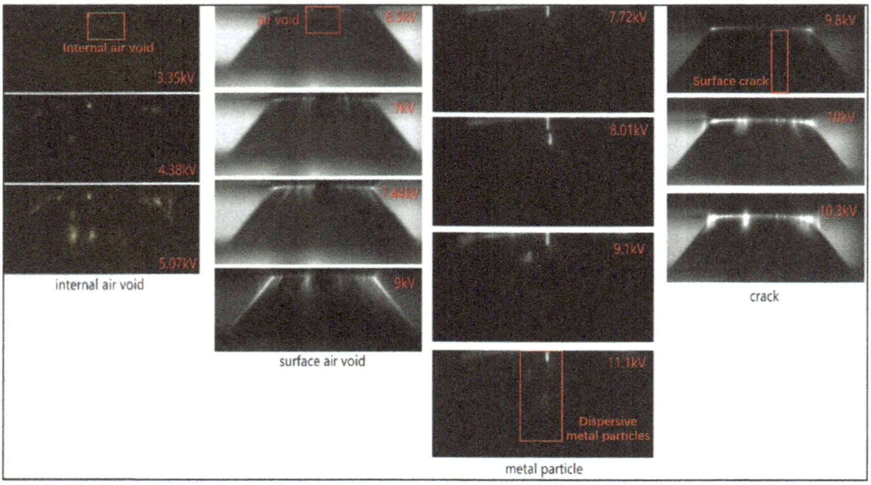

Fig. 5 ICCD photos at different voltages of different defects

4 Discussion

4.1 Analysis of the PD Check Signals Using the Algorithm of CNN

CNN (Convolutional Neural Network) is capable to avoid the complicated feature extraction and data reconstruction process in traditional recognition algorithms [15, 16]. At the beginning, input the calibrated data to operations and compare the output results with the calibration. Meanwhile the gradient descent algorithm is applied to continuously modify the network parameters to make the output closer to the calibration, thereby the mapping relationship from input to output could be confirmed (Fig. 6).

A total of 1000 PD signals of 4 types of defects are used as training data. After 2000 times of training, the neural network's recognition success rate of training data has increased from 21.81% to a stable above 98.74%. After this, the test set data with

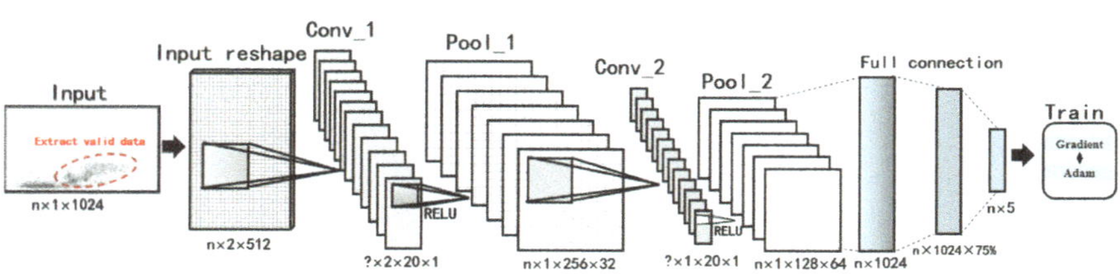

Fig. 6 Recognition pattern of CNN

Table 6 The separated results of the test set

Type of defect	Recognition ratio (%)
Metal particle	25.83
Internal air void	23.33
Surface air void	24.17
Crack defect	26.67

30 waveforms for each defect containing 120 sets of waveforms is imported into the network. The test results obtained are shown in the Table 6.

4.2 *Analysis of the UHF Signals*

In order to analyze UHF signals, discrete Fourier transform (DFT) conversion is carried out. The results of 4 defects are given in Fig. 7 respectively. The input signals of UHF method have been attenuated several times, besides the amplitude is strongly influenced by the voltage level of the gird. Comparing all these characteristics of each defect, one feature in common is that all of them own two main peaks, of which the higher one is at around 250 MHz while the lower one is at around 350 MHz. Conflicted with our prediction, only the amplitude of signals of the metal particles

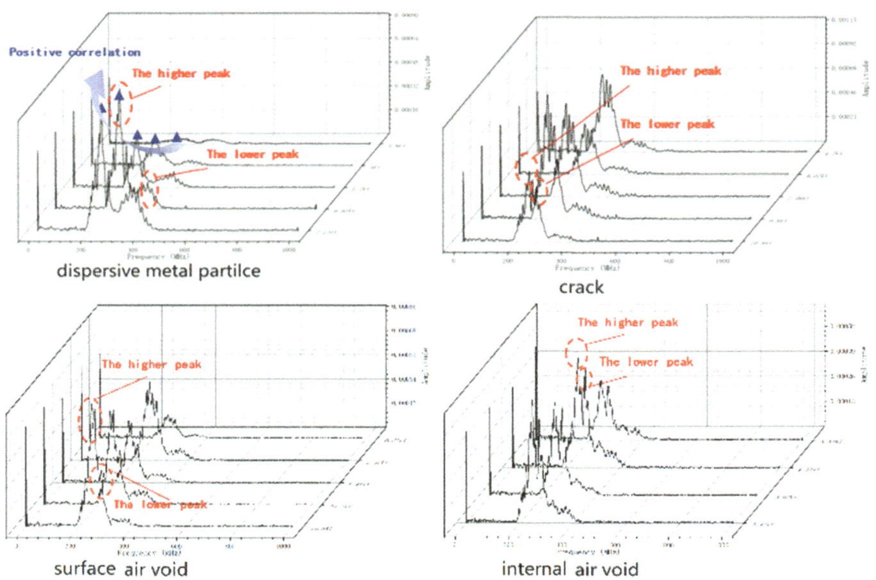

Fig. 7 Amplitude frequency characteristic of UHF signals of linear metal particle

represents an obvious positive correlation tendency with the applied voltage. This is quite unique and beyond all expectations.

5 Conclusion

In this study, the characteristics of the PD creeping development of 4 types of GIS epoxy insulation defects were studied, and finally reached the following conclusions. The discharge signals of 4 types of defects are able to be recognized by the convolutional neural network, and a high recognition accuracy is achieved with an appropriate learning rate. In addition, we also found that no matter what types of defects, the frequency of the UHF signals is at around 250 MHz. Except for the metal particles defect, the UHF signals of the other defects do not change significantly. Only when the applied voltage was about to reach the threshold voltage, the UHF signal had a significant mutation, i.e., the so-called "0–1" phenomenon of discharge in SF_6.

Acknowledgements This work was supported in part by the State Grid Shaanxi Electric Power Research Institute.

References

1. Hasegawa, T., et al. 1997. Development of insulation structure and enhancement of insulation reliability of 500 kV DC GIS. *12*(1): 194–202.
2. Kobayashi, S. and A. Horide. 1992. Development and field test evaluation of optical current and voltage transformers for gas insulated switchgear. *7*(2): 815–821.
3. Sabot, A. et al. 1996. *GIS Insulation Co-Ordination: On-Site Tests and Dielectric Diagnostic Techniques. A utility point of view.*
4. Boeck, W. and E. Al. 1998. *Insulation Co-Ordination of GIS, Return of Experience, On Site Tests and Diagnostic Techniques.*
5. Shang, Y., et al. 2017. Mechanical fault diagnosis system based on acoustic feature analysis in gas insulated switchgear. In *2017 1st International Conference on Electrical Materials and Power Equipment (ICEMPE).*
6. Takahashi, M., et al. 1997. Optical current transformer for gas insulated switchgear using silica optical fiber. *IEEE Transactions on power DELIVERY* 12 (4): 1422–1427.
7. Tang, J., et al. 2012. Partial discharge recognition through an analysis of SF6 decomposition products part 1: decomposition characteristics of SF6 under four different partial discharges. *19*(1): 29–36.
8. Bian, C. and S.B.J.H.V.A. Chen. 2010. *Study on Anti-Interference Technique used in PD Detection for GIS based on pulse current waveshap.*
9. Wang, F., et al. 2002. Insulator surface charge accumulation under impulse voltage. *IEEE Transactions on Dielectrics and Electrical Insulation* 11 (5): 847–854.
10. Si, W.R., et al. 2010. Investigation of a comprehensive identification method used in acoustic detection system for GIS. *IEEE Transactions on Dielectrics and Electrical Insulation* 17 (3): 721–732.
11. Song, B., et al. 2018. A new optical method of partial discharge distant positioning in GIS. In *2018 IEEE Electrical Insulation Conference (EIC).*

12. Ren, M., M. Dong, and J.J.E. Liu. 2016. Statistical Analysis of Partial Discharges in SF6 Gas via Optical Detection in Various Spectral Ranges. *9*(3).
13. Ding, W., et al.. 2016. *Decomposition Characteristics of SFunder Creeping Discharge on Solid Insulator*
14. Iwabuchi, H., et al. 2013. Influence of tiny metal particles on charge accumulation phenomena of GIS model spacer in high-pressure SF6 gas. *20*(5): 1895–1901.
15. Zhang, A.A., et al. 2020. Recognition of partial discharge of cable accessories based on convolutional neural network with small data set. *Compel-the International Journal for Computation and Mathematics in Electrical and Electronic Engineering* 39 (2): 431–446.
16. Peng, X.S., et al. 2019. A convolutional neural network-based deep learning methodology for recognition of partial discharge patterns from high-voltage cables. *IEEE Transactions on Power Delivery* 34 (4): 1460–1469.

A Survey of Emergency Self-Running Power Supply Schemes for Rail Transit Vehicles

Yutong Zhu, Tengfei Qiu, Jiamin Gao, Shuichang Li, and Lijun Diao

Abstract When the power supply fails, the rail transit vehicles stay in place and wait for rescue, which not only has great hidden danger, but also affects the efficiency of train operation. In order to solve this problem, we could equip the train with an energy storage system that would enable the train to run to the nearest stop. According to the installation position of the energy storage system, the emergency self-running power supply scheme can be divided into vehicle type and ground type. According to the output voltage level of the energy storage system, the emergency self-running power supply scheme can be divided into low-voltage, medium-voltage and high-voltage. In this paper, the research status and hot issues of emergency self-running power supply scheme for rail transit vehicles are summarized, and the characteristics and shortcomings of each technical scheme are analyzed. Finally, the future research direction and hotspot of emergency self-running power supply scheme is pointed out.

Keywords Rail transit vehicles · Emergency Self-Running · Energy storage · Power supply

1 Introduction

Rail transit vehicles are mainly powered through catenary or third rail, and have strong dependence on them. During the operation of rail transit vehicles, the train power may sometime be cut off due to bad weather, power system or subsystem failure and so on, the train can only stay in place and wait for rescue. The power and

Y. Zhu
Locomotive and Car Research Institute, China Academy of Railway Sciences Corporation Limited, Beijing, China

Y. Zhu · T. Qiu · S. Li
Beijing Zongheng Electro-Mechanical Technology Co., Ltd, Beijing, China

J. Gao · L. Diao (✉)
Beijing Jiaotong University, Beijing, China
e-mail: Ljdiao@bjtu.edu.cn

W. Chen et al. (eds.), *The Proceedings of the 9th Frontier Academic Forum of Electrical Engineering*, Lecture Notes in Electrical Engineering 743,
https://doi.org/10.1007/978-981-33-6609-1_14

capacity of the emergency power supply can only provide short-term ventilation and emergency lighting power in emergency situations, it will also have an impact on the operation and scheduling of trains. If the train can achieve the emergency traction through the onboard or ground energy storage system, forcing the stop of the train to run to the nearest stop, it can effectively solve the problem of emergency treatment caused by the invalid electrification. When the power supply is normal, the energy storage system can be charged through the traction network, and the regenerative braking energy of the train can be absorbed and stored [1, 2].

2 Traction Power Supply System for Rail Transit Vehicles

A typical traction and auxiliary power supply system for rail transit vehicles is shown in Fig. 1. Traction power supply system mainly consists of DC link, traction converter, traction motor, et al. [3]. It powers the traction system to ensure the operation of the train. The auxiliary power supply system is composed of auxiliary converter, charger, storage battery, et al. [4]. It supplies power to all electric equipment except traction power system. DC110 V batteries are fitted to power the control system and other DC loads. When the power failure occurs, the DC110 V storage battery releases the stored energy to realize the functions of emergency ventilation and emergency lighting.

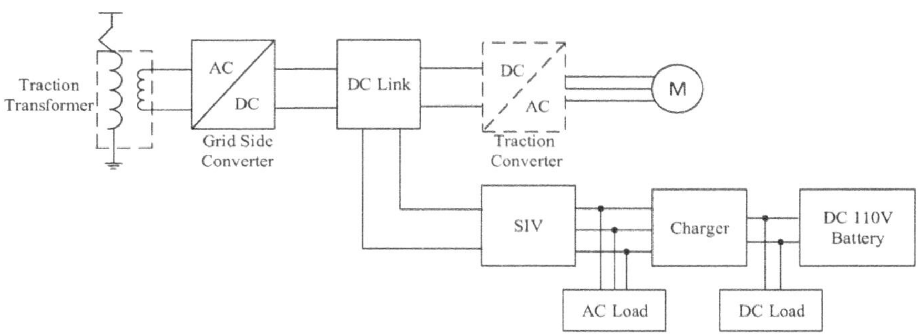

Fig. 1 Traction and auxiliary power supply system for rail transit vehicles

3 Onboard Low-Voltage Energy Storage Power Supply Scheme

3.1 Direct Power Supply

Since the speed of emergency self-running is relatively low, the traction motor does not need to reach rated power and torque, so the onboard low-voltage 110 V battery pack can be directly used to supply power for the DC link. Reference [5] established an electro-thermal-mechanical coupling model for traction system based on this scheme and proposed an optimization method of efficiency and energy.

This power supply scheme has the advantages of simple structure and high reliability, but in order to reach the driving power of traction motor, it is necessary to enlarge the capacity of battery pack [6], that is to say, to add more batteries, which will take up a certain amount of train space. In addition, the power of traction motor and the speed and distance of emergency self-running is limited.

3.2 Power Supply Via Boost Converter

Direct battery power though can achieve a certain degree of emergency running, the battery output voltage is low, the speed and distance of emergency self-running are limited to a certain extent. This problem can be solved by boosting the 110 V low voltage storage battery and supplying it to the traction inverter.

Reference [7] suggest that multiple batteries can be connected in series by adding contact switches, but this method increases the complexity of system wiring, the range of boosting voltage is also limited. Using converter boost is a better solution, Ref. [8] proposed to increase the output voltage of the battery through the DC/DC boost converter. Reference [9] proposed to design the auxiliary converter and charger into a bidirectional work mode. Under emergency conditions, the DC 110 V battery pack will supply power to the auxiliary converter through the charger, and the auxiliary converter will work in reverse to transmit power to the DC link, the traction inverter takes electricity from the DC bus to drive the traction motor. This scheme can achieve the emergency traction with the same voltage as the original traction power supply system of the train. However, the charger has to work in two directions, the power of the charger when working in reverse is greater than the actual demand, this leads to an increase in the size and weight of the charger, and also increases the unreliability of the charger.

Compared with the direct power supply scheme, the DC 110 V battery can provide higher DC bus voltage after boosting, so that the traction motor can obtain higher working voltage and output greater torque, but it is also limited by the capacity of the battery. In addition, due to the emergency lighting and control power also comes from DC 110 V battery, a reasonable energy management strategy is needed in order to ensure its uninterrupted power supply [10].

4 Onboard Medium-Voltage Energy Storage Power Supply Scheme

In order to achieve higher output power and greater walking distance, additional energy storage system is a desirable option. In Ref. [11], an extra onboard medium-voltage energy storage system is proposed to provide power supply through power-frequency auxiliary converter under emergency conditions, as shown in Fig. 2. In Ref. [12], it is also proposed that power supply can be realized through high frequency isolated DC/DC converter, as shown in Fig. 3.

The above two power supply schemes provide power through the auxiliary converter or part of the circuit in the auxiliary power supply system. Although they can maintain the high DC voltage supply of the traction system during the emergency

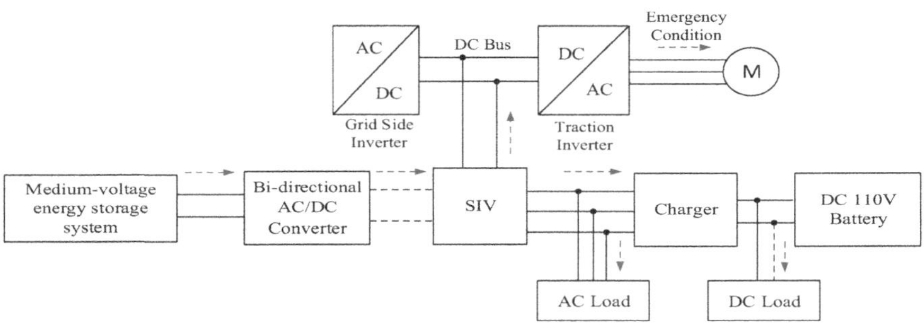

Fig. 2 Schematic diagram of onboard medium-voltage power supply scheme by power frequency SIV

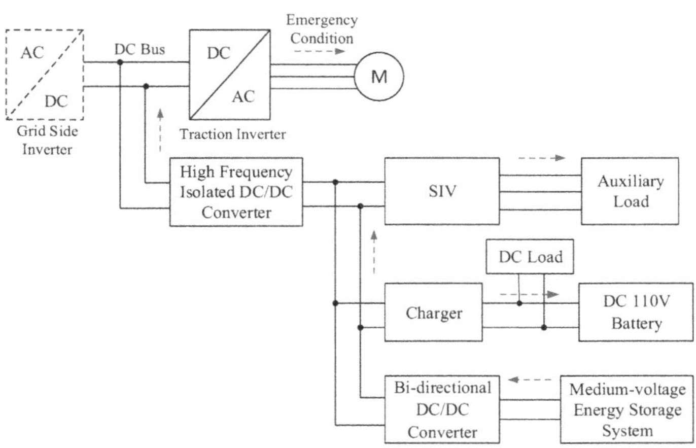

Fig. 3 Schematic diagram of onboard medium-voltage power supply scheme by high frequency isolated DC/DC converter

conditions, it is difficult to design and control the auxiliary converter, which reduces the reliability of the auxiliary converter, and the power of the emergency traction is limited by the power level of the auxiliary converter. Based on this, a power supply scheme of adding a charger to charge the extra medium voltage energy storage system and discharging the energy storage system directly to the DC bus is adopted. The typical case is Beijing-Zhangjiakou high-speed railway, with the self-running ability of 30 km/h speed on 5 km 5‰ uphill road and 15 km straight road [13]. Similar to the scheme, the emergency traction scheme for Beijing metro line 16 with high voltage battery power supply in Ref. [14] is proposed.

5 Onboard High-Voltage Energy Storage Power Supply Scheme

Although the extra medium-voltage energy storage system can provide higher running power and the DC bus voltage can basically meet the demand of low-speed running of the train under emergency conditions, its power is still limited by the voltage of the energy storage system, and has limited capacity to absorb electrical braking. Based on this, the use of extra onboard high-voltage energy storage system becomes a higher performance scheme. The high-voltage energy storage system is connected to the DC bus through a bi-directional DC/DC converter [15], so that the DC bus voltage during emergency self-running is the same as when it works normally, it also avoids the influence of emergency traction on the control of power consumption, lighting and emergency ventilation power supply.

There are three main types of high-voltage energy storage systems: battery-powered [16], supercapacitor-powered [17] and battery and supercapacitor hybrid power supply. Supercapacitors have the advantages of high specific power, long cycle life, wide operating temperature range [18], which can quickly carry out high-power charge and discharge, but the energy density is lower than that of battery, and the unit voltage is too low to be used as an onboard energy storage element. The specific energy of the battery is high, the working voltage is stable, and the technology is mature. However, large current charging and discharging has a negative impact on the life and efficiency of the battery [19]. Therefore, the combination of super capacitor and battery is an effective solution to the power drive of rail transit vehicles. In terms of battery selection, lithium-ion battery is often used as emergency traction power supply [20] because of its advantages of no memory effect, no pollution, small self-discharge rate and long cycle life [21]. The topology of hybrid energy storage system composed of supercapacitor and battery can be divided into series and parallel. References [22] connect the battery and supercapacitor through DC/DC converter in series. Reference [23] connects the super capacitor and the battery in parallel. In addition to the common battery and supercapacitor power supply, a fuel cell system power supply scheme is proposed in Ref. [24].

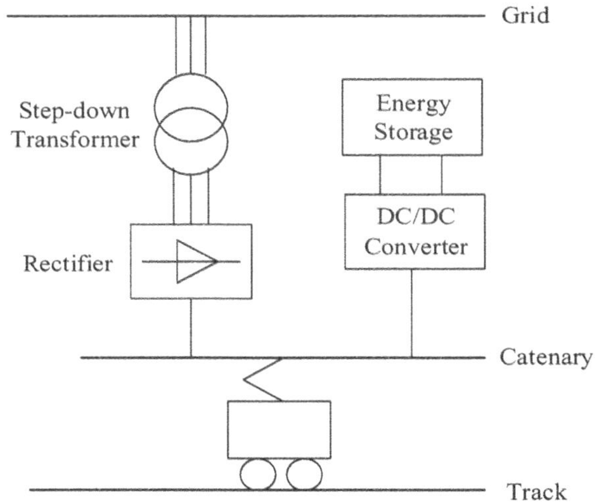

Fig. 4 Schematic diagram of ground energy storage and power supply scheme

6 Ground Energy Storage Power Supply Scheme

The above-mentioned emergency self-running power supply schemes are all based on the installation of energy storage systems on vehicles. With the increasing passenger flow density, the investment will continue to grow, however, the ground energy storage power supply scheme [25, 26] can save investment by installing the energy storage and absorption device on the ground, as shown in the Fig. 4. But the current path of the ground energy storage system when train braking or pulling is far away, the line voltage drop is large, and the electric energy loss is larger than the on-board energy storage system. The charging and discharging control of energy storage system is studied in Ref. [27]. A control method based on load state and time-varying parameters is proposed.

7 Analysis and Discussion

Based on the above analysis of various power supply schemes for rail transit vehicles. Table 1 lists the characteristics and shortcomings of each scheme.

The onboard low-voltage energy storage power supply scheme uses the original 110 V battery of the train to supply power directly to the traction motor or after boosting voltage. The direct power supply scheme is simple in structure and does not need to reform the train, but the output voltage of the battery is low, only suitable for low-speed and short-distance emergency self-running occasions. In contrast, the performance of the traction motor can be improved by boosting the voltage of the low-voltage energy storage system. The on-board medium-voltage energy storage power supply scheme adds an extra medium-voltage energy storage system to the vehicle,

Table 1 Comparison of emergency self-running power supply schemes

Scheme name		Costs	Reliability	Storage energy	Implementation difficulty	Emergency self-running performance
Onboard low-voltage energy storage	Direct power supply	Lowest	High	Least	No need to modify the existing system, the easiest to implement	Suitable for the lowest speed and short distance
	Series boost power supply	Higher	Higher	Less	Need to add contactor, easy to implement	High DC voltage, suitable for short distance occasions
	Low-voltage converter boost power supply	A slightly higher	High	Less	Need to increase boost chopper, easier to implement	DC bus can reach rated voltage, suitable for short distance occasions
	Boost power supply of bi-directional auxiliary converter	High	Low	Less	Need to transform the original auxiliary converter and charger to make it work bidirectionally, more difficult to achieve	DC bus can reach rated voltage, suitable for short distance occasions
Onboard medium-voltage energy storage	Bi-directional SIV power supply	Higher	Lower	More	Need to add bidirectional AC/DC converter, more difficult to achieve	DC bus can reach rated voltage, suitable for high-speed, longer distance occasions
	Bidirectional high frequency isolated DC/DC power supply	Higher	Lower	More	Need to add bidirectional DC/DC converter, more difficult to achieve	DC bus can reach rated voltage, suitable for high-speed, longer distance occasions

(continued)

Table 1 (continued)

Scheme name		Costs	Reliability	Storage energy	Implementation difficulty	Emergency self-running performance
	Direct power supply	Higher	High	More	Need to add an additional charger, more difficult to achieve	DC bus can reach rated voltage, suitable for high-speed, longer distance occasions
Onboard high-voltage energy storage		High	High	Many	Need to add bi-directional DC/DC converter and high voltage energy storage device, easier to implement	DC bus can reach rated voltage, suitable for high-speed, long-distance occasions
Ground storage		Highest	Low	Most	Need to install bi-directional DC/DC converters, energy storage devices and other equipment on the ground, difficult to achieve	DC bus can reach rated voltage, suitable for high-speed, long-distance occasions

with higher output voltage. The traction motor can work under rated voltage, and the speed and distance of emergency self-running are increased to a certain extent. The output voltage of the energy storage system is further improved by the on-board high-voltage energy storage power supply scheme, which requires the addition of bi-directional DC/DC converter, and has simple structure and high reliability. The high-voltage energy storage system can be composed of battery, supercapacitor, mixed battery and supercapacitor, fuel cell, etc. The combination of supercapacitor and battery hybrid energy storage system can integrate the advantages of both and drive the vehicle better, which is a hot issue in the future. In addition, due to the voltage difference between DC bus and DC110V battery, there is a problem that the low-voltage side current is too large in voltage conversion. The references [28, 29] have made relevant research on this problem. The energy storage system is installed on the ground in the scheme of ground power supply, and the structure is complex. At present, relevant research is less, and further theoretical research is needed.

8 Conclusion

In this paper, the emergency self-running power supply scheme for rail transit vehicles is summarized, and the current research situation is summarized from four aspects: low-voltage, medium-voltage, high-voltage and ground, the future research direction of emergency self-running power supply is discussed. At present, there are still many basic theories and key technologies that need to be further studied about emergency self-running power supply technology. For example, the energy management strategy of energy storage system, the combination of supercapacitor and battery, the realization of boost converter and the scheme of ground energy storage, these will become the hot issues in the future.

Acknowledgements We would like to thank the participants and companies who take part in this study. This research was carried out within the Science and Technology Research and Development Plan Project of China Railway Corporation (P2018G047).

References

1. Takahashi, H., Y. Kume, K. Honda, et al. 2016. The function development for emergency train running by battery application in traction power supply system. In *IEEE International Power Electronics and Motion Control Conference (PEMC)*, pp. 1137–1140. Varna.
2. Hayashiya, H. et al., Possibility of energy saving by introducing energy conversion and energy storage technologies in traction power supply system. In *15th European Conference on Power Electronics and Applications (EPE)*, pp. 1–8. Lille.
3. Larsson, P.L., and C. Gerster. 2007. Integrated propulsion and auxiliary supply systems for multi-system operation. In *European Conference on Power Electronics and Applications*, pp. 1–8. Aalborg.
4. Zheng, Huaxi, Gao Jilei, Zheng Qionglin. 2010. Comparison and analysis of auxiliary power supply system of high-speed EMU in China. *Electric Drive, Journal 403*: 53–59 (in Chinese).
5. Wang, K., Q. Ge, and Y. Li. 2014. Analysis and optimized control of emergency traction by storage battery for urban rail transit vehicle. In *17th International Conference on Electrical Machines and Systems (ICEMS)*, pp. 175–179. Hangzhou.
6. H. Takahashi, Y. Kume, K. Honda, H. Kawatsu, J. Kaminishi, and Y. Shimizu.: Development of emergency self-running train for energy utilization in Stationary Energy Storage System. In *17th European Conference on Power Electronics and Applications (EPE'15 ECCE-Europe)*, pp. 1–8. Geneva.
7. Cao, Chunwei, Sun Jian, Chen Xinghu. 2017. Study on emergency traction of subway train battery. *Dual-Use Technologies and Products, Journal 8*:54–55 (in Chinese).
8. Yang, Li, Hu Jinxin. 2019. Emergency traction improvement scheme of vehicle battery. *Locomotives and rolling stock, Journal 4*:32–34, 46 (in Chinese).
9. Guangwei, Xu, Liu Jiancheng, Chen Dong, et al. 2019. Battery-powered emergency traction and emergency air-conditioning refrigeration system for EMU. *Locomotive Electric Drive, Journal* 5: 116–120. (in Chinese).
10. Motapon, S.N., L.A. Dessaint, and K. Al-Haddad. 2013. A comparative study of energy management schemes for a fuel-cell hybrid emergency power system of more-electric aircraft. *IEEE Transactions on Industrial Electronics, Journal* 61 (3): 1320–1334.
11. Li, Jun, Yang Ying, Wu Jian, et al. 2018. *A New Auxiliary Power Supply System for Emergency Running of EMU and its Method*. CN107878228A (in Chinese).

12. Charmaine, Xu, Jin Wenbin, Qiu Junjie, et al. 2019. *An Emergency Self-Walking Traction System for Motor Vehicles*. CN209191940U (in Chinese).
13. Kong, Yushu, and Yu Mingyuan. 2020. Composition and function analysis of emergency self-running system of intelligent EMU of Jing Zhang high-speed railway. *Urban Rail Transit Research, Journal 2302*:71–74 (in Chinese).
14. Deng, Wenhao. 2016. Analysis of Line 16, Beijing Subway Traction Scheme. *Modern Urban Rail Transit, Journal 6*:20–23 (in Chinese).
15. Diao, Lijun, Liu Zhigang, Zhang Gang, et al. 2013. *Traction Converter for Hybrid Electric Multiple Mode EMU*. CN201310253122.7 (in Chinese).
16. Du, Yufeng, Wang Xiaoyong. 2011. Type selection design of traction battery for Singapore electric power storage battery dual energy engineering vehicle. *Electric Locomotives and urban rail vehicles, Journal 343*:28–30 (in Chinese).
17. Ma, Yingchun, Wang Yinan. 2019. Analysis of the application of ultracapacitor in emergency traction of subway. *Shandong Industrial Technology, Journal 1*:58 (in Chinese).
18. Conway, B.E. 1999. *Electrochemical Supercapacitors: Scientific Fundamentals and Technological Applications*. NewYork: Kluwer Academic Publishers.
19. Nima, Ghaviha, Yang Yu. 2018. Summary of application of energy storage system in railway transportation. *Foreign Rolling Stock, Journal 554*:8–11.
20. Matsuda, Y., T. Nakajima, Y. Ohzawa, et al. 2011. Safety improvement of lithium ion batteries by organo-fluorine compounds. *Journal of Fluorine Chemistry* 132: 1174–1181.
21. Teshima, M., and H. Takahashi. 2014. Lithium ion battery application in traction power supply system. In *International Power Electronics Conference (IPEC-Hiroshima 2014—ECCE ASIA)*, pp. 1068–1072. Hiroshima.
22. Huang, Hai, Guo Wanlu, Chen Shengcai. 2018. A hybrid power system circuit for rail transit based on energy storage element. *Technology and Markets, Journal 257*:83–84 (in Chinese).
23. Wang, Hugao, Qu Haiyang, Chen Zhongjie. 2006. Design and research of energy storage power applied to emergency traction of metro vehicles. *Electric Locomotives and urban rail vehicles, Journal 391*:50–53 (in Chinese).
24. Li, Ming, Han Guopeng, Zhang Qiumin, et al. 2018. *Emergency traction power supply system, method and trams*. CN108790885A (in Chinese).
25. Diao, Lijun, Li Weijie, Liu Bo, et al. 2019. *The railway electrification system's energy coordination system*. CN110492507A (in Chinese).
26. Wang, Yue, Qin Bin. 2018. Research on suppressing pressure fluctuation of traction network by ground lithium battery energy storage system. *New Industrialization, Journal 810*:1–5. (in Chinese).
27. Capasso, A., G. Ghilardi, and G. G. Buffarini. 2010. Bologna—Florence high speed railway line: MV emergency traction power supply, operating conditions and PQ issues. *Electrical Systems for Aircraft, Railway and Ship Propulsion, Bologna, Journal*, pp. 1–5.
28. Hong, Zhenhai, Gao Wei, Yu Kaifeng, et al. 2018. *High-power Bi-directional Charger for Metro Vehicles*. CN208247990U (in Chinese).
29. Shu, Junqing, Zhao Linchong, Yang Bin, et al. 2016. *The utility model relates to an on-board isolation step-up power supply device for emergency traction*. CN205674982U (in Chinese).

Overview of Intelligent Train Service System Design for Passenger Service

Jiao Zhang, Miaomiao Huo, and Yali Wang

Abstract The intelligent train service system refers to the train as the core, the use of general advanced technology to empower various train systems or equipment, realize train self-perception, decision-making, and self-optimization, realize train intelligent service, and comprehensively improve service quality. This article designs an intelligent train service system for passenger services, and combines the current industry development concepts and trends to explain the technical development route of the intelligent train service system, and designs around the purpose of serving passengers, service drivers, and service platform operators. The three-tier structure of the system, the design of the vehicle system network architecture and the internal data transmission mode of the vehicle system, provide a certain theoretical basis for the subsequent development of this field.

Keywords Intelligent train service system · Three-layer structure · The network architecture · The internal data transmission mode · The subsequent development

1 Introduction

The intelligent train service system refers to taking the train as the core and using general advanced technology to empower each train system or equipment to realize train self-awareness, self-decision and self-optimization, realize train intelligent service, and comprehensively improve service quality [1–4].

The technical development route of smart trains can be summarized as:

(1) On the basis of the existing vehicle platform, build a train sensor network to give the train the ability to sense its own environment. (2) Promote the integration and resource integration of the internal subsystems of the train, build an integrated data transmission and control platform, provide a basis for the transmission, processing

J. Zhang (✉) · M. Huo · Y. Wang
1Beijing Subway Operation Technology Centre, Beijing, China
e-mail: zhangjiao2099@163.com

W. Chen et al. (eds.), *The Proceedings of the 9th Frontier Academic Forum of Electrical Engineering*, Lecture Notes in Electrical Engineering 743,
https://doi.org/10.1007/978-981-33-6609-1_15

and analysis of massive data, and increase the degree of train intensification. (3) Based on massive data of multiple specialties and multiple subsystems, carry out data governance, integration and mining, so that the train has the ability of self-awareness, self-decision and self-optimization.

The development of intelligent train service system will promote the formation of new industrial clusters such as rail transit network communication, environment awareness, artificial intelligence, etc. [5–8]. The rapid reform of the rail transit industry will provide the impetus to bring new products, new services, and new formats to satisfy the people's travel Demand, improve the level of operations, improve urban management, promote industrial upgrading and accelerate the transformation of social and economic development will play a key role.

The intelligent train service system is directly oriented to passengers. Through advanced technologies such as multi-sensor fusion and integrated release of new multimedia information, the level of intelligent control is improved, and more intelligent services are provided for passengers from the aspects of travel terminals and automatic adjustment of the state of the cars to enhance passenger travel experience. At the same time, it promotes the track operation to actively seek reform and innovation in track operation management and service quality, starting from the passenger experience, and responding in a refined manner to the passenger's full travel chain process service needs.

2 System Structure

For the purpose of serving passengers, service drivers, and service platform operators, in the system design process, the intelligent service system is designed into three levels, namely the vehicle layer, the platform layer, and the central cloud platform. The entire intelligent service system is based on this the system architecture carries out follow-up work for the work objective (Fig. 1).

(1) Vehicle Intelligent Service Platform (VISP): Based on integrated control, it integrates various service terminals to realize train intelligent information service, environment perception and personnel behavior detection to empower passengers and drivers.
(2) Platform Intelligent Service Platform (SISP): It is responsible for obtaining the information of the congestion, train temperature and time of the arriving trains through data interaction with VISS, and providing intelligent waiting services for platform passengers.
(3) Central Intelligent Service Platform (CIMP): It is responsible for obtaining information on train car air quality, driver behavior detection results, abnormal behavior of passengers in the car through data interaction with VISS, and monitoring and recording the line operation.

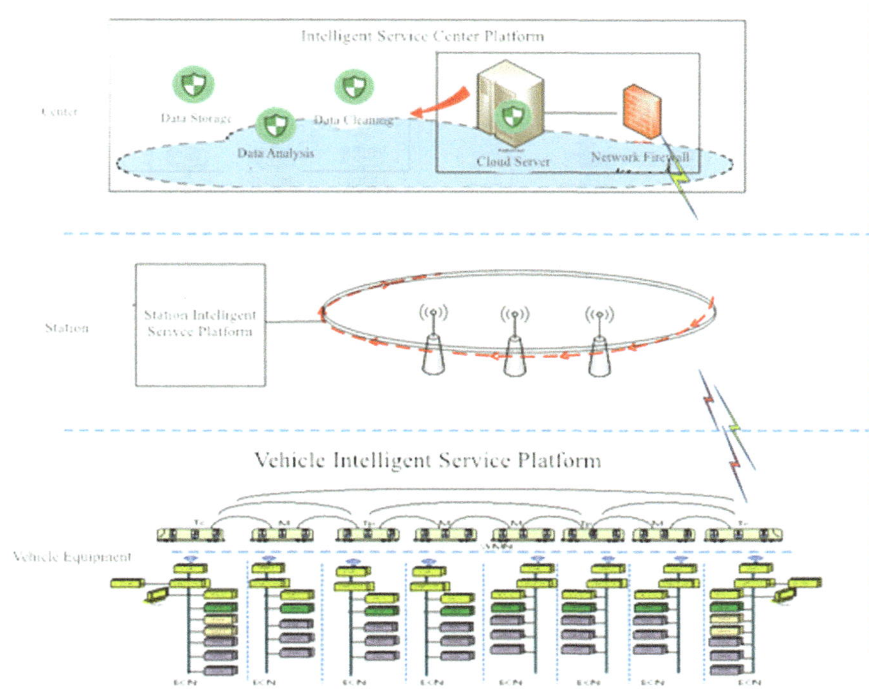

Fig. 1 System structure diagram

3 Vehicle System Network Architecture

The network architecture can be divided into train-level and vehicle-level networks according to its hierarchy.

Train level: The vehicle is set up with an on-board integrated control platform (VICP), which is responsible for summarizing the entire train information and interacting with platform equipment through the vehicle-ground wireless module (V2WU). The train-level wireless self-organizing network (WMN) is built through the vehicle-mounted wireless unit (VWU) to realize the data interaction between the ICCU and the intelligent vehicle control unit (IVCU) of other cars.

Vehicle-level: build an Ethernet marshalling network (ECN) in each car, each vehicle is equipped with an intelligent train service system on-board control unit (IVCU, abbreviated as on-board control unit), and connect to subsystem devices on the train marshalling network through the Ethernet interface Including: driver behavior detection module (DBTU), cabin environment detection unit (EDU), cabin abnormal behavior detection unit (ABDU), cabin congestion detection unit (CDU), passenger information service unit (PIU), etc. (Fig. 2).

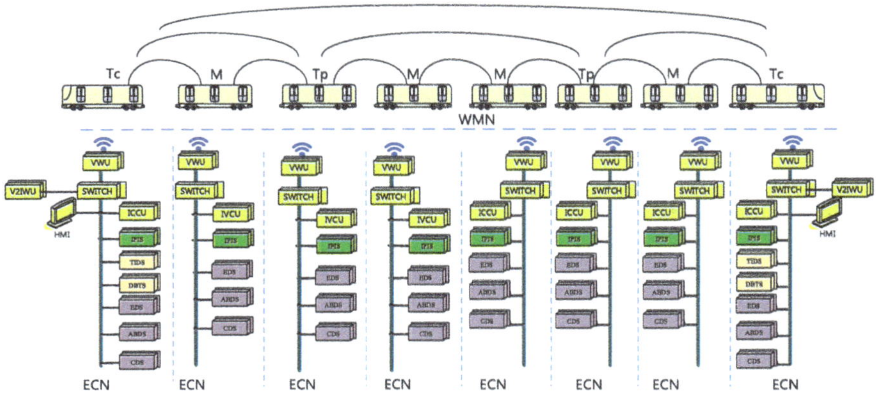

Fig. 2 Vehicle system network architecture

4 In-Vehicle System Data Transmission

(1) Vehicle-mounted intelligent train system and platform intelligent service system

The vehicle-mounted intelligent service system (VISS) communicates with the station intelligent service system (SISS) through a wireless interface. VISS sends information about the congestion and temperature of different cars to the SISS to realize the intelligent waiting guidance service for passengers on the platform.

(2) Vehicle-mounted intelligent train system and central integrated monitoring system

The vehicle-mounted intelligent service system (VISS) and the central integrated monitoring system (CIMS) communicate through a wireless interface, and VISS sends the air quality and driver behavior test results of different cars to CIMS.

5 Conclusion

This article describes the technical development route of the intelligent train service system in combination with the current industry development tem architecture carries out follow-up concepts and trends, and carries out relevant design of the passenger train-oriented intelligent train service system, which provides a certain basis for subsequent research in this field. It can integrate new technologies such as Internet of Things technology, intelligent perception technology, intelligent decision-making technology, and big data to intelligently upgrade ordinary trains, research and develop intelligent train service systems, and provide passengers with better service. The development of intelligent train service systems is The rail transportation industry can play a leading and leading role.

Acknowledgements This work was supported by National Key R&D Program of China (2016YFB1200402).

References

1. Zhang, Qiuliang, and Fang Kai. 2013. Design and implementation of train passenger transportation service management information system. *Railway Computer Application* 22 (2): 17–20.
2. Yang Guoyuan. 2016. *Research on the Railway Passenger Transportation Management Information System and Key Technologie.*
3. Tianyun, Shi. 2005. A general framework of information system of high-speed passenger railways. *Journal of Transportation Systems Engineering and Information Technology* 5 (92–497): 106.
4. Zhang, Xia, Zhao Yu. 2004. Study of railway passenger integrated service information system. *Journal of Transportation Systems Engineering and Information Technology 4*(2):64–67.
5. Fu, Di, Lv XiaoJun, Wang Jing, Yang Fan. 2008. Study on railway passenger service system for passenger station. *Railway Computer Application 17*(11):12–15.
6. Zhang Jingwen. 2010. Thoughts on measures of optimizing and increasing railway passenger traffic service. *Railway Transport And Economy 32*(3):47–48.
7. Hu, Haifeng, Zhou Donghong, Shao Xiaofeng. 2009. A study on the problems and solutions in China railway passenger service system. *Automation Panorama 26*(11):74–77.
8. Yang, Ke, Cui Bingmou. 2004. Analysis on the demand of comprehensive information for railway passenger transport service. *Railway Transport And Economy 26*(12):40–42.

Relevant Research on Image Recognition Content in the Intelligent Train Service System of a Subway Train

Jiao Zhang, Yujie Li, and Miaomiao Huo

Abstract The detection of the degree of congestion in the carriage is based on the evaluation of the number of passengers in the carriage. The target detection algorithm based on the deep neural network can accurately identify the number of passengers in the carriage and evaluate the degree of congestion in the carriage. This paper conducts related researches on the functions and uses of image recognition in an intelligent train service system in a subway train, including congestion detection research based on image recognition, passenger abnormal behavior analysis research based on image recognition, and driver fatigue based on image recognition The three pieces of content were detected and related principle analysis was carried out. This article expounds the principles of the above three pieces of content, and provides a certain research foundation for subsequent research in this field.

Keywords Image recognition · Intelligent train service system · Passenger abnormal behavior · Driver fatigue testing

1 Introduction

The detection of the degree of congestion in the car is based on the evaluation of the number of passengers in the car. The passenger perception camera can capture sufficient image information, and the calculation module performs target detection on the collected image information by constructing a multi-layer neural network, and evaluates the crowdedness of the car according to the number of detected passengers and a prescribed threshold.

In conventional target detection, a large number of candidate frames need to be produced to satisfy a high enough intersection ratio with each truth value [1–3]. At the same time, the candidate box does not match the true value, which is not conducive to the classification task of the bounding box. In order to solve the shortcomings of the candidate frame, a set of key points will be selected to detect the target method,

J. Zhang (✉) · Y. Li · M. Huo
Beijing Mass Transit Railway Operation Corporation LTD, Beijing Subway Operation Technology Centre, Beijing, China
e-mail: zhangjiao2099@163.com

W. Chen et al. (eds.), *The Proceedings of the 9th Frontier Academic Forum of Electrical Engineering*, Lecture Notes in Electrical Engineering 743,
https://doi.org/10.1007/978-981-33-6609-1_16

and then the candidate frame is removed, but due to its lack of reference to the target global information, it still has limitations. At the same time, each target consists of a set of key points. The algorithm is sensitive to the bounding box, and it is impossible to determine which set of key points belong to the same target. The environment of subway cars is special, the space is relatively narrow, the range of passenger distribution is wide, the personnel are concentrated and there will be more occlusion, which brings many difficulties to the detection of passenger targets [4–6]. Therefore, the calculation module adopts a new type of target detection algorithm to realize the target recognition for passengers in the area.

The new algorithm adds a function to observe the visual pattern of each candidate area on the basis of the traditional target detection algorithm, and then can judge the correctness of each bounding box. By adding a key point to explore the information of the middle area (approximate geometric center), the innovation of this article is that if the predicted bounding box has a higher intersection ratio with the true value, the probability that the center key point predicts the same category is higher, vice versa. Therefore, during inference, when a bounding box is generated through a set of key points, we continue to observe whether a key point with the same category falls into the center of the area, that is, using three points to represent the target.

At the same time, in order to better detect the key point of the center, namely the corner point, two methods are proposed to enrich the information of the center point and the corner point.

2 Two Methods

2.1 Center Pooling

The branch used to predict the key points of the center is helpful for the center to obtain more central areas of the target, and thus the central area of the proposal can be more easily perceived. This method is realized by taking the maximum value of the sum of the horizontal and vertical response values at the center position. The center of an object does not necessarily contain strong semantic information that can be easily distinguished from other categories. For example, a person's head contains strong semantic information that is easy to distinguish from other categories, but its center is often located in the middle of the person. center pooling extracts the maximum values of the horizontal and vertical directions of the center point and adds them to provide the center point with information outside the location. This operation gives the central point an opportunity to obtain semantic information that is easier to distinguish from other categories. Center pooling can be achieved by a combination of corner pooling in different directions. The maximum value in a horizontal direction can be achieved by connecting left pooling and right pooling in series. Similarly, the maximum value in a vertical direction can be achieved by connecting top pooling and bottom pooling in series.

2.2 *Cascade Corner Pooling*

Add the function of original corner pooling to sense internal information. The maximum value of the sum of the response values within the target object and the boundary direction in the feature map is combined to predict the corner point. Generally, the corners are located outside the object, and the location does not contain the semantic information of the associated object, which brings difficulties to the detection of the corners. The traditional approach is called corner pooling. It extracts the maximum value of the object boundary and adds it. This method can only provide the semantic information of the edge of the associated object, and it is difficult to extract the semantic information of the richer object. The cascade corner pooling first extracts the maximum value of the object boundary, and then continues to extract the maximum value at the maximum value of the boundary (in the direction of the dotted line in the figure), and adds it to the maximum value of the boundary, so as to provide a richer feature for the corner points Related object semantic information. Experiments have confirmed that this method is more stable in the presence of feature-level noise, which helps to improve the accuracy and recall rate.

In order to remove the incorrect bounding box, use the position of the detected center point and sort it according to the following process:

(1) Select the top k key points according to their scores.
(2) Remap the center keypoint to the input map according to the corresponding offset.
(3) Define a central area for each bounding box and ensure that there are central key points in the central area.

At the same time, ensure that the category of the point is consistent with the category of the bounding box. If the central key point is detected in the central area, the score of the bounding box is updated with the average of the scores of the upper left corner, the lower right corner and the central point, and the bounding box is saved. If no center point is detected, the bounding box is removed. Finally, the target accurate bounding box is obtained.

In summary, the target detection algorithm based on deep neural network can accurately identify the number of passengers in the car and evaluate the degree of congestion in the car.

3 Analysis of Abnormal Passenger Behavior Based on Image Recognition

Abnormal behavior is divided into two steps: detection of key points of multiple human bodies in the carriage and recognition of abnormal behavior actions. Generally speaking, human behavior recognition has multiple modalities, such as appearance, depth, optical flow, and body bones. Among these modalities, dynamic human bones can often complement other modalities and convey important information [7–9]. Through human key point detection, human bones can be identified, and dynamic bone modalities can be established. Reanalyzing its action patterns can achieve human behavior recognition. The general process of human joint point recognition is that the passenger compartment camera first captures the cabin image, and then the key point detector in the abnormal behavior perception algorithm of the calculation module will identify and mark the body features, helping the body tracking algorithm to understand the different angles The performance of each pose is presented in the form of a 3D color stickman. The recognition process is closely related to the camera system and computing power. The computing module's abnormal behavior perception algorithm is configured with a multi-thread module to speed up processing.

The algorithm mainly includes the following steps:

(1) Input an image, extract the features through the convolution network, get a set of feature maps, and then divide it into two branches, use CNN to extract part confidence maps and part affinity fields respectively;
(2) After obtaining the two pieces of information, use bipartite matching (even matching) in graph theory to find the part association and connect the joint points of the same person. Due to the vector nature of PAF, the generated even matching is very correct. Merged into a person's overall framework.
(3) Seeking multi-person parsing based on PAFs—Converting multi-person parsing problems into graphs problems—Hungarian algorithm.

In general, the abnormal behavior perception algorithm is equivalent to the upgraded version of the body tracking technology used in previous somatosensory games, but compared with the 20 key points tracked by Microsoft Kinect, the abnormal behavior perception algorithm is much more detailed. The same action, Kinect Perceiving that a person is raising their hands, the abnormal behavior perception algorithm can observe that the person is actually pointing something with his finger. In terms of facial tracking, the entire head in Kinect is just a point, and the eyebrows, eyes, nose, and mouth of the calculation module's abnormal behavior perception algorithm can be depicted by dozens of key points, and the expression can be recognized.

Here are a few examples of different scenarios:

Abnormal behavior perception algorithm is used to extract key points of human body and generate skeleton feature map, that is, behavior detection principle diagram. After extracting the human bones, the 2D coordinates of each human bone in each frame are used to represent the bone sequence. After that, a spatio-temporal map is generated based on the features of the bone sequence, and a spatio-temporal map of the bone sequence is constructed in two steps. In the first step, the edges between frames represent the temporal relationship of the corresponding bone points of the human body; in the second step, within each frame, a space graph is constructed according to the natural skeleton connection relationship of the human body. Finally, the spatio-temporal map is input into the ST-GCN network for behavior prediction. Predict the bone spatio-temporal map through the ST-GCN network and recognize human movements.

4 Driver Fatigue Detection Based on Image Recognition

The fatigue detection is completed through the camera and fatigue detection control unit installed on the driver's driving platform.

(1) Using computer vision technology, first preprocess the image, use the color space skin color model for coarse face positioning, and successively reduce the eye area according to the characteristics of the face; finally, the edge information is used to locate and close the eye measuring. Taking into account the correlation between the frames of the video image sequence, the linear motion prediction method is used to track the human eye, complete the doze detection, find the abnormal sound alarm in time, and send the recognition result to the integrated control unit.
(2) Use deep convolutional neural network to detect the driver's face and nose in sequence, use local two-bit features and random forest classifier training to generate an online nose detector, in order to correct the optical flow tracker during the nose tracking process Drift error and other parameters; finally, analyze the change of the boundary gradient of the mouth area under the nose, and combine the confidence of the nose tracker, the lateral movement of the face and other information to judge the completion of breathing detection. The result is sent to the integrated control unit.

5 Conclusion

In this paper, the three principles of image recognition-based congestion detection, image recognition-based passenger abnormal behavior analysis and image-based driver fatigue detection in the intelligent train service system of a subway train are analyzed and explained. These three parts are analyzed and explained, which provides a certain research basis for the subsequent research in this field.

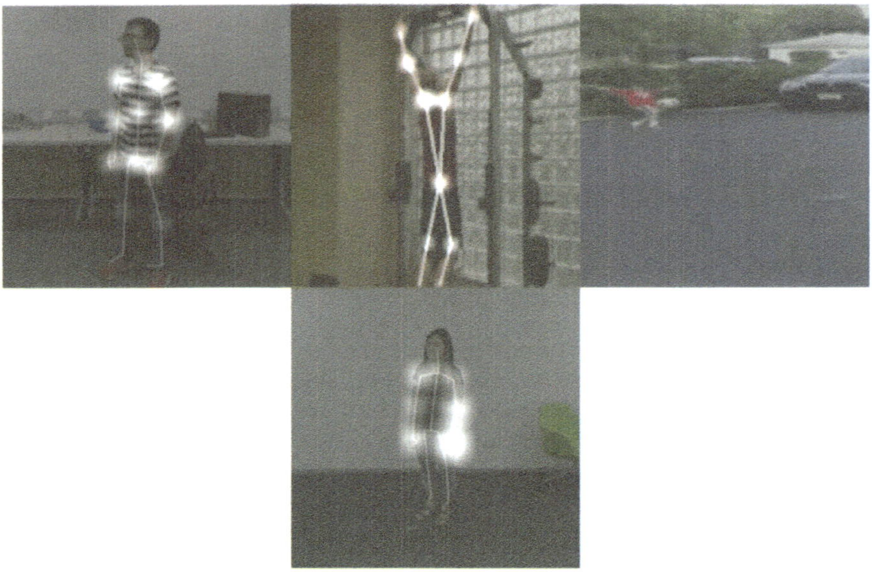

Fig.1 Examples of behavior detection application scenarios

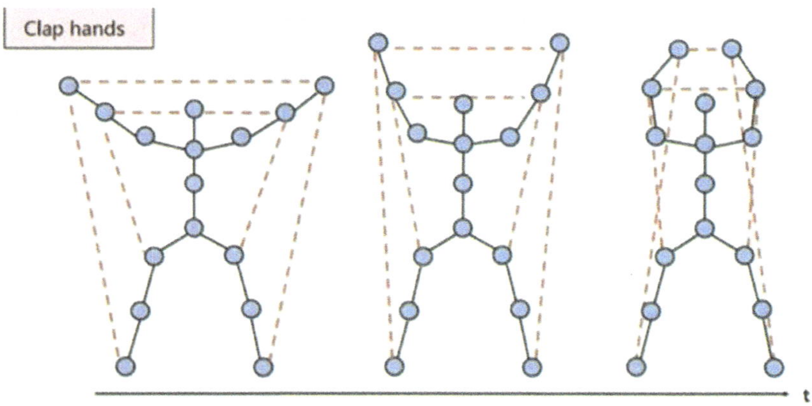

Fig.2 Schematic diagram of behavior detection

Acknowledgements This work was supported by National Key R&D Program of China (2016YFB1200402).

References

1. Li, Qing, and Wei Xueyun. 2020. SAR ATR based on shunt convolutional neural network. *Chinese Journal of Radio Science 35*(3):364–371.
2. Jiang, Manxing, Zhao Tongzhou, and Wu Zejun. 2020. Application of convolution neural network based on target shape in ships and warships classification. *Journal of Wuhan Institute of Technology 42*(2):213–217.
3. Wang, Hang, Chen Xiao, Tian Shengzhao, and Chen Duanbing. 2020. SAR image recognition based on few-shot learning. *Computer Science 47*(5):124–128.
4. Zhao, Jiangping, and Wang Yao. . Unsafe behavior recognition based on image recognition technology. *Safety and Environmental Engineering 27*(1):158–165.
5. Peng, Shuangqi, Xu Qiang, Li Huajin, Zheng Guang. 2019. Grain size distribution analysis of landslide deposits with reliable image identification. *Journal of Engineering Geology 27*(6):1290–1301.
6. Zhu Fengxia. 2020. License plate recognition technology based on neural network and image recognition. *Electronic Design Engineering 28*(2):130–133,138.
7. Fan, Di, and Ju Zhiyong. 2020. Method of small sample image recognition based on prototype network. *Computer and Modernization* (3):103–107.
8. Tian, Zhongke, Chen Chengjun, Li Dongnian, Zhao Zhengxu, and Hong Jun. 2020. Part recognition and assembly monitoring based on depth images. *Computer Integrated Manufacturing Systems 26*(2):300–311.
9. Zhang, Ke, and Hou Jie. 2020. A new image recognition method in improved convolution neural network. *Science Technology and Engineering 20*(1):252–257.

Vehicle-Ground Communication and Positioning Technology Based on Ultra-Wideband Technology in an Intelligent Train Service System of a Subway Train

Jiao Zhang, Yujie Li, and Miaomiao Huo

Abstract Ultra-wideband wireless communication is currently the fastest-growing and most dynamic technology, which has the advantages of high anti-interference performance, high transmission rate, large space capacity, and low power consumption. This article describes the vehicle-ground communication and positioning technology of ultra-wideband wireless communication (UWB) in an intelligent train service system in a subway train. UWB technology is used to complete the transmission of positioning information and related link information while participating in positioning. Wired The communication synchronization network integrated with the wireless can effectively reduce the cost and the redundancy of the network system, and provide a certain research foundation for subsequent research in this field.

Keywords Vehicle-ground communication and positioning technology · Ultra-wideband wireless communication (UWB) · Intelligent train service system • transmission of positioning information · Wired and wireless communication synchronization network

1 Introduction

The core of a subway train's intelligent train is an integrated platform. The platform supports the integration, access and control of various application logics through microservices and middleware technology, and provides basic services for various applications, including location information services and vehicle locations. Communication service, UWB wireless communication, as one of the most rapidly developing and most dynamic technologies, has the advantages of high anti-interference performance, high transmission rate, large space capacity, low power consumption, etc. It can meet the positioning and communication of intelligent train development

J. Zhang (✉) · Y. Li · M. Huo
Beijing Mass Transit Railway Operation Corporation LTD, Beijing Subway Operation Technology Centre, Beijing, China
e-mail: zhangjiao2099@163.com

W. Chen et al. (eds.), *The Proceedings of the 9th Frontier Academic Forum of Electrical Engineering*, Lecture Notes in Electrical Engineering 743,
https://doi.org/10.1007/978-981-33-6609-1_17

Demand: On the one hand, it provides accurate positioning services for trains, and on the other hand, it can be used as a communication method to achieve two-way train-ground communication [1–4].

Adopt a communication system that combines UWB communication network with a new generation of dedicated communication equipment [5–7], and use UWB communication base stations and positioning base stations deployed in the positioning network to complete the transmission of positioning information and related link information while participating in positioning [8–10]. The wireless synchronous communication synchronization network can effectively reduce the cost and reduce the redundancy of the network system. To ensure the communication rate and communication distance in the communication process, a new generation of dedicated communication equipment is used to improve the overall transmission efficiency of the system.

2 Principle

UWB is a carrier-free communication technology. It uses non-sinusoidal narrow pulses in the nanosecond to microsecond range to transmit data, and uses the time-domain pulses of the starting and landing points to directly modulate. Ultra-wideband transmission puts the modulation information process in a single It is performed on a very wide frequency band, and the duration of this process is used to determine the frequency range occupied by the bandwidth. We refer to −10 dB relative bandwidth exceeding 25%, or −10 dB absolute bandwidth exceeding 500 MHz, which is called ultra-wideband (UWB).

Absolute bandwidth refers to the difference between the upper cutoff frequency and the lower cutoff frequency corresponding to a roll-off point on both sides of the maximum value of the signal power spectrum, and the relative bandwidth refers to the ratio of the absolute bandwidth to the center frequency.

The spectrum range used by UWB in this article is 3774–4243.2 MHz, and the highest power spectral density of the signal transmitted by the base station is −41.3 dBm/MHz. Because the UWB communication rate is closely related to the transmission distance. In less than 10 m, the UWB signal transmission rate can reach up to 1 Gbps, but as the distance increases, the communication rate declines rapidly. When the communication distance is greater than 35 m, the maximum communication rate can reach 6.8 Mbps. Considering the need to meet the vehicle-to-ground communication capability of the vehicle in the moving state with an average speed of 80 km/h during operation, the communication capability meets the transmission rate requirement of 250 Kbps, and the communication bandwidth is set to 2 M.

In the system described in this article, the total bandwidth of the ultra-wideband system is 6 M. Since this system adopts the way of communication and positioning shared channels, in order to ensure the positioning while taking into account the communication capabilities, for this reason, the equipment bandwidth is allocated according to 2:1 The communication bandwidth is designed to be 2 M.

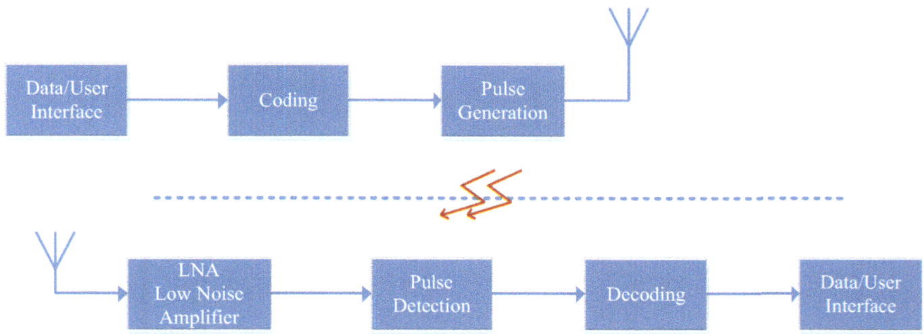

Fig. 1 UWB communication principle

Since the beam angle range of the base station antenna design is ±75°, considering the positioning accuracy and the probability of communication connectivity, the protection cut-off angle is set to ≤10°. According to the construction site, the base station is generally installed below 3 m, and the radio frequency coverage radius of a single base station is about 50 m. In order to ensure the positioning accuracy, it is deployed at a distance of 30 and 50 m under different environments of the platform and the interval (the actual distance will be slightly adjusted due to the field environment during the construction process), and combined with the verification of the previous positioning function, comprehensively considering the cost and positioning accuracy, It is optimal to deploy at intervals of 30 and 50 m. The communication principle of UWB system can be seen in Fig. 1.

UWB technology has the following features: high transmission rate, short communication distance; small size, low power consumption; good system coexistence, high communication density; strong anti-interference; extremely high positioning accuracy and strong anti-multipath capability.

The principle of UWB high-precision positioning technology based on bidirectional time measurement can be seen in Fig. 2.

Compared with the outdoor, the indoor space is narrow and the occlusion is serious. The multipath interference of the signal propagation is very large. The ultra-wideband (UWB) signal bandwidth is wide (500 MHz left and right), the signal is a pulse-type narrow signal (ns level), when the UWB signal is used indoors, the first path signal can be well distinguished, and then rely on a certain layout of the indoor positioning base station, Basically, the signal that reaches the user on the first path is a direct link, and the signal strength is the strongest, and other path signals that arrive subsequently are ignored directly. Therefore, an indoor positioning system with a certain scale capacity can be established with positioning accuracy ≤20 cm, free roaming.

Based on UWB's two-way ranging indoor positioning algorithm for positioning, the algorithm does not require high-precision time synchronization between the transceiver and can effectively measure the accurate distance between the transceiver.

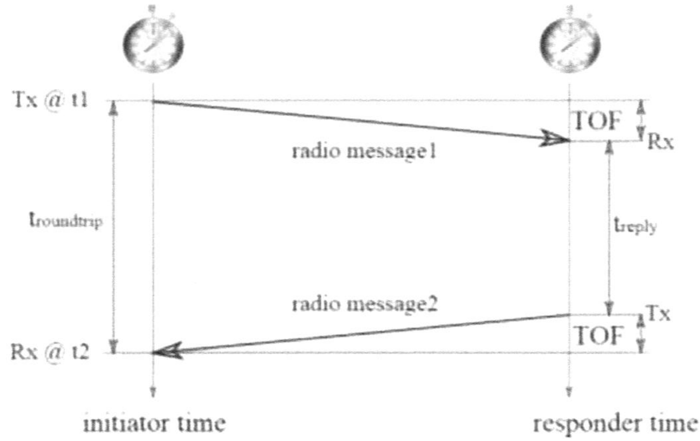

Fig. 2 Schematic diagram of UWB two-way ranging

Basic principle: The user sends a positioning request and records the signal transmission time t1. After receiving the request, the anchor feeds back to the user a signal. The time difference treplay is loaded into the signal and sent to the user. The user records the time t2 when the feedback information is received. The distance between the user and the anchor is $(t_2—t_1-t_{replay}) \cdot C/2$, where C is the speed of electromagnetic wave propagation.

3 System Structure

At the train level, an on-board Tag wireless unit is added to access the train integrated platform. Provide communication and positioning services for trains.

At the off-track level, an integrated positioning and communication base station is deployed as an interface device for vehicle wireless unit access and positioning.

At the station level, deploy a location and a communication server, and connect to the station's integrated platform to achieve data connection between the train and the station (Fig. 3).

Through the station-trackside-train link, the vehicle-ground communication information is transparently forwarded, and the data flow is as follows (Figs. 4 and 5).

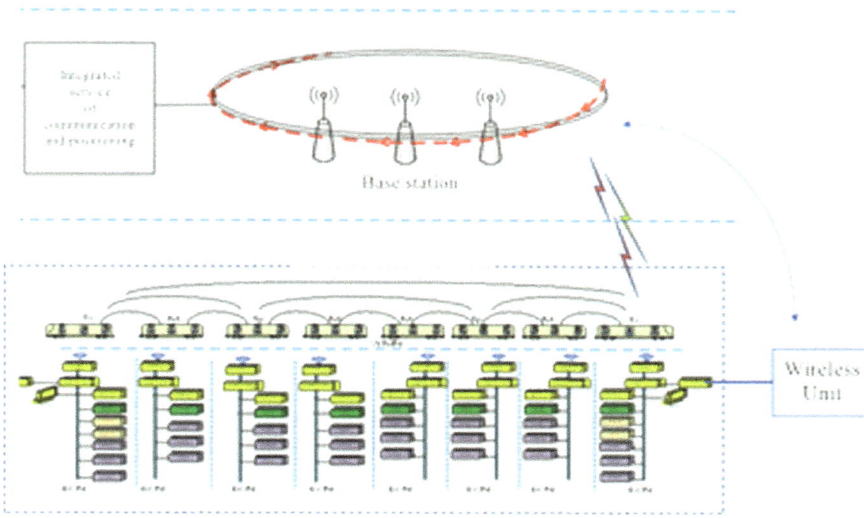

Fig. 3 System architecture

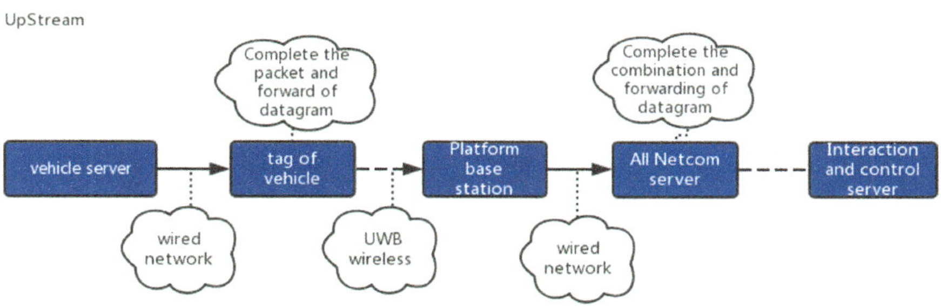

Fig. 4 Vehicle-to-ground data flow

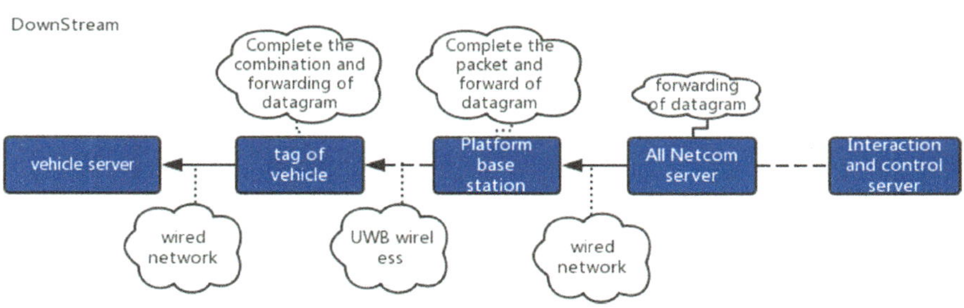

Fig. 5 Ground-to-vehicle data flow

4 Functions

(1) Train positioning

Urban rail transit positioning is one-dimensional positioning. According to the operating scenario, a positioning map is established to establish the correspondence between the base station and the actual location. Based on the ranging and positioning technology of wireless pulse arrival time difference, the train UWB terminal realizes the active positioning of the train through communication with the ground UWB base station.

(2) Car-ground wireless communication

By constructing a sensor network, the train detects information such as the degree of congestion inside the car. Before arriving at the station, the train platform sends information such as train location and congestion to the station platform through UWB wireless communication to meet the display needs of the station's guidance screen, including the display of train position and the display of the congestion of different train compartments.

5 Conclusion

This article analyzes and describes the vehicle-ground communication and positioning technology of ultra-wideband wireless communication (UWB) in the intelligent train service system of a subway train, and explains its positioning principles and communication links. While completing positioning, it transmits related link information, and uses a wired and wireless communication synchronization network to effectively reduce costs and reduce the redundancy of the network system, which can provide a certain research basis for subsequent research in this field.

Acknowledgements This work was supported by National Key R&D Program of China (2016YFB1200402).

References

1. Zhang, Baojun, Tian Qi, Wang Heng, Chen Xi. 2020. UWB indoor positioning algorithm based on CNN and online learning. *Chinese Journal of Sensors and Actuators 33*(4):511–516.
2. Zhou, Lin, Li Xiaoming, Jiang Xianzhi. 2020. Adaptive UWB/DR indoor co-localization approach based on UKF. *Computer Systems & Applications 29*(5):175–181.
3. Yan, Jiahui, Zhang Xiaoming, Liu Jun, Zhu Menglong, Qi Jinfan, Gao Lizhen. 2020. Design of sports field positioning system based on UWB. *Navigation Positioning and Timing 7*(1):98–103.

4. Yang, Sheng, Hu Zhigang, Zhao Qile, Zhang Huichao. 2020. Principle and precision verification of UWB positioning based on TOF. *Journal of Geodesy and Geodynamics 40*(3):247–251.
5. Liu, Yu, Chen Ziran, Gu Hongyang, Fu Lele, Lu Yongle, Li Xinghai. 2019. UWB assisted inertial positioning method based on fault-tolerant decision tree. *Piezoelectrics & Acoustooptics 41*(6):897–903.
6. Chen, Hao, He Jie, Ma Kai, Huang Qin. 2019. Application of UWB precise positioning based on Kalman filter for power field operation safety. *Electronic Design Engineering 27*(24):103–107.
7. Xu, Qingkun, Song Zhongyue. 2019. Design of positioning system of autonomous following robot based on UWB technology. *Computer Engineering and Design 40*(11):3337–3341.
8. Shi, Liuqiang, Fu Jiangtao. 2019. UWB-based positioning system design. *Electronic Design Engineering 27*(15):161–165.
9. Shi, Ce, Chen Shanxue, Li Fangwei. 2020. Narrow band interference suppression of time reversal UWB communication system. *Systems Engineering and Electronics 42*(4):948–953.
10. Gu, Huidong, Jiao Liangbao, Zhou Jian. 2020. Indoor high-precision positioning system design based on UWB. *Software Guide 19*(4):159–163.

Research on Intelligent Security Defense System Based on Full Scenario Protection of Smart Grid

Shijun Zhang, Jing Zeng, Ji Lai, and Shuo Li

Abstract This paper aims at the deficiencies of the network security protection system of the smart power grid, and studies how to improve the network security defense system of the power grid, to achieve comprehensive network security protection. Focus on terminal, host and boundary protection, this paper proposes a set of full-scenario protection intelligent security defense system, aiming at ensuring the safe and stable operation of the power grid, strengthening the foundation of network security, improving the quality of security protection management, and enhancing the level of network security protection.

Keywords Network security · Smart grid · Intelligent · Defense

1 Introduction

In recent years, the smart grid has developed rapidly, with the scale of the network constantly expanding and the application level of the network constantly improving [1]. It has become a huge force to promote social progress and economic development. With the increasing development of the network, the threat of network security is increasingly serious, accompanied by network attacks, and other hacker activities are more frequent and constantly upgraded [2]. With the continuous development of information construction, the original security situational awareness and monitoring and early warning have been unable to meet the requirements of smart grid security protection fully. It is urgent to take measures to establish a unified, safety monitoring mechanism, strengthen the linkage disposal capacity, and optimize the process of monitoring and early warning, to build an intelligent, three-dimensional, comprehensive security defense system [3].

They are combining with the present situation in the construction of the smart grid security, aiming at the shortcomings of the network security protection [4], this paper put forward the intelligent security defense system based on the full scenario

S. Zhang (✉) · J. Zeng · J. Lai · S. Li
State Grid Jibei Information & Telecommunication Company, No. 32 Zaolin Front Street, Xicheng District, Beijing, China
e-mail: 760044312@qq.com

W. Chen et al. (eds.), *The Proceedings of the 9th Frontier Academic Forum of Electrical Engineering*, Lecture Notes in Electrical Engineering 743,
https://doi.org/10.1007/978-981-33-6609-1_18

protection research direction. By taking measures in the terminal, the host, border protection, and other aspects build an intelligent security defense system, strengthen centralized security monitoring, event linkage analysis, intelligent threat processing ability, realize rapid response for network security threats.

2 Background

2.1 Current Situation of Network Security Protection

With the gradual development of smart grid construction, information security has become an important foundation for the safe and stable operation of the smart grid, and the construction of unified, reasonable, and safe information facilities is an important guarantee for the smart grid [4]. As shown in Fig. 1, the smart grid is under threat.

At present, network security protection mainly depends on the power grid company has deployed network security systems and equipment. Existing security equipment generally covers border security, terminal security, application security, data security, and another level, and has basic capabilities such as intrusion prevention, situation analysis, and host compliance detection [5]. However, there's still a part of the lack of protection.

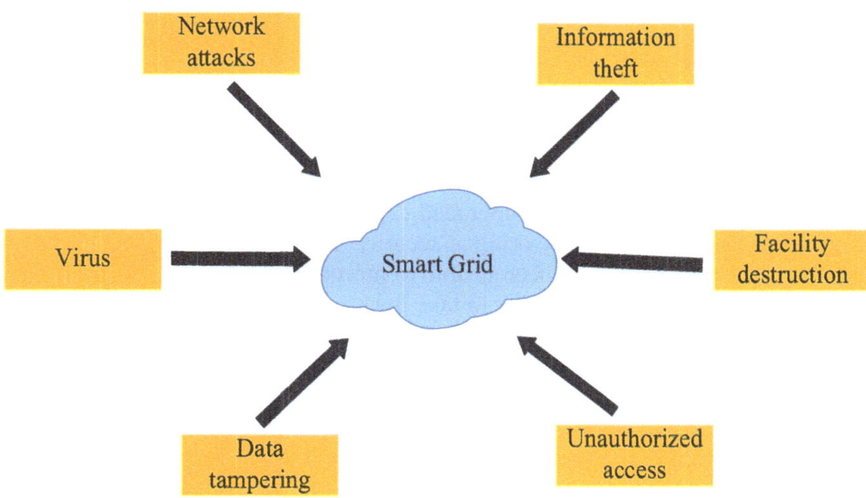

Fig. 1 The smart grid is under threat

2.2 *The Design of Intelligent Security Defense Institution*

Based on the understanding of the current security protection management situation of the smart grid, the construction of the full-scenario intelligent security defense system is proposed from terminal control, host security, boundary management, platform construction, safe operation, and other aspects. Figure 2 is the design drawing of the full-scenario intelligent security defense system.

The construction of full-scenario intelligent security defense system can realize the unified control of terminals and devices, improve the intelligent handling capacity of network security incidents, improve the quality of security management, improve the overall level of network security protection of smart grid, and ensure the stable operation of network security and information system.

3 Intelligent Security Defense Measures

The full-scenario intelligent security defense system takes measures in the three directions of terminal, host, and border. It supplements and strengthens advanced technical security facilities for network security, establishes security defense methods, strengthens network security defense capabilities, and improves the level of intelligent security defense in all scenarios.

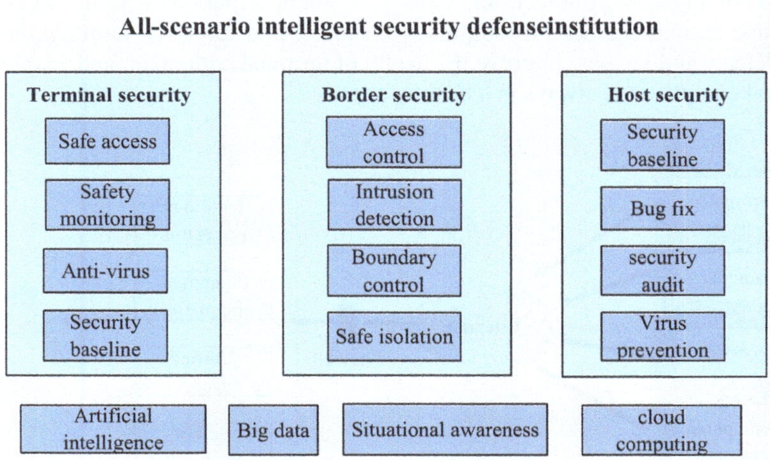

Fig. 2 The design of intelligent security defense institution for the full scenario

3.1 Security Defense Measures for Terminals

According to the management and control principle of "ontology protection, equipment compliance, unified access, security visibility," by means of terminal access, terminal security monitoring, antivirus, baseline monitoring, and other technology, through collect antivirus data, traffic threat analysis data, terminal access data, and baseline compliance data on the terminal equipment side realize comprehensive monitoring analysis from the terminal ontology, network analysis layer to global monitoring layer, and achieve comprehensive visual analysis and early warning for terminal equipment. Through the refined management and control concept, the terminal equipment can be fully protected, and the identification, ontology, behavior standardization, risk quantification, analysis intelligence, and management visualization of the general terminal equipment can be realized (Fig. 3).

Identification. Accurate device visibility is a cornerstone of safety practices [6]. All online devices are scanned and deeply identified by the terminal access system to obtain the device name, device type, property rights unit, IP address, Mac address, network, and other information of the terminal, and use intelligent identification and unique algorithms to generate unique codes for each terminal device. According to the acquired asset ledger information, it can be associated with security threat warning, and locate and deal with the equipment according to the equipment code, to facilitate the control of the security threat situation of the terminal equipment.

Proprioception. The higher the health of the equipment body, the stronger the ability to resist external threats. Through research and application of baseline scanning, vulnerability scanning, virus scanning, and other technologies, regular scans are used to obtain ontology defects and virus infection information on terminal devices. Then use various methods to eliminate, disinfect, and reinforce terminal devices with defects and viruses. Improve the health of terminal equipment and ensure that terminal equipment is always in a healthy state.

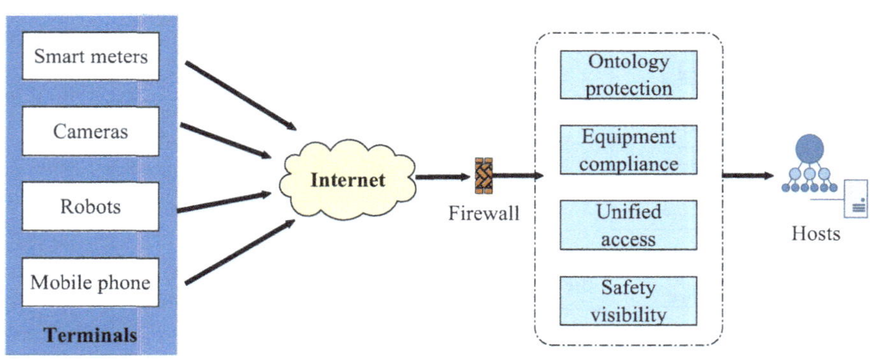

Fig. 3 Security defense measures for terminals

Standardization of behavior. Many terminal devices are close to external users, such as smart electricity meters and payment machines, which are vulnerable to abnormal behaviors or social attacks [7]. Through technologies such as terminal access, terminal flow analysis, and behavior analysis audit, collect and monitor some abnormal behaviors of terminal devices or abnormal data generated by the abnormal behavior. Moreover, according to the abnormal data made some abnormal behavior generates some special monitoring alarms to ensure the standardization of terminal equipment usage behavior.

Risk quantification. Through fine-grained statistics and detailed log statements, we can have a general view of the security situation of the whole network terminal, including the situation of high-risk vulnerability repair, virus detection and killing trend, the classification of external threat level, to achieve quantitative observation, efficient management and comprehensive monitoring of the risk of the whole network terminal. Adopt big data analysis technology to continuously collect user and equipment environmental data, user business access behavior data, and conduct risk modeling and correlation analysis to measure potential security risks [8].

Intellectualization of analysis. Collect scattered terminal behavior data, antivirus data, equipment asset data, and external attack data in a unified manner to form a behavior portrait centered on terminal-business-behavior [9]. Use big data analysis technology and multi-dimensional analysis technology to conduct correlation analysis from the perspective of threats, and obtain the details of threats such as victim targets, attackers, attack methods, vulnerabilities, malicious samples, and propagation methods in time. According to the analyzed threat results, through the threat verification, the authenticity and impact of the threat are judged.

Management visualization. The visualization of asset management realizes asset import, asset labeling, and asset portfolio, through the extraction of alarm information from multiple systems such as the security access system and antivirus system. Moreover, according to the risks covered by its warning, it realizes the correlation between assets and risks and forms a complete asset library. Risk visualization takes the dimension of threatening the entire life cycle and visualizes threats through a variety of threat detection methods. Automatically notify and issue alarms according to the set actions, which improves the efficiency of daily operations.

3.2 Security Defense Measures for the Host

For the host, vulnerabilities, virus prevention, resource control, security baseline, and security audit have been adopted to ensure the security and credibility of the host. However, there are still some deficiencies, mainly lack means for virus protection, security audit, and virtualized resource monitoring. By enhancing virus prevention capabilities, security auditing capabilities, and virtualization security protection capabilities, the host's threat detection capabilities are enhanced, and the host security protection link in the intelligent defense system is consolidated (Fig. 4).

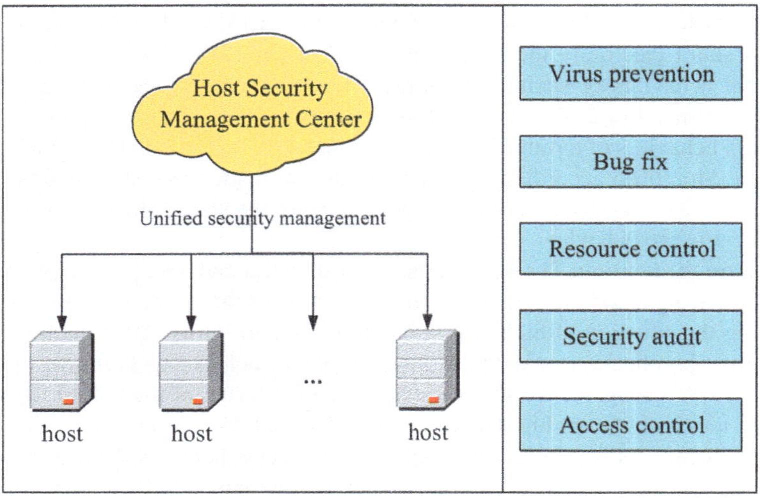

Fig. 4 Security defense measures for the host

Improve virus prevention capabilities. Host viruses are latent, and they are usually hidden in the host to spread and spread through certain channels [10]. The longer the hiding time, the more extensive the range of influence, and the greater the destructiveness it brings. Therefore, it is extremely important to improve host-virus prevention capabilities. By deploying a protection mechanism against viruses and malicious code, to prevent the intrusion of host viruses, realize the detection and blocking of host viruses, improve the host's antivirus capabilities, and enhance the capabilities of malware detection and detection.

Strengthen security audits. According to the actual needs of power grid operation, through log audit and operation and maintenance audit measures, strengthen the control of host security audit. Through the deployment of a log audit system, the log of the host is collected, stored, and analyzed in a unified manner. Generate audit reports and perform detection and analysis on the audit reports to ensure host security. By deploying the operation and maintenance audit system, the operation and maintenance operations are re-certified, and the operation and maintenance operations are recorded and audited to ensure the safety of the host operation and maintenance operations.

Explore the direction of virtualization protection. Explore the direction of host virtualization security protection, and study the feasibility of cloud virtualization antivirus, virtualized firewall, virtual patching, and other technologies for communication security in the virtual cloud environment. By deploying a cloud platform, effective monitoring of east–west traffic between virtualized resources can prevent attackers from moving laterally between servers and ensure the security of virtualized cloud hosts.

3.3 Design of the Controller Protection Measures Against Network Boundaries

At present, there are seven types of network boundaries in smart grids, including Internet boundaries, information internal and external network boundaries, information internal network third-party boundaries, information internal network vertical boundaries, information internal and external user domain boundaries, information internal network horizontal domains, and others [11]. Security devices such as firewalls, intrusion detection, antivirus walls, and secure access gateways have been deployed by relevant protection requirements, and have access control, intrusion detection, and security isolation capabilities. However, there are still some shortcomings in border protection. The network structure has unclear areas, and the network border protection equipment is partially missing and outdated.

In response to the lack of border protection, continue to refine border management, supplement border security protection measures, and focus on the improvement of third-party network border security protection, and comprehensively enhance the company's security protection capabilities. Design from two aspects of precision protection and secure operation and maintenance automation supplemented with basic security equipment such as IDS, IPS, WAF, and firewall, and strengthened the construction of technical capabilities such as unknown threat detection, situational awareness, and automatic verification tools for security equipment configuration. Third-party network boundary security protection capabilities have been improved, boundary management and control have been refined, protection measures have been added, and situation monitoring has been strengthened.

4 Discussion

4.1 Launch Typical Applications and Build a Defense Platform

With the continuous development of advanced continuous attack technology, network security attack is faster, more professional means, more covert process, more complex technology, network security personnel, and organizations are increasingly demanding. It is necessary to have the ability of rapid information sharing, alarm correlation analysis, timely handling of events, etc., to meet the demand of early warning analysis for scalability and flexibility. In the face of frequent network attacks, massive attack data, and endless attack means, network security personnel rely on traditional processing means, has been unable to achieve rapid response to network attacks.

In response to the needs of network security under the new situation, big data, artificial intelligence, and other technologies are used to build a full-scenario intelligent security defense platform. By summarizing the daily work experience of network security, combining security defense measures of terminals, hosts, and borders, carry out typical scenario applications, and complete automatic distribution and intelligent matching of network security incident classification. Realize automatic monitoring, analysis, and disposal of known attacks and threats, manual disposal of unknown attacks and threats, and improve the level of network security intelligent defense.

4.2 A Normal Operating Mechanism Based on a Full-Scenario Intelligent Security Defense Platform

Based on the full-scenario intelligent security defense platform combined with border security equipment, establish a normalized security operation mechanism to carry out daily network security monitoring, operation process and responsibility management, system equipment planning and design management, daily operation and maintenance, network security monitoring, security analysis response, reinsurance, and other safety operations.

In terms of security operations, it will make full use of the security events monitored by the full-scenario intelligent security defense platform. It is focusing on the various stage of the system life cycle, detailed operational content, and requirements. Utilize situational awareness, big data, artificial intelligence, and other technologies to fully enhance protective intelligence ability, carry out various safety operations in a normal way.

5 Conclusion

Based on the current status of smart grid security construction, the deficiencies of smart grid security defense, this paper integrates new information technology, initially proposes a smart security defense system based on full-scenario protection. Since the construction of this system involves a large number of new and improved security technical defense measures, it is a huge project. This article is only based on the understanding of the current management situation and proposes a system construction idea, which will be gradually followed by intelligent security defense according to the construction idea. The system was implemented and improved.

Acknowledgements This work was supported by the Science and Technology Project of State Grid Jibei Power Company Limited (No. 52018E190075).

References

1. All, F., A.R. Al-Ali, R. Al-Dalky, et al. 2012. Smart grid security: threats, vulnerabilities, and solutions. *International Journal of Smart Grid & Clean Energy*.
2. Mrabet, Z.E., N. Kaabouch, H.E. Ghazi, et al. 2018. Cyber-security in smart grid: Survey and challenges. *Computers & Electrical Engineering*: S0045790617313423.
3. Tan, S., D. De, W.Z. Song, et al. 2017. Survey of security advances in smart grid: A data driven approach. *IEEE Communications Surveys & Tutorials* 1: 1–1.
4. Huang, B., and Y. Yao. 2016. Failure mode analysis on machining center based on possibility theory.*Proceedings of the 5th international conference on electrical engineering and automatic control [Lecture Notes in Electrical Engineering]* 367: 627–636. https://doi.org/10.1007/978-3-662-48768-6
5. Lee, S., J. Bong, S. Shin, et al. 2014. A security mechanism of Smart Grid AMI network through smart device mutual authentication. In *International Conference on Information Networking*. IEEE.
6. Wu, J., K. Ota, M. Dong, et al. 2016. Big Data analysis-based security situational awareness for smart grid. *IEEE Transactions on Big Data*: 408–417.
7. Anzalchi, A., and A. Sarwat. 2015. A survey on security assessment of metering infrastructure in smart grid systems. In *Southeastcon*. IEEE.
8. Min, B., and V. Varadharajan. 2014. Design and analysis of security attacks against critical smart grid infrastructures. In *2014 19th International Conference on Engineering of Complex Computer Systems (ICECCS)*. IEEE.
9. Fan, Xinxin, and Guang Gong. 2013. Security challenges in smart-grid metering and control systems. *Technology Innovation Management Review* 3 (7): 42–49.
10. Amin, S.M. 2012. Smart grid security, privacy, and resilient architectures: Opportunities and challenges. *IEEE Power & Energy Society General Meeting*. IEEE.
11. Fangfang, Dang, Fajia Ji, Lei Zhu. 2018. A brief talk on information security basic management work of intelligent electric grid security defense system. In *International Conference on Machinery*. ACM.

Online PMSM Stator Interturn Fault Detection Aided by Phase Angle Diagnosis of Zero-sequence Components

Yecheng Zhang and Gai Liu

Abstract Accurate diagnosis of slight inter-turn short circuit fault (ISCF) of permanent magnet synchronous motor (PMSM) has always been a difficult problem in the field of electrical engineering. Aiming at the problem of ineffective ISCF diagnosis, a method on Phase angle diagnosis of zero-sequence components is proposed. In this method, a calculable expression of the fault indicator was derived in a wye-connected winding PMSM with ISCF, being for both slight and severe ISCF cases. Using the derived expression, ISCF were detected, as well as the degree and the location of the ISCF. Simulation results demonstrated that the proposed method not only has the capacity of diagnosing the ISCF, but also assessing the faulty severity and identifying the fault location under the different conditions of the rotor speed and load torque.

Keywords PMSM · Interturn short-circuit fault · Phase angle diagnosis · Faulty severity and location

1 Introduction

Permanent magnet synchronous motor (PMSM) is increasingly applied in household appliances, electric vehicles and wind turbines because of its high power density and efficiency [1, 2]. However, the reliability and safety of these machines remain one of the most vital aspects for PMSM applications. Consequently, to avoid PMSM failures which may bring about huge economic losses and even human casualties [3, 4], the online monitoring and fault diagnosis methods of PMSM have been receiving considerably increasing attention.

Electric Power Research Institute (EPRI) [5] had made several investigations on the reliability of industrial motors and concluded that the stator winding faults

Y. Zhang (✉)
School of Electrical Engineering, Jiangsu University, No. 301 Xuefu Road, Zhenjiang, Jiangsu, China
e-mail: 15050859371@139.com

Y. Zhang · G. Liu
Jingjiang College of Jiangsu University, Zhenjiang 212013, China

W. Chen et al. (eds.), *The Proceedings of the 9th Frontier Academic Forum of Electrical Engineering*, Lecture Notes in Electrical Engineering 743,
https://doi.org/10.1007/978-981-33-6609-1_19

of motors account for about 21–37% of the motor failures. Meanwhile, it is well recognized that the interturn short-circuit fault (ISCF), that the insulation degradation in several turns of a stator winding within one phase, is one of the most common stator winding faults [6, 7]. This fault generates a large circulating current in the faulted turns [8] which can cause further insulation failures by forming a localized heat spot and ultimately leading to a complete failure of the winding as a phase-to-phase or phase-ground fault. Therefore, it is quite necessary for rapid detection of the ISCF, which can prevent serious damages from happening and improve the reliability of the PMSM.

There are three main methods that have been employed for ISCF diagnosis technology during recent years. One method is the knowledge-based approach [3], using artificial tools such as a fuzzy control system [4], pattern recognition and artificial neural networks for the purpose of process monitoring. This approach provides various training algorithms for a rapid and precise simulation of the motor, and analysis of all possible interactions between predicted values. However, it is very difficult to collect sufficient amount of real-time data under different conditions.

The second method is the motor current signature analysis (MCSA) approach [6, 7]. This method is highly effective and simple for rapid diagnosis of ISCF since it is well developed for PMSM and offers online monitoring without additional sensors for diagnosis [6]. Although MCSA has more advantages in the application of ISCF diagnosis, it still faces several obstacles to overcome. Firstly, previous studies focused mainly on the frequency spectral analysis. The stator current harmonic components due to interturn fault in a PMSM are superimposed with those that appear in a healthy PMSM [8]. Another vital drawback of spectral analysis is that there is a strong correlation between the fault signatures and the structures of motors, which may display different results for the same fault conditions of PMSM with different winding configurations [9]. These will lead to people's indetermination and misjudging.

The third method is the motor voltage signature analysis (MVSA) approach. This method is capable of detecting a minor ISCF occurring due to the voltage signal which can change abruptly in inductive circuit. In view of recent voltage-based methods, the Negative-Sequence Components Diagnosis (NSCD) and the Zero-Sequence Components Diagnosis (ZSCD) become the mainstream. NSCD has the ability to detect the ISCF occurrence and the fault severity [10], but fails to judge the fault phase, while ZSCD has the ability to judge the fault phase [8], but has worse performance in detecting the ISCF occurrence and the fault severity than NSCD [10].

In order to solve the above problems, the phase angle diagnosis (PAD) method based on Zero-Sequence Components is introduced. PAD aims to achieve two goals. The first goal is to determine a fault indicator that is not influenced by the change of the rotor speed with all degree of ISCF. Through the fault voltage analysis, the relationship is confirmed that in the case of slight ISCFs, the amplitude of the fault current is proportional to the rotor speed, while the amplitude is constant under the severe ISCF condition [10]. PAD can effectively judge whether the ISCF severity is slight or severe before choosing the proper fault indicator. The second goal is to make the right judgment of the fault phase with different values of the contact resistance in the short path. A calculable expression according to the contact resistance for PAD is proposed. The proposed computation method shows that the fault phase can be correctly determined regardless of the contact resistance change in the short path.

The remainder of this paper is organized as follows. Section II describes the ISCF model equations of a PMSM with Wye-connected winding. Section III provides details about PAD. Section IV presents simulation results to validate the proposed method. Finally, Section V concludes this paper.

2 PMSM Dynamic Model with ISCF

2.1 ISCF Stator Model

The ISCF may originate large circulating currents in the shorted turns. The severity of ISCF mainly depends on the two terms, the contact impedance of ISCF r_f and the short circuit ratio $\mu = n/N_C N_1$, where n is the number of ISCF turns, N_C is the number of coils per phase and N_1 is the turns number in one coil [2]. The one term r_f limits the current in the short circuit winding when ISCF occurs. The greater r_f, the closer to health state in which r_f is considered infinite. If $r_f = 0$, it means a full short circuit forming a huge loop current in the fault winding. The other term μ increases the voltage of short circuit winding and expands the impact scope of the ISCF. If $\mu = 1$, it means inter-coil short circuit. When the ISCF becomes more serious, the r_f is getting close to zero and μ increases rapidly. Figure 1 illustrates an equivalent model of a wye-connected winding PMSM with an ISCF in phase A. If ISCF occurs in phase B or C, the model can still be used with the change of the fault phase mark.

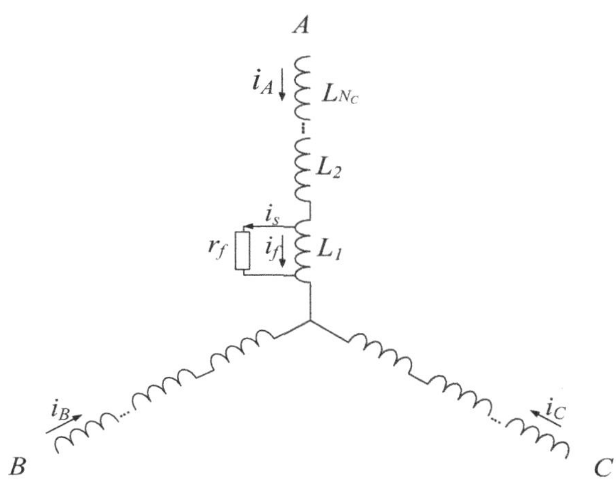

Fig. 1 An equivalent model of a wye-connected winding PMSM with an ISCF in phase A

2.2 ISCF Detailed Analysis

In a healthy 3-Phase PMSM, it is well known that there are balanced resistance, self-inductance and mutual-inductance in three phases. Hence, the healthy resistance inductance, self-inductance and mutual-inductance can be set as follows:

$$\begin{cases} R_A = R_B = R_C = R_S \\ L_A = L_B = L_C = L_S \\ M_{AB} = M_{BC} = M_{AC} = M_S \end{cases} \tag{1}$$

Additionally, in the wye-connected PMSM, the following condition is always satisfied:

$$i_A + i_B + i_C = 0 \tag{2}$$

When the ISCF occurs, it can be seen that the current i_A in the short circulating path is divided into two different components. One flowing in the short path is the short circuit current i_s, while the other one flowing in the shorted turns is the fault current i_f, which influences much wider than i_s as shown in Fig. 2. The current i_A has the following description:

$$i_A = i_s + i_f \tag{3}$$

In the case of PMSM with ISCF, the equation of three-phase voltage generated by the inverter is expressed as follows:

$$V_A R_{Ah} i_A + L_{ah}\frac{di_A}{dt} + e_{Ah} + M_{Ah-Af}\frac{di_f}{dt} + M_{A-B}\frac{di_B}{dt} + M_{A-C}\frac{di_C}{dt} + R_{Af} i_f$$

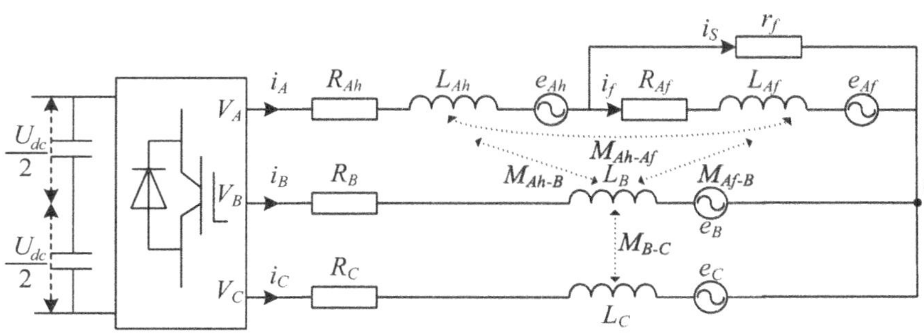

Fig. 2 The detailed wye-connected PMSM with an artificial neutral point using three high resistors

$$+ L_{af}\frac{di_f}{dt} + M_{Ah-Af}\frac{di_A}{dt} + e_{Af} + V_0 \tag{4}$$

$$\begin{aligned} V_B = R_B i_B + L_B\frac{di_B}{dt} + M_{Ah-B}\frac{di_A}{dt} \\ + M_{Af-B}\frac{di_f}{dt} + M_{B-C}\frac{di_C}{dt} + e_B + V_0 \end{aligned} \tag{5}$$

$$\begin{aligned} V_C = R_C i_C + L_C\frac{di_C}{dt} + M_{Ah-C}\frac{di_A}{dt} \\ + M_{Af-C}\frac{di_f}{dt} + M_{B-C}\frac{di_B}{dt} + e_C + V_0 \end{aligned} \tag{6}$$

Meanwhile, the equation of the fault voltage in the shorted turns is expressed as follows:

$$\begin{aligned} V_f = r_f i_s = R_{Af} i_f + L_{Af}\frac{di_f}{dt} \\ + M_{Ah-Af}\frac{di_A}{dt} + M_{Af-B}\frac{di_B}{dt} + M_{Af-C}\frac{di_C}{dt} + e_{Af} \end{aligned} \tag{7}$$

where R_{Ah} and L_{Ah} is the resistance and self-inductance of remaining healthy turns, while R_{Af} and L_{Af} is the resistance and self-inductance of shorted turns, and e_{Af} and e_{Ah} is the back-EMF voltage of phase A in shorted turns and in healthy turns, respectively. According to [8, 10], they are satisfied as follows:

$$\begin{cases} R_{Ah} = (1-\mu)R_S \\ L_{Ah} = (1-\mu)^2 L_S \\ R_{Af} = \mu R_S \\ L_{Af} = \mu^2 L_S \\ e_{Af} = \mu e_A \\ e_{Ah} = (1-\mu)e_A \end{cases} \tag{8}$$

According to Formulas (4), (5), (6), (7) there are also three kinds of mutual-inductances in the fault phase *A*. The first is the mutual-inductance $M_{Ah\text{-}Af}$ between the shorted turns and the remaining normal winding of phase *A*. The second is $M_{Af\text{-}B}$ and $M_{Af\text{-}C}$, the mutual-inductance between the shorted turns and healthy winding of phase *B and C* respectively. The last is the mutual-inductance between the remaining normal winding of phase *A* and healthy winding of phase *B and C*. Referring to [8, 10], they can be expressed as follows:

$$\begin{cases} M_{Ah-Af} = \mu(1-\mu)L_S \\ M_{Af-B} = M_{Af-C} = \mu M_S \\ M_{Ah-B} = M_{Ah-C} = (1-\mu)M_S \end{cases} \tag{9}$$

Substituting (3), (8) and (9) into the Formula (7) yields:

$$r_f i_s + \mu R_s i_s + \mu^2 L_s \frac{di_s}{dt} = \mu R_s i_A + \mu L_s \cdot \frac{di_A}{dt} + \mu M_s \frac{di_B}{dt} + \mu M_s \frac{di_C}{dt} + \mu e_A \quad (10)$$

Substituting (2) into (10) yields:

$$r_f i_s + \mu R_s i_s + \mu^2 L_s \frac{di_s}{dt} = \mu R_s i_A + \mu (L_s - M_s) \frac{di_A}{dt} + \mu e_A \quad (11)$$

When ISCF fault occurs, μ is in the interval (0,1], we can substitute all variables into vectors as follows:

$$\left(\frac{r_f}{\mu} + R_s + j\omega_e \mu L_s\right)\overrightarrow{I_s} = [R_s + j\omega_e (L_s - M_s)]\overrightarrow{I_A} + \overrightarrow{e_A} \quad (12)$$

Setting $\left(\frac{r_f}{\mu} + R_s + j\omega_e \mu L_s\right)\overrightarrow{I_s} = \overrightarrow{V_K}$, where ω_e is the electrical angular velocity, it yields that $\overrightarrow{V_K} = [R_s + j\omega_e (L_s - M_s)]\overrightarrow{I_A} + \overrightarrow{e_A}$ according to (12). The relationship among $\overrightarrow{V_K}$, $\overrightarrow{I_s}$ and $\overrightarrow{I_A}$ is shown in Fig. 3. It can be seen that the angle between $\overrightarrow{V_K}$ and $\overrightarrow{I_s}$ is α, and the angle between $\overrightarrow{I_s}$ and $\overrightarrow{I_A}$ is θ.

According to (4) and (10), we obtain that

$$i_s = \frac{\mu (V_A - V_0)}{\mu (1 - \mu) R_s + r_f} \quad (13)$$

In (13), the last term V_0 is always much smaller than V_A, therefore, (13) can be simplified as

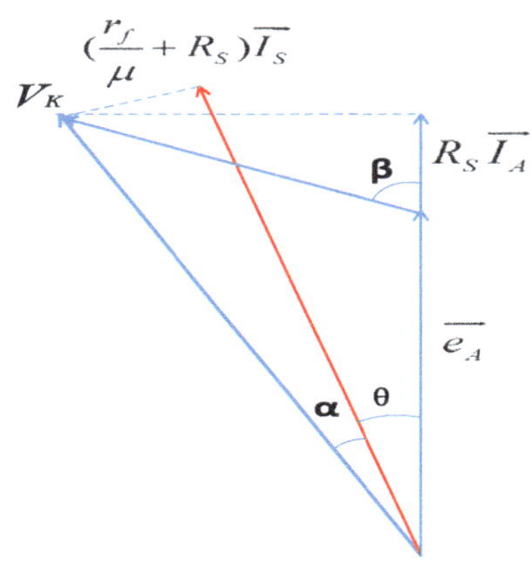

Fig. 3 The angle relationship between faulty phase current and short circuit current

$$i_s \approx \frac{\mu V_A}{\mu(1-\mu)R_s + r_f} \tag{14}$$

It can be seen from (14) that the initial phase angle of the fault current i_s is approximately equal to that of the phase voltage V_A when μ is not equal to 0. As the circuit of PMSM is inductive, the initial phase angle of V_A is ahead of that of i_A, hence, the initial phase angle of i_s is ahead of that of i_A. The angle between i_s and i_A can be defined as $\theta = \theta_{i_s} - \theta_{i_A}$, where θ_{i_s} and θ_{i_A} are the initial phase angle of i_s and i_A, respectively.

According to (4), (5) and (6), we obtain that

$$\begin{aligned} V_0 = {} & \frac{V_A + V_B + V_C}{3} + \frac{1}{3}\mu R_s i_s \\ & + \frac{1}{3}(L_s + 2M_s)\frac{di_s}{dt} - \frac{e_A + e_B + e_C}{3} \end{aligned} \tag{15}$$

A three-phase balanced resistor network is adopted, it results in

$$\begin{aligned} V_0 &= \frac{1}{3}\mu R_s i_s + \frac{1}{3}(L_s + 2M_s)\frac{di_s}{dt} - \frac{e_A + e_B + e_C}{3} \\ &= V_1 \sin\left(\omega_e t + \theta_{V_1}\right) - \frac{e_A + e_B + e_C}{3} \end{aligned} \tag{16}$$

where V_1 and θ_{V_1} are the amplitude and initial phase angle of the fundamental component in the V_0, respectively.

Since i_s cannot be measured directly for motor in use, θ_{i_s} is expressed as follows using (17):

$$\theta_{i_s} = \theta_{V_1} - \varphi \tag{17}$$

where $\varphi = tan^{-1}(\omega_e(L_s + 2M_s)/R_s)$.

Therefore,θ can be calculated easily using (18) as follows:

$$\theta = \theta_{V_1} - \varphi - \theta_{i_A} \tag{18}$$

Supposing γ is the angle between $\overrightarrow{V_K}$ and $\overrightarrow{I_A}$, there is the following angle relations:

$$\gamma = \alpha + \theta \tag{19}$$

γ can also be calculated by (20):

$$\gamma = \tan^{-1}\frac{\omega_e(L_s - M_s)}{R_s + \left|\frac{\overrightarrow{e_A}}{\overrightarrow{I_A}}\right|} \tag{20}$$

According to (17), (18) and (19), α will be calculated as follows:

$$\alpha = \gamma - \theta = \tan^{-1}\frac{\omega_e(L_s - M_s)}{R_s + \left|\frac{\vec{e_A}}{\vec{I_A}}\right|} - \theta_{V_1} + \tan^{-1}(\omega_e(L_s + 2M_s)/R_s) + \theta_{i_A} \tag{21}$$

It can be seen from Fig. 3 that $\alpha = \tan^{-1}\frac{\omega_e L_s}{\frac{R_s}{\mu} + \frac{r_f}{\mu^2}}$.

Setting

$$\mathrm{F}(\mu, r_f) = \frac{R_s}{\mu} + \frac{r_f}{\mu^2}, \text{ then } \alpha = \tan^{-1}\frac{\omega_e L_s}{\mathrm{F}(\mu, r_f)} \tag{22}$$

In the case of slight ISCF, r_f is much larger than zero and μ is much smaller than 1, $\mathrm{F}(\mu, r_f)$ is mainly dependent on the $\frac{r_f}{\mu^2}$. Because r_f is very large (even nearly infinite in the healthy state) in this period, $\mathrm{F}(\mu, r_f)$ is also very large relative to ωL_s,$\alpha \approx 0$ and $\theta = \gamma - \alpha \approx \gamma$. However, when ISCF becomes severe, r_f decreases dramatically even close to 0 and μ increases rapidly, $\mathrm{F}(\mu, r_f)$ is mainly dependent on the $\frac{R_s}{\mu}$, and α grows a big step. In the case of severe ISCF, $\alpha \approx \gamma$ at last and $\theta = \gamma - \alpha \approx 0$.

3 Fault Location and Degree Indicator

3.1 Proposed Fault Indicator

Base on the above analysis, it can be judged the ISCF degree by θ, meanwhile, the fault phase can also be determined by θ when ISCF occurs in the corresponding phase. θ_A, θ_B and θ_C are defined as θ when ISCF occurs in Phase A, B and C, respectively. Therefore, when the severe ISCF occurs, the fault phase can be judged as follows:

(1) If ISCF occurs in the phase A, θ_A, θ_B and θ_C are close to 0, $2\pi/3$, and $-2\pi/3$, respectively;
(2) If ISCF occurs in the phase B, θ_A, θ_B and θ_C are close to $-2\pi/3$, 0, and $2\pi/3$, respectively;
(3) If ISCF occurs in the phase C, θ_A, θ_B and θ_C are close to $2\pi/3$, $-2\pi/3$, and 0, respectively.

When the slight ISCF occurs, the fault phase can be judged as follows:

(1) If ISCF occurs in the phase A, θ_A, θ_B and θ_C are close to γ_A, $\gamma_A + 2\pi/3$, and $\gamma_A - 2\pi/3$, respectively;

(2) If ISCF occurs in the phase B, θ_A, θ_B and θ_C are close to $\gamma_B - 2\pi/3$, γ_B, and $\gamma_B + 2\pi/3$, respectively;
(3) If ISCF occurs in the phase C, θ_A, θ_B and θ_C are close to $\gamma_C + 2\pi/3$, $\gamma_C - 2\pi/3$, and γ_C, respectively.

where $\gamma_B = \tan^{-1}\frac{\omega_e(L_s - M_s)}{R_s + \left|\frac{\vec{e_B}}{\vec{I_B}}\right|}$ and $\gamma_C = \tan^{-1}\frac{\omega_e(L_s - M_s)}{R_s + \left|\frac{\vec{e_C}}{\vec{I_C}}\right|}$ according to (20).

In PMSM, $\vec{e_A}$, $\vec{e_B}$ and $\vec{e_C}$ are proportional to ω_e, meanwhile, as the value of $\left|\frac{\vec{e_A}}{\vec{I_A}}\right|$, $\left|\frac{\vec{e_B}}{\vec{I_B}}\right|$, $\left|\frac{\vec{e_C}}{\vec{I_C}}\right|$ is much larger than R_s, $R_s + \left|\frac{\vec{e_A}}{\vec{I_A}}\right|$, $R_s + \left|\frac{\vec{e_B}}{\vec{I_B}}\right|$, $R_s + \left|\frac{\vec{e_C}}{\vec{I_C}}\right|$ are approximately proportional to ω_e, therefore, γ_A, γ_B, γ_C are almost 3 constants. When ISCF occurs, the healthy motor balance among $\vec{I_A}$, $\vec{I_B}$ and $\vec{I_C}$ is destroyed, γ_A, γ_B, γ_C seems a little difference between each other.

3.2 *Calculation of the Proposed Fault Indicator*

In order to achieve online diagnosis, the defined fault indicator should be extracted in real time. If the high-order harmonic components of V_0 are neglected, V_0 can be expressed as follows:

$$V_0 = V_1\sin(\omega_e t + \theta_{V_1}) + V_3\sin(3\omega_e t + \theta_{V_3}) \tag{23}$$

where θ_{V_3} is initial phase angle of the third harmonic component in the V_0.

Then (23) can be transformed into a stationary orthogonal reference frame, and the voltage components are defined as:

$$\begin{cases} V_d = \left(V_1\sin(\omega_e t + \theta_{V_1}) + V_3\sin(3\omega_e t + \theta_{V_3})\right)\cos(\omega_e t) \\ V_q = \left(V_1\sin(\omega_e t + \theta_{V_1}) + V_3\sin(3\omega_e t + \theta_{V_3})\right)\sin(\omega_e t) \end{cases} \tag{24}$$

By using trigonometric identities, (24) can be expressed as:

$$\begin{cases} V_d = \frac{1}{2}\left(V_1\sin(2\omega_e t + \theta_{V_1}\right) + \frac{1}{2}V_1\sin(\theta_{V_1}) + \frac{1}{2}V_3\sin(4\omega_e t + \theta_{V_3}) + \frac{1}{2}V_3\sin(2\omega_e t + \theta_{V_3}) \\ V_q = -\frac{1}{2}\left(V_1\cos(2\omega_e t + \theta_{V_1}\right) + \frac{1}{2}V_1\cos(\theta_{V_1}) - \frac{1}{2}V_3\cos(4\omega_e t + \theta_{V_3}) + \frac{1}{2}V_3\cos(2\omega_e t + \theta_{V_3}) \end{cases} \tag{25}$$

Then, filtering the harmonic components from the voltage signals defined by (25), yields

$$\begin{cases} V_{dl} = \frac{1}{2}V_1\sin(\theta_{V_1}) \\ V_{ql} = \frac{1}{2}V_1\cos(\theta_{V_1}) \end{cases} \tag{26}$$

where V_{dl} and V_{ql} are the DC voltage components in V_d and V_q, respectively. It can be seen from (26) that V_{dl} and V_{ql} have a direct relationship with the amplitude and initial phase angle of the fundamental component in the V_0. Finally, the initial phase angle can be calculated as:

$$\theta_{V_1} = \tan^{-1}\left(\frac{V_{dl}}{V_{ql}}\right) \tag{27}$$

According to (18), the fault location and degree indicator $\theta_{A(BC)}$ can be obtained as:

$$\begin{aligned}\theta_{A(BC)} = \theta_{V_1} - \varphi - \theta_{i_{A(BC)}} = \tan^{-1}\left(\frac{V_{dl}}{V_{ql}}\right) \\ - \tan^{-1}(\omega_e(L_s + 2M_s)/R_s) - \theta_{i_{A(BC)}}\end{aligned} \tag{28}$$

Base on the above analysis, θ_A, θ_B and θ_C not only can determine the fault phase, but also can indicate the fault degree.

4 Simulation Validation

4.1 Configuration of Simulation

The proposed method is verified in a motor signal acquisition simulation platform. The signals were collected at the sampling frequency of 10 kHz. The degree of ISCF was changed by varying values of μ and r_f in Tables 1, respectively. Therefore, a total of 3 cases of the ISCF were tested, including the values of Tables 1. The slightest ISCF was when $\mu = 0.05$ and $\boldsymbol{r}_f = 4\ \Omega$, and the most severe ISCF was when $\mu = 0.1$ and $\boldsymbol{r}_f = 0.\ 1\ \Omega$.

Simulations were performed using healthy and faulty PMSM under various operating conditions as follows:

(1) ***Steady-State Condition***
Figure 4 shows the simulation using the fault indicator when the rotor speed remained unvaried at 1500 r/min under the constant load torque of 4 Nm. The first fault is introduced at t = 5 s with $\mu = 0.05$, in which $r_f = 4\ \Omega$ is connected in series with the wire induced by short-circuit inductance of the stator phase A, where r_f is used to mimic the fault contact resistant. It can be seen from

Table 1 ISCF Situations with μ and r_f

	Slight	Medium	Severe
μ	0.05	0.1	0.1
r_f (Ω)	4	1	0.1

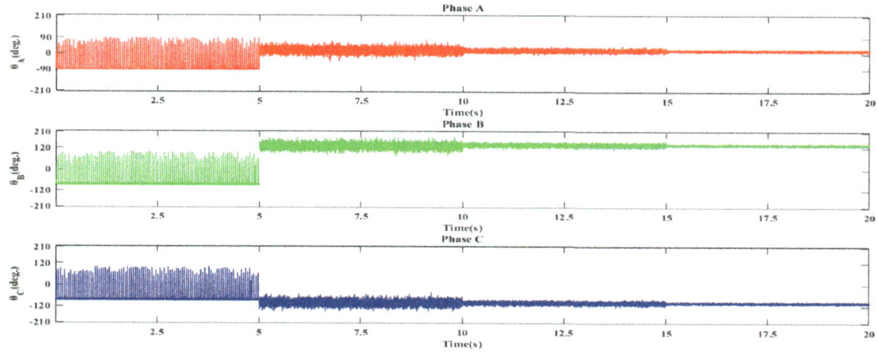

Fig. 4 ISCF indicator effects from slight to severe faults

Fig. 4 that the fault indicator θ_A,θ_B and θ_C are all changed and close to $\pi/9$, $-5\pi/9$, and $7\pi/9$, respectively. At t = 10 s, a more serious fault occurs in phases A (μ = 0.1 and r_f = 1 Ω), In this case, θ_A, θ_B and θ_C are close to $\pi/18$, $-11\pi/18$, and $13\pi/18$, respectively. At t = 15 s, μ remained unchanged and r_f was decreased to 0.1 Ω,θ_A, θ_B and θ_C were approximately equal to $1\pi/36$, $-11\pi/36$, and $13\pi/36$, respectively.

(2) ***Dynamic Condition***
In this case, ISCF faults were introduced under the stable fault condition of r_f = 1 Ω and μ = 0.1. Figure 5 shows the simulation when the rotor speed varied at 300, 500, 1000, and 1500 r/min under the constant load torque of 4 Nm, while Fig. 6 reveals the simulation when the rotor speed fixed at 1500 r/min under two different constant load torque of 2 and 4 Nm. It can be seen from Fig. 5

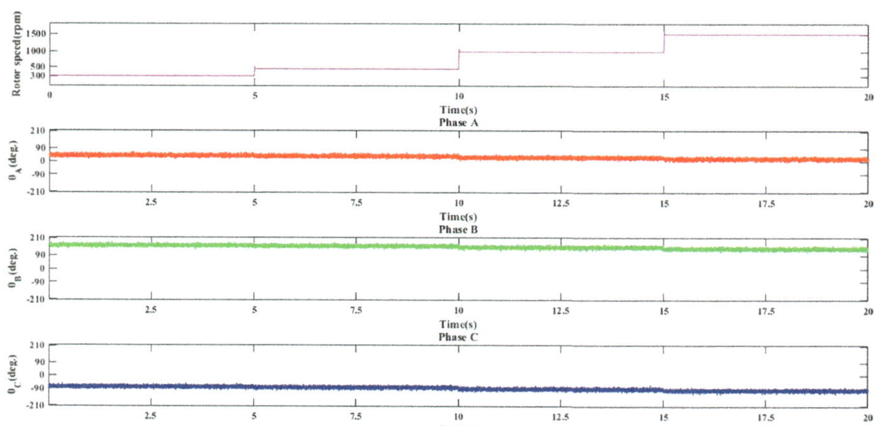

Fig. 5 ISCF indicator effects at different rotor speed

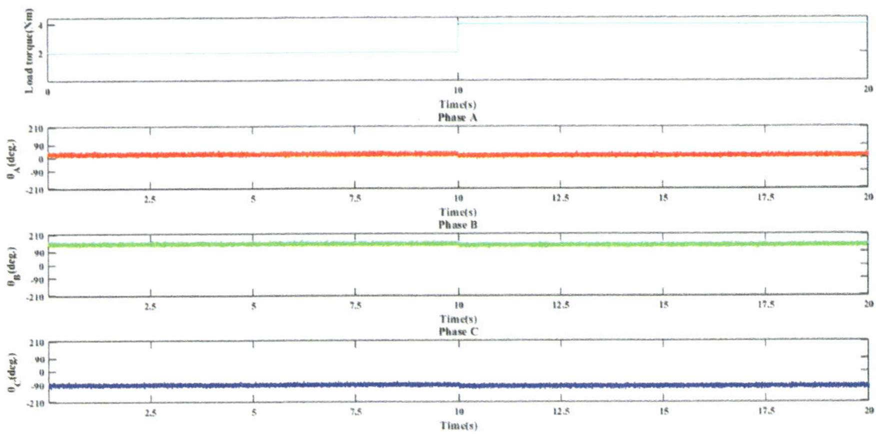

Fig. 6 ISCF indicator effects at different load torque

that θ_A, θ_B and θ_C fluctuated and varied a little with the increase of rotor speed, changing 10 degree from 300 to 1500 r/min. In Fig. 6, the values of the fault indicator do not change when the load torque varies because the rotor speed is fixed.

4.2 Simulation Results

(1) ***Varying ISCF degree condition***
The fault indicator is effectively detected through the difference of the fault indicator between the non-fault and fault states when the ISCF occurs. The indicator clearly shows the faulty phase under different ISCF degree conditions. Meanwhile, with ISCF degree increasing, the indicator makes corresponding changes which is in clear agreement with results discussed in Sect. 3.1.

(2) ***Varying rotor Speed condition***
Although there is a little change with the increase of the rotor speed, it is within the controllable range and the fault phase can still be easily detected. Meanwhile, compensation algorithm can be introduced to make corresponding adjustments for better appearance in ISCF degree detection.

(3) ***Varying load torque condition***
It can be observed that the varying load torque has little influence in the ISCF detection by the indicator, being in good agreement with the theoretical analysis.

(4) ***Comparison with other methods***
Compared with the traditional method [8, 9], this ISCF detection method not only has less measurement parameters and data to be collected, but also can detect fault phase and fault severity at the same time. In [10], the NSC method cannot separate the fault phase and its indicator is invalid when a serious ISCF

occurs as well, whereas the proposed method in this paper clearly separates the fault phase no matter the ISCF is slight or serious.

5 Conclusion

This paper proposed a fault location and degree diagnosis method of ISCF by using the PAD in PMSMs operating in various rotor speeds and load torque. A calculable expression of the fault indicator was derived in a wye-connected winding PMSM with ISCF, being for both slight and severe ISCF cases. Using the derived expression, ISCF were detected, as well as the degree and the location of the ISCF. Simulation results demonstrated that the proposed method not only has the capacity of diagnosing the ISCF, but also assessing the faulty severity and identifying the fault location under the different conditions of the rotor speed and load torque.

Acknowledgements This work was financially supported by the National Natural Science Foundation of China [grant No. 61673195], the Priority Academic Program Development of Jiangsu Higher Education Institutions (PAPD) and the Natural Science Research of Jiangsu Higher Education Institutions of China (CN) [grant No. 18KJB510008].

References

1. Zhu, Z.Q., and D. Howe. 2007. Electrical machines and drives for electric, hybrid, and fuel cell vehicles. *Proceedings of the IEEE* 95 (4): 746–765.
2. Zhang, Y., G. Liu, W. Zhao, H. Zhou, Q. Chen and M. Wei. On-line diagnosis of slight interturn short-circuit fault for a low-speed permanent magnet synchronous motor. *IEEE Trans. Transportation Electrification*. https://ieeexplore.ieee.org/document/9082017.
3. Nyanteh, Y., C. Edrington, S. Srivastava, and D. Cartes. 2013. Application of artificial intelligence to real-time fault detection in permanent-magnet synchronous machines. *IEEE Transactions on Industry Applications* 49 (3): 1205–1214.
4. Gao, Z., C. Cecati, and S.X. Ding. 2015. A survey of fault diagnosis and fault-tolerant techniques—Part II: Fault diagnosis with knowledge-based and hybrid/active approaches. *IEEE Transactions on Industrial Electronics* 62 (6): 3768–3774.
5. *Improved motors for utility applications, volume 1: Industry assessment study: Update and analysis*. Electric Power Research Institute, Palo Alto, CA, USA, Tech. Rep. EL-4286-V1, Oct. 1982.
6. Jung, J.-H., J.-J. Lee, and B.-H. Kwon. 2006. Online diagnosis of induction motors using MCSA. *IEEE Transactions on Industrial Electronics* 53 (6): 1842–1852.
7. Kim, K.H. 2011. Simple online fault detecting scheme for short-circuited turn in a PMSM through current harmonic monitoring. *IEEE Transactions on Industrial Electronics* 58 (6): 2565–2568.
8. Hang, J., J. Zhang, M. Cheng, and J. Huang. 2015. Online interturn fault diagnosis of permanent magnet synchronous machine using zero-sequence components. *IEEE Transactions on Power Electronics* 30 (12): 6731–6741.

9. Garcia, P., F. Briz, M.W. Degner, and A.B. Diez. 2004. Diagnostics of induction machines using the zero sequence voltage, vol. 1. In *2004 IEEE Industry Applications Conference: 39th IAS Annual Meeting*.
10. Jeong, H., S. Moon, and S.W. Kim. 2017. An early stage interturn fault diagnosis of PMSMs by using negative-sequence components. *IEEE Transactions on Industrial Electronics* 64 (7): 5701–5708.

A Study on the Center Controller of Distributed Power Supply Based on PLC Micro-grid

Ping Chen, Qinfei Sun, Zhao Wang, Xianglong Li, and Qingzhu Wan

Abstract In recent years, wind power generation, photovoltaic power generation, battery and other emerging energy sources have attracted great attention, research results are quite abundant, so microgrid and distributed energy have been highly concerned. Under the existing research, the new energy sources including wind power generation, solar power generation and battery will have a certain impact on the system due to the uncertainty of power generation. this topic to ensure the stability of microgrid and improve the power quality, based on programmable logic controller PLC, control strategy using hierarchical control mode, communication using Modbus-TCP based industrial ethernet, can realize microgrid grid-connected mode operation, islanding mode operation, fast switching between grid-connected operation and islanding operation mode, energy management scheduling based on energy storage SOC, real-time power balance, energy scheduling among multiple distributed generation systems, etc.

Keywords PLC · Microgrid · Central controller · Control strategy

1 Introduction

With the rapid development of the world economy and science and technology, the demand for energy is becoming higher and higher, and the new energy has gradually entered the people's vision, and the proposal of microgrid is to solve the problem of the utilization of distributed power supply. Because the microgrid with solar photovoltaic power generation, wind power generation, battery and other new energy sources will have a great impact on the distribution network when the system is connected to the grid, such as harmonics, voltage instability, large frequency

P. Chen · Q. Sun · Z. Wang · X. Li
State Grid Beijing Electric Power Research Institute, No. 30 Nan San Huan Zhong Lu, Feng Tai District, Beijing, China

Q. Wan (✉)
School of Electrical and Control Engineering, North China University of Technology, No. 5 Jin Yuan Zhuang, Shijingshan District, Beijing 100144, China
e-mail: wanqz@ncut.edu.cn

W. Chen et al. (eds.), *The Proceedings of the 9th Frontier Academic Forum of Electrical Engineering*, Lecture Notes in Electrical Engineering 743,
https://doi.org/10.1007/978-981-33-6609-1_20

deviation and so on [1]. Therefore, for the control system of microgrid, we need to study it deeply and thoroughly in order to solve these problems. The system consists of 2 sets of new energy photovoltaic power generation system, 1 two-way energy storage converter, 1 set of microgrid energy storage power station and 1 set of visual management and monitoring system. The system architecture diagram is shown in Fig. 1. In this paper, two new energy generation systems are used to simulate many users of new energy generation system, and a set of energy storage system is built in the residential area. The programmable logic controller P LC (N80-MDBS model) is used as the central controller of microgrid to establish a set of control system to realize the fast switching of microgrid grid-connected operation, isolated island operation, grid-connected operation and isolated island operation, microgrid energy management scheduling based on energy storage SOC, real-time power balance, energy scheduling among multiple distributed generation systems, etc.

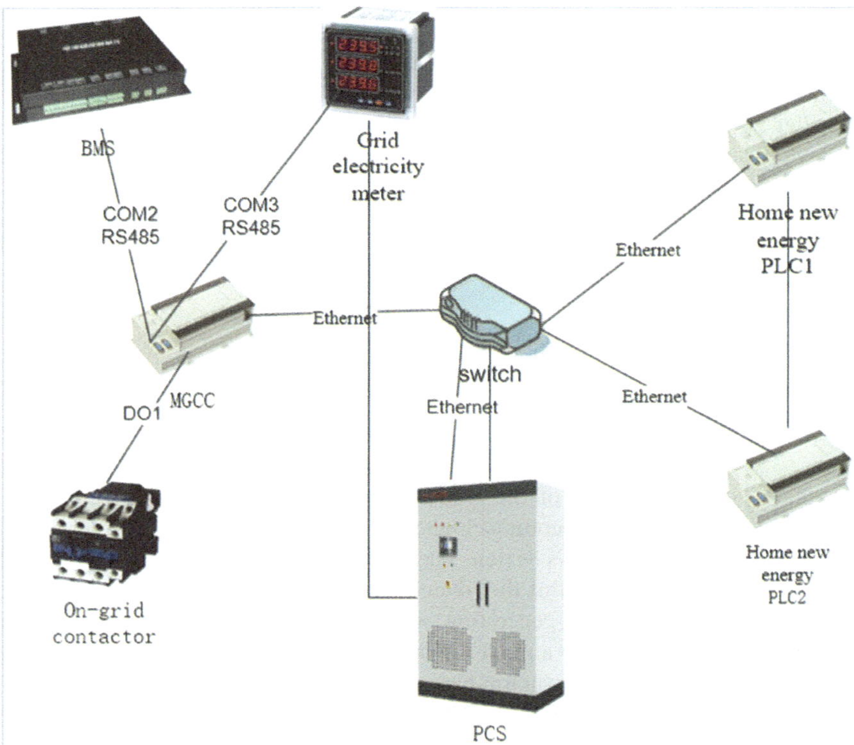

Fig. 1 MGCC network topology

2 Control Strategy

The system adopts distributed network communication topology based on Modbus-TCP. In the end, the real-time is easy to expand, and the energy management scheduling of microgrid accepts the real-time scheduling MGCC microgrid center controller.The MGCC of the middle layer controls the whole microgrid according to the instructions of the upper management to ensure the stability of the microgrid voltage and frequency and the switching off-grid; the lower layer executive layer is mainly the new energy system of each household, responsible for the start and stop of the bottom inverter and electric heating load switching, execute the MGCC instructions and upload the local operation status.

2.1 Microgrid Control Strategy

The control objects MGCC the microgrid central controller are: BMS (lithium battery energy storage system), PCS (bidirectional energy storage converter), home new energy system (2), power grid meter, grid-connected contactor. Among them, BMS and MGCC communicate through serial port 2, Communications agreements are R S485; Electricity meter and MGCC communication through serial 3, Communications protocols are also RS485; The PLC and PCS of the MGCC and the household new energy generation system 1 are communicated via switches through Ethernet; Grid-connected contactor for I/O control. The network topology is shown in Fig. 1.

2.1.1 Grid-Connected Operation Strategy

The grid-connected operation is mainly involved in the energy dispatching, which can be transported to the internal electric heating load and energy storage unit of the microgrid by using the power grid. Because of the large capacity of the large power network and the difficulty of changing the frequency and voltage of the large power network, the system can be supplied with rigid frequency and voltage. So all distributed power sources of the system work in constant power control (P/Q) mode when connected to the grid [2].

P/Q control is mainly used to make distributed power equal to its reference power [3], That is, when the terminal voltage and frequency of the grid connected by the grid-connected inverter of the distributed power supply fluctuate within the allowable deviation, the power generated by the distributed power supply remains stable, and the control principle is shown in Fig. 2.

within the allowable deviation (f) of the frequency by adjusting the droop characteristic curve of the frequency ($f_{\min} \leq f \leq f_{\max}$) When fluctuating, the active power emitted by the distributed power supply is equal to its given value. By adjusting the voltage sag characteristic curve, within the allowable deviation (u) of the voltage

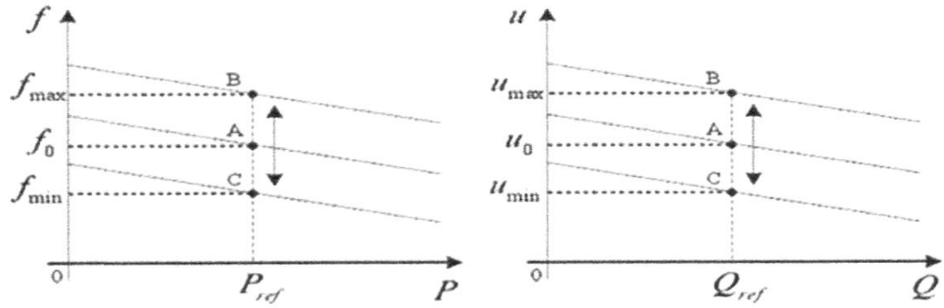

(a) Frequency droop characteristics (b) Voltage droop characteristics

Fig. 2 PQ Control schematic diagram

$(u_{\min} \leq u \leq u_{\max})$ When fluctuating, the reactive power emitted by the distributed power supply is equal to its given value [4].

If the power exchange between the microgrid and the external large power grid is determined according to the given value, the power flowing through the PCC should be monitored [5]. By MGCC to adjust the switching power so that the actual switching power is the same as the switching power. The formula for calculating the deviation between the actual switching power and the planned value is shown in formula (1).

$$\Delta \mathrm{P(t)} = \mathrm{P_{pcc}(t)} - \mathrm{P_{plan}(t)} \tag{1}$$

Formula: $P_{plan}(t)$ represents the active power plan value, P, of the t time exchanged between the external large network and the microgrid$_{pcc}$t represents the active power at the common connection point (PCC) at the t time.

When ΔP (t) > 0, the power value of the actual flow through the PCC is less than the set value, so it is necessary for the microgrid center controller to increase the power generation of the distributed power supply, or to remove the non-important electric heating load in the microgrid;

When ΔP (t) > 0, the power value of the actual flow through the PCC is greater than the set value, so it is necessary for the microgrid center controller to reduce the generation power of the distributed power supply and to put in part of the electric heating load.

Its control strategy flow chart is shown in Fig. 3.

MGCC read power P of 2 new home energy systems via communication P_1 and P_2, P_1 and P_2 For the signed number, the positive representation system needs to feed to the power grid, and the negative representation system needs to take electricity from the power grid.

When the system is powered on, it is connected to the grid. After PCS operation, the power loss of the system and the power of the new energy source of the home are calculated, and the power difference is calculated as the $P_{Difference}$, $P_{Difference} = P_1 + P_2 + P_{Damage}$, $P_{Difference}$ For signed numbers, P_{Damage} For the system itself consumption, for the positive number.

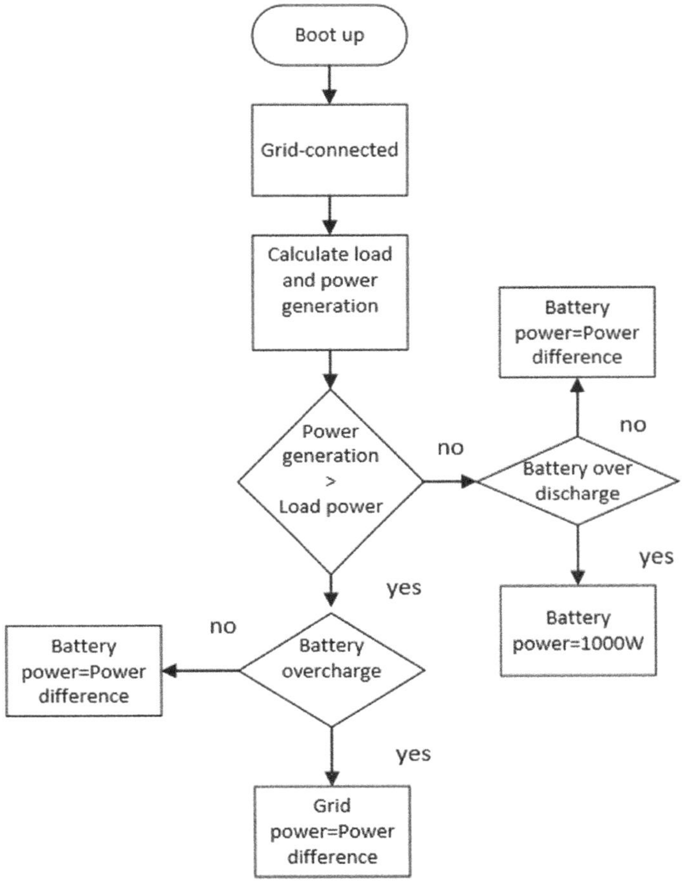

Fig. 3 Flowchart MGCC grid-connected operation

$P_{Difference}$ When > 0(feed): the system generates more power than electricity, it needs to charge the battery or power grid, at this time it needs to judge the state of the battery, when the battery is not overcharged, the power PCS the communication is set to $P_{Difference}$ When the battery is overcharged, the excess electricity is sent to the power grid.

$P_{Difference}$ When < 0(need to take electricity): the system generation is less than electricity, need to take electricity from the battery or power grid, at this time need to judge the state of the battery, when the battery is not overplay, the power PCS the communication setting is $P_{Difference}$ When the battery is overcharged, the power of the communication PCS is 1000 W, the state is constant power charging and the battery is maintained.

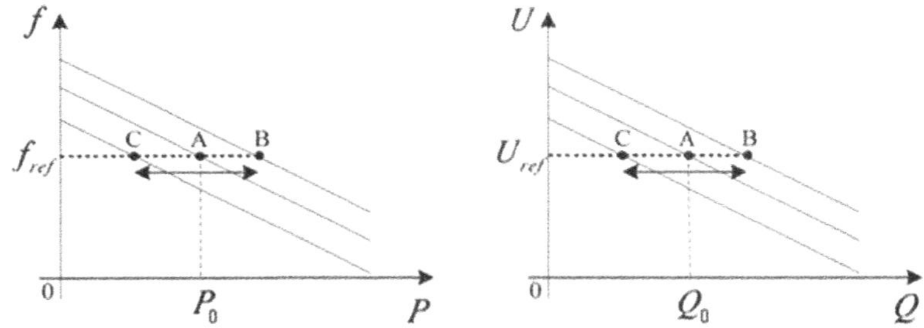

Fig. 4 Constant voltage constant frequency control principle

2.1.2 Off-Grid Operation Strategy

The control mode PCS bidirectional energy storage converter adopts constant voltage and constant frequency (U/f) control mode. U/f control mode is used to ensure that the voltage amplitude and system frequency of the grid side connected with the inverter remain unchanged when the power emitted by the distributed power source changes [4], Its control principle is shown in Fig. 4.

By increasing or reducing the active power emitted by the distributed power supply, the frèquency of the system is kept at a given value, and the voltage amplitude of the system is kept at a given value by increasing or reducing the reactive power emitted by the distributed power supply.

When the external large power network fails, the microgrid needs to cut off the connection with the external power network and leave the network. At this time, the microgrid itself can maintain the energy balance. It can change the power of distributed power supply, charge and discharge power of battery, switching electric heating load and so on, so as to ensure the operation stability in the isolated island mode of microgrid [6].

Calculate the loss of the system and the new energy power of the home, and calculate the P of the power difference when the system is off the grid $P_{Difference}$, $P_{Difference} = P_1 + P_2 + P_{Damage}$, $P_{Difference}$. For signed numbers, P_{Damage}. The consumption of the system itself is positive.

$P_{Difference}$ When > 0(feed): the system generates more electricity than electricity, it needs to charge the battery with excess power. At this time, it is necessary to judge the state of the battery. When the battery is not overcharged, the system remains intact; when the battery is overcharged, the P is closed Electricity generation >P_{Load} The generation of new energy systems in households.

$P_{Difference}$ < 0(need to take electricity): the system generation is less than electricity, need to take electricity from the battery, at this time need to judge the state of the battery, when the battery is not overplay, the system remains intact; when the battery is overplay, Communication controls home new energy systems 1 and 2 and PCS shutdown to protect batteries.

Its control policy flow chart is shown in Fig. 5.

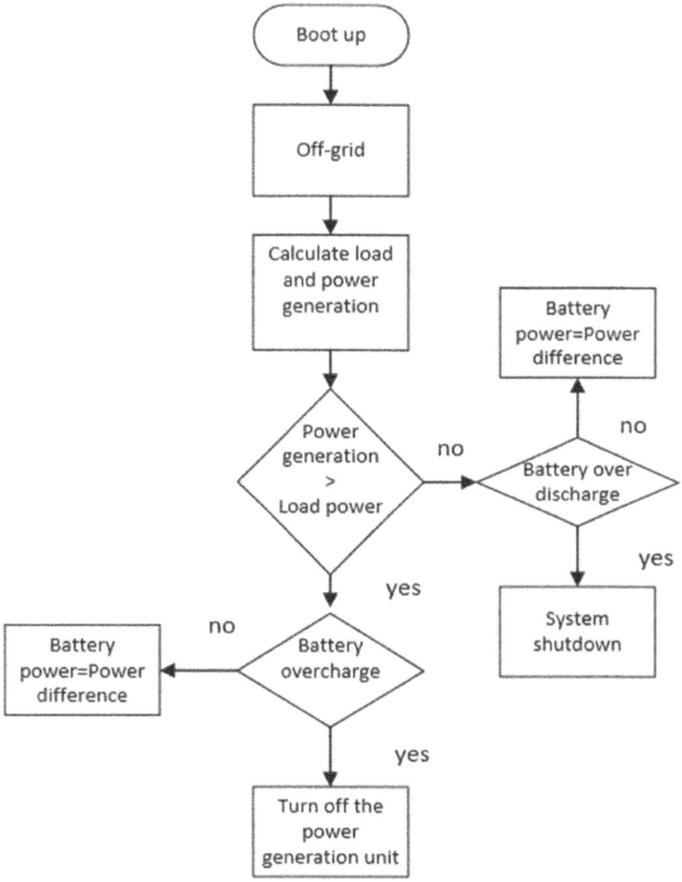

Fig. 5 Flowchart of the PLC program

2.2 *Control Strategy of 2 Household New Energy Generation System*

The control objects PLC home new energy generation system are: inverter, power meter (electric heating load, photovoltaic, grid), electric heating load, grid-connected contactor, main control MGCC.The PLC serial port 2 communicates with the inverter through the RS485 bus to collect the inverter data and control the inverter start and stop. The serial port 3 of the PLC is collected by the data of the RS485 bus and the power meter (electric heating load, photovoltaic, power grid), etc.PLC switch the electric heating load size, heat dissipation fan and grid-connected contactor through the digital output port. The PLC is connected to the main control MGCC through

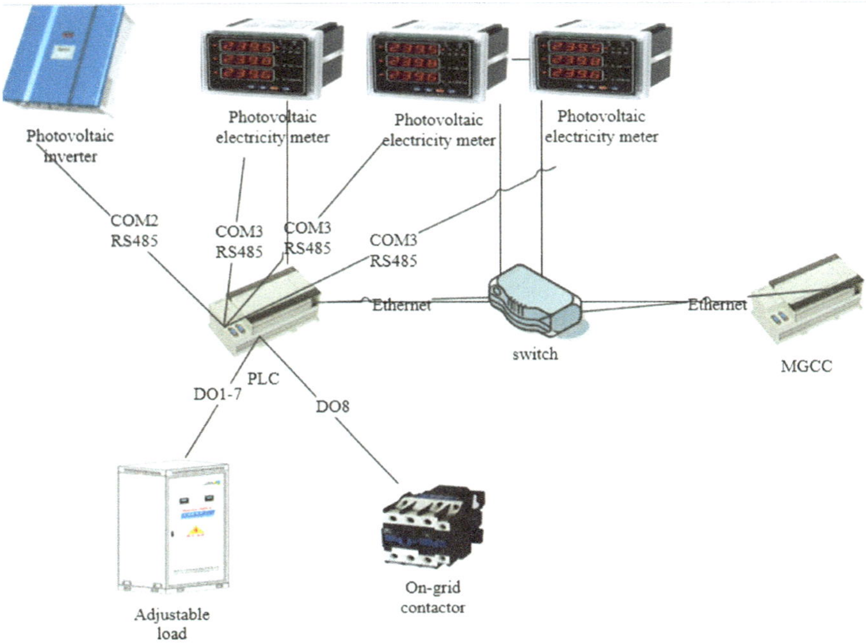

Fig. 6 Topology of new energy generation system network

Ethernet to receive and process commands in real time. The topology of the home new energy generation system is shown in Fig. 6.

After the system is turned on, the grid-connected contactor automatically absorbs and closes, receives the boot instruction when standby, judges whether the bus has electricity, the bus does not have electricity to continue standby, when the bus has electricity, the communication controls the inverter to start, the system runs. After receiving the shutdown instruction or MGCC the remote shutdown instruction, the communication control inverter stops and the system enters the standby state.

3 Design of Visual Monitoring Interface

The monitoring interface is mainly realized by Kingview software. In this system, I mainly designed the login interface, home page, subsystem of the detailed interface and history curve, alarm window and other interfaces.

The home page is the core of the whole system, as shown in Fig. 7, you can enter any sub-item in the home page, and the energy dynamic information display of the whole system and the state setting display of the system. Click on the corresponding button on the right side of the corresponding parallel, off-grid, on and off operation,

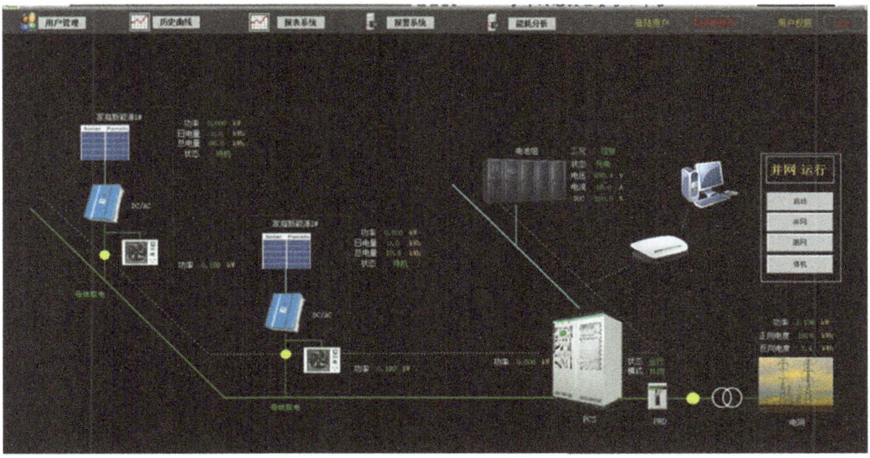

Fig. 7 Home page of upper computer monitoring system

while clicking on the corresponding area can enter the corresponding subsystem to view more detailed data. For example, click on the home new energy area to enter the subsystem interface shown in Fig. 8.

The new energy electronic system interface, as shown in Fig. 8, can view the detailed data of the new energy generation system, such as the voltage, current and other states of photovoltaic, inverter and so on. And it can be started and stopped control.

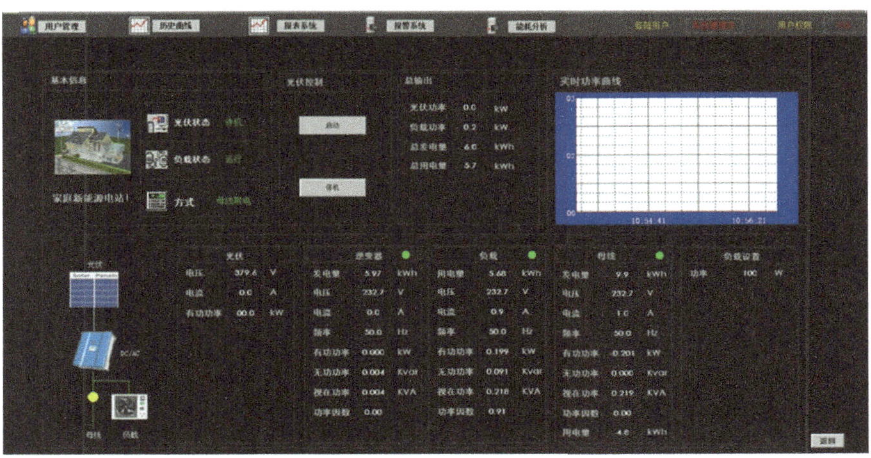

Fig. 8 Page of energy generation system for upper PC monitoring system

Table 1 Grid-connected power balance verification

S. No.	Photovoltaic one power generation (kw)	electric heating load one power (kw)	Photovoltaic two power generation (kw)	Electric heating load two power (kw)	Energy storage power (kw)	Grid power (kw)
1	1.829	0.63	0.369	1.711	−0.236	0.048
2	2.539	0.625	1.483	1.166	−2.038	0.018
3	0.366	1.704	1.114	1.164	−1.511	−0.017

Table 2 Off-grid power balance verification

Serial number	Photovoltaic one power generation (kw)	Electric heating load one power (kw)	Photovoltaic two power generation (kw)	Electric heating load two power (kw)	Energy storage power (kw)
1	1.445	1.066	0.37	1.084	−0.351
2	1.468	0.279	1.834	0.28	2.637
3	0.734	1.578	0.358	1.076	−1.517

4 Results and Analysis

4.1 Validation of Power Balance

Through the above data analysis (when the battery is rechargeable and dischargeable), it can be concluded that (Table 1).

The sum of PV power and electric heating load power ≈ energy storage power, so when connected to the grid, the power of the whole system is in dynamic balance (Table 2).

According to the above data analysis, the photovoltaic power and electric heating load power and ≈ energy storage power are obtained, so the power of the whole system is in dynamic balance when off-grid.

4.2 Verification of Frequency and Voltage Stability of 4 Systems

When the system is running, the frequency and voltage amplitude of the inverter, load and bus, and the frequency and voltage amplitude of the PCS output are observed through the monitoring interface. It is found that the frequency can be kept in the range of 50 ± 0.5 Hz and the voltage can be kept in the range of 5% and the voltage amplitude is stable.

5 Conclusions

Taking PLC as the central controller of microgrid, it communicates with home new energy generation system, BMS energy storage system, PCS bidirectional energy storage converter system, power meter of power grid, controls grid-connected contactor, realizes grid-connected operation, off-grid operation, parallel off-grid switching, energy management scheduling based on energy storage SOC, real-time power balance, energy scheduling among multiple distributed generation systems, and ensures the stability of voltage and frequency of the system in operation. With Kingview monitoring system to monitor the operation of the whole system to achieve visual operation.

Acknowledgements The authors acknowledge financial supported by Scientific Research Foundation of State Grid Corporation of China (5202011 8000M) and also Collaborative Innovation Center of Key Power Energy-Saving Technologies in Beijing,China(PXM2018_014212_000015_4_4).

The authors also thank the referees for the detailed review on the technical correctness and presentation quality.

References

1. Li Jinying, and Xin Ai. 2016. Hierarchical control strategy of microgrid with multiple distributed power sources. *Ningxia Electric Power* 3: 16–21.
2. Wang, Chengshan, Zhen Wu, Peng Li. 2014. Research on key technologies of microgrid. *Journal of Electrotechnical Technology* 2: 1–12.
3. Jing, Xiao. 2013. *Modeling of inverter-type distributed power generation based on DIgSILENT and its impact on distribution network*. Beijing: Beijing Jiaotong University.
4. Yang, Zhangang. 2010. *Microgrid experimental system research [EB/OL]*. https://www.tceic.com/l8h5818jhh4il7ki4glkj0ig.html.
5. Zhang Yang, Wei Wang, and Xianwei Li. 2011. Research on centralized control strategy of microgrid based on active power shortage. *Power System Protection and Control* 23: 106–111.
6. *Document space. Microelectronics review report [EB/OL]*. https://max.book118.com/html/2016/0229/36374065.shtm, 2016-03-06.

An Edge Calculation and Analysis System for State Perception and Operation and Maintenance of High-voltage Cable Outdoor Terminals

Youxiang Yan and Shuhong Wang

Abstract There are two types of traditional monitoring of outdoor transmission cables terminals. One is state tests on outdoor transmission cables terminals that are put into operation on a regular basis. But the outdoor cable terminal usually cannot be tested on time; The other way is to regularly take and analyze the thermal imaging of the outdoor cable terminal in daily operation and maintenance, however there is a monitoring blank period. It is necessary to use real-time online monitoring to monitor the operating status of the outdoor cable terminal. Therefore, a system for state perception and operation and maintenance of high-voltage cable outdoor terminals is developed based on the application of multiple big data methods such as edge computing and cloud storage, the loss of labor and material for cable operation and maintenance could be reduced greatly. It has preliminarily solved the problems of data blanking, historical data cannot be traced, operation and maintenance detection fails to meet the standard, automatically monitored in real time cannot be done, and operation and maintenance personnel cannot understand the situation in the first time when there is an alarm.

Keywords State perception · High-voltage · Cable terminals

1 Introduction

With the development of the urban economy in China, 110 kV and above transmission cables have become an important part of the urban power grid [1]. By the end of December 2019, the total length of 66–500 kV cabe lines in operation of State Grid Corporation of China has exceeded 20,000 km. There are approximately 30,000 outdoor transmission cables terminals running on the line. Hence it is very important to ensure the safe operation of so many outdoor transmission cables terminals. The outdoor cable terminal [2] is a very important device in the cable line, and it is an important node between the cable line and the overhead line or other

Y. Yan · S. Wang (✉)
School of Electrical Engineering, Xi'an Jiaotong University, Xi'an 710000, Shaanxi, China
e-mail: shwang@mail.xjtu.edu.cn

W. Chen et al. (eds.), *The Proceedings of the 9th Frontier Academic Forum of Electrical Engineering*, Lecture Notes in Electrical Engineering 743,
https://doi.org/10.1007/978-981-33-6609-1_21

equipment. The construction objectives of ubiquitous power Internet of things (IOT) [3–5] including: the visualization construction of power transmission and transformation equipment [6], the visualization monitoring [7] of important equipment body and environmental equipment, the application of intelligent operation and inspection control platform, the realization of intelligent image recognition [8] and early warning, unified monitoring and display.

The traditional monitoring of the outdoor transmission cable terminal is mainly to conduct off-line state test on the outdoor terminal transmission cable which is put into operation on a regular basis, but the main equipment must be shut down for the test. Sometimes, the main equipment, especially the high-voltage equipment, cannot be shut down due to the limitation of operation mode, which causes the outdoor terminal of the transmission cable to be unable to test on time. At present, the method of cable operation and maintenance is to regularly photograph and analyze the thermal imaging in daily operation and maintenance. The current problems include: (1) During the long-term operation, Outdoor cables terminals bear power frequency voltage and impulse voltage for a long time; (2) The long-term effect of harsh environment will make the connecting clamp bolt aging, even overheating dam-age; (3) Due to the environment impact, the surface of the outdoor cable terminal will be polluted, and flashover occurs in severe cases. In addition, the construction quality problems and the long-term power frequency voltage on the outdoor cables terminals enlarged the internal defects of the equipment, which may lead to the explosion of the cable terminals; (4) Daily operation and maintenance cannot achieve real-time monitoring, which result in a monitoring blank period; (5) It is impossible to grasp the status of the terminal in real time, and only increase the operation and maintenance time and cost of personnel; (6) Unable to know the emergency in time, resulting in missing the golden time for dealing with problems quickly; (7) There is no review and summary of historical abnormal problems.

2 Solution

2.1 Introduction of Cable Terminal System

Cable terminal system is committed to solve the problem of real-time monitoring and prevention of cables. The new operation and maintenance service support system is constructed by using web and WeChat. The system changes from the traditional single point mode to the new mode of “field + centralized control center + mobile team”, providing multiple, real-time and reliable security guarantee. At the same time, the system can realize 7 × 24 h continuous real-time monitoring, and eliminate the information collection blind area of traditional periodic manual duty inspection and paper record. As the result, the passive mechanical operation and maintenance method is changed to the active elastic operation and maintenance method, and the efficiency is improved. And for the emergency situation, the stable SMS early warning form is adopted to timely feedback the alarm to the operation and maintenance personnel. This system also supports the persistent storage of data, and understand the abnormal history of each cable terminal with one-button, which effectively solve the analysis and problem induction of historical anomaly. It supports the monitoring of key areas of cable terminals, and realizes temperature difference comparison of key areas, single area and different area. It also supports one-button real-time monitoring of real time operation temperature and surrounding environment for remote cable.

2.2 *Principle of Cable Terminal System*

The cable terminal system adopts high sensitivity and high precision infra-red thermal imaging camera, and then the temperature of each pixel can be obtained. After setting the regional coordinates, temperature threshold and other configuration information, the maximum temperature and minimum temperature of each area is obtained from the segmentation image and data analysis, and judge whether there is alarm. Then classifying and processing the temperature data and report it to the IOT platform. The IOT platform stores these messages temporarily and sends them to the server. According to the type of message received by the server, there are two processing methods including normal message processing steps and abnormal message processing steps.

2.3 *Software Development of Cable Terminal System*

2.3.1 System Structure

See Fig. 1.

2.3.2 Temperature Measurement Technology of Thermal Imaging Camera

The technology of converting the temperature distribution of the object surface into visible image and displaying the temperature distribution of the object surface in different colors is called infrared thermal imaging technology, and this electronic device is called infrared thermal imager. However, compared with the visible image, the infrared thermography lacks hierarchy and stereo sense due to its weak signal. Therefore, in order to judge the infra-red thermal distribution field of the measured target more effectively, some auxiliary measures are often used to increase the practical functions of the instrument, such as image brightness, contrast control, correcting, etc.

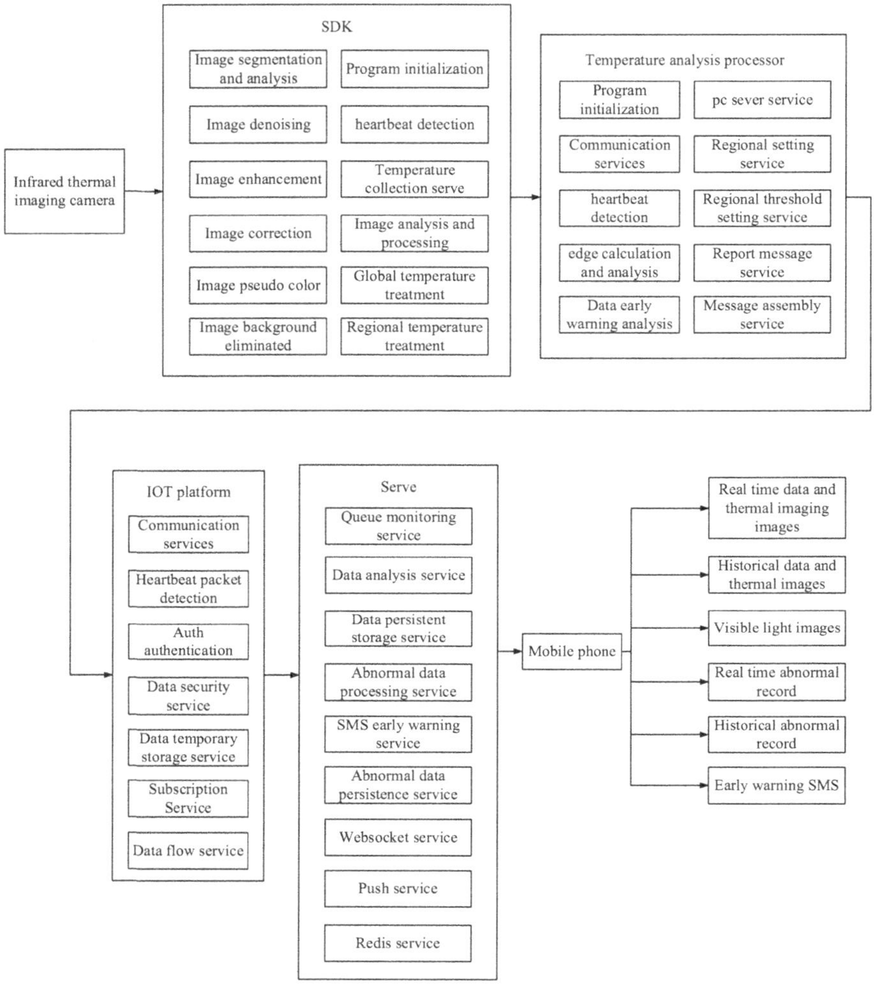

Fig. 1 System structure

All objects above absolute zero (−273 °C) emit infrared radiation. The infrared imager uses the infrared detector and objective lens to receive the infrared radiation energy of the measured target and reflect it to the photosensitive elements of the infrared detector, so as to obtain the infrared thermal image. In other words, the infrared thermal imager transforms the invisible infrared energy emitted by the object into the visible thermal image. Then the whole temperature distribution of the measured target can be observed by the thermal image. Finally, studying the heating of the target, so as to carry out the next work.

2.3.3 Realization of Infrared Thermal Imaging App

Infrared thermal imaging app is a further improvement of infrared thermal imaging system. It mainly includes the following function points: (a) Monitoring key areas based on infrared thermal imaging technology; (b) The over temperature threshold is set for the monitoring area; (c) Temperature difference threshold setting for single or different areas of monitoring area; (d) Rendering real-time thermal images and collecting the maximum and mini-mum temperature values of important areas; (e) Real time alarm according to overheating.

The infrared thermal imaging app (server side) selects the region and sets the corresponding overheating threshold and temperature difference threshold in the monitored area, so that the system will focus on the real-time monitoring of the selected region. Then the app will transmit the above data to the acquisition and edge computing analysis server and through TCP. Further operation is done by this server.

2.3.4 Implementation Principle and Data Forwarding of the Acquisition and Edge Computing Analysis Server

The 64 bit Linux system is adopted in the acquisition and edge computing analysis server. As an intermediate processor, it has the function of reporting data messages and collecting data from the infrared thermal imaging camera. The specific functions are as follows.

(1) Interacting with app server

The server side of the app transfers the setting area coordinates and relevant threshold setting, and obtain the collected data according to the coordinates. After calculating and analyzing the collected data obtained by edge calculation, the collected thermal images and the analyzed data are returned to the server side of the app through TCP for subsequent operation.

(2) Analysis, processing and data forwarding of edge computing data

The edge computing data is calculated by the second data correction algorithm to obtain more accurate temperature value. Then compared with the overheating threshold and temperature difference threshold, getting the normal data and abnormal data. Assembling the above data into the message in the specified format and reporting the message to the IOT platform.

2.3.5 Data Flow of IOT Platform

(1) Heartbeat detection and offline equipment processing: The IOT platform will send heartbeat packets to the acquisition and edge computing analysis server to ensure that the processor is online. When the IOT platform detects that the

acquisition and edge computing and analysis server is not online, it will realize the re-connection at the frequency interval of 30, 60 and 90 s. At the same time of re-connection, the IOT platform will send a message to the cloud server to inform the offline message of the specific equipment, so that the cloud server can make follow-up operation for this situation.

(2) Data flow: Through reading the subscription service and consumption group configured by the server, the IOT platform sends the received message to the designated consumption group for the server to read.

2.3.6 Real-Time Data Push by Broadcast Private Broadcasting

The server collects the data scripts issued by the IOT platform, and the array will be parsed and put into the specified private broadcast. The process is as follows: (a) The server needs to support WebSocket protocol and allow clients to establish WeSocket connections; (b) Connecting WebSocket. After the connection is successful, the client gets the unique identifier; (c) The client connection is successful, indicating that the client subscribes to the specified channel and will receive the message of the channel; (d) The client registers the listening events of the channel it subscribes to; (e) After the server completes the specified function, the program informs the webSocket server with the information of the specified channel name and event name; (f) The WebSocket server pushes the specified event in the form of broadcast to all clients registered to listen to this channel.

2.3.7 The Control Strategy of Real-Time Data Persistent Storage Service

Data storage can be divided into two directions: (1) Redis cache; (2) Insert database storage. Redis cache is used for frequently used mobile data rendering, which effectively solves the rendering efficiency and speed; Database storage is used for persistent storage for historical data review, historical temperature curve rendering, dis-play of historical temperature data, and analysis and summary of historical abnormal data.

2.3.8 Intelligent Monthly Report Analysis of Large Amount of Data

The system takes the frequency of collecting data once every 10 min set by general users as the reference. Then, the amount of acquisition records in a month is 4320. The system will make monthly report analysis for 4320 data according to the collected data value combined with the abnormal alarm situation in this month.

2.3.9 One-Button Temporary Collection Service

One-button temporary collection service is a function developed to solve the operation and maintenance personnel's expectation of real-time understanding and tracking the operation of remote cable terminals. The function supports real-time return: real-time visible image, real-time infrared thermal imaging and corresponding regional collection temperature of infrared thermal imaging (Fig. 2).

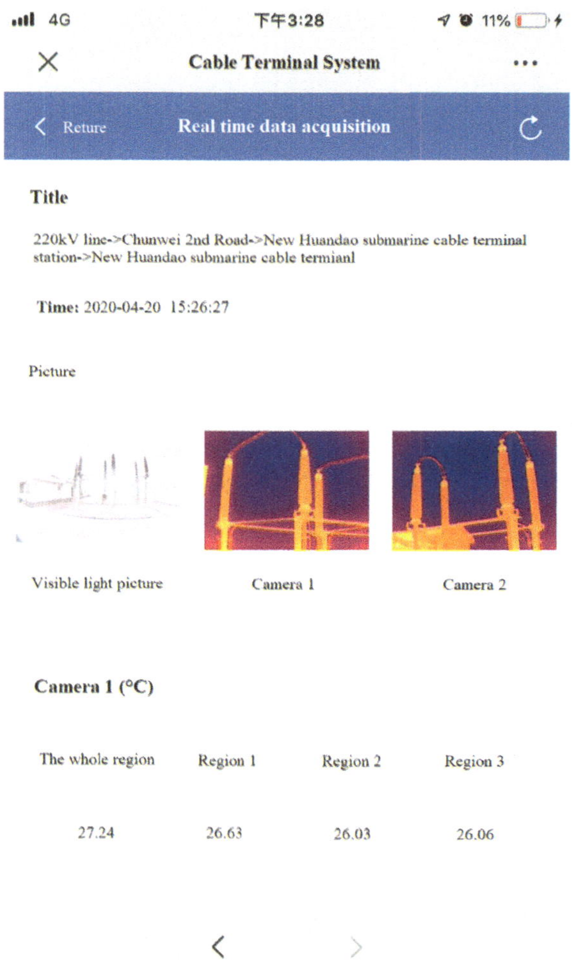

Fig. 2 Mobile phone app of acquisition system

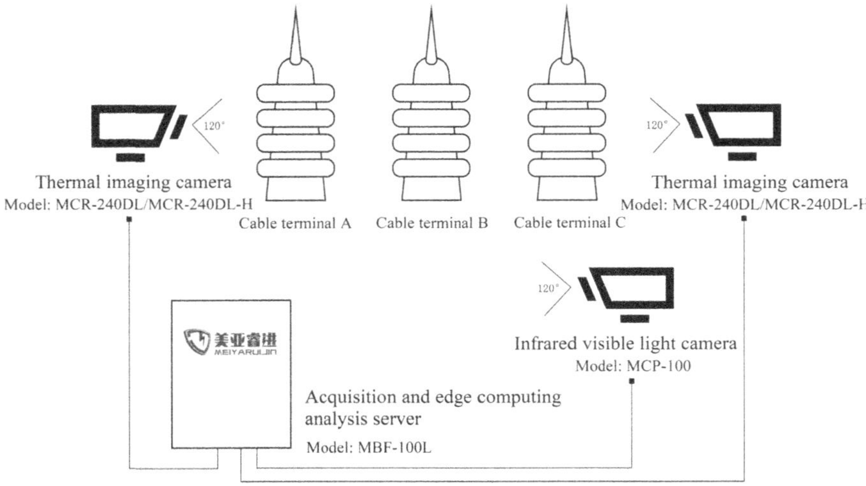

Fig. 3 Site installation diagram

2.3.10 Intelligent Early Warning Service

The system will send messages to the mobile phones of one or more operation and maintenance personnel in real time through the script of parsing abnormal data, and the SMS template will describe the specific cable terminal line, site, threshold, alarm temperature value and other information in detail.

3 Implementation and Analysis of Practical Problems

3.1 Installation Plane Figure

See Fig. 3.

3.2 Analysis of Practical Problems

From the above abnormal record, the following information can be obtained: Abnormal equipment, Abnormal time, Abnormal type, Abnormal temperature, Threshold, and summarized abnormal content. At the same time, the situation of the scene can be clearly understood through the visible light pictures as well as the fire can be known in time, and the location of abnormal temperature can be analyzed through thermal imaging (Figs. 4 and 5).

Abnormal equipment	#1 collector
Abnormal time	2020-04-08 06:38:50
Abnormal type	Temperature difference
Abnormal temperature	26.05 °C
Temperature difference threshold	25 °C

Abnormal title

Site: Chunwei 2nd Road-New Huandao submarine cable terminal station-New Huandao submarine cable terminal equipment: [#1 collector] abnormal temperature

Abnormal content

Site: Chunwei 2nd Road-New Huandao submarine cable terminal station-New Huandao submarine cable terminal equipment:[#1 collector] Temperature difference threshold: 25°C Alarm temperature: 26.05°C

Fig. 4 Example of cable terminal heating problem

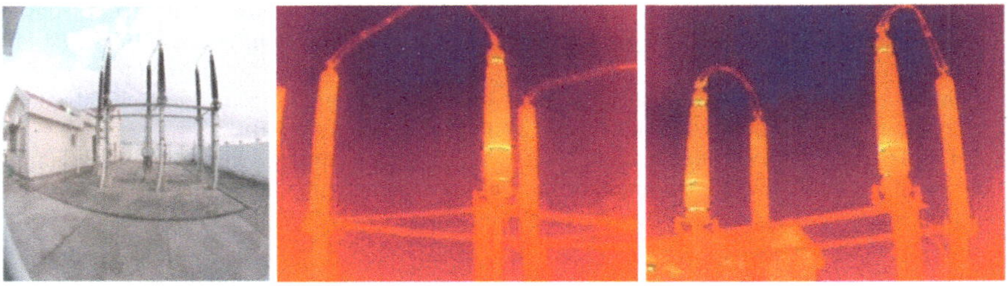

(a) visible light (b) Phase A (c) Phase B and C

Fig. 5 Pictures taken by visible light camera and thermal imaging camera

4 Conclusion

Based on the application of multiple big data methods such as edge computing and cloud storage, a system is developed for state perception and operation and maintenance of high-voltage cable outdoor terminals. The loss of labor and material for cable operation and maintenance could be reduced greatly. It has preliminarily solved the problems of data blanking, historical data cannot be traced, operation and maintenance detection fails to meet the standard, automatically monitored in real time cannot be done, and operation and maintenance personnel cannot under-stand the situation in the first time when there is an alarm. It provides a feasible example in the intelligent operation and maintenance of cables, laying the foundation for the future. At the same time, the system preliminarily achieves intelligent real-time monitoring, which effectively reduces the time of operation and maintenance and the problem of the detection unable to meet the standard. It meets the intelligent management needs of the cable department for cable operation.

References

1. Michael, S. 2015. China expects to spend $300 Billion by 2020 to improve electric grid, distribution system. *International Environment Reporter: Reference File* 38 (18): 1159–1159.
2. Huang, J., J. Huang, E. Xiang, et al. 2020. Study on insulation performance of EPR cable terminal containing metal particles. *IOP Conference Series: Earth and Environmental Science* 431 (1): 012003.
3. Addo, I.D., S.I. Ahamed, S.S. Yau, et al. 2014. A reference architecture for improving security and privacy in internet of things applications. In *2014 IEEE third international conference on mobile services: 2014 IEEE 3rd international conference on mobile services*, pp. 108–115. USA.
4. Wibowo, F.W. 2020. Wireless communication design of internet of things based on FPGA and WiFi module. *Journal of Physics: Conference Series* 1577 (1): 012035.
5. Zakaria, M., Kajan Ejub, Muhammad Asim, et al. 2019. Open challenges in vetting the internet of things. *Internet Technology Letters* 2(5).
6. Chao, Z., Z. Chao, Q. Jiafeng, et al. 2020. A method for preprocessing state data of power transmission and transformation equipment. *Journal of Physics: Conference Series* 1584 (1): 012059.
7. Jiang, K., L. Du, Y. Wang, et al. 2019. A smart overvoltage monitoring and hierarchical pattern recognizing system for power grid with HTS cables. *Electronics* 8 (10): 1194.
8. Selivanova, Z.M., D.S. Kurenkov, et al. 2020. Algorithmic support for solution of classification problems and image recognition of the studied objects by the intelligent information and measuring system. *Journal of Physics: Conference Series* 1441 (1): 012079.

Design and Testing of Flexible Security and Stability Control in Industrial Power Grid

Zaixin Yang, Jun Tao, and Chen Gao

Abstract Analyzing the security and stability control of the industrial power grid under large disturbance is of significant importance for industrial production. In this study, a system for flexible security and stability control (FSSC) is designed with active power emergency control and frequency-voltage coordination function. Fast transfer performance, in terms of industrial comprehensive load model, is analyzed by comparison. A system-level dynamic test platform based on real-time digital simulator (RTDS) is used. The results show that not only the function and operation of the flexible stability controller and fast transfer are verified, but also the industrial system response and transient characteristics are demonstrated stability. The FSSC system ensures the stability of the second defense line for the industrial power grid and provides technical support for engineering applications.

Keywords Flexible security and stability control (FSSC) · Real-time digital simulator (RTDS) · Fast transfer · System-level dynamic test

1 Introduction

With the continuous development of new energy clusters, long-distance, large-capacity transmission of UHV AC and DC technology, the problem of stability damage to the industrial system under disturbances has drawn increasing attention. The security and stability control measures mainly based on rigid cutting generator, cutting new energy line, and load cutting methods need to be improved. With the changes in the forms of power generation and sales in China's high energy-consuming industries, The motor accounts for a large proportion of the total load, and the rotating reserve capacity of the generator is insufficient. At the same time, a weak connection interface between the industrial power grid and the external large power grid results in insufficient shock resistance. After isolated grids, the risk of large-scale chain faults in the industrial power grid increases.

Z. Yang (✉) · J. Tao · C. Gao
Inner Mongolia Enterprise Key Laboratory of Smart Grid Simulation of Electrical Power System, Inner Mongolia Power Research Institute, Hohhot 010020, China
e-mail: zaixin.yang@foxmail.com

W. Chen et al. (eds.), *The Proceedings of the 9th Frontier Academic Forum of Electrical Engineering*, Lecture Notes in Electrical Engineering 743,
https://doi.org/10.1007/978-981-33-6609-1_22

The load model has certain influence on the transient stability calculation, low-frequency oscillation calculation and voltage stability calculation of the power system. Optimal load control in industrial sector combined with renewable generator and energy storage is a hot topic in recent years [1]. A fast converging short-term load forecasting method [2] and frequency control strategy considering load characteristics [3, 4] are studied. Therefore, the accuracy of load modeling directly affects industrial power generation and frequency stability. Industrial loads have a high proportion of motors, so the fast transfer characteristics of different voltage levels require in-depth research.

With the vertical multi-hierarchy and horizontal multi-level penetration of regional security and stability control systems, it has become a trend to deploy stability control systems for large industrial power grids [5]. The conventional regional security and stability control strategy rarely considers the uninterrupted power supply of the internal load of the power plant when shedding generator or cutting the grid connection of the power plant. Therefore, frequency stability technologies are applied from the aspect of isolated grid control [6]. Consequently, many studies have been applied from the aspect of isolated grid control. A frequency and voltage regulation with no steady-state error is evaluated and validated in [7]. Moreover, hardware-in-the-loop testing is also a typical feature of system-level test, which is used in power system simulation and performance evaluation of protection equipment [8, 9].

In this paper, a security and stability controller with flexible regulation of voltage and frequency is studied from the perspective of industrial power grid. The paper is organized as follows: In Sect. 2, the FSSC which function consist of flexible stability control and fast transfer coordination control is presented. In Sect. 3, the details of the system-level dynamic testing for industrial power grid are given and tested. Finally, the conclusions are presented in Sect. 4.

2 Strategy Design of Flexible Security and Stability Control System

Stability control function of the industrial grid is different from the large power grid, which mainly maintains the stability of the connection interface and various industrial production areas. Therefore, the strategy is designed for master station control, flexible stability control, and fast transfer coordination control, as shown in Fig. 1.

The function of the master station control is mainly to collect the electrical quantities and breakers position of the connection lines, and then the operation mode of the industrial grid is judged. At the moment of turning to an isolated grid, according to the interface power before the fault, the generators reduced power or the loads to be cut off are calculated. Furthermore, the control target is communicated with the flexible stability controller and fast transfer coordination controller.

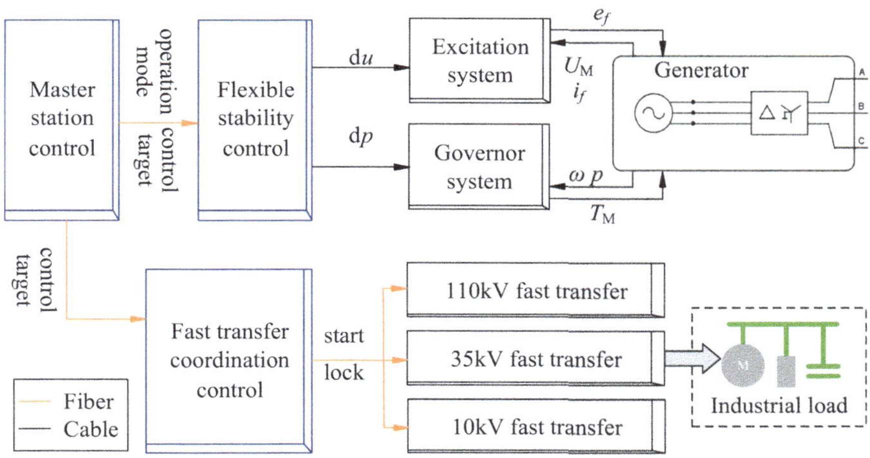

Fig. 1 Design block of flexible security and stability

2.1 *Flexible Stability Control*

The flexible stability controller is designed to perform connection interface voltage-frequency control and active power emergency control. The voltage loop monitors the bias voltage of the interface bus, and forms an excitation signal du, which is sent to the automatic voltage regulator (AVR). The frequency loop monitors the deviation frequency of the bus after the isolated grid, and forms a power signal dp, which is sent to governor. The power loop dynamically adjusts the output of multiple generators according to the control target of the master station.

The stability controller can flexibly and quickly adjust the voltage and power according to the controller PID and other parameters. The fast secondary frequency control of the generator is realized. The control effect is shown in Fig. 2.

A test showed that an N − 2 fault is occurred in the connection line, forming an isolated grid. The down-grid power before the accident was 410 MW. Without taking stability-control measures, the 220 kV bus frequency drops below 48.5 Hz, and the minimum value of the bus voltage is 0.93 p.u. According to the technical specification of grid-related protection [10], industrial grids cannot operate securely and stably at the time of isolated. Using flexible stability controller, the connection-interface frequency is controlled at 50.02–49.82 Hz, and the maximum overvoltage is 1.015 p.u. Considering the influence of boilers and turbines, it is difficult to continuously and steadily increase the active power by 40 MW within 1 s during actual operation. Therefore, through FSSC methods, timely and accurate load cutting measures are used, which can reliably guarantee the stable operation of the industrial power grid.

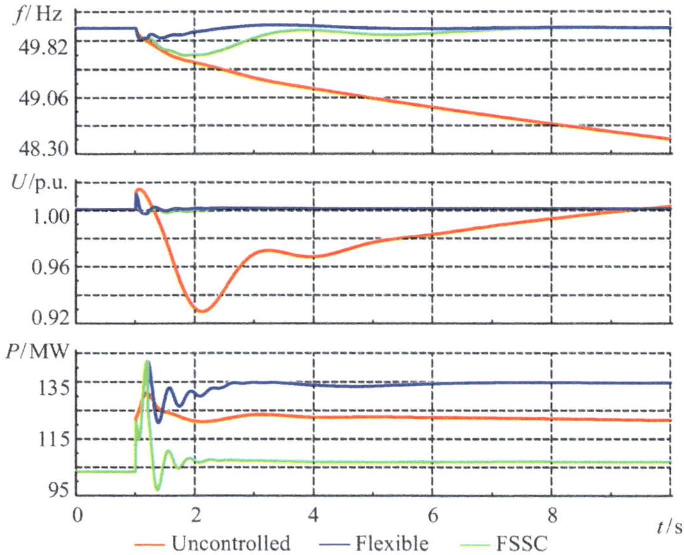

Fig. 2 Comparison of bus voltage, frequency and generator power in different control modes

2.2 *Fast Transfer Coordination Control*

According to the closing conditions of the backup breaker, the suitable transfer methods for industrial power grids are fast switch and synchronization capture. The frequencies and phases of the buses at different voltage levels are shown in Figs. 3 and 5. Figure 4 shows that the bus frequency changes are consistent. Figure 4 shows that 110 kV bus, 35 kV bus with 70% motors, 35 kV bus less than 30% combined loads and 35 kV bus with 90% motors.

Fig. 3 Frequency difference between blackout bus and backup bus

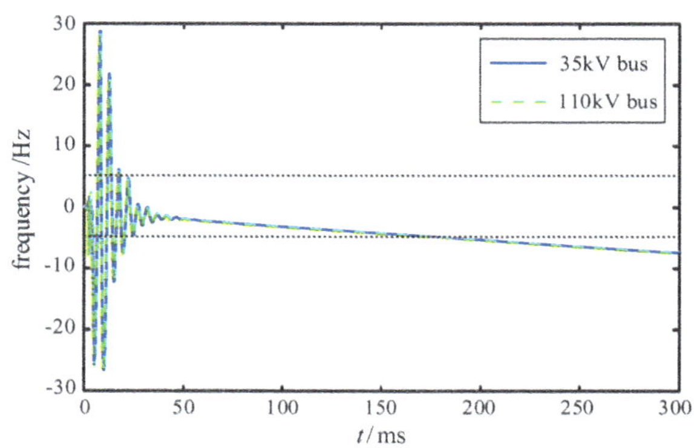

Fig. 4 Phase difference between blackout bus and backup bus

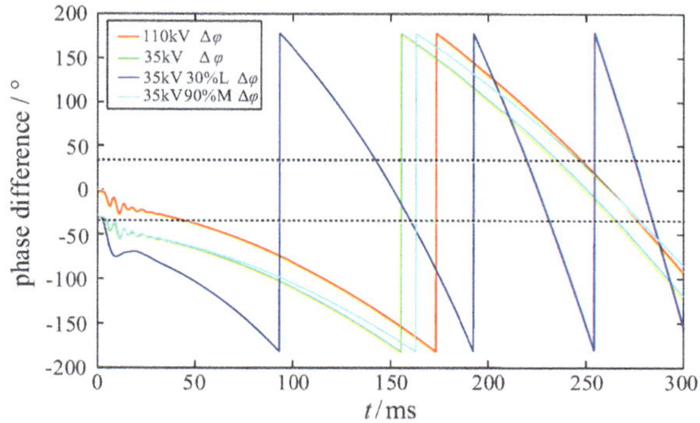

Fig. 5 Control effect of generator trip in isolated grid: **a** No. 3 generator P under FSSC power control; **b** 220 kV bus frequency and FSSC excitation adjustment value; **c** generator grid-connected breaker signal and all load breaker signals

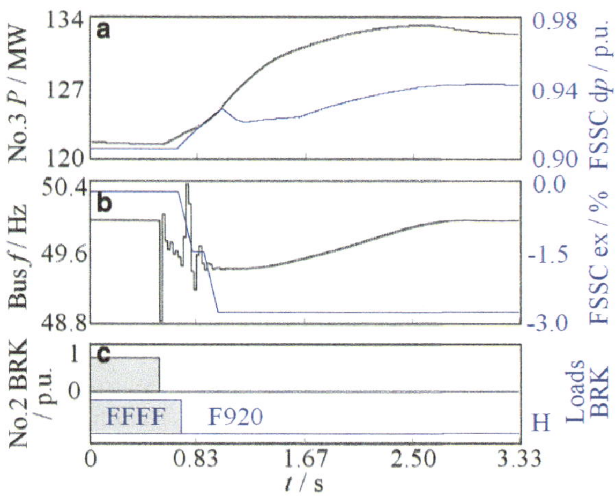

The fast switch condition is that the frequency difference is within 1.5 Hz, the phase difference is less than 30°, and the closing is rapid within 50 ms. Therefore, the fast switch success rate of 110 kV bus is higher than that of 35 kV bus. The synchronization capture condition is that at the first two synchronization points, the frequency difference is within 5 Hz, the phase difference is less than 20°, and the closing time is less than 200 ms. Therefore, the synchronization capture success rate of 35 kV light-loads bus is higher than that of 35 kV heavy-loads bus.

In order to eliminate the risk for fast transfer failure of industrial load after power failure, multi-level fast transfer coordination control is adopted. 110, 35 and 10 kV production buses are equipped with stability control devices. After a fault occurs and bus blackout, the stability controller sends a start-up command to this level of fast transfer and the next voltage level of fast transfer at the same time. Applied to engineering practice, the time difference between the two-stage fast transfer start signals is 8 ms, which effectively improves the transfer success rate and ensures the reliability of industrial power supply.

3 System-Level Dynamic Testing for Industrial Power Grid

A RTDS model based on the actual industrial power grid is built. The rated active power of the generator is 135 MW. The communication interface with external power grid is two 220 kV lines. 110 kV and below buses are divided into running modes. The industrial loads during routine production are 220–300 MW, which included 6 constant- impedance loads and 10 motors.

According to the dynamic test standard [11], the control strategy of the device is checked in closed loop. During the test, not only the function and operation of the devices are verified, but also the industrial system response and transient characteristics are demonstrated stability. Take the test items of isolated grid mode a generator trip as an example, as shown in Fig. 5.

Before the fault, No. 2 generator P = 118 MW, No. 3 generator P = 122 MW. The 16 loads are put into operation, that is, loads breaker status is FFFF. The No. 2 tripped at 0.53 s, and the bus frequency decreased accordingly. FSSC is activated and implemented control measures. The active power deviation dp is generated at 0.63 s for fast secondary frequency control. At the same time, a rapid reduction of the excitation is triggered in order to suppress the increase of the bus overvoltage. The loads breaker is opened at 0.70 s according to the load cutting command issued by the stability control device.

According to the under-cut strategy, the chemical loads are the first priority, and the constant impedance loads are cut first and then the motors are cut. In the chemical production area, each motor is 20 MW and the constant impedance is 2 MW. Therefore, according to the breaker state F920, a total of 108 MW loads are cut off. Under the governor system and FSSC control, No. 3 generator increases 10 MW active power and runs stably after 3 s. Therefore, tests show that FSSC control can restore the voltage, power, and frequency stability of industrial power grids.

4 Conclusions

In this paper, a flexible security and stability control system for industrial power grid is designed and tested. The FSSC architecture is demonstrated and the strategy of flexible stability controller and fast transfer device are tested and proved effective. The system-level test platform for industrial power grid is designed and dynamic tests are fully functionally verified. It avoids the serious consequences caused by system instability and can better support security and stability of the industrial power grid.

Acknowledgements The work described in this paper was fully supported by Natural Science Foundation of Inner Mongolia Autonomous Region of china (2021SHZR1483).

References

1. Gholian, A., M. Hamed, and H. Yingbo. 2015. Optimal industrial load control in smart grid. *IEEE Transactions on Smart Grid* 7: 2305–2316.
2. Ahmad, A., N. Javaid, M. Guizani, N. Alrajeh, and Z. Khan. 2016. An accurate and fast converging short-term load forecasting model for industrial applications in a smart grid. *IEEE Transactions on Industrial Informatics* 13: 2587–2596.
3. Lou, Y., R. Jiang, F. Qian, et al. 2019. Frequency control strategy after receiving-end power grid splitting considering load characteristics. *Power System Technology* 43: 213–220 (in Chinese).
4. Xu, J., Y. Chen, S. Liao, et al. 2019. Demand side industrial load control for local utilization of wind power in isolated grids. *Applied Energy* 243: 47–56.
5. Luo, J., X. Dong, X. Cui, et al. 2018. Discussion on reliability of large scale security and stability control system. *Power System Protection and Control* 46: 65–72 (in Chinese).
6. Zhang, M., Z. Du, N. Li, et al. 2012. Control strategies of frequency stability for islanding high-voltage microgrids. *Proceedings of the CSEE* 32: 20–26.
7. Farrokhabadi, M., C.A. Cañizares, et al. 2015. Frequency control in isolated microgrids through voltage regulation. *IEEE Transactions on Smart Grid* 5: 1185–1194.
8. Yang, Z., Y. Wang, L. Xing, et al. 2020. Relay protection simulation and testing of online setting value modification based on RTDS. *IEEE Access* 8: 4693–4699.
9. Alam, M., H. Leite, N. Silva, et al. 2017. Performance evaluation of distance protection of transmission lines connected with VSC-HVDC system using closed-loop test in RTDS. *Electric Power Systems Research* 152: 168–183.
10. Technical specification for grid-related protection of large generating unit, Standard DL/T 1309-2013, China Electricity Council, 2013 (in Chinese).
11. The dynamic test of the power system protective products, Standard GB 26864, China Electricity Council, 2011 (in Chinese).

Research of Electrical Test System and Simulation of Universal Converter Products Applied in DC Distribution Network

Zhilong Hu, Zimeng Xu, Xiaoling Yu, Congwei Tong, Teng Zhang, Ben Wang, Xidong Huang, and Weigong Qin

Abstract In recent years, the issues of energy depletion and power supply quality have received increasing attention. This has led to the emergence and continuous development of DC distribution network technology and renewable energy technology with unique advantages. At this stage, the state of the power grid is still based on a mature AC power grid as the main form. This makes a large number of applications of various converter products a foreseeable trend. In order to ensure the safe and reliable operation of such products, the research on its test technology is particularly important. This paper analyzes and studies the electrical test requirements and test methods of converter products, and proposes a technical scheme for the electrical test system of converter products. It is based on three full/half-bridge hybrid MMC converters. Through the parameter adjustment of the test system and the switching of multiple operating modes, the performance verification of different types of large-capacity converter products can be achieved with a small amount of active power consumption. PSCAD software simulation results also show that it can meet the electrical performance test requirements of universal converter products with single and dual power flow directions.

Keywords Converter · Full/half-bridge hybrid MMC · Electrical test · Active power loss

1 Introduction

As the continuous development of modern society, the demand for energy continues to rise exponentially. With continuous resource development, the accompanying environmental pollution and climate change issues will continue to plague us. At the same time, the gradual depletion of traditional energy is also a problem that mankind has

Z. Hu · Z. Xu (✉) · X. Yu · C. Tong · T. Zhang · B. Wang · X. Huang · W. Qin
Xi'an High Voltage Apparatus Research Institute Co., Ltd., Xi'an 710077, China
e-mail: xuzimeng@xihari.com

Z. Hu
e-mail: huzhilong@xihari.com

W. Chen et al. (eds.), *The Proceedings of the 9th Frontier Academic Forum of Electrical Engineering*, Lecture Notes in Electrical Engineering 743,
https://doi.org/10.1007/978-981-33-6609-1_23

to face. Vigorously develop renewable energy, based on the current energy development trend, will be an effective response. On the other hand, with the continuous precision of electricity demand, the importance of power supply quality has become the top priority of the development of power supply technology [1–6].

The DC distribution network technology, with its unique advantages, has also been valued and studied. Compared with AC distribution network, its advantages are mainly embodied in saving line corridors, low cost; high transmission efficiency, low line loss; high quality of power supply. At the same time, since most of the renewable energy and energy storage devices are DC ports, large-scale access can be realized flexibly and conveniently based on DC distribution network technology. The status quo of our national grid, on the one hand, is the demand for a large number of renewable energy access and the continuous development of DC distribution network technology, on the other hand, the mature AC transmission mode of the existing grid, which makes various converter products in the distribution A large number of applications in the Internet have become a foreseeable trend. As an important way to ensure product quality and operational reliability, the research on its test technology is particularly important [7, 8].

2 Topology of Converter Electrical Tests Circuit Based on Full/half Bridge Hybrid MMC Technology

2.1 Basic Principle of Electrical Tests of Converter

As a connection point device, the converter is used to connect different power grids or electrical equipment. Its operational performance, safety and reliability should be the focus of research and attention. In terms of operating performance, the performance of the converter under rated conditions is as important as the performance under abnormal conditions such as load fluctuations. Therefore, it is necessary to use a suitable test system to verify its performance.

The port external characteristics of the converter product can be expressed as the power side and the load side. The basic principle of the electrical test is to connect the test power supply to its power source side and the test load to its load side. Adjust the characteristics of the test power supply and test load to make the tested converter operate in specific working state to verify its performance under different working conditions. A basic structure of the electrical test circuit of the converter product is shown in Fig. 1.

In addition to realizing the basic function test of the converter, the electrical test circuit should also meet the general requirements, that is, it can test different types of converter products, such as AC/DC converters, DC/DC converters and test converters with a larger parameter range.

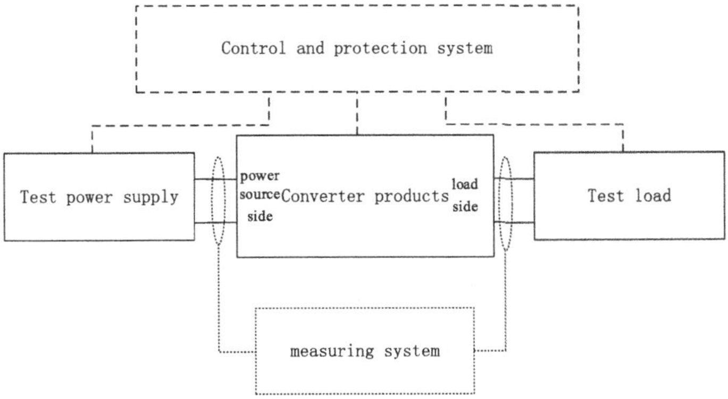

Fig. 1 Schematic diagram of electrical test of converter product

2.2 The Overall Scheme of the Test System

Based on the full/half-bridge hybrid MMC technology, this paper proposes a converter electrical test loop topology, as shown in Fig. 2. It is mainly composed of converter MMC1, converter MMC2, converter MMC3, test transformer T1, test transformer T2. MMC1, MMC2 and MMC3 are full/half-bridge hybrid structure MMC type converters. MMC3 can be controlled to operate in single-phase mode or three-phase mode.

T1 is connected to the MMC1 AC terminal. T2 is connected to the MMC2 AC terminal. T1 and T2 are connected to the AC grid from the same point. The DC terminal of MMC2 is directly connected to a DC terminal of the test object. According to the test requirements, the DC side of MMC1 can be directly connected to the other

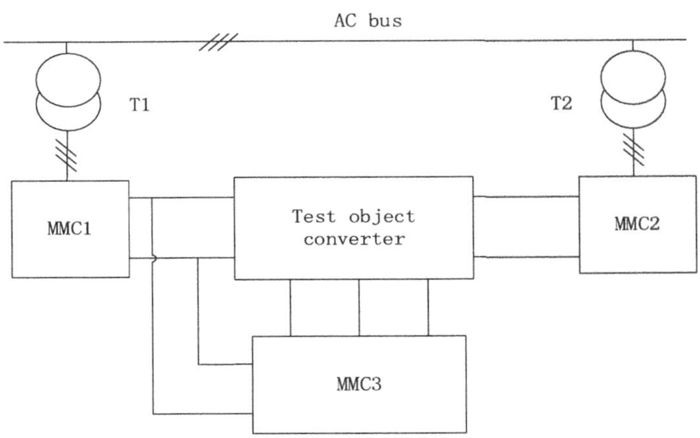

Fig. 2 Test circuit topology diagram

DC terminal of the test object or connected to the DC terminal of MMC3, and the AC terminal of MMC3 is connected to the AC terminal of the test object.

The MMC1, MMC2, MMC3 can realize the function of test power source and test load through the flexible control of rectifier state and inverter state.

2.3 Test System Solutions for Different Sample Types

According to the different application scenarios of the converters, the test object of the test system mainly includes DC/DC converter, DC/AC and AC/DC converter. Among them, the electrical test of the DC/AC converter and the AC/DC converter differs only in the focus of the input/output side, and the implementation methods are generally the same. In either mode, the test power supply and test load can be replaced with each other to meet the test requirements of dual-power flow direction converter products.

Mode1: DC/DC converter Test.

In this mode, the test object is DC/DC converter product. Use MMC1 as the test power source which connects to the input side of the test object, while MMC2 as the test load which connects to the output side of the test object.

In this condition, constant voltage control is adopted by MMC1 to provide a stable and controllable DC test voltage for the test object. The MMC2 uses constant power control or constant voltage control according to the requirements of the test object. It is used to match the characteristics and control methods of different test objects, and control the test power, test current and test voltage in real time during the test. The test implementation is shown in Fig. 3.

Mode 2: DC/AC converter and AC/DC converter mode.

The test products in this mode can be divided into two types according to the AC side interface mode: three-phase converter and single-phase converter. The test implementation is shown in Fig. 4.

When the test object is a three-phase converter. The MMC1 DC terminal is connected to the DC terminal of the auxiliary converter MMC3, the MMC3 AC

Fig. 3 DC/DC converter test implementation

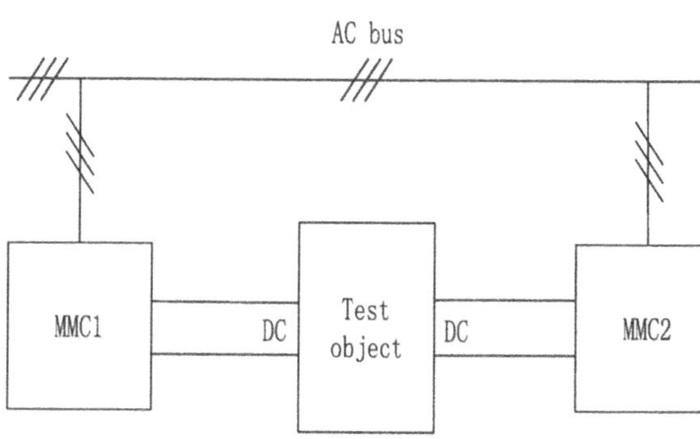

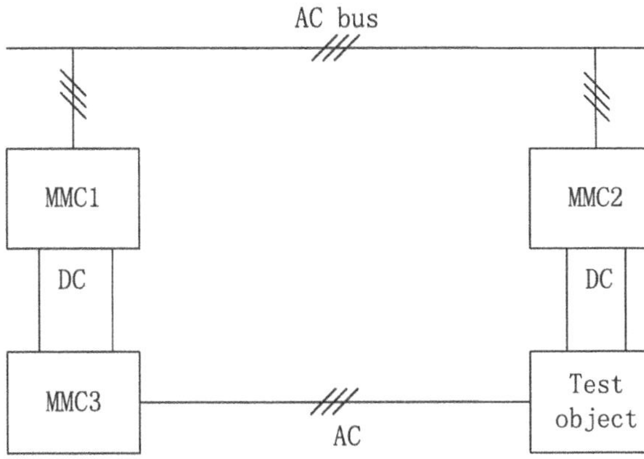

Fig. 4 DC/AC converter (three-phase) test implementation

terminal is used as the test power supply to connect to the AC terminal of the test object, and the MMC2 DC terminal is used as the test load to connect to the DC terminal of the test object.

When the test product is a single-phase converter, its implementation is similar to the three-phase converter. Use MMC1 DC terminal to connect to the DC terminal of the auxiliary converter MMC3. Use MMC3 AC side single-phase mode as the test power supply to connect to the AC terminal of the test object, and MMC2 DC terminal as the test load to connect to the DC terminal of the test object.

In this condition, the test power source adopts constant voltage control to provide a stable and controllable test voltage for the test object. The test load uses constant power control or constant voltage control. It is used to match the characteristics and control methods of different test objects, and control the test power, test current and test voltage in real time during the test.

2.4 *Characteristics of the Test System*

2.4.1 Diversification of Test Functions

The test system described in this article can meet the test requirements of various types and larger parameter ranges through different modes. Through the combination of different working conditions of three MMC converters, it can meet the test requirements including DC/DC converters, three-phase DC/AC converters, and single-phase DC/AC converters. At the same time, the flexible control of the power reversal based on the MMC converter can meet the test requirements of the unidirectional power flow direction and the bidirectional power flow direction [9]. Through the DC voltage adjustment characteristics of the full/half-bridge hybrid MMC converter and the active power control function, the DC voltage and test active power can be flexibly controlled in a wide range to meet the test requirements of various products with different parameter ranges.

2.4.2 Flexible Control of Test Power Supply and Test Load

This article uses the full/half bridge hybrid MMC converter MMC1, MMC2 and MMC3 to work together as the test power source and test load for the test system.

The advantages of this scheme mainly include:

(1) The full/half bridge hybrid MMC converter is used as the DC power output, and the real-time adjustment of the output voltage in a wide range can be realized through the converter control strategy. Real-time adjustment and reversal of test power source can also be achieved. It has the advantages of high control precision and fast response time [10, 11].
(2) The adjustment of the output DC voltage is completely realized by the converter, and the output voltage of the converter transformer does not change. That is, when the transformer is involved in voltage adjustment, a huge loss in actual output power due to the decrease in output voltage is avoided [12, 13].
(3) The test system adopts the test power source and test load connected to the grid system from the same grid connection point, and a special test control strategy is designed. The test power source and the test load are mutually compensated, and the power circulation of the test power source-the test object-the test load-the test power source is realized.

3 Research on Digital Simulation Based on PSCAD

3.1 Simulation Model Establishment

3.1.1 The Overall Structure of the Simulation System

A simulation system based on PSCAD is built according to the above-mentioned converter electrical test system scheme. The overall structure is mainly composed of converter modules MMC1, MMC2, MMC3, MMC4 and DC/DC modules, among which the converter modules all adopt full/half-bridge hybrid MMC converters. The DC/DC module is an equivalent DC/DC converter model, and MMC4 is a DC/AC converter model. MMC3 and MMC4 can be controlled to operate in three-phase mode and single-phase mode to verify the test functions of three-phase converter and single-phase converter respectively. The test object converter model adopts different control strategies to simulate the control function and output characteristics of the test converter. The simulation model structure is shown in Fig. 5.

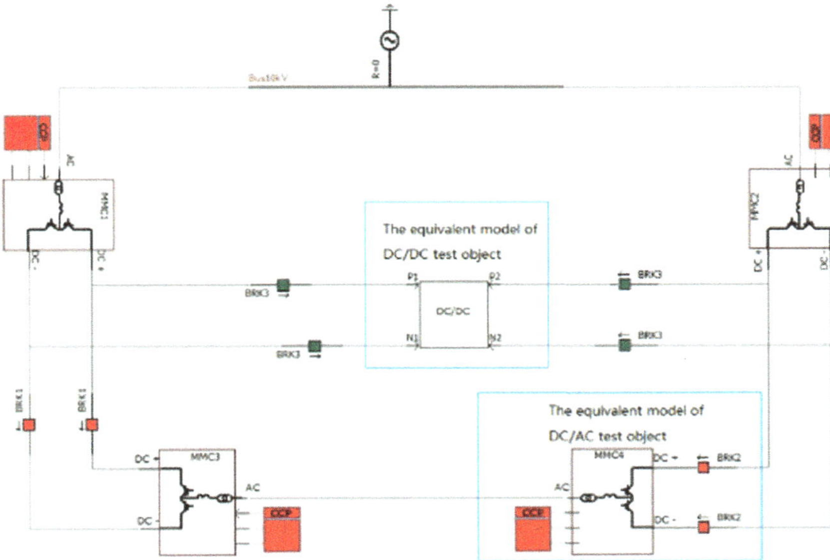

Fig. 5 Overall structure of the simulation model

In this model, the MMC2 AC terminal is connected to the AC grid, and the DC terminal is connected to the DC side of the test object. It can be used as a rectifier to provide the required adjustable DC voltage (3–20 kV) for the test object, or it can operate as an inverter. The DC sides of MMC1 and MMC3 adopt a back-to-back structure. The AC side of MMC1 is connected to the AC grid, and the AC side of MMC3 is connected to the AC side of the test object. MMC3 can be used as an inverter to provide AC voltage (1.5–10 kV) to the test object. The AC power grid uses an ideal voltage source of 10 kV. MMC1, MMC2, and MMC3 are all auxiliary converters, and different control strategies are used to simulate the actual operating conditions of the test object.

3.1.2 Model of Key Equipment MMCs Converter

In this model, there are 4 MMC converter models, namely MMC1, MMC2, MMC3 and MMC4. All are three-phase converter structure. MMC3 and MMC4 can operate in single-phase mode and three-phase mode according to requirements.

In the model of the main structure of the three-phase and single-phase MMC converter, each phase is composed of a module, and the input of each module contains the modulation voltage of the upper and lower bridge arms and the system unlock/lock command, and the output is the upper and lower bridge arms The sum of the bridge arm current and the bridge arm module voltage. The output is the sum of the bridge

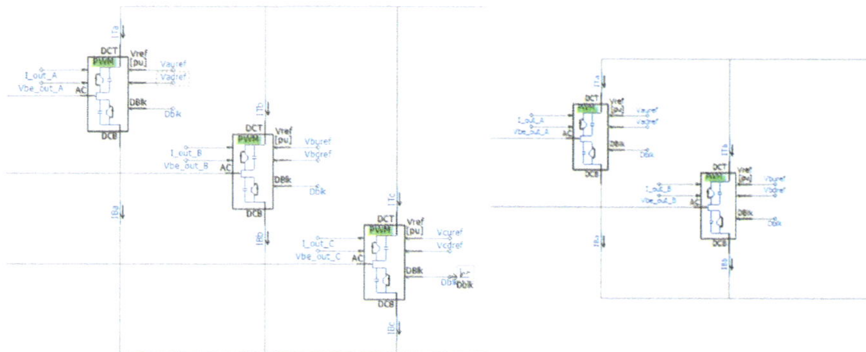

Fig. 6 Three-phase and single-phase MMC converter simulation model

arm currents of the upper and lower bridge arms and the sum of the bridge arm module voltages of the upper and lower bridge arms.

Each module is composed of upper and lower bridge arms. Each bridge arm contains 2 half-bridge sub-modules, 12 full-bridge sub-modules, bridge arm reactor L and bridge arm equivalent resistance. The full and half bridge sub-modules receive the IGBT trigger command generated by the control, and feedback the sub-module capacitor voltage to the control for capacitor voltage balance control. As shown in Fig. 6.

3.1.3 Test System Control Function Design

The control function of the test system is mainly realized by the control and protection function of each converter. The control and protection of the converter mainly includes circulating current suppression, level approximation control, and capacitor voltage balance control. Among them, the circulating current suppression control calculates the circulating current component in the bridge arm through the current of the bridge arm, and adopts the current closed-loop control based on dq transformation according to the circulating current component, and the modulation wave compensation amount generated by the control is superimposed on the modulation wave issued by the control protection.

According to the characteristics of different test modes, each MMC converter in the simulation system can be set to the corresponding control mode:

The MMC1 converter control mode is fixedly set to constant DC voltage control, which provides a stable and controllable DC voltage to MMC 3 or DC/DC test object through back-to-back.

MMC2 converter has two control modes, namely constant DC voltage control and constant power control, and it can work in rectification or inverter mode by back-to-back with MMC4.The MMC2 and MMC4 control modes are mutually coordinated. If MMC2 selects the constant DC voltage control mode, then MMC4 can select the constant AC voltage and frequency control or the constant power control. If MMC2 chooses constant power control, MMC4 can only choose constant DC voltage control.

MMC3 converter has two control modes, namely constant AC voltage and frequency control and constant power control. The MMC3 AC side is connected to the MMC4 AC side. If the MMC3 selects the constant AC voltage and frequency control mode, the MMC4 can choose the constant power control or the constant DC voltage control mode. If MMC3 selects constant power control, MMC4 can only select constant AC voltage frequency control mode.

MMC4 is the sample converter. There are three control modes: constant power control, constant AC voltage and frequency control, and constant DC voltage control. MMC4 control mode needs to cooperate with MMC2 and MMC3.

3.2 *Simulation Verification of Typical Operating Conditions*

(1) **DC/DC test object mode**
Adjust the parameters of the test object to control its DC side voltage to 20 kV and power to 12 MW. Wait for it to run stably, and then control it to produce an overload condition of 1.2 times the rated load with a duration of 0.5 s. After the overload is over, it runs stably at 12 MW again, and the simulation waveform is shown in Figs. 7 and8. In the whole test process, the maximum value of the overall active power loss of the test circuit measured does not exceed 2.5 MW, as shown in Fig. 8.

(2) **DC/AC (three-phase) test object mode**
Adjust the parameters of the test object to control its DC side voltage to 20 kV, AC side voltage to 10 kV and power to 12 MW. Wait for it to run stably, and then control it to produce an overload condition of 1.2 times the rated load with a duration of 0.5 s. After the overload is over, it runs stably at 12 MW again, and the simulation waveform is shown in Figs. 9 and10. In the whole test process, the maximum value of the overall active power loss of the test circuit measured does not exceed 1.2 MW, as shown in Fig. 10.

Keep the main wiring status of the simulation system unchanged. Adjust the parameters of the test object to control its DC side voltage to 20 kV, AC side voltage to 10 kV and power to −12 MW. Wait for it to run stably, and then control it to produce an overload condition of 1.2 times the rated load with a duration of 0.5 s. After the overload is over, it runs stably at −12 MW again, and the simulation waveform is shown in Figs. 11 and 12. In the whole test process, the maximum value of the overall active power loss of the test circuit measured does not exceed 0.8 MW, as shown in Fig. 12.

(3) **DC/AC (single -phase) test object mode**
Adjust the parameters of the test object to control its DC side voltage to 20 kV, AC side voltage to 10 kV and power to 4 MW. Wait for it to run stably, and

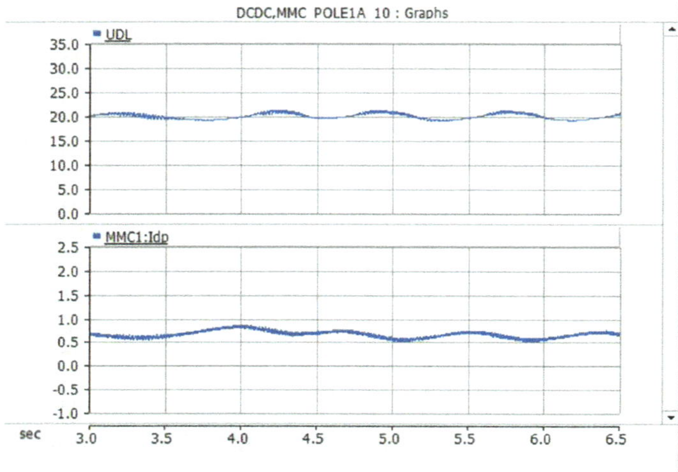

(a)Voltage and current on the input side of the test object

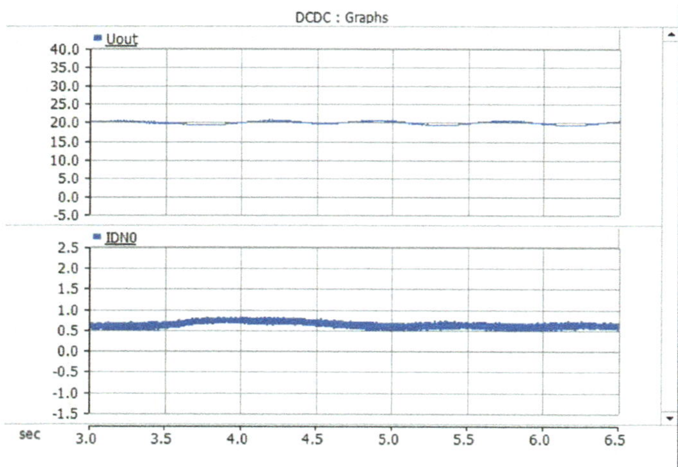

(b) Voltage and current on the input side of the test object

Fig. 7 Voltage and current of the test object

then control it to produce an overload condition of 1.2 times the rated load with a duration of 1 s. After the overload is over, it runs stably at 4 MW again, and the simulation waveform is shown in Figs. 13 and14. In the whole test process, the maximum value of the overall active power loss of the test circuit measured does not exceed 0.5 MW, as shown in Fig. 14.

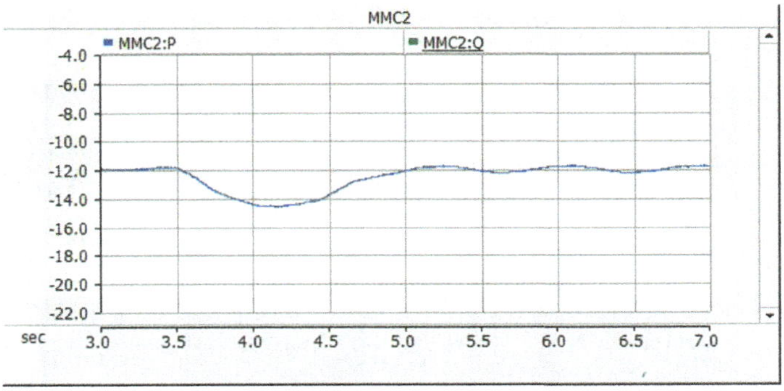

(a) Active power of the test object

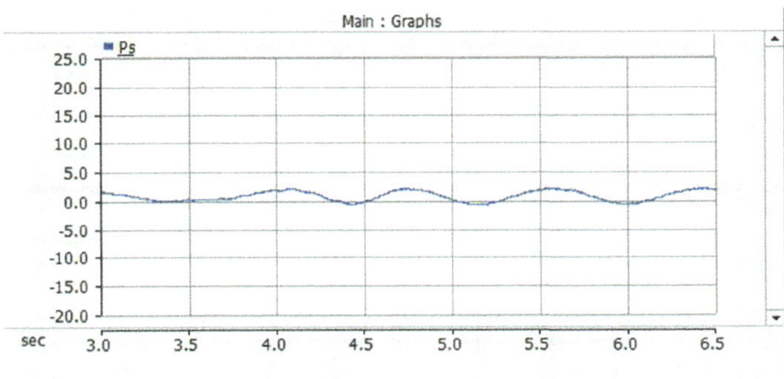

(b) Active power loss of test circuit

Fig. 8 Active power of the test

Keep the main wiring status of the simulation system unchanged. Adjust the parameters of the test object to control its DC side voltage to 20 kV, AC side voltage to 10 kV and power to −4 MW. Wait for it to run stably, and then control it to produce an overload condition of 1.2 times the rated load with a duration of 1 s. After the overload is over, it runs stably at −4 MW again, and the simulation waveform is shown in Figs. 15 and16. In the whole test process, the maximum value of the overall active power loss of the test circuit measured does not exceed 0.8 MW, as shown in Fig. 16.

4 Conclusion

The simulation results show that the test system proposed in this paper is based on a full/half-bridge hybrid MMC converter. Through the coordination of different control strategies and operation modes of three MMC converters, various types of test power

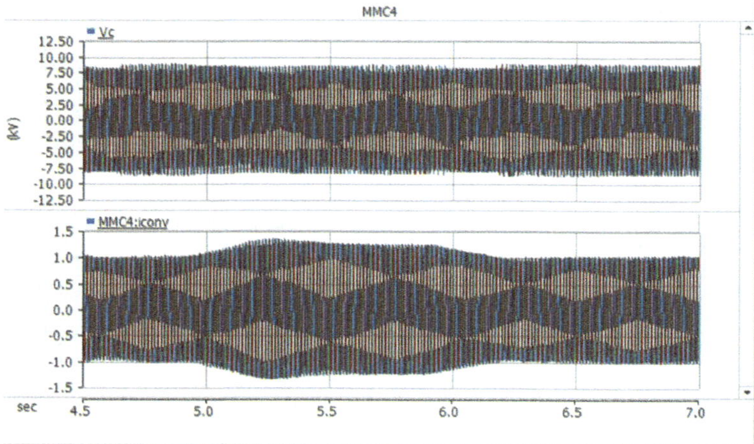

(a)Voltage and current on the AC side of the test object

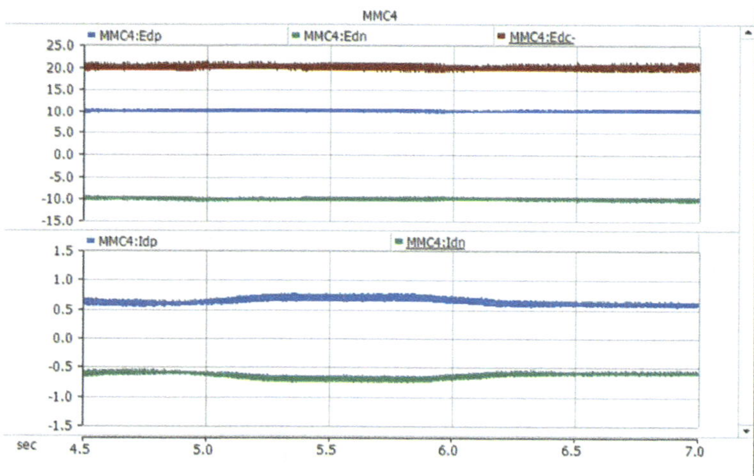

(b) Voltage and current on the DC side of the test object

Fig. 9 Voltage and current of the test object

supply functions and test load functions are realized. It can meet the electrical test requirements of DC/DC converters, DC/AC converters and AC/DC converters with multiple types and large-capacity converter products while the overall test circuit consumes less active power. At the same time, based on the real-time control of the MMC converter power, the test system scheme can flexibly realize the real-time exchange of test power and test load to meet the test requirements of dual-power flow direction products. And it can verify the performance of the test product under rated conditions and load fluctuation conditions.

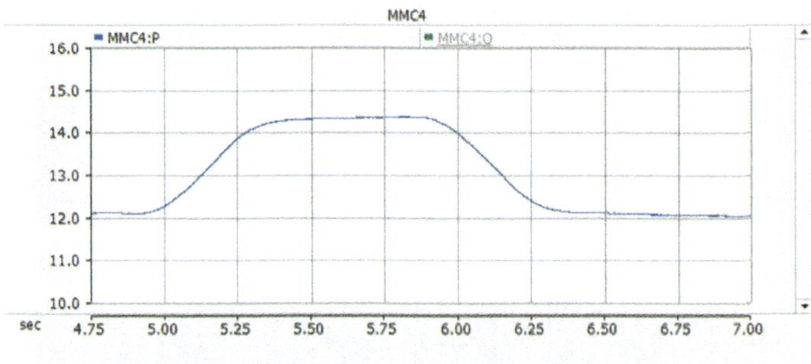

(a) Active power of the test object

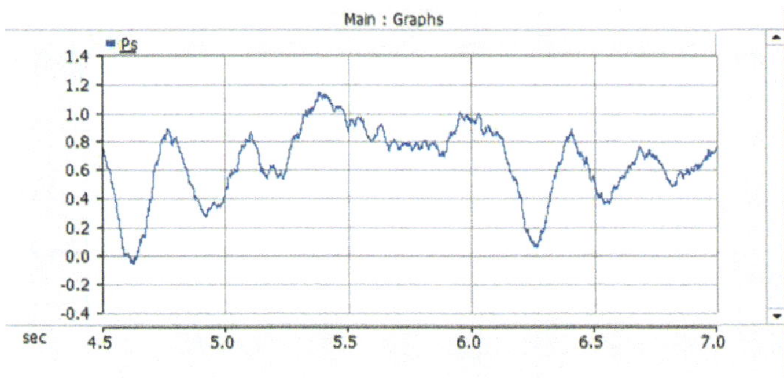

(b) Active power loss of test circuit

Fig. 10 Active power of the test

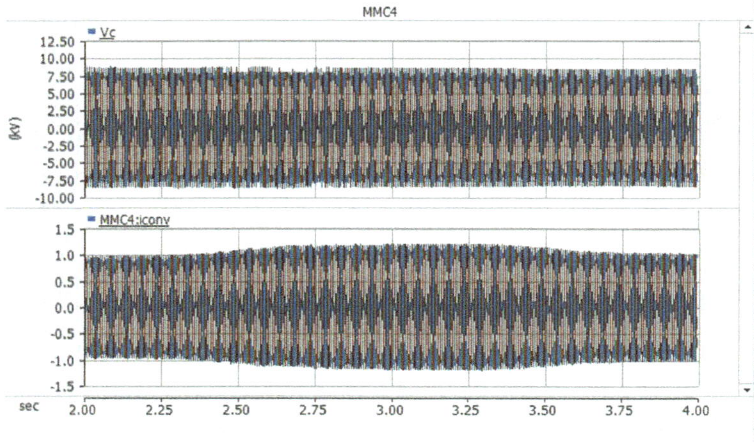

(a)Voltage and current on the AC side of the test object

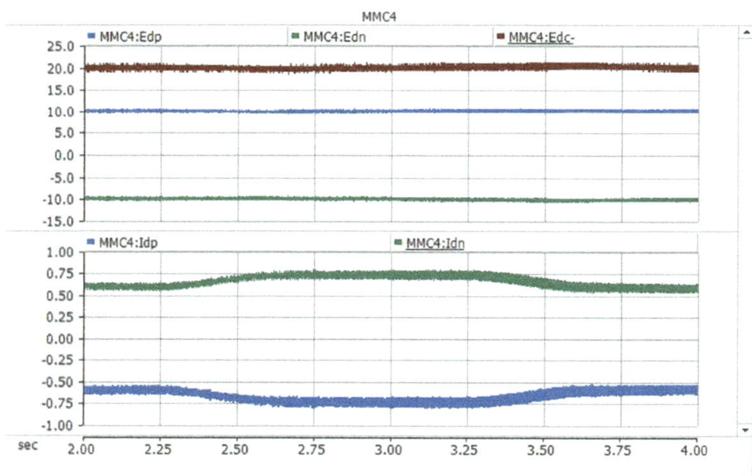

(b) Voltage and current on the DC of the test object

Fig. 11 Voltage and current of the test object

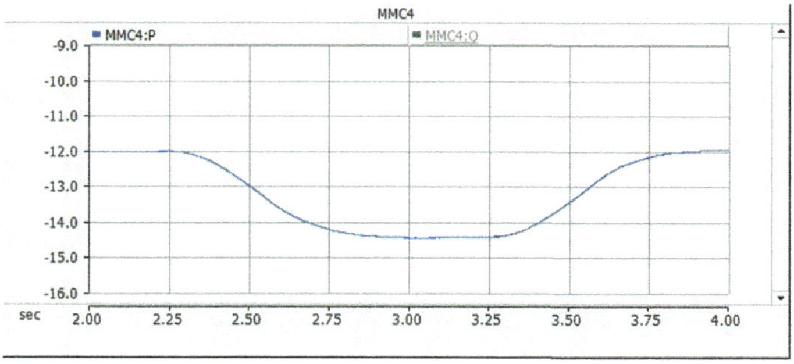

(a) Active power of the test object

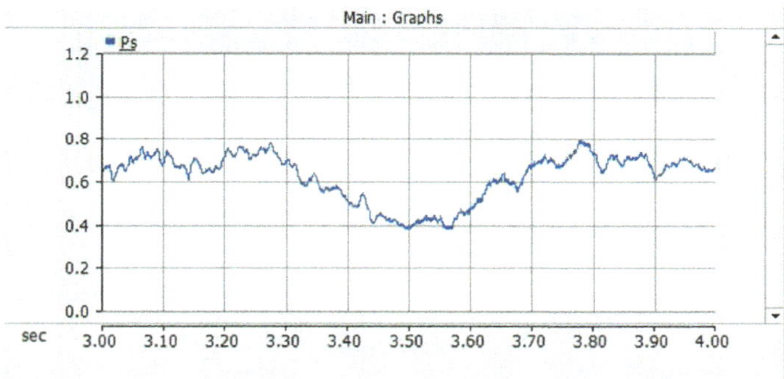

(b) Active power loss of test circuit

Fig. 12 Active power of the test

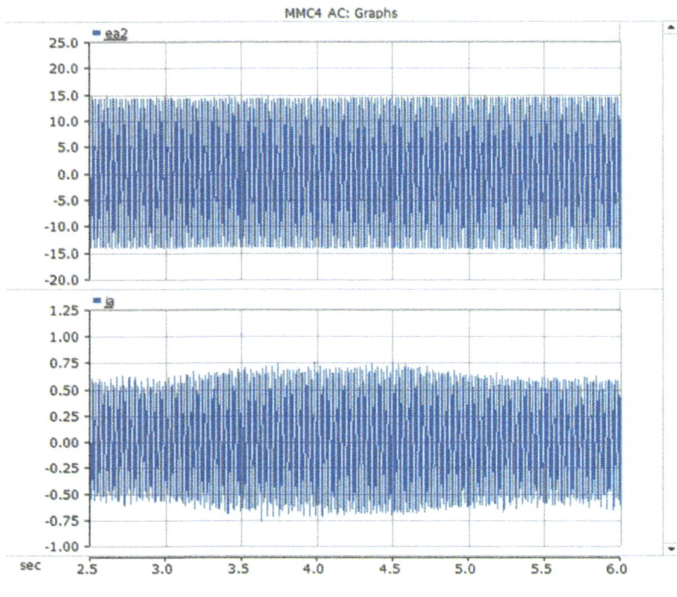

(a) Voltage and current on the AC side of the test object

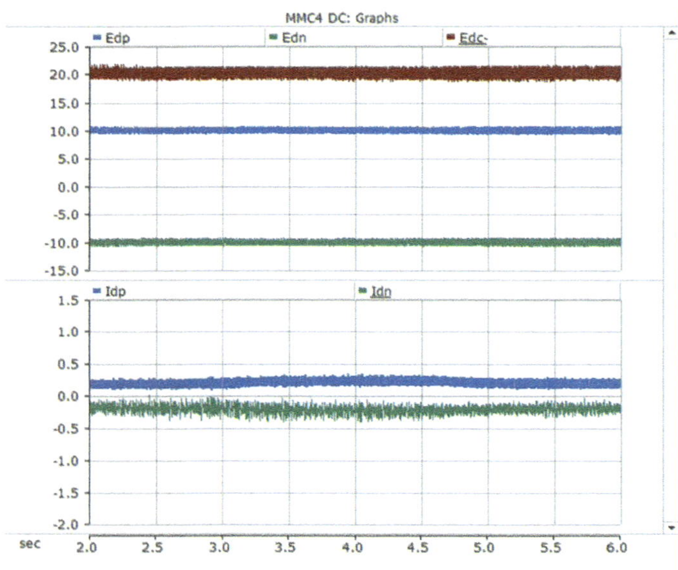

(b) Voltage and current on the DC side of the test object

Fig. 13 Voltage and current of the test object

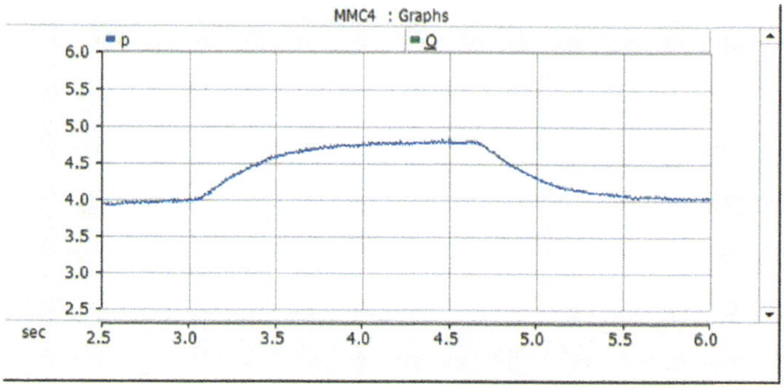

(a) Active power of the test object

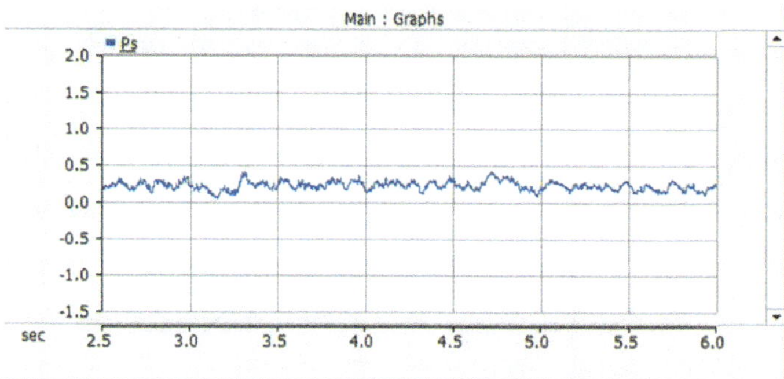

(b) Active power loss of test circuit

Fig. 14 Active power of the test

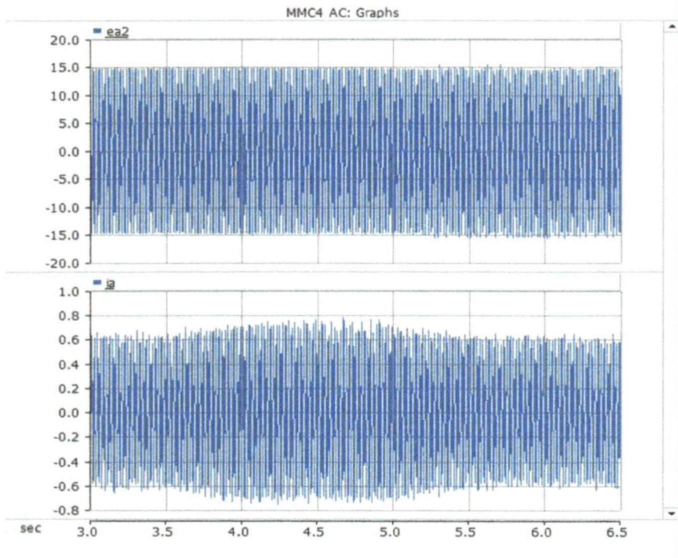

(a)Voltage and current on the AC side of the test object

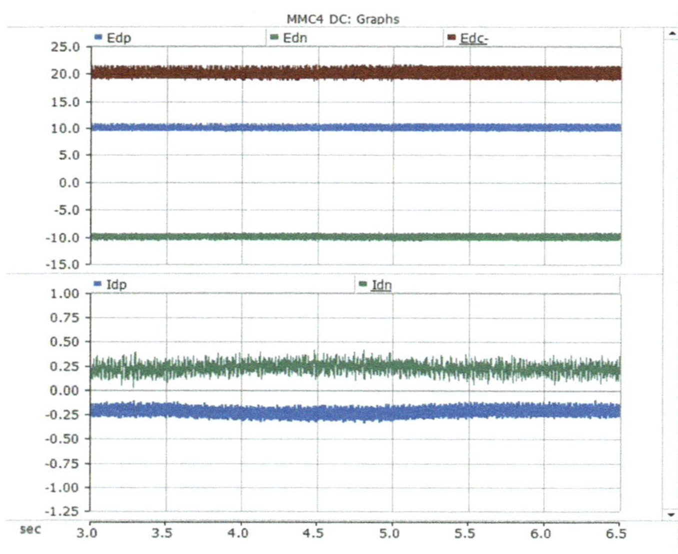

(b) Voltage and current on the DC side of the test object

Fig. 15 Voltage and current of the test object

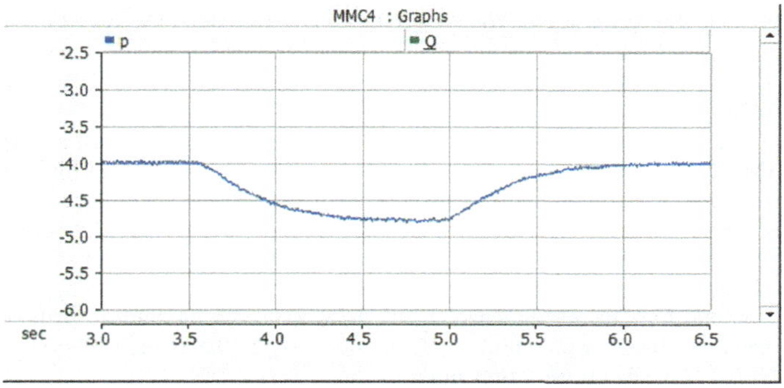

(a) Active power of the test object

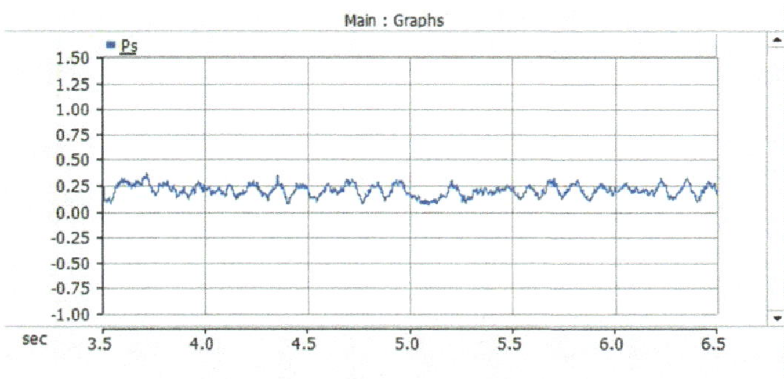

(b) Active power loss of test circuit

Fig. 16 Active power of the test

References

1. Liu, Zhenya. 2016. Research of global clean energy resource and power grid interconnection. *Proceedings of the CSEE* 36(19): 5103–5110 (in Chinese).
2. Wang, Mingfu, Huahua Wu, Linhua Yang, et al. 2020. Development status and prospects of the energy Internet in electricity markets. *POWER DSM* 22(2): 1–7 (in Chinese).
3. Sheng, Wanxing, Qing Duan, Ying Liang, et al. 2015. Research of power distribution and application grid structure and equipment for future energy internet. *Proceedings of the CSEE* 35(15): 3760–3769 (in Chinese).
4. Pu, Tianjiao, Kewen Liu, Naishi Chen, et al. 2015. Design of ADN based urban energy internet architecture and its technological issues. *Proceedings of the CSEE* 35(14): 3511–3521 (in Chinese).
5. Zeng, Ming, Yongqi Yang, Yuanfei Li, et al. 2016. The preliminary research for key operation mode and technologies of electrical power system with renewable energy sources under energy internet. *Proceedings of the CSEE* 36(3): 681–691 (in Chinese).
6. Cao, Junwei, Mingbo Yang, Dehua Zhang, et al. 2014. Energy internet: An infrastructure for cyber-energy integration. *Southern Power System Technology* 8(4): 1–10 (in Chinese).
7. Song, Qiang, Biao Zhao, Wenhua Liu, et al. 2013. An overview of research on smart DC distribution power network. *Proceedings of the CSEE* 33(25): 9–19 (in Chinese).

8. Liu, Ruihuang, Jinggang Yang, Yongyong Jia, et al. 2019. engineering application of DC transformer in medium voltage DC distribution network. *Autumation of Electric Power System* 43(23): 131–140 (in Chinese).
9. GB/T 13422. 2013. *Semiconductor converters-electrical test methods*.
10. Sun, Li, Feng Hu, Chenglian Ma, et al. 2017. Research on carrier phase shift modulation mode of the MMC inverter. *Power System and Clean Energy* 33(10): 56–62 (in Chinese).
11. Sun, Changjiang, Jianwen Zhang, Xu Cai, et al. 2016. Current-fed operation of isolated MMC-based DC transformer. *Proceedings of the CSEE* 36(7): 1977–1986 (in Chinese).
12. Qiang, Song, and Rao Hong. 2015. *Analysis and design of the converters in VSC-HVDC system*. Beijing: Tsinghua University Press (in Chinese).
13. Lu, Shi, Jialong Li, Chengyong Zhao, et al. 2018. Research on control method of voltage balance and fault clearance of hybrid MMC composed of D-HBSM and P-FBSM. *Proceedings of the CSEE* 38(21): 6411–6419 (in Chinese).

Study on Aging Characteristics and Products of Liquid Insulating Medium Under Partial Discharge

Zhihao Chen, Zhichun Qiu, Yongqiang Fu, and Ruobing Zhang

Abstract The accident happened on cable termination often starts from partial discharge which will cause electrical aging of insulating silicone oil. Therefore, the research on the electrical aging characteristics and products of the insulating silicone oil under partial discharge is significant. In this paper, a dielectric barrier discharge with pin plate electrodes is used to simulate partial discharges under real conditions. The electrical characteristics of the silicone oil and the products formed before and after the discharge are analyzed. It is found that prolonged partial discharges will significantly alter the dielectric properties of insulating silicone oils. The dielectric loss factor and the relative permittivity will increase and the volume resistivity will be greatly reduced. At the same time, gel-like substances polymerized by Si–O bonds and gases such as hydrogen and carbon monoxide were produced. Finally, changes at the molecular level were analyzed on the basis of the products and their formation and electrical aging properties were explained. In addition, the gas product characteristics were analyzed using the three-ratio method.

Keywords Insulating silicon oil · Partial discharge · Dielectric properties · Products

1 Instruction

With increasing urbanization, underground high-voltage power cable equipment has become a major transmission method for saving urban space. In the past ten years, the average growth rate of power cable laying length has exceeded 15% [1]. However, some problems still exist in the use of high-voltage power cables, including equipment manufacturing problems, human problems of improper installation and maintenance, over-voltage and overload operation of cables, and insulation problems caused by external forces [2–5]. In addition to human factors (e.g. construction problems),

Z. Chen · Z. Qiu · Y. Fu · R. Zhang (✉)
Tsinghua Shenzhen International Graduate School, Tsinghua University, Shenzhen, China
e-mail: zhangrb@sz.tsinghua.edu.cn

W. Chen et al. (eds.), *The Proceedings of the 9th Frontier Academic Forum of Electrical Engineering*, Lecture Notes in Electrical Engineering 743,
https://doi.org/10.1007/978-981-33-6609-1_24

the proportion of failures and accidents caused by cable terminals and their accessories has exceeded 20% [6]. According to the simulation of related literature, the electric field distribution at the cable terminals is very nonuniform due to the complex internal structure of the terminals [7, 8]. And the semi-conductive areas at the cable terminals and joints are subject to strong electrical stresses. To solve these problems, the oil-filled cable terminals are equipped with a stress cone between the edge of the metal sheath inside the terminal and the outer surface of the added winding insulation, and low-viscosity insulating silicone oil is used as the coolant and insulating medium to fill it. Currently, the use of oil-immersed terminals is relatively high due to more mature technology and long experience with them. Therefore, it is of great significance to study the aging characteristics of low-viscosity insulating silicone fluid. In recent years, several accidents and failures have occurred due to the aging of the insulating silicone oil in oil-filled terminals.

The causes of terminal failures of Gas Insulated Switchgear (GIS) have been analyzed by relevant studies combined with actual engineering cases [9–15]. Among them, the main reasons for the insulation problem are the mismatching of cable accessories and cable pretreatment dimensions, poor fit between the stress cone and the cable surface, uneven surface of the stress cone, and insufficient insulation between the stress cone and the epoxy casing. Problems such as not being full of insulating medium between insulation defects can occur, easily lead to the occurrence of discharges, causing silicone oil aging and insulation performance degradation (Fig. 1). And the silicone oil in the GIS equipment will leak because of the deviation of the conductor and the terminal installation, the insufficient pre-tightening force of the end bolts, and the external load. Silicone oil will be exposed to the air and age quickly. Preliminary research has been done on the initial discharge voltage, gas and solid products of silicone oil when partial discharge occurs in insulating oil [16, 17]. However, the electrical performance before and after aging has not been further studied. Partial discharge is relatively unobvious, and its direct impact is relatively weak, so there is no more specific and systematic research in this area. However,

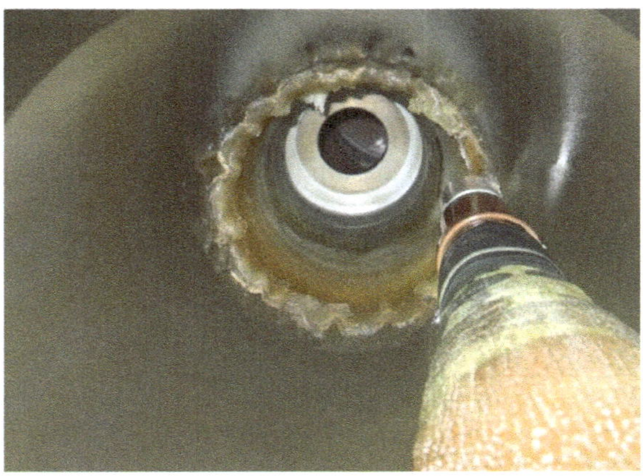

Fig. 1 Appearance diagram of gel polymer in aging terminal

the effect of partial discharge on the electrical parameters of silicone oil is likely to become cause of serious accidents for a certain length of time. Therefore, the study on the influence of the electrical properties silicone oil under partial discharge has strong practical significance and lays the foundation for the analysis of the causes of related accidents.

In this paper, we simulate the weak electrical aging process of silicone oil using the form of dielectric barrier discharge, measure the electrical parameters and physicochemical parameters of silicone oil after aging, collect and analyze the dissolved gas components and other aging products generated in silicone oil, and study the rules of silicone oil aging products and the microscopic changes of silicone oil aging process under partial discharge.

2 Introduction to Test Samples and Test Methods

In this paper, low viscosity TR50 silicone insulating fluids from Wacker Germany were chosen as the sample, which are the most common type used in cable GIS terminals in power systems [18]. TR50 is a completely synthetic, high-purity, low-viscosity silicone oil that is not corrosive and volatile. It is a colorless, odorless, and transparent non-polar liquid. The chemical composition of TR50 is mainly polydimethylsiloxane. Its molecular formula is shown in Fig. 2.

In the test, 150 mL of crude silicone oil and anaerobic treated oil were respectively poured into the single needle-plate electrode oil cup for treatment. In order to simulate the partial discharge condition, a piece of quartz glass was added between the needle electrode and plate electrode to block the discharge, so that the discharge between the needle and plate electrode could not form a complete discharge channel. The thickness of the quartz glass sheet is 3 mm, the distance between the needle electrode and the plate electrode is 6 mm, and a voltage of 24 kV is applied to the needle electrode. For crude oil, the top of the oil cup is connected to the outside air through a vent hole. For anaerobic treated oil, the air in the upper layer of the oil cup will be removed by nitrogen substitution. And a pressure gauge to detect the internal pressure is connected to the top vent. The experimental circuit consists of a voltage regulator, a test transformer, a protective resistor, a needle-plate electrode oil cup, a high frequency voltage test device and a partial discharge measurement circuit. The partial discharge measurement loop is composed of high-voltage coupling capacitors,

Fig. 2 Polydimethylsiloxane

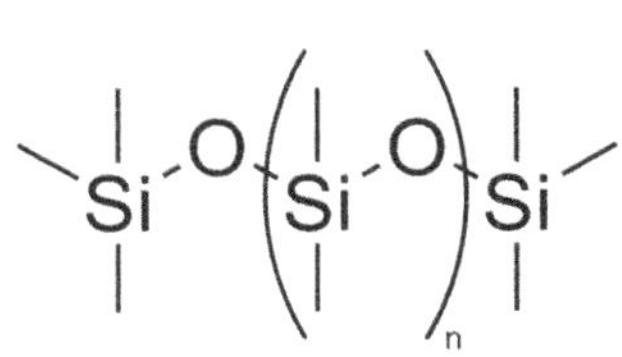

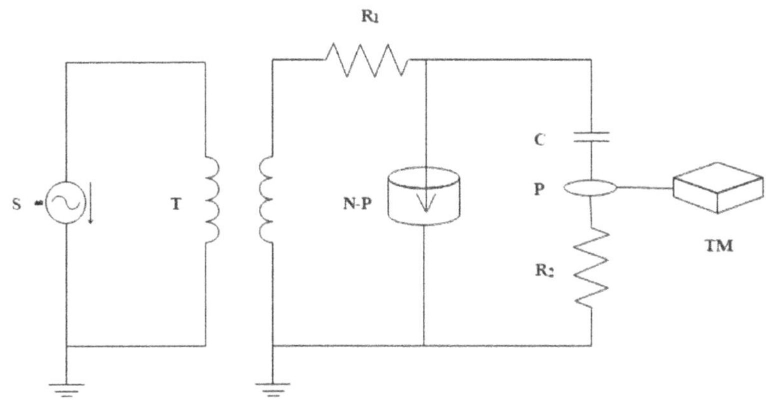

Fig. 3 Experiment circuit schematic. From left to right, S is power frequency AC voltage source, T is transformer, R_1 is protection resistor, N-P is the needle-plate electrode oil cup, C is high-voltage coupling capacitors, R_2 is non-inductive resistors and P is high voltage probe, TM is terminal (oscilloscope or acquisition card)

non-inductive resistors and high-frequency CT, as shown in Fig. 3. The design of the measuring circuit refers to the standard GB/T 7354-2003.

Dielectric loss factor, volume resistivity and relative permittivity are detected by insulating oil dielectric loss tester. The kinematic viscosity of silicone oil is measured by Pinger's viscometer with references to the standard HG/T 2363-1992. The acid value is measured using a basic burette, with alkali blue 6B as an indicator and KOH solution as a titrant for strong alkali titration of weak acids. The gas products were extracted by the headspace gas extraction method in the standard GB/T 17623-1998, and then analyzed by gas chromatograph. The chromatographic column is a special column for transformer oil analysis and a GS-GASPRO capillary column [19]. The impurities generated after aging are analyzed by a Fourier infrared spectrometer.

3 Experiment Result

3.1 Discharge Phenomenon and Discharge Volume

The dielectric barrier discharge has a very fine white filament discharge channels, as shown in Fig. 4, which can be observed directly by the naked eyes only under black background and in dim light conditions. The adjacent discharge time intervals between two discharges is relatively long. After the discharge lasts for several days, no obvious changes can be observed with the naked eye in the silicone oil sample. A knocking sound could be heard on the quartz glass sheet during discharge. We speculate that it's the sound of the stream that develops on the needle electrode hitting the glass. It is worth mentioning that during the test, no bubbles were observed to

Fig. 4 Discharge phenomenon

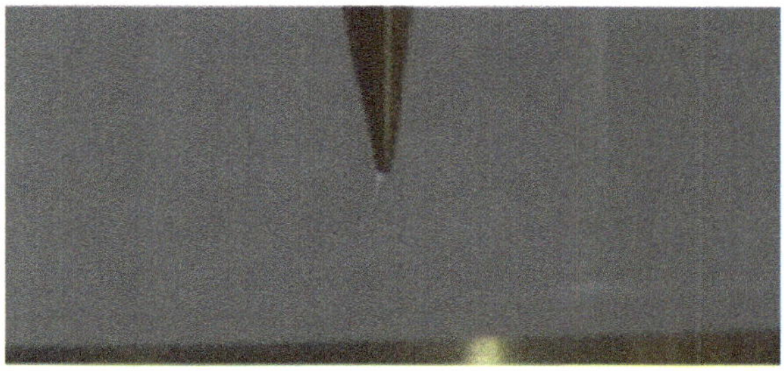

be generated and escaped from the liquid surface. This is because the discharge energy of the dielectric barrier discharge is very weak, and the rate of gas products generated during discharge is not enough to make the gas generated to form bubbles. So, it can be assumed that the gas products generated by dielectric barrier discharge are completely dissolved in silicone oil.

The partial discharges are relatively very weak and the amount of each discharge is small. Figure 5 shows the voltage waveform on the inductive resistor in the RC circuit during partial discharge. Due to the small partial discharge, there is no obvious change in the voltage waveform of the needle electrode during discharge, so only the partial discharge pulse can be seen on the filtered inductive resistor. The amplitude of the discharge pulse is usually only a few tens of volts in the inductive resistor. In this order of magnitude, in addition to the discharge pulse, a power frequency sinusoidal signal that has been reduced by a thousand times can also be seen.

Due to the small amount of partial discharge, the temperature in the processing chamber was always around 27 °C room temperature during the experiment.

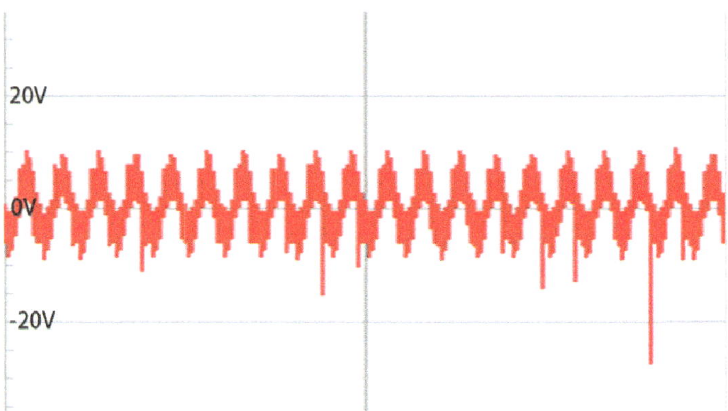

Fig. 5 Voltage waveform on non-inductive resistor

Table 1 Silicone oil parameter change after DBD

Parameters	Test groups		
	Crude oil	P0	P1
Test duration/h	0	118	155
Cumulative discharge/10^{10}pC	0	2.34	2.09
Dielectric dissipation factor%	0.003	0.008	0.008
Relative permittivity	2.532	2.535	2.541
Volume resistivity/$10^{11}\,\Omega$ m	90.2	18.73	20.00
Kinematic viscosity/mm^2 s^{-1}	118.82	118.69	119.36
Acid number/mgKOH g^{-1}	0.0039	0.0071	0.0341

3.2 Changes in Electrical Parameters of Silicone Oil

The parameter changes of silicone oil before and after dielectric barrier discharge under anaerobic and aerobic conditions are shown in Table 1, where P0 is the result under anaerobic conditions, and P1 under aerobic conditions. We can see that the parameters of silicone oil have changed significantly after a longer period of discharge treatment. For the insulation related parameters, the changes under anaerobic and aerobic conditions are basically the same: the dielectric loss tangent value has increased substantially, while the volume resistivity has decreased to a large extent. In the anaerobic test P0, the kinematic viscosity of silicone oil almost did not change. And in the aerobic test P1, there was a slight increase in kinematic viscosity. We speculate that this is because the oxygen in the air participates in the aging process, causing cross-linking between silicone oil molecules and enhancing the force between molecules. At the same time, the acid value of silicone oil does not change significantly under anaerobic conditions but varies greatly under aerobic conditions.

3.3 Gas Products

As shown, Fig. 6a is the gas generation product of the dielectric barrier discharge of silicone oil in the absence of oxygen. The two gases with the largest production volume are hydrogen and carbon monoxide, respectively, methane and carbon dioxide production volumes are small, while ethane, ethylene and acetylene production volumes are extremely small, almost undetectable. The reason is that the energy density of DBD is low, affecting the generation of ethylene and acetylene. On the other hand, however, the generation of C_2 hydrocarbons is an indication of relatively high energy density, which is consistent with the characteristics of DBD, that is, during DBD, there is high energy density near the needle-plate electrode and low electric field strength and low energy density in most other regions.

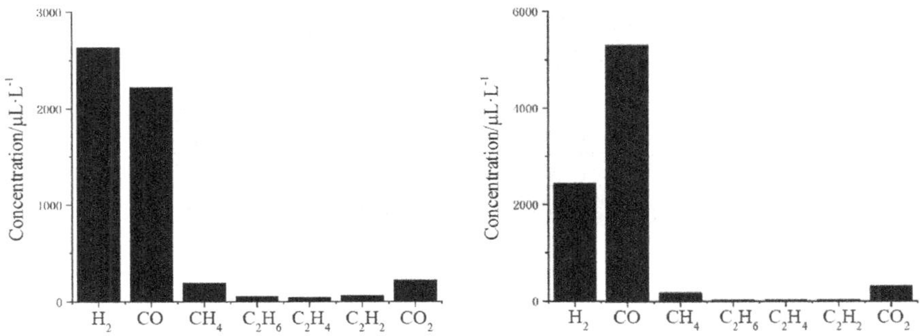

(a) Gas products under anaerobic condition (b) Gas products under aerobic condition

Fig. 6 Gas products

Figure 6b shows the gas products of crude oil during experiment under aerobic condition. The largest production is still carbon monoxide and hydrogen, but under the influence of oxygen, the carbon monoxide production is higher than that of hydrogen under anaerobic conditions. The amount of hydrogen produced in the presence of oxygen does not differ much from that in the absence of oxygen. For the two gas products with relatively low generation, methane generation was slightly lower than in the anaerobic test, while carbon dioxide generation increased somewhat. Ethane, ethylene, acetylene and anaerobic conditions were similar, with very low production.

Combined with the results of the gas products, it can be concluded that the molecular level changes in the dielectric barrier discharge include the breakage of C-H bond in the Si-CH3 structure, resulting in free hydrogen atoms; at the same time, the Si-CH3 bond will also be broken, resulting in free CH3* groups; in addition, the Si–O bond will also be broken in the dielectric barrier discharge, resulting in free O oxidized CH3* groups. The oxidation of CH3* groups also occurs more frequently under aerobic conditions. The free CH2* and CH* groups produced during the dielectric barrier discharge are rare.

3.4 Gel Polymer Analysis Results

After the end of discharge aging under oxygen enriched condition, the silicone oil remained virtually unchanged in appearance, but there was a gel polymer attachment at the needle electrode as shown in Fig. 7, similar to the gel at the actual accident site. And we were not able to collect any gel polymer in the experiment under lean oxygen condition.

The results of the gel analysis using Fourier infrared spectrometer are shown in Fig. 8. It can be seen that there are elastic vibration absorption peaks of CH_3 in the range of 3000–2800 cm^{-1}, deformation vibration absorption peaks of CH_3 in the

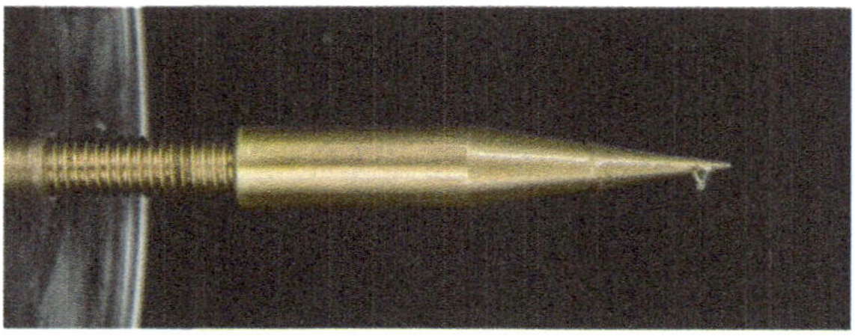

Fig. 7 Gel polymer

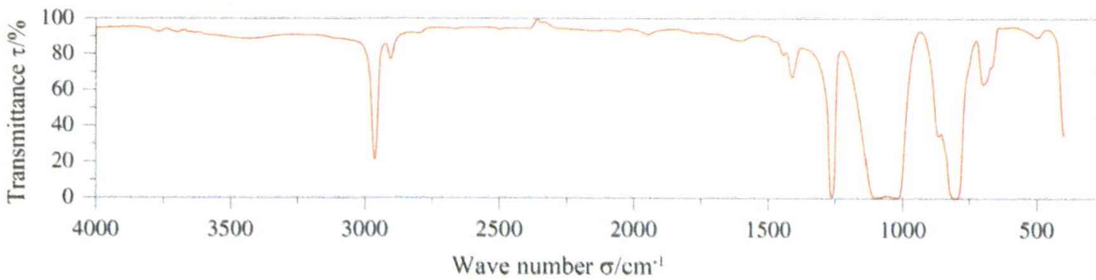

Fig. 8 Gel polymer Fourier infrared spectrum

range of 1300–1200 cm^{-1}, swing vibration absorption peaks of CH_3 near 800 cm^{-1}, absorption peak of Si-HR group near 2300 cm^{-1} and absorption peak of Si–C in the range of 700–600 cm^{-1}. Especially for the absorption peaks corresponding to Si–O bonds in the wavelength range of 1130–1000 cm^{-1}, the width of the peaks and the splitting of the peaks can be distinguished, although the intensity of the absorption peaks of the gel polymer is larger due to the insufficient doping of potassium bromide.

The above results indicate an increase in the degree of polymerization and chain length. The increase in Si–O bonds at the end of the silicone oil is probably due to the polymerization of Si–O bonds at the end of the long chains or the formation of cyclodimethylsiloxane units. The gradual increase of the viscosity of the silicone oil from a liquid to a gel polymer is caused by the polymerization of the Si–O bonds into long chains or even a network structure.

4 Discussion

4.1 Reaction Principle Discussion

It is generally believed that the main chemical reaction under weak discharge is caused by electron excitation and electron impact with energy less than 10 eV [20]. These reactions can lead to the breaking of C–H, Si–C and Si–O bonds, with C–H bond energies of 338 kJ/mol, Si–C bond energies of 347 kJ/mol and Si–O bond energies of 460 kJ/mol. According to the results of the gel polymer infrared spectroscopy

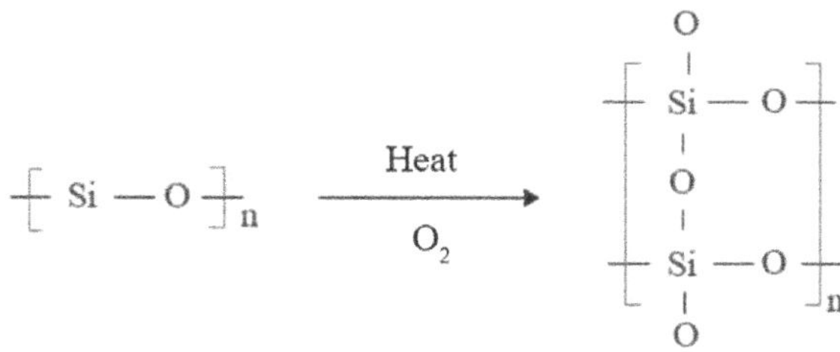

Fig. 9 Structural change of silicone oil under partial discharge

and gas products, it can be concluded that in DBD, the most important change in the molecular level is the breakage of the Si–C bond and the C-H bond, which will produce a large number of CH_3* groups and free hydrogen ions. In addition, the Si–O bond will also be broken and free O will be generated, so that CH_3* will be oxidized under lean oxygen condition to generate CO and CO_2. Under oxygen enriched conditions, more CH_3* groups are oxidized to generate COOH groups, which leads to an increase in the acid value of silicone oil. During the DBD process, due to the high local energy density, smaller amounts of free CH2* and CH* groups are also produced, resulting in the formation of ethylene and ethane.

The characteristics of infrared spectra show that the sample polymerization degree and chain length increase with the participation of oxygen. The polymerization of Si–O bonds at the end of the long chain or the formation of cyclodimethylsiloxane units may cause the increase of Si–O bonds. And the gradual increase in viscosity of silicone oil from liquid to gel is caused by the polymerization of Si–O bonds in silicone oil into long chains or even network structures. It can be concluded that the production of gel polymer in the experiment and at the accident site are similar, that is, after a long period of weak partial discharge, some of the Si–C bonds in the silicone oil will be broken, and the remaining substances will undergo oxidative polymerization with oxygen in the air to form a network of high polymer. The structural changes are shown in Fig. 9. Therefore, when the high energy density of partial discharge leads to a large number of Si–O bond breaks, the gel polymer will be difficult to form in large quantities.

4.2 Comparative Analysis of On-Site Aging Silicone Oil

In this section, the gas product results of aged silicone oil in the test and at the engineering site will be compared and analyzed. Table 2 shows the analysis results of gas products in some problematic GIS terminals at the accident site. In addition, the silicone oil inside the four terminals of 1B, 1C, 2A, and 2B all have yellowing phenomenon, and there are different amounts of gels and adhesions inside. Among them, there are only a small amount of adhesions in phase 1B, more gels are formed in phases 1C and 2A, and a large amount of gel polymers and adhesions are found in

Table 2 Gas products of on-site aging silicone oil

Samples	CH_4	C_2H_4	C_2H_6	C_2H_2	H2	CO
Crude oil	3.25	0.81	0.29	0.00	2.46	2.65
X-1B	801.7	1.08	150.6	0.00	3993	194.0
X-1C	726.8	1.38	2059	0.00	9741	34.61
X-2A	914.7	0.89	2264	0.00	10,388	27.01
X-2B	695.3	42.56	2573	21	8560	18.11

2B. It has been diagnosed that partial discharges of different levels occurred in these four terminals for a long time.

The gas generation characteristics in the DBD test are consistent with the project site. The CH_4/H2 ratio is close to the results of the partial discharge test, and the generation of acetylene is extremely low, with only a small amount of generation in phase 2B. The hydrocarbon substance most generated in the silicone oil inside the phase 1B terminal is methane, and this result is consistent with the test, but with the difference that there is a large amount of ethane generation in phases 1C, 2A and 2B. At the same time, CO generation was relatively low. Comparing the phenomena at the accident site with those in the test, it can be found that the common phenomena of these groups at the accident site are that the silicone oil turns yellow and there is less gel polymer in phase 1B, while more gels or adhesions are generated in phases 1C, 2A and 2B. But the silicone oil hardly changes color in the partial discharge test and the amount of gels generated is also small. It can be concluded that the local discharge energy in 1C, 2A and 2B phases was smaller, so the Si–O bond could not be broken, but the Si–C bond was still broken, thus more gels were generated by Si–O polymerization in the near oxygen-free environment. At the same time many free CH_3* groups were not oxidized due to lack of O, thus generating a large amount of ethane and a relatively small amount of CO. Whereas the energy of the 1B phase and the DBD used to imitate the partial discharge during the test is relatively large, so that there is less polymer and more CO.

From the above analysis, the aging phenomenon of silicone oil under partial discharge at the engineering site is consistent with the conclusions obtained from the test. The overall level of aging of silicone oil can be characterized by the amount of CH_4 and H2 generated. The aging energy density can be reflected by the amount of ethane, CO and gel polymer generated. And the local aging energy can be reflected by ethylene, acetylene and other products.

4.3 Analysis of the Judgment Results of Gas Products by Three Ratio Method

The three-ratio method is commonly used to determine the type of fault in transformer oil. It is based on the thermodynamics and is summarized by a large amount of

Table 3 The code obtained by the diagnosis of gas products by the three ratio method

	C_2H_2/C_2H_4	$CH_4/H2$	C_2H_4/C_2H_6
Anaerobic	1	1	0
Aerobic	1	0	0

engineering practice experience, which can more accurately judge the type of fault that occurs in the transformer. The three-ratio analysis of the gas products obtained in the experiments in this paper can get the codes shown in Table 3.

The fault types obtained according to this code are all arc discharges, which are different from the results in this article. Therefore, it can be considered that the three-ratio method is not suitable for the diagnosis of silicon oil partial discharges.

5 Conclusion

(1) The main components of the gas production products of silicon oil under partial discharge are hydrogen and carbon monoxide, and the minor components are methane and carbon dioxide. The production of ethane, ethylene and acetylene is very small.

(2) Partial discharge will cause the break of the C–H, Si–C and Si–O bonds in the silicone oil, resulting in CH_3* groups and free hydrogen and oxygen. In an oxygen-free environment, the CH_3* group will be oxidized by the free O formed by the cleavage of Si–O to produce CO and a small amount of CO_2; in an oxygen environment, CH_3* will be oxidized to form a COOH group, leading to an increase in the acid value of silicone oil.

(3) Under the local high temperature caused by the weak partial discharge due to the insulation defect, some of the Si–C bonds in the silicone oil will be broken, and the remaining substances will undergo oxidative polymerization with the oxygen in the air to form gel polymer. The gel polymer will be difficult to form when the partial discharge energy density is high enough to cause a large number of ruptures of the Si–O bond.

(4) The overall extent of silicone oil aging can be characterized by the amount of CH_4 and H2 generated. The aging energy density can be reflected by the amount of ethane, CO and gel polymer generated. And the local aging energy can be reflected by ethylene, acetylene and other products.

(5) The three-ratio method to judge the fault type of transformer oil is not suitable for the diagnosis of silicon oil partial discharges.

References

1. Zhou, Yuanxiang, Jiankang Zhao, Rui Liu, et al. 2014. Key technical analysis and prospect of high voltage and extra-high voltage power cable. *High Voltage Engineering* 40 (9): 2593–2612 (in Chinese).
2. Li, Lulu, Jing Yong, Liqiang Zeng, et al. 2018. On-line monitoring of insulation overall aging for cross-bonded cables based on system power disturbances. *Transactions of China Electrotechnical Society* 33 (14): 3396–3405 (in Chinese).
3. Huang, Guanglei, Zhe Li, Fengyuan Yang, et al. 2019. Experimental research on leakage current of DC cross linked polyethylene cable. *Transactions of China Electrotechnical Society* 34 (1): 192–201 (in Chinese).
4. Xie, Shengyi, Fan Yang, Xin Huang, et al. 2020. Air gap detection and analysis of XLPE cable insulation based on terahertz time domain spectroscopy. *Transactions of China Electrotechnical Society* 35 (12): 2698–2707 (in Chinese).
5. Wang, Chuanxu. 2014. The high voltage cable fault and its condition detection technique. *Electrical Engineering* 15 (09): 70–73 (in Chinese).
6. Xun, Jun. 2004. Analysis of cable fault in construction and operation. *High Voltage Engineering* z1: 89–93 (in Chinese).
7. Liu, Hui, Qinghe Shen, Yufeng Chen, et al. 2014. Research on corona interference in partial discharge detection on GIS cable end. *Electrical Measurement and Instrumentation* 51 (12): 44–49 (in Chinese).
8. Wang, Shaohua, Ye Ziqiang, Bingxiao Mei, et al. 2011. Causes of power cable faults and test method study. *Electrotechnics Electric* 5: 48–52 (in Chinese).
9. Li, Zhonghua, and Xia Zhang. 2018. Effect of nonlinear properties of stress cone reinforced insulation on electric field distribution of HVDC cable terminal. *Journal of Electrical Engineering* 11: 37–43 (in Chinese).
10. Hu, Peng, Chenggang Li, and Dabing Chen. 2017. Cause analysis and countermeasure study of cracking accident of cable GIS terminal epoxy casing. *Electric Power Engineering Technology* 36 (1): 102–105 (in Chinese).
11. Huang, Jiankai, and Jiangang Zhao. 2018. Analysis of a GIS terminal connector failure of 110 kV cable. *Metallurgical Power* 225 (11): 5–8 (in Chinese).
12. Ren, Zhigang, Guang Cong, and Huachun Li. 2014. Combined diagnosis of partial discharge for GIS cable terminal. *High Voltage Apparatus* 50 (3): 44–49 (in Chinese).
13. Su, Jiahua. 2013. Analysis and treatment of a high voltage test abnormality of a 110kV cable GIS terminal. *Ability and Wisdom* 7: 274–275 (in Chinese).
14. Wu, Ke, Yunsong Qiao, Xiong Gang, et al. Failure analysis of partial discharge of coaxial cable. *Electrical Engineering* 21(6): 83–87 (in Chinese).
15. Cao, Jingying, Xianguang Zha, Jie Chen, et al. 2018. Fault and simulation analysis for the terminal of 220 kV cable line. *Electric Power Engineering Technology* 37 (6): 151–155 (in Chinese).
16. Loiselle, L., I. Fofana, J. Sabau, et al. 2015. Comparative studies of the stability of various fluids under electrical discharge and thermal stresses. *IEEE Transactions on Dielectrics Electrical Insulation* 22 (5): 2491–2499.
17. Schmidt, W.F., and K. Yoshino. 2015. Ion mobilities in non-polar dielectric liquids: Silicone oils. *IEEE Transactions on Dielectrics and Electrical Insulation* 22 (5): 2424–2427.
18. R. Zhang, J. Wu, G. Du and Z. Qiu. 2018. Study on the aging properties of silicone fluid in GIS cable terminations. In 2018 12th International Conference on the Properties and Applications of Dielectric Materials (ICPADM). IEEE, Xi'an, pp. 429-434. doi: 10.1109/ICPADM.2018.8401029

19. R. Zhang, Z. Qiu, J. Wu, X. Li and S. Wang. 2018. Study on Gaseous Products in the Aging Process of Silicone Oil in Cable Terminals. In 2018 IEEE Conference on Electrical Insulation and Dielectric Phenomena (CEIDP). IEEE, Cancun, pp. 394-397. doi: 10.1109/CEIDP.2018.8544876
20. L. Wang, S.Y. Liu, C. Xu, et al. 2016. Direct conversion of methanol to n-C_4H_{10} and H_2 in a dielectric barrier discharge reactor. *Green Chemistry* 18 (20): 5658–5666.

Wireless Synchronous Transmission of Power and Signal Based on Electric-Field Coupling

Xianmin Mu, Yan Liu, Jiwei Guo, and Xiyou Chen

Abstract Synchronous transmission of signal is as important as wireless power transmission. In order to solve the problem of mutual interference between power and signal in the wireless synchronous transmission system of power and signal with shared channel, this paper proposes a new electric field-coupled power and signal transfer system using common mode and differential mode separation circuits. Based on the principle of transmitting power in common mode and transmitting signal in differential mode, this paper establishes the circuit models of power transmission channel and signal transmission channel, and analyzes the working performance of the system in detail according to the circuit theory. At the same time, the design principles of the system parameters are given in this paper. Finally, simulation and experiment verify the feasibility of the proposed system, and the effect of plate capacitance change on system performance is studied.

Keywords Electric-field coupling · Wireless power transmission · Synchronous transmission of power and signal · Mutual interference · Plate capacitance change

1 Introduction

At present, wireless power transfer (WPT) technology has attracted widespread attention from scholars all over the world [1–3]. Electric-filed coupled power transfer (ECPT) technology has been applied in some fields [4] based on the ability to transmit power through metals [5, 6]. In many applications, it is necessary to realize wireless synchronous transmission of power and signal [7–9], for example, the robot rotary joint and underwater wireless power transmission system [10], sensor networks, RFID, mobile devices [11] and so on.

X. Mu · Y. Liu (✉) · J. Guo · X. Chen
School of Electrical Engineering, Dalian University of Technology, Dalian, China
e-mail: 1686977832@qq.com

X. Mu
e-mail: muxm@dlut.edu.cn

W. Chen et al. (eds.), *The Proceedings of the 9th Frontier Academic Forum of Electrical Engineering*, Lecture Notes in Electrical Engineering 743,
https://doi.org/10.1007/978-981-33-6609-1_25

There are two main ways to realize the synchronous transmission of power and signal. One is to transmit power and signal by adding signal transmission channels [12, 13]. The other is to use the shared channel to achieve synchronous transmission of both [14, 15]. In this paper, a new type of electric-field coupled power and signal transfer (ECPST) system using shared channel to transmit both is designed by adopting common mode and differential mode separation circuits.

2 Model Establishment and Principle Analysis

The ECPST system circuit proposed in this paper is shown in Fig. 1.

This system transfers power in common mode and transfers signal in differential mode. In which C_1, C_2 and C_3 are used to transmit power simultaneously; C_1 and C_2 build the signal channel to transmit signals; coupling inductors L_{p1}, L_{s1} and L_{p2}, L_{s2} separate the receiving and transmitting circuit in signal transfer channel from the power transfer channel, which suppressing the interference of power channel to signal channel.

According to the superposition principle of linear circuits, analyze the working conditions of the circuit when the power source and the signal source work independently.

2.1 Circuit Model of Power Transfer Channel

When the power source works alone, the circuit of the power transfer channel is shown in Fig. 2.

The equivalent topology of the power transfer channel is shown in Fig. 3 based on the T decoupling equivalent principle of the coupling inductor and the principle of equivalent transformation of the Δ connection and the Y connection of impedance.

The input impedance Zpin of the system circuit can be expressed as:

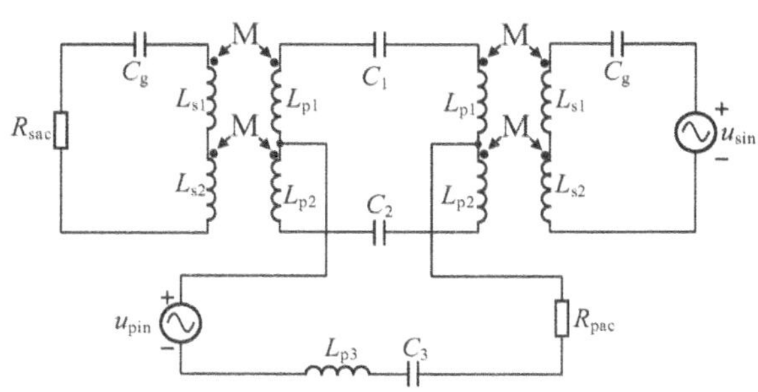

Fig. 1 Circuit diagram of ECPST system

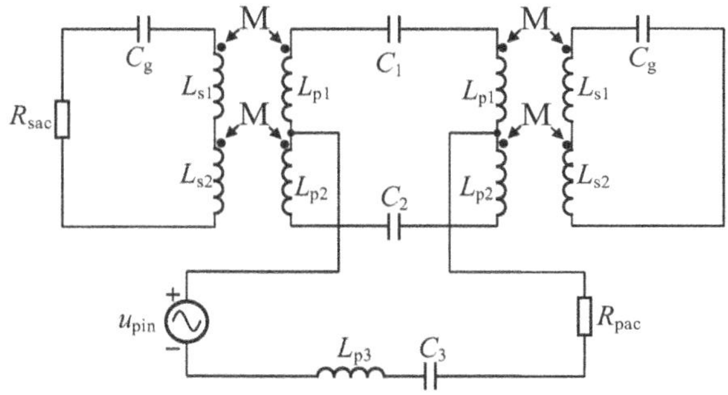

Fig. 2 Circuit diagram of the power transfer channel

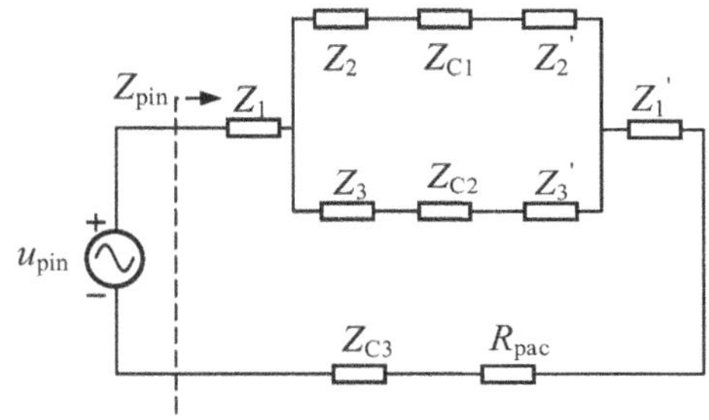

Fig. 3 Equivalent topology of the power transfer channel

$$Z_{pin} = Z_1 + (Z_2 + Z_{C1} + Z_2')/(Z_3 + Z_{C2} + Z_3') + Z_1' + Z_{C3} + R_{pac} \quad (1)$$

Based on (1), the voltage gain expression of power transfer channel is expressed as (2). Where, U_{Rpac} is the voltage across the load R_{pac}.

$$G_{R_{pac}} = \left|\frac{\dot{U}_{R_{pac}}}{\dot{U}_{pin}}\right| = \left|\frac{R_{pac}}{Z_{pin}}\right| \quad (2)$$

The further equivalent topology is obtained based on the principle of equivalent transformation of the Δ connection and the Y connection of impedance.

Considering the complexity of the circuit topology, the circuit is divided into four parts as shown in Fig. 4, and the impedance expression of each part as shown as (3):

$$\begin{cases} Z_{ps1} = Z_d + Z_e \\ Z_{ps2} = \frac{Z_{ps1} \times (Z_b + Z_c)}{Z_{ps1} + Z_b + Z_c} \\ Z_{ps3} = Z_a + Z_{ps2} \end{cases} \quad (3)$$

According to Kirchhoffs voltage law and Kirchhoff's current laws, the transfer func-tion of each part is derived. Where, U_{Zc}, U_{Ze}, $U_{\text{Zb+Zc}}$, $U_{\text{Zd+Ze}}$ and U_{Rsac} are the

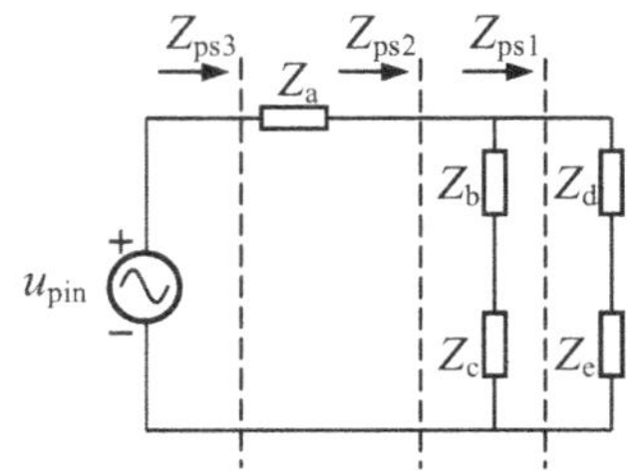

Fig. 4 Equivalent topology of the power transfer channel

voltage across the impedances Z_c, Z_e, $Z_b + Z_c$, $Z_d + Z_e$ and load resistance Rsac, respectively.

$$\begin{cases} G_{ps1} = \frac{\dot{I}_{pin}}{\dot{U}_{pin}} = \frac{1}{Z_{ps3}} & G_{ps4} = \frac{\dot{U}_{Z_e}}{\dot{U}_{Z_d+Z_e}} = \frac{Z_e}{Z_d+Z_e} \\ G_{ps2} = \frac{\dot{U}_{Z_b+Z_c}}{\dot{I}_{pin}} = Z_{ps2} & G_{ps5} = \frac{\dot{U}_{Z_d+Z_e}}{\dot{U}_{Z_b+Z_c}} = 1 \\ G_{ps3} = \frac{\dot{U}_{Z_c}}{\dot{U}_{Z_b+Z_c}} = \frac{Z_c}{Z_b+Z_c} & G_{ps6} = \frac{\dot{U}_{Rsac}}{\dot{U}_{Z_c}-\dot{U}_{Z_e}} = \frac{R_{sac}}{Z_s+R_{sac}} \end{cases} \tag{4}$$

Based on (3) and (4), the interference voltage gain expression of the power transfer channel to the signal transfer channel is expressed as:

$$G_{P_{Rsac}} = \left|\frac{\dot{U}_{R_{sac}}}{\dot{U}_{pin}}\right| = \left|\prod_{i=1,2,3,6} G_{psi} - \prod_{j=1,2,4,5,6} G_{psj}\right| \tag{5}$$

2.2 Circuit Model of Signal Transfer Channel

When the signal source works alone, the circuit of the power transfer channel is shown in Fig. 5.

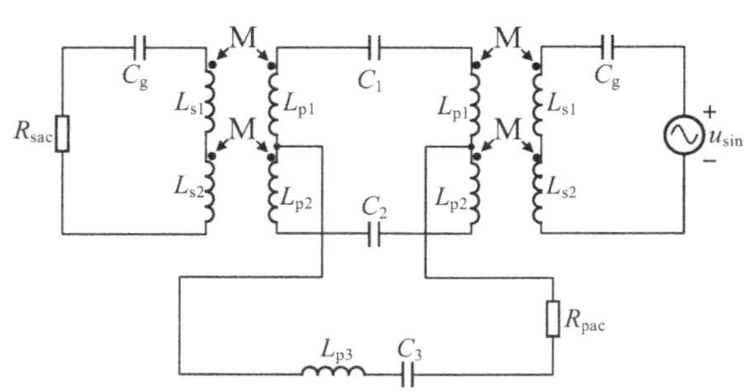

Fig. 5 Circuit diagram of the signal transfer channel

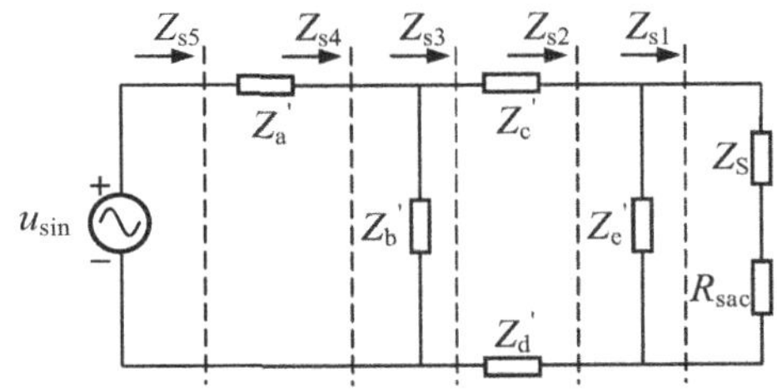

Fig. 6 Equivalent topology of the signal transfer channel

The equivalent topology of the power transfer channel is shown in Fig. 6 based on the T decoupling equivalent principle of the coupling inductor and the principle of equivalent transformation of the Δ connection and the Y connection of impedance. Considering the complexity of the circuit topology, the circuit is divided into six parts as shown in Fig. 6, and the impedance expression of each part as shown as (6):

$$\begin{cases} Z_{s1} = Z_s + R_{sac} & Z_{s4} = \frac{Z_{s3}\times Z'_b}{Z_{s3}+Z'_b} \\ Z_{s2} = \frac{Z_{s1}\times Z'_e}{Z_{s1}+Z'_e} & Z_{s5} = Z'_a + Z_{s4} \\ Z_{s3} = Z'_c + Z'_d + Z_{s2} & \end{cases} \tag{6}$$

According to Kirchhoffs voltage law and Kirchhoff's current laws, the transfer func-tion of each part is derived. Where, $U_{Zb'}$, $U_{Ze'}$, and U_{Rsac} are the voltage across the im-pedances $Z_{b'}$, $Z_{e'}$ and load resistance R_{sac}, respectively; $I_{Zc'}$ and I_{Rsac} are the current flowing through the impedances $Z_{c'}$ and load resistance R_{sac}, respectively.

$$\begin{cases} G_{s1} = \frac{\dot{I}_{sin}}{\dot{U}_{sin}} = \frac{1}{Z_{s5}} & G_{s4} = \frac{\dot{U}_{Z'_e}}{\dot{I}_{Z'_c}} = Z_{s2} \\ G_{s2} = \frac{\dot{U}_{Z'_b}}{\dot{I}_{sin}} = Z_{s4} & G_{s5} = \frac{\dot{I}_{Rsac}}{\dot{U}_{Z'_e}} = \frac{1}{Z_{s1}} \\ G_{s3} = \frac{\dot{I}_{Z'_c}}{\dot{U}_{Z'_b}} = \frac{1}{Z_{s3}} & G_{s6} = \frac{\dot{U}_{Rsac}}{\dot{I}_{Rsac}} = R_{sac} \end{cases} \tag{7}$$

Based on (6) and (7), the voltage gain expression of signal transfer channel is expressed as:

$$G_{R_{sac}} = \left|\frac{\dot{U}_{R_{sac}}}{\dot{U}_{pin}}\right| = \left|\prod_{i=1,2,3,4,5,6} G_{si}\right| \tag{8}$$

3 Parameters Design

Considering the symmetry and simplification of the structure, make the plate coupling capacitance equal, that is, $C_1 = C_2$, and let the coupling inductance in signal transfer channel with the same mutual inductance coefficient M satisfy the equation $L_{p1}/L_{s1} = L_{p2}/L_{s2}$. In order to improve the transmitted power and efficiency of the power transfer channel, the compensation inductors L_{p1}, L_{p2} and the plate coupling capacitors C_1, C_2 and C_3 satisfy the resonance relationship shown as (9), where ω_p is the angular frequency of u_{pin}.

$$\begin{cases} \omega_p^2(2L_{p1})C_1 = 1 \\ \omega_p^2(2L_{p2})C_2 = 1 \\ \omega_p^2 L_{p3}C_3 = 1 \end{cases} \tag{9}$$

In order to ensure the accuracy and anti-interference of signal transfer channel, the inductance L_{s1}, L_{s2} and the compensation capacitor C_g satisfy the resonance relationship shown as (10).

$$\omega_s^2(L_{s2} + L_{s1} - 2M)C_g = 1 \tag{10}$$

4 Simulation and Experiment

The simulation and experimental circuit are constructed according to Figs. 2 and 5. Table 1 lists the circuit parameters for the prototype and simulation.

Table 1 Parameters of ECPST system

Parameters	Value
U_{pin}/V	50
f_p/Hz	10^6
C_1, C_2/pF	300
C_3/pF	600
L_{pi}/uH	42
R_{pac}/Ω	50
U_{sin}/V	5
f_s/Hz	10^7
L_{si}/uH	42
M/uH	41.58
C_g/pF	300
R_{sac}/Ω	1000

The simulation waveforms is shown in Fig. 7. It can be seen that the voltage gain G_{Rpac} is 1 and G_{PRsac} is $2 * 10^{-14}$ which value is very small and negligible.

In order to verify the correctness of the scheme, a prototype was constructed. The experimental waveforms is shown in Fig. 8. It can be seen that the voltage gain G_{Rpac} is 0.94 and G_{PRsac} is 0.03.

Figures 9, 10 and 11 show the change curves of the voltage gain G_{Rpac}, G_{PRsac} and G_{Rsac} with coupling capacitors C_1 and C_2 changing. As shown in Fig. 9, it

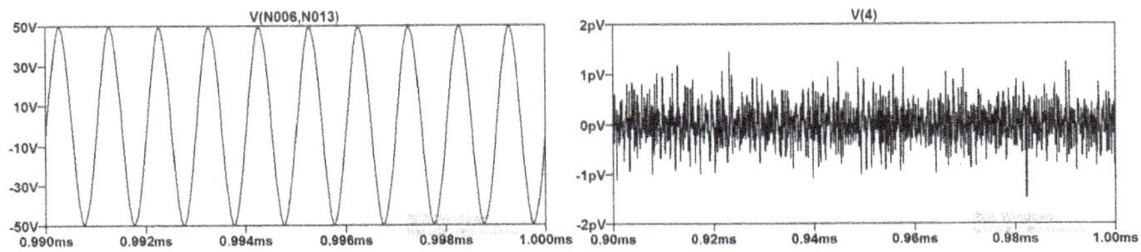

Fig. 7 U_{Rpac} and U_{PRsac} of the power transfer channel

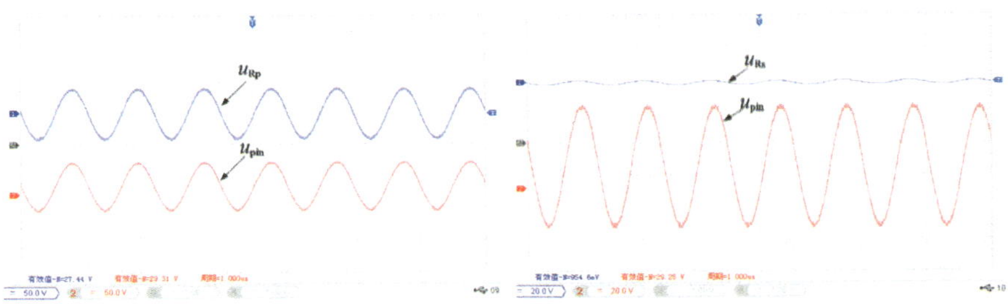

Fig. 8 U_{Rpac} and U_{PRsac} of the power transfer channel

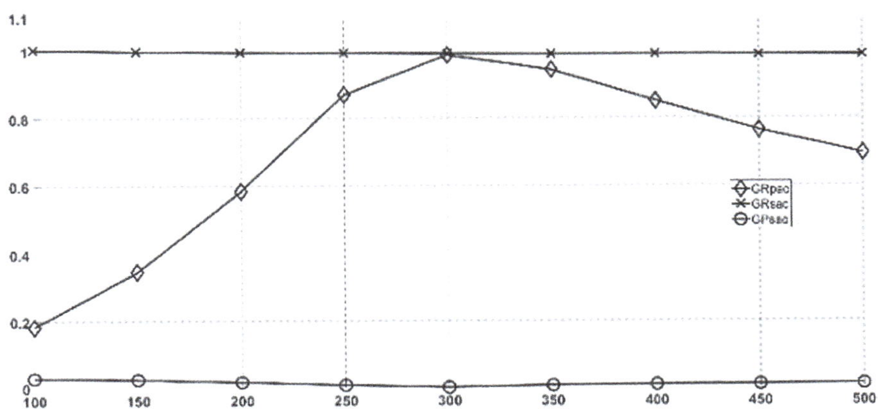

Fig. 9 Voltage gain varies with C_1

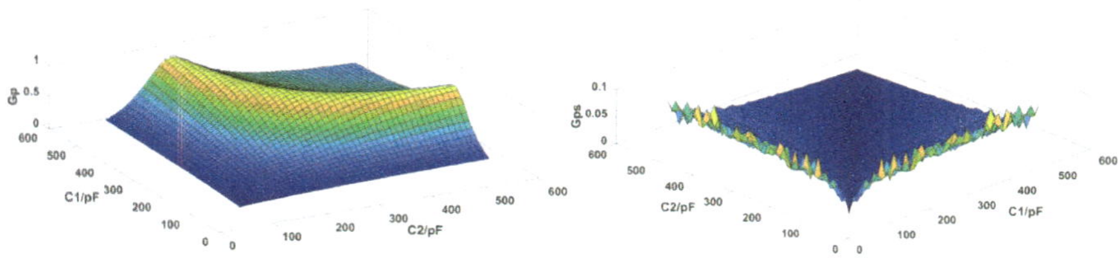

Fig. 10 The relationship between G_{Rsac}, G_{PRsac} and C_1, C_2

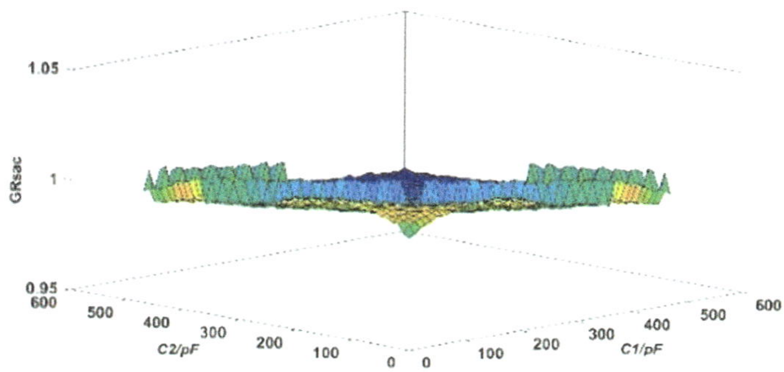

Fig. 11 The relationship between G_{Rsac} and C_1, C_2

has an influence on the power transfer channel and signal transfer channel with the coupling capacitor C_1 changing in the range of 33.3–160%, and the impact on the power transfer channel is more obvious.

As shown in Fig. 10, it has an influence on the power transfer channel and signal transfer channel with the coupling capacitors C_1 and C_2 both changing in the range of 26.7–200%. It can be seen that the power channel has no interference to the signal channel when the changes of C_1 and C_2 are the same, and the energy transfer channel may still achieve good transmission effects when the changes of C_1 and C_2 are different. Figure 11 shows that it has no effect on signal transfer channel no matter whether C_1 and C_2 change the same.

5 Conclusion

This paper designs a new type ECPST system by adopting the principle of common mode transmission power and differential mode transmission signal which solves the problem of mutual interference between power and signal through reasonable design of circuit parameters.

The simulation and experimental results show that the synchronous transmission of power and signal can be realized with pF coupling capacitor, and verify the feasibility of the proposed system. In addition, the output voltage gains also change with the change of the plate capacitance C_1 and C_2, so the influence of plate capacitance changes on the system performance will be deeply discussed later.

References

1. Xie, L., Y. Shi, et al. 2013. Wireless power transfer and applications to sensor networks. *IEEE Wireless Communications* 20 (4): 140–145.
2. Xingming, Fan, Gao Linlin, et al. 2019. Overview of research status and application of wireless power transmission technology. *Transactions of China Electrotechnical Society* 034 (007): 1353–1380.
3. Moon, S., B. Kim, S. Cho, et al. 2014. Analysis and design of a wireless power transfer system with an intermediate coil for high efficiency. *IEEE Transactions on Industrial Electronics* 61 (11): 5861–5870.
4. Su, Yugang, Shiyun Xie, Aiguo Hu, et al. 2015. Transmission property analysis of electric-field coupled wireless power transfer system with LCL resonant network. *Transactions of China Electrotechnical Society* 30(19): 55–60 (in Chinese).
5. Lu, F., H. Zhang, H. Hofmann, et al. 2015. A double-sided LCLC-compensated capacitive power transfer system for electric vehicle charging. *IEEE Transactions on Power Electronics* 30 (11): 6011–6014.
6. Cao, Lingling, Qianhong Chen, Xiaoyong Ren, et al. 2012. Review of the efficient wireless power transmission technique for electric vehicles. *Transactions of China Electrotechnical Society* 27(8): 1–13 (in Chinese).
7. Yujin, Jang, Jung Kyu Han, Shin Young Cho, et al. 2016. Wireless power and data transfer system for smart bridge sensors. In *Applied Power Electronics Conference & Exposition*. IEEE.
8. Tu, Zhangjie, Xiongzhu Bu, Miaomiao Xu. 2016. Research on synchronous transmission technology of wireless power and signal based on magnetic coupling resonance. *Electronic Measurement Technology* 263(03): 38–41 (in Chinese).
9. Yong, Zhu, Yuenan Wu. 2017. A survey on SWIPT system based on coupled inductor. *Hans Journal of Wireless Communications* 007(001): P.7–16 (in Chinese).
10. Hu, A.P., C. Liu, H.L. Li. 2008. A novel contactless battery charging system for soccer playing robot. In *International Conference on Mechatronics and Machine Vision in Practice*, pp. 646–650.
11. Xueliang, Huang, Tan Linlin, Chen Zhong, et al. 2013. Review and research progress on wireless power transfer technology. *Transactions of China Electrotechnical Society* 28 (10): 1–11 (in Chinese).
12. Dionigi, M., and M. Mongiardo. 2012. A novel resonator for simultaneous wireless power transfer and near field magnetic communications. In *Microwave Symposium Digest*. IEEE.
13. Wang, G., P. Wang, Y. Tang, et al. 2012. Analysis of dual band power and data telemetry for biomedical implants. *IEEE Trans Biomed Circuits System* 6 (3): 208–215.
14. Su, Y.G., W. Zhou, A.P. Hu, et al. 2017. Full-duplex communication on the shared channel of a capacitively coupled power transfer system. *IEEE Transactions on Power Electronics PE* 32 (4): 3229–3239.
15. Qingxin, Yang, Li. Yang, Yin Jianbin, et al. 2017. Wireless synchronous transmission of power and information based on ASK in WPT via coupled magnetic resonances. *Transactions of China Electrotechnical Society* 32 (016): 153–161 (in Chinese).

Weak Fault Diagnosis Method for Subway Bearing Based on Multi-scale Class Distance

Yin Tian

Abstract According to the separability criterion of overall scale Class Distance there will be a fault cluster indiscernibility problem when this diagnosis is used in the process of subway bearing weak fault diagnosis. Aiming at this problem, this paper presents a diagnosis method based on Multi-Scale Class Distance (MSCD). In the basis of traditional criterion which focused on the classes distance within single overall scale class, we use the distance measure of multi-scale class as the separability criterion, extract the class of fault from the sample and calculate the euclidean distance between classes. Then classify the sample into several sub-samples, and use these sub-samples as a whole sample to carry on the separability criterion calculation of the next scale. Repeat this process until the multi-scale classification is finished and obtain the sub-samples with best separability. We can get the vibration signals of bearing weak fault in different working conditions from the subway gearbox test bench, and use the method proposed by this paper to carry on weak fault diagnosis. The result shows that this method can achieve the classification of bearing weak fault. Comparing with the traditional PCA classification result based on single scale, this method can obtain a fault classification result with high separability.

Keywords Multi-scale class distance · Subway bearing · Weak fault diagnosis · Principal component analysis

1 Introduction

The working condition of subway gearbox bearing is quite severe. The rapid changes of load generated by the frequent start-ups of the train can make the weak faults of the gearbox bearing develop into a significant failure in a short time, which cause great damage to the operating safety of subway train. In that way, carrying weak fault diagnosis of subway bearing is very necessary [1]. The subway gearbox has a quiet complicated structure. It contains lots of rotating parts which makes the signals

Y. Tian (✉)
CRRC Academy, No. 5 Building, 2nd Phase of Noble Center, East ROAD of Beijing Auto Museum, Fengtai District, Beijing, China
e-mail: tianyin@crrcgc.cc

W. Chen et al. (eds.), *The Proceedings of the 9th Frontier Academic Forum of Electrical Engineering*, Lecture Notes in Electrical Engineering 743,
https://doi.org/10.1007/978-981-33-6609-1_26

we get contain a large amount of noises. Under the interference of noise, the fault feature signal can be strongly coupled, attenuated or modulated, which significantly increase the difficulty to diagnose a weak fault.

Cluster analysis is a multivariate statistical analysis method which can classify several samples quantitatively [2]. In the field of subway gearbox fault diagnosis, it's a common way to analyze the weak fault signal in a cluster way and classify the fault signal into several subsets. Carry out a series of time domain, frequency domain and statistic transformation to obtain high-dimensional characteristic parameters which can present some intrinsic characteristics of the signal [3–5]. Evaluate the effect of the parameters based on distance measure, information measure and dependence measure. Using methods like principle component analysis [6], independent component analysis [7] and manifold learning [8] to reduce the dimension of high-dimensional characteristic parameters, and extract those who have strong correlation with cluster and weak correlation with characteristic. In that way, the classification of feature fault is succeed. The key in this process is the feature evaluation criteria, in which as the measure parameter for measure separation, measure difference and the identification ability of the measure, the distance measure is used widely in fault subset classification [9]. But in fact the traditional criterion which based on the measure of the distance between classed is the average distance of each types of eigenvectors [10]. The calculation of such criterion is simple but flawed. When the distance between classes is small and the distance within a class is big, the value of this criterion is large. A large criterion value represent the fault subset with high separability, but in fact at this moment the separability of the subset is not that high. In that way a cluster indiscernibility problem will occur.

Aiming at this problem, experts proposed some solutions. Zai et al. introduce a separability criterion called Kernel Distance Measure [11]. This method calculate the distance of clusters in kernel space, obtain effective classification result and reduce the running time of algorithm. Li Dong et al. compare and analyze evaluation standards like distance measure and information measure, clarify the advantages and disadvantages of each standards [12].

This paper introduce the concept of multi-scale in the basis of traditional class distance. The method introduced in this paper starts from the overall cluster scale, calculate the Euclidean distance set by set, classify the class of fault according to the separability criterion of distance measure. This method take into account of class distance and distance in single class at the same time, promote the separability of fault class successfully.

2 Multiscale Class Distance

2.1 Separability Criterion of Multiscale Class Distance

The separability criterion of class distance in single class can be defined as follow:

$$J_d(x) = \frac{1}{n_i n_j} \sum_{k=1}^{n_i} \sum_{l=1}^{n_j} \delta(x_k^{(i)}, x_l^{(j)}) \tag{1}$$

where, c is the number of class; n_i, n_j represent the number of samples in class ω_i and class ω_j respectively. $x_k^{(i)}$ and $x_l^{(j)}$ represent the D-dimension eigenvectors in class ω_i and class ω_j; $\delta(x_k^{(i)}, x_l^{(j)})$ is the distance between vector $x_k^{(i)}$ and $x_l^{(j)}$.

From formula (1) we can see that the inner thoughts of $J_d(\chi)$ is to calculate the space distance between each eigenvectors and the average distance of vectors under one overall scale, then evaluate the separability through the space distance gotten in single scale. Since this criterion is a overall scale criterion, it's criterion standard can be written as: the larger the $J_d(\chi)$ value, the bigger the overall distance of each classify-able class under overall scale, the better the separability of each class [13]. View $\delta(x_k^{(i)}, x_l^{(j)})$ as Euclidean distance, then it can be written as:

$$\delta(x_k^{(i)}, x_l^{(j)}) = (x_k^{(i)} - x_l^{(j)})^T (x_k^{(i)} - x_l^{(j)}) \tag{2}$$

Which is

$$\begin{aligned} J_d(x) = tr(S_b + S_w) &= \sum_{i=1}^{c} (m_i - m)^T (m_i - m) \\ &+ \sum_{i=1}^{c} \frac{1}{n_i} \sum_{k=1}^{n_i} (x_k^i - m_i)^T (x_k^i - m_i) \end{aligned} \tag{3}$$

where,

$$m_i = \frac{1}{n_i} \sum_{k=1}^{n_i} x_k^{(i)}$$

$$m = \sum_{i=1}^{c} m_i$$

In the above formula, S_b is the divergence matrix between classes; m_i is mean vector of sample class i; m is the overall mean vector of all sets of samples; S_w is the divergence matrix within class, represent in class distance measure. In the traditional criterion standard, when we use $J_d(\chi)$ in single overall scale as separability criterion, the track of S_b is big and the track of S_w is small, in that case when the value of $J_d(\chi)$ is big, the results are more ideal. Since this criterion is a kind of overall average algorithm, when the track of S_w is big, the track of S_b is small, the value of $J_d(\chi)$ will be quiet big too. Theoretically, a bigger $J_d(\chi)$ value means the sample to be processed has better separability, but actually, a small distance between classes means that it's difficult to distinguish the classification result. In other words, the sample do not have a good separability. At this time if we still use the $J_d(\chi)$ in single overall scale as the separability criterion, there will be a wrong classification result.

In view of the fact that only the average class distance under a single overall scale is considered, This paper proposes a multi-scale class distance criterion $J_{MSD}(\chi)$ based on multi-scale ideas. $J_{MSD}(\chi)$ solve the problem from the multi-scale point of view. It first calculate the separability criterion under initial scale α. α scale can be written as:

$$J_{MSD}^{\alpha}(x) = \frac{1}{2^{\alpha}} \sum_{i=1}^{n_a} \sum_{j=1}^{n_b} d(x_i, x_j) \quad x_i \in \omega_a, x_j \in \omega_b \tag{4}$$

where α is the scale of initial sample. It's value represent the sample number in the set for unsorted sample. x_i, x_j represent the initial D-dimension eigenvector in ω_i class and ω_j class. x_i, x_j can be gotten through extract some initial feature of sample; $d(x_i, x_j)$ is the vector distance between x_i, x_j, can be represented by Euclidean distance, S - order Minkowski measure, Chebychev distance or squared distance. In this paper we choose Euclidean distance as distance measure, so the $d(x_i, x_j)$ can be written as:

$$d(x_i, x_j) = \|x_i - x_j\| = (x_i - x_j)^T (x_i - x_j) \tag{5}$$

that is

$$\begin{aligned} J_{MSD}^{\alpha}(x) &= \frac{1}{2^{\alpha}} \sum_{i=1}^{n_a} \sum_{j=1}^{n_b} (x_i - x_j)^T (x_i - x_j) \\ &= \frac{1}{2^{\alpha}} \left[\sum_{i=1}^{c} (m_i^{\alpha} - m^{\alpha})^T (m_i^{\alpha} - m^{\alpha}) + \sum_{i=1}^{c} \frac{1}{n_i} \sum_{k=1}^{n_i} (x_k^i - m_i^{\alpha})^T (x_k^i - m_i^{\alpha}) \right] \end{aligned} \tag{6}$$

where, m_i^{α} is the mean vector of sample set class i in α scale;m^{α} is the mean vector of all different sample sets in α scale.

Through the calculation of separability criterion in initial scale α, the classification result under this scale can be gotten. Since $J_{MSD}^{\alpha}(x)$ focus on the class distance in specific scale, in average process it can avoid the the case where the distance in the class is too small and the distance between classes is too large. Thus, this method can get classification result with good separability in first scale.

β scale:

When the classification result in scale α is gotten, then we can move to the next β scale to continue classification process. β represent the number of samples to be sorted. The classification for scale β is based on the one for scale α, which means that further classification is done in the initial scale classification results, the corresponding selection standard of separability criterion will be changed accordingly. Let ω_α be the classification result in scale α, then the separability criterion under scale β can be written as:

$$J_{MSD}^{\beta}(x) = \frac{1}{2^{\beta}} \sum_{i=1}^{n_a} \sum_{j=1}^{n_b} (x_i - x_j)^T (x_i - x_j) \quad x_i, x_j \in \omega_\alpha \tag{7}$$

Equation (7) can be used to classify ω_α in next scale, so as to obtain a high separability classification set ω_β.

After the classification under scale β, the original sample set has been able to get a classification result with high separability. If the classification result is not very well, the further classification based on scale γ can be carried on according to Eq. (7). The separability criterion under this scale can be written as:

$$J_{MSD}^{\gamma}(x) = \frac{1}{2^{\gamma}} \sum_{i=1}^{n_a} \sum_{j=1}^{n_b} (x_i - x_j)^T (x_i - x_j) \quad x_i, x_j \in \omega_\beta \tag{8}$$

2.2 *Classification of Fault Subset*

After determining the multi-scale class distance separability criterion, the dimension reduction operation based on this criterion can be taken for this group of classification data. For reduction operation, this paper choose PCA as dimension reduction algorithm. The PCA is often used for dimension reduction of high dimensional data. It can transform raw data into a group of representations who is linearly independent in all dimensions, and this algorithm can extract the main feature component of the data. In the process of dimension reduction, PCA can make the total variance of the sample matrix unchanged, make sure the transformed features are arranged in descending order of statistical variance. Meanwhile, PCA can ensure that each sample is kept independent with each other, and can order the eigenvalues in descending order according to their variance contribution rate. Here we introduce the separability criterion $J_{MSD}(x)$ which we have mentioned above. The data separability will better with the increase of $J_{MSD}(x)$ value, vice versa.

In order to effectively classify the fault subsets, the original features need to be extracted from the original fault sample. Features that are often used in rolling bearing fault diagnosis include time domain characteristics, frequency domain characteristics, and time–frequency domain joint characteristics [14]. Common initial features include time domain characteristics, frequency domain characteristics and spectral vectors, in which time domain characteristics include mean, root mean square, square root, absolute mean, skew, kurtosis, variance, minimum, maximum, peak and peak,

Table 1 Dimensions of the different initial feature vectors

Type of feature vectors	Dimensions of feature vectors
Time domain characteristics	16
Frequency domain characteristics	13
Spectral vector	5120 (FFT points)

waveform index, peak index, pulse index, margin index, skew Indicators, kurtosis indicator, and frequency domain characteristics include mean frequency, frequency standard deviation, spectral dispersion, frequency center, rms frequency and main band position. The spectral vector is the spectral distribution characteristic of the signal. The dimensions of the different initial feature vectors of each group are shown in Table 1.

After collecting the fault vibration signal, extract initial features from the signal and obtain different eigenvectors. Use the multi-scale distance criterion as the standard, and reduce the dimension of high-dimensional eigenvectors through PCA algorithm. At last get a classification result with high separability. Figure 1 is the fault classification algorithm flow chart.

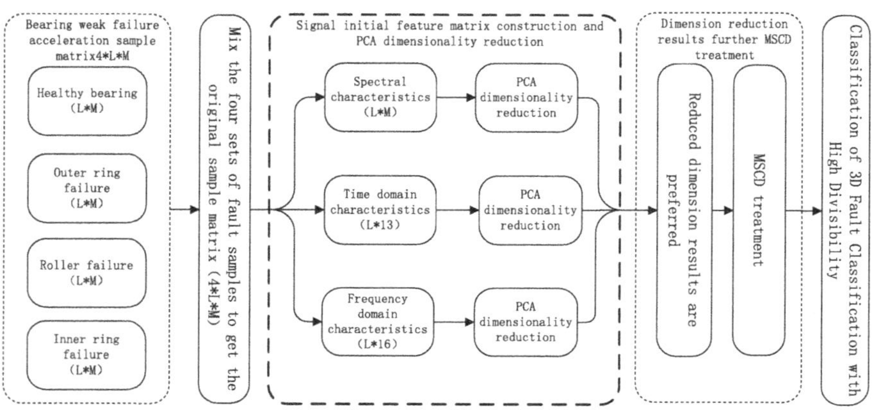

Fig. 1 Algorithm flow chart

3 Experimental Verification

3.1 Weak Fault Vibration Signal Acquisition

In order to verify the correctness of the above theory, we collected the fault data of the subway bearing to verify. In this paper we use the subway gearbox failure simulation test bench produced by Spectra Quest Company to collect data. This gearbox is designed according to the real subway gearbox with a zoom ratio of 1:3. Different fault sample can be obtained through changing the assembly methods of the components. It can fully simulate the vibration characteristics of subway gearbox bearing in real working conditions. The shape of the test bench is showed in Fig. 2.

In this paper, we choose SQI608A11-10F type ICP accelerometer to acquire vibration signal. This sensor can collect the vibration acceleration signal range from 0.5 to 10,000 Hz, with a working range of $\pm 490\ \text{m/s}^2$. This sensor can acquire vibration signal accurately, and make sure the signal is reliable and true. When bearing failure occurs, the high frequency vibration signal created by the fault will spread out in the form of hemispherical waves. The center of the wave will be the vibration impact point. Since the frequency and energy loss are proportional in the process of propagation, the energy efficiency of contact interface in single-node is about 22.17%. Thus, in order to reduce the energy attenuation during signal acquisition, we choose the top of the bearing as the best measuring point [15]. Put accelerometer in the top of bearing, collect the vibration signal in this direction as the bearing vibration signal. The actual arrangement of the sensor is shown in Fig. 3.

Fig. 2 Gearbox failure test bench

Fig. 3 The arrangement of ICP acceleration sensor

The text subject is NU 202 ECP type cylindrical roller bearing. In order to simulate weak failure and real failure points, we process grooves in outer-ring, rolling-part and inner-ring respectively to generate periodic impact signal. The real condition of bearings is showed in Fig. 4.

In this experiment, in order to make the weak fault signal more weaker, to highlight the significance of weak signal research, we remove the Bearing load.

Describe the sampling theorem in a mathematical way [16]:

$$s(t) = \sum_{-\infty}^{\infty} s(kT) \frac{\sin \pi (t/T - k)}{\pi (t/T - k)} \tag{9}$$

Normal bearing | Outer-ring failure | Rolling-part failure | Inner-ring failure

Fig. 4 The bearings used in this paper

where, $s(t)$ is the sampling signal, k is an integer, and t is sampling time.

Through the analysis of Eq. (9) we can see that in order to ensure the integrity of the sampling information, the sampling frequency needs to be very high. Normally the sampling frequency is 5–10 times higher than the frequency to be analyzed. In this paper, the input shaft is rotated at 30 Hz, so the characteristic frequency of outer-ring failure, rolling part fault and the inner-ring fault are 128.4, 70.9 and 201.6 Hz. Since the main components of rolling bearing vibration signal are base frequency and 2 times frequency, in this paper the sampling frequency is 5 kHz, which ensures the completeness of the effective frequency acquisition in the vibration data acquisition process. In this paper, 30 sets of sample data are collected for each bearing condition. The length of each sample is 1 s. Under a sampling frequency of 5 K, matrix of each set sample is 30 * 5120 dimension, the full sample matrix is 120 * 5120 dimension.

3.2 Initial Feature Sample Set Construction

After getting the full sample matrix, in order to obtain the time domain and frequency domain distribution of each group, the initial FFT processing is performed. From Fig. 5 we can see that we can't distinguish fault types directly from the time domain and spectrum distribution. The characteristic frequency of each fault is calculated according to the bearing structure and input speed, which is

$$F_{BPFO} = 128.4\,\text{Hz} \tag{10}$$

$$F_{BSF} = 70.95\,\text{Hz} \tag{11}$$

$$F_{BPFI} = 201.59\,\text{Hz} \tag{12}$$

The sub-band energy statistics are performed for each fault characteristic frequency. In real time, the movement between raceway and rolling part is not pure rolling. This movement is with a certain random sliding [17, 18], which will result in a 1–2% random error between the actual fault impact interval and the theoretical fault impact interval [19]. For energy statistics accuracy, in this paper, we take the plus or minus two percent of fault characteristic frequency and its double frequency as the frequency range of fault sub-band, which is

$$f \in \left[0.98 \times f_{fcf}, 1.02 \times f_{fcf}\right] \cup \left[0.98 \times 2f_{fcf}, 1.02 \times 2f_{fcf}\right] \tag{13}$$

Use Eq. (13) to calculate the energy ratio of the fault sub-band. The result is showed in Fig. 6.

From the Fig. 6 we can see that the energy ratio of each fault sub-band is extremely small throughout the energy distribution. The energy ratio in the inner-ring is the

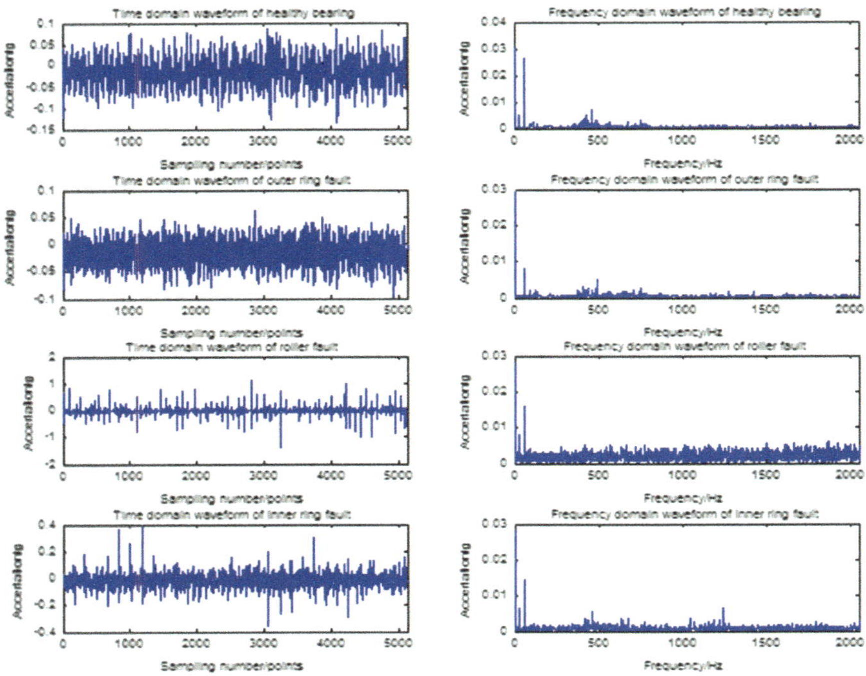

Fig. 5 Sample time–frequency waveform

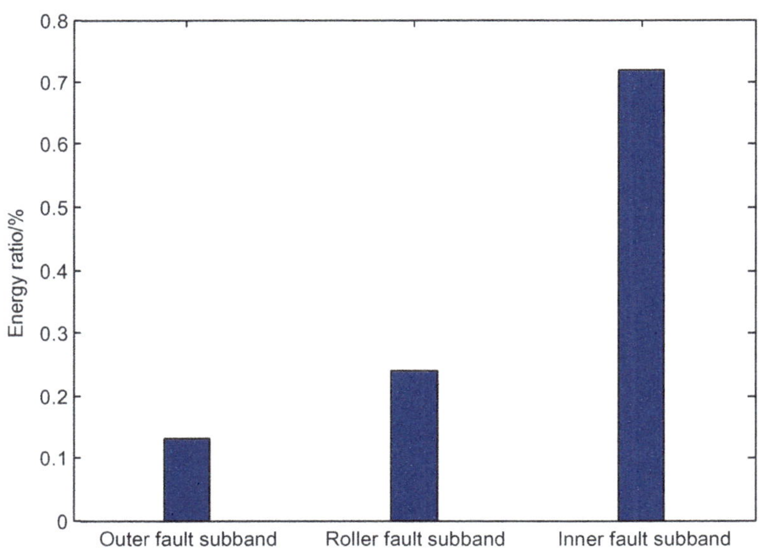

Fig. 6 Energy ratio of the fault sub-band

Table 2 Parts of the eigenvector value

	Normal bearing	outer ring failure	Rolling part failure	Inner ring failure
Mean/g	−0.01500	−0.014907	−0.013997	−0.014203
Root mean square/g	0.028448	0.018940	0.086067	0.037611
Absolute average/g	0.022798	0.014846	0.037791	0.026137
Skew/g^3	7.7024e−07	−4.0969e−07	1.8042e−05	1.4762e−05
Kurtosis /g^4	2.0046e−06	4.2822e−07	0.0033529	2.7704e−05
Variance /g^2	0.00080930	0.00035873	0.0074076	0.0014146
Maximum/g	0.10198	0.077468	1.1385	0.40330
Minimum/g	−0.12232	−0.082748	−1.3806	−0.33980
Mean frequency/Hz	1.2262	1.0737	5.3819	2.1577
Standard deviation/g	4.946	2.9594	10.951	4.6512
Frequency center/Hz	1176.9	1150.2	1370.8	1290.2
Root mean square frequency/Hz	1418.9	1383.6	1556.3	1507.1

highest one but that number is lower then 0.8%. That means the vibration signal we got is weak fault signal.

Carry out the initial feature extraction of each vibration signal. These feature include domain characteristics, frequency domain characteristics and spectral characteristics. The characteristics of each condition are showed in Table 2.

After getting the initial eigenvector of each sample, the multi-scaleclass distance criterion can be used to reduce the dimension of each high dimensional eigenvector. First we choose those initial vectors who represent time domain and frequency domain characteristics to carry out dimension reduction operation. The result is showed in Fig. 7.

From the Fig. 7 we can see that when use time domain and frequency domain characteristics as the initial sample set, the PCA dimension reduction result is not very well. Each fault class can not be distinguished with each other. Thus, then we choose spectral vector as the initial sample set to carry out PCA dimension reduction.

From Fig. 8 we can see that after traditional PCA dimension reduction the classification result of different characteristic parameters is not very well. Although the classification result of those who use spectral vector as initial characteristics is much better but this result still have the problem of the distance in class is too large and the distance between classes is too small. In that case, we choose spectral vector to carry out PCA dimension reduction operation and then use multi-scale class distance for further operation.

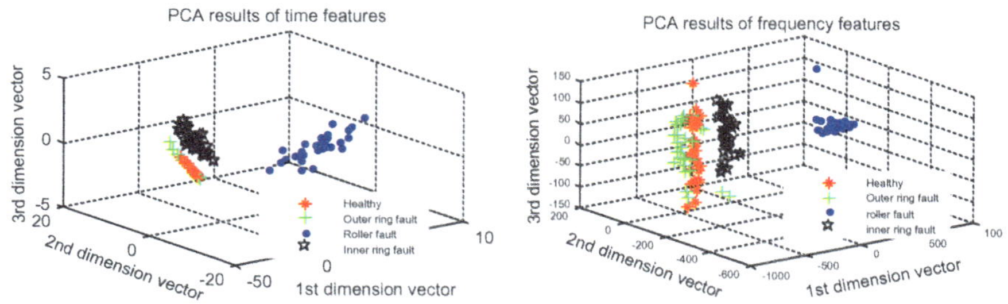

a) PCA dimension reduction result for time domain characteristic

b) PCA dimension reduction result for frequency domain characteristic

Fig. 7 PCA characteristic dimension reduction result for time domain and frequency domain

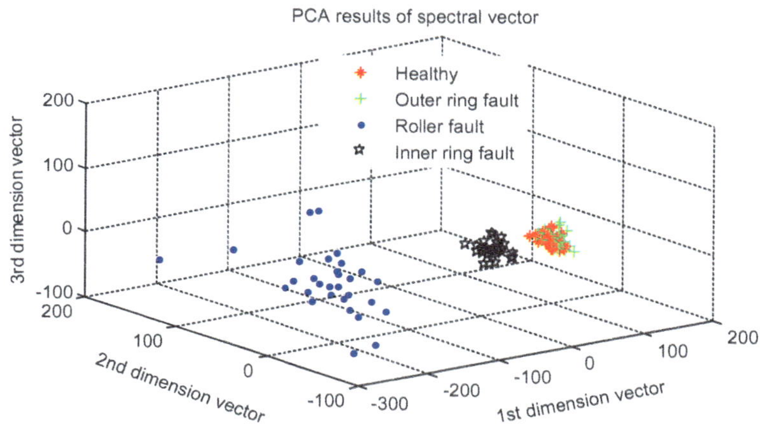

Fig. 8 Characteristics choose based on multi-scale class distance

When use spectral vector as initial characteristics, the sample matrix is 120 * 5120 dimensions. After PCA-MSCD operation, a 2-dimension classification result can be gotten. The result is showed in Fig. 9.

Apparently there are 4 fault subset classes of different conditions in Fig. 9. This is consistent with the default sample data type, proving the correctness of the PCA-MSCD. Comparing Figs. 8 and 9, we can see that after MSCD operation, the separability between the four fault classes has been greatly improved, especially the result between normal bearing and outer-ring fault shows high separability. Achieving the target of the distance between classes is large and the distance within the class is small. This result prove the efficiency of the method proposed in this paper.

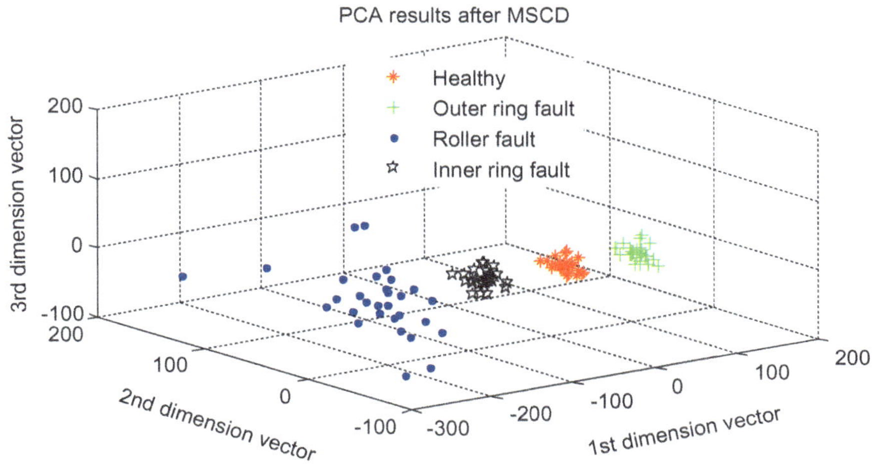

Fig. 9 PCA-MSCD result

4 Conclusion

In this paper we propose MSCD-PCA sample classification method. And apply this method to the classification of weak fault of subway gearbox, acquire fault classification results with high separability successfully. Aiming at the defect of single overall scale class distance criterion (the distance in class is too large and the distance between classes is too small), this paper introduce the multi-scale analysis thought. Combine with PCA dimension reduction effect and base on the overall scale, this method can extract sample matrix feature, reduce the dimension of high dimensional data, and get a preliminary result. Base on the result, the classification operation in the next scale can be done. Repeat the classification processes until a high separability result can be gotten. Use this method in the classification of subway gearbox weak fault. Comparing the result with the one based on traditional class distance classification, we can see this method can improve the classification result significantly.

References

1. Cui, L., D. Mo, and W. Na. 2014. Application of sparse signal decomposition using dual-BP in gear-box weak fault diagnosis. *Chinese Journal of Scientific Instrument* 35(11): 2633–2640.
2. Everitt, B., and T. Hothorn. 1980. Cluster analysis. *Quality & Quantity* 14 (1): 75–100.
3. Li, M., Q. Yang, S. Meng, et al. 2014. Vibration signal feature extraction and diagnosis of diesel engine faults with the combination of multi-characteristic parameters. *Automotive Engineering* 36 (4): 438–442.

4. Cui, L., N. Wu, W. Wang, et al. 2014. Sensor-based vibration signal feature extraction using an improved composite dictionary matching pursuit algorithm. *Sensors* 14 (9): 16715–16739.
5. Peng, Z., F. Chu, and Y. He. 2002. Vibration signal analysis and feature extraction based on reassigned wavelet scalogram. *Journal of Sound & Vibration* 253(5): 1087–1100.
6. Candès, E.J., X. Li, et al. 2009. Robust principal component analysis. *Journal of the ACM* 58(3): 1–73.
7. Hyvarinen, A. 1999. Fast and robust fixed-point algorithms for independent component analysis. *IEEE Transactions on Neural Networks* 10(3): 626–634.
8. Zhang, Z., J. Wang, and H. Zha. 2012. Adaptive manifold learning. *IEEE Transactions on Pattern Analysis & Machine Intelligence* 34 (2): 253–265.
9. Everitt, B.S., G. Dunn, B.S. Everitt, et al. 2011. *Cluster analysis.* Wiley.
10. Wu, X., Y. Zhou. 2012. Cluster analysis and display of target feature sensitivity evaluation method based on clustering analysis and geometry. *Control and Decision* 27(6): 915–918.
11. Cai, Zheyuan, Jianguo Yu, Xianpeng Li, et al. 2010. Feature selection algorithm based on kernel distance measure. *Pattern Recognition and Artificial Intelligence* 23 (2): 236–238.
12. Xu, Yao, Xiaodan Wang, and Yuxi Zhang. 2012. A survey of feature selection. *Control and Decision* 27 (2): 162–165.
13. Yeung, D.Y., H. Chang, and G. Dai. 2007. Learning the kernel matrix by maximizing a KFD-based class separability criterion. *Pattern Recognition* 40 (7): 2021–2028.
14. Patricia, Henriquez, Jesus B. Alonso, Miguel A. Ferrer, etc. 2014. Review of automatic fault diagnosis systems using audio and vibration signals. *IEEE Transactions on Systems Man and Cybernetics: Systems* 44(05): 642–652.
15. Wang, Bin, Jian-yu Zhang, Li-xin Gao. 2012. Dynamic response analysis for rolling bearing vibration measuring point. *Journal of Vibration and Shock* 31(19): 166–168.
16. Benedetto, J.J., J.S. Paulo, and G. Ferreira. 2001. Modern sampling theory: mathematics and applications. Birkhauser Boston, Boston.
17. Randall, R.B., J. Antoni, and S. Chobsaard. 2001. The relationship between spectral correlation and envelope analysis in the diagnostics of bearing faults and other cyclostationary machine signals. *Mechanical Systems and Signal Processing* 15 (5): 945–962.
18. Antoni, J., and R.B. Randall. 2002. Differential diagnosis of gear and bearing faults. *Journal of Vibration and Acoustics* 124 (2): 165–171.
19. Randall, R.B., and J. Antoni. 2011. Rolling element bearing diagnostics-A tutorial. *Mechanical Systems and Signal Processing* 25 (2): 485–520.

Research on Face Recognition Algorithms in the Context of Power Internet of Things

Qionglan Na, Dan Su, Yixi Yang, Jing Lou, Jia Wu, and Jing Zeng

Abstract According to the unified arrangement of the State Grid Corporation of China, combined with full service 2022 Zhangjiakou "Zero Carbon Winter Olympics" and the State Grid Corporation of China " Three types two networks" of the actual work (Ran et al. in Application of self-service analysis tool for data management application, pp 243–249 (2020) [1]), the construction of the power of things put forward higher intelligent requirements, artificial intelligence is the future of scientific and technological progress. Facial recognition technology is the key, and face recognition technology is the leader in artificial intelligence. This paper introduces the fundamental techniques of face recognition, starting from the strategic goal of "Three Types, Two Networks, World Class" of State Grid Corporation of China, and It gives different application directions according to the actual power system and finally puts forward suggestions and prospects for the development of power intelligence.

Keywords Artificial intelligence · Face recognition · Zero carbon winter Olympics · Power IoT

1 Introduction

With the rapid development of Internet technology, the industrial revolution is accelerating, information is spreading rapidly, and personal information protection has received unprecedented attention. Face recognition technology has natural advantages in non-contact, high recognition rate, non-mandatory, etc., making it easier to integrate with professional technology, so it has attracted the attention of experts and scholars from all walks of life. At the same time, under the background of the development strategy of "three types and two networks" of State Grid Corporation

Q. Na (✉) · D. Su · J. Lou · J. Wu · J. Zeng
State Grid Jibei Information and Telecommunication Company, Beijing 100053, China
e-mail: 81885883@qq.com

Y. Yang
State Grid Information and Telecommunication Branch, Beijing 100761, China

W. Chen et al. (eds.), *The Proceedings of the 9th Frontier Academic Forum of Electrical Engineering*, Lecture Notes in Electrical Engineering 743,
https://doi.org/10.1007/978-981-33-6609-1_27

of China [2], "big data, cloud computing, Internet of Things, mobile computing, and mobile computing" have been widely used in the engineering field in recent years [3, 4]. Application, artificial intelligence," has been rapidly developed as the underlying technology and is widely used in various professional fields of companies. The main research idea of this paper is to organically integrate face recognition technology with the power dispatch communication system and its business, and extend it to existing. The power dispatch information and communication service system.

2 Background

2.1 State Grid Corporation of China's "Three Types, Two Networks, World Class" Strategic Target

In 2019, State Grid Co., Ltd. put forward the strategic goal of "three-type two-grid, world-class." The goal is to establish a "hub, platform, and shared" enterprise and achieve "a strong, smart grid and ubiquitous power Internet of Things [5]." Achieve the integration of the three streams of "energy flow, business flow, and data flow" to form a world-class energy Internet enterprise that is safe, reliable, efficient, and high-quality [6].

2.2 Creating a Zero-Carbon Winter Olympics Provides a Significant Historical Opportunity

Around the creation of fresh, green winter Olympic particular area power grid, the ubiquitous power IoT comprehensive coverage, real-time connection energy power "cloud net end" all chain business, will cause the transformation of the IoT application architecture and model, through physical information convergence and the new generation of power system interpenetration and Intelligent interaction, realizing real-time online connection and integrated development of people, machines and things in all aspects of energy and power production and consumption [7]. Information and Communication Branch of State Grid Jibei Electric Power Company Limited researches face recognition technology and gradually builds support for the power grid's whole business. The company is committed to promoting the development of new energy internet businesses and the realization of zero carbon emissions from energy consumption in the core area of the Olympic Games [8].

2.3 Natural Advantages of Facial Recognition Technology in Power Systems

The operation and management of the power grid involves a large amount of data and requires the collection and analysis of various data types, which provides a perfect scenario for the Application of artificial intelligence technologies such as big data analytics, speech recognition, image recognition, and so on.

Artificial intelligence technology is the direction of development for the whole society, and State Grid Corporation Limited has also issued guidelines on artificial intelligence. Face recognition technology is at the forefront of artificial intelligence, which is the key to planning the future of State Grid Corporation Limited.

3 Introduction to Face Recognition Technology

In the past five years, technologies such as artificial intelligence, big data, and cloud computing have developed rapidly, and biometrics technology that uses face recognition as an application has also been significantly developed. More and more face recognition technologies are used in industrial production, and It has been widely used in daily life. At the same time, with the demand for biometrics in the field of power information and communication, more and more power companies are devoting their energy to relevant localized research and development. With the construction of the power Internet of Things, The application scenarios of face recognition are also continually digging [9].

3.1 Facial Recognition Technology Development History

Facial recognition technology shifted from neurological and cognitive science to engineering in the 1950s, and facial recognition technology in the 1960s The Application of facial recognition technology gradually entered the field of engineering. At that time, face recognition technology was mainly concerned with the study of face projection using geometric structures, and the study of facial features—attributes and the direct physical relationship between these features to realize the grounding of face recognition technology. Up to now, face recognition technology has also been commonly used in various industries for decades and has had good results (Fig. 1).

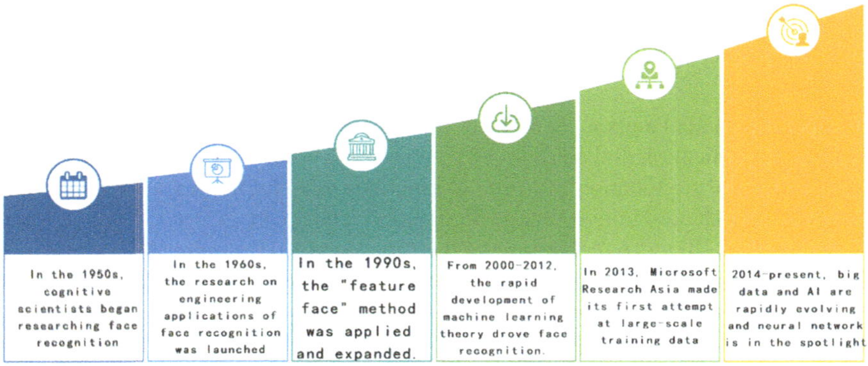

Fig. 1 The development of face recognition

3.2 *Face Recognition Technology Principle*

Facial recognition technology refers to the summarization and conclusion of facial features to achieve recognizable results, and then judge the identified person's Biometric technology for identity. Acquires a stream of images or video from a camera or camcorder and detects images or video containing facial features in the detected images or video, which is tracked and detected, is a technology called face recognition.

3.3 *Facial Recognition Technology Process*

The development of facial recognition is advancing by leaps and bounds in the field of biometrics, mainly because of the advantages in various aspects such as convenience and accuracy. Specifically, it includes features such as non-contact, non-intrusive, perfect hardware foundation and fast and convenient acquisition, and good scalability. Generally speaking, the face recognition system includes four parts, which are: face image acquisition and detection, face image pre-processing, and face detection system. Four aspects of image feature extraction and matching and recognition are covered (Fig. 2).

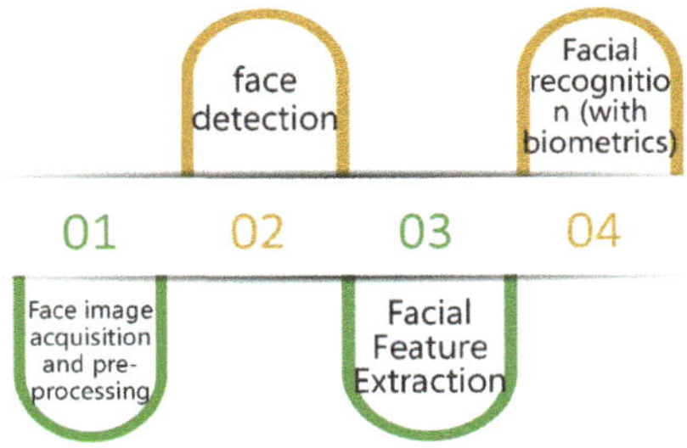

Fig. 2 Face recognition technology flow

3.4 Face Recognition Technology Industry Chain

The face recognition industry chain consists of upstream, midstream, and downstream layers, with upstream being the underlying layer, including AI chips, algorithms, and the Data collection; midstream includes technologies such as face recognition of graphic images, face recognition of captured video, and face feature database matching. In the field, the companies in this field are mainly embedded manufacturers and provide a package of solutions for the downstream. The downstream mainly aims to solve various application scenarios, such as the intelligent security field.

3.5 Face Recognition Technology Implementation Basis

With the popularization of big data and Internet of Things technologies, the number of PCs, clients, mobile terminals, etc. that can access the Internet all over the world is increasing. The structured and unstructured data generated are also increasing geometrically. The more data you have, The higher the efficiency of machine learning and deep learning, as the number of data increases, the more problems the machine can solve.

Due to the full Application of GPUs in deep learning computing, the efficiency of deep learning algorithms processed by GPUs is much higher than that of CPUs, and more and more calculations enter the cloud. Cloud computing reduces the difficulty of implementing in-depth learning calculations and reduces users. Cost of learning. The GPU adopts the traditional computing model based on the data center architecture. It has a faster machine learning training framework. By using GPU chips to build clusters, it is possible to iterate the network faster and make deep learning processing faster and more accurate.

With the increase of big data and industrial applications, the demand for in-depth learning algorithm research has also increased, such as the use of open-source frameworks such as Berkeley's Caffe, Google's TensorFlow and Torch, which has led to the rapid development of deep learning technology, enterprises, and research. Organizations can deploy deep learning applications faster while also providing more tools for open source (Table 1).

Table 1 Face recognition industry-related policies

Time	Policy	Related content
March 2017	Government Report 2017	It was pointed out that we should accelerate the cultivation and growth of emerging industries, including artificial intelligence
July 2017	State Council Circular on the Issuance of a New Generation of Artificial Intelligence Development Plan	By 2020, AI technologies and applications will be in step with the world's advanced level, and the scale of AI core industries will exceed 1,500. Billion, driving the scale of related industries to over 1 trillion yuan
December 2017	Three-year Action Plan to Promote the Development of a New Generation of Artificial Intelligence Industry (2018–2020)	From promoting industrial development, combined with "Made in China 2025", the relevant tasks of the "Development Plan for a New Generation of Artificial Intelligence" have been refined and implemented, with the deep integration of information technology and manufacturing technology as the mainline, with the industrialization and integrated Application of a new generation of artificial intelligence technology as the focus, to promote the deep integration of artificial intelligence and the real economy
November 2018	2018 Government Report	We will make more significant and more robust emerging industrial clusters, implement big data development actions, strengthen the research and development and Application of new-generation artificial intelligence, and promote "Internet Plus" in various fields such as medical care, elderly care, education, culture, and sports
November 2018	Work plan for unveiling the list of critical tasks for innovation in the next-generation artificial intelligence industry	Deploy smart products, core foundations, smart manufacturing, support systems, and other critical mission directions, solicit and select a group of enterprises, research institutes, etc. that master key core technologies of artificial intelligence, strong innovation capabilities, and excellent development potential, to carry out "discovery" Strive to achieve breakthroughs in iconic technologies, products, and services

(continued)

Table 1 (continued)

Time	Policy	Related content
March 2019	2019 Government Report	Upgrade artificial intelligence to "Smart + ," build an industrial internet platform, expand "Smart + ," and empower the transformation and upgrading of the manufacturing industry

4 Research Method

4.1 Template Matching Algorithm

The method determines whether the target face is in the face library by using the vector distance of the face-protected image location in the captured image, Whether the detected faces are similar. For example, facial feature datasets include factors such as nose, eyes, face shape, etc., and comparing features to determine if they are similar is face recognition, The basic approach of technology. This method requires less data volume, higher recognition efficiency and faster recognition speed, and lower configuration requirements for the recognition system.

4.2 Feature Face Algorithm

The algorithm is based on the PCA method, i.e., PCA principal element analysis, a template matching-based optimization algorithm with high accuracy and efficiency and is one of the underlying algorithms for face recognition.

4.3 Adaboost Face Detection Algorithm

Face detection is the starting step of the algorithm in the system to find and label faces in the input image. The face detection algorithm is implemented based on the cascade classifier trained by Adaboost algorithm, The Adaboost algorithm is an adaptive classification method, where a sample that has been misclassified in the previous classifier will proceed to the next one. Classifier for adaptive training. The algorithm is characterized by the rapid calculation of Haar-like features using an "integral image," which is then calculated from the Haar. The poorly composed classifiers are combined into better classifiers. Finally, multiple robust classifiers are cascaded to form stronger classifiers for Face detection.

In order to solve this problem, the following improvement methods are proposed here: setting a limit condition for the update of the weight value, and setting a threshold value in each round of the algorithm update W_m [10].

$$W_m = \left(\sum_{i=1}^{N} W_{m,i}, i\right)/N \tag{1}$$

$$\begin{aligned} e_m &= P(G_m(x_i) \neq y_i) \\ &= \sum_{i=1}^{N} W_{mi} I(G_m(x_i) \neq y_i) \end{aligned} \tag{2}$$

where: $W_{m,i}$ is the weight of the algorithm in the mth iteration; W_m is the threshold of the mth round, and its size is the value of the round The mean of the weights of all samples trained (N is the number of samples). From Eq. (2), $G_m(x)$ the error e_m on the training set $G_m(x)$ is the sum of the misclassified samples.

$$\beta = \frac{1}{2} \log \frac{1 - e_m}{e_m} \tag{3}$$

$$W_{m+1,i} = \begin{cases} W_{m,i}\beta, \; If G_m(x_i) = y_i \\ W_{m,i}\beta^{-1}, \; If G_m(x_i) \neq y_i, \quad \text{And} W_{m,i} < W_m \\ W_{m,i}, \; If G_m(x_i) \neq y_i, \quad \text{And} W_{m,i} > W_m \end{cases} \tag{4}$$

where β is the weight of the primary classifier in the final classifier, from Eq. (3), $\beta > 0$ for $e_m < 1/2$, which β increases as e_m decreases and the smaller the classification error, the smaller the primary classifier will be in the final classification. The more useful the classification is. From Eq. (4). The weight of the next iteration only increases when the weighting factor obtained from the iterative operation is less than the threshold set in this round. Otherwise, it remains unchanged. In this way, the performance degradation of the algorithm due to the increasing sample weight can be avoided to a certain extent.

5 Discussion

5.1 Face Recognition Access Control

Users who use the computer room need to submit the work order and face information through the work order system. Once the person in charge of the work order system passes the approval of the person in charge, it will be automatically sent to the face

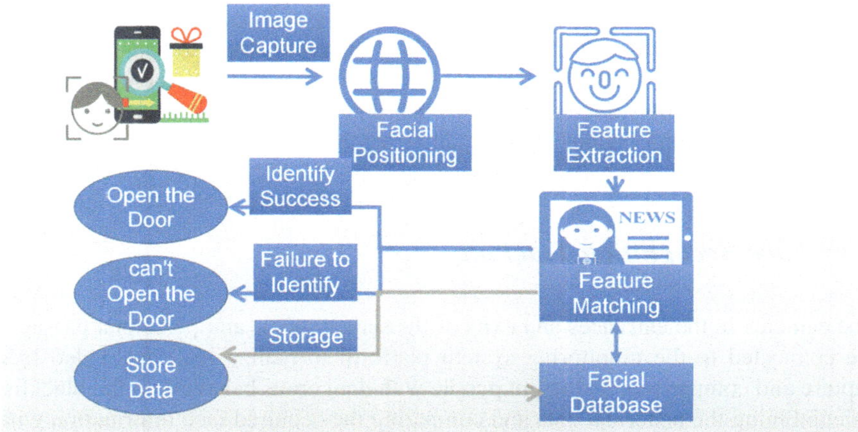

Fig. 3 Flow chart of face recognition access control in dispatching room

recognition access control system that needs to use the machine room to complete face authentication. You can enter the machine room by swiping face recognition. After you leave the machine room, the authority is automatically revoked. Every time you enter, you need to submit a ticket application online.

In addition to the technical personnel completing the work order tasks, other personnel must apply for the signature to the person in charge of the relevant management department before entering the computer room. Other personnel collect face information at the entrance of the computer room, enter the computer room after double verification of face recognition and card swipe. After passing the face recognition verification, the work permit must be swiped before the controller can start the access control door. The face information must be consistent with the work permit information, and the access authority is automatically cancelled after the staff leaves the computer room. You need to reapply and collect facial information before entering the next time.

After the visitor completes identification through the facial recognition system, the system issues a command to open the electric lock and records the time of opening. When the visitor leaves the server room, the facial recognition system needs to identify the visitor at the inner door and open the electric lock only after confirming his identity. Let go and revoke the visitor's face privileges, as well as record the time out (Fig. 3).

5.2 Face Recognition Monitoring

Establish a monitoring center in the dispatching computer room, use high-definition face recognition cameras to confirm the identity of the people in and out of each computer room and monitor their behavior, so that you can clearly understand the operation of personnel in various scenarios in the computer room, and deal with the situation in time. Thus, the operation standardization of the control room is strengthened, the safety and efficiency of the room management are improved, and the system operation is stable and safe.

5.3 Face Recognition Tracking

The cameras in the entrances and exits of the server rooms and personnel passages are connected to the monitoring system platform to realize real-time video face capture and comparison, and to put people with dangerous behavior on the blacklist by combining the historical data and comparing the acquired face information with the blacklist, to discover suspicious people in a useful and timely manner and take relevant measures quickly, and display the map trajectory of suspicious people on the monitoring screen so that that suspicious people can be thoroughly investigated afterward. The area where they go can be shown visually, providing strong support for post-event forensics and loophole analysis.

5.4 Staffing Statistics

Leaves the server room. Depending on the content of the work performed by the person entering the server room, the person can customize the application time in the server room, but can not exceed the maximum time limit specified. If the personnel in the server room exceed the requested operating hours, they are considered strand-ed, and the monitoring system will issue an alert to the security personnel in the control room. Those who are stranded should re-submit their request for the server room's use within the specified time, or they should leave the server room immediately. If the specified time is exceeded, their face information will be automatically entered into the blacklist system.

5.5 Detention of Staff in the Engine Room

The access control system automatically records when a server room user enters and leaves the server room. Depending on the content of the work performed by the person entering the server room, the person can customize the application time in the server room, but can not exceed the maximum time limit specified. If the personnel in the server room exceed the requested operating hours, they are considered stranded, and the monitoring system will issue an alert to the security personnel in the control room. Those who are stranded should re-submit their request for the server room's use within the specified time, or they should leave the server room immediately. If the specified time is exceeded, their face information will be automatically entered into the blacklist system.

6 Conclusion

In this paper, we analyze the "Three Types, Two Networks, World Class" strategy of State Grid Corporation of China, and discuss the significant history of the Zero Carbon Winter Olympics. The introduction elaborates on the principles of face recognition technology and industrial structure, focusing on template matching algorithms, feature face algorithms, and the Adaboost face detection algorithm. Finally, the Application of the face recognition algorithm in the field of power IoT is introduced. With the rapid development of innovative technologies such as deep learning, big data, and artificial neural networks, a new and dynamic era of artificial intelligence has arrived. Promoting the Application of face recognition technology in the construction of power IoT to build intelligent dispatching capabilities is the future trend, and AI. The challenges of technology to the power IoT.

Acknowledgements This work is supported by the Science and Technology Project of State Grid Jibei Power Company Limited (No. 52018E190075).

References

1. Ran, R., Z. Lei, D. Zhou, et al. 2020. Application of self-service analysis tool for data management application. *The International Conference on Cyber Security Intelligence and Analytics*: 243–249.
2. Wu, S., G. Ji, Q. Duan, et al. 2019. Development strategy of energy internet industry for power grid enterprises. In *International Conference on Frontier Computing*, pp. 1284–1293.
3. Xu, Z., K.-K.R. Choo, A. Dehghantanha, et al. 2019. Cyber security intelligence and analytics, 928. Springer.
4. Wu, J., L. Zhang, H.J.M. Lv. 2020. The concept layer, feature layer and landing layer of ubiquitous internet of things in power system 768(6): 062039.
5. Li, N., and K. Zheng. 2020. Regional energy internet construction framework and key tasks 446(2): 022034.
6. Shuming, S., Z. Yongjia, X. Yongjin, et al. 2019. Research on standard system architecture of metrology technology for energy interconnection. *Journal of Physics: Conference Series*: 042072.
7. Tian, W., X. Li, and F. Shang. 2019. Design scheme of electric IoT wireless private network. In *2019 6th International Conference on Systems and Informatics (ICSAI)*, pp. 314–318.
8. Siwu, L., X. Linzhi, M. Xiaodi, et al. 2019. Integrated energy service platform under the umbrella of ubiquitous power internet of things. *Journal of Physics: Conference Series*: 012052.

9. Xue, Y., Y. Zheng, and S. Rahman. 2019. *Proceedings of PURPLE MOUNTAIN FORUM 2019-international forum on smart grid protection and control: Volume II*, 585. Springer.
10. Liu, X., and J. Liu. 2015. Spam filtering based on AdaBoost and active learning methods. In *Electronic engineering and information science: Proceedings of the international conference of electronic engineering and information science 2015 (ICEEIS 2015)*, pp. 409, 17–18 Jan 2015, Harbin, China.

A Quadra-Layer Direct Speed SMPC Strategy for PMSM Rotor Position

Yao Wei, Mengyuan Li, Yanjun Wei, and Hanhong Qi

Abstract Sequential model predictive control (SMPC) is an effective method to eliminate the weighting factors when two objectives or variables need to be controlled at the same time. However, a problem is that the cost function with multiple objectives or variables cannot operate on the sequential structure. A quadra-layer direct speed SMPC (Q-DS-SMPC) method for permanent magnet synchronous machines (PMSMs) rotor position is proposed in this paper. The rotor position control system with multiple control objectives and a switching vector table with 14 vectors have been applied to select the effective vectors in each model predictive control (MPC) process. Moreover, according to the tracking performances of rotor position and the importance of objectives, a sort of divided MPCs and a group of extracted values of cost function for limitation have been decided. Besides, a multiple layers SMPC principle and a selecting method for the cost functions in each divided MPC have been discussed in this paper. Simulation and experimental results corroborate the benefits of control performances and demonstrate the correctness of the theories and analyses.

Keywords DS-SMPC · Multiple objectives · PMSM · Switching vector selecting processes

---

Y. Wei · M. Li · Y. Wei · H. Qi (✉)
College of Electrical Engineering, Yanshan University, Qinhuangdao 066004, China
e-mail: hhqi@ysu.edu.cn

Y. Wei
e-mail: weiyao@stumail.ysu.edu.cn

M. Li
e-mail: 1253424371@outlook.com

Y. Wei
e-mail: yjwei@ysu.edu.cn

W. Chen et al. (eds.), *The Proceedings of the 9th Frontier Academic Forum of Electrical Engineering*, Lecture Notes in Electrical Engineering 743,
https://doi.org/10.1007/978-981-33-6609-1_28

1 Introduction

The MPC has been widely researched in the realm of power electronic and power transmission such as inverter, rectifier, matrix converter and motor according to different applications and environments [1, 2]. A high-performance control system with high tracking precise and fast dynamic response need to operate for motor system. One effective improved method is the adjustment of the cost function. The MPC strategy for the PMSM system can be divided into predictive current control (PCC), predictive speed control (PSC) and predictive torque control (PTC) according to controlling objectives of the cost function [3, 4].

The structure of MPC with PMSM has been improved as different advanced structures such as the sequential MPC (SMPC). The SMPC structure divides the cost function with multiple objectives into a group of cost functions and each of cost function has a single objective. The MPC compensation block is inserted into the sequential structure and the predicted variables are at time $k + 2$ [5]. SMPC with two vectors has been applied in the PMSM system without any weighting factors which are the essential factors of the cost function in MPC [6]. This method removes the online or offline tuning processes of weighting factors and an improvement realizes high-quality performance based on PTC [7]. Moreover, a SMPC with three vectors has been applied in the first layer of sequential structure and the performances have been further improved [8, 9]. An expanded switching table including three zero-vectors and 18 nonzero-vectors has been used to select the switching vector in a grid-connected three-level circuit, and a cascaded optimal switching sequence MPC (OSS-MPC) method which changes the outer MPC into a sequence structure has been used to control the circuit. This method reduces the burden of hardware and optimally controls both the grid currents and the capacitor voltages [10].

A quadra-layer direct-speed SMPC (Q-DS-SMPC) for PMSM rotor position has been proposed and the principle of multiple layers sequential structure of SMPC has been analyzed in this paper. Four objectives including three tracking objectives and one limitation of rotor position control with Q-DS-SMPC are divided into four single calculation processes of MPC, and each of the process includes one objective. A comparison between three typical sorts of objectives is displayed the performances of the Q-DS-SMPC method with different sorts and different extracted values of the cost function which participate the processes of limitation objective. The proposed method can provide advantages and theoretical contributions including sequential method of multiple-layer SMPC, determined location of extracted cost function values of limitation objective and the sort of objectives in SMPC, and comparison between typical three different sorts. According to the servo stiffness, the parameter of position controller and the integral of time-weighted absolute value of the errors (ITAEs) in tracking processes of the references with different amplitudes, frequencies and initial phases, the principles of Q-DS-SMPC and the performances of proposed method are verified by simulation and experimental results.

2 Multi-layer Structure of SMPC

2.1 Multi-layer Sequential Model Prediction Control

The predictive horizon of conventional MPC is selected as one sampling time T_s and the objectives including tracking objectives and variable limitations. The objectives and limitations are summarized as a cost function which can be expressed as:

$$\min J(k+1) = \sum_{i=1}^{P} q_i \left\| \mathbf{x}^*(k+1) - \mathbf{x}(k+1) \right\|^2 + \sum_{j=1}^{M} r_j \Delta \mathbf{u} \tag{1}$$

where $\boldsymbol{x}^*(\mathrm{k}+1)$ and $\boldsymbol{x}(\mathrm{k}+1)$ are the vectors of state variables at time $\mathrm{k}+1$ and $\Delta\boldsymbol{u}$ is the vector of variable limitations. q_i and r_j are weighting factors, and P and M are the number of tracking objectives and limitations respectively.

The sequential structure divides the cost function in (1) into multiple cost functions with single objective. Each layer of the multiple layers SMPC is regrouped into some MPCs including the primary MPC with main objective and some other MPCs with secondary objectives. The primary objective always selects a tracking objective as the main objective and can be expressed as:

$$\min J_{pr}(k+1) = \left\| x_1^*(k+1) - x_1(k+1) \right\|^2 \tag{2}$$

where x_1 is the main variable in vector $\boldsymbol{x}$. The limitations of variable always select as other MPCs because a linear or a nonlinear function has calculated based on a group of known cost function values. The other MPCs can be expressed as:

$$\min J_{se}(k+1) = \left\| x_2^*(k+1) - x_2(k+1) \right\|^2 \quad \text{or} \quad \min J_{se}(k+1) = \Delta u \tag{3}$$

where x_2 is another variable in vector $\boldsymbol{x}$ and Δu is a limitation in vector $\Delta\boldsymbol{u}$.

A compensation block has been used in SMPC and the state variables sending into the primary MPC are at time $\mathrm{k}+1$. Therefore, the cost functions in (2) and (3) are at time $\mathrm{k}+2$ and can be expressed as follow:

$$\begin{cases} \min J_{pr}(k+2) = \left\| x_1^*(k+2) - x_1(k+2) \right\|^2 \\ \min J_{se}(k+2) = \left\| x_2^*(k+2) - x_2(k+2) \right\|^2 \quad \text{or} \quad \min J_{se}(k+2) = \Delta u \end{cases} \tag{4}$$

2.2 Switching Vector Selection

The number of switching vectors is limited in actual because of limited switches number. In the sequential structure, the angle of effective vectors in the primary MPC need to less than 180° and the system operates toward the respected operating state for the main objective. Therefore, conventional switching vectors SMPC with 8 vectors of inverter only can be selected the maximum 3 effective vectors in the primary MPC, and the whole SMPC only can include the maximum 3 divided MPCs.

Multiple objectives in the multi-layer SMPC need a group of extra switching vectors to expand the switching table which is listed in Table 1 printed in Fig. 1b, where the symbol X means that no switch closed in corresponding bridge arm. The switching table of inverter can be extended into 14 vectors including 2 zero vectors and 12 non-zero vectors. As shown in the figures, the maximum numbers of the objectives of the SMPC and the effective vectors in the primary MPC are increased to 6.

Table 1 Switching table of three-phase inverter with extra-switching states

Symbol	Voltage vector
1X0	$V_9 = V_{dc}/2 + j\ \text{sqrt}(3)V_{dc}/6$
X10	$V_{10} = -j\ \text{sqrt}(3)V_{dc}/3$
01X	$V_{11} = -V_{dc}/2 + j\ \text{sqrt}(3)V_{dc}/6$
0X1	$V_{12} = -V_{dc}/2 - j\ \text{sqrt}(3)V_{dc}/6$
X01	$V_{13} = -j\ \text{sqrt}(3)V_{dc}/6$
10X	$V_{14} = V_{dc}/2 - j\ \text{sqrt}(3)V_{dc}/6$

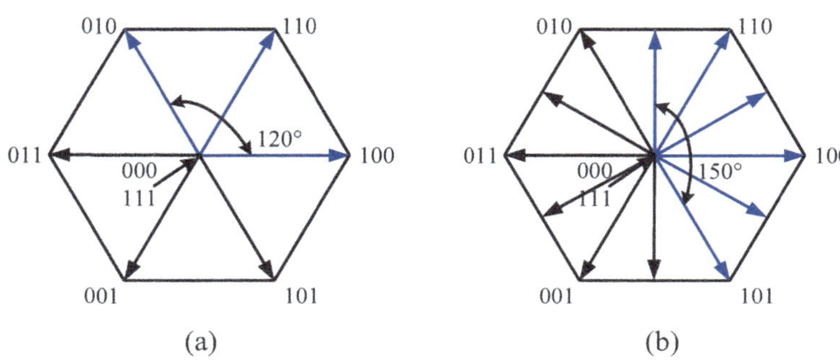

Fig. 1 The sketch of effective vectors in the first layer of SMPC and multi-layer SMPC. **a** Conventional SMPC, **b** Multi-layer SMPC

3 Multi-layer Structure of SMPC

3.1 Quadra-Layer DS-SMPC Strategy

The discrete-time PMSM model is generated by the two-step Euler interpolation method. Based on the discrete-time model, the objectives of PMSM rotor position system with predictive speed control are listed as speed reference tracking, smooth behavior of the electrical torque, torque by ampere optimization and current magnitude limitation. These four objectives can be summarized as follows:

$$J = \lambda_{\omega}\left(\omega_r^*(k+1) - \omega_r(k+1)\right)^2 + \lambda_i(i_{sd}(k+1))^2 + \lambda_{if}\left(i_{sqf}(k+1)\right)^2 + \hat{f}\left(i_{sd}(k+1), i_{sq}(k+1)\right) \tag{5}$$

where λ_{ω}, λ_i and λ_{if} are weighting factors and the $i_{sqf}(k+1)$ is the filtered value of the i_{sq} current. The last term of function in (5) is a nonlinear function for limiting the amplitude of the stator currents and is defined as:

$$\hat{f}\left(i_{sd}(k+1), i_{sq}(k+1)\right) = \begin{cases} \infty \text{ if } |i_{sd}(k+1)| > i\text{max or } \left|i_{sq}(k+1)\right| > i\text{max} \\ 0 \text{ if } |i_{sd}(k+1)| \le i\text{max and } \left|i_{sq}(k+1)\right| \le i\text{max} \end{cases} \tag{6}$$

where i_{max} is the value of the maximum allowed stator current magnitude.

Considering the importance of the objectives, the cost functions of DS-SMPC with four objectives can be divided into 4 functions and each of them can realize single objective. Based on the delay compensation block, the predictive variables at time $k + 2$ are expressed as:

$$J_{pr} = \left(\omega_r^*(k+2) - \omega_r(k+2)\right)^2 \tag{7}$$

$$J_{se1} = (i_{sd}(k+2))^2 \tag{8}$$

$$J_{se2} = \left(i_{sqf}(k+2)\right)^2 \tag{9}$$

$$J_{se3} = \hat{f}\left(i_{sd}(k+2), i_{sq}(k+2)\right) \tag{10}$$

and the numbers of effective vectors in each MPCs are 6, 4, 2 and 1 respectively.

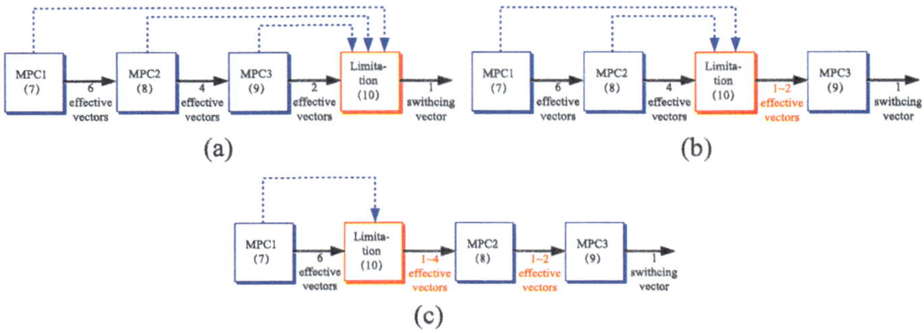

Fig. 2 Three typical possible sorting options of DS-SMPC. **a** Sort1, **b** Sort2, **c** Sort3

3.2 Extra Sort Selecting Method

The sort of cost functions of the DS-SMPC in (7)–(10) is according to the importance of the objectives. Besides, these secondary objectives can be sorted arbitrarily, and there are six sorting options of the last three MPCs.

The cost function (10) needs to be calculated based on a group of known values of cost function. Therefore, the groups of cost function value before the MPC with (10) can be used in the calculating processes of the MPC. Some typical possible sorting options are summarized in Fig. 2 and the cost function values in the MPC with (10) are replaced by a dotted line in blue. The numbers of effective vectors after limitation are unknown.

4 Simulation and Experimental Results

4.1 Simulation Results and Performances Analysis

A simulation environment has been built in MATLAB/simulink software and the parameters of PMSM are summarized in Tabble 2. The structure of DS-SMPC with the sort based on importance for PMSM rotor position system is shown in Fig. 3.

Table 2 Main PMSM parameters

Parameters	Values	Parameters	Values
Number of pole pairs p	4	Magnet flux linkage ψ_m	0.175 Wb
Stator induction L_s	0.835 mH	Rotor inertia J	0.0008 kg.m2
Stator resistance R_s	2.360Ω	Friction coefficient B	0.0008 N.m.s

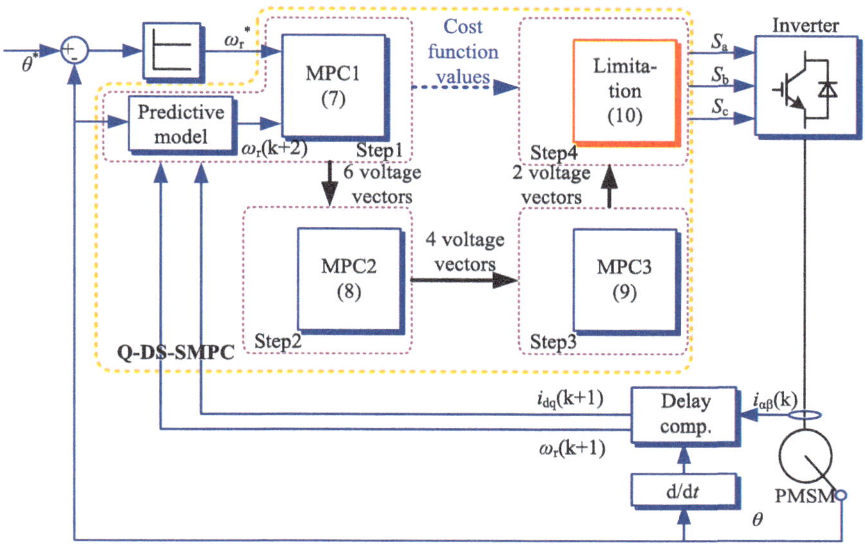

Fig. 3 The structure of Q-DS-SMPC for PMSM rotor position

4.1.1 Tracking Performances of Q-DS-SMPC

One sinusoidal position reference signal with amplitude 10 rad, frequency 5 Hz and 0 initial phases had been used to verify the dynamics of system. The tracking waveforms of rotor position are shown in Fig. 4a, b. ITAE is an important index reflecting the

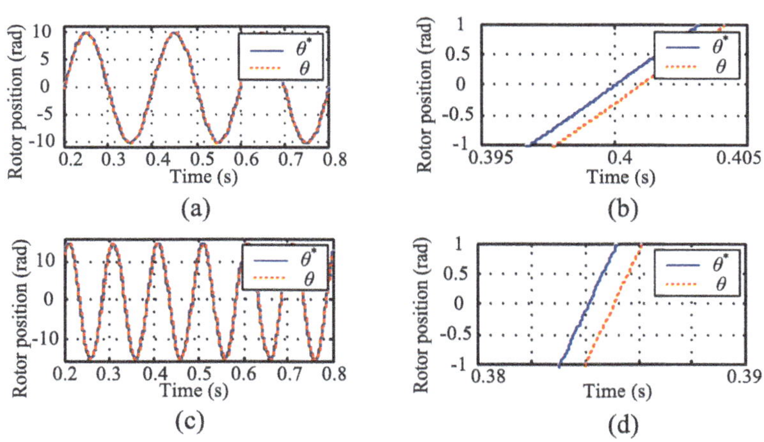

Fig. 4 Waveforms of rotor position. **a** Global waves for the first reference, **b** Enlarged waves for the first reference, **c** global waves for the second reference, **d** enlarged waves for the second reference

comprehensive performance of system, and the ITAEs of rotor position and speed in 0.8 s were 0.06129 and 1.613 respectively.

Another sinusoidal position reference signal with amplitude 15 rad, frequency 10 Hz and 13 rad initial phase had been used, and the simulation results are shown in Fig. 4c, d. The ITAE of rotor position and speed were 0.1865 and 2.264 respectively. Moreover, an overshoot with 6.2267% happened at the first cycle of tracking process due to the larger slope of the first cycle.

Moreover, a required cost function values changed structure and a location of limitation block changed structure of Q-DS-SMPC are shown in Fig. 5, and the simulation results are shown in Fig. 6. The ITAEs of rotor position and speed were 0.1443 and 52.4 in 0.8 s respectively for the sort 1 with the first reference. The ITAEs were increased because the ripples of rotor position and speed were increased sharply. The tracking performances of the sort2 are shown in Fig. 6c, d, and the ITAEs of rotor position and speed were 0.1361 and 39.16 in 0.8 s respectively. The ripple of the

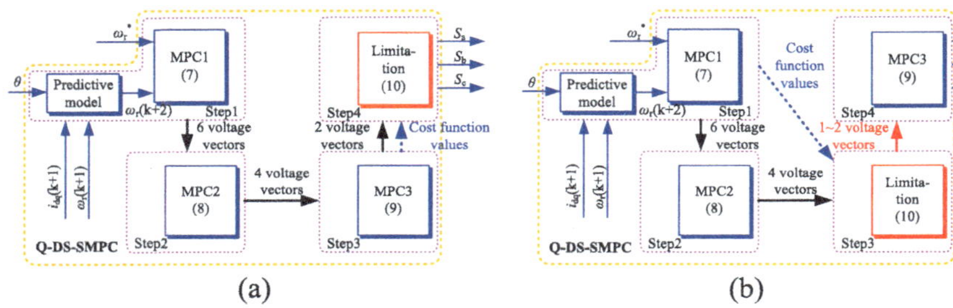

Fig. 5 Changed structures of Q-DS-SMPC. **a** Sort1, **b** Sort2

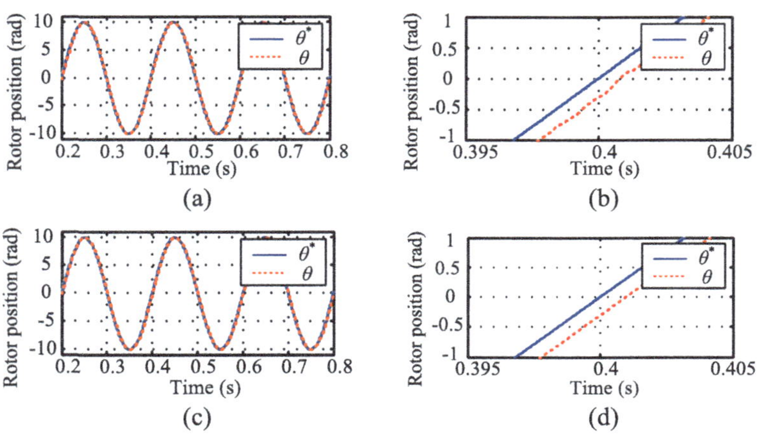

Fig. 6 Waveforms of rotor position. **a** Global waves for sort 1, **b** enlarged waves for sort 1, **c** global waves for sort 2, **d** enlarged waves for sort 2

Table 3 ITAE of rotor position and speed in simulations

Structure	References	ITAE of rotor position	ITAE of speed
Initial sort	Amplitude 10 rad, frequency 5 Hz and 0 initial phases	0.06129	1.1613
Sort1		0.1443	52.40
Sort2		0.1361	39.16
Initial sort	Amplitude 15 rad, frequency 10 Hz and 13 rad initial phase	0.1865	2.264
Sort1		0.2101	19.25
Sort2		0.1959	6.695

system had been decreased. However, the ITAEs were increased because the rotor position in first cycle was not following closely.

Thus, different extracted cost function values for the limitation block could affect the ripple and tracking performance mainly because of the cost function values under different operating states, and different locations of the limitation block affect the tracking performance mainly. The ITAEs for different sorts have been listed in Table 3. Besides, the burden of hardware and software with the changed structures are increased because the number of cost function values and the number of effective vectors in limitation MPC need to be detected and involved in the calculating processes. Therefore, comparing with the results of three sorts, the limitation block is suitable to extract the cost function values from the MPC with primary objective and suitable to locate at the end of the DS-SMPC.

4.1.2 Proportional Parameter of Position Controller

A proportional parameter needs to be tuned in the rotor position controller according to the performances of system including ITAEs of rotor position, ITAEs of speed, overshoots, settling times, static errors and ripples in the tracking processes of the responses. The performances are shown in Fig. 7. As shown in the figures, the parameter in proportion controller had a suitable range of value from 200 to 500 which all performances were appropriate.

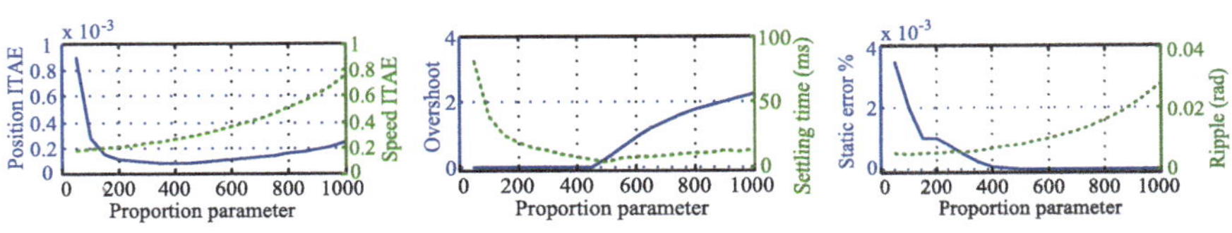

Fig. 7 Performances of proportional parameter of position controller. **a** ITAEs, **b** overshoot and settling time, **c** static error and ripple

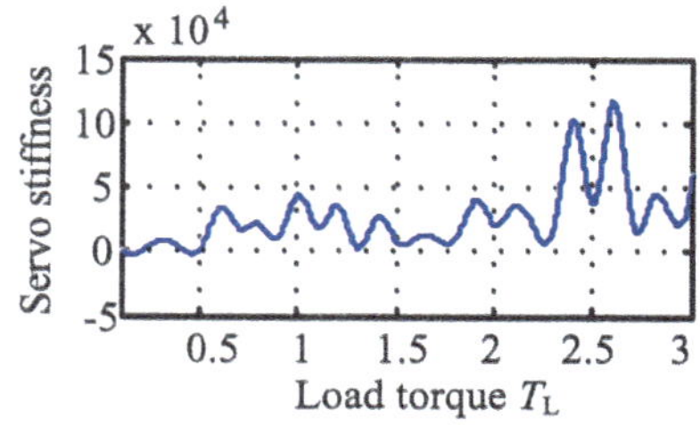

Fig. 8 Servo stiffness of Q-DS-SMPC

4.1.3 Servo Stiffness Test

The servo stiffness is an important index for the PMSM rotor position control. The simulation waveform of the servo stiffness is shown in Fig. 8. As shown in the figure, the average value of servo stiffness of Q-DS-SMPC with load torque range during 0.1–3 N.m is 28,021.32, and the standard deviation of servo stiffness is 26,691.91. The servo stiffness has upward trend with the load torque T_L increasing.

4.2 Experimental Results

An experimental setup has been built including a main circuit and a control circuit mainly. An inverter with IGBTs (FGL35N120FTD) and a 1 kW PMSM (INOVANCE ISMH2-10C30CD) with an incremental encoder (INOVANCE EI34H) are selected as main circuit, and a DSP (TMS320F2812) and a CPLD (EPM240T100I5N) are combined as control circuit. The DS-SMPC algorithm was written into DSP and the system was controlled. The reference signal in experiments is changed as a sinusoidal wave with frequency 0.5 Hz and amplitude 4 rad (equals 229.183°). The experimental waveforms of rotor positions with the initial sort are shown in Fig. 9a, b. The delay time of rotor position is about 72 ms.

Moreover, the experimental results and there enlarged waveforms of the rotor position with the sort 1 and sort 2 are shown in Fig. 9c–f. As shown in figures, these waveforms have similar performances. The delay times of the sort 1 and sort 2 are not improved obviously. However, the obvious ripples of the sort 1 in simulation were almost flooded by noise at other positions.

A sinusoidal reference signal with lower frequency and larger amplitude of 0.2 Hz and 8 rad (equals 458.366°) had been used into the experimental processes. The experimental waveforms of rotor position and reference with different sorts and different extracted cost function values are shown in Fig. 10a–c respectively. The system could track the reference successfully, and the performances of the different sorts conformed to the analyses above-mentioned.

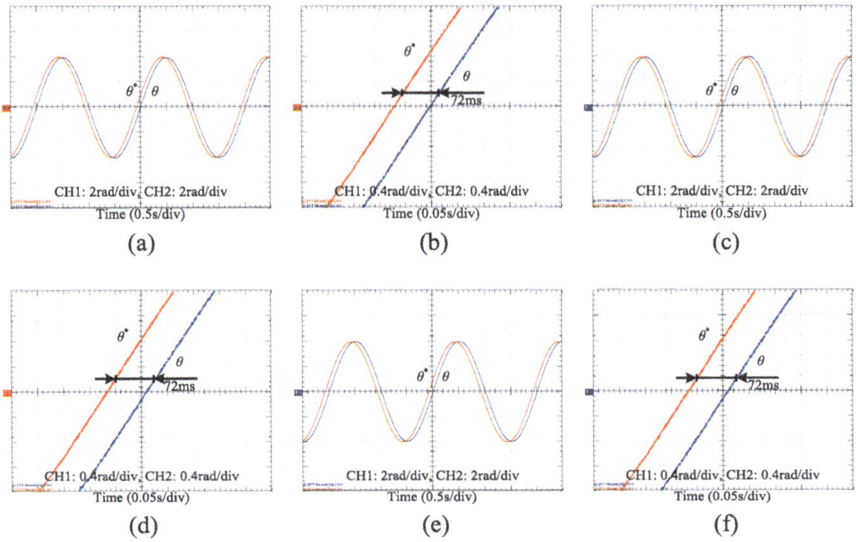

Fig. 9 Experimental waveforms of Q-DS-SMPC for PMSM rotor position system. **a** rotor position for initial sort, **b** enlarged rotor position for initial sort, **c** rotor position for sort1, **d** enlarged rotor position for sort1, **e** rotor position for sort2, **f** enlarged rotor position for sort2

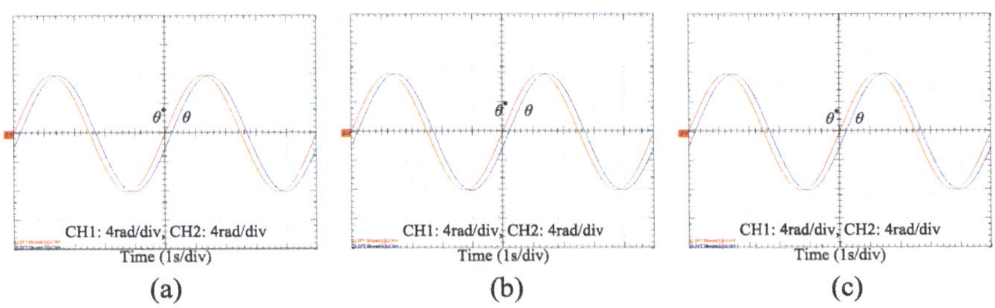

Fig. 10 Experimental waveforms of Q-DS-SMPC for PMSM rotor position system lower frequency and larger amplitude reference. **a** Initial sort, **b** Sort1, **c** Sort2

5 Conclusions

This paper has presented a quadra-layer sequential model predictive control for PMSM rotor position. The principle of multiple layers sequential structure is discussed. The cost functions of proposed method realize objectives without any weighting factors and tuning processes. The performances of different proportional parameters in position controller and servo stiffness are analyzed by results with same sampling times.

Three typical sorting opinions with different sorts and extracted values of cost function are analyzed from all possible sorting opinions. Comparing with three typical sorting opinions, the ITAEs of rotor position are decreased about 57.5 and 55.0% with lower reference frequency and decreased about 11.2 and 4.80% with higher reference frequency. The results show that the objective importance is the key of the sort and the cost function values for limitation are extracted from the primary MPC including the primary objective. Furthermore, the multiple-layer SMPC is limited by the number of switching vector. The main circuit can be improved as a three-level or a multi-level circuit including dozens possible switching vectors. Therefore, more difficulty objectives can be summarized and realized into cost function.

References

1. Geyer, T. 2016. *Model predictive control of high power converters and industrial drives*, 1st ed. Chichedter, U.K.: Wiley.
2. Acuna, P., C.A. Rojas, R. Baidya, et al. 2019. On the Impact of transients on multistep model predictive control for medium-voltage drives. *IEEE Transactions on Power Electron* 34 (9): 8342–8355.
3. Wang, J., F. Wang, Z. Zhang, et al. 2017. Design and implementation of disturbance compensation-based enhanced robust finite control set predictive torque control for induction motor systems. *IEEE Transactions on Industrial Informatics* 13 (5): 2645–2656.
4. Wei, Y., Y. Wei, Y. Sun, et al. 2020. Prediction horizons optimized nonlinear predictive control for PMSM position system. *IEEE Transactions on Industrial Electronics* 67 (11): 9153–9163.
5. Cortes, P., J. Rodriguez, C. Silva, et al. 2012. Delay compensation in model predictive current control of a three-phase inverter. *IEEE Transactions on Industrial Electronics* 59 (2): 1323–1325.
6. Zhang, Y., and H. Yang. 2016. Two-Vector-based model predictive torque control without weighting factors for induction motor drives. *IEEE Transactions on Power Electronics* 31 (2): 1381–1390.
7. Norambuena, M., J. Rodriguez, Z. Zhang, et al. 2019. A very simple strategy for high-quality performance of AC Machines Using Model Predictive Control. *IEEE Transactions on Power Electronics* 34 (1): 794–800.
8. Zhang, Y., B. Zhang, H. Yang, et al. 2019. Generalized sequential model predictive control of IM drives with field-weakening ability. *IEEE Transactions on Power Electronics* 34 (9): 8944–8955.
9. Wang, X., and D. Sun. 2017. Three-Vector-based low-complexity model predictive direct power control strategy for doubly fed induction generators. *IEEE Transactions on Power Electronics* 32 (1): 773–782.
10. Andrés, M., C.D. Roberto, P.A. Ricardo, et al. 2019. Computationally efficient cascaded optimal switching sequence MPC for grid-connected three-level NPC converters. *IEEE Transactions on Power Electronics* 34 (12): 12464–12475.

A Parameters Tuning Method of LADRC Based on Reference Value Filtered Two-Degree-of-Freedom for PMSM Current Control

Yao Wei, Yening Sun, Yanjun Wei, and Hanhong Qi

Abstract Active-disturbance-rejection-control (ADRC) is a kind of control method which can obtain good dynamic and static performances. In order to effectively solve the problem of parameter tuning in ADRC in permanent magnet synchronous motor (PMSM) speed control system and maintain performance, an analysis of motor speed control principle have been made. Based on linear-active-disturbance-rejection-control (LADRC), a parameter tuning method of reference value filtered two-degree-of-freedom (2-DoF) control is proposed for LADRC current controller in speed control PMSM bypass changing and optimizing the structure of controller. The simulation and experimental results show that the stability and rapidity of the system are maintained and the parameters could be optimized by decoupling of system performance, and the correctness of the method for PMSM speed control is verified.

Keywords PMSM · LADRC · Reference value filter structure · 2-DoF control

1 Introduction

ADRC becomes a hotspot to apply to replace PI controller in the servo control system recently, due to its simple realization, strong resisting disturbance, higher dynamics and accuracy. The basic principle of ADRC is that all of the internal and external disturbances are extracted and eliminated actively [1]. The ADRC can be designed by the order of the controlled plant and applied into the control system.

Y. Wei · Y. Sun · Y. Wei · H. Qi (✉)
College of Electrical Engineering, Yanshan University, Qinhuangdao 066004, China
e-mail: hhqi@ysu.edu.cn

Y. Wei
e-mail: weiyao@stumail.ysu.edu.cn

Y. Sun
e-mail: sunyening@stumail.ysu.edu.cn

Y. Wei
e-mail: yjwei@ysu.edu.cn

W. Chen et al. (eds.), *The Proceedings of the 9th Frontier Academic Forum of Electrical Engineering*, Lecture Notes in Electrical Engineering 743,
https://doi.org/10.1007/978-981-33-6609-1_29

The PMSM model is simplified as a 2-order system and the stability is analyzed by the describing function method [2]. Based on the 2-order PMSM model, a load observer and a compensator are combined with the model to resist external disturbance and enlarged model error due to two-order derivations [3]. The PMSM order can be further simplified as a 1-order system and designed the tracking-derivator by the rotor displacement to improve performances [4]. According to the large coverage of ADRC parameters, these parameters are always tuned by results of many simulation and experimental results in industry because of nonlinear structure. Based on the frequency-domain tuning method of PI controller, the bandwidth of controller is combined with the amplitude frequency characteristic and a group of parameters of ADRC with suitable stability are obtained [5]. Moreover, a repetitive control is inserted to the ADRC structure to decrease the parameter sensitivities and to further enhance robustness [6]. The least square support vector machine (LSSVM) can also be inserted to the ADRC structure to train the optimal regression model and to obtain improving parameters with increasing calculation burdens [7, 8]. A reduced order ADRC and a high-order extended state observer (ESO) with feedforward and feedback loops are applied to the PMSM control system respectively to simplify requiring parameters and sensitivities of parameters [9]. Based on the complex frequency-domain, the parameters of LADRC structure change into some bandwidth functions and the complex frequency-domain, and an approximation method and a relationship analysis method between the dynamics and parameters are applied to tuning parameters of 2-order LADRC respectively to further simplify the tuning processes [10].

The IMC structure can be converted to the reference value filtered 2-DoF with two controllers based on the internal model of the plant and applied into PMSM control system to maintain robustness and to decouple reference tracking performance and resisting disturbance performance based on the 2-DoF Smith estimated controller. A stator current LADRC tuning method for the PMSM speed control system is proposed in this paper. The structure of LADRC is converted to the reference value filtered 2-DoF structure and tuned by the IMC complex frequency-domain method. The system with parameters by the proposed method has advantages including better dynamics and steady performances. The correctness and effectiveness are verifies by the simulation and experimental results, and the performances could be improved by the decoupled 2-DoF structure.

2 Basic Principle

2.1 LADRC Basic Principle

LADRC has two main parts including proportional-derivation (PD) control combination and linear ESO (LESO) where the PD control combination is combined by the PD controller and the linear SEF (LSEF). For n-order control system, the LESO can be expressed as:

$$\begin{cases} \dot{z} = \boldsymbol{A}z + \boldsymbol{B}u + \boldsymbol{L}(y - \hat{y}) \\ \hat{y} = \boldsymbol{C}z \end{cases} \tag{1}$$

where

$$\boldsymbol{A} = \begin{bmatrix} 0 \; 1 \; 0 \cdots 0 \\ 0 \; 0 \; 1 \cdots 0 \\ \vdots \; \vdots \; \vdots \; \ddots \; \vdots \\ 0 \; 0 \; 0 \cdots 1 \\ 0 \; 0 \; 0 \cdots 0 \end{bmatrix} \quad \boldsymbol{B} = \begin{bmatrix} 0 \\ 0 \\ \vdots \\ b_0 \\ 0 \end{bmatrix} \quad \boldsymbol{C} = \begin{bmatrix} 1 \\ 0 \\ \vdots \\ 0 \\ 0 \end{bmatrix}^{\mathrm{T}} \tag{2}$$

z is the observing vectors of the state vectors $\boldsymbol{x}$, and $\boldsymbol{L}$ is the error feedback control gain matrix.

The control system poles can be adjusted as the multiple poles to realize parameterization design, i.e. characteristic equation satisfying $\lambda(s) = (s + \omega_o)^{\mathrm{n}}$, where ω_o is the observer bandwidth. The error feedback gain matrix $\boldsymbol{L}$ is only related with ω_o which can be rewritten as β_i in the discretization system. The LESO and LSEF in the discretization system can be expressed as:

$$\begin{cases} z_1' = z_2 - \beta_1(z_1 - y) \\ z_2' = z_3 - \beta_2(z_1 - y) \\ \vdots \\ z_{n-1}' = z_n - \beta_{n-1}(z_1 - y) \\ z_n' = z_{n+1} - \beta_n(z_1 - y) + bu \\ z_{n+1}' = -\beta_{n+1}(z_1 - y) \end{cases} \tag{3}$$

$$u = \frac{1}{b_E}\left(-z_{n+1} + \sum_{i=1}^{n} p_{\mathrm{i}}\left(r^{(i-1)} - z_i\right)\right) \tag{4}$$

where variables p_i and β_i are positive adjustable values, and r is the reference value. The structure of PMSM with stator current LADRCs is shown in Fig. 1.

3 LADRC Design

3.1 Stator Current LADRC Design Method

The decoupled PMSM stator current can be expressed as:

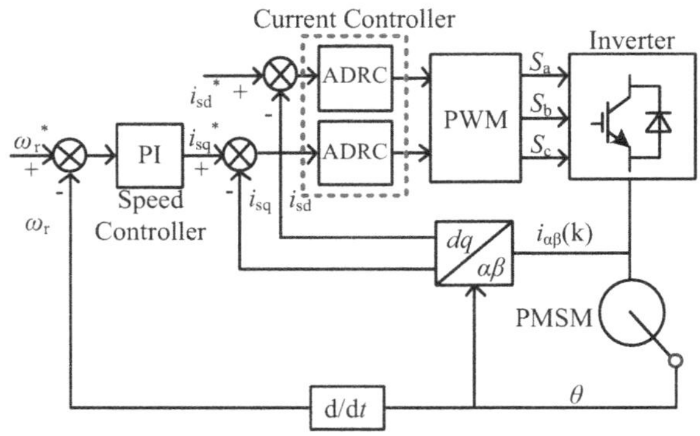

Fig. 1 Structure of LADRC current controller for PMSM speed control

$$\frac{d}{dt}\begin{bmatrix} i_{sd} \\ i_{sq} \end{bmatrix} = \begin{bmatrix} -\frac{R_s}{L_s} & 0 \\ 0 & -\frac{R_s}{L_s} \end{bmatrix}\begin{bmatrix} i_{sd} \\ i_{sq} \end{bmatrix} + \frac{1}{L_s}\begin{bmatrix} u'_{sq} \\ u'_{sq} \end{bmatrix} \tag{5}$$

and the stator voltage satisfying:

$$\begin{cases} u_{sd} = u'_{sq} - \omega_r L_s i_{sq} \\ u_{sq} = u'_{sq} + \omega_r(\psi_m + L_s i_{sd}) \end{cases} \tag{6}$$

where u_{sd} and u_{sq} are the d-axis and q-axis components of stator voltage, i_{sd} and i_{sq} are the components of stator current. L_s and R_s are stator induction and resistance, ω_r is the angular velocity, and ψ_m is the magnet flux linkage, u_{sd}' and u_{sq}' are the d-axis and q-axis voltage components of the armature winding impedances. The controlled plant can be expressed as following expression which can be seen as a 1-order inertia system.

$$G_p(s) = \frac{1}{L_s s + R_s} \tag{7}$$

The structure of 1-order LADRC is shown in Fig. 2a, and the LESO and LSEF can be expressed as:

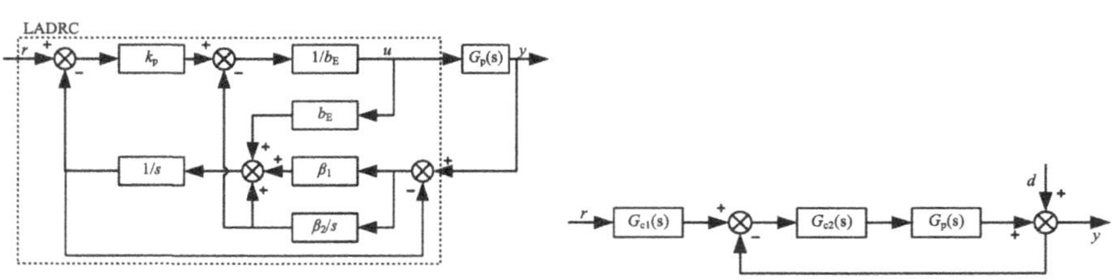

(a) Structure of 1-order LADRC

(b) Structure of IMC-2DoF

Fig. 2 Structure of current controller

$$\begin{cases} z_1' = z_2 - \beta_1(z_1 - y) + bu \\ z_2' = -\beta_2(z_1 - y) \end{cases} \tag{8}$$

$$u = \frac{1}{b_E}\left(-z_2 + k_p(r - z_1)\right) \tag{9}$$

Assuming variable $\beta_1 \gg 0$, $\beta_2 \gg 0$ and $\beta_2^2 \gg \beta_1$, the structure in Fig. 2a can be adjusted into the reference value filtered 2-DoF structure which is shown in Fig. 2b, and two controllers are expressed as:

$$G_{c1}(s) = \frac{\frac{\beta_1}{\beta_2}s + 1}{\frac{\beta_2 + \beta_1 k_p}{\beta_2 k_p}s + 1} \tag{10}$$

$$G_{c2}(s) = \frac{2\beta_1}{\beta_2}A\left(1 + \left(\frac{\beta_1 + 2\frac{\beta_2}{\beta_1}}{2\beta_2}\right)s + \frac{1}{\left(\frac{2\beta_1}{\beta_2}\right)s}\right) \cdot \frac{1}{\left(\frac{\beta_1\beta_2 + 2\beta_2 k_p + \beta_1^2 k_p + \beta_1 k_p}{\beta_2 k_p(\beta_1 + k_p)}\right)s + 1} \tag{11}$$

where

$$A = \frac{k_p \beta_2}{b_E(\beta_1 + k_p)} \tag{12}$$

Based on (10) and the IMC structure tuning method, defines intermediate variables k_1, k_2, k_3 and T_F as:

$$k_1 = \frac{2\beta_1}{\beta_2}A = \frac{T + \frac{T_s}{2}}{K(\lambda_1 + T_s)} \tag{13}$$

$$k_2 = \frac{2\beta_1}{\beta_2} = T + \frac{T_s}{2} \tag{14}$$

$$k_3 = \frac{\beta_1 + 2\frac{\beta_2}{\beta_1}}{2\beta_2} = \frac{\frac{T_s}{2}T}{T + \frac{T_s}{2}} \tag{15}$$

$$T_F = \frac{\beta_1\beta_2 + 2\beta_2 k_p + \beta_1^2 k_p + \beta_1 k_p}{\beta_2 k_p(\beta_1 + k_p)} = \frac{T_s\lambda_1}{2(T_s + \lambda_1)} \tag{16}$$

where K is the plant proportional coefficient, T is the plant integral coefficient, T_s is the sampling period and λ_1 is an adjustable coefficient. Combining (13)–(16), there four functions and four requiring variables, and some groups of certain values for the requiring variables can be obtained.

Substituting (7) into (13)–(16), the intermediate variables can be expressed as:

$$k_1 = \frac{2\beta_1}{\beta_2} A = \frac{\frac{L_s}{R_s} + \frac{T_s}{2}}{\frac{1}{R_s}(\lambda_1 + T_s)} \tag{17}$$

$$k_2 = \frac{2\beta_1}{\beta_2} = \frac{L_s}{R_s} + \frac{T_s}{2} \tag{18}$$

$$k_3 = \frac{\beta_1 + 2\frac{\beta_2}{\beta_1}}{2\beta_2} = \frac{\frac{1}{2}\frac{L_s}{R_s}T_s}{\frac{T_s}{2} + \frac{L_s}{R_s}} \tag{19}$$

$$T_F = \frac{\beta_1\beta_2 + 2\beta_2 k_p + \beta_1^2 k_p + \beta_1 k_p}{\beta_2 k_p(\beta_1 + k_p)} = \frac{T_s\lambda_1}{2(\lambda_1 + T_s)} \tag{20}$$

and the bandwidth β_1 and β_2 of the discrete-time observer are:

$$\beta_1 = \left| \frac{2\left(\frac{T_s}{2} + \frac{L_s}{R_s}\right)}{T_s\frac{L_s}{R_s} - \frac{1}{2}\left(\frac{T_s}{2} + \frac{L_s}{R_s}\right)^2} \right| \tag{21}$$

$$\beta_2 = \left| \frac{4}{T_s\frac{L_s}{R_s} - \frac{1}{2}\left(\frac{T_s}{2} + \frac{L_s}{R_s}\right)^2} \right| \tag{22}$$

$$k_p = \left| \frac{-\left(2\beta_2 + \beta_1^2 + \beta_1 - \frac{T_s\lambda_1}{2(\lambda_1 + T_s)}\beta_1\beta_2\right) \pm \sqrt{\left(2\beta_2 + \beta_1^2 + \beta_1 - \frac{T_s\lambda_1}{2(\lambda_1 + T_s)}\beta_1\beta_2\right)^2 - \frac{2T_s\lambda_1}{\lambda_1 + T_s}\beta_1\beta_2^2}}{\frac{T_s\lambda_1}{\lambda_1 + T_s}\beta_2} \right| \tag{23}$$

$$b_E = \left| \frac{k_p\beta_2(\lambda_1 + T_s)}{(\beta_1 + k_p)\left(L_s + \frac{1}{2}R_s T_s\right)} \right| \tag{24}$$

Substituting (21) and (22) into (23) and (24), two groups of effective values of the variable b_E and k_p can be obtained if the variables are satisfied a condition:

$$\left(2\beta_2 + \beta_1^2 + \beta_1 - \frac{T_s\lambda_1}{2(\lambda_1 + T_s)}\beta_1\beta_2\right)^2 \geq \frac{2T_s\lambda_1}{\lambda_1 + T_s}\beta_1\beta_2^2 \tag{25}$$

and the tracking performance and disturbance resisting performance are decoupled by the variable b_E to further improve control performances.

3.2 Robustness and Stability Analysis

The actual controlled plant is defined as:

$$G_{\mathrm{p}}(s) = G_{\mathrm{p0}}(s) + \Delta P_{\mathrm{a}}(s) \tag{26}$$

where $\Delta P_a(s)$ is the uncertain disturbance between the actual controlled plant $G_p(s)$ and the plant model $G_{p0}(s)$. When s closes to zero, the output $u(0)$ of LADRC can be expressed as:

$$u(0) = G_{\mathrm{p0}}^{-1}(0) \tag{27}$$

According to the transfer function in (5), the system satisfies $|r(0)u(0)| < 1$ during random frequency ranges, and a stable robustness can be obtained when the LADRC applies the obtained parameters.

The close-loop transfer function of the system can be expressed as:

$$D(s) = a_4 s^4 + a_3 s^3 + a_2 s^2 + a_1 s + a_0 \tag{28}$$

where

$$a_4 = T_{\mathrm{F}} \frac{L_{\mathrm{s}}}{R_{\mathrm{s}}} \frac{\beta_2 + \beta_1 k_{\mathrm{p}}}{\beta_2 k_{\mathrm{p}}} \tag{29}$$

$$a_3 = T_{\mathrm{F}} \frac{L_{\mathrm{s}}}{R_{\mathrm{s}}} + \left(T_{\mathrm{F}} + \frac{L_{\mathrm{s}} + k_1 k_3}{R_{\mathrm{s}}} \right) \frac{\beta_2 + \beta_1 k_{\mathrm{p}}}{\beta_2 k_{\mathrm{p}}} \tag{30}$$

$$a_2 = T_{\mathrm{F}} + \frac{L_{\mathrm{s}} + k_1 k_3}{R_{\mathrm{s}}} + \left(1 + \frac{k_1}{R_{\mathrm{s}}} \right) \frac{\beta_2 + \beta_1 k_{\mathrm{p}}}{\beta_2 k_{\mathrm{p}}} \tag{31}$$

$$a_1 = \frac{k_1}{R_{\mathrm{s}}} + 1 + \frac{k_1}{R_{\mathrm{s}} k_2} \frac{\beta_2 + \beta_1 k_{\mathrm{p}}}{\beta_2 k_{\mathrm{p}}} \tag{32}$$

$$a_0 = \frac{k_1}{R_{\mathrm{s}} k_2} \tag{33}$$

Based on the Rous-Hurwitz law and obtained parameters above-mentioned, a_0, a_1, a_2, a_3 and a_4 are positive values which are satisfying $a_1 a_2 > a_0 a_3$ and $a_3(a_1 a_2 - a_0 a_3) > a_1^2 a_4$. All of the poles are located in the left-half of complex plane, and the stator current LADRC system is stable.

4 Simulation and Experimental Results

A PMSM speed control system with stator current LADRCs and obtained parameters is built in the MATLAB/simulink software. The selected basic PMSM parameters are listed in Table 1, the LADRC parameters of the current controller are determined by the PMSM.

Table 1 Table captions should be placed above the tables

Parameters	Values	Parameters	Values
Number of pole pairs p	4	Magnet flux linkage ψ_m	0.175 Wb
Stator induction L_s	0.835 mH	Rotor inertia J	0.0008 kg.m2
Stator resistance R_s	2.360 Ω	Friction coefficient B	0.0008 N.m.s

Two groups of LADRC parameters are obtained and noted as β_1, k_{p1}, b_{E1} and β_2, k_{p2}, b_{E2} respectively. The step responses of LSEO and LADRC with two groups of parameters are shown in Figs. 3 and 4 respectively, and the z_2 signal is narrowed 8000 times to observe easily. The system can track the step reference and operate into steady state in about 2 ms, and the LESO also can observe the total disturbance quickly and stably. For the decoupled 2-DoF structure, a rated torque is inserted to the shaft at 5 ms, and the 0.1, 1, 5 and 10 times b_E are tested and combined with the first group of LADRC parameters. As shown the testing results and the related right y-axis in Fig. 3, all of the waveforms have similar resisting disturbance performances and different reference tracking performances by different b_E values, and the performances are decoupled.

Bode figures of the LADRC with tuning method above-mentioned and the PI controllers with frequency domain tuning method are shown in Fig. 5. The right

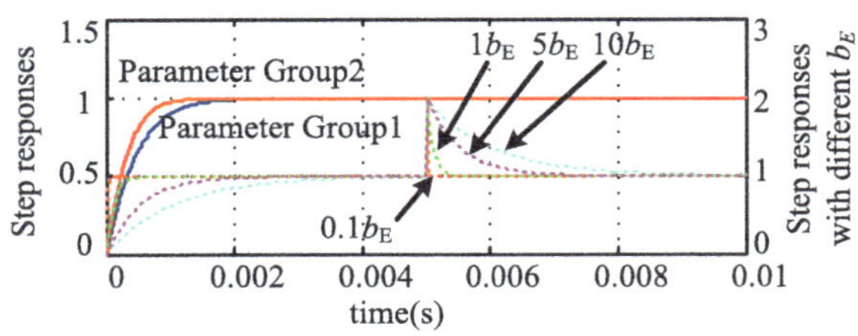

Fig. 3 Waveforms of step response for current LADRC

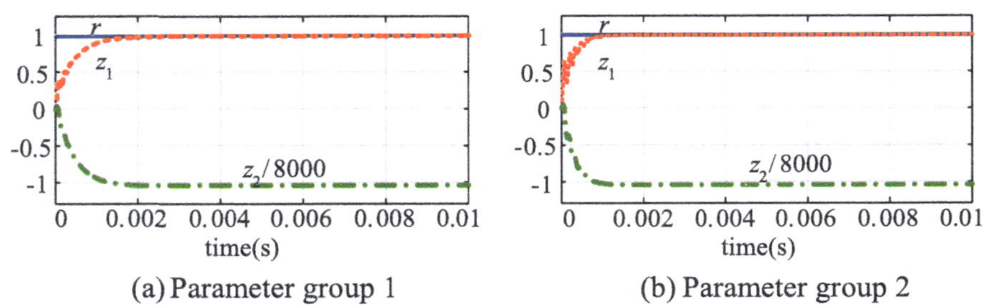

Fig. 4 Waveforms of step response for current LADRC

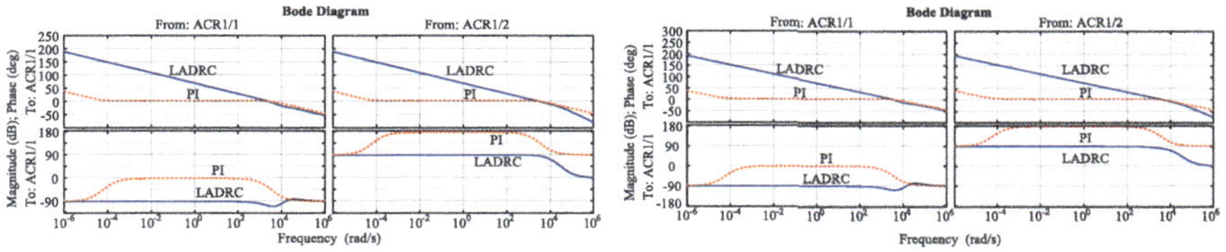

(a) Parameter group 1

(b) Parameter group 2

Fig. 5 Bode diagrams for current LADRC

Fig. 6 Waveforms of step response for current LADRC

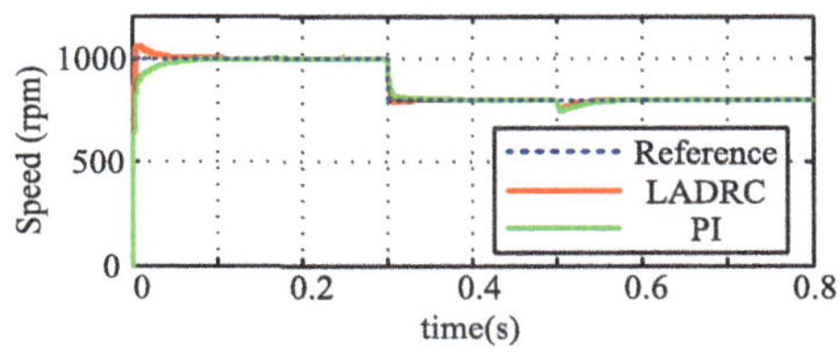

side is the output from disturbance, and left side is the output from input. The bodes with two groups of obtaining parameters have similar performances. Comparing with the PI controllers, the LADRC stability and resisting disturbance performance are enhanced because of the higher magnitudes at low-frequency part, higher slopes at high-frequency part and -20 dB/dec slope at 0 dB. The systems with two types of controllers have similar rapidity due to similar shear frequencies.

The initial speed reference is settled as 1000 rpm at start and changed as 800 rpm to ensure tracking reference performance, and a 3 N.m load torque is inserted on the shaft at 0.5 s to test resisting disturbance performance. The parameter group 2 with better comprehensive performances is selected, and the parameters of PI controller are tuned by the frequency-domain method which obtains $P_s = 0.0578$, $I_s = 2.78$. The speed simulation results are shown in Fig. 6. As shown in the figures, the overshoot σ_1 and the settling time t_{s1} are 6.25% and 0.08 s at start operating state, the overshoot σ_2 and the settling time t_{s2} are 1.82% and 0.04 s at 0.3 s, and the overshoot σ_3 and the settling time t_{s3} are 5.96% and 0.05 s at 0.5 s respectively. ITAE reflects the comprehensive performance during whole operating process. ITAEs of the system with stator current LADRC and PI controller in 0.8 s are 0.9761 and 1.927 respectively mainly because the overshoot σ_3 is increased to 6.94% for PI controllers.

A 1 kW PMSM with an incensement encoder is selected as the controlled plant and the basic parameters are same with Table 1, and the controller is selected as a DSP with TMS320F2812PGFA to obtain experimental platform. Based on the obtaining parameters above-mentioned, the experimental waveforms at steady state are shown in Fig. 7 and the experimental waveforms at transient state are shown in Fig. 8. The actual speed can track the reference and resist the disturbance about 120 ms and 750 ms respectively, and the obtaining parameters of LADRC can realize stably control.

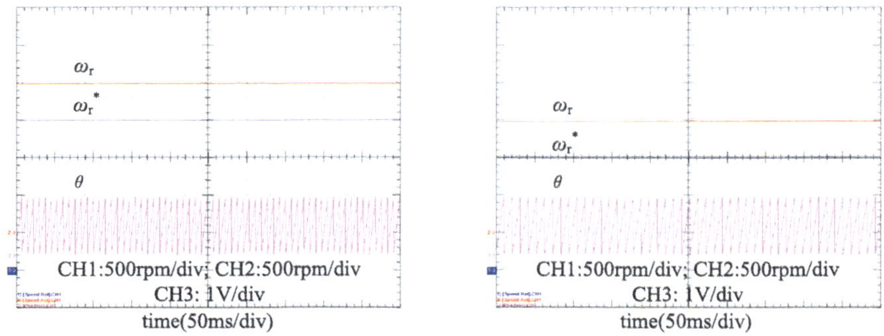

(a) 2000rpm (b) 1500rpm

Fig. 7 Experimental waveforms of LADRC current controller for PMSM speed control in steady state

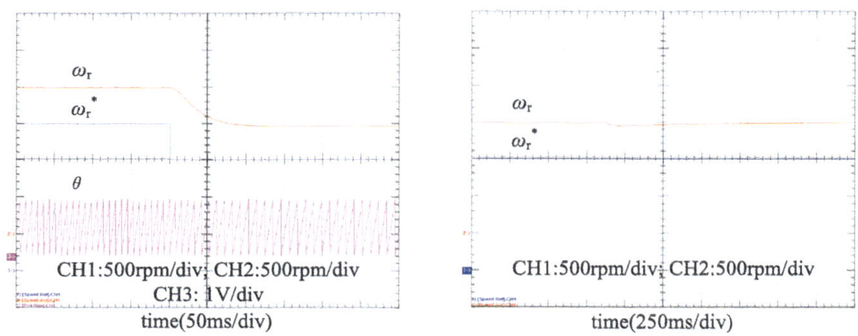

(a) Reference change (b) Load torque change

Fig. 8 Experimental waveforms of LADRC current controller for PMSM speed control in transient state

5 Conclusions

A reference value filtered 2-DoF structure tuning method for the PMSM speed control system with stator current LADRCs are proposed in this paper. Comparing with the PI controller by the frequency-domain method, the proposed method has advantages including better stability and resisting disturbance performance which are verified by the simulations. The overshoot and settling time of LADRC are decreased obviously, and the ITAE is decreased about 49.347% under same disturbance condition. Excepting effective reference tracking performance and resisting disturbance performance of the system, these performances can be decoupled by the parameter b_E, and further improved by the adjustable value to improve system performances.

References

1. Han, J. 2008. *Active Disturbance Rejection Control Technique—the technique for estimating and compensating the uncertainties*. Beijing: National Defense Industry Press Co, Ltd., 243–263.
2. Zhu, J., Q. Ge, K. Sun, et al. 2020. Traction-system research of high-speed maglev based on active disturbance rejection control. *Transactions of China Electrotechnical Society* 35 (5): 1065–1074.
3. Zuo, Y., J. Zhang, C. Liu, et al. 2017. A modified adaptive disturbance rejection controller for permanent magnetic synchronous motor speed-regulation system with time-varying input. *Transactions of China Electrotechnical Society* 32 (2): 161–170.
4. Liu, C., G. Luo, W. Tu, et al. 2017. Servo systems with double closed-loops based on active disturbance rejection controllers. *Proceedings of the CSEE* 37 (23): 7032–7039.
5. Ma, Y., Y. Zhou, H. Liu, et al. 2017. A local control method of multi-machine low frequency oscillation based on active disturbance rejection control. *Proceedings of the CSEE* 37 (5): 1360–1372.
6. Du, C., Z. Yin, Y. Li, et al. 2017. Research on repetitive-auto disturbance rejection control strategy for induction motors baesd on vector control. *Transactions of China Electrotechnical Society* 32 (19): 85–93.
7. He, Y., Z. Wang, and S. Zheng. 2016. Decoupling control of high speed permanent magnet synchronous motor based on online least square support vector machine inverse system method. *Proceedings of the CSEE* 36 (20): 5639–5646.
8. Zhao, J., M. Hua, and T. Liu. 2019. Cooperative optimization and fault-tolerant control method of multi-disk permanent magnet synchronous motor for electric vehicles. *Proceedings of the CSEE* 39 (2): 386–394.
9. Zuo, Y., H. Fu, C. Liu, et al. 2016. A novel two-degree-of-freedom controller for permanent magnet synchronous motor speed-regulation system. *Transactions of China Electrotechnical Society* 31 (17): 140–146.
10. Gao, Z. 2013. On the foundation of active disturbance rejection control. *Control Theory & Applications* 30 (12): 1498–1507.

Data Augmentation and Class Based Model Evaluation for Load Disaggregation Based on Deep Learning

Bo Li, Yandi Li, Changyuan Liang, Weifeng Su, and Zhe XuanYuan

Abstract Load disaggregation aims to disaggregate the power usage of individual appliance based only on aggregated power monitoring data. Applications such as appliance monitoring, appliance-based power metering can be realized based on this technique. In recent years, research has been done to apply deep learning to load disaggregation and initial results can be found in literature. However, for all machine learning based models, the class imbalance problem or the skewed distribution of values in training datasets, has long been recognized as a severe problem dragging down the overall accuracy of the model trained. This is also a common problem in current public datasets for load disaggregation research due to natural imbalance of appliance on/off time. In this work, we address the data imbalance problem by proposing a novel data augmentation method that combines both original data sequences and selected data sequences. Moreover, we propose a more fair and natural evaluation method to compare the models' performance under imbalance datasets. Extensive empirical study shows that the proposed method can achieve 53.48–83.81% performance enhancement over the current state-of-the-art model.

Keywords Load disaggregation · Deep learning · Imbalance dataset · Data augmentation · Model evaluation

---

B. Li
Electric Power Research Institute of Yunnan Power Grid, Yunnan 650000, China
e-mail: 705097939@qq.com

Y. Li · C. Liang · W. Su · Z. XuanYuan (✉)
Beijing Normal University-Hong Kong Baptist University United International College, Zhuhai 519000, China
e-mail: zhexuanyuan@uic.edu.cn

Y. Li
e-mail: yandili@uic.edu.cn

C. Liang
e-mail: m730026058@uic.edu.cn

W. Su
e-mail: wfsu@uic.edu.cn

W. Chen et al. (eds.), *The Proceedings of the 9th Frontier Academic Forum of Electrical Engineering*, Lecture Notes in Electrical Engineering 743,
https://doi.org/10.1007/978-981-33-6609-1_30

1 Introduction

Electrical energy monitoring can be classified into Intrusive Load Monitoring (ILM) and Non-Intrusive Load Monitoring (NILM) [1, 2]. Comparing with ILM, which depends on metering of each single appliance, NILM is characterised by low cost as well as easy equipment installation and maintenance due to only a single, whole-home smart meter required [3]. However, NILM needs the research on how to disaggregate the gross electrical power consumption of multiple appliances into individual ones, which is known as load disaggregation. In mid 1980s, Hart's research on non-intrusive load decomposition opens the prelude to research in this field [4, 5]. The research of NILM is usually based on the high and low frequency sampling data of the active power of electrical appliances, and the hidden Markov model (HMM) is used to predict the power of a single electrical appliance. For example, Makonin et al. proposed a load identification algorithm based on sparse hidden Markov chain [6]. Furthermore, a non-intrusive decomposition method based on affinity propagation and genetic algorithm optimization was proposed in [7].

With the development of deep learning techniques, it has been widely used in image processing, speech recognition and other fields. At the same time, the development of smart power grids has produced exponentially increasing data volume, which has laid a good foundation for the application of deep learning in power systems. Paper [8] proposed an improved recurrent neural network (RNN) to detect the sensor fault of the chiller and achieved better detection efficiency than traditional methods. Paper [9] uses convolutional neural network (CNN) to solve the problem of DC XLPE cable partial discharge pattern recognition, and realizes the self-extraction of key features of strong random signals, which greatly improves the accuracy of recognition compared with previous methods. In the meantime, deep learning techniques applied to NILM are also developing rapidly. Research [10] proposed a deep belief network (DBN) for non-invasive power load identification based on a deep neural network architecture. The denoising autoencoding algorithm based on CNN was originally used by Kelly for non-invasive load decomposition [11], using the UK-DALE data set to realize neural network self-extracting useful features for learning [12]. Zhang et al. proposed a "sequence-to-point" model, which uses a more complex deep CNN, also based on the UK-DALE data set, and the prediction results are better than Kelly [13].

For all power load decomposition algorithms based on machine learning, the data set used for training and verification largely determines the performance of the algorithm. At present, there are several data sets collected based on real home or commercial building usage scenarios in the world, such as UK-DALE in the United Kingdom, REDD in the United States, AMPDs in Canada, and iAWE in India, etc. [14]. These data sets generally include the electrical characteristic data of the total electric meter and the sub-sockets in several homes or office spaces collected in time series in several consecutive time periods. Although the amount of data and types of appliances in these data sets is considerable, there is a common problem, which is the distribution imbalance of appliance's on and off states.

Aiming at the problem of unbalanced data in training data set used in previous deep learning models, we propose a method to augment the training data set, mixing real time series data and manually extracted fragment data of appliance's on states according to a certain ratio. Then use this new generated data set to train the deep learning model. Experiments results show that this training method helps to improve the accuracy of load disaggregation. Also, we propose an improved load disaggregation algorithm evaluation indicator, which makes the comparison between different algorithms more scientific and meet the realistic application requirement.

The rest of this paper is structured as follows: Sect. 2 mainly introduces the basic principles of CNN and its latest developments in NILM. Section 3 summarizes the constitution of existing training data sets and proposes innovative training set augmentation methods. Section 4 proposes a load disaggregation algorithm evaluation method that is more in line with actual application requirements. Section 5 provides the results and makes a comparison with baseline models. Section 6 presents the conclusions for the paper.

2 Load Disaggregation Algorithms Based on CNN

2.1 CNN and Its Application in Load Disaggregation

With its unique advantages, CNN has developed into one of the most widely used models in the field of deep learning, showing significantly better performance than traditional machine learning algorithms especially in image recognition and signal processing tasks [16–18].

The convolutional layer uses a sliding window to traverse all the neuron regions of the upper level by pre-defined local perceptual matrices of a certain size, and each local perceptual domain corresponds to a fundamental structure of features. The biggest advantage of CNN is it makes the model invariant to displacement, scaling and other forms of distortion.

The input data in the NILM task is the sequential total power value of the time, which can be input into one-dimensional CNN model for training. Regarding a certain length of sequential power data as a one-dimensional static picture, we can move a sliding window with a fixed width along the time axis at a certain pace, known as "step", and the intercepted power data segment in the window after each movement is used as the model input [11]. CNN can discover useful features automatically during the training process, thereby avoiding the lack of core features due to insufficient prior knowledge, especially advantageous to complex application scenarios. Figure 1 is a typical NILM process flow. The raw power data is used for model training after pre-processing, and then the new power data is disaggregated by the model with the trained parameters, and finally the model is evaluated by a suitable evaluation system.

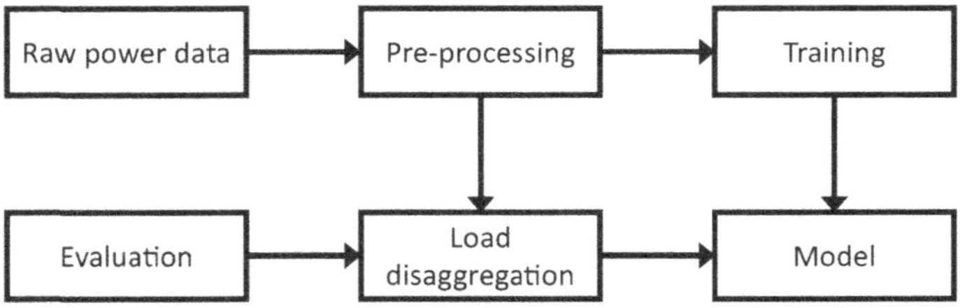

Fig. 1 NILM basic process flow

Fig. 2 DAE model architecture (conv. refers to the convolutional layer, dense refers to the fully connected layer, length and units refers to the number of nodes, and Filters refers to the filters in the convolutional layer)

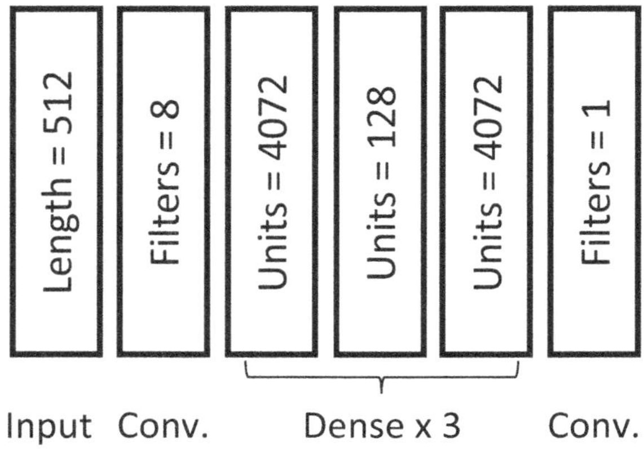

2.2 *Denoising Autoencoder Model*

Kelly et al. [11] regard load disaggregation as a "noise reduction" task, where the target power of the appliance is analogous to the "clean" main signal, while the power generated by other appliances at the same time represents the background "noise". The goal is to separate the target power from the background "noise" power.

Denoising Autoencoder (DAE) concept can be exploited to build a deep neural network [19]. Convolutional layers are set as the first and last layer of the network, and extract useful intermediate features by its invariance to displacement, scaling and distortion. For example, a feature of 1000 watts step change is likely to be a useful feature no matter where it occurs in the sequential power data segment. The specific architecture of DAE model is shown in Fig. 2.

2.3 *Seq2Point Model*

The input and output of the DAE model are all sequence data of the same length. This type of model is usually called a "sequence-to-sequence (seq2seq)" model. Zhang

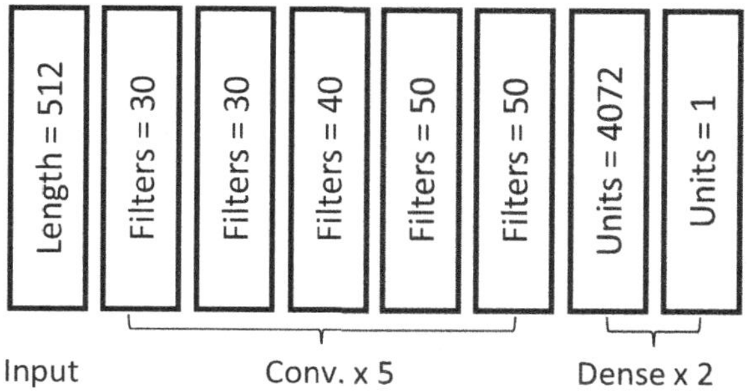

Fig. 3 Seq2point model architecture

et al. [13] proposed a corresponding "sequence-to-point (seq2point)" model. Its core idea is to input the aggregated power sequence data of all appliances, but output the predicted single point power value at the midpoint position of the input sequence of the target appliance in order to achieve better prediction accuracy. The architecture of this model is shown in Fig. 3.

Both DAE and Seq2Point are machine learning models based on CNN. Their structure is representative and produces good performance in terms of NILM in recent years. However, due to the aforementioned imbalance in the training data of the model, there is room for improvement in the prediction performance. The work in this paper confirms this conclusion through innovative experimental methods and detailed experimental data.

3 Data Set Processing

3.1 Data Set

Unbalanced data sets widely exist in the real world, and the unbalanced data will affect the bias of the model trained with it, making the model more inclined to predict outputs in line with the most categories or states in the data set. Previously, the imbalance of data set has received more attention in the classification tasks of machine learning [15, 20], but in the regression problem such as load disaggregation, there are relatively few discussions. One of the reasons is that in the regression problem, data is continuous value, making it more challenging in the definition and optimization of "imbalance". However, the common imbalance problem in the load disaggregation data set has greatly affected the performance of machine learning algorithms in such applications, and should be paid attention to.

We use the public data set UK-DALE (UK domestic appliance-level electricity) as the model training and testing data set [12]. This data set collects power consumption records of dozens of appliances over five local households in the UK, of which the duration of house 2–4 ranges from 2013 to 2014 for several months, and house 1 has the longest span from November 2012 to April 2017. The collected data includes the sequential power data of the whole-house meter and of each single appliance.

We mainly consider four common appliances in UK-DALE: kettle, washing machine, microwave oven and dishwasher. These four electrical appliances consume a large proportion of electrical energy in general households. and also show high exposure in the research using UK-DALE, which makes it convenient for comparison. In actual scenarios, different households often use different brands or models of the same kind of appliance, of which the power characteristics are different. Based on this fact, this research pays more attention to the generalisation ability of the model in different households. Since only house 1 and house 2 cover all four kinds of appliances [12], to test the generalisation ability of the disaggregation model, we select house 1, 3, 4, 5 as the source of training data with house 2 for testing.

Since most appliances do not work continuously, their working hours are much shorter than non-working hours. Figure 4 shows the power curves of a kettle and a washing machine in a random period of time. The figure clearly shows that the working time of appliances accounts for relatively low ratios, and this phenomenon is more serious in a longer time span. Table 1 lists the proportion of the number of random sampling points in the working state to the corresponding total sampling points, of the four appliances respectively. The working time ratio is less than 5%, which has severe imbalance.

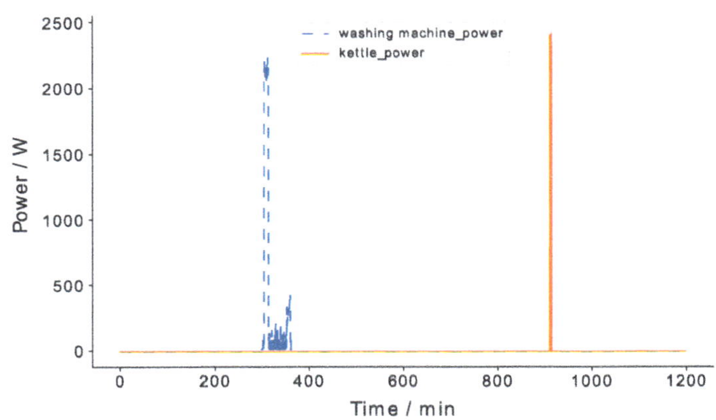

Fig. 4 Kettle and washing machine power samples within a period of time

Table 1 Activation period ratio of 4 appliances within 3 months

Appliance	Kettle (%)	Washing machine (%)	Microwave (%)	Dish washer (%)
Proportion of sampling points in working state (3 months)	0.67	4.85	4.83	2.60

3.2 Training Set Augmentation

For unbalanced data sets, an effective method is to over-sample the minority categories while under-sampling the majority categories [21]. However, the method of under-sampling the majority of categories may cause the missing of some potentially important information related to the majority of categories which exists in the discarded samples [20]. In order to avoid losing important information of most categories, the idea of over-sampling minority categories is introduced in this research where the impact of training data sets with different working time ratios on model performance is thoroughly explored.

The training set is constituted of same-length power value sequences. We refer to the power value sequence, including entire or partial working state of target appliance, as "on-state sequence", and otherwise "off-state sequence". And we also refer to the ratio of the number of on-state sequence in the training set as "on-state sequence ratio". With this definition, the extent of balance in training set can be characterised by on-state sequence ratio. Then, we will conduct experimental investigations on the influence of on-state sequence ratio on the model performance. The specific implementation method is to first pre-define three types of data sets (see Table 2), then the new training sets will be generated based on these definitions. In Table 2, "Time Series Set" contains a long continuous natural time period of power value sequences with information on the usages of appliance, which can theoretically help the model acquire a certain degree of users' appliance usage habits. "Extraction Set" contain sequences artificially extracted from the raw power data set, only 50% of which are on-state sequences of the target appliance. On this basis, "Mixture Set" concept is proposed to generate the new training sets in our experiments, the specific groups and parameters (take washing machine as example) of which are exhibited in Table 3.

Table 2 Datasets names and definition

Name	Definition
Time series set	A data set formed by dividing the raw time series power data into same-length sequences and arranging them in temporal sequential order
Extraction set	A data set consisting of sequences artificially extracted from the raw power data set, 50% of which are on-state sequences of the target appliance while the remaining 50% are off-state ones with no working state part
Mixture set	A data set formed by sampling sequences from time series set and extraction set following certain probabilities

Table 3 Imbalanced dataset parameters

Group	Data set constitution	On-state sequence ratio (washing machine as example) (%)
1	100% Time series set	9.75
2	75% Time series set + 25% extraction set	20.08
3	50% Time series set + 50% extraction set	29.77
4	25% Time series set + 75% extraction set	39.89
5	100% Extraction set	49.72

Table 4 Pre- and post-processing parameters

Unit: watt	Kettle	Washing machine	Microwave	Dish washer
Input standard deviation	991.90	1008.37	1012.63	928.60
Target standard deviation	332.02	493.69	159.15	618.67
On-state power threshold	1000.00	5.00	5.00	5.00
Off-state power assignment	0.00	0.00	0.00	0.00

3.3 Pre-processing

The power sequence data input by the model is normalised by Z-score method. and the mean and variance used are derived from the statistics of data set, see Table 4. The normalisation equations are shown below, where x_{in} is the input power value; x_{total} is the whole-house power value at the same time; $\bar{x}$ is mean value of whole-house power; σ_{total} is the input standard deviation in Table 4; x_{target} is the target power value after pre-processing; $x_{\text{GroundTruth}}$ is the real power value of target appliance; σ_{target} is the target standard deviation in Table 4.

$$x_{\text{in}} = \frac{x_{\text{total}} - \bar{x}}{\sigma_{\text{total}}} \tag{1}$$

$$x_{\text{target}} = \frac{x_{\text{GroundTruth}}}{\sigma_{\text{target}}} \tag{2}$$

3.4 Post-processing of Zero-Assignment

The model has a certain probability of regressing the "off-state" predicted value to a non-zero value, which is usually expressed as a low-amplitude value near zero, as shown in Fig. 5. In reality "off state" occupies most proportion for most kinds of appliances, such mispredictions will lead to an increase in the overall prediction error and have a great impact on the results. Aiming at this problem, we propose a

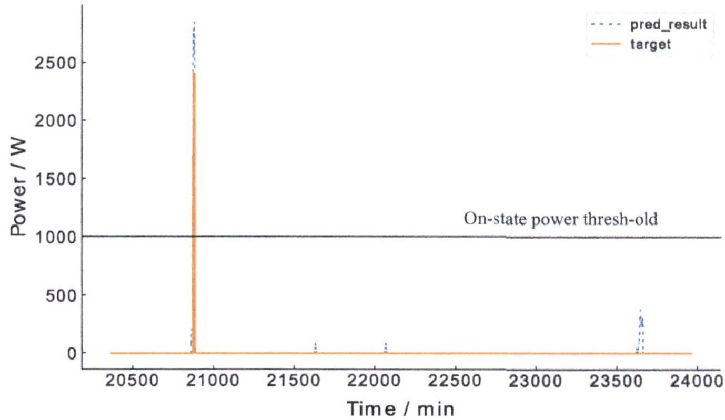

Fig. 5 Kettle's ground truth power and model predicted power profiles over a period of time

post-processing method, named Zero-Assignment, specially dealing with "off-state" predicted power values. Since the predicted "off state" of discontinuous working-mode appliances by our model generally corresponds to apparently consecutive close-to-zero power values, which is significantly different from the power level of predicted "on state" (mostly ranging from tens to several kilowatts), the threshold method is used to classify the prediction results to either "off state" or "on state. We use "on-state power threshold" in Table 4 as classification threshold value, then the part below the threshold denoted as "off-state area" and is set to zero value to eliminate the mis-predicted power consumption in off-state area, reducing the overall error level.

4 Class-Based Evaluation System

4.1 Evaluation Indicator

The main way to measure the performance of the model is to quantify the degree of fitting the predicted power waveform by the model to the real waveform. Using the mean absolute error (MAE) between the ground truth power value and the corresponding predicted value, the calculation formula is as follows, where x_t is the target power value at time t; $\hat{x}$ is the predicted power at time t; T is the total number of time steps.

$$\mathrm{MAE} = \frac{1}{T}\sum_{t=0}^{T}\left|\hat{x}_t - x_t\right| \tag{3}$$

Table 5 Comparison among washing machine models trained by temporal sequential data

Training set	Model	Test set span (house 2)	MAE
Time series set	Kelly's seq2seq	Unknown time span	163.468
	Zhang's seq2point	Unknown time span	12.663
	Zhang's seq2seq	Unknown time span	10.153
	Seq2point + dataset augmentation (this paper)	2013.03.01–2013.03.10	10.6188
	Naïve zero predictor	2013.03.01–2013.03.10	11.995
	Seq2point + dataset augmentation (this paper)	2013.03–2013.05	5.2584
	Naïve zero predictor	2013.03–2013.05	6.63

The power data is time series data. The larger the MAE between the ground truth power value of the sampling point and the corresponding predicted power value is, the larger the gap between the model prediction result and the real result. Therefore, MAE can effectively measure model performance.

4.2 Drawback of Evaluation Method in Seq2point Research

However, the evaluation method in most NILM research directly using MAE might not be reasonable enough. Table 5 contains the performance of seven load disaggregation models with different architectures (the target appliance is a washing machine for example) on Time Series Set of house 2 in UK-DALE data set during different time spans. The model test data of Kelly and Zhang is quoted from the article [13]. The output of naïve zero predictor is always 0. The other two models are both trained by us, using the test data sets of the corresponding time spans in Table 5.

Comparing the test results of those active models with the naïve zero predictor, the model with the method of augmenting data set proposed by us can get the best performance, but at the same time it is found that the test results of the naïve zero predictor are even significantly better than the Kelly's model, so the overall MAE with the test set of Time Series Set does not indicate the model's ability to predict on-state power values of appliances. This phenomenon stems from the unbalanced working-state distribution of the target appliance in the test data set, and the off state dominates the distribution.

4.3 Class-Based Evaluation Method

To solve the problem of unbalance in test data set aforementioned in Sec. 4.2, we proposed a class-based evaluation method, measuring performance for different working states respectively. Specifically, we use pre-defined "On-state power threshold" value in Table 4 to classify every sampling point into "on-state" class and "off-state" class first, then evaluate two classes separately by the evaluation indicator (like MAE). This method implements individual evaluation for each class, including the working-state power prediction that people pay more attention to, which contributes to more objective and reasonable evaluation of model performance in terms of extensively existing unbalanced distribution in electricity data sets.

For example, Fig. 5 shows a randomly intercepted power sequence containing the on-state and off-state of a kettle. The working state threshold is 1000 watts. The sampling points, whose power values are lower than the threshold, are classified as off-state points, while the ones above the threshold are regarded as on-state points, thus the sequence is divided into two parts and compared with the model-predicted values respectively.

5 Experiments and Results Analysis

5.1 Experiment Steps

We implemented the following steps for experiments:

1. Model construction: use deep learning framework Keras to build seq2point model [13]. The input of model each time is power value sequence with length of 599 sampling points, and the output is the power value of target appliance at the time step same as the centre point of input sequence. The specific model structure is depicted as Fig. 3.
2. Training set generation: take washing machines, kettles, microwaves and dish washers from house 1, 3, 4, 5 in UK-DALE as the target appliances with 1/6 Hz sampling rate, and generate experimental training data set groups in Table 3.
3. Data pre-processing: normalise input data by subtracting mean value and getting divided by standard deviation, and normalise target data by only getting divided by standard deviation. Related parameters can be referred to in Table 4.
4. Model Training: use every group of training set to train the same sep2point model separately, with around 1.28×10^7 input sequences for each group. Adopt mini-batch method with 64 sequences per batch. Choose MAE as the loss function. Set the early stop after successive 5×10^4 batches of no better MAE.
5. Test set selection: choose house 2 low-frequency time series data in UK-DALE as test set with 1/6 Hz sampling rate and the time span of six months from Mar to Jun 2020.
6. Post-processing: employ two methods to process the predicted power values. First, directly use the predicted values for evaluation. Second, apply Zero-Assignment introduced in Sect. 3.4 to the predicted values to obtain the processed results for evaluation.

7. Model test: use two methods to evaluate the results on the test set with MAE indicator. First, test the performance on the over-all test set data which consists of much more off-state power value points and less on-state points. Second, use class-based evaluation method described in Sect. 4.3 to measure the performance.
8. Experimental facilities: utilise Sever with CPU of Intel Xeon E5-2620 v4 and GPU of GeForce GTX TITAN X for model training and testing.

5.2 Results and Analysis

Impact of on-state sequence ratio. The performance of the corresponding models of four appliances on the test set is shown in Table 6. For the other three types of appliances except washing machine, the model trained on the 100% Time Series Set performs significantly worse than the other four models using more balanced training sets in both the overall evaluations and the class-based evaluations. The four more balanced training sets (with on-state sequence ratio higher than or equal to 25% in Table 6) make the MAE improved by more than 50%, proving the proper

Table 6 Evaluation results of models trained by datasets with diverse on-state sequence ratio

Training set		100% T	75% T 25% E	50% T 50% E	25% T 75% E	100% E
Kettle	On Ratio	16.66%	24.77%	32.98%	41.47%	49.58%
	MAE (all)	58.51	7.87	6.35	8.94	8.76
	MAE (on)	2919.07	481.58	472.72	607.88	626.39
	MAE (off)	33.44	3.72	2.26	3.69	3.35
Washing machine	On Ratio	9.75%	20.08%	29.77%	39.89%	49.72%
	MAE (all)	5.26	26.30	12.63	14.94	15.28
	MAE (on)	574.83	396.82	267.40	282.93	308.37
	MAE (off)	1.69	23.97	11.03	13.26	13.44
Microwave	On Ratio	20.57%	25.90%	35.02%	41.78%	49.59%
	MAE (all)	17.38	7.34	6.99	5.45	5.50
	MAE (on)	746.46	359.77	317.57	290.08	338.30
	MAE (off)	11.69	4.59	4.58	3.53	1.43
Dish washer	On Ratio	5.14%	16.29%	27.47%	38.60%	49.80%
	MAE (all)	97.45	24.52	26.77	20.71	16.02
	MAE (on)	1300.12	466.06	511.04	273.19	319.57
	MAE (off)	65.86	12.92	14.06	14.08	8.05

T—Time series set, E—Extraction set, On Ratio—on-state sequence ratio, MAE (all) stands for MAE on whole test set, MAE (on) stands for MAE of on-state points (on-state MAE), MAE (off) stands for MAE of off-state points (off-state MAE)

balance of data distribution in electrical power training set can significantly improves the load-disaggregation model performance. For washing machine, the evaluation performance of the models with four more balanced training sets in the on-state parts is apparently higher than the that of the model with 100% Time Series Set, where the MAE is improved by more than 30%.

Performance in the on-state parts is normally of more importance than the off-state parts, thus we pay more attention to and mainly discuss the on-state performance in this section. For the same appliance, 100% Time Series Set corresponds to the worst performance of the model, because this group of data sets have the severest imbalance with the largest proportion of off-state sequences, which makes it difficult for the model to learn the characteristics of the on-state power behaviour during the training process. As the proportion of on-state sequences of the training set increases, the performance of the model improves, but to a certain extent, it shows a downward trend. Because large proportion of the on-state sequence affects the learning of the power behaviour of those distracting appliances in the off-state duration of the target appliance. Also, excessive duplication of the minority-state data in training set would overfit the model, and then undermine the generalisation ability on the unseen data set [20].

The best training set composition for different appliances are not always the same. For example, for kettle and washing machine, the group of 50% Time Series Set plus 50% Extraction Set contributes the lowest(best) MAE, while the group of 25% Time Series Set plus 75% Extraction Set works best for microwave and dishwasher. Figure 6 shows the changing trend of the on-state MAE with the on-state sequence ratio. For these four kinds of appliances in UK-DALE, the optimal on-state sequence ratios in training set are located between 30 and 45%, and exceeding this range leads to performance degradation to varying degrees. In Fig. 6, the MAE corresponding to the optimal on-state sequence ratios are all considerably better than that corresponding to 100% Time Series Set, of which the increase ratios are 83.8% for kettle, 53.48% for washing machine, 61.14% for microwave and 78.99% for dish washer.

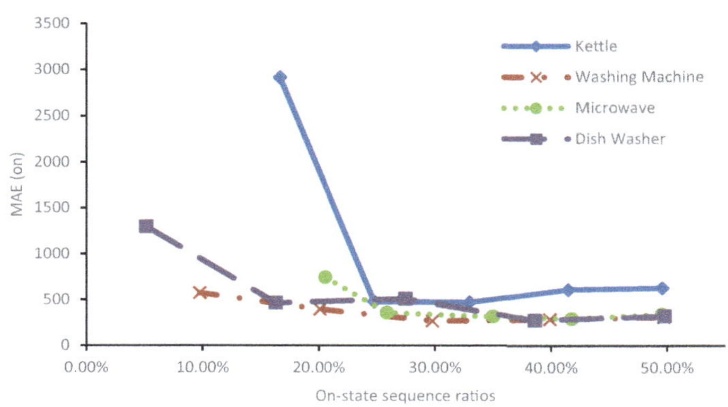

Fig. 6 On-state MAE versus on-state sequence ratio of training sets

Results of Zero-Assignment Post-processing. Zero-Assignment post-processing is essentially setting the low-amplitude power values generated by misprediction in the off-state duration to zeroes, thus mainly focusing on off-state performance. Table 7 shows the evaluation results after Zero-Assignment processing, where the on-state MAE keep comparable to those before processing, but the off-state MAE decrease significantly than before, which also results in the decrease of MAE on whole test set. We choose the best group of each kind of appliance, then post-process their prediction results by Zero-Assignment method and compare the performance in Fig. 7. The increase ratio in MAE ranges from 11 to 67%, having proved the effectiveness of Zero-Assignment post-processing method.

In practical applications, due to the high proportion of appliances' off-state time, the accumulative amount of false power mis-predicted by the model is considerable, severely affecting the prediction of target appliances' long-term electricity consumption which is obtained by accumulating predicted power within the period. With Zero-Assignment, this problem can be greatly improved.

Table 7 Evaluation results of models trained by datasets with diverse on-state sequence ratio after "zero-assignment" post-processing

Training set (zero-assignment)		100% T	75% T 25% E	50% T 50% E	25% T 75% E	100% E
Kettle	On Ratio	16.66%	24.77%	32.98%	41.47%	49.58%
	MAE (all)	26.04	5.12	4.97	6.02	6.30
	MAE (on)	2915.65	486.97	479.52	608.36	633.68
	MAE (off)	0.35	0.84	0.75	0.67	0.72
Washing machine	On Ratio	9.75%	20.08%	29.77%	39.89%	49.72%
	MAE (all)	3.64	25.04	11.42	8.21	4.68
	MAE (on)	573.21	395.49	265.38	278.11	303.89
	MAE (off)	0.07	22.71	9.83	6.52	2.80
Microwave	On Ratio	20.57%	25.90%	35.02%	41.78%	49.59%
	MAE (all)	5.86	5.07	4.40	4.59	4.57
	MAE (on)	758.15	355.09	318.66	289.10	337.37
	MAE (off)	0.00	2.35	1.95	2.37	1.97
Dish washer	On Ratio	5.14%	16.29%	27.47%	38.60%	49.80%
	MAE (all)	97.45	19.13	22.36	17.93	11.78
	MAE (on)	1300.12	464.65	509.16	271.47	317.05
	MAE (off)	65.86	7.87	9.57	11.27	3.77

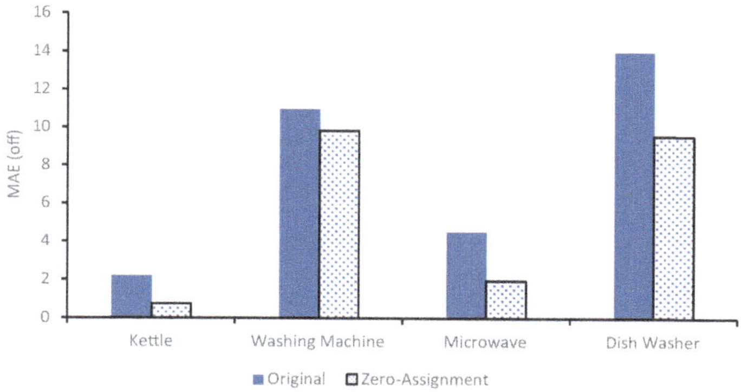

Fig. 7 Comparison between pre- and post- zero-assignment processing

6 Conclusion

In this paper, we described the commonly existing unbalance problem of data sets in the field of loading disaggregation based on machine learning. Then we proposed a training set augmentation method to improve it. Due to the demand for more scientific measurement of model performance in load disaggregation, we also proposed a class-based evaluation system. The experimental results indicated increases of 53.48–83.81% on the performance of the current popular NILM model structure. Moreover, we proposed a post-processing method named Zero-Assignment to improve the off-state misprediction.

References

1. Yu, Y., B. Liu, and W. Luan. 2013. Nonintrusive residential load monitoring and decomposition technology. *Southern Power System Technology* 7 (4): 1–5. (in Chinese).
2. Cheng, X., L. Li, H. Wu, et al. 2016. A survey of the research on non-intrusive load monitoring and disaggregation. *Power System Technology* 40 (10): 3108–3117. (in Chinese).
3. Liu, Y., and M. Chen. 2014. A review of nonintrusive load monitoring and its application in commercial building. In *The 4th Annual IEEE International Conference on Cyber Technology in Automation, Control and Intelligent*, 623–629. IEEE.
4. Hart, G.W. 1985. Prototype nonintrusive appliance load monitor. MIT Energy Laboratory Technical Report, and Electric Power Research Institute Technical Report.
5. Hart, G.W. 1992. Nonintrusive appliance load monitoring. *Proceedings of the IEEE* 80 (12): 1870–1891.
6. Makonin, S., F. Popowich, I.V. Bajić, et al. 2015. Exploiting HMM sparsity to perform online real-time nonintrusive load monitoring. *IEEE Transactions on Smart Grid* 7 (6): 2575–2585.
7. Xu, Q., O. Lou., A. Zheng, et al. 2018. A non-intrusive load decomposition method based on affinity propagation and genetic algorithm optimization. *Transactions of China Electrotechnical Society* 33 (16): 3868–3878. (in Chinese).

8. Li, D., H. Yin, B. Zheng, et al. 2015. Application of improved LSTM method in sensor fault detection of the chiller. *Transactions of China Electrotechnical Society* 34 (11): 2324–2332. (in Chinese).
9. Zhu, Y., Y. Xu, X. Chen, et al. 2020. Pattern recognition of partial discharges in DC XLPE cables based on convolutional neural network. *Transactions of China Electrotechnical Society* 35 (3): 659–668. (in Chinese).
10. Xu, C., K. Chen, J. Ma, et al. 2019. Recognition of power loads based on deep belief network. *Transactions of China Electrotechnical Society* 34 (19): 4135–4142. (in Chinese).
11. Kelly, J., and W. Knottenbelt. 2015. Neural NILM: Deep neural networks applied to energy disaggregation. In *Proceedings of the 2nd ACM International Conference on Embedded Systems for Energy-Efficient Built Environments*, 55–64. ACM.
12. Kelly, J., and W. Knottenbelt. 2015. The UK-DALE dataset, domestic appliance-level electricity demand and whole-house demand from five UK homes. *Scientific data* 2 (1): 1–14.
13. Zhang, C., M. Zhong, Z. Wang, et al. 2018. Sequence-to-point learning with neural networks for non-intrusive load monitoring. In *Thirty-Second AAAI Conference on Artificial Intelligence*.
14. Faustine, A., N.H. Mvungi, S. Kaijage, et al. 2017. A survey on non-intrusive load monitoring methodies and techniques for energy disaggregation problem. arXiv preprint arXiv: 1703.00785.
15. Branco, P.O., L. Torgo, and R.P. Ribeiro. 2017. SMOGN: A pre-processing approach for imbalanced regression. *Proceedings of Machine Learning Research* 1: 1–15.
16. Lauer, F., C.Y. Suen, and G. Bloch. 2007. A trainable feature extractor for handwritten digit recognition. *Pattern Recognition* 40 (6): 1816–1824.
17. Tivive, F.H.C., and A. Bouzerdown. 2005. An eye feature detector based on convolutional neural network. In *International Symposium on Signal Processing & Its Applications*. IEEE.
18. Szarvas, M., A. Yoshizawa, M. Yamamoto, et al. 2005. Pedestrian detection with convolutional neural network. In *Intelligent Vehicles Symposium*, 224–229.
19. Vincent, P., H. Larochelle, Y. Bengio, et al. 2008. Extracting and composing robust features with denoising autoencoders. In *Proceedings of the 25th International Conference on Machine Learning*, 1096–1103. ACM.
20. He, H., and E.A. Garcia. 2009. Learning from imbalanced data. *IEEE Transactions on Knowledge and Data Engineering* 21 (9): 1263–1284.
21. Chawla, N.V., K.W. Bowyer, L.O. Hall, et al. 2002. SMOTE: Synthetic minority over-sampling technique. *Journal of Artificial Intelligence Research* 16: 321–357.

Method for Restoring Distribution Rule of DC Ground Potential Under HVDC System

Shangmao Hu, Kunquan Li, Gang Liu, Hansheng Cai, Lei Jia, and Ruifang Li

Abstract The ground potential distribution of the High Voltage Direct Current (HVDC) transmission system during ground loop operation is a prerequisite for accurately assessing the DC bias levels of AC and DC grid transformers, track circuits, buried metals, and oil and gas pipelines, and being affected by the ground electrode current. The size of the ESP near the ground electrode of the DC transmission system is affected by the DC system; one is affected by the size, shape and embedding method of the ground electrode; the second depends on the spatial distribution of the earth resistivity. The complex geo-logical environment makes the electrical structure of the earth complex and changeable. Therefore, it is particularly important to establish a model of the earth resistance that simulates the actual electrical structure of the earth, model the earth electric field near the DC ground electrode, and find out the factors that can affect the ESP distribution. In this paper, a mathematical algorithm is used to derive the mathematical formula of the ground potential of the composite earth model, and then this analytical formula is applied to the practical engineering problems of ESP distribution in South China. Through the calculation results of different models, the simulation model that reflects the ideal of South China is identified.

Keywords HVDC · Transmission system · CDEGS · Geoelectric structure · Grounding model · Reduction algorithm · DC bias

1 Introduction

HVDC transmission has the characteristics of long transmission distance, large transmission capacity, and low loss, which can make up for the national situation of the reverse distribution of China's energy resources and demand. All the ground loop operation modes of the HVDC system have current flowing to the ground through the ground connection. Among them, the single-stage earth return line operation mode and the same polarity operation mode have the largest ground current, which

---

S. Hu · K. Li (✉) · G. Liu · H. Cai · L. Jia · R. Li
High Voltage Research Institute, CSG Electric Power Research Institute, Guangzhou, China
e-mail: andy_lau_china@163.com

W. Chen et al. (eds.), *The Proceedings of the 9th Frontier Academic Forum of Electrical Engineering*, Lecture Notes in Electrical Engineering 743,
https://doi.org/10.1007/978-981-33-6609-1_31

can reach 3 KA [1–3]. The direct current is injected into the ground through the ground electrode, forming a ground constant current field in the soil, and forming a potential distribution (ESP, Earth Surface Potential) on the ground surface. The greater the current, the greater the ESP. Theoretically, the ground with zero potential is no longer zero under the strong current of the ground electrode. Shaped space area. The non-zero potential of the ground surface will cause many negative effects, such as corrosion of metal pipes, DC bias of transformers, human and animal safety, and radio interference. In order to analyze the severity of the above-mentioned negative effects, it is necessary to consider various parameters of the earth and accurately establish a model of the earth electric field in the DC-to-earth proximity area. It is expected that the ESP distribution is as accurate as possible. This article focuses on the distribution of the surface potential of the compound earth model, uses the mirror image method and electromagnetic wave theory to derive the calculation algorithm and calculation formula of the compound earth model, and applies this formula to the earth model in southern China. Modeling of the Battery Based ESS.

2 HVDC System and Establishment of Earth Model

2.1 *Structure of the System*

When the DC transmission line is operated in a single maximum loop, the current takes the ground as a channel, and a potential will be generated on the surface of the earth. The operation structure of the DC transmission single maximum loop [4] is shown in Fig. 1.

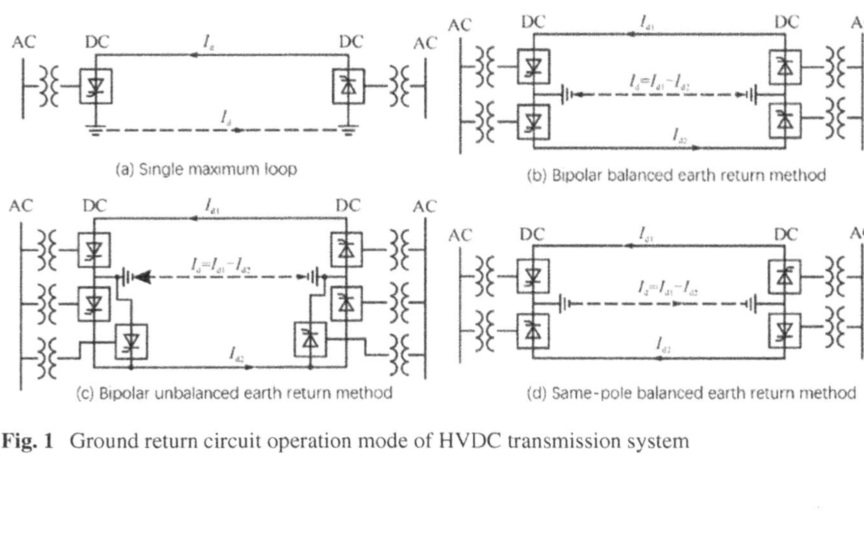

Fig. 1 Ground return circuit operation mode of HVDC transmission system

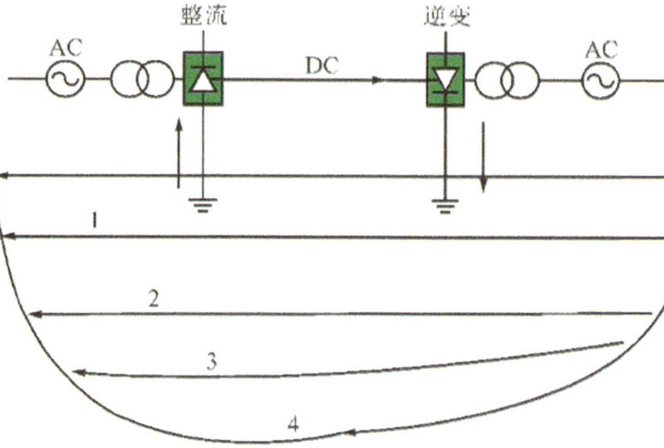

Fig. 2 Model of the complex multi-layer soil

When DC current flows into the ground and flows from one ground electrode to another, it will gradually flow from the upper soil to the deep soil. Due to the continuity of the current, the current actually flows through the soil with different resistivities near the ground electrode [5–7]. Figure 2 is a complex multi-layer soil earth model, with a current of I at the DC grounding electrode entering the ground. When considering the large-scale ground potential distribution, the current is regarded as a point current source, and the current source is not included. Equation, then the field equation in the area surrounding the point current is a Poisson equation (Fig. 3).

2.2 *Earth Resistance Model*

The earth is an important part of the HVDC transmission system, because the unipolar current and bipolar unbalanced or asymmetric current in the system all flow into the earth [7–13]. Because the resistivity of the earth is affected by many factors, such as geographic environment and climate environment, many variables need to be considered when quantifying the resistivity. As shown in Table 1, the earth resistivity measured by the electromagnetic detection method indicates the depth of each layer of the earth and the corresponding soil resistivity Ω·m.

The resistivity shown in Table 1 is divided horizontally against the ground to obtain the approximate range of the soil resistivity of each layer. It must be combined with the soil resistivity of the specific study area [13–15]. For example, the vertical layered resistance of the ocean must be considered in southern China. The Sichuan and Sichuan regions must consider the resistivity of these multiple vertical layers in the mountain. The calculation accuracy of the surface potential increases with the fineness of the spatial resistivity we divide. In addition to the horizontal and vertical

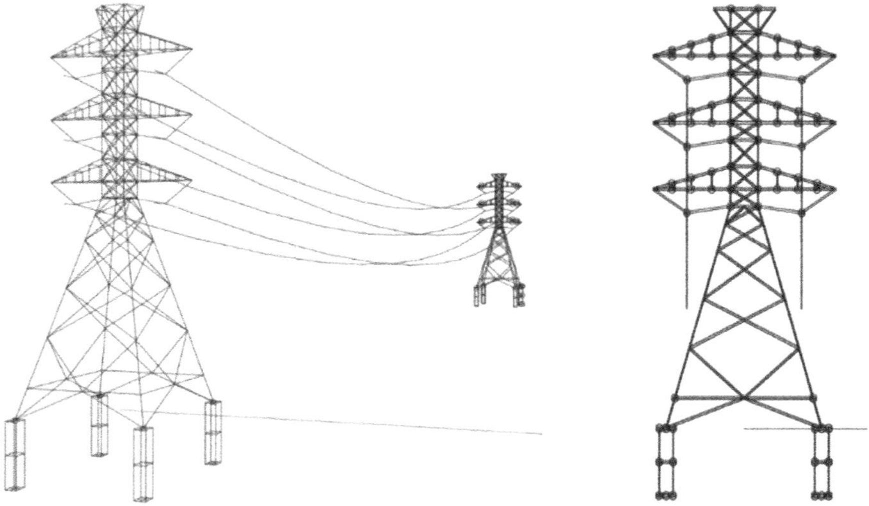

Fig. 3 HVDC transmission system model

Table 1 Address structure table

Address layer name	Parameters	Value
1-Shallow deposit	100	300
2-deep layer	200	10,000
3-bedrock layer	15,000	10,000
4-Mantle to earth	150	∞
5-Ocean	0.5	∞

plane divisions mentioned above, there are common slope and cylinder divisions. Of course, the use of planes to divide the earth resistivity is suitable for a large range of space. For a small space, the current causes physical or chemical changes in the soil near the ground electrode, thereby changing the resistivity. This change law Must be characterized by functional relationships or other methods.

3 Theoretical Calculation of Uniform Medium ESP

3.1 Ground Current Changes with Depth

The current density J_x in the x direction parallel to the ground surface at any point M (0, y, z) on the perpendicular bisector of the connection line between the two ground electrodes. Calculate the current flowing through the ground electrode into the ground and out of the soil with a depth of h or more:

$$\frac{I_h}{I} = \frac{2}{\pi} \arctan \frac{2h}{D} \tag{1}$$

where I is the current injected into the grounding point; D is the straight-line distance between the two grounding electrodes of the receiving end; θ is the angle between the connection between point M and the current entry point and the positive direction of the x-axis.

According to the calculation formula (1), the curve of the percentage of the current flowing through the earth in the ground above the depth of h on the center line of the two grounding electrodes in Fig. 4 as a function of h/D changes. It can be seen that in the uniform ground, 70% of the ground current flows in the ground with the same depth as the ground electrode spacing, and 30% of the ground current flows in the deep ground larger than the ground distance. This indicates that a large amount of ground current will circulate in the deep earth. In the UHV DC transmission project, the electrical properties of the earth and its structure are complicated at a distance of several thousand kilometers. The earth resistivity model near the ground electrode is established to calculate the ESP distribution accurately, and the deep earth resistivity should be taken into account when modeling. And the survey of deep earth resistivity should be carried out.

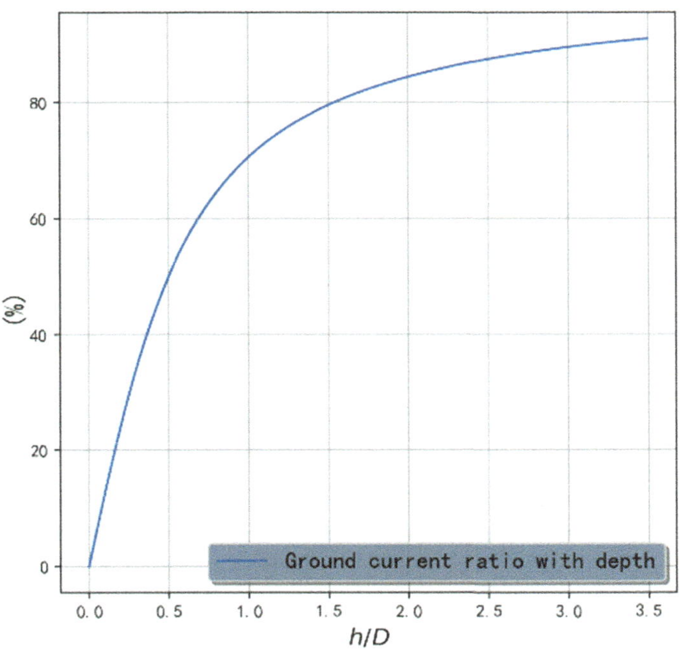

Fig. 4 Percentage of ground current with depth

3.2 Infinite Uniform Earth Model

The point power in an infinite uniform medium is shown in the Fig. 5.

According to the current continuity theorem, the calculation formula of potential distribution in the medium $\phi(x, y, z)$ is obtained:

$$\Phi(x, y, z) = \frac{\rho i}{4\pi \sqrt{(x - x_o)^2 + (y - y_o)^2 + (z - z_o)^2}} \quad (2)$$

where (x_0, y_0, z_0) is the coordinate of the point power supply; (x, y, z) is the coordinate of any point; ρ is the resistivity of the medium; i is the point power supply Current. We use python's three-dimensional drawing module to visualize this formula. The ESP distribution diagram of Fig. 6 is obtained. (Note: The distance scale is meters).

Then we will set up a single power supply based on the above high-voltage transmission model to simulate the ESP distribution of the ground electrode. The ESP distribution shown in Fig. 7 is obtained.

Comparing the above two figures, we can see that the theoretical calculation and simulation model results are almost the same, which shows the feasibility of the theoretical algorithm.

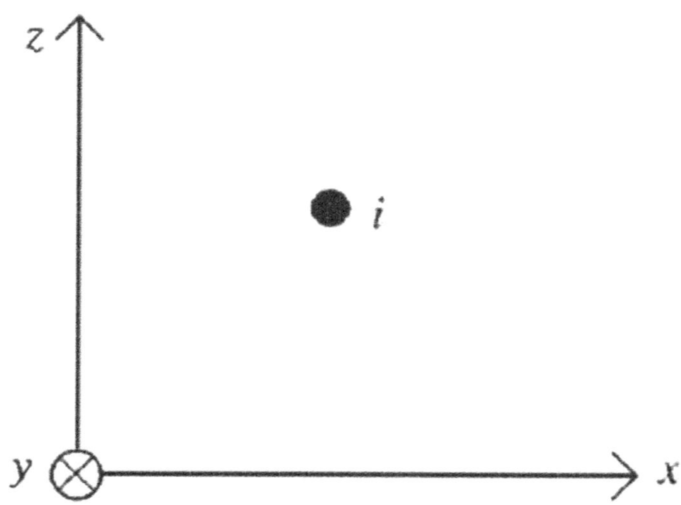

Fig. 5 Wireless uniform medium power supply

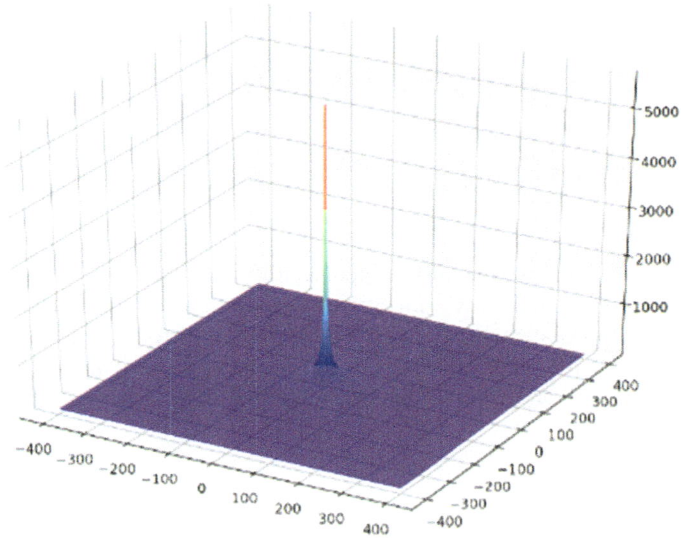

Fig. 6 ESP calculated value of homogeneous medium

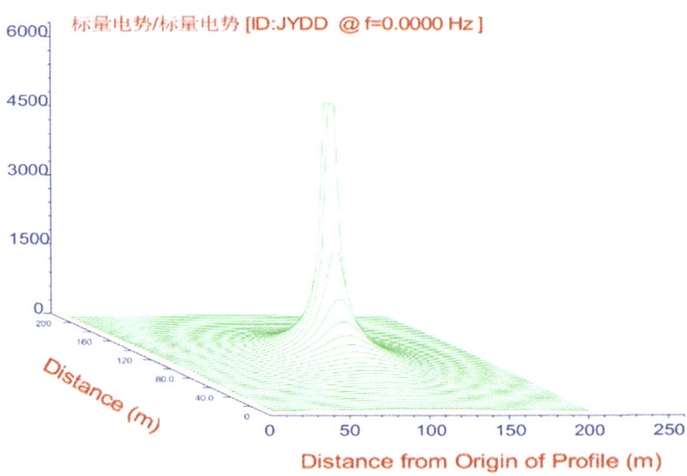

Fig. 7 ESP simulation values of homogeneous media

3.3 Theoretical Calculation of ESP of Compound Soil Model

$$\Phi_H(x, y, z) = \sum_{j=0}^{\infty} \Gamma_4^j [\Phi(x, y, z + 2jB + b) + \Phi(x, y, z - 2jB - b)] + \sum_{j=0}^{\infty} \Gamma_4^j [\Phi(x, y, z - 2jB + b) + \Phi(x, y, z + 2jB - b)] \quad (3)$$

According to the above-mentioned geoelectric field equations, boundary conditions, and current continuity, we can derive the analytical formula of the ESP distribution of the composite geological model by using the mirror image method and the law of wave refraction and reflection. Visualize the formula. The ESP distribution diagram of Figs. 8, 9 and 10 is obtained. (Note: The distance scale is kilo-meters).

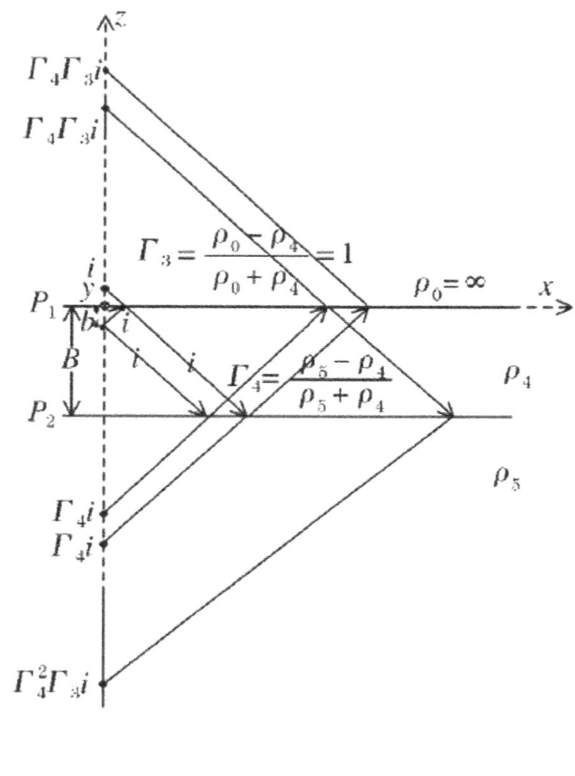

Fig. 8 Two-layer horizontal layered earth model

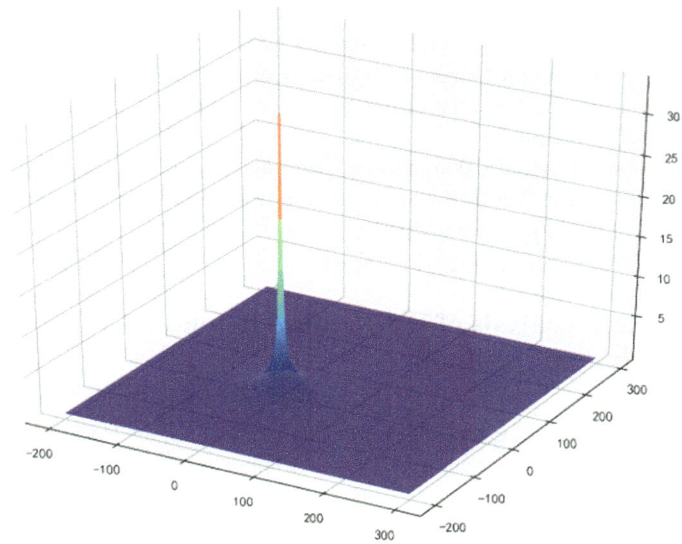

Fig. 9 Three-layer vertical layered earth model ESP distribution

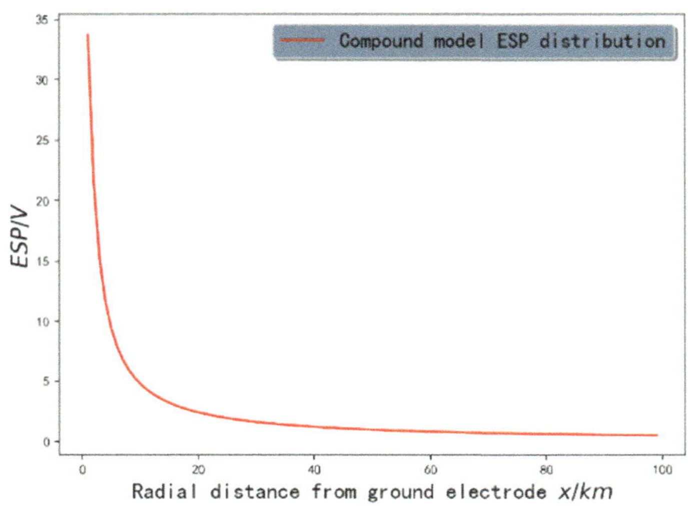

Fig. 10 ESP distribution of the two-layer horizontal layered earth model

3.4 In Conclusion

According to the above simple model to compound model, we can get the following conclusions:

(1) The maximum value of the ground potential appears on the surface of the ground electrode. The ground current of 3000 A can reach 100 to 200 V, and the size of the potential has no obvious relationship with the selected model, indicating that the potential of the ground electrode surface is not affected by the model selection;

(2) The ground potential calculated by all models is sharply reduced from 0 to 20 km, indicating that the location of the ground electrode is preferably 20 km away from the city or a transformer above 110 kV.

4 Simulation Analysis of DC Ground Potential Under HVDC System

In the module HIFREQ of the simulation software CDEGS, the HVDC transmission model shown in Fig. 3 is created, and a geological model of the coastal area in southern China is constructed based on Table 1. This system is a 220 kV overhead power transmission system. The three-phase three-circuit lines with vertical and parallel conductors are arranged at a phase interval of 10 m and a distance of 10 m. The natural sag of the conductor is considered. To construct a high-voltage transmission tower model, detailed parameter definitions need to be made according to the transmission lines of the high-voltage transmission system. Through simulation modeling and analysis, we obtain the surface potential distribution near the ground electrode. At this time, the simulation shows that the ground potential attenuation on the observation line is shown in Fig. 11. It can be seen that in the high-voltage DC transmission system of 220 kV voltage level, ESP attenuates sharply within 500 m of the ground electrode, and ESP almost drops to zero outside 500 m.

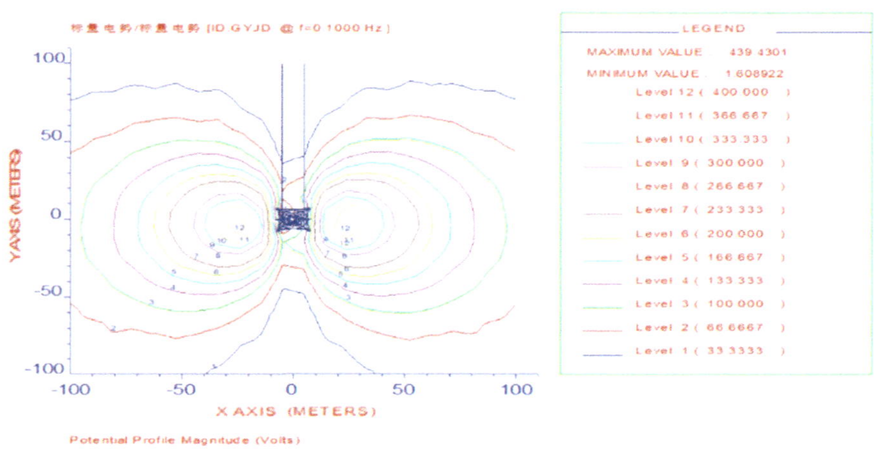

Fig. 11 ESP distribution under high-voltage transmission system

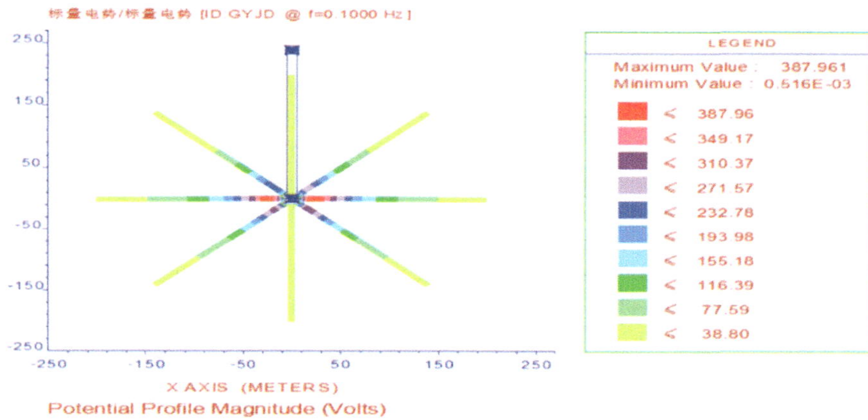

Fig. 12 ESP distribution of observation lines in 8 directions

In order to compare the attenuation rules of ESP in various directions in detail, we have drawn 8 observation lines, which are distributed in a Mi-shape, as shown in Fig. 12.

And get the ESP distribution on each observation line. It can be seen that the ESP amplitude along the direction of the high-voltage transmission corridor is extremely low, with a maximum value of 20 V and a sharp decay to zero. The ESP in the direction of the vertical corridor is the largest, and it can reach nearly 400 V at the outer boundary of the tower. After that, it attenuates sharply, and it attenuates to less than 20 V outside the range of 200 m, and to less than 10 V at 300 m. In the direction of 45° to the transmission corridor, the attenuation law is consistent with the vertical direction, but the maximum amplitude is less than 300 V.

In summary, we can get the following conclusions: When the 220 kV HVDC is running normally and normally, the distribution and attenuation of ESP in different directions are different; the left and right are symmetrically distributed, and the closer to the transmission line, the weaker the ESP.

5 Conclusion and Outlook

In this paper, the calculation method of the surface potential in the composite earth model is derived using the composite mirror image method and the reflection and refraction laws of electromagnetic waves. Comparing the potential curves, the ideal calculation model to obtain the potential distribution in the coastal area of South China is a three-layer ocean composite earth model, which not only ensures the accuracy of the calculation, but also saves the memory of the computer. Combining the simulation results of the compound earth model with ocean in South China, the following suggestions and prospects are proposed.

(1) Transformers or metal pipes located at neutral points in coastal areas should pay attention to the operation of the ground loop of the HVDC system. Due to the low impedance of the sea, the city near the ocean has a low potential. The large potential difference makes the DC in the area larger, causing more serious metal corrosion or DC bias of the transformer and other adverse effects;
(2) Replacing complex summation formulas with polynomials can greatly simplify the calculation of the ideal model surface potential. After finding the ideal model, you can find the potential difference between any two points by fitting the function. This potential difference can be used to determine whether the step voltage and contact voltage are within the safety value range, and assist the location of the HVDC ground electrode.

Acknowledgements This work was supported by project (ZBKJXM20180644) of Electric Power Research Institute, CSG.

References

1. Liu, Jiming, and Xiao Dong. 2006. Analysis of effects of DC earth pole on earth surface Potential distributions. *High Voltage Engineering* 32 (9). (in Chinese).
2. Zhuohong, Pan, Lu Zhang, et al. 2012. Simulation and analysis of earth surface potential distribution in horizontal multi-layer soil. *High Voltage Engineering* 38 (1): 116–123. (in Chinese).
3. Liu, Qu, Li Licheng, et al. 2007. DC current distribution in HVDC systems of monopolar operation with ground retrun in complex soil structure. *Proceedings of the CSEE* 27 (36): 8–13. (in Chinese).
4. Ishiguro, F., F. Tanaka, S. Tanaka, et al. 1986. Coordinated stabilizing control of exciter turbine and braking resistor. *IEEE Power Engineering Review* 6 (08): 25–26.
5. Tang Xinling, Jiang jianguo, et al. 2013. Analysis of earth surface potential distribution based on finite element. *Modem Electric Power* 30 (3): 65–68. (in Chinese).
6. Guoping, Zou, Yao Hui, He Wenlin, et al. 2016. Xiluodu-West Zhejiang 800kv UHV DC transmission project receiving end Network DC bias treatment. *High Voltage Technology* 42 (2): 543–550. (in Chinese).
7. Southey, R.D., M. Siahrang, S. Fortin, et al. 2015. Using fall-of-potential measurements to im-prove deep soil resistivity estimates. *IEEE Transactions on Industry Applications* 51 (6): 5023–5029.
8. Dai, Zhihui, Ge Hongbo, Yan Siqi, Wang Zengping, Chen Xi. 2018. Failure Analysis of flexible DC distribution network. *Transactions of China Electrotechnical Society* 33 (08): 1863–1874. (in Chinese).
9. Zhang, Qiushi, Wang Linong, Fang Yaqi, Liu Kai, Gao Jiachen. 2019. Simulation analysis method of gap discharge characteristics of equipotential personnel-tower tower structure. *Transactions of China Electrotechnical Society* 34 (11). (in Chinese).
10. Yingxi, Liu. 2017. *Research on the dynamic distribution and leakage monitoring of urban rail transit stray current*. Xuzhou: China University of Mining and Technology. (in Chinese).
11. Zhiwen, Zhang, Lei Shijie, Zhai Chengda, and Yi Meisheng. 2019. Characteristic analysis of Absorption and shunt Capacitor commutation Converter in HVDC Transmission system. *Transactions of China Electrotechnical Society* 34 (S2): 684–691. (in Chinese).

12. Yongbin, Zhang, Nie Minglin, Zhang Junpeng, Song Hui, Chi Mingwu, and Yi Qianyu. 2019. Characteristic analysis of absorption and shunt capacitor commutation converter in HVDC transmission system. *Transactions of China Electrotechnical Society* 34 (S2): 684–691. (in Chinese).
13. Chen, Le, Peng Yongquan, Lin Xiangning, Li Zhengtian, Jin Neng. 2019. Fuzzy planning of distribution network frame considering the uncertainty of distributed power. *Transactions of China Electrotechnical Society* 34 (S1): 258–263. (in Chinese).
14. Yufei, Teng, Li Xiaopeng, Lin Sheng, and Li Shilong. 2019. Research on the adaptability of the impedance monitoring system for the grounding pole of the UHV DC system. *Trans-actions of China Electrotechnical Society* 34 (19): 4154–4161. (in Chinese).
15. Xiaowei, Wang, Wu Gao Jie, Song Guobing Lei, and Wei Yanfang. 2019. High-impedance ground fault detection method for flexible DC switching network. *Transactions of China Electrotechnical Society* 34 (13): 2806–2819. (in Chinese).

Research on Speed Control Method of Twelve-phase Permanent Magnet Synchronous Motor

Mingqing Yao, Hongwei Ma, and Jingpan Ren

Abstract The mathematical model of multi-phase permanent magnet synchronous motor (PMSM) can be summed up as a single winding and multi winding models in MT coordinate system. This paper studies the speed regulation method of 12-phase PMSM, and two speed control strategies of multi-phase vector and multi-phase multi-vector are realized. According to the vector control mode of "id = 0", the air gap flux observer speed regulating system is constructed, and the multi-phase vector control strategy of air gap flux-oriented control 12 phase PMSM is realized. In addition, the multi-phase multi-vector control strategy of two modulation methods of "space vector pulse width modulation (SVPWM)" and "capacitor voltage balanced diode clamped type (NPC) three-level inverter" are adopted to motor speed regulation. On this basis, the operation of the motor under phase failure is analyzed and verified. The simulation results show that the system has good control and operation performance based on the above two control strategies. At the same time, the proposed control strategies can be extended to the model of "n" phase PMSM with the same mode.

Keywords Twelve-phase PMSM · Multiphase vector control · Multi-phase Multi-vector control · SVPWM · NPC three-level inverter hybrid

1 Introduction

Multi-pole multi-phase motors have the characteristics of large capacity, large torque, high power, wide range speed regulation, small size, light weight, safe and reliable structure. It is mostly used in ship electric propulsion due to its high torque density, small pulsation and strong fault-tolerant, it has attracted many scholars research interest of multi-phase motors [1]. This paper takes 12-phase PMSM as an example, explores the mathematical model and equivalent analysis of multi-phase PMSM [2]. Literature [3] and Literature [4] obtained the mathematical model of the 12-phase synchronous motor in a, b, c coordinate system, d, q coordinate system and MT

M. Yao (✉) · H. Ma · J. Ren
School of Automation, Beijing Institute of Technology, Beijing, China
e-mail: mhw@bit.edu.cn

W. Chen et al. (eds.), *The Proceedings of the 9th Frontier Academic Forum of Electrical Engineering*, Lecture Notes in Electrical Engineering 743,
https://doi.org/10.1007/978-981-33-6609-1_32

coordinate system of the air-gap flux orientation through in-depth analysis of the twelve-phase synchronous motor. On this basis, through the simulation modeling of 12-phase PMSM, the multi-phase vector and multi-phase multi-vector control method is designed and implemented.

2 Theoretical Analysis

The 12-phase PMSM studied in this paper is composed of four sets of three-phase windings whose neutral points are independent of each other with an electrical angle of 15°. The mutual positions of the four sets of windings are shown in Fig. 1 below. And make the following assumptions: the four sets of windings of Y_1, Y_2, Y_3, and Y_4 are symmetrically distributed in space, which are sequentially 15° apart from each other, and the air gap magnetic field is sinusoidal in space. The saturation, hysteresis, eddy current and skin effect of conductor are not considered. The influence of temperature on motor parameters is ignored. The inner surface of the stator is smooth, ignoring the effect of cogging, that is, the influence of the tooth slot and the ventilation slot is not considered.

2.1 Mathematical Model of Symmetrical 12-Phase (4Y Shift 15°) PMSM

According to the principle of power invariance [5, 6], the transformation matrix from the 12-phase stationary coordinate system to the d, q rotating coordinate system is obtained as follows:

$$C_{dq}^{12} = \sqrt{\frac{2}{12}}\begin{bmatrix} \cos\theta & \cos\left(\theta - \frac{2\pi}{12}\right) & \cdots & \cos\left[\theta - 11 * \frac{2\pi}{12}\right] \\ -\sin\theta & -\sin\left(\theta - \frac{2\pi}{12}\right) & \cdots & -\sin\left[\theta - 11 * \frac{2\pi}{12}\right] \end{bmatrix} \tag{1}$$

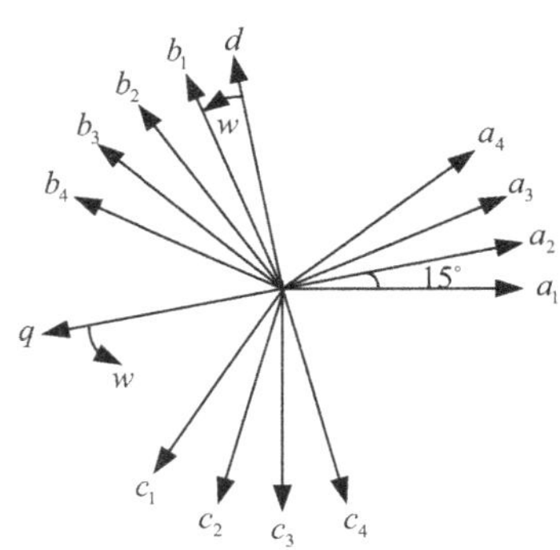

Fig. 1 Relative position of each phase winding of 12-phase PMSM

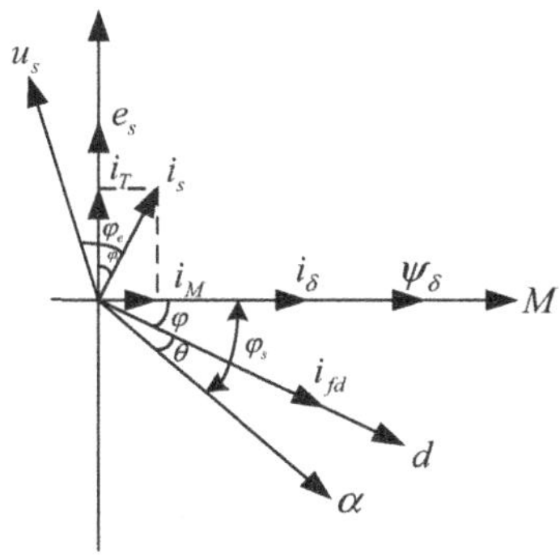

Fig. 2 PMSM motor current, flux linkage vector illustration

The corresponding inverse transformation matrix is as follows:

$$C_{12}^{dq} = \sqrt{\frac{2}{12}} \begin{bmatrix} \cos\theta & -\sin\theta \\ \cos\left(\theta - \frac{2\pi}{12}\right) & -\sin\left(\theta - \frac{2\pi}{12}\right) \\ \vdots & \vdots \\ \cos\left[\theta - 11 * \frac{2\pi}{12}\right] & -\sin\left[\theta - 11 * \frac{2\pi}{12}\right] \end{bmatrix} \tag{2}$$

Figure 2 is the corresponding phasor diagram in the MT coordinate system. The rotation transformation matrix from the d, q rotating coordinate system to the MT coordinate system is shown as follows:

2.2 The Mathematical Model in D, Q and M, T Coordinates

After the above analysis, the 12-phase winding can be regarded as four Y-connected three-phase symmetric windings with the same stator and rotor. The transformation matrix is as follows:

$$C_{dq}^{abc} = \frac{2}{3} \begin{bmatrix} \cos[\theta - (i-1)\cdot 15^\circ] & -\sin[\theta - (i-1)\cdot 15^\circ] \\ \cos[\theta - 120^\circ - (i-1)\cdot 15^\circ] & -\sin[\theta - 120^\circ - (i-1)\cdot 15^\circ] \\ \cos[\theta + 120^\circ - (i-1)\cdot 15^\circ] & -\sin[\theta + 120^\circ - (i-1)\cdot 15^\circ] \end{bmatrix}^T$$
$$i = 1, 2, 3, 4 \tag{3}$$

From the analysis of the model, the torque equation of the motor can be obtained as follows:

$$T_d = \frac{1}{4}\psi_\delta \sum_{i=1}^{4} i_{Ti} \tag{4}$$

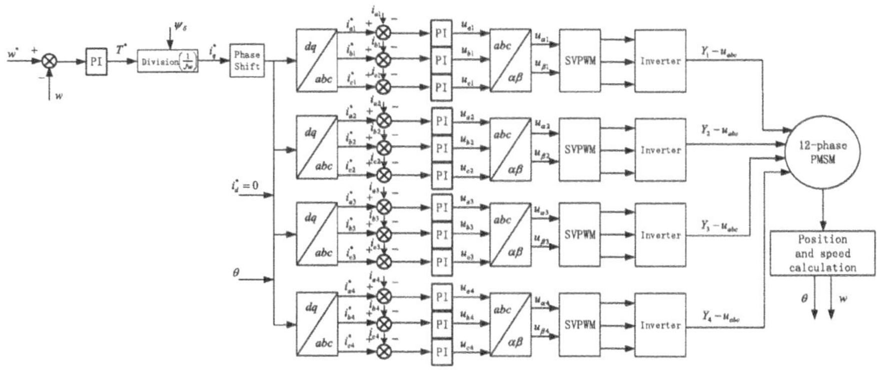

Fig. 3 12-phase PMSM multi-phase vector control block diagram

$\sum_{i=1}^{4} i_{Ti}$ is the sum of the torque components of the four Y-connected windings.

3 Control Strategy of 12-phase PMSM

3.1 *Multi-phase Vector Control Method*

If the air-gap flux linkage remains constant, the torque is proportional to the torque component of the stator current. The speed of the 12-phase permanent magnet synchronous motor can be controlled by the torque component of the stator current. In this case, the flux-oriented control principle of the 12-phase PMSM, that is, the multi-phase stator current variable is decomposed into a magnetization component and a torque component through vector transformation, and these two variables can be controlled independently. The control strategy of the 12-phase permanent magnet synchronous motor explained above is the control block diagram shown in Fig. 3.

3.2 *Multi-phase Multi-vector Control Method*

If the torque component and magnetization component of the four three-phase windings have the same magnitude, the composition of the motor torque will reach the maximum, and the control efficiency will also reach the optimal state. The corresponding vector control block diagram is shown in the Fig. 4. Figure 5 is the control method of NPC three-level inverter [7, 8] with capacitor voltage balance.

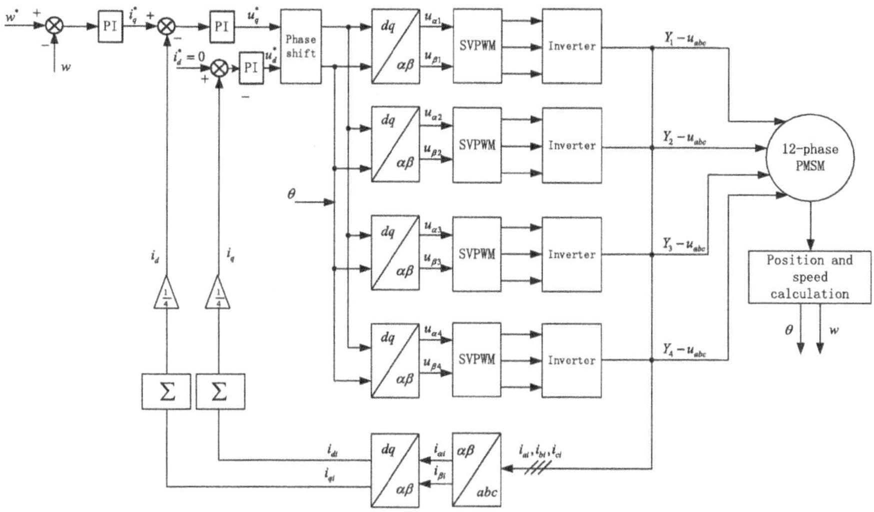

Fig. 4 12-phase permanent magnet synchronous motor multi-phase multi-vector (SVPWM) control block diagram

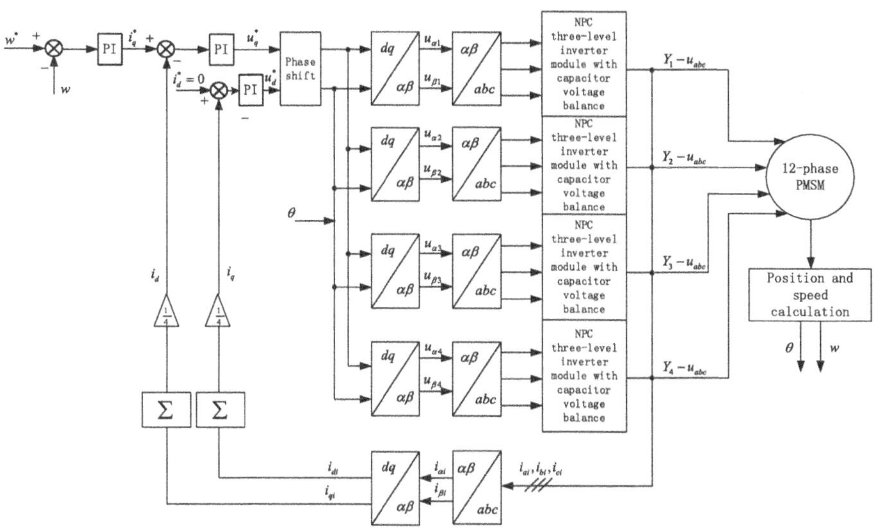

Fig. 5 Multi-phase multi-vector control block diagram of NPC three-level inverter with capacitor voltage balance

4 Simulation Example Verification

In order to verify the correctness and feasibility of the above theoretical analysis, this article takes a built-in twelve-phase PMSM as an example. The detailed parameters are as follows: $L_d = 8.065e\text{-}5H$, $L_q = 11.35e\text{-}5H$, $R = 5.11\ m\Omega$, $\Psi_\delta = 69.32\ mWb$, $p = 2$, $U_d = 311\ V$ and Simulation time $t = 0.2\ s$.

4.1 *Analysis of Operating Conditions Under Different Control Strategies*

Figure 6 is the use of a multi-phase vector control strategy to adjust the speed of 12-phase PMSM. When the speed command is 0.1 s, it changes from 200 to 500 rpm. It can be seen from the speed diagram of the motor that the motor can reach the given speed in a short time with small overshoot.

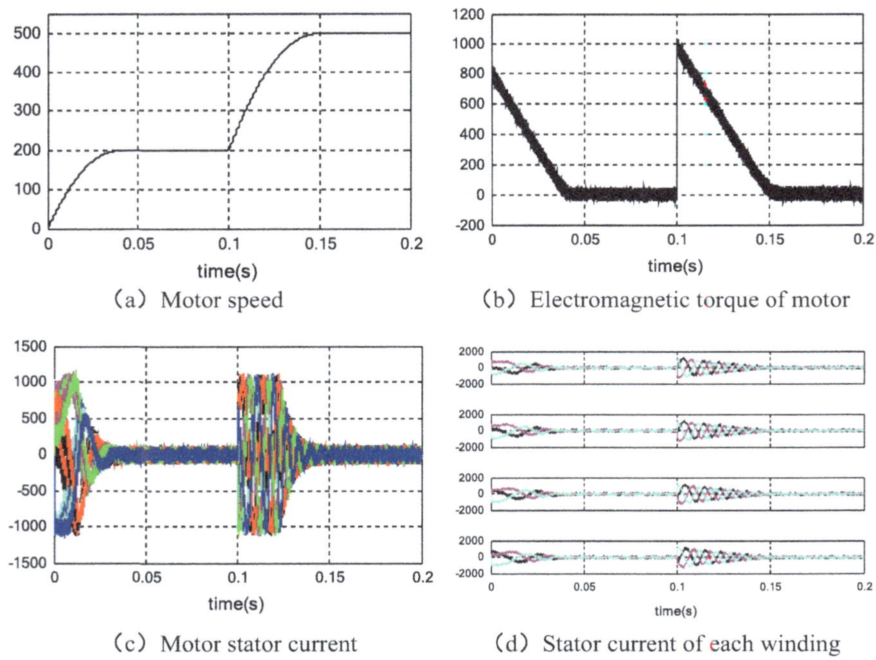

(a) Motor speed

(b) Electromagnetic torque of motor

(c) Motor stator current

(d) Stator current of each winding

Fig. 6 Multiphase vector control 12-phase PMSM

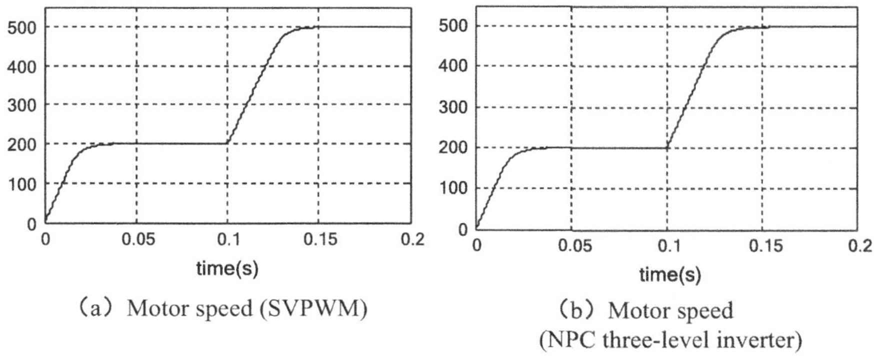

Fig. 7 Multi-phase multi-vector control 12-phase PMSM

Figure 7 are two different modulation methods under the multi-phase multi-vector control strategy. Compared with the multi-phase vector control strategy, the adjustment of this method is simpler and faster. This also reflects the feasibility of two speed regulation from the side.

4.2 Analysis of Motor Phase Loss Operation Under Different Control Strategies

For "Y_1, Y_2, Y_3, Y_4" four sets of symmetrical Y-connected three-phase windings, there are four operating modes when they are put into the cut-out operation. The change of stator winding structure is shown in Fig. 8.

As is shown in Figs. 9 and 10, using the current limit criterion method [9, 10], when the motor is open in a certain phase winding, the speed of the motor can always reach the given speed stably without large fluctuation. This is because the load parameters of the multi-phase power supply and the polyphase circuit are balanced and symmetric. When one or more port units are damaged, it can still maintain most of the rated values under balanced excitation.

5 Conclusion

In this paper, by analyzing the structure characteristics of 12-phase PMSM, the control strategy of multi-phase vector and multi-phase vector is designed and realized. The simulation results show that the system has good control and operation performance under these two control strategies. In addition, on the basis of these two control strategies, the phase loss of 12-phase PMSM is analyzed. Through simulation, it is found that under these two control strategies, the system can still be reliable

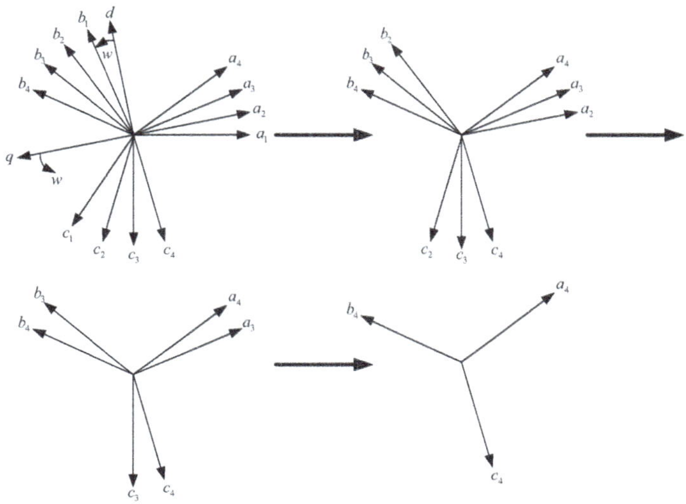

Fig. 8 12-phase PMSM symmetrical phase loss operation stator winding structure changes

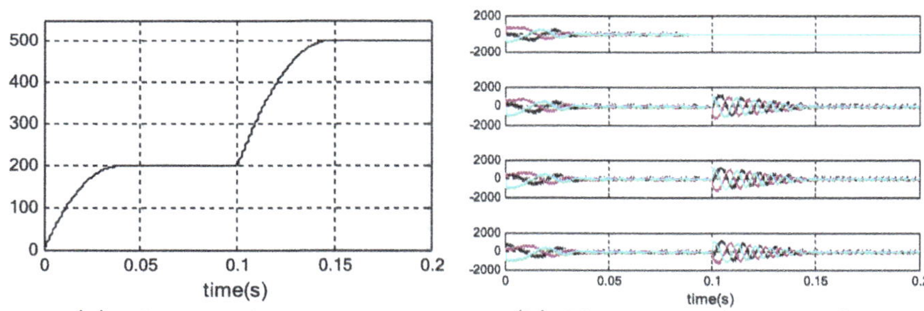

(a) Motor speed (b) Electromagnetic torque of motor

Fig. 9 Open circuit of one phase of the motor under Multi-phase vector control strategy

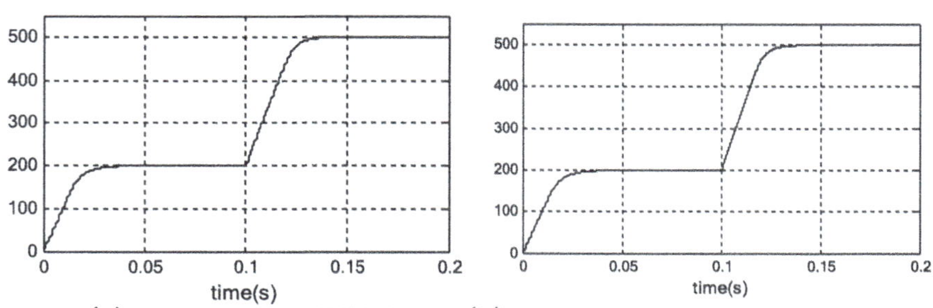

(a) Motor speed (SVPWM) (b) Motor speed (NPC three-level inverter)

Fig. 10 Open circuit of one phase of the motor under Multi-phase Multi-vector control operation when one phase of the motor is out of phase, which provides a theoretical basis and simulation platform for further analysis of the operation characteristics of the system.

Acknowledgements Project Supported by the National Key Research and Development Program of China (2018YFB0905500).

References

1. Lin, Hua, Zou Yunping, and He Bi. 2006. The vector control strategies for multiphase synchronous motor drive systems. Montreal, Quebec, Canada: IEEE ISIE 9–12 July 2006.
2. Ma, Xiaoliang. 2004. *Powerful cycloconverter speed regulation and vector control technology*. Beijing: China Machine Press.
3. Zhaohui, Zhuang, Xiong Youlun, and Ma. Ting. 2001. Review, current situation and prospect of multi-phase induction motor frequency conversion speed regulation system. *Electric Drive* 2: 3–7. (in Chinese).
4. Zhihao, Ye, Zhang Xiaofeng, Li Hongjiang, Yu. Fei, and Du Chengdong. 2004. Simulation analysis of 12-phase synchronous motor model and speed control. *Journal of Wuhan University of Technology (Transportation Science and Engineering Edition)* 05: 671–674. (in Chinese).
5. Hua, Lin, and Zou Yunping. 2004. Mathematical model of twelve-phase synchronous motor with air gap flux linkage orientation control. *Small and Medium Motors* 02: 21–27. (in Chinese).
6. Hua, L., Z. Yunping, and H. Bi. 2006. The vector control strategies for multiphase synchronous motor drive systems. In *2006 IEEE International Symposium on Industrial Electronics*, Montreal, Que., 2205–2210.
7. Ma, Weiming, Hu An, and Dezhi Liu. 2000. Stability of a syn- chronous generator with diode-bridge rectifier and back-EMF load. *IEEE Transactions on Energy Conversion* 15: 458–463.
8. Yongdong Li, Xiao Xi, and Gao Yue. 2005. Large-capacity multilevel converter: principle, control and application. Science Press. (in Chinese)..
9. Feng, Yu., Cheng Ming, Tian Zhujie, and Zhang Wei. 2019. Control strategy for active phase loss operation of nine-phase flux switching permanent magnet motor. *Journal of Electrical Engineering and Technology* 34 (08): 1626–1635. (in Chinese).
10. Song, Baiyang. 2019. Research on symmetrical phase loss and steady state operation of nine-phase induction motor. Qingdao University, (in Chinese).

HTS Conductor for Superconducting Fault Current Limiting Transformer

Kang Qiangqiang, Wang Yinshun, Ma Siming, Wang Yueyin, Wei Defu, and Tie Guo

Abstract With the increasing power grid capacity and voltage level, the fault current of the system is getting higher, which will have a great impact on the economy of the region and the safety of the power system. In order to solve this problem, the researchers have developed a variety of superconducting fault current limiting devices based on the characteristics of the second-generation (2G) high-temperature superconducting (HTS) tapes with high resistivity after quenching. Among them, the superconducting fault current limiting transformer (SFCLT) has the dual functions of a transformer and a fault current limiter. In this paper, an HTS conductor for superconducting current limiting transformer is proposed. Based on the short circuit fault current of the secondary winding of 120 kVA HTS transformer, the specific structural parameters of the conductor are given. On the basis of the parameters, combined with the numerical analysis of conductor's circuit and thermal equations, the critical current and current limiting characteristics of the conductor are obtained by modeling and simulation. The simulation results show that the conductor can limit the fault current well, and it is a scheme for the superconducting current limiting transformer winding.

K. Qiangqiang (✉) · W. Yinshun · M. Siming · W. Yueyin
State Key Laboratory of Alternate Electrical Power System with Renewable Energy Sources, North China Electric Power University, Beijing 102206, China
e-mail: 15176260171@163.com

W. Yinshun
e-mail: yswang@ncepu.edu.cn

M. Siming
e-mail: 15931132503@qq.com

W. Yueyin
e-mail: wangyueyin1@163.com

W. Defu · T. Guo
Electric Power Research Institute of State Grid Liaoning Electric Co Ltd, Shenyang 110006, Liaoning, China
e-mail: Wdf_dky@163.com

T. Guo
e-mail: guotie9999@163.com

W. Chen et al. (eds.), *The Proceedings of the 9th Frontier Academic Forum of Electrical Engineering*, Lecture Notes in Electrical Engineering 743,
https://doi.org/10.1007/978-981-33-6609-1_33

Keywords Characteristic of the fault current limiting · HTS conductor · Superconducting fault current limiting transformer (SFCLT)

1 Introduction

With the rapid growth of population and economy, the ever-increasing grid capacity and voltage level will generate large fault currents that exceed the capacity of the circuit breaker, which in turn will have a great impact on the economy and the safety of the power system in the region. Therefore, the development of power equipment with fault current limiting function is very necessary to improve the safety and reliability of the power system.

Because the second-generation (2G) high-temperature superconducting (HTS) tapes have high resistivity after quenching, the researchers proposed and produced some superconducting power devices with the function of limiting fault current. American Super Power Company produced and tested the resistive superconducting fault current limiter based on YBCO 2G HTS tapes in 2009 [1]. The U.S. Department of Homeland Security, American Superconducting Corporation and Consolidated Edison jointly funded the demonstration projects of HTS cables with fault current limiting function [2]. In 2014, the New Energy and Industrial Technology Development Organization (NEDO) of Japan and the Institute of Superconducting Science and Systems, Kyushu University, Fukuoka, Japan, designed and assembled a three-phase 66 kV/6.9 kV, 20 MVA capacity fault current limiting HTS transformer [3]. In the above equipment, the HTS fault current limiting transformer not only has the function of a transformer, but also has the function of a fault current limiter. In normal operation, the superconducting fault current limiting transformer (SFCLT) works in a superconducting state, with large capacity, low loss and high efficiency. When the fault current exceeds the critical current of the HTS conductor, the overall resistance of the transformer rapidly becomes larger to limit the fault current [4].

At present, researches on SFCLT is only limited to the principle prototype [5, 6]. Based on 2G HTS tapes, a variety of superconducting conductor structures have been proposed internationally, such as CORC (Conductor On Round Core) conductors, TSTC (Twisted Stacked-Tapes Cable) conductors and quasi-isotropic conductors. CORC conductor has the advantages of good flexibility, high mechanical strength, high current density, large capacity and low AC losses. In 2017, Advanced Conductor Technologies LLC designed commercial CORC conductors with diameters of 7 mm and 3 mm [7]. In 2019, with the support of the U.S. Navy, the company designed, fabricated and tested a fault current limiting CORC conductor with a critical current of 1.1 kA. [8]. CORC is suitable for superconducting fault current limiting transformer windings due to its compact structure and flexibility.

This paper proposes a new type of CORC-like conductor for SFCLT winding. Since stainless steel has a higher resistivity than copper at 77 K,we choose the stainless steel former and stainless steel reinforced REBCO superconducting tapes.

In this paper, the winding for SFCLT is designed, and its fault current limiting characteristics are numerically analysed by the circuit and thermal equation.

2 Design Method of Conductor Structure

The basic structure of the conductor is shown in Fig. 1. The superconducting tape is helically wound on a circular section former with a certain pitch. The superconducting layer can be one or more layers according to the winding requirements. In order to balance the electric field between adjacent superconducting layers, a semiconducting carbon paper layer is placed between adjacent superconducting layers.

Because the resistivity of stainless steel is two orders of magnitude higher than copper at 77 K, the current limiting effect of stainless steel is better. We use twisted strands of stainless steel wire as the former. For superconducting tapes, we used stainless steel reinforced 2G HTS tapes.

2.1 Former

The former is located in the innermost layer of the conductor and plays the role of supporting the entire conductor. For conventional CORC conductors, the former uses round cross-section copper stranded wire. When the fault current exceeds the conductor's critical current, the former can play a role in shunting [9], using its own current capacity to shunt the fault current and protect the CORC conductor. Because the resistivity of stainless steel at 77 K is two orders of magnitude higher than that of copper, using stainless steel round cross-section stranded wire as the former can achieve better current limiting and protection effects. Since the superconducting wire substrate is stainless steel (Hastelloy), it has approximately the same resistivity as the

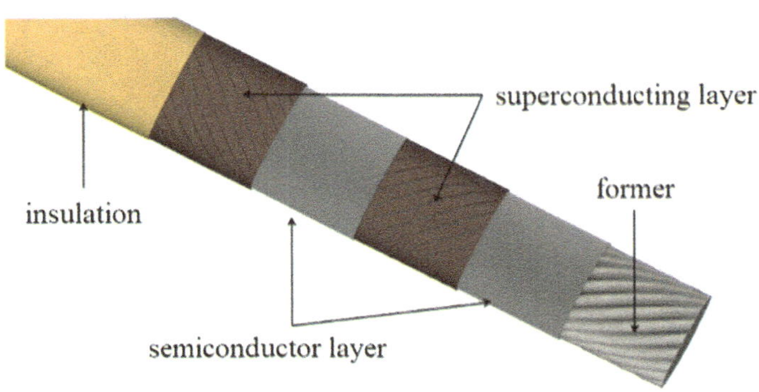

Fig. 1 Schematic geometrical configuration of the HTS conductor

stainless-steel former. The fault current will evenly flow through the stainless steel former and the superconducting tapes. The heating and the temperature is uniform, so that local overheating and damage to the conductor will not generated. In order to avoid the excessive temperature caused by the heating of the former, the selection of the cross-section of the former must be reasonably designed.

Generally, a short circuit fault occurs in the system, and the circuit breaker or switch will start in a short time (usually on the order of seconds). In such a short time, the short-circuit current flows through the conductor, and the heat is too late to spread. As a conservative design, adiabatic approximation can be used [10],

$$\int_{0}^{\Delta t_{\mathrm{F}}} \frac{\rho(T)}{A^2} I_{\mathrm{F}}^2 \mathrm{d}t = \int_{T_{\mathrm{op}}}^{T_{\mathrm{P}}} \gamma C(T) \mathrm{d}T \tag{1}$$

where A is the stainless steel former section, γ is the density of stainless steel, C is the specific heat capacity of the stainless steel frame, T is the former's temperature, ρ is the resistance of the stainless steel, Δt_{F} is the breaking time of the circuit breaker, I_{F} is the fault current, T_{op} is the temperature during normal operation, and T_{p} is the highest temperature that the conductor can withstand. Since the cross-section of the superconducting wire is much smaller than that of the former, it is ignored during the fault and all the fault current flows through the former. Solving formula (1), the cross-sectional area of the stainless-steel former can be determined.

2.2 Superconducting Layers

The number of tapes in the conductor layer is determined by the rated current of the transformer winding. The number of conductor layers n should be an even number as much as possible.

Suppose the critical current of a single superconducting tape is I_{c}, and I_{N} is the rated current of the winding, then the total number of tapes N_0 is

$$N_0 = \frac{\sqrt{2} I_{\mathrm{N}}}{I_{\mathrm{c}}}(1 + \beta) \tag{2}$$

where β is the safety margin, which can be determined according to different needs.

According to formula (1), the cross-sectional area A of the former and its radius r_{f} is obtained. Assuming the thickness of the superconducting tape is d and the thickness of the semiconductor carbon paper layer is h,

$$r_{\mathrm{i}} = \frac{r_{\mathrm{f}} + h \cdot i + d \cdot i}{2} \tag{3}$$

Fig. 2 Schematic view of i-th layer in conductor

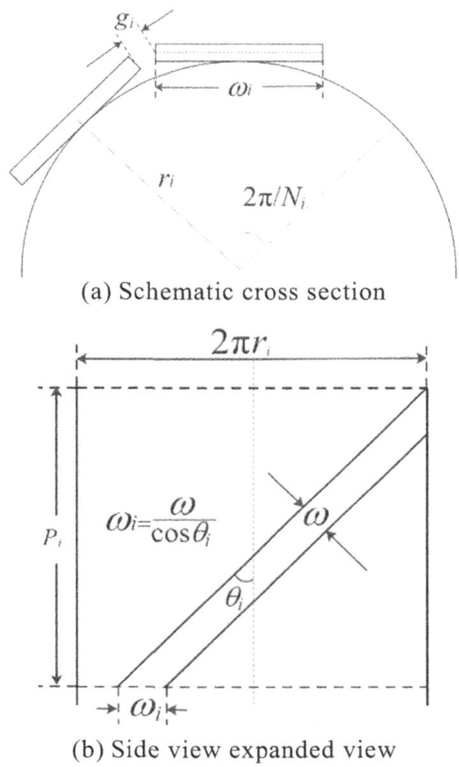

where i is 1, 2, 3 …, n, r_i is the center radius of the i-th conductor layer, as shown in Fig. 2.

The superconducting tape is spirally wound on the stainless-steel former with a certain length pitch, and the relationship between the pitch and the winding angle is shown in Fig. 2 [11],

$$P_i = \frac{2\pi r_i}{\tan\theta_i} \tag{4}$$

where P_i is the length pitch of the i-th conductor layer strip, and θ_i is the winding angle of the i-th conductor layer tapes.

According to the literature [11] and Fig. 2, the relationship between the number of superconducting tapes N_i in each layer, the gap g_i between adjacent tapes, the tape width ω and the center radius r_{ii} of the conductor layer is as follows,

$$r_{ii} \cdot \tan\left(\frac{\pi}{N_i}\right) = \frac{1}{2\cos(\theta_i)}\left(\frac{g_i}{\cos\left(\frac{\pi}{N_i}\right)} + \omega\right) \tag{5}$$

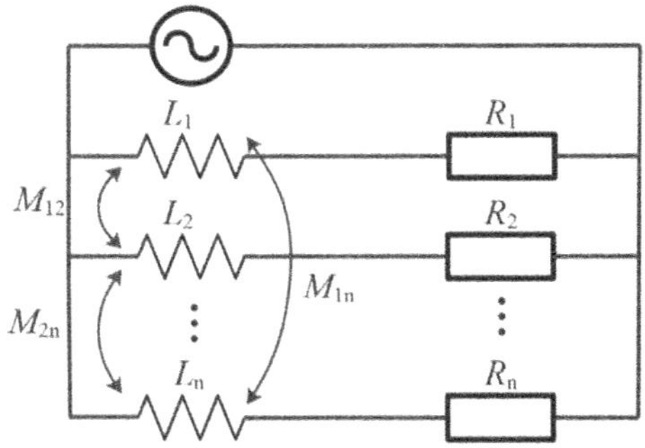

Fig. 3 Equivalent circuit model of HTS conductor

Taking into account the mechanical properties of the superconducting tape at low temperatures, g_i also needs to meet the following conditions [12],

$$g_i \geq \frac{\varepsilon_t}{N_i} 2\pi r_i \left(1 + \frac{P_i^2}{(2\pi r_i)^2}\right) \tag{6}$$

where ε_t is the free thermal shrinkage rate of the superconducting tape.

The simplified circuit model of the conductor is shown in Fig. 3 [13], where R_i represents the contact resistance of each layer of the superconducting layer, L_i represents the self-inductance of each layer of the superconducting layer and M_{ij} (j = 1, 2, …, n) represents the mutual inductance between the *i*-th and *j*-th layers of the superconducting layer. According to this circuit model,

$$jw \begin{bmatrix} L_1 & M_{12} & \cdots & M_{1n} \\ M_{21} & L_2 & \cdots & M_{2n} \\ \vdots & \vdots & \ddots & \vdots \\ M_{n1} & M_{n2} & \cdots & L_n \end{bmatrix} \begin{bmatrix} I_1 \\ I_2 \\ \vdots \\ I_n \end{bmatrix} = \begin{bmatrix} V_1 \\ V_2 \\ \vdots \\ Vn \end{bmatrix} \tag{7}$$

where ω is the angular frequency, I_i is the current of the *i*-th layer of the conductor layer and V_i is the voltage of the *i*-th layer of the conductor layer. Since the joint resistance of each layer is much smaller than the number of inductive reactance of each layer, it is ignored in formula (7) [14, 15]. According to formula (7), it can be seen that the current distribution of the conductor is mainly determined by the inductance and mutual inductance of each layer. The calculation formulas for the inductance and mutual inductance of each layer of the conductor are as follows [16],

$$L_i = \frac{\mu_0 \pi r_i^2}{P_i^2} + \frac{\mu_0 \ln(D/r_i)}{2\pi} \tag{8}$$

$$M_{ij} = \alpha_i \alpha_j \frac{\mu_0 \pi r_i^2}{P_i P_j} + \frac{\mu_0 \ln(D/r_j)}{2\pi} \tag{9}$$

where D is the radius of the loop where the magnetic field outside the conductor section is approximately equal to zero, and α_i and α_j are constants with values $+1$ or -1, depending on the winding direction of the superconducting tapes of each layer.

Since the conductor is a multilayer structure, the current distribution of each layer will be nonuniform. The outer layer's current is obviously greater than the inner layer's, which will weaken the current carrying capacity of the conductor and increase the AC losses [17]. In order to make the current distribution of each layer uniform and reduce the AC losses, it is made uniform by adjusting the winding pitch of each layer and the winding direction of each layer. This is the principle of current sharing [18–20]. According to current sharing conditions,

$$I_1 = I_2 = \cdots = I_n \tag{10}$$

$$V_1 = V_2 = \cdots = V_n \tag{11}$$

In order to maintain the overall tightness of the conductor and prevent the current-carrying capacity from degrading due to the stress caused by the cold shrinkage of the conductor from room temperature to low temperature, the winding angle θ of the superconducting tape should also meet the following condition ($r < R$) [21],

$$\sin^{-1}\left(\sqrt{\frac{\varepsilon_t - \varepsilon_s - \varepsilon_p}{\varepsilon_r - \varepsilon_p}}\right) \le \theta \le \sin^{-1}\left(\sqrt{\frac{r_i}{R_c}}\right) \tag{12}$$

where ε_s, ε_p, and ε_r are the critical tensile strain force, the pitch change rate and the radial heat shrinkage rate of the former, respectively, and R_c is the critical radius of the superconducting tape.

Solving Eqs. (7)–(12), the length pitch of each layer of the superconducting layer can be obtained.

2.3 Analysis of the Magnetic Field

After determining the various parameters of the conductor superconducting layer, the magnetic field of the conductor can be analysed. As shown in Fig. 4, the magnetic field generated by the superconducting layer can be divided into the circumferential magnetic field $B_{i\theta}$, distributed along the outer surface of each layer, and the axial magnetic field B_{iZ}, distributed along the axial direction of each layer. The calculation formula is [22]

$$B_{iZ} = \mu_0\left(\sum_{k=i+1}^{n} \alpha_k \frac{I_k}{P_k} + \frac{\alpha_i I_i}{2P_i}\right) \tag{13}$$

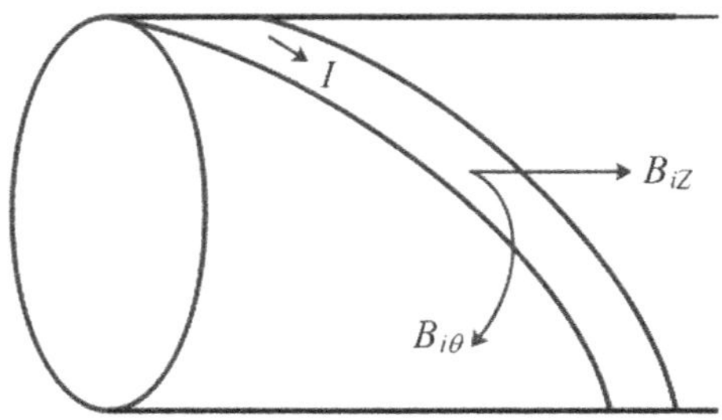

Fig. 4 Schematic view of magnetic field in i-th layer of conductor

$$B_{i\theta} = \mu_0\left(\frac{1}{2\pi r_i}\sum_{k=1}^{i-1}\alpha_k I_k + \frac{1}{2}\frac{I_i}{2\pi r_{io}}\alpha_i\right) \tag{14}$$

$$B_i = \sqrt{B_{i\theta}^2 + B_{iZ}^2} \tag{15}$$

where r_{io} represents the outer radius of the *i*-th layer of the superconducting layer.

3 Results of Conductor Design

3.1 Conductor Parameters

In order to explain the conductor design method in more detail, the corresponding SFCLT parameters are given here. The rated capacity of the transformer is 125 kVA, the rated voltage is 6 kV/400 V, and the rated current is 20.8 A/312.5 A. The conductor can be used as the secondary winding of the transformer. The stainless steel reinforced REBCO superconducting tape, produced by Shanghai Creative Superconducting Company, is selected. The main parameters of the superconducting tape are shown in Table 1. The outermost layer of the conductor is an insulating layer, and polyimide or PPLP can be selected.

Table 1 Main parameters of stainless steel reinforced REBCO tapes

Parameters	Value
Width/mm	3
Thickness/mm	0.2
Critical current under self-field I_{c0}/A (@77 K)	28.5
Critical tensile stress/Mpa	400
Critical bending radius/mm	20
Critical tensile strain/%	0.3
Room temperature resistivity/(Ω·mm^2/m)	0.847

Table 2 Parameters of HTS conductor

Parameters	Value
Former radius r_f/mm	3
Number of superconducting layers n	4
Center radius of each superconducting layer/mm	3.2, 3.5, 3.8, 4.1
Winding direction of each superconducting layer α	1, 1, −1, −1
Pitch of each superconducting layer P/mm	55, 208, 65, 30
Winding angle of each superconducting layer θ/(°)	20.4, 6.1, 20.3, 41.5
Number of tapes per layer of superconducting layer/piece	4, 4, 4, 4
Room temperature resistance per kilometer/Ω	21.3

when calculating the cross-sectional area of the former, considering that the breaking time of the existing distribution network circuit breaker ranges from 20 to 200 ms, Δt_F is 100 ms. I_F is the fault current of 3125 A, which is 10 times the rated current. T_{op} is 77 K. T_p is the melting point temperature of solder, 433 K (160 °C). According to the formula (1), the cross-sectional diameter of the former is 23.6 mm^2. Due to the small radius of the conductor, the winding angle of the superconducting layer will slightly exceed the required range of formula (12) when the current sharing condition is met. Therefore, starting from the purpose of achieving a better current sharing effect and reducing AC losses, the values of r_f and θ can be selected appropriately. The conductor parameters are shown in Table 2.

3.2 Critical Current

In this part, a three-dimensional finite element model of the conductor is established, as shown in Fig. 5.

In order to simplify the calculation, only the stainless steel reinforced REBCO superconducting tapes are considered in the model [23]. The critical current of the 2G HTS tape is generally described by the following formula:

$$I_c(B) = I_{c0}\left(1 + \frac{1}{B_0}\sqrt{\lambda^{-2}B_{\parallel}^2 + B_{\perp}^2}\right)^{-\tau} \tag{16}$$

where $B_{\parallel}$ and $B_{\perp}$ are the parallel and perpendicular components of the magnetic field on the surface of the superconducting tape, respectively. B_0 is 20 mT, and λ

Fig. 5 FE model of the HTS conductor

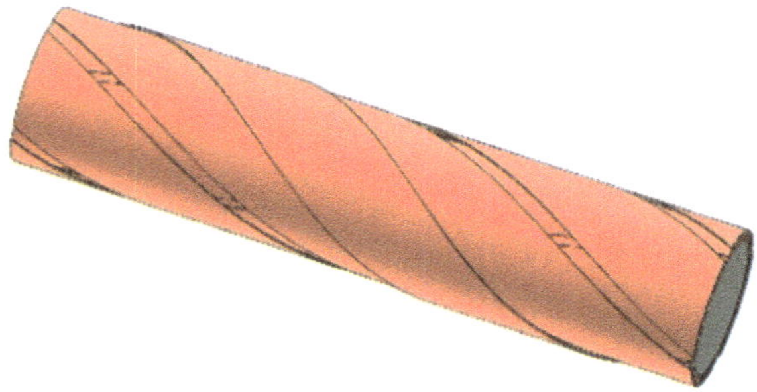

and τ are adjustable parameters. By applying a transport current to the conductor, the magnetic field distribution of can be calculated. According to the magnetic field distribution of the conductor, the critical current at this time can be calculated. When the applied transport current is equal to the calculated critical current, the critical current at this time is defined as the critical current under the conductor's self-field. Figure 6 shows the distribution of the magnetic field and the normalized current density of the conductor when the transport current is 700 A.

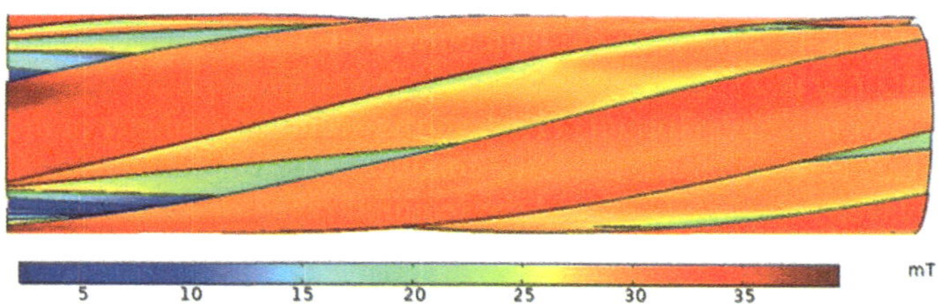

(a) Distribution of magnetic field

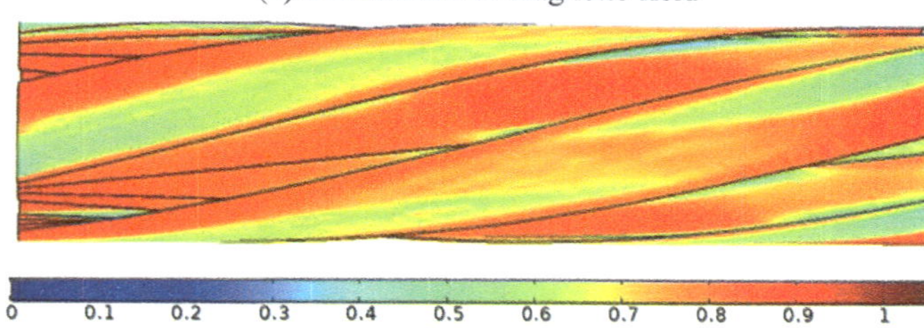

(b) Normalized current density distribution

Fig. 6 Magnetic field distribution and critical current density's distribution of the conductor

4 Simulation and Result Analysis

4.1 Simulation

The conductor designed in this article is oriented to SFCLT, so its current limiting characteristic is a very important indicator. We assume that J and J_c are the current density and the critical current density of the superconducting tape, respectively. In order to simplify the calculation, suppose the resistivity of the superconducting tape in the superconducting state is 0. When $3J_c \leq |J|$, the resistivity of the superconducting tape at this time is the resistivity of the stainless-steel reinforcement layer. Combined with the E-J characteristics of superconducting tapes [24, 25],

$$\rho_{sc} = \begin{cases} 0 & |J| < J_c \\ \frac{E_0}{|J_c|}\left(\frac{|J|}{J_c} - 1\right)^n & J_c \leq |J| < 3J_c \\ \rho_{ss} & 3J_c \leq |J| \end{cases} \tag{17}$$

where ρ_{sc} is the resistivity of the superconducting tapes, ρ_{ss} is the resistivity of the stainless steel, E_0 is the calculated reference voltage, and n is the value of n in the quench change of the superconducting tape.

In order to obtain the current limiting characteristics of the conductor, we introduce it into a simple power system. Its simplified circuit diagram is shown in Fig. 7.

We assume that the superconducting conductor has longitudinal uniformity, from which we can obtain the circuit equation and heat equation of the system [26],

$$V = (L_{sc} \cdot l)\frac{dI}{dt} + (R_{L1} + R_{L2} + R_{sc} \cdot l)I \tag{18}$$

$$C_0(T)\frac{dT}{dt} = (R_{sc} \cdot l)I^2 - Q_{LN} \tag{19}$$

where V is the system voltage, l is the length of the conductor, R_{L1} and R_{L2} are the resistance of the loads, C_0 is the heat capacity of the conductor, and Q_{LN} is the heat exchanged by the outermost part of the conductor with liquid nitrogen. As shown in

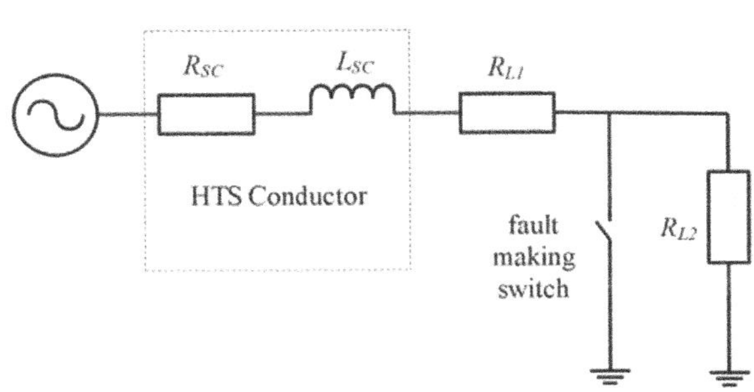

Fig. 7 Simulation circuit

Fig. 7, we control the load R_{L2} to connect and disconnect from the circuit by opening and closing the fault switch, and simulate the short-circuit fault of the power system. Then we obtain the current limiting characteristics of the conductor.

4.2 Result Analysis

According to the requirements of the corresponding transformer, the system voltage is set to 400 V. The prospective fault current I_{pro} is set to 10 times the winding rated current I_N, which is 3125 A. The breaking time of the circuit breaker is selected as 100 ms. The short-circuit current fault is cleared within 5 power frequency cycles after occurrence. The length l of the conductor can be determined according to different current-limiting requirements and the specifications of the transformer, here is 20 m. According to the aforementioned circuit and thermal equation model, the current through the conductor, the temperature and resistance of the conductor can be obtained when a short-circuit fault occurs. Figure 8 shows the changes in the current waveform, the temperature and the resistance of the conductor during a short-circuit fault.

It can be seen from Fig. 8a that the effective value I_{lim} of the current flowing through the conductor is limited to about 1000 A during the first cycle of the short-circuit fault, which is about 30% of the expected fault current I_{pro}. This shows that the conductor effectively reduces the short-circuit current. Secondly, we can also get that the overall current changes little during the fault. From Fig. 8b, it can be seen that the temperature of the conductor has not change much, about 10 K. The value of the resistance is very small and the change trend is similar to the temperature change trend. Its change during the fault is not obvious, which corresponds to the inconspicuous current change shown in Fig. 8a.

5 Conclusion

A new type of CORC-like conductor for HTS fault current limiting transformers is proposed. The conductor is made of stainless steel former and stainless steel reinforced REBCO superconducting tapes. The conductor structure with fault current limiting function is conceptually designed. Based on the short-circuit fault current of the secondary winding of a 120 kVA HTS transformer, the specific parameters of the conductor are given, and the relevant modeling and simulation of the conductor are carried out. By simulating the short-circuit fault, the current-limiting characteristics of the conductor are obtained. The simulation results show that the conductor has a good effect of limiting the short-circuit current, and it is a design scheme for the SFCLT winding.

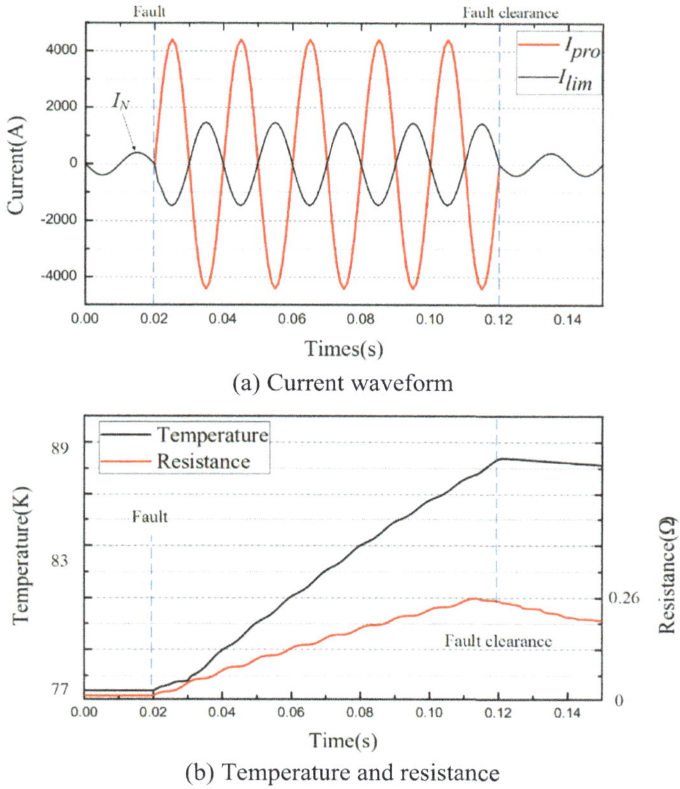

Fig. 8 Current limitation characteristics

Acknowledgements This work was supported in part by State Grid under Grant DG71-19-004.

References

1. Neumueller, H., W. Schmidt, and H. Kraemer. 2009. Development of resistive fault current limiters based on YBCO coated conductors. *IEEE Transactions on Applied Superconductivity* 19 (3): 1950–1955.
2. Maguire, J., D. Folts, and J. Yuan. 2011. Status quo and progress of hydra HTS cable with fault current limiting for con edison grids. *Power System and Clean Energy* 27 (5): 01–04.
3. Iwakuma, M., K. Sakaki, A. Tomioka. 2015. Development of a 3φ-66/6.9 kV-2 MVA REBCO superconducting transformer. *IEEE Transactions on Applied Superconductivity* 25 (3): 1–6.
4. Okubo, H., C. Kurupakon, and S. Ito. 2007. High-Tc superconducting fault current limiting transformer (HTc-SFCLT) with 2G coated conductors. *IEEE Transactions on Applied Superconductivity* 17 (2): 1768–1771.

5. Kojima, H., M. Kotari, and T. Kito. 2011. Current limiting and recovery characteristics of 2 MVA class superconducting fault current limiting transformer (SFCLT). *IEEE Transactions on Applied Superconductivity* 21 (3): 1401–1404.
6. Hellmann, S., M. Abplanalp, and L. Hofstetter. 2017. Manufacturing of a 1-MVA-class superconducting fault current limiting transformer with recovery-under-load capabilities. *IEEE Transactions Applied Superconductivity* 27 (4): 1–5.
7. Weiss, J., T. Mulder, K. Ten. 2017. Introduction of CORC@ wires: highly flexible, round high-temperature superconducting wires for magnet and power transmission applications. *Superconductor Science and Technology* 30 (1): 4002.
8. van der Laan, D.C., J.D. Weiss, C.H. Kim. 2018. Development of CORC® cables for helium gas cooled power transmission and fault current limiting applications. *Superconductor Science and Technology* 31 (8): 5011.
9. Victor, E.S., S.V. Vitaly, and V.R. Alexander. 2009. 30 m HTS power cable development and witness sample test. *IEEE Transaction on Applied Superconductivity* 19 (3): 1702–1705.
10. Yibo, Z. 2015. *Research on key technology of 220 kV could dielectric high temperature superconducting cable*. Beijing: North China Electric Power University. (in chinese).
11. Satoshi, F., K. Ryurchi, and O. Jun. 2006. Numerical analysis of AC losses characteristics of cable conductor assembled by HTS tapes in polygonal arrangement. *IEEE Transactions on Applied Superconductivity* 16 (2): 143–146.
12. Yinshun, W., Z. Fengyuan, G. Zhiyuan. 2009. Development of a high-temperature superconducting bus conductor with large current capacity. *Superconductor Science and Technology* 22 (5): 055018 (5 pp).
13. Grilli, F., S. Stavrev, and B. Dutoit. 2004. Numerical analysis of the effects of the magnetic self-field on the transport properties of a multilayer HTS cable. *IEEE Transactions on Applied Superconductivity* 14 (1): 94–102.
14. Jian, L., L. Liangzhen. 2000. Layer-current distribution of conductor for HTS AC transmission cable. *Advanced Technology of Electric Engineering and Energy* 2: 9–12+18) (in chinese).
15. Jiahui, Z., L. Huifeng, C. Xiaoyu. 2016. Test and analysis on the DC and AC I-V characteristics of high temperature superconducting cable. *Transactions of China Electrotechnical Societ* 31 (24): 120–125(2016) (in chinese).
16. Yinshun, W., L. Hongwei, and Z. Huiyuan. 2010. A conceptual design for HTS cable with large current capacity using co-axial configurations. *IEEE Transactions on Applied Superconductivity* 20 (3): 1263–1267.
17. Kruger, O.S., C. Traholt, and A. Kuhle. 1999. Loss and inductance investigation in a 4-layer superconducting prototype cable conductor. *IEEE Transactions on Applied Superconductivity* 9 (2): 833–836.
18. Mukoyama, S., K. Miyoshi, and H. Tsubouti. 1997. 50-m long HTS conductor for power cable. *IEEE Transactions on Applied Superconductivity* 7 (2): 1069–1072.
19. Mukoyama, S., K. Miyoshi, and H. Tsubouti. 1999. Uniform current distribution conductor of HTS power cable with variable tape-winding pitches. *IEEE Transactions on Applied Superconductivity* 9 (2): 1269–1272.
20. Sytnikov, V.E., P.I. Dolgosheev, and G.G. Svalov. 1998. Influence of the multilayer HTS-cable conductor design on current distribution. *Physica C: Superconductivity* 310: 387–397.
21. Yubao, L., L. Liangzhen, Z. Shaoping. 1999. Development of a 1000 A class bismuth-based HTS model cable. *Chinese Journal of Low Temperature Physics* 21 (2): 43–19 (1999) (in chinese).
22. Ji-kwang, L., and C. Gueesoo. 2001. Magnetization loss calculation in superconducting power transmission cable. *Cryogenics* 41 (3): 157–161.
23. Jiuke, W., W. Yinshun, W. Yueyin. Study on critical of a flexible quasi-isotropic HTS conductor with high engineering current density. *Fusion Engineering and Design* 160 (11) 1820.
24. Jun, G., Z. Jiahui, F. Jing. 2018. Analysis of transient resistance characteristics for resistive type high temperature superconducting fault current limiter. *Transactions of China Electrotechnical Society* 33 (9): 2130–2138 (in chinese).

25. Xiaoyu, C., Z. Jiahui, F. Jing. 2016. Simulation and analysis on the operational characteristics of the power transmission lines with a 110 kV/3 kA high temperature superconducting cable in a meshed grid. *Transactions of China Electrotechnical Society* 31 (16): 7–14 (in chinese).
26. Kojima, H., F. Kato, and N. Hayakawa. 2012. Superconducting fault current limiting cable (SFCLC) with current limitation and recovery function. *Physics Procedia* 36: 1296–1300.

Assisted Diagnosis of Real-Virtual Twin Space for Data Insufficiency

Jiajun Duan, Yigang He, and Xiaoxin Wu

Abstract This paper proposes a virtual-real twin spatial fusion theory to deal with the problem of insufficient data for intelligent diagnosis of power equipment. First, build a transformer winding fault experiment platform, and use Frequency Response Analysis (FRA) to obtain the measured sample set that contains different fault locations, fault types and severity of the transformer to form an accurate physical space. Then, use COMSOL and matlab to build a transformer digital space model, correspondingly set the fault, obtain the simulation sample set, and form a fuzzy mirror space. Then, the sample data in the two spaces is pre-processed by feature extraction and fused with each other to obtain the virtual and real twin spaces. The fused samples are used as auxiliary training samples of intelligent fault diagnosis network. Finally, based on the proposed method, a variety of intelligent diagnostic networks are applied to diagnose the measured dataset, verifying that the proposed method can effectively improve the diagnosis and positioning effects of small sample data.

Keywords Real-Virtual twin space · Transformer · Fault diagnosis · Frequency response analysis (FRA) · Long-Short-Term memory network (LSTM) · Convolutional neural networks (CNN)

1 Introduction

Researchers have studied a variety of adaptive, high-accuracy intelligent fault diagnosis methods [1, 2]. The current intelligent fault diagnosis capabilities for power equipment are still weak. Most researches are looking for more effective feature extraction and fault diagnosis methods [3, 4] to improve diagnostic capabilities. Literature [5] pointed out that the most critical problem affecting the effectiveness of intelligent diagnosis methods is the insufficient size of training data, especially for power equipment. The lack of fault samples of power equipment severely limits the in-depth study of its intelligent diagnosis methods [6]. However, it is difficult to

J. Duan · Y. He (✉) · X. Wu
Wuhan University, 430072 Wuhan, China
e-mail: yghe1221@whu.edu.cn

W. Chen et al. (eds.), *The Proceedings of the 9th Frontier Academic Forum of Electrical Engineering*, Lecture Notes in Electrical Engineering 743,
https://doi.org/10.1007/978-981-33-6609-1_34

achieve large-scale implementation of destructive tests for modern power equipment such as transformers to obtain fault samples required for diagnosis, whether from the perspective of economy or safety [5].

Aiming at the problem of fault diagnosis in the case of insufficient samples, the general approach is to expand the amount of training data through data enhancement [7]; some researchers try to reduce the number of training samples required for equipment fault diagnosis through transfer learning (TL) [8]; Other researchers have constructed simulation models of the equipment based on the parameters of the equipment to be diagnosed. For example, the winding structure of a power transformer is converted into circuit elements to obtain the frequency sweep response (FRA) trace of the device [9]. Then research on fault diagnosis based on the simulation model [10]. The diagnosis effect can be improved by continuously increasing the accuracy of the model; or the use of simulation and programs to establish a digital twin model of the equipment, so as to predict the fault state according to the input monitoring data [11]. Therefore, how to build a bridge between the digital space model and the actual equipment to effectively couple the two, so as to facilitate the use of the fault samples of the digital space model to assist the diagnosis of the actual equipment, is a direction worth exploring.

Transformers are one of the most critical power equipment. And the fault of the transformer is mainly manifested as a winding fault [12], and its safe and reliable operation is essential. In general, the current fault diagnosis research of transformer windings has the following problems:

(1) The use of measured data for fault diagnosis research is limited by the shortage of data, there are fewer diagnosis categories, and the diversity of fault severity is not fully considered. There are fewer studies on winding fault location, and the diagnosis methods are not intelligent enough, usually parameter analysis [13];
(2) The use of simulation models to study fault diagnosis methods basically does not consider measured data, so the effect of the final diagnosis model is not yet It can be seen that there is a lack of strong generalization ability in actual diagnosis [14];
(3) The lack of fault samples in the field of power equipment diagnosis is a common phenomenon, but there are few in-depth studies on how to deal with the diagnosis of small samples. No research has considered effective coupling of measured data and simulation data of power equipment to assist actual diagnosis.

Therefore, this article considers that power equipment generally has relatively mature simulation methods, and proposed a real-virtual twin space, which couples simulation data and measured data to complement each other, so as to give full play to each other's advantages, reasonably expand the data set, and further improve The intelligence and generalization capabilities of the diagnostic model provide a new idea for future researches on small-sample equipment diagnostics.

2 Analytical Model

This paper proposes a virtual-real twin space for solving the lack of data in fault diagnosis scenarios, effectively coupling simulation data and measured data, and applying it to transformer FRA intelligent fault diagnosis. The schematic diagram is shown in Fig. 1. Obtain the experimental data set and the simulation data set. The data sets of all their labels respectively constitute the precise entity space and the fuzzy mirror space. For brevity, they are denoted as R-Space and V-Space. Then space fusion is performed to form a virtual-real twin space (abbreviated as RV-Space) to aid diagnosis.

2.1 R/V Space Integration

Assuming that the transformer status label corresponding to the data is H_γ ($\gamma = 1, 2, 3, \ldots$), where γ represents the label number. The samples with label γ in R-Space are recorded as $\boldsymbol{SR_\gamma} = \{SR_{\gamma 1}; SR_{\gamma 2}; SR_{\gamma 3}; \ldots\}$, which is called a subset of R-Space, it contains a total of NR_γ samples. Similarly, the subset labeled γ in V-Space is denoted as $\boldsymbol{SV_\gamma} = \{SV_{\gamma 1}; SV_{\gamma 2}; SV_{\gamma 3}; \ldots\}$, which contains a total of NV_γ samples.

Next, feature fusion of V-space and R-space is required. For samples corresponding to the same label γ in the two spaces, generally, part or all of the SR_γ subspace and the SV_γ subspace are extracted and fused. Use $\ell^{(\lambda)}(A)$ to represent the matrix composed of λ samples drawn from space A ($A = \boldsymbol{SR_\gamma}$ or $\boldsymbol{SV_\gamma}$), and each sample is regarded as a column, then:

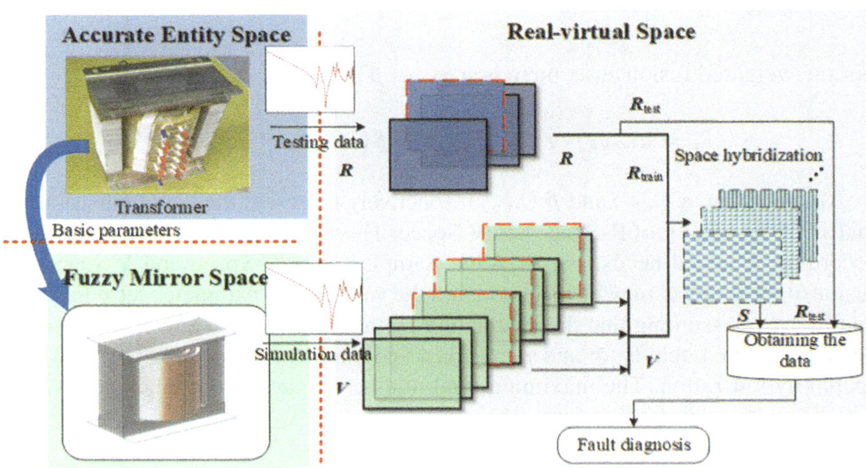

Fig. 1 Schematic diagram of the diagnosis process through the real-virtual digital twin space

$$\ell^{(\lambda)}(A) = \begin{bmatrix} x_{11} & x_{21} & \cdots & x_{\lambda 1} \\ x_{12} & x_{22} & \cdots & x_{\lambda 2} \\ \vdots & \vdots & \vdots & \vdots \\ x_{1N_w} & x_{2N_w} & \cdots & x_{\lambda N_w} \end{bmatrix} \tag{1}$$

Among them, 'x_{1N_w}' represents the N_w-th point of sample No. 1, and has $\lambda \leq NV_\gamma$ or $\lambda \leq NR_\gamma$.

The symbol '⊗' is used to indicate the feature fusion algorithm. The calculation process can be expressed by the fusion function $\varphi(R, V)$, which has: $\varphi(R, V) = R \otimes V$. R-Space and V-Space are merged to obtain the RV-Space, and its sample set is denoted as $\boldsymbol{S}$. In the sample set $\boldsymbol{S}$ after hybridization, the corresponding samples can be expressed by (2):

$$S_\gamma = \varphi\Big(\ell^{(\lambda_{R_\gamma})}(\mathcal{SR}_\gamma), \ell^{(\lambda_{V_\gamma})}(\mathcal{SV}_\gamma)\Big) = \ell^{(\lambda_{R_\gamma})}(\mathcal{SR}_\gamma) \otimes \ell^{(\lambda_{V_\gamma})}(\mathcal{SV}_\gamma) \tag{2}$$

The subset $\boldsymbol{S}_\gamma$ formed by the samples belonging to the label γ is:

$$\mathcal{S}_\gamma = \left\{ S_\gamma \middle| \begin{matrix} \lambda_{R\gamma} = 1, 2, \ldots, NR_\gamma \\ \lambda_{V\gamma} = 1, 2, \ldots, NV_\gamma \end{matrix}, \mathcal{P}(\rho) \right\} \tag{3}$$

where $\boldsymbol{\rho} = \{\rho_1, \rho_2, \ldots\}$ represents the parameter set of the fusion algorithm, $\mathcal{P}(\rho)$ is the parameter constraint equation.

2.2 Weighted Fusion

For the weighted fusion case, there is $\boldsymbol{\rho} = \{\boldsymbol{\alpha}, \boldsymbol{\beta}\}$:

$$S_\gamma = \boldsymbol{a}(\lambda_{R\gamma}) \cdot \ell^{(\lambda_{R_\gamma})}(SR_\gamma) + \boldsymbol{\beta}(\lambda_{V\gamma}) \cdot \ell^{(\lambda_{V_\gamma})}(SV_\gamma) \tag{4}$$

Among them, $\boldsymbol{\alpha}$ $(\lambda_{R\gamma})$ and $\boldsymbol{\beta}$ $(\lambda_{V\gamma})$ respectively represent the weight distribution matrix of the label γ of R-Space and V-Space. The sample size of $\boldsymbol{S}$ can be changed according to actual needs. For the same samples from R-Space and V-Space, the fusion algorithm '⊗' may be different, and the value of the parameter set $\boldsymbol{\rho}$ may also be different. Assuming that the calculation method of R-V space hybridization has been determined, that is, ⊗ and $\boldsymbol{\rho}$ are fixed and unique, RV-Space is obtained after spatial hybridization. The maximum total number of samples of label γ is NRV_γ, which can be calculated by formula (5):

$$NRV_\gamma = \sum_{\lambda_{V_\gamma}=1}^{NV_\gamma} C_{NV_\gamma}^{\lambda_{V_\gamma}} \cdot \sum_{\lambda_{R\gamma}=1}^{NR_\gamma} C_{NR_\gamma}^{\lambda_{R\gamma}}$$

$$= \left(C^{1}_{NV_{\gamma}} + C^{2}_{NV_{\gamma}} + \ldots + C^{NV_{\gamma}}_{NV_{\gamma}}\right)\left(C^{1}_{NR_{\gamma}} + C^{2}_{NR_{\gamma}} + \ldots + C^{NR_{\gamma}}_{NR_{\gamma}}\right)$$
$$= \left(2^{NV_{\gamma}} - 1\right)\left(2^{NR_{\gamma}} - 1\right) \quad (5)$$

3 Diagnostic Testing and Analysis

The diagnostic test process includes the construction of R-Space and V-Space, and space fusion to generate new data as an aid to the diagnosis of the original measured data. Finally, the intelligent fault diagnosis method is used for training and verification.

3.1 Data Acquisition and Method Execution

This article chooses our laboratory's customized single-phase transformer (TDG-3kVA, 220 V/36 V, 50/60 Hz) as the experimental test object. There are a total of 28 cascaded disc windings, with one tap for every two disc windings. A simulation model was constructed based on the transformer parameters, and three types of faults were set for the experimental equipment and the simulation model. Divide the equipment to be diagnosed into 7 areas, where the fault area is denoted as *i*, and then the combination of them is written as '*ji*' to denote the corresponding label H_{γ}, namely: $H_{\gamma} = \{$'00', '11', ..., 'ji', ..., '36', '37'$\}$, the serial number of the label is $\gamma = 1, 2, \ldots, 22$. The label of the normal state is '00'.

Sample acquisition of R-Space The specific fault setting method is shown in Fig. 2: Assuming that a certain part of the winding is faulty, connect the circuit devices (*L* or C) to the two taps closest to the fault location, or connect the two taps directly to simulate a short-circuit fault. Then perform sweep frequency analysis, set the frequency range to 1 kHz to 1 MHz, and the sampling interval to 1 kHz to obtain the FRA trace H_f.

Among them, $U_1\ (f)$ and $U_{test}\ (f)$ represent the voltage of the excitation terminal and the detection terminal respectively. Obtain failure samples of different tags.

Sample acquisition of V-Space (fault setting 10–30%) First, establish a finite element model based on the geometry and material parameters of the transformer to be diagnosed. In order to calculate the basic parameters such as capacitance between windings, capacitance to ground, self-inductance and mutual inductance. Next, the obtained parameters are substituted into the transformer equivalent circuit, and the equivalent circuit is constructed by referring to [12]. Compile a program through the node voltage matrix, simulate the frequency sweep response of the circuit, and obtain the fault type and fault area corresponding to the diagnostic labels. Corresponding

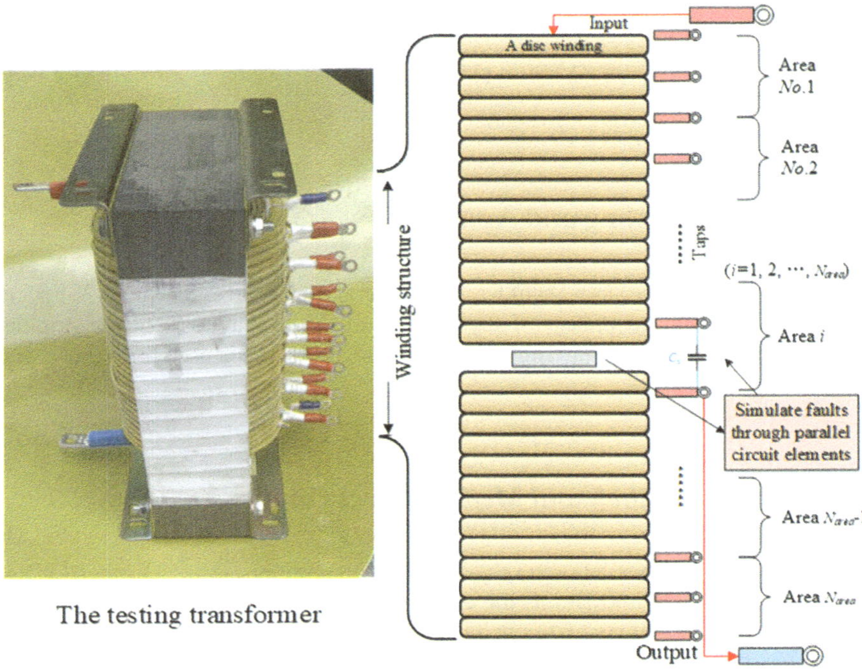

Fig. 2 Transformer winding fault simulation method

to the experimental fault simulation method, the fault is simulated by setting the abnormal components in the equivalent circuit to a parameter abnormality of ±(10–20%). Obtain simulation data sets of different fault degree/fault location/fault type. Every label contains 10 samples.

3.2 The Specific Implementation Process of Space Hybridization

Feature extraction of samples in space The simulation data set constitutes V-Space, and the experimental data set constitutes R-Space. For V-space, all the samples in it will be used for spatial hybridization calculations, which are collectively denoted as $\boldsymbol{V}$. As for the $\boldsymbol{R}$ space, 30% of the samples are randomly selected for spatial fusion and network training, and 70% are used to verify the effect, denoted as $\boldsymbol{R}_{\text{train}}$ and $\boldsymbol{R}_{\text{test}}$, respectively. It can be seen that the number of experimental data sets is extremely rare. In this section, only 3 samples per label are used for training to verify the effectiveness of the proposed method for small sample and multi-label diagnosis.

Feature hybridization and imaging Take the method described in the first section to fuse $\boldsymbol{V}$ and R_{train} to obtain RV-Space. The fusion method adopts the weighted average method, as shown in formula (4). In this paper, $\lambda_{R_\gamma}=\lambda_{V_\gamma}=1$ the parameter set $\boldsymbol{\rho} = \{\alpha, \beta\}$. Randomly select $\alpha \in [0.4, 0.6]$, then. According to formula (5), in the newly generated RV-Space after fusion, each fault label contains $\sum_{\lambda_{V_\gamma}=1}^{1} C_{NV_\gamma}^{\lambda_{V_\gamma}} \cdot \sum_{\lambda_{R_\gamma}=1}^{1} C_{NR_\gamma}^{\lambda_{R_\gamma}} = NV_\gamma \cdot NR_\gamma$ samples, which changes with the value of $\boldsymbol{R}_{\text{train}}/\boldsymbol{R}$.

The data imaging method can intuitively condense a large amount of data into a two-dimensional image, which effectively saves storage space while highlighting fault features, which is conducive to fault feature identification. Before further building a deep vision diagnosis network, this article directly corresponds the value to the color map to visualize the waveform data. So as to achieve the purpose of highlighting the features of data.

3.3 Fault Diagnosis Effects

Next, the training sample set S of the virtual and real subspace is used as a supplement to the training set $\boldsymbol{R}_{\text{train}}$, and the fault type and location are used as diagnostic labels, which are input into the deep convolutional neural network for training. Use $\boldsymbol{R}_{\text{test}}$ to identify and locate faults to verify the diagnosis effect. Take the virtual-real twin space assisted fault diagnosis in this paper (Training dataset = $\boldsymbol{S} + \boldsymbol{R}_{\text{train}}$). Based on the MobileNet-V2 pre-training network, the method of Fine-tune is adopted to freeze the parameters of the first 50% of the pre-training network, set $\boldsymbol{R}_{\text{train}}/\boldsymbol{R}$ = 30%, and input the training set for training. The final verification effect is shown in Table 1. In addition, Table 1 also plots the diagnosis verification process based on MobileNet-V2 (Training dataset = $\boldsymbol{R}_{\text{train}}$) that does not use the virtual-real twin space, directly uses the V-space data as the auxiliary training set of the deep migration learning model, and performs the diagnosis verification results (Training dataset = $\boldsymbol{V} + \boldsymbol{R}_{\text{train}}$), and PSO optimizes the diagnosis result of SVM. The R-Space data is the same for all diagnosis situations. The operating system is Intel i5 and a single GPU: GTX1660ti. The hyper-parameters and diagnostic accuracy of each method are shown in the second column of Table 1.

4 Conclusion

This paper proposes a virtual-real twin space theory that comprehensively considers equipment digital simulation models and physical equipment, and provides a new solution for the fault diagnosis scenarios of equipment with insufficient fault samples in practical applications, such as transformers. It can be seen from Table 1 that the diagnosis effect of the deep transfer learning method based on MobileNet-V2 is better

Table 1 Diagnostic settings and results

Diagnosing method	Hyper parameters	Training dataset	Diagnostic results (%)
CNN (MobileNet-V2)	Learning rate: 3×10^{-5} Mini batch size: 20 Epochs: 100 Drop period: 100 Drop factor: 0.2 Optimizer: Adam	$\boldsymbol{R}_{\text{train}}$	87.0
		$\boldsymbol{R}_{\text{train}} + \boldsymbol{V}$	89.6
		$\boldsymbol{R}_{\text{train}} + \boldsymbol{RV}$	92.2
Bi-LSTM	Number of hidden units per layer: 80 Mini batch size: 20 Layers: input → Bi-LSTM × 3 → FL → Softmax → Output Learning rate: 0.01 Drop factor: 0.5	$\boldsymbol{R}_{\text{train}}$	45.5
		$\boldsymbol{R}_{\text{train}} + \boldsymbol{V}$	61.0
		$\boldsymbol{R}_{\text{train}} + \boldsymbol{RV}$	70.1
PSO-SVM	Kernel function: Gaussian, linear, quadratic, cubic Particle speed limit: [−1, 1] PSO parameters C1 and C2: 1.5 Population size: 25	$\boldsymbol{R}_{\text{train}}$	83.1
		$\boldsymbol{R}_{\text{train}} + \boldsymbol{V}$	85.7
		$\boldsymbol{R}_{\text{train}} + \boldsymbol{RV}$	88.3

than the PSO optimized SVM (PSO-SVM), and better than LSTM. Based on the 22-label transformer fault diagnosis scenario, the average accuracy of the proposed method can reach 92.9%. After adopting the virtual-real twin space, compared with the traditional diagnosis result and the method of using simulation as auxiliary training data for diagnosis, its result is improved. It has a certain auxiliary effect on the problem of lack of samples in fault diagnosis of power equipment.

Acknowledgements This work was supported by the National Natural Science Foundation of China under Grant No. 51977153, 51977161, 51577046, the State Key Program of National Natural Science Foundation of China under Grant No. 51637004, the national key research and development plan "important scientific instruments and equipment development" Grant No. 2016YFF0102200, Equipment research project in advance Grant No. 41402040301.

References

1. Bazan, G.H., P.R. Scalassara, W. Endo, and A. Goedtel. 2020. Information theoretical measurements from induction motors under several load and voltage conditions for bearing faults classification. *IEEE Transactions on Industrial Informatics* 16 (6): 3640–3650.
2. Ruijin, Liao, Wang Youyuan, Liu Hang, et al. 2018. Research status of condition assessment method for power equipment. *High Voltage Engineering* 44 (11): 3454–3464 (in Chinese).
3. Enwen, Li, Wang Linong, Song Bin, Fang Yaqi. 2019. Parallel clustering analysis of dissolved gas analysis data based on chaotic sequences. *Transactions of China Electro-technical Society*

34 (24): 5104–5114 (in Chinese).

4. Bo, Qi, Zhang Peng, Rong Zhihai, et al. 2017. Validity assessment method of DGA sensors based on data driven and multiple criterion integration. *Power System Technology* 41 (11): 3662–3669 (in Chinese).
5. Ozcanli, A.K., F. Yaprakdal, and M. Baysal. 2020. Deep learning methods and applications for electrical power systems: A comprehensive review. *International Journal of Energy Research* 44 (9): 7136–7157.
6. Ren, Z., Y. Zhu, K. Yan, et al. 2020. A novel model with the ability of few-shot learning and quick updating for intelligent fault diagnosis. *Mechanical Systems and Signal Processing* 138 (1): 106608.
7. Hu, T., T. Tang, and M. Chen. 2019. Data simulation by resampling—A practical data augmentation algorithm for periodical signal analysis-based fault diagnosis. *IEEE Access* 7: 125133–125145.
8. Zhiwei, Yang, Shen Wei, Yang Fan, et al. 2019. Fault diagnosis model for distribution transformer migration learning considering multi-factor situation evolution. *Transactions of China Electro-technical Society* 34 (7): 1505–1515 (in Chinese).
9. Zhao, X.Z., C.G. Yao, A. Abu-Siada, and R.J. Liao. 2019. High frequency electric circuit modeling for transformer frequency response analysis studies. *International Journal of Electrical Power & Energy Systems* 111: 351–368.
10. Jahan, M.S., R. Keypour, H.R. Izadfar, and M.T. Keshavarzi. 2019. Detecting the location and severity of transformer winding deformation by a novel adaptive particle swarm optimization algorithm. *International Transactions on Electrical Energy Systems* 29 (1): e2666.
11. Venkatesan, S., K. Manickavasagam, N. Tengenkai, and N. Vijayalakshmi. 2019. Health monitoring and prognosis of electric vehicle motor using intelligent-digital twin. *IET Electric Power Applications* 13 (9): 1328–1335.
12. Yunpeng, Liu, Cheng Weihao, Hu Huan, et al. 2019. Method for classification of transformer winding deformation based on complex values of the frequency response and digital image processing technology. *High Voltage Engineering* 45 (3): 907–913 (in Chinese).
13. Zhixian, Zhang, Chen Weigen, Tang Sirui, et al. 2019. State feature extraction and abnormal state diagnosis of on-load tap-changers based on complementary lumped empirical mode decomposition and local abnormal factors. *Journal of Electrical Engineering and Technology* 21: 4508–4518. (in Chinese).
14. Ghanizadeh, A.J., and G.B. Gharehpetian. 2014. ANN and cross-correlation based features for discrimination between electrical and mechanical defects and their localization in transformer winding. *IEEE Transactions on Dielectrics and Electrical Insulation* 21 (5): 2374–2382.

Analysis of the Temperature Field of a High-Speed Permanent Magnet Motor

Wang Hao, Liang Deliang, Jia Shaofeng, Chu Shuaijun, and Liang Yongtao

Abstract When the high-speed motor is running, due to its ultra-high speed and the current frequency of thousands of hertz, the loss value of each part will increase greatly. High loss will inevitably lead to high temperature rise. Excessive temperature rise may cause problems such as motor insulation failure, permanent magnet demagnetization, and insufficient mechanical strength. This paper establishes a lumped parameter thermal network model of a 20 kW, 10 Wrpm surface-mount 2 pole 6 slot high speed permanent magnet motor, and conducts a finite element analysis, and analyzes the effect of a cooling method in the stator conductor. The final finite element analysis results show that the copper conductor temperature rise of this high-speed motor is 254.49 °C, and the maximum temperature rise of the motor is 258.71 °C. When the copper conductor has an internal cooling structure and cooling water is passed, the copper conductor temperature rise almost drops to 0 °C, and the maximum temperature rise of the motor has also dropped to 80.27 °C.

Keywords High-speed motor · Conductor water cooling · Lumped parameter thermal network · Finite element analysis

1 Introduction

The high-speed motor has the advantages of high efficiency and high energy density, and it has small size and high speed. Using a high-speed motor to form a direct drive system can eliminate the traditional speed-increasing gear structure. This can make the system more compact, reduce costs, improve system efficiency, and reduce maintenance time [1]. High-speed motors can be widely used in refrigerators or air-conditioning centrifugal compressors [2] and other fields. In addition, high-speed

W. Hao (✉) · L. Deliang · J. Shaofeng · C. Shuaijun · L. Yongtao
State Key Laboratory of Electrical Insulation and Power Equipment, Xi'an Jiaotong University, Xi'an 710049, China
e-mail: wanghao707@stu.xjtu.edu.cn

Xi'an Jiaotong University, Xi'an 710049, China

W. Chen et al. (eds.), *The Proceedings of the 9th Frontier Academic Forum of Electrical Engineering*, Lecture Notes in Electrical Engineering 743,
https://doi.org/10.1007/978-981-33-6609-1_35

motors also have very broad application prospects in the fields of electric vehicles, oilfield exploration, ships, and all-electric aircraft [3–6].

Literature [7] designed a 2.7 kW, 16 Wrpm ultra-speed high-speed permanent magnet synchronous motor for automotive internal combustion engine superchargers. Finite element analysis and prototype experiments have proved the possibility of electrification of automotive superchargers. Literature [8] designed and constructed a 5 kW, 24 Wrpm ultra-high speed test system, and finally reached a speed of 18 Wrpm in the experiment.

However, high-speed motors have ultra-high linear speed and current frequency, these factors have brought great hidden dangers to the safe operation of high-speed motors. The ultra-high current frequency makes the loss of the high-speed motor far greater than that of the ordinary motor. The small size makes it difficult to dissipate heat. The temperature of the motor will rise to an unbearable level, and the insulation of the motor will face greater challenges. Therefore, how to effectively cool high-speed motors has become a more important issue.

In this paper, a lumped parameter thermal network model of a 20 kW, 10 Wrpm surface mount 2-pole 6-slot high-speed permanent magnet motor is established, and the loss and heating of the motor are analyzed by finite element method. On this basis, this article proposes a cooling method in the stator conductor and analyzes its cooling effect.

2 Principle Analysis and Cooling Structure Inside the Conductor

2.1 *Winding Loss Analysis*

The winding loss of high-speed motor includes copper loss and eddy current loss, and the expression of copper loss is shown in Eq. (1). Since the electrical frequency and speed of high-speed motors are greatly increased compared with ordinary motors, the skin effect, proximity effect and the air gap magnetic field transformation caused by the high-speed rotation of the motor [9] lead to the eddy current loss in the windings. Equation (2) [10] is the eddy current loss formula of a single cylindrical copper conductor. This article uses a rectangular parallelepiped conductor, and Eq. (3) can be derived from Eqs. (2), and (3) is the eddy current loss formula of a single rectangular parallelepiped copper conductor.

$$P_{cu} = mI^2R \tag{1}$$

In formula (1), P_{cu} is the winding copper loss, and m is the number of winding phases.

$$P_e = \frac{\pi l d^4 B_{pk}^2 \omega^2}{32\rho} \quad (2)$$

In formula (2), P_e is the eddy current loss of the cylindrical copper conductor, l is the conductor length, d is the conductor diameter, B_{pk} is the peak magnetic density, ω is the magnetic field angular velocity, and ρ is the resistivity of copper.

$$P_{er} = \frac{b l d^3 B_{pk}^2 \omega^2}{8\rho} \quad (3)$$

In formula (3), P_{er} is the eddy current loss of the rectangular parallelepiped copper conductor, b is the conductor width, d is the conductor thickness, and the other parameters are the same as in formula (2).

2.2 Cooling Analysis

Air, hydrogen and water are commonly used motor cooling media, and due to the large heat capacity, low viscosity, and good fluidity of the water cooling media, the stator coil adopts water cooling to dissipate heat. The temperature difference between copper and water is generally about 1 °C, and the temperature difference between hydrogen-cooled conductor and hydrogen is as high as 30 °C, so the current density of water-cooled stator coils can be several times higher than that of air-cooled coils [11]. Equation (4) is the convection heat transfer formula.

$$\Phi = \alpha A \left| t_{\text{copper}} - t_{\text{water}} \right| \quad (4)$$

In the formula (4), Φ is the heat flow, α is the convective heat transfer coefficient, A is the heat exchange surface area, t_{copper} is the temperature of the copper conductor, and t_{water} is the temperature of the cooling water. It can be seen from Eq. (4) that for a certain contact area, when the temperature difference is constant, the larger the convective heat transfer coefficient, the faster the heat transfer. The natural convective heat transfer coefficient of air is only 5 ~ 25 W/m^2 °C, The forced convection heat transfer coefficient of water can reach 15,000 W/m^2 °C.

2.3 Thermal Network Analysis

The thermal network model includes elements such as heat source and thermal resistance. The heat source is the loss of each component of the motor. The thermal resistance during the heat transfer process will be discussed below. There are three basic forms of heat transfer, including heat conduction, heat convection and heat radiation [12]. Among them, the heat transfer inside the solid is mainly heat conduction, and the heat transfer process between the solid and liquid is mainly heat convection, and the heat radiation can be neglected when the extremely high temperature is not reached [13, 14].

The heat conduction process satisfies Fourier's law. When one-dimensional steady-state conduction is conducted, the mathematical expression of Fourier's law is shown in Eq. (5). According to Fourier's law, We can get the heat transfer per unit time of a solid with a thickness of δ and a cross-sectional area of A, as shown in Eq. (6).

$$\Phi = -\lambda A \frac{dt}{dx} \tag{5}$$

$$\Phi = \lambda A(t_1 - t_2) = \frac{\Delta t}{\delta/\lambda A} \tag{6}$$

In formulas (5) and (6), Φ is the amount of heat conduction through the cross-section per unit time, λ is the thermal conductivity, A is the cross-sectional area of the cross-section, and $\frac{dt}{dx}$ is the rate of change of temperature at a certain point in the flow direction of the heat, t_1 and t_2 are the temperatures of the two ends, and Δt is the temperature difference between the two ends. The $\delta/\lambda A$ in formula (6) is defined as the thermal resistance R during heat conduction.

$$R = \frac{\delta}{\lambda A} \tag{7}$$

The process of convective heat transfer between solid and liquid satisfies the Newtonian cooling formula, such as Eq. (8).

$$\Phi = \alpha A|t_s - t_w| = \frac{\Delta t}{1/\alpha A} \tag{8}$$

In formula (8), Φ is the heat transferred by heat convection per unit time, α is the convective heat transfer coefficient, A is the contact surface area, and t_s is the solid surface temperature. The $1/\alpha A$ in formula (8) is defined as the thermal resistance R during heat conduction.

$$R = \frac{1}{\alpha A} \tag{9}$$

In the process of heat conduction and thermal convection, the schematic diagram of the thermal resistance is shown in Fig. 1, where R_1 and R_3 are the thermal resistance during the thermal convection process, and R_2 is the thermal resistance during the thermal conduction process.

Fig. 1 Thermal resistance

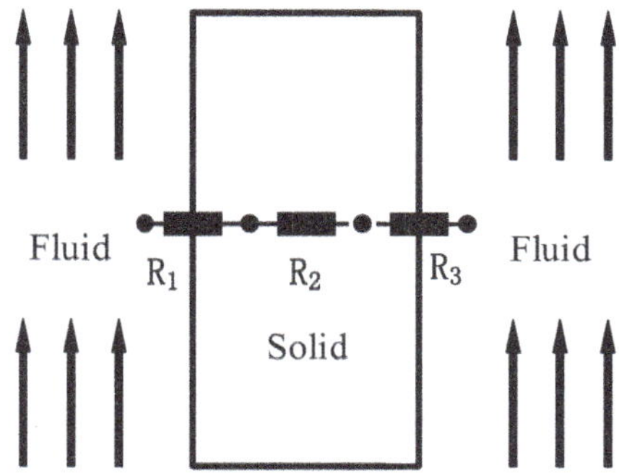

On the premise of ensuring the calculation accuracy, the entire motor system is reasonably simplified. After selecting the appropriate nodes, the thermal network model of the high-speed permanent magnet motor can be established [15]. Therefore, the radial equivalent thermal network of the high-speed motor with cooling channels can be obtained, as shown in Fig. 2.

When the motor temperature rise is constant and reaches the steady-state thermal balance, the equivalent nodes in each part of the motor satisfy the steady-state thermal balance equation, and the matrix form is shown in Eq. (11).

$$G = \frac{1}{R} \tag{10}$$

$$P = GT \tag{11}$$

In formula (11), P is the n × 1 heat source matrix, which represents the loss of each point, G is the n × n thermal conductivity matrix, and the inside is the thermal conductivity between the nodes, and the thermal conductivity is the reciprocal of the

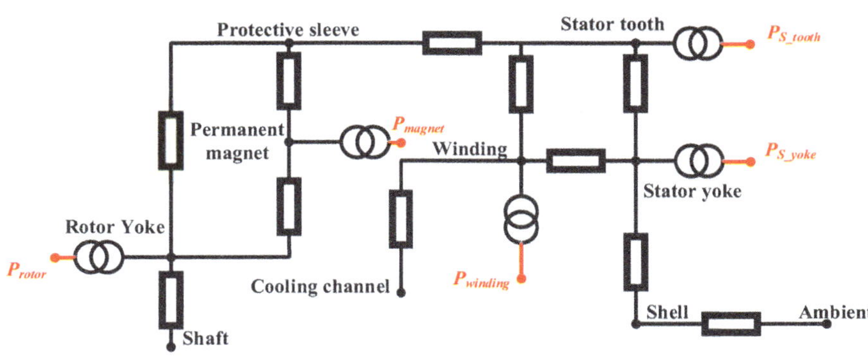

Fig. 2 Lumped parameter thermal network model

thermal resistance, and T is the temperature matrix of n × 1, and n is the number of nodes. For the lumped parameter thermal network established in this paper, n is 10.

2.4 Water Cooling Structure in Conductor

The water cooling structure in the stator conductor used in this paper is shown in Fig. 3, in which cooling water will flow. The cooling structure mainly includes a main channel and a plurality of convex auxiliary channels arranged along both sides of the main channel. This structure can make the cooling water and the copper conductor more fully contact, thereby more efficiently remove the heat generated by the copper conductor, thereby achieving the purpose of reducing temperature rise. In addition, since the current flowing in the conductor has a skin effect, opening a hole in the center of the conductor will not have much impact on the current density.

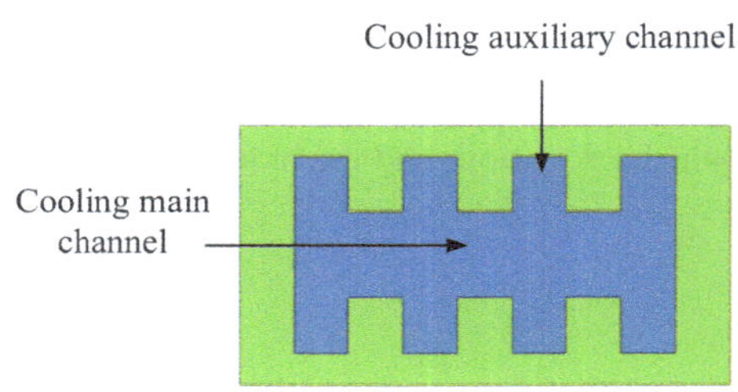

Fig. 3 Water cooling structure in stator conductor

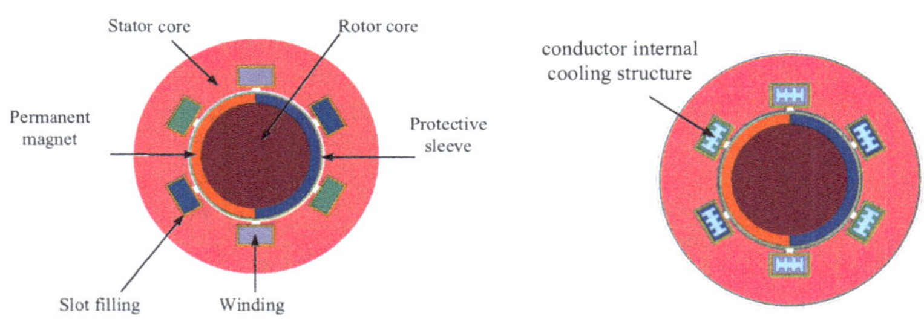

(a) Without conductor internal cooling structure (b) With conductor internal cooling structure

Fig. 4 Motor model

Table 1 The parameters of motor

Parameter	Value	Parameter	Value
Rated power/kW	20	Air gap thickness/mm	1
Rated voltage/V	380	Stator tooth width/mm	10
Rated speed/Wrpm	10	Groove depth/mm	7.5
Number of stator slots	6	Conductor length/mm	10
Number of rotor poles	2	Conductor width/mm	5.5
Permanent magnet thickness/mm	3	Rubber thickness/mm	0.5
Stator outer diameter/mm	72	main channel/mm*mm	7.777*1.795
Stator inner diameter/mm	40	auxiliary channel/mm*mm	1.196*1.111

Table 2 The material of motor

Parts	Material	Parts	Material
Stator and rotor	DW310_35	Winding	Copper
Permanent magnets	SmCo24	Slot filling	Rubber
Permanent magnet protective sleeve	Titanium		

3 Motor Model

The motor is a surface mount high-speed permanent magnet motor, and the permanent magnet protective sleeve is made of titanium. Figure 4a, b show the models of high-speed permanent magnet motors without water cooling in the conductor and water cooling in the conductor. These two schemes have the same parameters except for the conductor structure. The main parameters are shown in Tables 1 and 2.

4 Simulation Results

4.1 Motor Loss Distribution

The loss of each part of the motor directly determines the temperature rise and temperature distribution of the motor. Figure 5a, b show the loss distribution diagrams of high-speed permanent magnet motors without water cooling in the conductor and water cooling in the conductor, respectively. It can be seen from the figure that the positions of the two models with the greatest heat generation appear on the stator teeth.

Since the cooling channel in the copper conductor of the stator leads to the reduction of the cross-sectional area of the conductor and the increase of the current density, the heat generation of the largest heating point on the copper conductor increases to

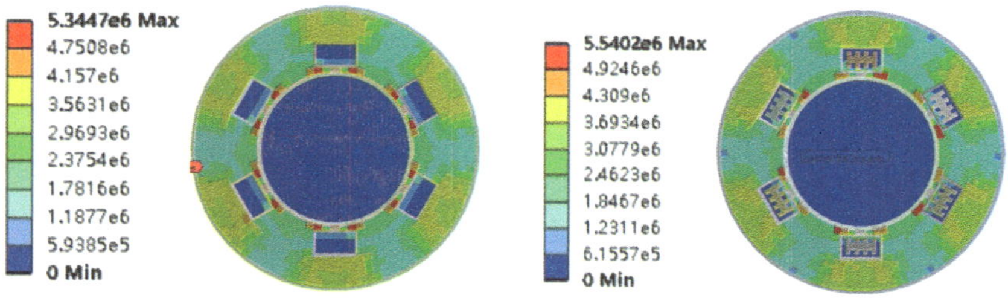

(a) Without conductor internal cooling structure (b) With conductor internal cooling structure

Fig. 5 The distribution of motor loss

Table 3 The copper conductor loss

Cooling method	Copper loss/W	The maximum heating point of the conductor heat/MW/m^3
No water cooling	102.463	1.524
With water cooling	114.262	4.114

2.699 times of the original value, but because the current in the copper conductor has a skin Effect, the hollowness of the conductor does not have much influence on its overall loss. The overall copper consumption of the latter is only increased by 10.522% over the former. The specific values of the winding copper loss and the maximum heating point of the copper conductor are shown in Table 3.

4.2 *Motor Temperature Distribution*

After introducing the motor losses, the motor temperature field can be analyzed. Figure 6a, b show the temperature distribution diagrams of high-speed permanent magnet motors without water cooling in the conductor and water cooling in the

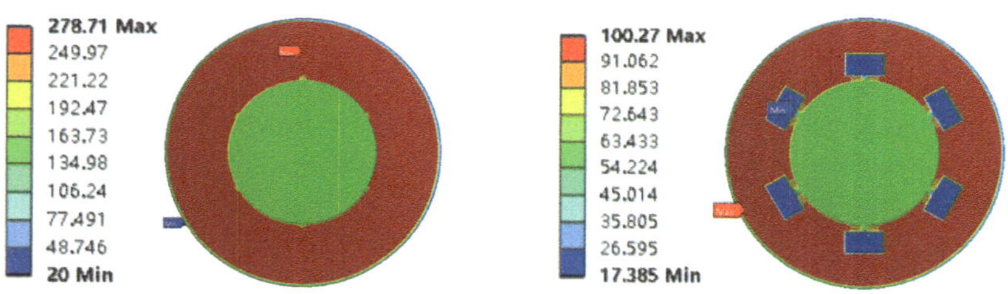

(a) Without conductor internal cooling structure (b) With conductor internal cooling structure

Fig. 6 The distribution of motor temperature

Table 4 The convection heat transfer coefficient

Parts	Convection heat transfer coefficient/W/m^2 °C
Motor housing	25
Air gap	66
Cooling channel	6000

Table 5 The maximum temperature rise data

Parts	Maximum temperature rise/°C	
	Without cooling	With cooling
conductor	254.49	0.20
stator	257.77	80.27
Slot filler	258.71	72.38
Rotor	140.34	40.91
Permanent magnets	140.50	41.03
protective case	140.53	41.04

consaductor, respectively. The initial temperature of the motor is set to 20 °C, the temperature of the cooling water is set to 20 °C, and the convective heat transfer coefficients of each interface are shown in Table 4.

Table 5 shows the comparison of the temperature rise data of each part of the motor before and after adding cooling structure to the conductor. It can be seen that the maximum temperature rise of the entire motor appears in the slot filling, reaching 258.71 °C, and the maximum temperature rise of the winding also reaches 254.49 °C. After adding internal water cooling to the copper conductor, the cooling effect of the cooling structure on the winding is very significant. Figure 7b, c show the temperature distribution diagram of a single copper conductor without internal water cooling and with internal water cooling respectively. Figure 7a shows the temperature change curve of the two conductor models with time. The temperature rise of the copper conductor is almost 0 °C. It can be seen that the cooling structure can efficiently take away the heat of the conductor, thereby reducing its temperature rise.

After the copper conductor is added to the internal water cooling, the temperature rise of other parts of the motor has also been greatly reduced. The temperature rise

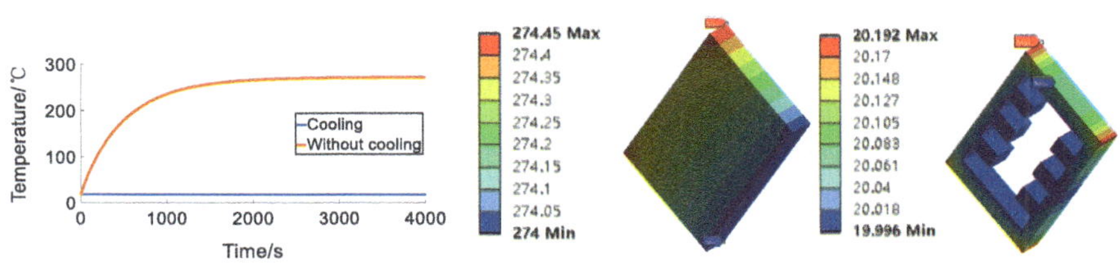

(a) The temperature change curve (b) Without cooling (c) With cooling

Fig. 7 The distribution of copper conductor

of the stator has dropped by 177.50 °C compared with the previous one, and the temperature rise of the rotor, permanent magnet and permanent magnet protective sleeve has all dropped by 100 °C around. Although the temperature rise has been greatly reduced, its value is still high, especially for the stator, which needs to be optimized in the next step.

5 Conclusion

This paper analyzes the loss in the copper conductor, and establishes a lumped parameter thermal network model of a 20 kW, 10 Wrpm surface mount 2-pole 6-slot high-speed permanent magnet synchronous motor, and conducts a finite element analysis of its temperature field distribution. A cooling method in the stator conductor is proposed and its cooling effect is analyzed. The final finite element analysis results show that the temperature rise of the copper conductor of the high-speed motor is 254.49 °C. When cooling water is passed into the copper conductor, the temperature rise of the copper conductor almost drops to 0 °C. The cooling effect of this cooling method on other parts of the motor is also obvious, and the maximum temperature rise of the whole motor falls to 80.27 °C. In the subsequent work, the cooling of the stator core will be further optimized.

Acknowledgements Supported by the National Natural Science Foundation of China (51737010, 51677144), the State Key Laboratory of Electrical Insulation and Power Equipment (EIPE19109), and the "Delta Power Electronics Science and Education Development Program".

References

1. Duan, C., H. Guo, W. Xing, W. Tian and J. Xu. 2018. Design and analysis of a 120 kW high-speed permanent magnet motor with a novel evaporative cooling configuration for centrifugal compressor. In *2018 21st International Conference on Electrical Machines and Systems (ICEMS)*, Jeju, 393–397. https://doi.org/10.23919/icems.2018.8549402.
2. Gonzalez, D. 2016. Performance validation of a high-speed permanent-magnet motor for centrifugal compressors. In *2016 XXII International Conference on Electrical Machines (ICEM)*, Lausanne, 545–549. https://doi.org/10.1109/icelmach.2016.7732579.
3. Jiang, C., M. Qiao, P. Zhu and Q. Zheng. 2018. Design and verification of high speed permanent magnet synchronous motor for electric car. In *2018 2nd IEEE Advanced Information Management, Communicates, Electronic and Automation Control Conference (IMCEC)*, Xi'an, 2371-2375. https://doi.org/10.1109/IMCEC.2018.8469398.
4. Tianyu, Wang, Wen Fuqiang, and Zhang Fengge. 2018. Analysis of multi-field coupling strength for mw high-speed permanent magnet machine. *Transactions of China Electrotechnical Society* 33 (19): 4508–4516. https://doi.org/10.19595/j.cnki.1000-6753.tces.171238. (in Chinese).
5. Weeber, K., C. Stephens, J. Vandam, et al. 2007. High-speed permanent-magnet motors for the oil & gas industry. In ASME (American Society of Mechanical Engineers) Turbo Expo, Montreal (CA). vol. 4, 1511–1520. pt. B; 20070514-17.

6. Shoujun, Song, Liu Weiguo, and Uwe Schaefer. 2010. Optimal control of a high speed switched reluctance starter/generator for the more/all electric aircraft. *Transactions of China Electrotechnical Society* 4: 44–52. (in Chinese).
7. Noguchi, T., Y. Takata, Y. Yamashita and S. Ibaraki. 2005. 160,000-r/min, 2.7-kW Electric Drive of Supercharger for Automobiles. In *2005 International Conference on Power Electronics and Drives Systems*, Kuala Lumpur, 1380–1385. https://doi.org/10.1109/peds.2005.1619904.
8. Oyama, J., T. Higuchi, T. Abe, K. Shigematsu, X. Yang and E. Matsuo. 2003. A trial production of small size ultra-high speed drive system. In *IEEE International Electric Machines and Drives Conference, IEMDC'03*, Madison, WI, USA, vol. 1, 31–36. https://doi.org/10.1109/iemdc.2003.1211239.
9. Popescu, Mircea, David G. Dorrell. 2013. Skin effect and proximity losses in high speed brushless permanent magnet motors. In *2013 IEEE Energy Conversion Congress and Exposition (ECCE 2013)*, vol. 5, 3520–3527. Denver, Colorado (US). 15–19 September 2013
10. Bang-cheng, H.A.N., P.E.N.G. Song, and H.E. Zan. 2020. Eddy current loss calculation and thermal analysis of high-speed motor winding in magnetically suspended control moment gyroscope. *Optics and Precision Engineering* 28 (1): 130–140. (in Chinese).
11. Shunian, Ding. 1992. Heating and cooling of large motors. (in Chinese).
12. Wenquan, Tao. 2019. *Heat Transfer*. Beijing: Higher Education Press (in Chinese).
13. Xiaofei, Wang, Dai Ying, and Luo Jian. 2019. Waterway design and temperature field analysis of vehicle permanent magnet synchronous motor based on fluid-solid coupling. *Transactions of China Electrotechnical Society* 34 (z1): 22–29. https://doi.org/10.19595/j.cnki.1000-6753.tces.L80816. (in Chinese).
14. Li, Yu. Research on performance analysis and design of high speed permanent magnet brushless DC Motor. Nanjing: School of Electrical Engineering, Southeast University (in Chinese).
15. Xintong, Zhang, Zhang Chengming, and Li Liyi. 2019. Split ratio optimization for permanent magnet synchronous machines considering winding temperature limitation. *Transactions of China Electrotechnical Society* 34 (9): 1886–1899. https://doi.org/10.19595/j.cnki.1000-6753.tces.180027. (in Chinese).

Review of the Application of Intelligent Optimization Algorithm for Design of Novel Electric Machines

Shang Yanzhe, Liang Deliang, Jia Shaofeng, Luo Yating, and Wang Hao

Abstract Compared with traditional optimization algorithm, intelligent optimization algorithm has many advantages, such as fast convergence speed, high precision and strong search effect in electric machine design, so it is favored by more and more researchers, and has attracted more academic research and engineering applications in recent years. In this paper, the introduction of the novel machines, the principle, the main application status and application prospect of intelligent optimization algorithm are discussed in detail, after comprehensively investigating the existing domestic and overseas literature. The application of the swarm intelligent algorithm, evolutionary optimization algorithm, surrogate-based optimization algorithm and other optimization algorithms in the single objective optimization and multi-objective optimization of the novel machine is comprehensively described and the future development direction of intelligent optimization algorithm is discussed.

Keywords Novel machine · Intelligent optimization algorithm · Machine design · Review

1 Introduction

With the continuous increase in demand for rail transit, new energy vehicles, wind power generation and other industrial equipment in recent years, the development concept of energy saving and emission reduction has been continuously strengthened, and the demand for high-efficiency, high-reliability, and high-cost-effective motors has become stronger. Therefore, how to further optimize the design of the motor and greatly improve the performance of the motor has been the focus of current research. At present, both the traditional structure motor and the new motor will have disadvantages such as large torque ripple and large loss.

On the other hand, the production of high-efficiency motors and their optimized design are very important to ensure energy conservation and reduce costs, and are also

S. Yanzhe (✉) · L. Deliang · J. Shaofeng · L. Yating · W. Hao
Xi'an Jiaotong University, Xi'an 710049, China
e-mail: syz0206@stu.xjtu.edu.cn

W. Chen et al. (eds.), *The Proceedings of the 9th Frontier Academic Forum of Electrical Engineering*, Lecture Notes in Electrical Engineering 743,
https://doi.org/10.1007/978-981-33-6609-1_36

of great benefit to extending the life of the motors. For the problem of motor design optimization, traditional optimization methods such as step acceleration method and Newton method are very complicated, and the optimization results are not satisfactory. There are more trials and longer duration. Therefore, they are facing more complex motor optimization. When designing problems, traditional optimization methods are not very effective.

In recent years, intelligent optimization algorithms have gradually been widely used in motor design optimization problems. Compared with traditional optimization methods, intelligent optimization algorithms have fast convergence speed, strong optimization effect, good optimization performance, and few test times and durations. Therefore, in motor design optimization, scholars at home and abroad gradually abandon traditional optimization methods. Instead, use or improve the intelligent optimization algorithm to optimize the target motor.

This article first briefly introduces the new motor, showing that the acquisition of its design formula is very cumbersome and complicated. Secondly, it classifies the intelligent optimization algorithms used in the motor design optimization problem and introduces the basic principles, advantages and disadvantages of each algorithm; then, the current intelligence The application status of optimization algorithm in motor design is explained; finally, the application prospect of intelligent optimization algorithm in motor design is analyzed and discussed.

2 Introduction of the Novel Machine Structure

At present, traditional motors such as DC motors, synchronous motors, induction motors, surface-mount permanent magnet motors are widely used in various industries, but with the rapid development of new energy vehicles, aerospace, robotics and other strategic emerging fields, motor operation The situation has become more and more complicated, and more stringent requirements have been put forward on the performance indicators of the motor. In addition to continuing to study the performance improvement of traditional motors, many scholars have also proposed other new working principles and new topology motors from the source of the innovation of the motor structure with advantages such as high power density, high output torque, wide speed range, high fault tolerance, high robustness, and high cost-effectiveness.

The current mainstream new motor types are shown in Fig. 1 [1].

Because the analytical design formulas of new motors are generally imperfect, most of them use finite element analysis, design, and optimization, which is time-consuming and laborious, so intelligent optimization algorithms must be used to optimize the design.

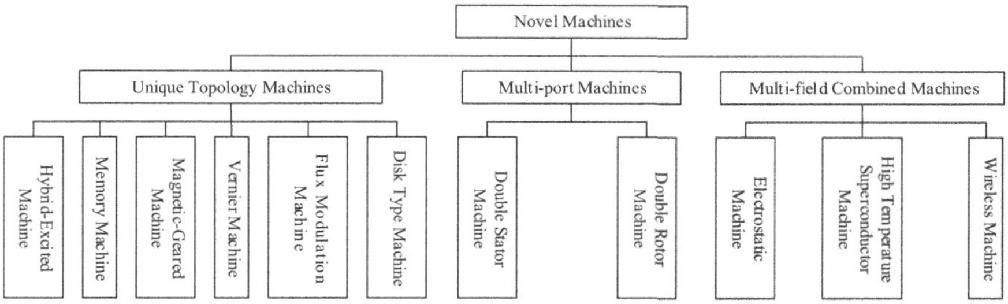

Fig. 1 Current types of mainstream novel machines

3 Introduction of Intelligent Optimization Algorithms

At present, intelligent optimization algorithms (IOA) used for motor design optimization can be divided into four categories: Swarm Intelligent Optimization Algorithms (SIOA), Evolutionary Optimization Algorithms (EOA), Surrogate Model Optimization Algorithms (SMOA), and Other Optimization Algorithms (OOA).

There are many types of SIOA. The main applications used in motor design optimization problems are Ant Colony Optimization Algorithm (ACOA), Particle Swarm Optimization Algorithm (PSOA), Bat Optimization Algorithm (BOA), etc.

EOA mainly include Genetic Algorithm (GA), Differential Evolution Algorithm (DEA), Backtracking Search Algorithm (BSA), etc.

SMOA mainly include Support Vector Machine Algorithm (SVMA), Kriging Method (KM), Response Surface Methodology (RSM), etc.

OOA are used in motor design optimization, mainly Taguchi Method (TM) and Simulated Annealing Algorithm (SAA).

The classification diagram of IOA is shown in Fig. 2. Due to space limitations, only one of the four types of intelligent optimization algorithms is introduced separately.

3.1 Swarm Intelligence Optimization Algorithm

The swarm intelligence optimization algorithm mainly simulates the habits of animal groups such as insect swarms and bird swarms. The group forages or hunts in a cooperative manner. Each individual in the group continuously changes the search

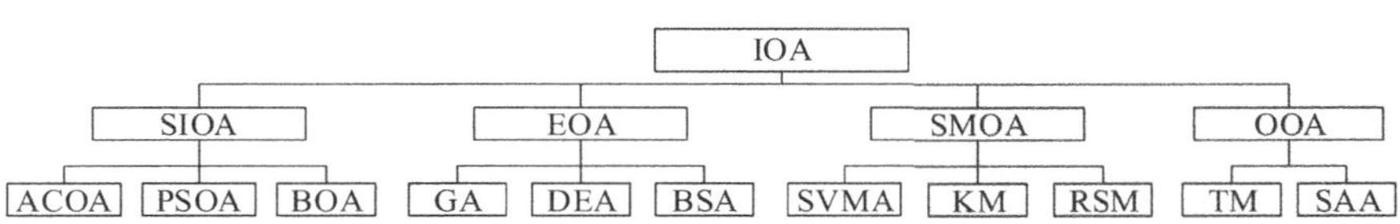

Fig. 2 Classification of intelligent optimization algorithms

target by learning from their own experience and experience of other individuals direction.

3.1.1 Particle Swarm Optimization Algorithm

Aiming at the continuous variable optimization problem, American scholars J. Kennedy and RC Eberhart proposed PSOA in 1995 based on the phenomenon of group biological activities such as bird foraging and fish migration. Algorithms have been widely used in the fields of function optimization and neural networks [2].

PSOA has a strong ability to optimize small data and runs faster, but when the number of iterations is large, it is easy to fall into a local optimum [3].

3.2 Evolutionary Optimization Algorithm

Evolutionary optimization algorithm is a kind of algorithm that abstracts biological evolution process and mechanism to solve a series of optimization problems.

3.2.1 Genetic Algorithm

GA was proposed by the American scholar J. H. Holland in 1975. GA is an abstraction of genetic selection and the theory of survival of the fittest. The optimal solution is finally obtained through the selection, crossover, and mutation of genetic operators. It has been widely used in function optimization, path planning, combinatorial optimization and other fields [4].

The operation steps of GA consist of coding, generating initial population, calculating fitness value, selecting parent, crossover operation and mutation operation [5].

GA performs a series of operations in the order of selection, crossover, and mutation, which increases the reproduction probability of individuals with higher fitness values, making the fitness value of offspring higher. As the number of evolution increases, the best individual can eventually be obtained, namely The optimal solution of the problem [6].

GA has the disadvantages of poor local search ability, slow convergence speed, easy to fall into local extreme points and population assimilation in the later stage.

3.3 Surrogate Model Optimization Algorithm

SMOA can establish a sample library of the relationship between motor structure parameters and output performance, and then perform regression fitting on the sample data to obtain a fast calculation model of the motor. It is usually optimized by combining with other intelligent optimization algorithms.

3.3.1 Kriging Method

KM is a statistical-based interpolation method named after South African mining engineer D. G. Krige, which aims to more accurately predict ore reserves. KM is a linear regression algorithm for spatial modeling and interpolation of the research target based on the covariance function. It can give the best linear unbiased estimate.

KM is divided into ordinary Kriging and general Kriging according to the adopted global trend function. Among them, the general Kriging model is widely used, and the general Kriging expression form is shown in formula (1) [7]:

$$Z^*(\boldsymbol{x}) = \sum_{i=1}^{n} \lambda_i Z(\boldsymbol{x}_i) \tag{1}$$

In the formula, $Z^*(\boldsymbol{x})$ represents the variable to be estimated; x represents the sample point; λ_i represents the weighting coefficient; $Z(\boldsymbol{x})$ represents the response value corresponding to the sample point.

Although KM can improve the estimation accuracy, it is easy to overlook some special points due to its smooth processing characteristics.

3.4 Other Optimization Algorithms

In addition to the above-mentioned SIOA, EOA, and SMOA, there are OOA used in motor design optimization. Taguchi method is mainly introduced here.

3.4.1 Taguchi Method

TM is a local optimization design method based on orthogonal experiment and signal-to-noise ratio technology proposed by Japanese scholar Genichi Taguchi. Different from other local optimization design methods, TM can optimize multiple goals. By establishing an orthogonal table, the optimal configuration of optimized variables can be obtained with the least number of experiments [8].

TM involves three main stages in the optimization design problem:

(1) Selecting performance characteristics, control factors and noise factors;
(2) Designing an orthogonal table and conduct experiments;
(3) Using mean and variance method to obtain the best parameter settings.

Like the inherent shortcomings of other local optimization methods, TM is highly dependent on the selection of the initial point, so global optimization ability is poor.

4 Application of Intelligent Optimization Algorithm

The optimization design goals of the motor mainly include improving electromagnetic torque, improving efficiency, reducing cost, reducing noise, reducing copper loss and iron loss, and restraining torque ripple. Whether it is single-objective optimization or multi-objective optimization, three requirements must be determined before optimization. Elements: optimization variables, objective functions and constraints.

With the increasing complexity of working conditions, the calls for environmental protection and green are becoming higher and higher, and the engineering requirements for motors are becoming more and more stringent. Therefore, only single-objective optimization of motors is usually not enough. Realizing multi-objective optimization of motors has become the mainstream.

4.1 *Multi-objective Optimization*

Compared with single-objective optimization, the multi-objective optimization process is more complicated, and its expression form is shown in formula (2).

$$\begin{cases} \min f_k(\S), k = 1, 2, \ldots, p \\ \text{s.t. } g_i(\S) \geq 0, i = 1, 2, \ldots, q \\ h_j(\S) = 0, j = q + 1, \ldots, m \\ \S \in S \subseteq \mathbf{R}^n \end{cases} \tag{2}$$

In the formula, p represents the number of optimization objective functions.

The various objectives of multi-objective optimization are often conflicting and restricting each other. One of the objective functions is optimal, which may cause the performance of the other objectives to decrease, that is, there may not be all objective functions at the solution x to achieve the maximum or minimum value. Therefore, the French economist V. Pareto called the optimal solution of the multi-objective optimization problem as the Pareto optimal solution in 1896, and proposed the Pareto solution set [9]. The Pareto optimal solution can make the overall optimization have the least objective conflict.

The multi-objective optimization process of the motor is shown in Fig. 3.

The motor design optimization method based on SIOA is widely used in the multi-objective optimization problem of motor design. On the basis of the traditional artificial bee colony, literature [10] proposed an optimization algorithm combining

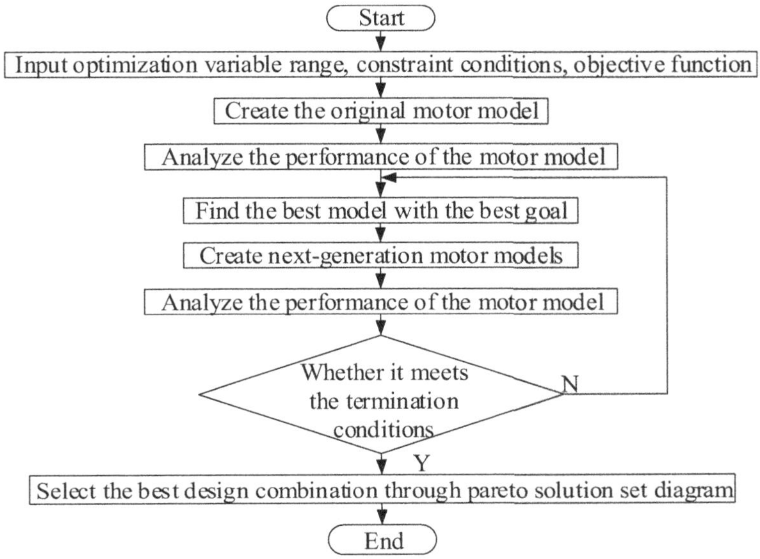

Fig. 3 Flow chart of machine multi-objective optimization

artificial bee colony algorithm and strength Pareto evolutionary algorithm to optimize the efficiency and torque ripple of surface-mounted permanent magnet motors for electric vehicles. The literature [11, 12] made different improvements to the traditional artificial bee colony algorithm to achieve multi-objective optimization of the target motor. Similarly, ACOA, PSOA, and BOA are also used in the multi-objective optimization design of brushless DC motors [13], double stator brushless doubly-fed motors [14] and self-starting permanent magnet synchronous motors [15].

At the same time, many scholars use the optimization algorithm with EOA as the core to realize the multi-objective optimization design of the motor. In the traditional DEA, the selection part is based on the fitness function, and the optimal model may be discarded in the optimization process, resulting in the slow convergence speed and poor final optimization performance. Therefore, literature [16] improves the traditional DEA by using the non-dominated sorting to find the Pareto optimal solution set, and adopts the optimal pre reservation strategy to ensure that the optimal model will not be discarded, and at the same time, the optimal pre reservation strategy is used to ensure that the optimal model will not be discarded At the end of the crossover part of the system, the new algorithm will generate a random integer between the variables and the number of variables, and the current individual variables cross with the corresponding variables. Through this improved DEA, the torque of the motor is improved, the induced voltage is reduced, the optimization time is greatly reduced, and the overall performance of the field modulated motor is improved.

In recent years, because TM can obtain the optimal solution through a small number of experiments, more and more studies tend to use TM or a hybrid optimization algorithm combined with TM to perform multi-objective optimization of the target motor, and achieve good results. Literature [17–20] uses TM to optimize the torque ripple, cogging torque, back-EMF harmonic rate and other objectives of permanent magnet synchronous motors.

5 Application Prospects

At present, for the optimization of motor design, multi-objective optimization of motor design is still the mainstream. Many scholars at home and abroad have adopted multiple optimization algorithms to achieve multi-objective optimization of motors.

At the same time, many improved algorithms are based on conventional optimization algorithms. The above introduces other optimization algorithms to improve optimization performance, and the multi-objective optimization algorithm used has three main characteristics. The first is to combine various optimization algorithms, synthesize the advantages of each algorithm and give full play to the advantages of each algorithm; the second is to reduce the number of optimization iterations as much as possible, and reduce the test time; the third is to improve the local search ability on the basis of ensuring the global optimization effect.

Literature [21–25] combined EOA with SMOA to realize multi-objective optimization of motor. Among them, literature [21] uses the combination of RSM method and GA algorithm to establish an analysis model of the motor through the RSM method, while the GA algorithm optimizes the model, and finally realizes the multi-objective optimization of the motor; literature [22] uses GA and KM to optimize the length of armature core and the shape of external teeth of the designed permanent magnet linear synchronous motor, in order to minimize the braking force with large end force and weaken the vibration and noise of the motor during operation; Reference [23] adopts the combination algorithm of GA and convolution neural network, which reduces the core loss and improves the operation performance of the motor; literature [24, 25] uses the combination method of SVMA and GA to optimize the studied motor, using SVMA establishes the non-parametric model of the motor, and then uses GA to optimize structural parameters of the motor to obtain the optimal parameter combination.

Literature [26] combines SIOA with EOA, and introduces a multi population GA, which uses multiple relatively independent populations to search the solution space, realizes information transmission between populations through migration operator, and uses manual selection operator to record the optimal individual of each population in each generation. Compared with the conventional GA algorithm, the algorithm has better performance in complex objective function It overcomes the disadvantages of poor local search ability.

Literature [27, 28] combined the three optimization algorithms, and the literature [27] used a hybrid optimization algorithm based on the nuclear extreme learning machine method, SAA and PSO. The non-parametric model of the resistance motor, SAA is introduced into PSOA to make up for the shortcomings of PSO that is easy to fall into the local optimum, so as to achieve faster model solving speed and better optimization effect; literature [28] adopts sensitivity analysis to divide the key parameters of the electric excitation double stator field modulation motor into layers of sensitivity, using GA to optimize the structural parameters of the strong sensitivity, and continuing to layer the structural parameters of the weak sensitivity, and using RSM and single parameter sweep method for optimization analysis respectively.

In literature [29], the method of combining GA algorithm with TM was used to re optimize the scheme optimized by GA algorithm, so as to realize multi-objective optimization of IPMSM.

6 Conclusion

This article systematically explains the classification of current mainstream new motors and briefly introduces the structure of several new motors. At the same time, it also introduces the classification, principle introduction, application status and prospects of intelligent optimization algorithms in current motor design optimization problems. The application of intelligent optimization algorithm in the multi-objective optimization of motors and the form and advantages of mixing multiple optimization algorithms are analyzed. In general, since the development of motor design optimization, the types of intelligent optimization algorithms have shown a spurt-like growth. The motor design optimization algorithms are relatively mature, and certain phased results have been achieved for optimizing motor performance. It is moving towards balancing global search and local search. Combining multiple optimization algorithms, comprehensive advantages, improving convergence accuracy and optimization speed, and reducing the number of iterations and trials. It can be expected that in the future hybrid vehicles, aerospace, industrial equipment, power generation and transmission and other socially focused application motor fields, intelligent optimization algorithms will have broad development space and application prospects, and will also drive more new motors R&D and production.

Acknowledgements Thanks to the National Natural Science Foundation of China (51737010, 51677144), the State Key Laboratory of Electrical Insulation and Power Equipment (EIPE19109), and the 'Delta Power Electronics Science and Education Development Program' for their support.

References

1. Liu, C. 2018. Emerging electric machines and drives—An overview. *IEEE Transactions on Energy Conversion* 33 (4): 2270–2280. https://doi.org/10.1109/TEC.2018.2852732.

2. Jingwei, Haung, Zhu Fuxi, and Kang Lishan. 2018. *Computational Intelligence.* Beijing: Science Press (in Chinese).
3. Xiping, Liu, Hu Weiping, Ding Weizhong, Xu Hui, and Zhang Yun. 2020. Research on multi-parameter identification method of permanent magnet synchronous motor. *Transactions of China Electrotechnical Society* 35 (06): 1198–1207 (in Chinese).
4. Qingshan, Xu, Lou Oudie, Zheng Aixia, and Liu Yujun. 2018. A non-intrusive load decomposition method based on affinity propagation and genetic algorithm optimization. *Transactions of China Electrotechnical Society* 33 (16): 3868–3878. (in Chinese).
5. Guobin, Yang. 2019. Optimization design of switched reluctance motor based on response surface method and genetic algorithm. Jiangxi University of Science and Technology (in Chinese).
6. Yao, Zhang, Du Rongfa, Zhao Xinchao, and Ye Qainjie. 2020. Motor noise optimization based on multi-objective genetic algorithm. *Small & Special Electrical Machines* 48 (04): 21–25 (in Chinese).
7. Yanli, Zhang, Xia Bin, Xie Dexin, and Bai Baodong. 2013. Optimum design of pole face of switched reluctance motor based on Kriging interpolation. *Electric Machines and Control* 17 (03): 56–61. (in Chinese).
8. Jinxin, Jia, Yang Xiangyu, and Cao Jianghua. 2013. Optimization design of interior permanent magnet motor based on Taguchi method. *Micromotors* 46 (06): 1–4. (in Chinese).
9. Xiaoshu, Ma, Li Yulong, and Yan Lang. 2010. Comparsion review of traditional multi-objective optimization methods and multi-objective genetic algorithm. *Electric Drive Automation* 32 (03): 48–50+53 (in Chinese).
10. Krasopoulos, C.T., I.P. Armouti and A. G. Kladas. 2016. Hybrid multi-objective optimization algorithm for PM motor design. In *2016 IEEE Conference on Electromagnetic Field Computation (CEFC)*, Miami, FL, 1–1. https://doi.org/10.1109/cefc.2016.7816014.
11. Das, P.P., and S. N. Mahato. 2016. Design optimization of a six-phase induction motor by Flower pollination and modified Artificial Bee Colony Algorithms. In *2016 IEEE Region 10 Conference (TENCON)*, Singapore, 3313–3317. https://doi.org/10.1109/TENCON.2016.7848665.
12. Heshan, Zhang, Deng Zhaoxiang, Tuo Jiying, Yu. Zhang, and Tao Shengchao. 2019. Multi-objective optimum design for in-wheel motor based on improved artificial bee colony algorithm. *Journal of Southwest Jiaotong University* 54 (04): 671–678. (in Chinese).
13. Umadevi, N., M. Balaji and V. Kamaraj. 2014. Design optimization of brushless DC motor using Particle Swarm Optimization. In *2014 IEEE 2nd International Conference on Electrical Energy Systems (ICEES)*, Chennai, 122–125. https://doi.org/10.1109/icees.2014.6924153.
14. Dandan, Ma. 2019. Rotor optimization design of dual-stator brushless doubly fed machine based on improved particle swarm optimization. Shenyang University of Technology (in Chinese).
15. Knypiński, Ł. 2017. Application of Bat Algorithm in the optimal design of line-start permanent magnet synchronous motor. In *2017 18th International Symposium on Electromagnetic Fields in Mechatronics, Electrical and Electronic Engineering (ISEF) Book of Abstracts*, Lodz, 1–2. https://doi.org/10.1109/ISEF.2017.8090697.
16. Pei, T., D. Li, R. Qu, M. R. Shah and P. Zhang, Multi-objective optimization algorithm of a magnetic field modulation motor based on advanced differential evolution. In *2017 20th International Conference on Electrical Machines and Systems (ICEMS)*, Sydney, NSW, 1–5. https://doi.org/10.1109/icems.2017.8056129.
17. Jianjun, Li, Wu Huang Kaisheng, and Huang Qu Ning. 2018. Multi-objective optimization of surface permanent magnet synchronous motor based on Taguchi method. *Small & Special Electrical Machines* 46 (07): 10–13. (in Chinese).
18. Yang, Lu, Miao Hong, Zeng Chengbi. 2019 Multi-objective optimization of surface-mounted and interior permanent magnet synchronous motor based on Taguchi method. *Micromotors* 52 (09): 1–5+24 (in Chinese).
19. Xu, J., L. Gao, L. Zeng and R. Pei. 2019. Optimum design of interior permanent magnet synchronous motor using Taguchi method. In 2019 IEEE Transportation Electrification

Conference and Expo (ITEC), Detroit, MI, USA, 1–4. https://doi.org/10.1109/ITEC.2019.8790592.
20. Yunqing, Pei, Yang Xiangyu, Cheng Xiaohua, Zhao Shiwei. 2018. Optimal design of a permanent magnet synchronous motor with novel structure rotor. *Micromotors* 51 (12): 1–5+47 (in Chinese).
21. Hasanien, H.M., A.S. Abd-Rabou, and S.M. Sakr. 2010. Design optimization of transverse flux linear motor for weight reduction and performance improvement using response surface methodology and Genetic algorithms. *IEEE Transactions on Energy Conversion* 25 (3): 598–605. https://doi.org/10.1109/TEC.2010.2050591.
22. Zhang, Y., Jingguo Yuan, D. Xie, In Sung Hwang and C. S. Koh. 2010. Shape optimization of a PMLSM using Kriging and genetic algorithm. In *2010 5th IEEE Conference on Industrial Electronics and Applications*, Taichung, 1496–1499. https://doi.org/10.1109/iciea.2010.5514796.
23. Doi, S., H. Sasaki and H. Igarashi. 2019. Multi-objective topology optimization of rotating machines using deep learning. *IEEE Transactions on Magnetics* 55 (6): 1–5. Art No. 7202605. https://doi.org/10.1109/tmag.2019.2899934.
24. Xiang, Q., Y. Sun and X. Zhang. 2011. Optimization design of the bearingless switched reluctance motor based on SVM and GA. In *Proceedings of the 30th Chinese Control Conference*, Yantai, 1472–1475.
25. Le, W., M. Lin, L. Jia, J. Ai, X. Fu and Z. Chen. 2019. Multi-objective optimization of an air-cored axial flux permanent magnet synchronous machine with segmented PMs based on support vector machine and genetic algorithm. In *2019 22nd International Conference on Electrical Machines and Systems (ICEMS)*, Harbin, China, 1–4. https://doi.org/10.1109/icems.2019.8922465.
26. Liyi, Li, Tang Yongbin, Liu Jiaxi, et al. 2013. Application of the multiple population genetic algorithm in optimum design of air-core permanent magnet linear synchronous motors. *Proceedings of the CSEE* 33 (15): 69–77+14 (in Chinese).
27. Yongzhi, Liu, Li Jie, Shan Chenglong, and Lin Bowen. 2019. Optimization design of switched reluctance motor based on kernel extreme learning machine and simulated annealing particle swarm optimization. *Journal of Electronic Measurement and Instrumentation* 33 (02): 148–153. (in Chinese).
28. Xianglin, Li, Li Jinyang, Yang Guangyong, Wang Yubin, and Feng Xingtian. 2020. Multi-objective optimization analysis of electric-excitation double-stator field-modulated machine. *Transactions of China Electrotechnical Society* 35 (05): 972–982. (in Chinese).
29. Aimeng, Wang, and Wen Yun. Optimal design of interior permanent magnet synchronous motor based on hybrid genetic algorithm. *Electric Machines & Control Application* 44 (03): 59–65+95 (in Chinese).

Magnetically Coupled Resonant Wireless Power Transmission Coil Temperature Rise and Parameter Optimization

Da Li, Xu-sheng Wu, Wei Gao, Jian-xin Gao, Yi-qin Lei, and Chun-yang An

Abstract Wireless power transmission has received extensive attention in the fields of aviation, aerospace, automobile, and new energy. With the increasing demand of power electronics, the power and transmission distance of wireless power transmission is increasing. The increase of transmission power leads to the increase of coil loss and the coil temperature rise. The temperature rise of the coil will cause problems such as material aging, insulation damage, and performance degradation, so the calculation and research of the temperature rise of the coupled coil are very important for the research of the wireless power transmission.

Keywords Magnetically coupled resonance · Coupling coil · Coil temperature rise · Parameter optimization

1 Introduction

As a safe and convenient charging method, wireless power transmission has deeply research foundation and broad application prospects in electric vehicles, rail transit, aerospace, industrial equipment, consumer electronics, underwater and other fields [1–4]. WPT can be divided into three types: microwave transmission, electromagnetic induction, and magnetic coupling resonance. Magnetic coupling wireless power transmission technology realizes energy conversion between electric field and magnetic field through high-quality coupler. It has long transmission distance and high biosafety. Among them, the high-power magnetic coupling resonance wireless power transmission has attracted extensive attention of scholars at home and abroad for its power advantages and broad application prospects, and could change the way people use electrical appliances [5, 6].

Wireless power transmission is a combination of electromagnetic field coupling control, power electronics technology, communication, and control technology. Therefore, the overall performance of wireless power transmission technology is

D. Li · X. Wu · W. Gao (✉) · J. Gao · Y. Lei · C. An
School of Electrical Engineering, Naval Engineering University, Wuhan 430032, China
e-mail: depkin@163.com

W. Chen et al. (eds.), *The Proceedings of the 9th Frontier Academic Forum of Electrical Engineering*, Lecture Notes in Electrical Engineering 743,
https://doi.org/10.1007/978-981-33-6609-1_37

not only related to the topology of the circuit, but also depends on the magnetic field distribution between coils. The optimization and design of the coupling coil is also the key to the design of the wireless power transmission system. Introducing WPT technology into different fields must face the adjustment and optimization of the coupler coil to adapt to different work requirements. Electric vehicle wireless charging has become a hot spot in the current research on electric vehicle charging points due to its high safety, strong operability, and space saving. The Beijing Jiaotong University team used the field-circuit coupling model to compare the effective transmission distance and offset performance of the circular coil and the square coil [7]; the Jilin University team used the three-coil method to achieve stronger load robustness in the wireless charging system and relatively low environmental electromagnetic radiation characteristics in wireless charging system [8]. Compared with static charging, dynamic wireless charging expands the charging occasions of electric vehicles and helps build a distributed charging network. In order to ensure the stable transmission of the transmitting and receiving power coils in the direction of travel, the Huazhong University of Science and Technology team proposed a segmented power coil design with an asymmetric orthogonal rectangular structure, which improves the fault tolerance of lateral offset and keeps it high. The coupling coefficient [9]. At the same time, the performance requirements of magnetic coupling coils are also different in different working environments such as rail transit, mines, and oceans [10–12].

Many experts and scholars have also conducted a series of in-depth studies to improve the performance of the coil. Tianjin Polytechnic University team used the field-circuit coupling model to calculate the spatial electromagnetic field of single-turn coils of different radii and the overall change trend of the circuit parameters, effectively avoiding blindness in design and testing [5]; Hunan University team aimed at the coupled resonance process The problem of low efficiency when the transmitting coil and the receiving coil are not coaxially transmitted. A dynamic parameter adjustment based on chaos optimization algorithm is proposed, which effectively improves the system efficiency [13]; Chongqing University has proposed new coils such as DD to improve the anti-offset ability of coils and the parameter characteristics of wireless power transmission system [14]. The research team of Beijing Jiaotong University aims at the tram parking rule with only longitudinal offset and normal distribution parking error. By improving the wireless transmission efficiency of the secondary coil, the transmission efficiency of the secondary coil is improved. At the same time, UOA of New Zealand and KAIST of South Korea respectively conducted relevant research and optimization design on the coupling mechanism of static wireless charging and dynamic wireless charging of electric vehicles [16–20].

At present, the coil optimization of high-power wireless power transmission system mainly includes three methods to improve the coil transmission efficiency: new coupling structure design, coil parameter optimization and different topology compensation structure. However, a considerable part of the electric energy consumed in the process of electric energy transmission is converted into heat energy, which causes the temperature of the coil to rise and the heat is radiated to the area around the coil. The heating of the coil is one of the main factors that lead to the aging of the coil material, the destruction of the insulation, and the shortening of the life of the coil. Improving the heating condition of the coil is of great practical significance for prolonging the service life of the coil. Therefore, this paper takes three kinds of common coupling coils as the research object, studies the temperature rise and optimizes the coil parameters in the transmission process.

2 Temperature Rise Calculation and Modeling

2.1 Model of Magnetic Coupling System

According to the coupled mode theory, a coupled system model composed of the transceiver coils in the wireless power transmission is established [21]:

$$\frac{da_1(t)}{dt} = -j\omega_1 a_1(t) + jK_{12}a_2(t) - \tau_1 a_1(t) + S(t) \tag{1}$$

$$\frac{da_2(t)}{dt} = -j\omega_2 a_2(t) + jK_{21}a_1(t) - (\tau_1 + \tau_2)a_2(t) \tag{2}$$

$a_1(t), a_2(t)$ is the mode amplitude integral frequency component of transceiver coil; $|a_1(t)|^2$, $|a_2(t)|^2$ is the energy stored in the receiving and transmitting coils respectively; ω_1, ω_2 indicate the angular frequency of the receiving and transmitting coils respectively; $K_{12} = \frac{\omega_1 M}{2\sqrt{L_1 L_2}}$, $K_{21} = \frac{\omega_2 M}{2\sqrt{L_1 L_2}}$ are the coupling coefficient of the system. $\tau_1 = \frac{\omega_1}{2Q_1}$, $\tau_2 = \frac{\omega_2}{2Q_2}$ are the loss coefficients of the transmitting and receiving coils; $\tau_L = \frac{\omega_L}{2Q_L}$ is the load factor; $S = Fe^{-jwt}$ is the excitation of Sine.

2.2 Power, Efficiency and Transient Response

Assuming that the inherent parameters of the resonant coil are the same, the power absorbed by the transmitting coil $P_1 = 2\tau_1|a_1(t)|^2$, receiving coil absorb power $P_2 = 2\tau_2|a_2(t)|^2$, load absorbed power $P_L = 2\tau_L|a_L(t)|^2$. The transmission efficiency of the system can be obtained as

$$\eta = \frac{P_L}{P_1 + P_2 + P_L} = \frac{\tau_L|a_2(t)|^2}{\tau_1|a_1(t)|^2 + \tau_2|a_2(t)|^2 + \tau_L|a_2(t)|^2} \tag{3}$$

According to Eqs. (1) and (2), the expression of the double-coil vibration in the coupled system can be obtained by solving:

$$a_1 = \frac{[\tau_2 - j(\omega_2 - \omega)]Fe^{-jwt}}{K^2 + \tau_1\tau_2 - (\omega_1 - \omega)(\omega_2 - \omega) + j[\tau_1((\omega_2 - \omega) + \tau_2(\omega_1 - \omega))]} \tag{4}$$

$$a_2 = \frac{jKFe^{-jwt}}{K^2 + \tau_1\tau_2 - (\omega_1 - \omega)(\omega_2 - \omega) + j[\tau_1((\omega_2 - \omega) + \tau_2(\omega_1 - \omega))]} \tag{5}$$

The system is in resonance state

$$\eta = \frac{(\tau_L/\tau_2)K^2/\tau_1\tau_2}{(1 + \tau_L/\tau_2)K^2/\tau_1\tau_2 + (1 + \tau_L/\tau_2)^2} \tag{6}$$

$$P_L = 2\tau_L \frac{K^2F^2}{\left(K^2 + \tau_1(\tau_2 + \tau_L)\right)^2} \tag{7}$$

According to the coupling model, the dual−coil coupling will cause the system to split the resonant frequency. The angular frequency of the transceiver coil is the same, the frequency difference between the two coils is $2|K|$.

Assuming t = 0, $a_1(0) = A, a_2(0) = 0$, ignore the excitation and load, the Eqs. (1) and (2) are subjected to a pull transformation and the initial value is taken into consideration after the pull transformation is performed to obtain the time domain value [22]

$$a_1(t) = A\left[\cos\left(\frac{t}{2}\sqrt{4K^2 - j(\tau_1 - \tau_2)^2}\right) + \frac{\tau_2 - \tau_1}{\sqrt{4K^2 - j(\tau_1 - \tau_2)^2}}\sin\left(\frac{t}{2}\sqrt{4K^2 - j(\tau_1 - \tau_2)^2}\right)\right]e^{-jwt - \frac{\tau_2+\tau_1}{2}t} \tag{8}$$

$$a_2(t) = A\left[\frac{j2K}{\sqrt{4K^2 - j(\tau_1 - \tau_2)^2}}\sin\left(\frac{t}{2}\sqrt{4K^2 - j(\tau_1 - \tau_2)^2}\right)\right]e^{-jwt - \frac{\tau_2+\tau_1}{2}t} \tag{9}$$

The loss system of the transmitting and receiving coil is the same (Fig. 1):

$$a_1(t) = A\cos(Kt)e^{-jw - \tau_1 t} \tag{10}$$

$$a_2(t) = A[j\sin(Kt)]e^{-jw - \tau_1 t} \tag{11}$$

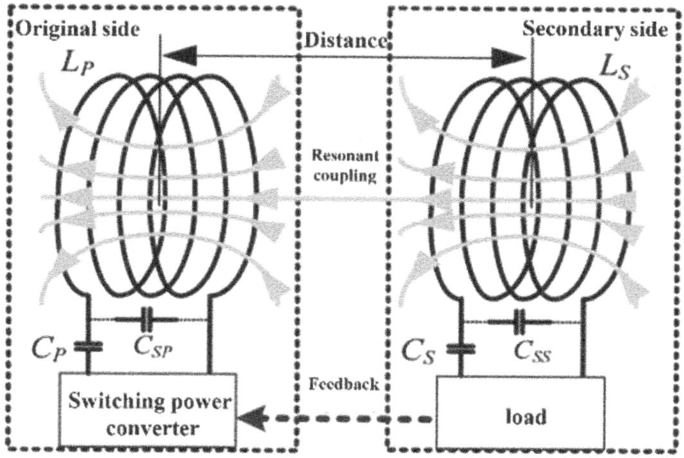

Fig. 1 Model of magnetic coupled resonance system

2.3 Calculation of Coil Temperature Rise

Joule heat will be produced in the process of wireless power transmission by magnetic coupling coil. The accumulation of Joule heat will lead to the change of coil temperature and affect the conductivity of materials. Therefore, it is necessary to establish a temperature field model to study the temperature rise in wireless power transmission system. The calculation formula of Joule heat is as follows:

$$Q = \int_0^t i^2 R\, dt \tag{12}$$

It can also be expressed as:

$$Q = \frac{1}{\gamma} J_{total} \times J_{total} \tag{13}$$

γ is the conductivity of the coil material, J_{total} is the coil current density. Considering that there may be high frequency resonance frequency and skin effect in the process of wireless power transmission, resulting in uneven distribution of coil current, the calculation formula of current density in this calculation model is as follows:

$$I_L = \int_S J_w ds \tag{14}$$

$$V_{CT} = 2\pi r \frac{J_w}{r_c} + 2\pi r \frac{\partial A}{\partial t} \tag{15}$$

I_L is the coil current in the equivalent circuit, S is the cross-sectional area of the wire coil. V_{CT} is the voltage across the single-turn conductor coil, r_c is the conductivity of the coil material. $2\pi r \frac{J_w}{r_c}$ is the coil resistance voltage, $2\pi r \frac{\partial A}{\partial t}$ is the induction voltage of the coil. $V_{coil} = \sum_{t=1}^{n} V_{CT}$, V_{coil} are the voltage across the coil, n is the number of coil turns.

The Joule heat generated by the wire coil can be obtained through the above equation, and the temperature rise and heat transfer process of the coil in the wireless power transmission need to be calculated through the heat conduction equation.

$$pC_p \frac{\partial T}{\partial t} + \nabla \cdot (-k\nabla T) = Q \tag{16}$$

p is the material density, C_p is the heat capacity of the material, K is the material thermal conductivity.

3 Solution of Temperature Rise of Wireless Power Transmission Coil

The resonant coil is the core of the wireless power transmission system. The resonant frequency of the system is determined by two resonant units; the transmission efficiency of the system is directly determined by the transmission coefficient between the coils, and the transmission status of the two coils directly affects the working status of the system. Spiral resonant coils can be roughly divided into cylindrical spirals and plane spirals. Since the spiral coil is axisymmetric in rotation, its calculation model can be simplified to a two-dimensional axisymmetric model. The modeling domain is separated by the infinite element domain, and the symmetry about the z = 0 plane is simulated by the ideal magnetic conductor boundary. Therefore, it is simplified to a half original domain (Fig. 2 and Table 1).

According to the two-dimensional axisymmetric model of the magnetic coupling coil, it can be found that the magnetic flux density between the single-layer spiral coil and the double-layer spiral coil is large, and the magnetic field strength decreases obviously in the diffusion area. The maximum flux density of single-layer spiral coil and cylinder spiral coil is 0.18 T, and that of double-layer spiral coil is 0.74 t. However, the minimum magnetic flux density of the cylindrical spiral coil is 2.26×10^{-4} T greater than the 9.95×10^{-5} T of the single-layer planar spiral coil, and that is greater than the minimum magnetic flux density of the double-layer planar spiral coil 3.2×10^{-5} T. Among the three different types of coils, the maximum current induction density of the cylindrical spiral coil is 6.65×10^6 A/m^2. At the same time, it is not difficult to find that there is a certain degree of skin effect on the surface of the metal coil through the current density map, but it has little effect on the wireless power transmission system at this frequency.

According to the temperature distribution of three different types of resonant coils in Fig. 3, the spiral coil is least affected by electromagnetic heat. The maximum temperature of cylindrical spiral coil is only 347 k after being electrified for one hour, while the maximum temperature of single-layer planar spiral coil is 390 k and that of double-layer planar spiral coil is 634 k. The temperature rise of single-layer spiral coil and double-layer plane spiral coil is concentrated on both sides of the receiving

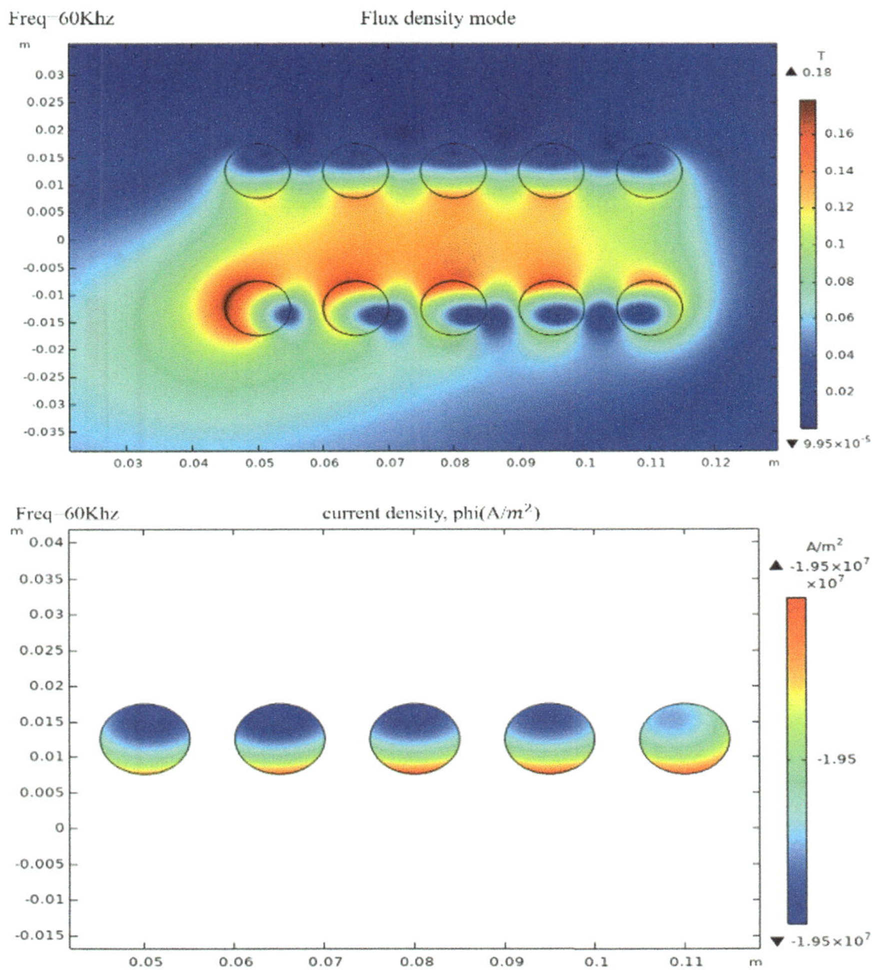

(a) Magnetic flux density and current density of single-layer planar spiral coil

Fig. 2 Magnetic flux density and current density distribution map of resonant coil

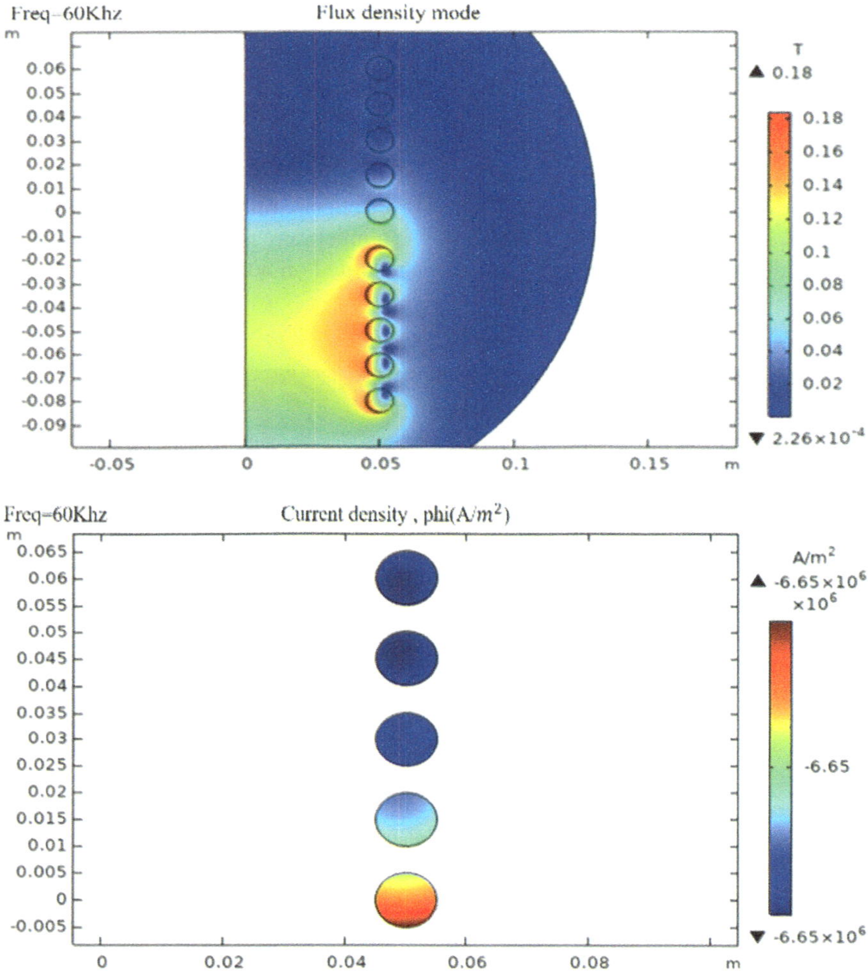

(b) Magnetic flux density and current density of cylindrical spiral coil

Fig. 2 (continued)

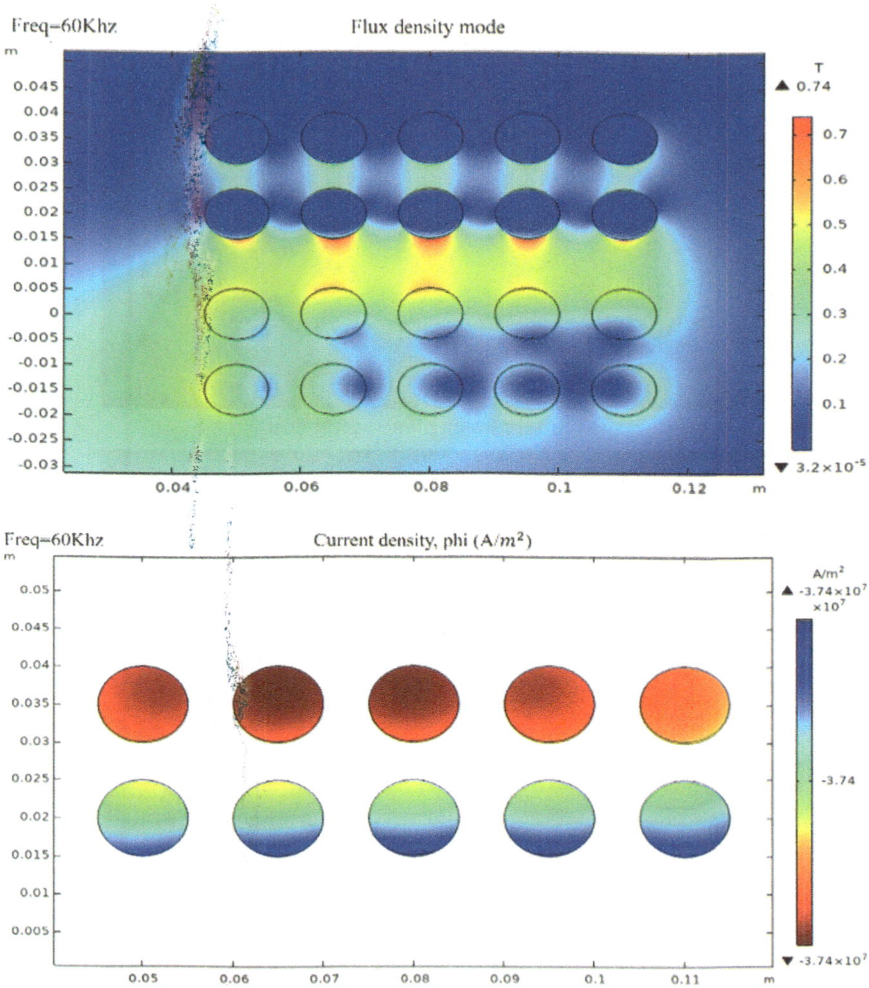

(c) Magnetic flux density and current density of double-layer planar spiral coil

Fig. 2 (continued)

Table 1 Coil parameter

Coil type/parameter	Single layer plane helix	Cylinder helix	Double plane helix
Number of turns	5	5	10
Coil radius (CM)	5–10	5	5–10
Wire radius (CM)	0.5	0.5	0.5

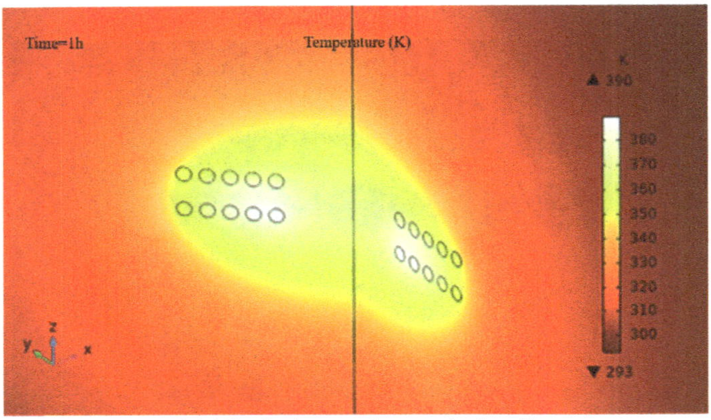

(a) Single layer plane spiral coil

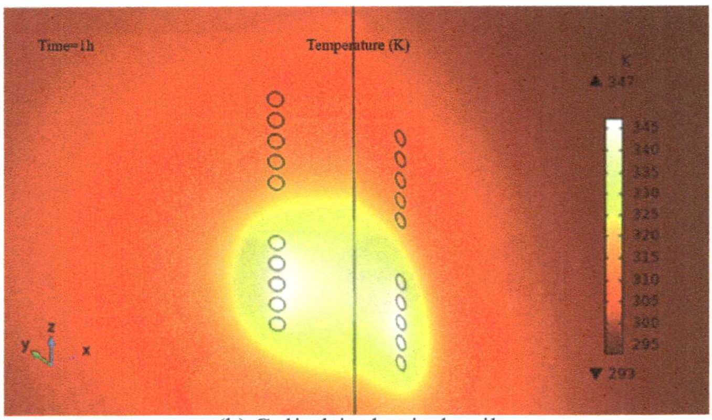

(b) Cylindrical spiral coil

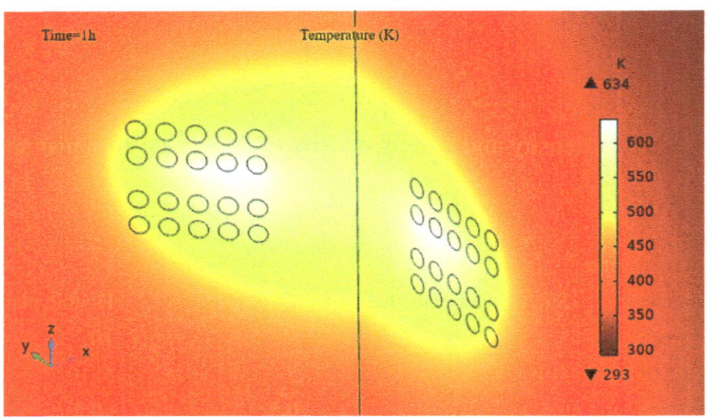

(c) Double-layer planar spiral coil

Fig. 3 Resonant coil temperature profile

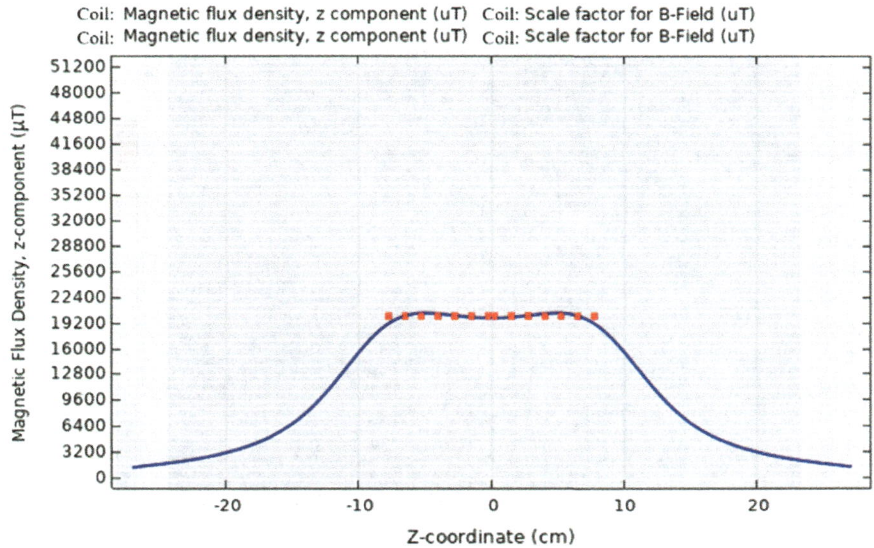

Fig. 4 Magnetic field optimization results

coil, while the temperature rise of the transmitting coil side of the cylinder spiral coil is significantly higher than that of the receiving coil side.

4 Coil Parameter Optimizations

From the above simulation effect diagram, the induced magnetic field of the cylindrical spiral coil is not uniformly distributed, so the uniform distribution of the magnetic field is achieved by optimizing the coil spacing. The Z component of the magnetic field intensity within the range of ±8 cm at the center line of the magnetic coupling coil is controlled to 20,000 μT to form a uniform magnetic field as shown in Fig. 4. The coil distribution after optimization is shown in Fig. 5. The optimization of the coil spacing here is helpful for the stable operation of wireless power transmission and the even distribution of coil heat, thereby increasing the service life of the magnetic coupling coil.

5 Conclusion

In this paper, the temperature rise effect of the commonly used planar and cylindrical spiral coils is analyzed and studied theoretically. By comparing the magnetic field distribution, induction intensity and space heat distribution of single-layer spiral coil,

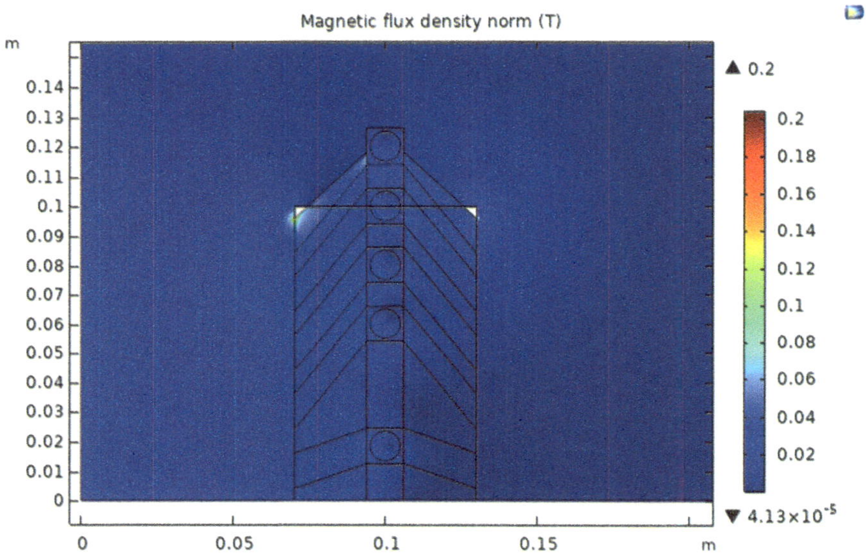

Fig. 5 Coil position after optimization

double-layer plane spiral coil and cylinder spiral coil, it is found that the temperature rise of cylinder spiral coil is the minimum under the same working condition, and its magnetic field intensity value is consistent with that of single-layer spiral coil. At the same time, aiming at the problem of uneven distribution of magnetic field in the cylindrical spiral coil, the coil spacing is optimized to realize the uniform distribution of magnetic field in the center of the cylinder, which is conducive to the heating uniformity of the coil and prolonging the working life of the magnetic coupling coil.

Acknowledgements This work was supported in part by the China National Science and Technology Support Program under Grant (51807197) and Funds for Creative Research Groups of Hubei under Grant (2018CFA008).

References

1. Zhengming, Zhao, Zhang Yiming, and Chen Kainan. 2013. New progress of magnetically coupled resonant radio energy transmission technology. *Chinese Journal of Electrical Engineering* 03: 26–38.
2. Xueliang, Huang, Tan Linlin, and Chen Zhong. 2013. Research and Application of radio energy Transmission Technology. *Journal of Electrical Technology* 28 (10): 1–11.
3. Garnica, J., R.A. Chinga, J. Lin. 2013. Wireless power transmission: From far field to near field. *Proceedings of the IEEE* 101 (6): 1321–1331. LNCS Homepage, http://www.springer.com/lncs. Last accessed 2016/11/21.

4. Jin, K., and W. Zhou. 2019. Wireless laser power transmission: A review of recent progress. *IEEE Transactions on Power Electronics* 34 (4): 3842–3859.
5. Xian, Zhang, Yang Qingxin, CUI Yulong. 2013 Design, optimization and validation of energy transmission coils in high-power radio energy transmission system. *Journal of Electrical Technology* 028 (010): 12–18.
6. Rozman, M., A. Ikpehai, and B. Adebisi. 2019. Smart wireless power transmission system for autonomous EV charging. *IEEE Access* 7: 112240–112248.
7. Hongyuan, Zhang. 2019. Research on coil type of magnetic coupling mechanism and topology interoperability of compensation network for electric vehicle wireless charging system.
8. Jian, Zhang. 2015. Design and efficiency optimization of wireless charging system for intermediate frequency magnetic resonance electric vehicles.
9. Jinbo, Zhao. 2016. Study on anti-offset and relay relay method of piece-based dynamic wireless charging.
10. Ruikun, Ma, Li Yong, and He Zhengyou. 2016. Research progress of Radio Energy Transmission technology and its application in rail transit. *Journal of Southwest Jiaotong University* 051 (003): 446–461.
11. Takatoshi, Gonda, Suguru. 2018. Effect of air space in waterproof sealed case containing transmitter and receiver of wireless power transfer in sea water. *IEEJ Transactions on Electronics Information & Systems*.
12. Zhao, D., E.J. Ding, and H. Xue. 2013. Multiple-input single-output wireless power transmission system for coal mine application. *Applied Mechanics and Materials* 462–463: 900–904.
13. Zhongqi, Li, Huang Shoudao, Yuan Xiaofang. 2017. Efficiency optimization of non-coaxial time-coupled resonant radio energy transmission system. *Journal of Electrical Technology* 2017 (8).
14. Chao, Hu. 2015. Study on energy efficiency characteristics and optimization methods of electromagnetic coupling mechanism for wireless power supply of electric vehicles.
15. Zhiyong, Zhao, Yang Zhongping, and Lin Fei. 2017. Coil optimization design of tram radio energy transmission system based on parking error law. *Chinese Journal of Electrical Engineering* S1: 196–203.
16. Zaheer, A., H. Hao, and G.A. Covic. 2015. Investigation of multiple decoupled coil primary pad topologies in lumped IPT systems for interoperable electric vehicle charging. *Power Electronics IEEE Transactions on* 30 (4): 1937–1955.
17. Elliott, G.A.J., S. Raabe, and G.A. Covic. 2010. Multiphase pickups for large lateral tolerance contactless power-transfer systems. *IEEE Transactions on Industrial Electronics* 57 (5): 1590–1598.
18. Budhia, M., Boys, J.T., Covic, G.A. 2013. Development of a single-sided flux magnetic coupler for electric vehicle IPT charging systems. *IEEE Transactions on Industrial Electronics* 60 (1): 318–328.
19. Choi, S.Y., B.W. Gu, and S.W. Lee. 2014. Generalized active EMF cancel methods for wireless electric vehicles. *IEEE Transactions on Power Electronics* 29 (11): 5770–5783.
20. Moon, S., B. Kim, and S. Cho. 2014. Analysis and design of a wireless power transfer system with an intermediate coil for high efficiency. *IEEE Transactions on Industrial Electronics* 61 (11): 5861–5870.
21. Kurs, André, A. Karalis, R. Moffatt. Wireless power transfer via strongly coupled magnetic resonances. *Science* 317 (5834): 83–86.
22. Chao, Zhang. 2010. Simulation of resonance coupled radio energy transmission system resonator 2010: 8–12 (in Chinese).

Effect of Sintering Temperature and Holding Time on Ionic Conductivity for $Li_{6.4}La_3Zr_{1.4}Ta_{0.6}O_{12}$ Electrolyte

Jie Gao, Weichang Guo, Hengrui Yang, Fei Shen, and Xiaogang Han

Abstract Garnet-type $Li_{6.4}La_3Zr_{1.4}Ta_{0.6}O_{12}$ (LLZTO) is considered as one of the most potential electrolytes for solid state lithium batteries. Sintering process is a critical step in the preparation of LLZTO. Herein, LLZTO solid-state electrolytes were prepared by high temperature solid-phase reaction with different sintering temperature from 1150 to 1250 °C and different sintering holding time from 1 min to 60 min. According to EIS, electrolytes with high ionic conductivity up to 8.1×10^{-4} S cm^{-1} were obtained by sintering at 1200 °C for 20 min. Besides, LLZTO pellets prepared in this experimental condition show high density reaching up to 97.6%. Li symmetric cell with the LLZTO sintered at 1200 °C for 20 min also exhibits stable cycle performance under current density of 0.25 mA cm^{-2} at room temperature.

Keywords $Li_{6.4}La_3Zr_{1.4}Ta_{0.6}O_{12}$ electrolyte · Sintering temperature · Sintering holding time · Ionic conductivity

1 Introduction

Lithium-ion batteries (LIBs) have been broadly applied to the fields of electronic devices, electric vehicles and large-scale energy storage [1]. However, conventional LIBs with flammable organic liquid electrolytes are prone to accidents such as combustion and even explosion [2–4]. To eliminate security hazards, inorganic solid-state electrolytes (SSEs) have been proposed to replace organic liquid electrolytes. SSEs are non-flammable and have higher mechanical strength to block the formation of lithium dendrites [5–8], leading to intrinsic safety of solid-state batteries (SSBs).

J. Gao · W. Guo · H. Yang · F. Shen · X. Han (✉)
State Key Laboratory of Electrical Insulation and Power Equipment, School of Electrical Engineering, Xian Jiaotong University, Xian, Shaanxi 710049, China
e-mail: xiaogang.han@xjtu.edu.cn

F. Shen
Xian Jiaotong University Suzhou Institute, Suzhou, Jiangsu 215123, China

X. Han
Key Laboratory of Smart Grid of Shaanxi Province, Xian, Shaanxi 710049, China

W. Chen et al. (eds.), *The Proceedings of the 9th Frontier Academic Forum of Electrical Engineering*, Lecture Notes in Electrical Engineering 743,
https://doi.org/10.1007/978-981-33-6609-1_38

Moreover, SSEs make it possible to use Li metal anode (6.49 g mol^{-1}, 3860 mAh g^{-1}, −3.040 V versus. standard hydrogen electrode) and high voltage cathode (up to 5 V), which allows for far greater energy density than commercial LIBs. Thus, it is generally recognized that traditional LIBs will be replaced by SSBs with inorganic solid-state electrolytes gradually.

As the core component of SSBs, solid electrolyte can be mainly categorized as sulphides and oxides [9, 10]. Sulphide-type SSEs possess high ionic conductivity ($\geq 10^{-3}$ S cm^{-1}), but poor compatibility against electrodes and high sensitivity to air, which hinders their application [11]. By contrast, oxide electrolytes show excellent stability towards Li anode and air, especially garnet-type $Li_{6.4}La_3Zr_{1.4}Ta_{0.6}O_{12}$(LLZTO), which exhibits comparable conductivity with organic liquid electrolytes [12]. Sintering process is considered as the key step during the preparation of LLZTO, drastically affecting the samples' crystal structure, density and conductivity. In this work, effects of different sintering temperature and sintering holding time on performance of LLZTO, especially the ionic conductivity, were studied. It was demonstrated that the optimal condition to prepare LLZTO electrolyte was sintering at 1200 °C for 20 min, and the obtained LLZTO pellets show high conductivity up to 8.1×10^{-4} S cm^{-1} as well as high density up to 97.6%. When sintered at higher temperature (1250 °C) or lower temperature (1150 °C) for from 1 min to 60 min, the conductivity of LLZTO pellets decreased in various degree.

2 Experimental Section

The preparation process of LLZTO pellets is illustrated in Fig. 1. LiOH (99.99%), La_2O_3 (99.99%), ZrO_2 (99.99%) and Ta_2O_5 (99.99%) were weighed according to mass based on the stoichiometric proportion. Among them, 15% wt excess of LiOH was added to compensate for the lithium volatilization loss during high-temperature sintering. These materials were mixed thoroughly by ball-milling with isopropanol as a dispersant at 300 rpm for 12 h and then dried and pressed into pellets. These pellets were calcined at 800 °C for 10 h in air and then triturated for another ball-milling. After dried in vacuum oven overnight, ultra-fine powder was obtained. The powder was pressed into green pellets with diameter of 12.5 mm under 700 MPa and sintered in different conditions to prepare LLZTO pellets.

The crystal structure of the sintered LLZTO powder obtained by grinding final LLZTO pellets was tested by X-ray diffraction (XRD, Bruker D2 Phaser) using Cu Kα radiation with 2θ in the range of 10–80° and a step size of 0.04°. The microstructure of pellet samples was observed by scanning electron microscope (SEM, Phenom ProX). The relative density of the LLZTO pellets was measured by Archimedes drainage. Electrochemical impedance spectroscopy (EIS, BioLogic SP-300) measurements were used to measure the ion conductivity indirectly with the frequency from 7 M to 1 Hz. Before EIS testing, a thin layer of gold was sputtered onto both sides of LLZTO pellets. The ionic conductivity was calculated by the formula $\sigma = L/(S*R)$, where L, S and R stand for thickness, area and resistance, respectively. Li symmetric

Fig. 1 Preparation process of LLZTO electrolyte

cells were assembled in an Ar-filled glovebox. LLZTO pellet was rubbed on molten Li at 230 °C till Li wetted the pellet's surface, and the flank of pellet was polished in case of short. Then the LLZTO pellet with Li was sandwiched between two pieces of stainless steel sheets and was sealed in Swagelok-type cell mold. The cycling performance of Li symmetric cell was carried out by a LAND battery tester.

3 Results and Discussion

LLZTO green pellets were sintered at 1150, 1200 and 1250 °C, and different holding time from 1, 20 and 60 min was used at each temperature. The XRD patterns of LLZTO after sintering in different conditions are shown in Fig. 2. It can be seen that the characteristic peaks of the nine samples all match well with the standard PDF card (# 80-0457). No other miscellaneous peaks are observed, which indicates that the crystalline phases of the samples all belong to garnet-type cubic phase. In other words, these nine combinations of temperature and time are all feasible to obtain LLZTO pellets with standard crystal phases.

Ionic conductivity is studied by the electrochemistry impedance spectroscopies (EIS). The EIS plots of LLZTO sintered in different temperature and time are presented in Fig. 3, in which the points of x-intercepts correspond to the resistance of LLZTO electrolytes, including both bulk and grain boundary [12]. According to the formula mentioned above, smaller impedance means higher conductivity. As presented in the EIS plots (Fig. 3), pellets sintered at 1200 °C exhibit the lowest impedance among three temperatures. By sintering at 1200 °C for 20 min, the highest

Fig. 2 XRD patterns of LLZTO sintered in different combinations of temperature and time

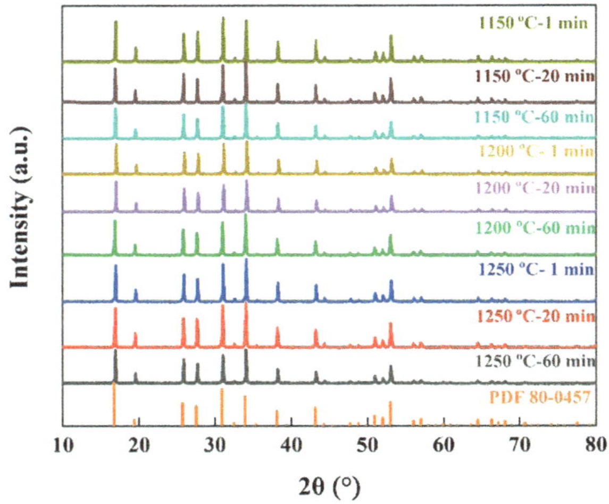

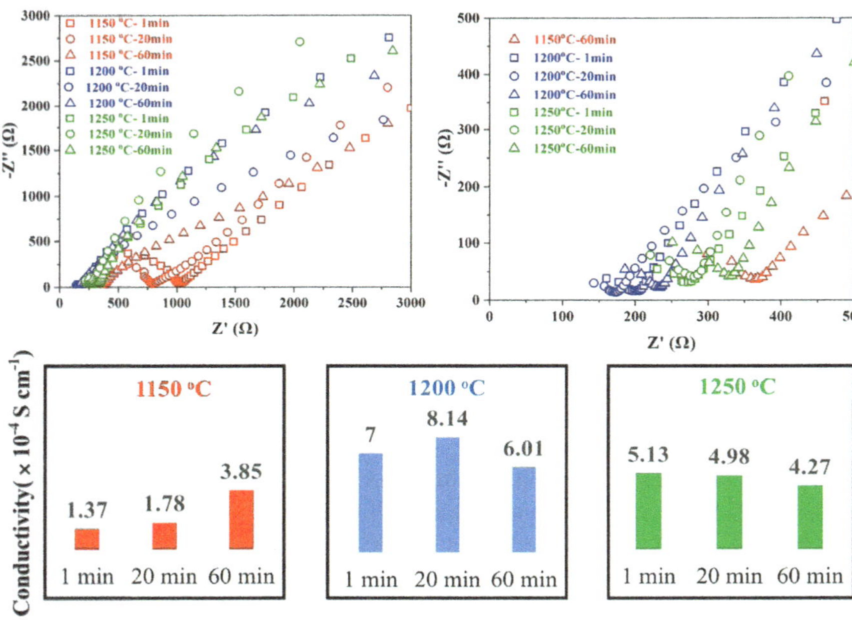

Fig. 3 EIS plots and conductivity bar graphs of LLZTO pellets sintered in different combinations of temperature and time

ionic conductivity up to 8.1×10^{-4} S cm^{-1} can be obtained. When the sintering holding time is extended to 60 min, the LLZTO pellet shows a lower conductivity (6.01×10^{-4} S cm^{-1}). Reducing time to 1 min also leads to a slight decrease in conductivity (7×10^{-4} S cm^{-1}). As shown in the bar graphs (Fig. 3), the conductivity of LLZTO pellets decreases gradually with the increase of time when the sintering temperature is 1250 °C. By contrast, higher conductivity of LLZTO pellets is obtained with the longer sintering holding time at 1150 °C.

The conductivity of LLZTO pellet is closely related to its microstructure [13, 14], so it is essential to study the microstructure evolution of LLZTO sintered in different conditions. Figures 4 and 5 show the cross-sectional SEM images of LLZTO pellets. As we can see in Fig. 4, the crystal grain size increases as the sintering holding time are raised at 1200 °C. Enough growth of grains is helpful for the connection at grain boundaries, leading to higher conductivity. However, longer sintering time at high temperature may give rise to lithium volatilization loss, which, in turn, causes a decrease of conductivity [15]. As the experimental results prove, 20 min is an optimal sintering holding time to obtain high conductivity when the sintering temperature is 1200 °C. Figure 5 shows morphological change of LLZTO when sintered for 20 min at different temperature. Smaller grain size and some pores can be observed in the SEM image of pellet sintered at 1150 °C for 20 min, which is in accord with the lower

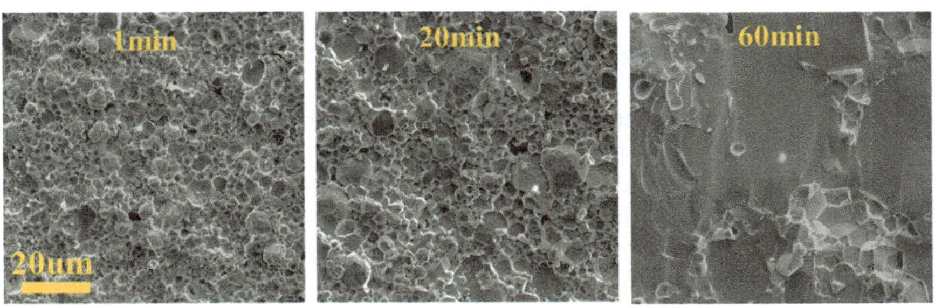

Fig. 4 Cross-sectional SEM images of LLZTO pellets sintered at 1200 °C for 1, 20 and 60 min

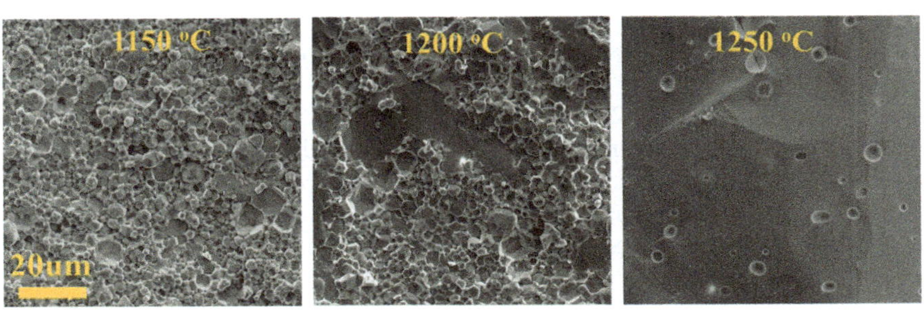

Fig. 5 Cross-sectional SEM images of LLZTO pellets sintered for 20 min at 1150, 1200 and 1250 °C

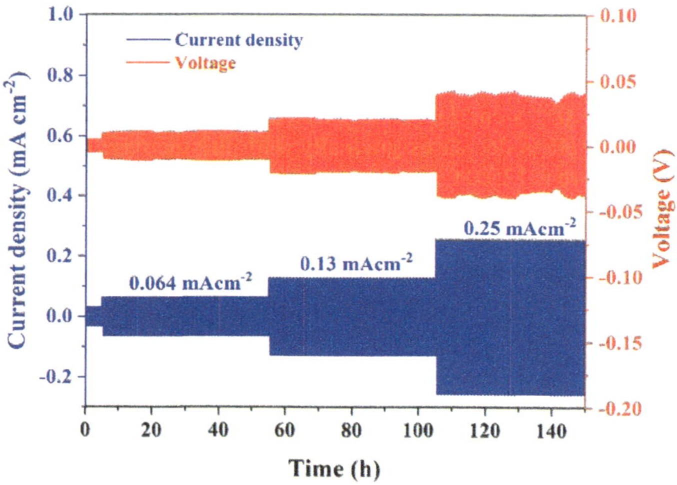

Fig. 6 Cycling performance of Li symmetric cell with LLZTO electrolyte sintered at 1200 °C for 20 min

conductivity. Although larger grain can be obtained by sintering at 1250 °C, severe lithium volatilization may become the dominant restrictive factor for conductivity. Thus, the combination of 1200 °C and 20 min is the best for improving conductivity of LLZTO electrolyte.

The LLZTO pellet sintered at 1200 °C for 20 min was used to assemble Li symmetric cell for cycling performance testing. The cell underwent a continuous stripping/plating cycling and each cycle lasted for 1 h. As shown in Fig. 6, the cell with LLZTO pellet performs stably, achieving at least 50 cycles at each current density (0.064, 0.13 and 0.25 mA cm^{-2}).

4 Conclusion

$Li_{6.4}La_3Zr_{1.4}Ta_{0.6}O_{12}$ solid-state electrolyte pellets were prepared by solid phase reaction. Samples sintered at different temperature (1150, 1200, 1250 °C) for different holding time (1, 20, 60 min) all show typical garnet-type cubic phase according to the XRD patterns. Through comparison, the combination of 1200 °C and 20 min is the optimal condition, under which the pellet exhibits high conductivity up to 8.1×10^{-4} S cm^{-1} with density of 97.6%. Li symmetric cell with this electrolyte also shows stable cycling under 0.25 mA cm^{-2}.

References

1. Doughty, D.H., P.C. Butler, A. Akhil, et al. 2010. Batteries for large-scale stationary electrical energy storage. *Electrochemical Society Interface* 19: 49–53.
2. Samson, A.J., K. Hofstetter, S. Bag, et al. 2019. A bird's-eye view of Li-stuffed garnet-type $Li_7La_3Zr_2O_{12}$ ceramic electrolytes for advanced all-solid-state Li batteries. *Energy & Environmental Science* 12 (10): 2957–2975.
3. Liu, Q., Z. Geng, C. Han, et al. 2018. Challenges and perspectives of garnet solid electrolytes for all solid-state lithium batteries. *Journal of Power Sources* 389: 120–134.
4. Zhang, B., R. Tan, L. Yang, et al. 2018. Mechanisms and properties of ion-transport in inorganic solid electrolytes. *Energy Storage Materials* 10: 139–159.
5. Ferraresi, G., M. El Kazzi, L. Czornomaz, et al. 2018. Electrochemical performance of all-solid-state li-ion batteries based on garnet electrolyte using silicon as a model electrode. *ACS Energy Letters* 3 (4): 1006–1012.
6. Han, F., J. Yue, C. Chen, et al. 2018. Interphase engineering enabled all-ceramic lithium battery. *Joule* 2 (3): 497–508.
7. Bachman, J.C., S. Muy, A. Grimaud, et al. 2016. Inorganic solid-state electrolytes for lithium batteries: mechanisms and properties governing ion conduction. *Chemical Reviews* 116 (1): 140–162.
8. Manthiram, A., X. Yu, and S. Wang. 2017. Lithium battery chemistries enabled by solid-state electrolytes. *Nature Reviews Materials* 2 (4): 16103.
9. Takada, K. 2018. Progress in solid electrolytes toward realizing solid-state lithium batteries. *Journal of Power Sources* 394: 74–85.
10. Shoji, M., E.J. Cheng, T. Kimura, et al. 2019. Recent progress for all solid state battery using sulfide and oxide solid electrolytes. *Journal of Physics. D. Applied Physics* 52 (10): 103001.
11. Kato, Y., S. Hori, T. Saito, et al. 2016. High-power all-solid-state batteries using sulfide superionic conductors. *Nature Energy* 1 (4): 16030.
12. Shen, F., W. Guo, D. Zeng, et al. 2020. A simple and highly efficient method toward high-density garnet-type LLZTO solid-state electrolyte. *ACS Applied Materials and Interfaces*.
13. Xue, W., Y. Yang, Q. Yang, et al. 2018. The effect of sintering process on lithium ionic conductivity of $Li_{6.4}Al_{0.2}La_3Zr_2O_{12}$ garnet produced by solid-state synthesis. *RSC Advances* 8 (24): 13083–13088.
14. Gong, Y., Z.-G. Liu, Y.-J. Jin, et al. 2019. Effect of sintering process on the microstructure and ionic conductivity of $Li_{7-x}La_3Zr_{2-x}Ta_xO_{12}$ ceramics. *Ceramics International* 45 (15): 18439–18444.
15. Huang, M., A. Dumon, Y. Shen, et al. 2013. Effect of Lithium content and sintering time on conductivity of $Li_7La_3Zr_2O_{12}$ as solid electrolyte. *Journal of the Chinese Ceramic Society* 41.

Research on Vibration and Noise Reduction of Motor Based on Negative Magnetostrictive Effect

Xin Zhang, Zihan Song, Wenbin Wang, and Yu Han

Abstract Noise suppression is a particularly important part in the design and application of motors. In some large motors, the magnetostrictive force makes a great contribution to vibration and noise. In response to this problem, this project proposes a method of drilling holes in the motor stator and filling negative magnetostrictive materials to reduce the vibration and noise of the motor, so that the negative magnetostrictive effect of the material to be filled and the positive magnetostrictive effect of the silicon steel sheet are combined. The resulting deformation is offset, thereby reducing the vibration and noise of the motor. This paper first establishes the electromagnetic-mechanical coupling numerical model of the motor, carries out finite element simulation calculation, and analyzes the vibration of the stator silicon steel sheet, finds the suitable filling position and size of the material to be filled, and makes the model for verification. In this paper, the negative magnetostrictive material nickel is used for simulation and actual verification experiments.

Keywords Magnetostrictive · Vibration and noise · Stator drilling · Finite element calculation

X. Zhang (✉) · Z. Song · W. Wang · Y. Han
Tianjin Key Laboratory of New Technology of Electrical and Electric Energy (Tiangong University), Tianjin 300387, China
e-mail: zhangxin@tiangong.edu.cn

Z. Song
e-mail: 1831045291@tiangong.edu.cn

W. Wang
e-mail: 1831045316@tiangong.edu.cn

Y. Han
e-mail: 1831045304@tiangong.edu.cn

W. Chen et al. (eds.), *The Proceedings of the 9th Frontier Academic Forum of Electrical Engineering*, Lecture Notes in Electrical Engineering 743,
https://doi.org/10.1007/978-981-33-6609-1_39

1 Introduction

Noise suppression is one of the urgent problems to be solved. In modern industry, motor has been widely used in various fields of national economy, and thus has higher and higher requirements on the performance of motor in all aspects.

The research on vibration and noise of motor originated in 1930s, and magnetostrictive effect was discovered by British scholars in 1842. In 2004, In reference [1]. American scholar Mohamed et al. Measured the effect of reverse magnetostriction on the magnetization characteristics of electrical steel, and calculated the magnetostrictive force using the principle of virtual work. It is shown in reference [2] that the calculated vibration of synchronous generator stator is different, depending on the vibration frequency considered. Due to magneto elastic coupling, the increase or decrease of vibration amplitude is about 20%. In reference [3], the magnetostrictive anisotropy and stress dependence of silicon steel sheets under alternating magnetization and DC bias were measured and analyzed. A magnetic finite element analysis method using magnetostrictive curves under DC bias was proposed.

In reference [4, 5], the magnetostrictive characteristics of electrical silicon steel sheet and the vibration noise of motor under different current are measured and analyzed. It is found that the test results are closer to the actual value when magnetostriction effect is considered. In reference [6], the contribution of magnetostrictive force and Maxwell force to vibration and noise of motor is compared through experiments. It is found that although the contribution of magnetostrictive force is less than that of Maxwell force, the research on magnetostrictive force is also very important. In reference [7], the magnetostrictive effect of non oriented silicon steel sheet in alternating magnetic field is analyzed, and the vibration of permanent magnet synchronous motor is tested experimentally. It is found that the vibration noise of the motor with magnetostriction effect is greater than that without magnetostriction effect.

The influence of magnetostrictive effect on stress of silicon steel sheet with or without insulating layer and the relationship between stress and strain in rolling direction and vertical direction of yoke and middle pillar of transformer core silicon steel sheet were measured in reference [8]. In reference [9], through the stator vibration response test of different types of motors, it was found that the main vibration frequency of the motor stator silicon steel sheet was twice and twice the power supply frequency, and the main vibration frequency was twice the power supply frequency. In reference [10], a magnetoelastic coupling model including magnetostriction was established to analyze the vibration of laminated core. Then, the numerical model was applied to measure the magnetization and MS curve. It was found that the deformation in the stator was not only caused by the magnetic resistance, but also by the magnetostrictive effect and harmonic wave, which was very important for the design of motors with lower vibration and noise make sense. In reference [11], the influence of electromagnetic force and magnetostrictive force on motor stator is considered, and the deformation of motor stator is expressed by equal force. It is found that electromagnetic force and magnetostrictive force will enhance and cancel each other under certain conditions. In reference [12], the relationship between stator vibration and magnetostriction was analyzed by establishing an analytical model of the motor stator. It was found that the vibration on the stator caused by magnetostrictive effect was directly proportional to the magnetostrictive coefficient of silicon steel sheet. References [13, 14] measured the vibration acceleration and stress distribution of the motor under different harmonic contents, and analyzed the influence of different harmonics on the vibration and noise of the motor.

In this paper, firstly, the electromagnetic mechanical coupling numerical model of the motor stator is established, and the simulation experiment of the negative magnetostrictive material filled with holes is carried out, and the stress change of the stator surface before and after the filling material is studied, and then the simulation is verified by experiments.

2 Mathematical Model

In order to test the influence of negative magnetostrictive material on the vibration and noise of the motor stator, the finite element simulation software COMSOL multiphysics is selected to establish the two-dimensional model, and the simulation experiment of the motor stator is carried out. The magnetic mechanical coupling numerical model can be expressed by the following formula:

$$[S][A] = [J_e] \tag{1}$$

$$[K][U] = [F] \tag{2}$$

where $\boldsymbol{K}$ is the mechanical stiffness matrix, $\boldsymbol{S}$ is the electromagnetic stiffness matrix, $\boldsymbol{J_e}$ is the Jacobian matrix of external current density, and $\boldsymbol{F}$ is the force. The vibration displacement $\boldsymbol{U}$ and vector magnetic potential $\boldsymbol{A}$ are obtained by solving.

In the finite element method, the solution equation of solid mechanics module is as follows:

$$\rho \frac{\partial^2 u}{\partial t^2} - \nabla \cdot \sigma = F \tag{3}$$

$$T - T_i = C(S - S_i) \tag{4}$$

where $\boldsymbol{\rho}$ is the density; $\boldsymbol{\mu}$ is the displacement vector; $\boldsymbol{F}$ is the load amount; $\boldsymbol{T_i}$ is the initial value of the given stress; $\boldsymbol{C}$ is the stiffness matrix; $\boldsymbol{S}$ is the strain; $\boldsymbol{S_i}$ is the initial value of the initial strain.

For the silicon steel sheet of motor stator, the stress formula of magnetostrictive surface is as follows.

The research equation of linear elastic material in solid mechanics is as follows:

$$\rho \frac{\partial^2 u}{\partial t^2} = \nabla \cdot (FS)^T + Fv \tag{5}$$

$$F = I + \nabla u \tag{6}$$

where $\boldsymbol{u}$ is the displacement vector, $\boldsymbol{F}$ is the load, $\boldsymbol{\rho}$ is the density, $\boldsymbol{S}$ is the strain, and $\boldsymbol{v}$ is the Poisson's ratio.

3 Simulation Results and Analysis

In this paper, nickel is used as the filling material, and based on the above mathematical model, multiple sets of finite element simulations are performed on the stator model of the permanent magnet synchronous motor, including the surface stress of the motor stator before the hole is filled with negative magnetostrictive material, and the motor The surface stress of the stator after punching and filling the negative magnetostrictive material. Then select several positions on the motor stator to compare the stress before and after the filling material. The selected measurement points have different distances from the filling position.

3.1 Initial Condition

The motor stator model is constructed in the finite element simulation software COMSOL Multiphysics. In order to reduce the influence of the rotor magnetic field on the stator, the motor rotor is removed in this experiment, which can reduce the magnetic flux leakage in the air gap, so that the magnetic flux is mainly concentrated on the stator, thereby reducing The electromagnetic force on the surface of the stator teeth of the small motor makes the main stress of the stator core of the motor magnetostrictive force. A three-phase alternating current is applied to the stator winding of the motor, the frequency is set to 50 Hz, various parameters of the magnetic field and solid mechanics field are set, and simulation experiments are carried out. The properties of nickel added in the experiment are shown in Table 1

3.2 Simulation Result

The left side of Fig. 1 shows the local stress distribution of stator before adding material, and the right side of Fig. 1 shows the local stress distribution of stator after

Table 1 Property value of Ni

Parameters	Value
Density kg/m^3	8902
Conductivity Ms/m	2088
Poisson's ratio nu	0.31
Young's modulus E/Pa	206×10^9
Relative permittivity ε	1
Saturation magnetization Ms/A m^{-1}	440,000
Saturation magnetostriction coefficient λs	-36×10^{-6}

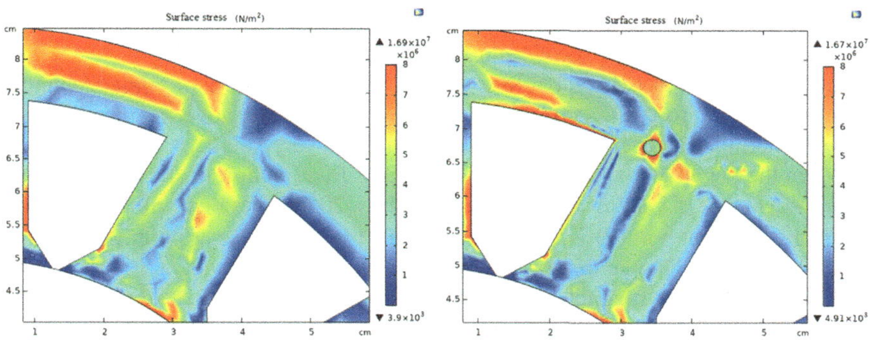

Fig. 1 Comparison of local stress before and after adding materials

adding material. Although the maximum surface stress is reduced from 1.69×10^7 to 1.67×10^7 N/m^2, the decrease is very small. However, the local red area is obviously reduced after the negative magnetostrictive material is filled.

Three measuring points A, B and C are selected. Point A is located in the area near the hole, while point B and point C are far away from the drilling position. From Table 2, we can see the stress change of the measuring point after adding negative magnetostrictive material into the hole.

Hole 3 is one of the advantages of filling negative magnetostrictive materials in this simulation. Hole 1 and hole 2 were randomly located (Fig. 2).

When the negative magnetostrictive material is filled in hole 3, the stress of point a decreases from 1×10^6 to 0.38×10^6 N/m^2 which decreases by 62%; the stress of

Table 2 Three point stress comparison

Parameter	A	B	C
Original stress N/m^2	1×10^6	1.03×10^6	0.85×10^6
Hole 1 add Ni N/m^2	0.8×10^6	1.04×10^6	0.85×10^6
Hole 2 add Ni N/m^2	1.18×10^6	1×10^6	0.85×10^6
Hole 3 add Ni N/m^2	0.38×10^6	1.02×10^6	0.83×10^6

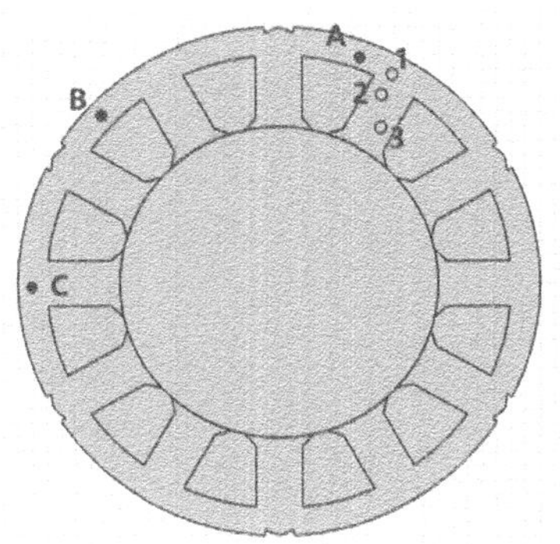

Fig. 2 Measuring point

point B changes from 1.03 × 10^6 to 1.02 × 10^6 N/m^2; the stress of point C changes from 0.85 × 10^6 to 0.83 × 10^6 N/m^2 after filling with negative magnetostrictive material, and the stress changes of point B and C are very small.

When the negative magnetostrictive material is filled in hole 1, the stress at point a becomes 0.8 × 10^6 N/m^2, which is reduced by 20% compared with the original stress; the stress at point B is 1.04 × 10^6 N/m^2, and the stress at point C is 0.85 × 10^6 N/m^2, and the change of stress in point B and C is very small.

When the negative magnetostrictive material is filled in hole 2, the stress at point a becomes 1.18 × 10^6 N/m^2, which is increased by 18% compared with the original stress; the stress at point B is 1 × 10^6 N/m^2, and that at point C is 0.85 × 10^6 N/m^2; the stress at point B and C does not change much before and after the negative magnetostrictive material is filled in hole 2.

From the above three groups of comparison, it can be found that after the negative magnetostrictive material is punched and filled, the stress and strain near the filling point are greatly affected, but with the increase of the distance, the influence on the longer distance is less; and only when the hole is drilled and filled with the negative magnetostrictive material in the appropriate position can the stress be reduced. However, the stress will increase when the negative magnetostrictive material is filled in the inappropriate position.

4 Experimental Results and Analysis

According to the simulation results, the experimental scheme was made. The stator surface stress of the motor without filling material and with negative magnetostrictive material was measured respectively, and several points were selected to compare the stress changes before and after the filling material.

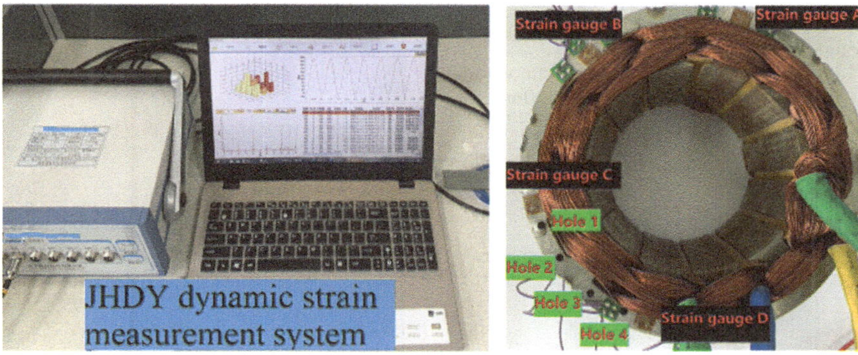

Fig. 3 Experimental equipment

4.1 Experimental Instrument and Preparation

Paste the strain gauges at 4 positions of the motor and name them A, B, C, D. The left side of strain gauge A has a small hole with a diameter of 1.2 mm, there is no hole around B, the right side of C is a hole with a diameter of 4 mm, and the left side is D. There are 4 mm holes on both sides. The 4 holes are numbered 1, 2, 3, and 4. Hole 1 is one of the suitable positions for adding negative magnetostrictive material (Fig. 3).

The frequency converter is used to generate the alternating rotating magnetic field by passing 50 Hz power frequency electricity into the motor winding. The strain of the strain gauge sticking position on the motor stator is measured by JHDY dynamic strain measurement system, and the data are saved in the computer.

4.2 Experimental Result

Tables 3 and 4 show the strain comparison of strain gauges B and C before and after the negative magnetostrictive material is filled in hole 1. After adding Ni into hole 1, the peak value of strain gauge B changes from 165.913 to 163.217 $\mu\varepsilon$, the variation is 1.6%, and the peak value of strain gauge C changes from 148.424 to 124.110 $\mu\varepsilon$, the variation is 16.3%. It can be seen that the distance between hole 1 and strain

Table 3 Strain contrast of strain gauge B

Parameter	Peak value ($\mu\varepsilon$)	Valley value ($\mu\varepsilon$)
Strain without Ni addition	165.913	−111.283
Strain of hole 1 after adding Ni	163.217	−108.878

Table 4 Strain contrast of strain gauge C

Parameter	Peak value (με)	Valley value (με)
Strain without Ni addition	148.424	−91.741
Strain of hole 1 after adding nickel	124.110	−96.699

Table 5 Strain gauge B, C and D after adding Ni to hole 1

Parameter	Peak value (με)	Valley value (με)
B	163.217	−108.878
C	124.110	−96.699
D	179.540	−96.341

gauge B is far, so the change is very small, while the strain gauge C is close to hole 1, so the change is large, which is the same as the simulation law.

Table 5 shows the strain of strain gauge D after adding nickel into hole 1, the strain of strain gauges B, C and D at hole 1 filled with negative magnetostrictive material is compared. It is found that the strain peak value of strain gauge C is 124.110 με, which is far less than 163.217 με of strain gauge B and 179.540 με of D. Therefore, it is concluded that when the negative magnetostrictive material is filled in the appropriate position of the motor stator, the strain near the filling material can be reduced, thus reducing the vibration noise of the motor.

5 Conclusion

In this paper, the electromagnetic-mechanical coupling numerical model is used to calculate and analyze the stator surface stress of the permanent magnet motor, and the appropriate drilling scheme is formulated through traversal scanning, and the experiment is verified on the actual motor. It is concluded that perforating holes and filling negative magnetostrictive materials at suitable positions of the motor stator can reduce the surface stress of the motor stator, thereby achieving the purpose of vibration reduction and noise reduction; perforating the motor stator and filling the negative magnetostrictive materials After the material is stretched, the influence on the surface stress of the motor is related to the distance of the punching position of the measuring point. With the increase of the distance between the measuring point and the filling position, the influence of the negative magnetostrictive material filled with holes decreases gradually, and the closer the distance, the greater the influence. This conclusion has important reference value for the design of low noise permanent magnet synchronous motor.

Acknowledgements This article is supported by the National Natural Science Foundation of China (51577131) "Research on Methods of Suppressing Motor Vibration and Noise Based on Negative Magnetostrictive Effect".

References

1. Mohammed, O.A., S. Liu, and N. Abed. 2004. Study of the inverse magnetostriction effect on machine deformation. In *SoutheastCon, 2004. Proceedings. IEEE*. IEEE.
2. Belahcen, A. 2005. Magnetoelastic coupling in rotating electrical machines. *IEEE Transactions on Magnetics* 41 (5): 1624–1627.
3. Zhang, Y., J. Wang, X. Sun, et al. 2014. Measurement and modeling of anisotropic magnetostriction characteristic of grain-oriented silicon steel sheet under DC bias. *IEEE Transactions on Magnetics* 50 (2): 1–1.
4. Xueyan, Han, Wu Zhang Zhe, and Chen Jian Shengnan. 2015. Study on vibration and noise of permanent magnet motor considering magnetostrictive effect. *New Electrical Energy Technology* 34 (01): 28–34. (in Chinese).
5. Xueyan, Han, Wu Zhang Zhe, and Chen Jian Shengnan. 2015. Influence of magnetostriction on vibration and noise of inverter fed permanent magnet motor. *Journal of motor and control* 19 (04): 1–6. (in Chinese).
6. Hao Qingliang, Hu, and Zhu Shaolin Yijun. 2011. Contribution analysis of magnetostriction in electromagnetic vibration of electric machines. *Application of Motor and Control* 38 (10): 31–35. (in Chinese).
7. Xu, Feng. 2014. Vibration research of variable frequency permanent magnet motor considering magnetostriction effect. *Hebei University of Technology* (in Chinese).
8. Lihua, Zhu. 2013. Research on the influence of magnetostrictive effect of laminated core on vibration and noise of transformer and AC motor. *Hebei University of Technology* (in Chinese).
9. Xin, Zhang, Xie Chaoqun, Zhu Lihua, et al. 2017. Numerical simulation and experiment of motor stress considering magnetostriction effect. *Acta Electrotechnics* 32 (S2): 50–55. (in Chinese).
10. Zhu, L., Q. Yang, R. Yan, et al. 2014. Magnetoelastic Numerical analysis of permanent magnet synchronous motor including magnetostriction effects and harmonics. *IEEE Transactions on Applied Superconductivity* 24 (3): 1–4.
11. Ghalamestani, Setareh Gorji, Lieven Vandevelde, and Jan A. A. Melkebeek. 2016. Magnetic forces and magnetostriction in rotating electrical machines. In *XXII International Conference on Electrical Machines*. IEEE.
12. Wu Shengnan, Yu., Tong Wenming Shenbo, and Tang Renyuan. 2019. Accurate analytical model of stator core vibration of radial flux motor caused by magnetostriction. *Acta electrotechnica Sinica* 34 (02): 226–235. (in Chinese).
13. Rongge, Yan, Liu Huaiwen, Bin Tong, Zhao Luna, and Zhou Jie. 2018 Study on the influence of harmonics on the vibration of induction motor. *Motor and Control Applications* 45 (01): 77–82 (in Chinese).
14. Tong, Ben, Chen long, Yan Rongge, Zhang Yujiao, and Yang Qingxin. 2019. Electromagnetic stress analysis of induction motor core considering the anisotropy of magnetization and magnetostriction. *Acta electrotechnics Sinica* 34 (01): 66–74 (in Chinese).

Research on the Influence of the Torsional Elastic Torque on Chaotic Characters of the Permanent Magnet Synchronous Motor System

Feng Zhang, Pengfei Wang, Xin Huang, and Xuelian Bai

Abstract The influence of the torsional elastic torque on chaotic characters of the permanent magnet synchronous motor system is studied. The state equations of the permanent magnet synchronous motor system including the torsional elastic torque, based on the *d-q* axis, are given. The stability of the system is analyzed by applying the Lyapunov's theorem on stability. The changes of the system's Lyapunov exponents with the torsional stiffness are investigated. Two torsional stiffnesses are selected. The time domain diagrams of the system and the system's phase diagrams in state space are given in these two cases. The system's sensitivity to the initial conditions are analyzed in these two cases, too. The results show that the permanent magnet synchronous motor system including the torsional elastic torque can be in a chaotic or hyperchaotic state under certain parameters.

Keywords Permanent magnet synchronous motor · Torsional stiffness · Chaos · Hyperchaos

1 Introduction

Since Lorenz discovered strange attractors in a three-dimensional autonomous system in 1963, chaotic systems are widely studied [1, 2].

Because the permanent magnet synchronous motor (PMSM) has the advantages of high efficiency and high-power density, it has been widely used in industrial productions, electric vehicles, robots and other fields [3–5].

Because the PMSM system is a nonlinear and strong coupling system, it may also be in a chaotic state. If the PMSM system can be in a chaotic state under certain conditions, it may be necessary to carry out further studies on the method to suppress the chaotic behavior. Therefore, it is meaningful to discuss whether the PMSM system can be in a chaotic state. Li et al. [6] studied chaotic characters of the PMSM system with uniform air gap based on the PMSM model on *d-q* axis. Jing et al. [7] studied

F. Zhang (✉) · P. Wang · X. Huang · X. Bai
Electric Engineering Section, Army Academy of Artillery and Air Defense, No. 451, Huang Shan Road, Shu Shan District, Hefei City, Anhui Province, China
e-mail: zf2006ger@aliyun.com

W. Chen et al. (eds.), *The Proceedings of the 9th Frontier Academic Forum of Electrical Engineering*, Lecture Notes in Electrical Engineering 743,
https://doi.org/10.1007/978-981-33-6609-1_40

further the case of non-uniform air gap. Some other studies [8–14] were also carried out further on the basis of Ref. [6].

However, all of these studies [6–14] did not considered the influence of the torsional elastic torque on chaotic characters of the PMSM system. So, we study that influence in this paper.

In addition, some parameter constraints in the PMSM system were relaxed in Ref. [6]. In order to avoid this shortcoming, parameter constraints in the PMSM system are considered more strictly in this paper.

In this paper, the state equations of the PMSM system including the torsional elastic torque, based on the *d-q* axis, are given. Then, the stability of the system is analyzed by applying the Lyapunov's theorem on stability. While the torsional stiffness changes in a certain range and other parameters of the system remain unchanged, the changes of the system's Lyapunov exponents with the torsional stiffness are investigated. Then, two torsional stiffnesses are selected. The time domain diagrams of the system and the system's phase diagrams in state space are given in these two cases. The system's sensitivity to the initial conditions are analyzed in these two cases, too. The results show that the PMSM system including the torsional elastic torque can be in a chaotic or hyperchaotic state under certain parameters.

2 System Model and Stability

The state equations of the PMSM system including the torsional elastic torque, based on the *d-q* axis, can be described as follows [15, 16]:

$$\begin{cases} \frac{di_d}{dt} = -\frac{R_s}{L_d} i_d + \frac{L_q}{L_d} i_q \omega + \frac{u_d}{L_d} \\ \frac{di_q}{dt} = -\frac{L_d}{L_q} i_d \omega - \frac{R_s}{L_q} i_q - \frac{\psi_{af}}{L_q} \omega + \frac{u_q}{L_q} \\ \frac{d\omega}{dt} = \frac{n_p^2 (L_d - L_q)}{J} i_d i_q + \frac{n_p^2 \psi_{af}}{J} i_q - \frac{D}{J} \omega - \frac{K}{J} \theta - \frac{n_p}{J} T_L \\ \frac{d\theta}{dt} = \omega \end{cases} \tag{1}$$

where i_d, i_q are the direct-axis and quadrature-axis current components, respectively; ω is the rotor electric velocity; θ is the rotor electric angle; L_d, L_q are the direct-axis and quadrature-axis stator inductors, respectively; R_s is the stator winding resistance; ψ_{af} is the permanent magnet exciting flux; n_p is the number of pole-pairs; J is the polar moment of inertia; D is the viscous damping coefficient; K is the torsional stiffness; T_L is the external load torque; u_d, u_q are the direct-axis and quadrature-axis voltage components, respectively; and i_d, i_q, ω, θ are the state variables.

Here, we consider T_L as a potential energy constant load, in which case T_L will keep constant no matter that the rotor electric velocity ω is positive, negative or zero. Furthermore, we consider that the potential energy constant load is connected with a spring. And we take the spring's equilibrium point as the starting position ($\theta = 0$). In addition, $K\theta / n_p$ represents the torsional elastic torque.

The formula (1) can be rewritten as follows:

$$\begin{cases} \frac{di_d}{dt} = -a_1 i_d + a_2 i_q \omega + b_1 \\ \frac{di_q}{dt} = -a_3 i_d \omega - a_4 i_q - a_5 \omega + b_2 \\ \frac{d\omega}{dt} = -a_6 i_d i_q + a_7 i_q - a_8 \omega - a_9 \theta - b_3 \\ \frac{d\theta}{dt} = \omega \end{cases} \tag{2}$$

where $a_1 = \frac{R_s}{L_d}, a_2 = \frac{L_q}{L_d}, a_3 = \frac{L_d}{L_q}, a_4 = \frac{R_s}{L_q}, a_5 = \frac{\psi_{af}}{L_q}, a_6 = -\frac{n_p^2(L_d - L_q)}{J}, a_7 = \frac{n_p^2 \psi_{af}}{J}$, $a_8 = \frac{D}{J}$, $a_9 = \frac{K}{J}$, $b_1 = \frac{u_d}{L_d}$, $b_2 = \frac{u_q}{L_q}$ and $b_3 = \frac{n_p}{J} T_L$.

Obviously, the formula (2) and the formula (1) are equivalent, but the form of formula (2) is more concise. In the later part of this section we will use the formula (2) to discuss the stability of the system.

Because the values of L_d, L_q, R_s, ψ_{af}, n_p and J are all positive, so the following relationships hold:

$$a_1 > 0, a_2 > 0, a_3 > 0, a_4 > 0, a_5 > 0, a_7 > 0 \tag{3}$$

The surface PMSM has the uniform air gap, and it means $L_d = L_q$ holds; the built-in PMSM has the non-uniform air gap, and it means $L_d < L_q$ holds. When the viscous damping torque is considered, $D > 0$ holds; when the viscous damping torque is not considered, $D = 0$ holds. When the torsional elastic torque is considered, $K > 0$ holds; when the torsional elastic torque is not considered, $K = 0$ holds. For the potential energy constant load considered in this paper, there is $T_L \geq 0$. So, the following relationships hold:

$$a_6 \geq 0, a_8 \geq 0, a_9 \geq 0, b_3 \geq 0 \tag{4}$$

Because the values of u_d and u_q can be positive, negative or zero, so b_1 and b_2 can also be positive, negative or zero.

If $a_9 > 0$ holds, a Lyapunov's potential function can be constructed as follows:

$$V = c_1 i_d^2 + c_2 i_q^2 + c_3 \omega^2 + c_4 \theta^2 \tag{5}$$

where

$$\begin{cases} c_1 = \frac{a_3 a_7 + a_5 a_6}{a_2} \\ c_2 = a_7 \\ c_3 = a_5 \\ c_4 = a_5 a_9 \end{cases} \tag{6}$$

It can be known from formulas (3), (4), (6) and the assumption of $a_9 > 0$:

$$c_1 > 0, c_2 > 0, c_3 > 0, c_4 > 0 \tag{7}$$

So, the potential function V is positive definite.

From formulas (2), (5) and (6), we can obtain:

$$\frac{dV}{dt} = -2\left(a_1c_1i_d^2 + a_4c_2i_q^2 + a_8c_3\omega^2 - b_1c_1i_d - b_2c_2i_q + b_3c_3\omega\right) \tag{8}$$

It can be known that dV/dt is not negative definite. According to the Lyapunov's theorem on stability, it cannot be proved that the system is stable by the potential function V shown in the formula (5). It should be noted that this does not mean that the system must be unstable, but only that the system may be unstable.

However, if the additional conditions

$$D > 0 \text{ and } u_d = u_q = T_L = 0 \tag{9}$$

hold, it means $a_8 > 0$ and $b_1 = b_2 = b_3 = 0$ hold, we can obtain:

$$\frac{dV}{dt} = -2\left(a_1c_1i_d^2 + a_4c_2i_q^2 + a_8c_3\omega^2\right) \tag{10}$$

is negative definite. According to the Lyapunov's theorem on stability, the system is stable.

If $a_9 = 0$ holds, we can obtain the same conclusions in a similar way, noting that formulas (2) and (5) will not contain θ any more in this case.

Therefore, we can know that if the additional conditions shown in the formula (9) hold, the PMSM system is stable regardless of whether the torsional elastic torque is considered, and if the additional conditions do not hold, the system may be unstable.

3 Changes of Lyapunov Exponents with *K*/*J*

The numerical analysis is of special importance to the study of chaos. To carry out the numerical analysis, some parameters of the PMSM system described in the formula (1) need be assigned.

There are 11 parameters in the formula (1), such as L_d, L_q, R_s, ψ_{af}, n_p, J, D, K, T_L, u_d and u_q. These parameters are related to many factors, including the running state of the PMSM, the transmission, the load, and the power supply, etc.

Because chaotic characters of a nonlinear system are very sensitive to parameters of the system, it is necessary to study chaotic characters of the system under different parameters. Since there are no reports on the influence of the torsional elastic torque on chaotic characters of the PMSM system, we study the case that the torsional

Table 1 The parameters of the PMSM system

Parameters	Value
D-axis inductor L_d/H	0.1425
Q-axis inductor L_q/H	0.1525
Stator winding resistance R_s/Ω	0.09
Permanent magnet exciting flux ψ_{af}/Wb	0.031
Number of pole-pairs n_p	1
Polar moment of inertia J/kg m^2	4.7×10^{-3}
Viscous damping coefficient D/N m s	0.00162
Torsional stiffness K/N m	[0–0.094]
External load torque T_L/N m	0.0282
D-axis voltage u_d/V	−28.5
Q-axis voltage u_q/V	0.915

stiffness changes in a certain range while other parameters of the system remain unchanged. The values of parameters in the formula (1) are shown in Table 1.

In this paper, we do not consider the control of the PMSM. So, we can set u_d and u_q as constant values. In addition, we set $L_d < L_q$. It means that the built-in PMSM is considered. Furthermore, if i_d is negative, the electromagnetic torque of the built-in PMSM can be improved. So, u_d of the built-in PMSM is usually taken as a negative value.

Under the parameters shown in Table 1, the value of K/J will change in the range of [0–20] s^{-2}. And, Lyapunov exponents of the PMSM system described in the formula (1) will change with K/J. The changes of the Lyapunov exponents with K/J are shown in Fig. 1.

Figure 1 shows that when the value of K/J changes in the range of [0–20] s^{-2}, LE1 is positive except for $K/J = 0$ (see Table 2); LE2 is positive under some K/J values and negative under other K/J values; LE3 and LE4 are always negative.

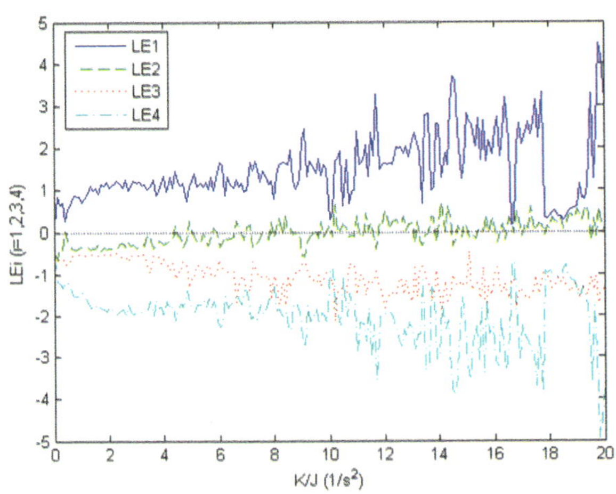

Fig. 1 The changes of the PMSM system's Lyapunov exponents (LE1-4) with K/J

Table 2 The Lyapunov exponents and the state types of the system under three K/J values

K/J (s^{-2})	LE1	LE2	LE2	LE3	LE4	State types
0	−0.028	−0.368	−0.368	−0.773	−0.396	Stable
9	2.2	−0.543	−0.543	−0.705	−2.517	Chaos
12	1.654	0.525	0.525	−1.795	−1.950	Hyperchaos

According to the theory on chaos [17], if LE3 and LE4 are always negative, when LE1 and LE2 are both negative, the system is in a stable state; when LE1 is positive and LE2 is negative, the system is in a chaotic state; when LE1 and LE2 are both positive, the system is in a hyperchaotic state.

Table 2 shows the values of Lyapunov exponents and the state types of the system when K/J takes three specific values.

In addition to judging whether a system is in a chaotic or hyperchaotic state by Lyapunov exponents of the system, it can also be judged by observing the time domain diagrams and the state space phase diagrams of the system. Furthermore, when a system is in a chaotic or hyperchaotic state, it will be sensitive to the initial conditions.

Since when $K/J = 9\,(s^{-2})$ and $K/J = 12\,(s^{-2})$, the system has similar time domain diagrams, state space phase diagrams and the sensitivity to the initial conditions, we will not show them when $K/J = 12\,(s^{-2})$ in Sects. 4 and 5 for conciseness.

4 Time Domain Diagrams and State Space Phase Diagrams Under Two *K/J* Values

The initial conditions for calculating the time domain diagrams and the state space phase diagrams of the system are $i_d(0) = -10\,A$, $i_q(0) = 1\,A$, $\omega(0) = 3\,rad$, $\theta(0) = 4\,rad/s$.

When $K/J = 0$, the time domain diagrams and the state space phase diagrams of the system are shown in Fig. 2.

When $K/J = 9$ (s^{-2}), the time domain diagrams and the state space phase diagrams of the system are shown in Fig. 3.

Figures 2 and 3 show that the PMSM system is in a stable state when $K/J = 0$, and a chaotic state when $K/J = 9$ (s^{-2}). Furthermore, Fig. 3b shows that when the PMSM system is in a chaotic state, its state space phase diagrams have some specific characters. And it means that the system may be a new chaotic system.

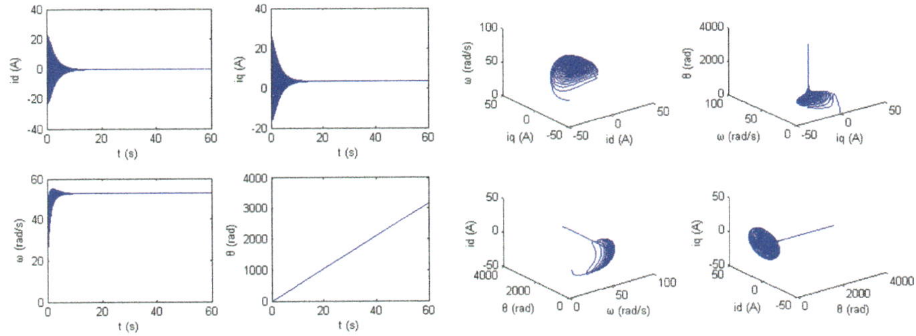

(a) The time domain diagrams (b) The state space phase diagrams

Fig. 2 The time domain diagrams and the state space phase diagrams when $K/J = 0$

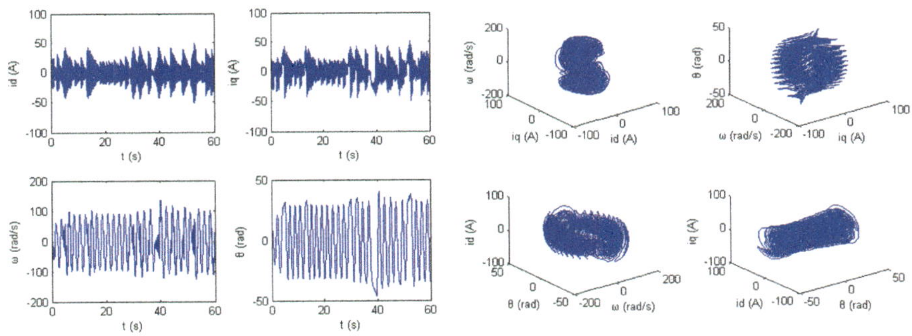

(a) The time domain diagrams (b) The state space phase diagrams

Fig. 3 The time domain diagrams and the state space phase diagrams when $K/J = 9\,(\mathrm{s}^{-2})$

5 Analysis of Sensitivity to Initial Conditions Under Two *K*/*J* Value

When a nonlinear system is in a chaotic or hyperchaotic state, it will show the sensitivity to the initial conditions. That is to say, a tiny change of the initial conditions will lead to an obvious change of the evolution trajectory of the system.

In order to judge whether the system has the sensitivity to the initial conditions, it is necessary to change slightly the initial conditions and recalculate the state variables of the system. The initial conditions are changed to $i_d(0) = -10.01$ A, $i_q(0) = 1.01$ A, $\omega(0) = 3.01$ rad, $\theta(0) = 4.01$ rad/s.

The differences of the state variables under the original initial conditions and the changed initial conditions are denoted as δi_d, δi_q, $\delta\omega$, $\delta\theta$. And, they are shown in Fig. 4.

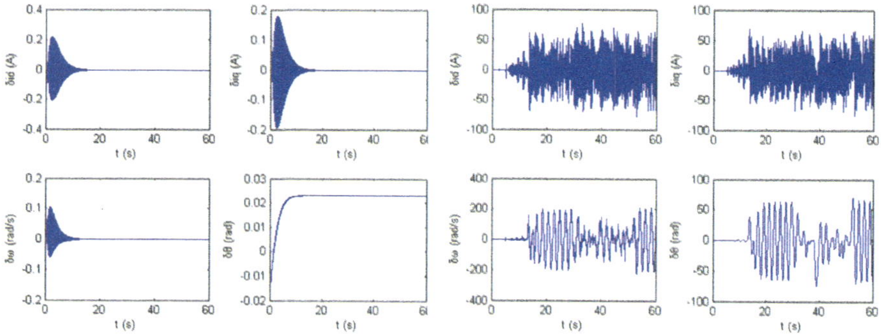

(a) When $K/J = 0$ (b) When $K/J = 9(\mathrm{s}^{-2})$

Fig. 4 The differences of state variables under two initial conditions when $K/J = 0$ and $K/J = 9$ (s^{-2})

Figure 4a shows that δi_d, δi_q, $\delta\omega$ all tend to zero, and $\delta\theta$ tends to 0.023 rad at about 19 s when $K/J = 0$. The tiny $\delta\theta$, that is 0.023 rad, can be ignored considering that θ is approximately 990 rad at about 19 s under the two initial conditions. Therefore, the system is not sensitive to the initial conditions when $K/J = 0$. And this is consistent with the conclusion that the system is in a stable state when $K/J = 0$.

In Fig. 4b, δi_d, δi_q, $\delta\omega$ and $\delta\theta$ do not tend to zero and show obvious fluctuations when $K/J = 9$ (s^{-2}). Therefore, the system is obviously sensitive to the initial conditions when $K/J = 9(\mathrm{s}^{-2})$. And this is consistent with the conclusion that the system is in a chaotic state when $K/J = 9$ (s^{-2}).

6 Conclusion

In this paper, the influence of the torsional elastic torque on chaotic characters of the PMSM system is studied. And, we can get the following main conclusions:

(1) If the additional conditions $D > 0$ and $u_d = u_q = T_L = 0$ hold, the PMSM system is stable regardless of whether the torsional elastic torque is considered, and if the additional conditions do not hold, the system may be unstable.
(2) The PMSM system can be in a chaotic or hyperchaotic state under certain parameters when the torsional elastic torque is considered.
(3) When the PMSM system is in a chaotic or hyperchaotic state, its state space phase diagrams have some specific characters. And it means that the system may be a new chaotic or hyperchaotic system.
(4) When the PMSM system is in a stable state, the system is not sensitive to the initial conditions. And when the PMSM system is in a chaotic or hyperchaotic state, the system is sensitive to the initial conditions.

References

1. Xu, Zhe, Chongxing Liu, and Tao Yang. 2010. Study on a new chaotic system with analysis and circuit experiment. *Acta Physica Sinica* 59 (1): 131–139. (in Chinese).
2. Zhang, Weiqiang, and Yangzheng Liu. 2011. Generation and circuit implementation of a unified hyperchaotic system. *Journal of Shandong University (Natural Science)* 46 (7): 30–34. (in Chinese).
3. Liu, Xiping, Weiping Hu, Weizhong Ding, et al. 2020. Resrarch on multi-parameter identification method of permanent magnet synchronous motor. *Transactions of China Electrotechnical Society* 35 (6): 1198–1207. (in Chinese).
4. Liu, Yubo, Xudong Wang, and Kai Zhou. 2020. Current deviation decoupling control with a sliding mode observer for permanent magnet synchronous motor. *Transactions of China Electrotechnical Society* 35 (8): 1642–1652. (in Chinese).
5. Li, Yuanjiang, Xin Dong, Haifeng Wei, et al. 2020. Sensorless compound PI control for surface permanent magnet synchronous motor speed regulation system. *Transactions of China Electrotechnical Society* 35 (10): 2119–2129. (in Chinese).
6. Li, Zhong, Bo Zhang, Lianfang Tian, et al. 1999. Strange attractors in permanent synchronous motors. In *IEEE 1999 International Conference on Power Electronics and Drive Systems*, Hong Kong, 150–155.
7. Jing, Zhujun, Chang Yu, and Guanrong Chen. 2004. Complex dynamics in a permanent-magnet synchronous motor model. *Chaos, Solitons and Fractals* 22 (4): 831–848.
8. Zhang, Bo, Zhong Li, Zongyuan Mao, et al. 2001. Analyzing chaotic phenomenon in permanent-magnet synchronous motors with Lyapunov exponent and capacity dimension. *Control Theory and Applications* 18 (4): 589–596. (in Chinese).
9. Zhang, Bo, Zhong Li, and Zongyuan Mao. 2002. Anti-control of its chaos and characteristics in the permanent-magnet synchronous motors. *Control Theory and Applications* 19 (4): 545–548. (in Chinese).
10. Wu, Zhongqiang, and Fuxiao Tan. 2006. Passivity control of permanent-magnet synchronous motors chaotic system. *Proceedings of the CSEE* 26 (18): 159–163. (in Chinese).
11. Li, Yunfeng, and Lixia Sun. 2015. Analysis of chaos and circuit implementation of permanent magnet synchronous motor system. *Micromotors* 48 (8): 61–82. (in Chinese).
12. Hou, Limin, Yong Li, and Zhao Sun. 2017. Chaotic control of PMSM based on nonsingular fast-terminal sliding mode. *Control Engineering of China* 24 (11): 2206–2210. (in Chinese).
13. Sun, Lixia, Sheng Lu, Zhenggeng Wen, et al. 2019. Analysis of chaotic motion mechanism of permanent magnet synchronous motors. *Electric Machines and Control* 23 (3): 97–104. (in Chinese).
14. Zhang, Neng, Yunfeng Jiang, and Cheng Cheng. 2019. Simulation of chaotic dynamics of permanent magnet synchronous motor under parameter constraints. *Computer Simulation* 36 (10): 223–348. (in Chinese).
15. Ruan, Yi, Ying Yang, and Boshi Chen. 2016. *Control Systems of Electric Drives-motion Control Systems*, 5th edn., 4237. Beijing: China Machine Press. (in Chinese).
16. Wang, Xiuhe, Yuping Sun, Guangyou Li, et al. 2013. *Electric machinery*, 2nd edn., 386–387. Beijing: China Machine Press. (in Chinese).
17. Liu, Xiaohe. *Nonlinear Circuits Theory*, 310. Beijing: China Machine Press. (in Chinese).

Development of 20 T Split Pulsed Magnet for Repetitive Pulses

Yiheng Li, Chengzhe Shu, Shan Jiang, and Tao Peng

Abstract A split pulsed magnet has been manufactured and tested. The magnet frame is made of G10 plate. Based on the consideration of the mechanical properties of the magnet, each layer of the coil is composed of 5 turns of wire. The light window is a fan shaped structure with a radial divergence angle of 30°. The bore of magnet is 10 mm and the height of the windows is 5 mm. The power supply is a 2.4 mF/10 kV miniature capacitor bank. The central field of the magnet reached 20.16 T at 4.3 kV. The final temperature is 134 K for repetitive pulses.

Keywords Split pulsed magnet · Repetitive pulsed magnetic field · Stress

1 Introduction

Split pulsed magnet has been used in X-ray diffraction experiment and neutron scattering experiment [1]. The structural diagram of the magnet is shown in Fig. 1. Compared to the traditional pulsed magnets, the split pulsed magnet has windows through which the X-rays and neutrons can travel. The center of the magnetic field area in the center can be observed during the experiment [2, 3].

Thus, spilt pulsed magnets with high field intensity have been developed in several high magnetic field laboratories. A 30 T split pulsed magnet was designed and manufactured at the National High Magnetic Field Laboratory (NHMFL) in 2005. Due to the defects of the manufacturing process, the magnet was damaged during the discharge process when the central field intensity was expected to reach 20 T [4]. A 38 T split pulsed magnet was developed at the University of Tokyo in 2006. Although rapid cooling technology was applied to the magnet, the coil heats was so severe that the cooling interval was still as long as 15 min. In 2009, a 16 T split magnet was developed by Bruker BioSpin AG. It was the first high field actively shielded fission magnet for neutron scattering experiments, mounted on a spallation neutron source in the United States [5]. In 2011, a 25 T split pulsed magnet with five-coil structure

Y. Li · C. Shu · S. Jiang · T. Peng (✉)
Wuhan National High Magnetic Field Center, Huazhong University of Science and Technology, No. 1037 Luo Yu Road, Hong Shan District, Wuhan, China
e-mail: pengtao@mail.hust.edu.cn

W. Chen et al. (eds.), *The Proceedings of the 9th Frontier Academic Forum of Electrical Engineering*, Lecture Notes in Electrical Engineering 743,
https://doi.org/10.1007/978-981-33-6609-1_41

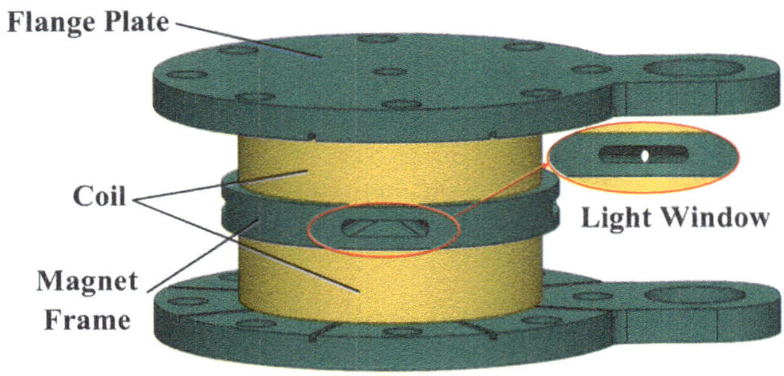

Fig. 1 Structure of the split pulsed magnet

was developed at the NHMFL. The working current of the magnet is as high as 40 kA, which requires a long cooling interval between every two pulses [6]. In 2014, a 31 T split pulsed magnet was developed at the Laboratoire National des Champs Magnétiques Intenses (LNCMI). The coil was composed of NbTi wire rein-forced copper wire and the cooling interval was 30 min [7]. A 10 T split pulsed magnet was developed at the Wuhan National High Magnetic Field Center (WHMFC) in 2018. Natural wind cooling technology was applied, which had low heat dissipation efficiency. The cooling interval of the magnet was more than 15 min. Therefore, it is necessary to develop a split pulsed magnet with high field strength and high cooling efficiency.

The biggest challenge in the development of the split pulsed magnet is that the huge axial electromagnetic force generated in the coil forces the light window to deform or even damage [8]. For instance, the axial electromagnetic force of a 25 T split magnet developed at the NHMFL is as high as 5762 kN [6]. Therefore, appropriate materials and structures must be used to overcome this challenge. At present, metal material is usually used as the frame, because of its good mechanical properties. However, due to the rapid change of the pulsed magnetic field, there will be a high eddy current in the metal frame. For instance, the eddy current in a 10 T split magnet frame developed at the WHMFC is as high as 25 kA. It leads to the increase of the power consumption, current, and the temperature [9–11]. Due to the poor mechanical properties of the non-metallic materials, the axial stress of the frame must be low as much as possible. It is necessary to analyze and optimize the electromagnetic, mechanical and heat dissipation in the magnet.

In this paper, a 20 T split pulsed magnet for repetitive pulses is designed based on the COMSOL Multiphysics commercial finite element analysis (FEA) software.

2 Magnet Design

Some parameters of the pulsed magnet need to be determined according to the experimental requirements. For instance, the size of light windows and the bore are determined by the neutron beam and samples. The magnetic field strength and the discharge frequency are expected to be as high as possible. However, high magnetic field strength means high current, which lead to the increase of internal stress and temperature rise. It is in conflict with the realization of high discharge frequency. Therefore, the influence of the field strength and discharge frequency should be balanced while optimizing the stress distribution and temperature rise.

In order to eliminate the eddy current of frame in pulses, G10 plate is selected to make the frame [12]. Therefore, it is necessary to reduce the axial electromagnetic force as much as possible because of the frame poor mechanical properties. That means the design of axial turns should not be too much. But too few turns will lead to the problems such as the difficulty in processing winding coil, heat dissipation problem and the uneven magnetic field in the center of the magnet. By analyzing the discharge process of the magnet, the parameters of the magnet are adjusted to achieve the optimal performance. Finally, the optimal stress distribution and heat dissipation design of the magnet are obtained. The main parameters of the magnet are listed in Table 1.

2.1 Electromagnetic Analysis

The frame plate is designed between two coils. The maximum magnetic field intensity inside the magnet is not located in the middle plane of the magnet [13]. The magnetic field distribution of the 1/2 magnet is shown in Fig. 2. The magnetic field intensity at

Table 1 Key parameters of the split pulsed magnet structure

Main part	Parameters	Value
Coil	Conductor	Cu
	Cross-section (mm × mm)	4.7 × 3.2
	Number of turns	9 × 5
	Inner diameter (mm)	10
	Outer diameter (mm)	76
	Inductance (μH)	100.62
Frame	Thickness (mm)	13
	Height of windows (mm)	5
	Number of windows	4
	Radial divergence angle of windows (°)	30

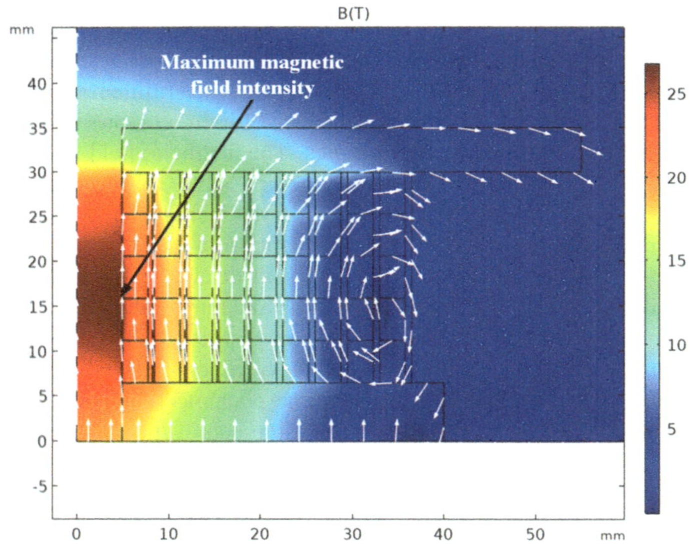

Fig. 2 Magnetic field distribution of the magnet

the center of the magnet can reach 20.23 T with the operating condition of 17.8 kA. The maximum magnetic field intensity is 26.94 T, which is located 15.4 mm away from the center. The uniformity of the magnetic field in the central hole of the magnet is better than $\pm 10\%$.

2.2 *Stress Analysis*

The conductor layers are reinforced with Zylon fiber. The thickness of interlayer filling is 0.5 mm. The ultimate tensile stress of Zylon fiber is 4 GPa [8]. The coil near the magnet frame is defined as the first turn, the coil near the flange plate is defined as the fifth turn. And the distribution of the von Mises stress of the coils is shown in Fig. 3. Where the von Mises stress of Zylon layers are higher than conductor layers. The results show that the maximum stress is 407.7 MPa, which is located at the fifth turn of the fourth layer of Zylon. It is far less than the tensile strength of Zylon fiber. With the current of 17.8 kA, the axial electromagnetic force of the upper and lower coils is 141.1 kN. In order to reduce the damage to the light windows, the magnet is designed with fan-shaped window and the radial divergence angle is 30°. The distribution of von Mises stress and deformation of the 1/4 frame in the process of the magnet being electrified is shown in Fig. 4. At the peak value of 20 T magnetic field, the maximum von Mises stress in the frame is 347.5 MPa, which is located at

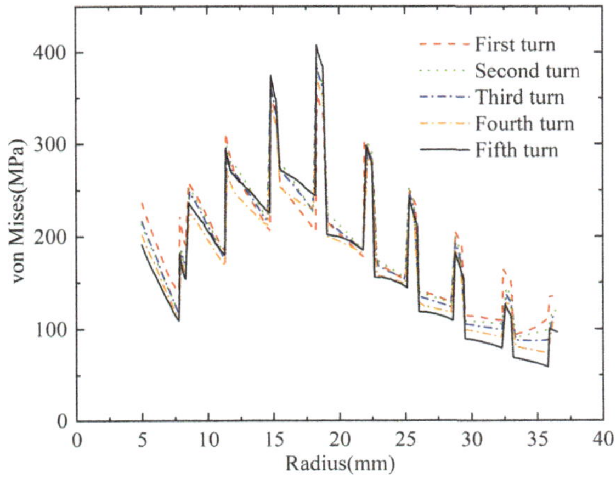

Fig. 3 Von Mises stress distribution of the magnet

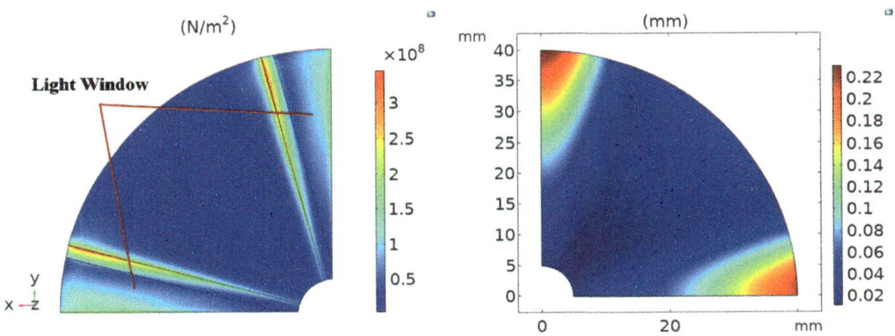

(a) Von Mises stress distribution (b) Deformation distribution

Fig. 4 Stress and deformation distribution of the magnet frame

the edge of both sides of the windows. The deformation is mainly concentrated in the outermost region of the windows, and the maximum deformation is only 0.22 mm. The mechanical strength meets the requirements and can ensure the high frequency discharge of the coil.

2.3 *Thermal Analysis*

The magnet is immersed in the liquid nitrogen. In order to improve the heat dissipation efficiency, the cooling channels are made on the flange plate as shown in Fig. 5. In the analysis of magnet temperature, it must take more than 50 pulses to stabilize

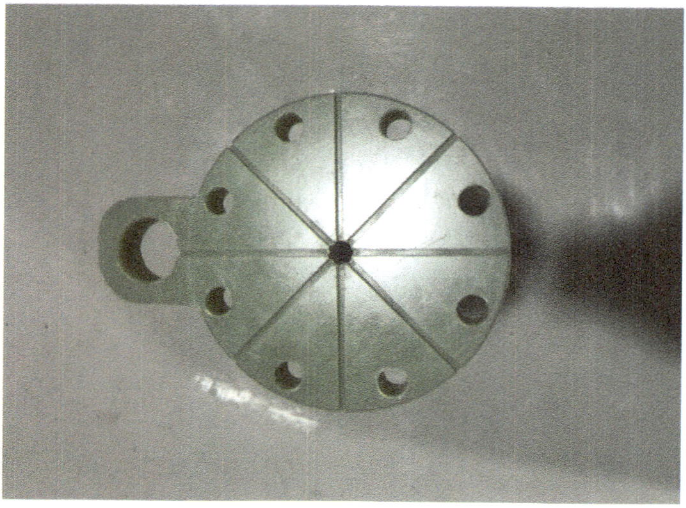

Fig. 5 Flange plate of the magnet

the final temperature when the initial temperature of the magnet is 77 K. In order to simplify the process, 140 K, which is close to the final temperature, is selected as the initial temperature of the magnet. While the discharge interval is 30 s, the average temperature variation of the magnet is shown in Fig. 6. The figure shows that the average temperature of the magnet is basically stable at about 140.5 K after the third discharge.

Fig. 6 Average temperature distribution of the magnet

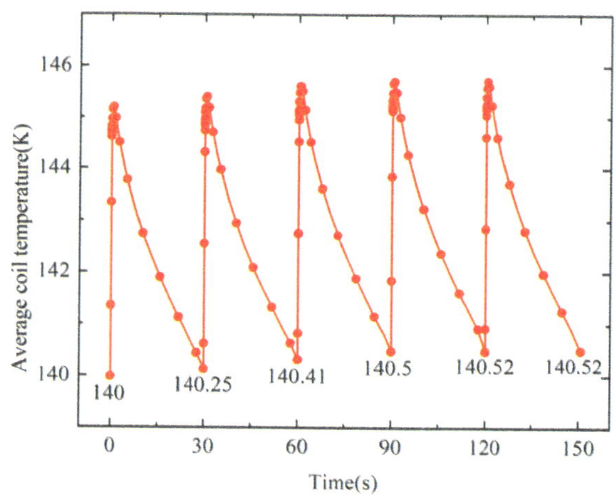

(a) Magnet after precooling (b) Magnet testing system

Fig. 7 Magnet test device

3 Test of the Magnet

The test power supply is a miniature capacitor bank. The capacitance is 2.4 mF and the maximum working voltage is 10 kV. The charge and discharge of the capacitor is controlled remotely by optical fiber, and the magnetic field is measured by pick-up coil. The magnet test device is shown in Fig. 7.

The voltage gradually increased from low to high, and the maximum is 4.3 kV. The peak value of the current is 17.21 kA, pulse width is 2 ms. The discharge interval is 5 min. The measured magnetic field waveform with different voltage is shown in Fig. 8.

In the repetitive pulse test, the discharge voltage of capacitor bank is configured as 4.3 kV, and the discharge interval is 30 s. The resistance value of magnet coil was recorded after every 5 pulses. The average temperature of magnet coil was calculated from the resistance. Average temperature of the magnet is shown in Fig. 9, the temperature is stable after 25 pulses, and the final temperature is about 134 K.

4 Conclusion

Based on the theory of material mechanics and heat transfer, the stress and temperature distribution of split pulsed magnet under repetitive pulses analyzed by COMSOL.

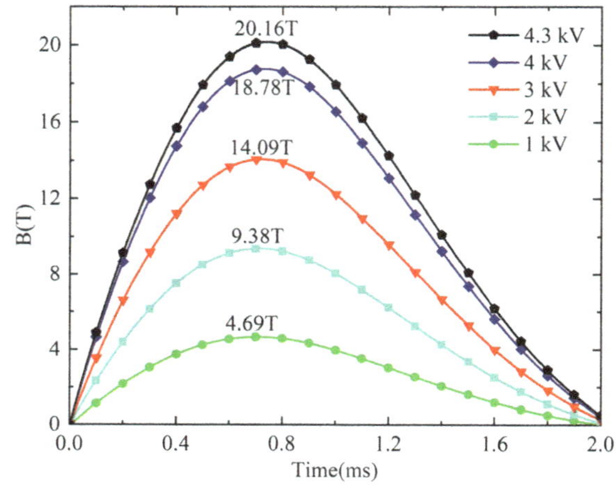

Fig. 8 The central magnetic field

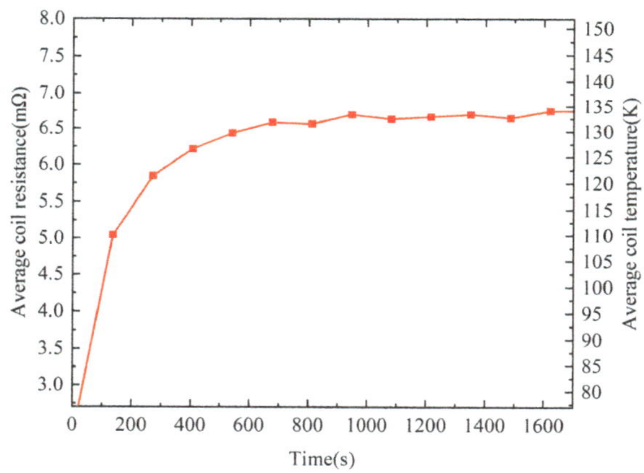

Fig. 9 Average temperature of the magnet under repetitive pulses

And the magnet has been manufactured and tested, which has a central aperture of 10 mm and an outer diameter of 76 mm. The maximum strength of the central magnetic field can reach 20.16 T under the pulse current with the peak value of 17.21 kA. Applied by the power supply of 2.4 mF/4.3 kV, the magnet can discharge repetitively at 30 s interval. It has certain reference significance for realizing higher magnetic field research of split pulsed magnet.

Acknowledgements This work was supported by the National Key Research and Development Program of China (2016YFA0401701), the National Natural Science Foundation of China (51677079, 51821005) and the Program for HUST Academic Frontier Youth Team (2017QYTD12).

References

1. Yongchao, Guo, Qian Xinxing, Jiang shili, et al. 2019. Design of cryostat for 13T superconducting magnet in neutron scattering experiments. *Cryogenics & Superconductivity* 47 (05): 22–26 + 32 (in Chinese).
2. Bird, M.D., S. Bole, J. Chen, et al. 2008. The powered scattering-magnet program at the NHMFL. *IEEE Transactions on Applied Superconductivity* 18 (2): 525–528.
3. Dai, Y., Q. Wang, H. Wang, et al. 2010. An 8 T superconducting split magnet system with large crossing warm bore. *IEEE Transactions on Applied Superconductivity* 20 (3): 608–611.
4. Bird, M.D., A.V. Gavrilin, S.R. Gundlach, et al. 2006. Design & testing of a repetitively pulsed magnet for neutron scattering. *IEEE Transactions on Applied Superconductivity* 16 (2): 1676–1679.
5. Gilardi, Raffaele. 2008. New high-field actively shielded split-coil magnet for neutron scattering. *Journal of Neutron Research* 16 (3): 93–96.
6. Toth, J., M.D. Bird, S. Bole, et al. 2012. Fabrication and assembly of the NHMFL 25 T resistive split magnet. *IEEE Transactions on Applied Superconductivity* 22 (3): 4301604.
7. Duc, F., X. Fabreges, T. Roth, et al. 2014. A 31T split-pair pulsed magnet for single crystal X-ray diffraction at low temperature. *Review of Scientific Instruments* 85 (5): 053905.
8. Peng, T., S. Wang, Y.D. Huang, et al. 2020. Study of the fatigue behavior of unidirectional zylon/epoxy composite used in pulsed magnets. *IEEE Transactions on Applied Superconductivity* 30 (4): 1–5.
9. Guojing, Dong, Liu Tao, and Li Qingmin. 2020. Numerical simulation for surface discharge of air-polyimide insulation under pulsed electrical stress. *Transactions of China Electrotechnical Society* 35 (09): 2006–2019. (in Chinese).
10. Zongtao, Niu, Zhang Cheng, Wang Ruixue, et al. 2016. Experimental study on the effect of the pulse repetition frequency on the characteristics of microsecond-pulse gliding discharges. *Transactions of China Electrotechnical Society* 31 (19): 191–198. (in Chinese).
11. Lv, Y., D. Xia, T. Peng, et al. 2017. Fast cooling pulsed magnet using distributed mini-gaps. *IEEE Transactions on Applied Superconductivity* 28 (3): 1–5.
12. Xiaofei, Wang, Dai Ying, and Luo Jian. 2019. Waterway design and temperature field analysis of vehicle permanent magnet synchronous motor based on fluid-solid coupling. *Transactions of China Electrotechnical Society* 34 (S1): 22–29. (in Chinese).
13. Tao, Peng, and Li Liang. 2010. Design and experiments of 65 T high field pulsed magnet. *High Power Laser and Particle Beams* 22 (01): 225–228. (in Chinese).

Comparative Study on COMSOL Finite Element Modeling Method of Ship Magnetic Field Based on Scalar Magnetic Potential and Vector Magnetic Potential

Kaisong Wang, Guohua Zhou, and Yuelin Liu

Abstract Obtaining the magnetic field spatial distribution of ships is necessary for magnetic silencing, and the numerical calculation of magnetic field is an effective method to obtain the magnetic field spatial distribution of ships. In this paper, the effectiveness of ship magnetic field calculation using COMSOL Multiphysics simulation platform is studied, such as the modeling steps and parameter setting methods, and the accuracy of magnetic field calculation based on scalar magnetic potential and vector magnetic potential is compared. Analytical verification examples of rotating ellipsoid shells and numerical examples of submarine engineering are provided. The results show that when using COMSOL for ship magnetic field calculation, the accuracy of vector magnetic potential method is slightly higher than that of scalar magnetic potential method under the same condition, and the modeling accuracy is better than 90%. It is helpful for ship magnetic field numerical calculation.

Keywords Ship magnetic field · Numerical calculation · COMSOL multiphysics · Finite element method · Simulation modeling

1 Introduction

Most of the modern ships are made of steel, and the ferromagnetic hull produces the ship magnetic field under the magnetization of the geomagnetic field. In order to improve the vitality and combat effectiveness of warships, it is necessary to carry out magnetic protection for modern warships [1]. The effective prediction and evaluation of the space magnetic field around the warship is a necessary prerequisite for the implementation of magnetic protection to the warship. At present, the prediction of ship magnetic field characteristics mainly includes empirical formula method, physical scale model method and numerical simulation method [2]. With the rapid development of computer software technology, numerical simulation calculation method has been widely used in warship magnetic field analysis and calculation, and is

K. Wang · G. Zhou (✉) · Y. Liu
Naval University of Engineering, No. 717 Liberation Avenue, Qiao Kou District Wuhan, China
e-mail: dandanqibing@126.com

W. Chen et al. (eds.), *The Proceedings of the 9th Frontier Academic Forum of Electrical Engineering*, Lecture Notes in Electrical Engineering 743,
https://doi.org/10.1007/978-981-33-6609-1_42

becoming more and more mature. Compared with the physical scale model method, the numerical simulation method not only saves the construction of the physical ship model, but also overcomes the disadvantages of poor flexibility, high cost and time-consuming of the physical ship model, and has the advantages of high speed, high efficiency, economy and versatility. Numerical simulation methods can be divided into four categories: equivalent source method, semi-analytical method, magnetic field integration method and magnetic field differentiation method. As a typical representative of magnetic field differential method, with the continuous development of numerical technology, finite element method has been widely used in many fields of electromagnetic numerical analysis [3], including the numerical calculation of ship magnetic field [4]. With the continuous promotion of finite element applications, a variety of commercial finite element software is becoming more and more mature, which provides a simulation platform for the numerical calculation of ship magnetic field, including OPERA, FLUX3D, TOSCA, ANSYS and COMSOL Multiphysics simulation software.

In this paper, the finite element simulation calculation of ship magnetic field is carried out by using COMSOL Multiphysics software as the simulation platform. Firstly, the software is introduced, and the basic calculation principle is analyzed, and then the concrete realization method of ship magnetic finite element simulation is described in detail, including geometric model modeling, material setting, interface selection, grid subdivision, calculation process and data post-processing. The accuracy of modeling and calculation based on scalar magnetic potential method and vector magnetic potential method is analyzed emphatically, and an analytical verification example of rotating ellipsoidal shell is designed. It has been successfully applied to the numerical prediction of the space magnetic field of a certain type of submarine.

2 Basic Theories

The magnetization of ship in geomagnetic field can be regarded as the quasi-static magnetization process of ferromagnetic material under the action of weak magnetic field, and the numerical calculation of ship magnetic field can be classified into the research category of open-domain magnetostatic field [5]. Based on the variational principle, the finite element method discretizes the variational problem so as to solve its approximate numerical solution. The discrete solution process includes field partition, function interpolation, element analysis, global synthesis, boundary condition treatment and final equation solution. The key of the finite element method is to find out the variational problems corresponding to the boundary value problems, and with the development of the technology of solving large sparse equations, the sparse matrix formed by the finite element method in the process of solving boundary value problems is easy to solve. When solving the open-domain magnetostatic modeling problem by the finite element method, the region boundary needs to be specially treated, while the COMSOL multi-physical field simulation software can deal with some equivalent boundary conditions when solving the finite element method [6], which is convenient for the finite element method to study the open-domain magnetostatic modeling problem.

The problem of ship magnetism calculation in magnetic protection can be reduced to the magnetization of ferromagnetic objects in a uniform static background magnetic field, and then the corresponding interface equation is solved by the finite element method. The ship magnetic calculation in this paper belongs to the three-dimensional open-domain magnetostatic problem, so the professional ACDC module in the electromagnetic part should be used accordingly. The electromagnetic field analysis in COMSOL is based on Maxwell equations [7], and different equations are selected according to the types of problems analyzed. The ACDC module has an encapsulated preset multi-physical field modeling interface, which can solve steady-state and time-varying problems for different physical fields. Aiming at the open-domain magnetostatic problem, the scalar magnetic potential method using "Magnetic field, no current" (mfnc) interface and the vector magnetic potential method using "Magnetic field" (mf) interface are commonly used in COMSOL finite element calculation. This paper will analyze from the comparison of two different modeling processes. At the same time, because the thin shell structure is the main structural feature of the ship model, the ship magnetic modeling in this paper also involves the treatment of the thin layer boundary problem [8]. In both mfnc and mf interfaces, the thin layer problem can be solved by adding "Magnetic shielding" condition, the thin shell of warship can be transformed into boundary surface, and the modeling and calculation of complex thin-walled structure ship hull can be realized by setting parameters such as thickness. The accuracy of using COMSOL software to solve the magnetic field of thin shell sphere has been verified by analyzing the example of thin shell sphere in Ref. [9].

The scalar magnetic potential method solves the magnetic flux conservation equation by defining the magnetic scalar potential formula, and the partial differential equation of magnetic potential field V_m is:

$$\nabla \bullet \mu_0 \mu_r(-\nabla V_m + H_b) = 0 \tag{1}$$

where H_b is the background field, and B field is calculated according to H field. The "Magnetic shielding" condition is defined as:

$$n \bullet (B_1 - B_2) = \nabla_T d_s B_T \tag{2}$$

$$H_T = -\nabla_T V_m + H_{b,T} \tag{3}$$

Among them, B_1 is the internal magnetic field, B_2 is the external magnetic field, the subscript T is the tangential vector, n is the normal vector, and d_s is the thin layer surface thickness. The condition of "External magnetic flux density" is defined as:

$$n \bullet B = n \bullet B_b \tag{4}$$

The vector magnetic potential method solves the electromagnetic induction equation by defining the magnetic vector potential formula, and the partial differential equation of magnetic vector potential A is:

$$\nabla \times \mu_0^{-1} \mu_r^{-1} \nabla \times (A + A_b) = 0 \tag{5}$$

$$H = \mu_0^{-1} \mu_r^{-1} B \tag{6}$$

where A_b is the background field, and B can also be used to specify the background magnetic field. The "Magnetic shielding" condition is defined as:

$$n \times (H_1 - H_2) = n \times H_T \tag{7}$$

$$B_T = n \times \frac{A_1 - A_2}{d_s} + B_{b,T} \tag{8}$$

where A_1 is the internal magnetic potential of the shell, A_2 is the external magnetic potential of the shell, similarly, the subscript T represents the tangential vector, n represents the normal vector, and d_s represents the surface thickness of the shell. The condition of "External vector magnetic potential" is defined as:

$$n \bullet A = n \bullet A_b \tag{9}$$

Aiming at the thin layer problem of warship shell, after adding "Magnetic shielding" condition, mfnc interface and mf interface need to add "External magnetic flux density" and "External vector magnetic potential".

3 Numerical Modeling Method and Steps

For the finite element solution of the ship open-domain static magnetic field problem, this paper puts forward two modeling schemes based on the COMSOL simulation platform, that is, the scalar potential method with mfnc interface and the vector potential method with mf interface. In order to facilitate comparison and analysis, except for the different settings of physical field excitation and boundary conditions, the other parameters and settings remain unchanged, so as to compare and analyze the modeling process of the two schemes.

3.1 Scalar Magnetic Potential Method Using Mfnc Interface

Step 1, Global parameter definition. Setting global parameters is convenient for calling parameters in the whole modeling process, and also for the adjustment of the structural parameters of the built-in CAD geometric modeling. In addition to setting the parameters required for geometric modeling, the geomagnetic field (gB) is defined as −4e−5T, and the relative permeability (μ_0) is defined as 150.

Step 2, Geometric modeling. COMSOL has its own CAD modeling tools, which can directly carry out geometric modeling on the software platform, and can also link with third-party software synchronously through CAD import module and Livelink interface. When using third-party software to import models, built-in CAD module can import various CAD files, grids or STL files, and also include the model files created by COMSOL software itself. In order to simplify the calculation, the ship body is properly simplified, and the geometric modeling of the hull is completed by using the CAD tool of the simulation platform. The ellipsoid model with analytical solution and the simplified submarine model are established respectively, which are located at the origin of the coordinate system and the bow points to the positive direction of the x axis. Then, according to the finite element calculation principle, it is necessary to establish an air domain surrounding the whole hull as the solution domain of the whole model calculation. For two geometric models, ellipsoid and submarine, a truncated boundary of 400 m × 200 m × 200 m cuboid is established, which is also located at the origin of coordinate system.

Step 3, Material definition. A rich material library is built in the software, and the material properties can be defined by users according to their needs. Here, only air materials and hull materials need to be added. The Air "domain" uses the air material "Air" built in the material library, and the hull material is added with user-defined materials. Select the corresponding hull "domain" and set the relative permeability (μ_0) to 150. According to the constitutive equation, when using scalar magnetic potential method, the relative permeability μ of Air material should be defined as 1, where "Air" in the material library will be automatically defined, and the "Domain" can be directly solved.

Step 4, Set magnetic field excitation and boundary conditions. It can be added by a preset 3D model in the Model Wizard, or it can be defined by selecting an empty model. First, select the mfnc interface corresponding to scalar bit method, and then set the magnetic field excitation and boundary conditions in the interface. The formula of "Reduced field" is used when adding magnetic field excitation in mfnc interface. In the formula, the geomagnetic field component (x, y or z, only one component is defined at a time) is $gB/mu0\_const$. This formula only solves the field disturbance in the physical field, which is convenient for solving the induced magnetic field component of submarine model, and also avoids the calculation error of the software when solving the mixed field. When setting the condition of "Magnetic shielding", first select all hull boundaries, then change the hull domain into boundary, and then set the hull surface thickness d_s to 10 mm, and the relative permeability μ_r of the hull can also be customized in the condition of "Magnetic shielding". In addition to the condition of "Magnetic shielding", it is necessary to add the condition of "External magnetic flux density" instead of "Magnetic insulation" to act on the boundary of magnetic field finite element truncation.

Step 5, Grid subdivision. For the complex model of ship, in order to improve the subdivision efficiency, the grid with tetrahedral elements is generally selected. The software platform comes with a powerful automatic grid subdivision tool, that is, you can select the predefined "Physical field control grid" for grid subdivision, or you can select the "User control grid" to customize grid subdivision according to user requirements. Here, the preset "Physical field control grid" is selected for automatic grid subdivision, and the grid accuracy is selected as "Finer". The number of units of ellipsoid model is 64,481, and that of submarine model is 64,944.

Step 6, Solve and calculate. Both methods add a "Steady-state" solution type in the research step, use the default steady-state solver, and then run "Calculate" to solve it. By default, "Convergence diagram" and "Three-dimensional drawing group" are generated.

Step 7, Post-processing. COMSOL is rich in post-processing tools, which can analyze and output various data, pictures, curves and animations according to users' requirements, thus realizing visual simulation of ship magnetic field graphics. Here, with the help of "Three-dimensional drawing group" and "One-dimensional drawing group", the data results of the model solution are constructed with three-dimensional graphics and one-dimensional curves, and then some calculated data are exported for comparative analysis.

3.2 Vector Magnetic Potential Method Using Mf Interface

The modeling steps of the vector magnetic potential method are similar to the scalar magnetic potential method, which is also divided into seven steps. The main difference is the magnetic field excitation and boundary conditions in the 4th step, and the other steps are basically the same.

In the 4th step, mf interface corresponding to vector magnetic potential method is selected first, then magnetic field excitation and boundary conditions used in the interface are set. The formula of "Reduced field" is also used when adding magnetic field excitation. The difference is that the geomagnetic field component (x, y or z, only one component is defined at a time) can be set as gB by a simpler definition method of "Uniform magnetic flux density". When setting boundary conditions, the condition of "Magnetic shielding" is also used. First, select all hull "Boundaries", change the hull "Domain" into "Boundaries", then set the hull surface thickness d_s to 10 mm, and select "From material" or "User-defined" relative permeability μ_r, with the difference that "External vector magnetic potential" is used instead of "Magnetic insulation" to define the boundary conditions of the truncated surface.

In addition to Step 4, in Step 3, the air material "Air" needs to define relative dielectric constant ε and conductivity σ besides μ. In this paper, ε is set as 1 and σ is set as 0 S/m because the current is not considered.

4 Example Analysis

In this paper, the vertical components of the induced magnetic field of the submarine model established by the two methods under the same geomagnetic environment and different depths are simulated, and the accuracy of the calculation is verified by an example of an analytical solution to the ellipsoidal shell, and then the vertical component of the induced magnetic field Z_{iz} is taken as an example to analyze the numerical results of the submarine model.

4.1 Analytical Example of an Ellipsoidal Shell

The major axis of outer surface of the ellipsoidal shell is 50 m, and the minor axis of outer surface is 5 m. The shell thickness d_s is set to 10 mm, relative permeability μ_r is 150, truncated boundary is a cuboid with 400 m × 200 m × 200 m, and the number of grid subdivision units is 64,481, as shown in Fig. 1.

Take 3 measuring lines, $y = 0$ under keel and $y = 4$, $y = -4$ under port and starboard, respectively, and set a step length of 5 m from $x = -200$ to $x = 200$ on each line. There are totally 81 measuring points in a line, and the measuring depth is 1.0 B (Beam) as the standard. According to the derivation of analytical formulas of magnetic field of ellipsoid and ellipsoid shell in Ref. [10] and the specific application case of calculating additional magnetic field of a hollow rotating ellipsoid at a point directly below the center in Ref. [11], the analytical values of all points on 3 measuring

Fig. 1 Grid subdivision diagram of an ellipsoidal shell model

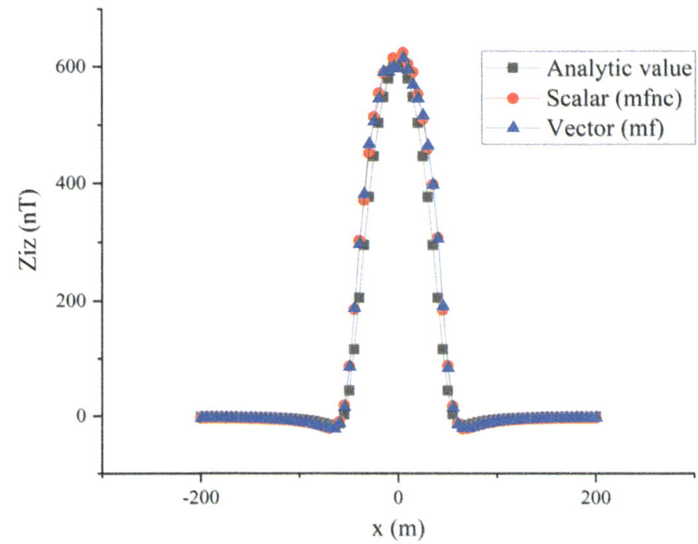

Fig. 2 Ziz of standard measuring depth (1.0 B) under keel line

lines under vertical magnetization are solved, and then compared with the simulation results of COMSOL platform.

Figure 2 shows the deviation between the Z_{iz} and the analytical value of the measurement points under the keel by the two methods. After calculation, the maximum relative error of scalar magnetic potential method is 10%, and that of vector magnetic potential method is 9.2%. Figure 3 shows the deviation of magnetic field values at measuring points below starboard line. After calculation, the maximum relative error of scalar magnetic potential method under starboard side is 9.2%, and that of vector magnetic potential method under starboard side is 9.1%. And Fig. 4 shows that the error of scalar magnetic potential method under port side is 9.5%, and the error of vector magnetic potential method is 9.1%, as shown below.

To sum up, COMSOL finite element simulation modeling can control the error within 10% when solving the open-domain magnetostatic problem of ships with ellipsoidal thin shells. Under the same conditions, the vector magnetic potential modeling method with mf interface has a slightly higher solution accuracy. It is suggested that the vector potential method should be preferred when solving the open-domain magnetostatic problem.

4.2 An Example of a Submarine

The submarine model is left and right symmetrical, and its geometric structure is similar to that of ellipsoidal shell model. The model boat is 100 m in length (x: −

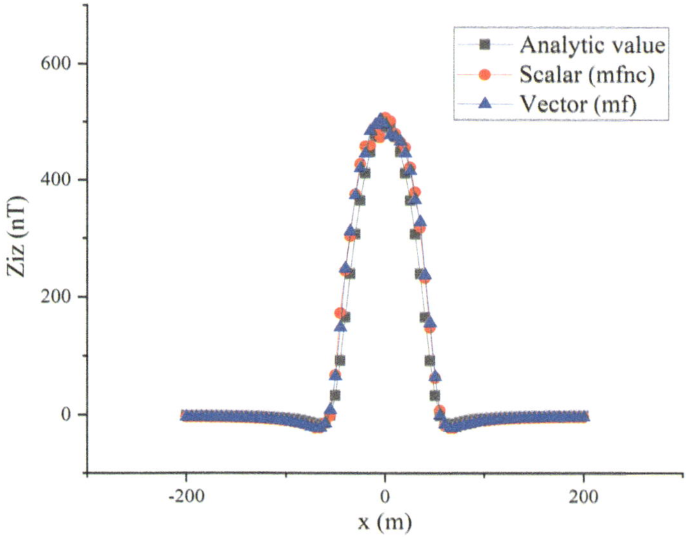

Fig. 3 Ziz of standard measuring depth (1.0 B) under starboard line

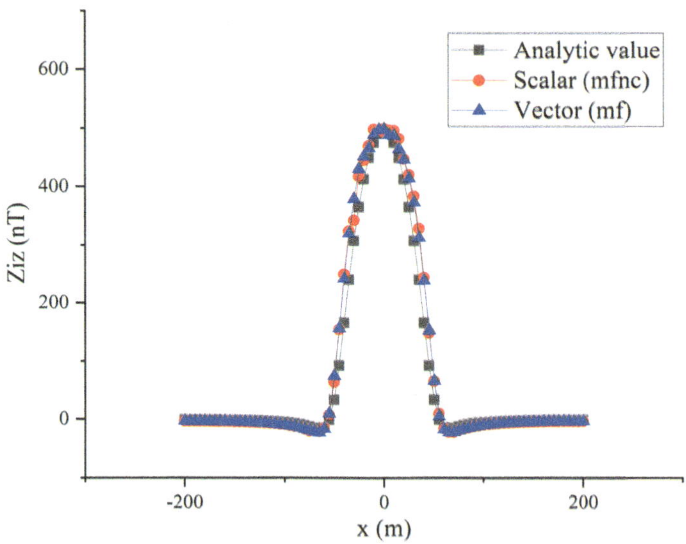

Fig. 4 Ziz of standard measuring depth (1.0 B) under port line

50 ~ 50), 10 m in width (y: −5 ~ 5) and 13.5 m in total height (z: −5 ~ 8.5), in which the main hull height is 10 m(z: −5 ~ 5) and the turret height is 3.5 m (z: 5 ~ 8.5), which is measured at a standard depth of 1 times the beam (1.0 B) of the ship. In the marine environment, since the permeability of seawater and air is close to 1, and the relative permeability of submarine is far greater than 1, it can be considered that submarine is magnetized by constant geomagnetic field in uniform medium with relative permeability of 1. The "Air" preset in the material library is still selected for air domain materials, and then the shell thickness d_s is set to 10 mm, the relative permeability μ_r is set to 150, the cut-off boundary is still a cuboid of 400 m × 200 m × 200 m. Finally, the number of grid subdivision units is 64,944, as shown in Fig. 5.

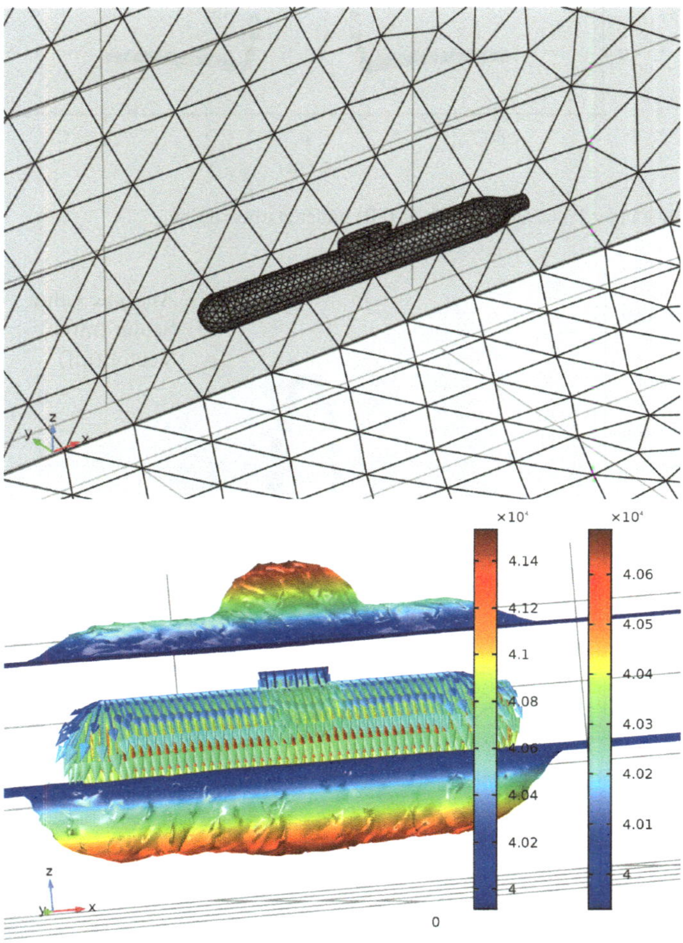

Fig. 5 Grid subdivision and 3-D magnetic field distribution diagrams of a submarine model

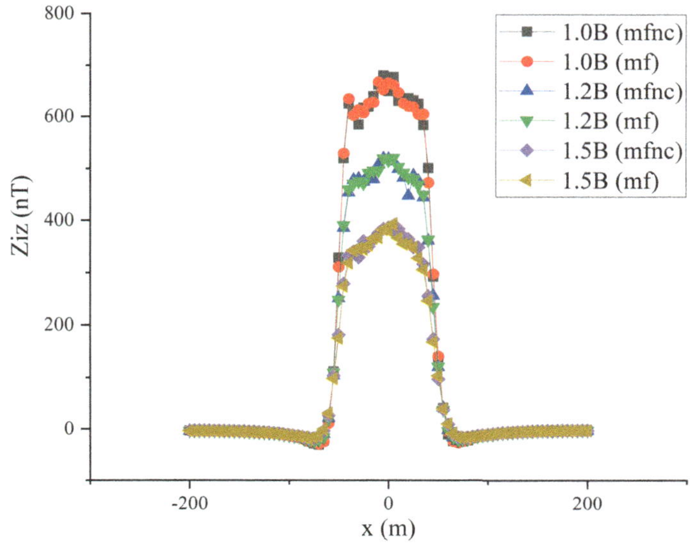

Fig. 6 Ziz of different measuring depths under keel line

Take 3 measuring lines: $y = 0$ under keel, $y = 4$ and $y = -4$ under port and starboard respectively, and set a step length of 5 m for each line from $x = -200$ to $x = 200$, with a total of 81 measuring points in a line. The simulation results are shown as follows.

Figure 6 and 7 are numerical simulation curves of the submarine model with two different modeling methods at the keel line and port line measuring points at three different measuring depths of 1.0 B, 1.2 B and 1.5 B, respectively. As shown in the figures, under the magnetization of z component of 40,000 nT geomagnetic field, the peak value of Z_{iz} at keel line measuring points is 680 nT, and the peak value of Z_{iz} at starboard measuring points is 561 nT. The magnetic anomaly points of keel curve are obviously more than those of port and starboard curves, and the degree of local unevenness is greater. At the same time, the Z_{iz} curve of the starboard line measuring points is close to the port line. According to the calculation, the relative root mean square error of Z_{iz} value between the port and starboard lines under the two methods is less than 4%, and the fitting degree is quite high. In addition, the calculated value of vector magnetic potential method is slightly smaller than the calculated value of scalar magnetic potential on the whole, and the error is smaller. With the increase of measurement depth, the magnetic field value will be attenuated continuously, and the magnetic anomaly points caused by local unevenness will be attenuated first. It has been verified that the Z_{iz} value of each side line measurement point under the two methods has been attenuated to around 0 when the measurement depth exceeds 2 times the ship beam (2.0 B). Therefore, as for the ship's magnetic model, the effective calculation depth under the hull should be controlled within the range of 2 times the ship beam (2.0 B).

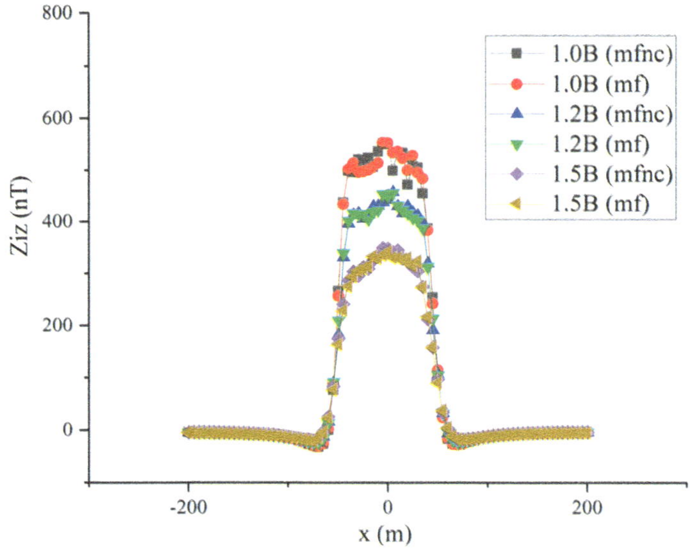

Fig. 7 Ziz of different measuring depths under port line

5 Conclusion

Taking COMSOL Multiphysics based on finite element method as the calculation platform, this paper studies the problem of ship open-domain static magnetic field modeling. Through the comparison of two different modeling methods of scalar magnetic potential and vector magnetic potential and the analysis of numerical examples, it is verified the applicability and accuracy of COMSOL finite element software for ship magnetic field numerical calculation. The modeling accuracy of vector magnetic potential method is slightly better, and the accuracy is better than 90%. It provides a certain reference for solving the problem of three-dimensional complex ship magnetic calculation.

Acknowledgements This work was supported in part by the National Natural Science Foundation of China under Grant (51107145, 51377165).

References

1. Chunsheng, Lin, Gong Shenguang. 2007. *Ship Physical Field*, 233–256. Beijing: Weapon Industry Press. (in Chinese).
2. Xingliang, Jiang, Hou Ledong, Han Xingbo, et al. 2020. Numerical simulation of torsional characteristics of transmission line conductors covered with ice. *Transactions of China Electrotechnical Society* 35 (08): 1818–1826. (in Chinese).

3. Fei, Peng, Yang Wenying, and Zhai Guofu. 2019. Parallel static magnetic field finite element method based on optimized transmission line method. *Transactions of China Electrotechnical Society* 34 (13): 2716–2725. (in Chinese).
4. Chengbao, Guo, Zhou Weichang. 2017. Numerical calculation and verification of magnetic characteristics of ship degaussing windings. *Journal of Military Technology* 38 (10): 1988–1994. (in Chinese).
5. Chao, Zuo, Geng Pan, Xia Jianchao, et al. 2015. Research on static and magnetic calculation of ship open space based on finite element boundary element method. *Ship Power Technology* 35 (04): 26–29. (in Chinese).
6. Chengying, Fan. 2019. Design method of rotating magnetic field based on Helmholtz coil and COMSOL finite element simulation. *Science and Technology and Innovation* 06: 46–48. (in Chinese).
7. Cizhang, Feng, and Ma. Xikui. 2000. *Introduction to Engineering Electromagnetic Field*. Beijing: Higher Education Press. (in Chinese).
8. Jankowski, P., and M. Woloszyn. 2018. Applying of thin plate boundary condition in analysis of ship's magnetic field. *Compel-the International Journal for Computation and Mathematics in Electrical and Electronic Engineering* 37 (5): 1609–1617.
9. Xiaoyong, Sun, Zhang Qi, Pan Mengchun, et al. 2017. Simulation and analysis of target magnetic characteristics based on COMSOL multiphysics. *China Testing* 43 (01): 122–126. (in Chinese).
10. Huaiyun, Peng, Wang Yuanxin, Pan Weiyan, et al. 2017. Inductive field produced by iron long rotating ellipsoid shell in uniform and constant magnetic field. *Journal of Electronics and Information Technology* 39 (05): 1250–1255. (in Chinese).
11. Yaozhong, Zhou, Zhang Guoyou. 2004. *Analysis and Calculation of Ship Magnetic Field*, 89–92. Beijing: National Defense Industry Press. (in Chinese).

Approximate Modeling of Gear Torque of Permanent Magnet Synchronous Motor Based on Improved Latin Hypercube Sampling

Xuerong Ye, Liqin Wu, Chengzhi Sun, Lin Wang, and Jun Zhang

Abstract Aim at there are lot of influence factors of the cogging torque of permanent magnet synchronous motor, the orders of magnitude are difference, the approximation modeling error is large, this paper put forward a kind of strong applicability approximate modeling method, can be used for all kinds of motor, namely the isometric cross complementary sampling method, improve the Latin hypercube sampling, improves the approximate model of permanent magnet synchronous motor cogging torque accuracy. In this paper, main influencing factors of cogging torque are determined. Firstly, the initial approximation model was established by the traditional Latin hypercube sampling and Kriging method, and the value range of the input parameters of the sample points whose approximate calculation error exceeded a certain threshold in the sensitivity analysis sample points was obtained. The training set was expanded based on the isometric cross-sampling method, and the calculation accuracy of the approximate modeling was improved. Taking the permanent magnet synchronous motor (PMSM) cogging torque as an example, the method presented in this paper is used for approximate modeling to achieve better calculation accuracy and verify the effectiveness of the method, which lays a foundation for the consistency optimization of PMSM cogging torque.

Keywords Permanent magnet synchronous motor · Approximate modeling · Cogging torque

1 Introduction

With its high efficiency and low torque ripple, PMSM is widely used in high-precision servo equipment such as turntables, robots and CNC machine tools [1]. With the wide application of permanent magnet synchronous motor, the requirements on the

X. Ye (✉) · L. Wu · C. Sun
Harbin Institute of Technology, Harbin 150001, China
e-mail: xuelai1981@163.com

L. Wang · J. Zhang
Guangzhou CNC Equipment Co., LTD, Guangzhou 510000, China

W. Chen et al. (eds.), *The Proceedings of the 9th Frontier Academic Forum of Electrical Engineering*, Lecture Notes in Electrical Engineering 743,
https://doi.org/10.1007/978-981-33-6609-1_43

control accuracy and noise and vibration of batch permanent magnet synchronous motor are increasingly improved [2]. As one of the main influencing factors of motor control accuracy and vibration and noise, the consistency of gear torque has become a research hotspot.

In the actual production process of permanent magnet synchronous motor, noise factors such as various size, shape and position tolerances and material attribute deviations cause the characteristic dispersion among mass-produced motor products [3–5]. Many influencing factors are coupled to each other, which makes it difficult to improve the cogging torque consistency of batch PMSM.

It is necessary to carry out batch modeling for permanent magnet synchronous motor (PMSM) to optimize the cogging torque consistency. Finite element modeling [6–8] is a common method, but the biggest problem of finite element method is that it takes a long time to calculate and is difficult to be directly used in the optimization process. In order to improve the calculation efficiency, some scholars proposed the precise subdomain model (ESM) [9–15] to improve the modeling efficiency. However, since ESM is established on the basis of the ideal motor model, the influence of machining deviation on the expression is not considered, and the magnetic permeability property of soft magnetic material is assumed to be infinite, and the influence of magnetic saturation on the model accuracy is not considered [16, 17]. When literature [2] introduced it into robust design, noise factor was taken into account and saturation coefficient was introduced. ESM method was adopted in combination with saturation coefficient to correct ESM. However, the influence of machining errors on the model is still not considered, so this method is not suitable for batch modeling of motor structure optimization considering machining dispersion. Some scholars used response surface method, Kriging method and radial basis method [18–20] to establish the input-output relationship model of the research object. Literature [4] USES the center composite design method to establish the approximate model between the structure size of the motor and the efficiency of the motor. However, as there are many influencing factors of the tooth torque, it takes a huge amount of work to use the center composite design method for modeling. [21–23] based on Latin hypercube sampling [24, 25] before and after the motor output torque, stator phase frequency, the approximate model of stator inner stress, but in view of the motor cogging torque is low, this kind of order of magnitude of big span, and the output of the input parameters more features, by using this method is difficult to get accurate approximation model of motor cogging torque.

In this paper, the cogging torque of a 10-pole 12-slot tabbed permanent magnet synchronous motor (PMSM) is studied, the mechanism of cogging torque is systematically revealed, the influence of rotor error is separated, and the main influencing factors are determined. In Latin hypercube approximation modeling, on the basis of Kriging method is used to get the cogging torque of initial approximation model, the approximate calculation was carried out on the sensitivity analysis of sample points, and get the error is bigger sample points corresponding to the input parameters, and puts forward a method of equidistant sampling at the intersection of complementary set of training samples, implements the less sample cases, achieve higher prediction precision, so as to solve multiple input, low output of the cogging torque of permanent magnet synchronous motor approximation error modeling problems. Finally, an example of permanent magnet synchronous motor (PMSM) cogging torque is used for approximate modeling to verify the effectiveness of the proposed method.

2 The Approximate Modeling Scheme for the Fast Calculation of Gear Torque of Permanent Magnet Synchronous Motor (PMSM)

Firstly, the finite element model of PMSM is established to obtain the cogging torque of PMSM, which is the output characteristic of PMSM. The main influencing factors of the PMSM's cogging torque were analyzed, and the key factors that had the most significant influence on the cogging torque of PMSM were selected in combination with the ranking of process manufacturability, design space limitation and sensitivity.

Secondly, within the allowable range of each key parameter, the Latin hypercube sampling method is adopted to establish the input parameters of the initial sample point Simulation output characteristic response. The relationship between input parameters and output characteristic responses $Y = (y_1, y_2, y_3, \ldots, y_n)^T$ is established by using Kriging method. $Y = f(X)$ As an initial approximation model.

Then the sensitivity is calculated to analyze the simulation results of the sample points $(y_1, y_2, y_3, \ldots, y_n)^T$ and the error of approximate calculation results $(\widetilde{y}_1, \widetilde{y}_2, \widetilde{y}_3, \ldots, \widetilde{y}_n)^T$ is $err(y_n \sim \widetilde{y}_n)$. Thus, the sample error of each input varies are $err(y_n \sim \widetilde{y}_n)$ separately, and the value range of each input corresponding to the sample with large error is obtained $\max err(y_n \sim \widetilde{y}_n)$. Thus, the sample error when each input changes separately is obtained, and the sample with large error is obtained. Corresponding value range of each input, such as the sample whose error exceeds a certain threshold, its input is A, B, The current corresponding interval is $A = [a_{\sigma 1}, a_{\sigma 2}]$, $B = [b_{\sigma 3}, b_{\sigma 4}]$, Then add sample points to the input method based on equidistant intersection respectively, and for the sample whose error does not exceed the threshold, the input is C, D. All center values of C, D were taken and combined as supplementary sample points. The supplementary samples are added to the original training set to form the improved training set $X' = (x_1, x_2, x_3, \ldots, x_n, x_s)^T$. In order to prevent sample agglomeration, the value rules of isometric cross supplementary sampling are shown in Formula (1):

$$\begin{aligned}
x_s &= [A, B, C, D, \ldots] \\
A &= [a_{\sigma 1}, a_{\sigma 1} + (a_{\sigma 2} - a_{\sigma 1})/n_s, a_{\sigma 1} + 2(a_{\sigma 2} - a_{\sigma 1})/n_s, \\
&\quad \ldots, a_{\sigma 2}, a_{mr}, \ldots, a_{mr}]^T \\
B &= [b_{mr}, \ldots, b_{mr}, b_{\sigma 3}, b_{\sigma 3} + (b_{\sigma 4} - b_{\sigma 3})/n_s, \\
&\quad b_{\sigma 3} + 2(b_{\sigma 4} - b_{\sigma 3})/n_s, \ldots, b_{\sigma 4},]^T \\
C &= [c_m, \ldots, c_m]^T
\end{aligned}$$

$$D = [d_m, \ldots, d_m]^T \tag{1}$$

where, x_s is the supplementary sampling sample matrix, $A, B, C, D, \ldots$ is the input of the approximate model, $a_{\sigma 1}, a_{\sigma 2}, b_{\sigma 3}, b_{\sigma 4}$ is the level value of A, B initial approximate calculation error of the input parameter, a_{mr}, b_{mr}, is the A, B random sampling within ±3% of the input center value, c_m, d_m, is the center value of C, D.

Finally, the extended training set is used to carry out Kriging approximate modeling, and the improved approximate model is obtained, which improves the precision of the approximate model of the motor and lays a foundation for the optimal design of the motor. The flow chart of improved Latin hypercube approximate modeling method is shown in Fig. 1.

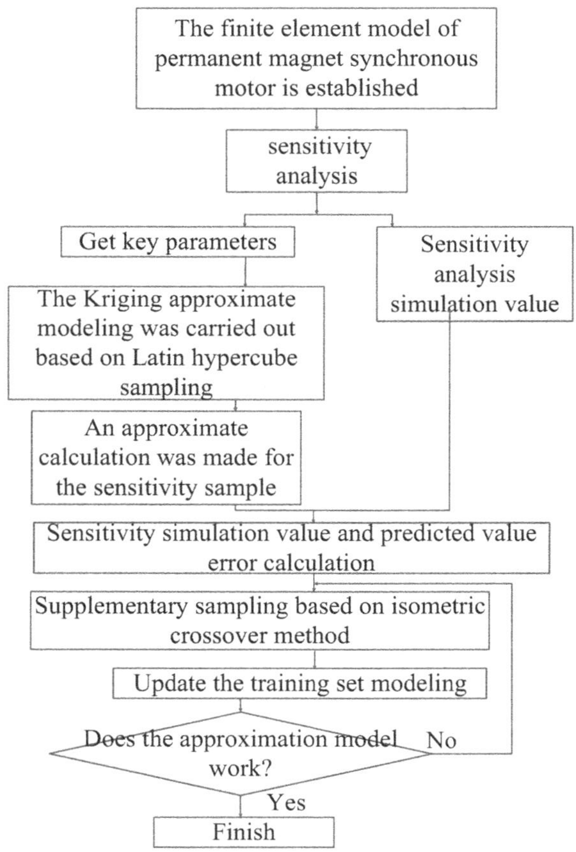

Fig. 1 Flow chart of improved Latin hypercube approximation modeling method

3 Analysis of Key Influencing Factors of Permanent Magnet Synchronous Motor (PMSM) Cogging Torque

3.1 Simulation Analysis of Cogging Torque

In this paper, the cogging torque of permanent magnet synchronous motor (PMSM) is taken as the research object. Firstly, finite element simulation is carried out. Its structure diagram is shown in Fig. 2, and the initial size parameters are shown in Table 1. A 10-pole, 12-slot tabular permanent magnet synchronous motor is studied. Its operating state is no load, the set speed is 5r/min, the sampling time is 2.4 s, and the sampling frequency is 100 Hz. It simulates the actual collection of the motor's

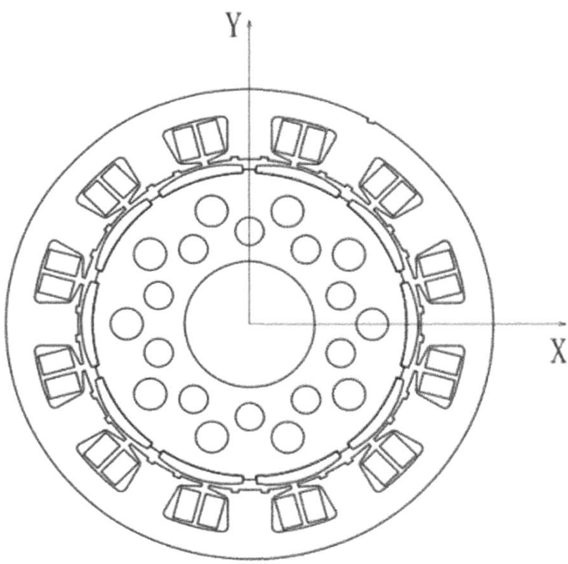

Fig. 2 Structure diagram of permanent magnet synchronous motor

Table 1 Initial dimension parameters of permanent magnet synchronous motor

Parameter	Value
stator outer diameter/mm	135
stator inner diameter/mm	95
width of air gap/mm	0.5
rotor outer diameter/mm	87
The rotor diameter/mm	36
Shaft diameter/mm	36
Magnetic tile pole-arc eccentricity/mm	18
Fixed rotor axial height/mm	55

cogging torque. Based on ANSYS, a two-dimensional finite element model of the motor is established to solve the electromagnetic field of the motor and output parameters such as flux density, inductance and flux linkage, etc., to obtain the inductance waveform of the motor, flux linkage waveform and cogging torque waveform are shown in Figs. 2, 3 and 4 (Fig. 5).

The data of one electric period of the measured cogging torque was compared with that of the simulation. The average value of the measured one-peak of the cogging torque was 0.048771 Nm, and the average value of the simulated one-peak of the cogging torque was 0.043759 Nm, with an error of 10.27%. Figure 6 shows the waveform of the simulated and measured cogging torque.

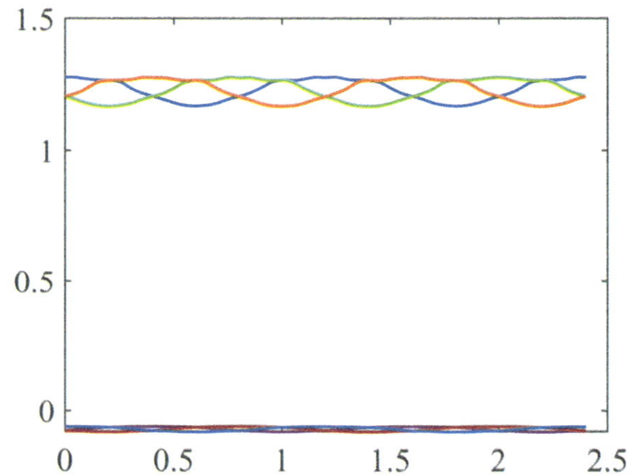

Fig. 3 Simulation of motor inductance

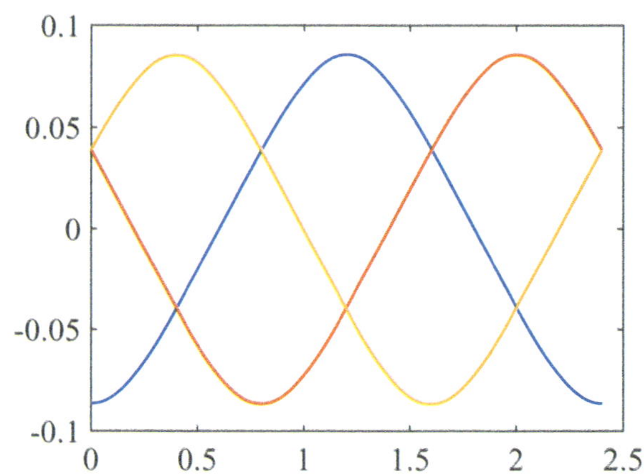

Fig. 4 Simulation of flux linkage

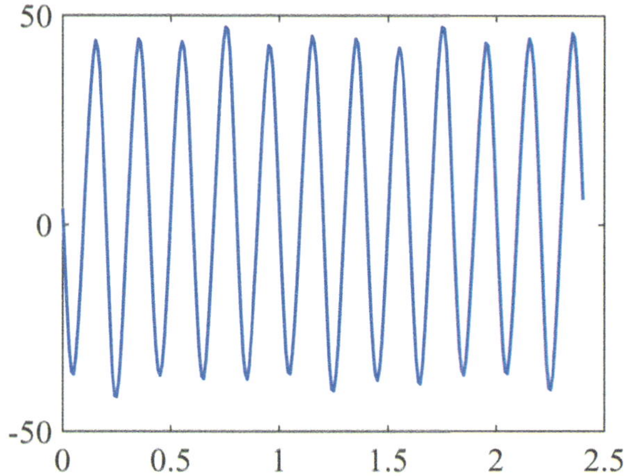

Fig. 5 Simulation of cogging torque waveform

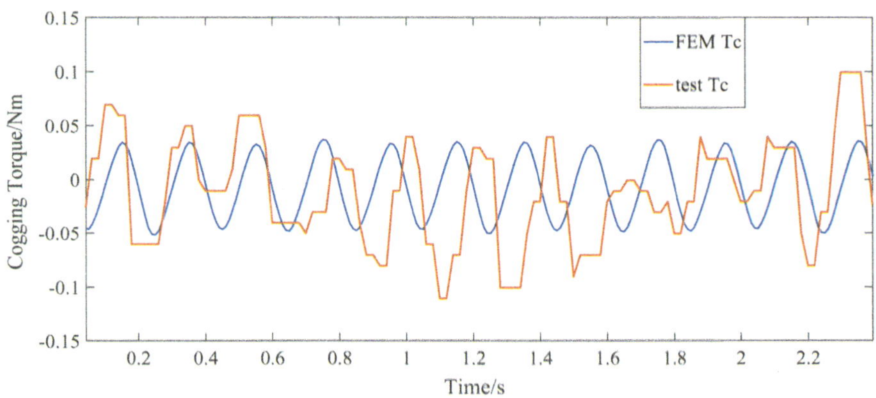

Fig. 6 Comparison of simulated cogging torque

3.2 *Sensitivity Analysis of Influencing Factors of Cogging Torque*

The cogging torque fluctuation, which affects the performance of the motor in many aspects, is widely concerned. Therefore, the cogging torque fluctuation is taken as the characterization of the cogging torque and studied. Noise factors such as size and shape tolerance, installation error and material property deviation in the process of motor production will cause the change of cogging torque volatility of PMSM. The influencing factors of the cogging torque fluctuation include rotor side, air gap and stator side. The rotor side noise factor includes the material properties such as the installation position of the magnetic tile, the coercivity and the relative permeability, the deviation of the polar arc eccentricity and the external dimension, the outside

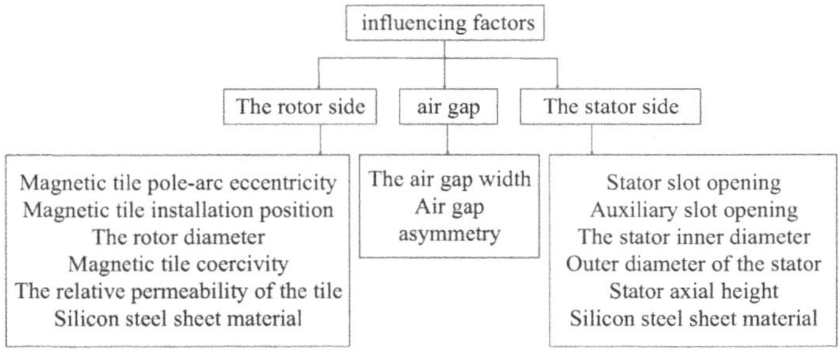

Fig. 7 Influence factors of cogging torque fluctuation rate of motor

diameter of the rotor and the dimension deviation of the rotor, etc. Air gap noise factors include air gap length deviation, static and dynamic eccentricity caused by air gap asymmetry; The stator side noise factors include the width of the slot opening, the inner and outer diameter of the stator, the width of the auxiliary slot opening, the non-uniformity of the stator core material and the overlaying coefficient of the core. The influencing factors of the motor cogging torque volatility are shown in Fig. 7.

The Taguchi experiment allows multiple control factors and optimization objectives to be in the same test to ensure that the level of each control factor is given equal weight. According to the dimension chain of the motor, the influence degree of each parameter on the variation of the cogging torque fluctuation is analyzed by comparing the cogging torque fluctuation under the condition that each dimension parameter changes $\pm 10\%$ at the same time. Each parameter takes the three horizontal values of center value and upper and lower limit value respectively. In the motor design, the selection of the main motor geometric parameters include: magnetic tile pole arc eccentricity, rotor outer diameter, relative permeability of magnetic tile, stator slot opening width, auxiliary slot opening width, stator outer diameter, fixed rotor axial height, stator inner diameter, magnetic tile force. Sensitivity analysis results of different control factor deviations are shown in Fig. 8, and range analysis is shown in Table 2.

According to the above sensitivity analysis results, factors K1, K2, K4, K6, K7 and K8 that have the most significant influence on the gear torque volatility of PMSM are obtained. K6 is limited by the design space, and K8 is the material property, which is not controllable in the processing, and these two parameters do not participate in the design. The other four parameters, namely, pole arc eccentricity, rotor outer diameter, stator slot opening width and stator inner diameter, are the design parameters that have the most significant influence on the cogging torque fluctuation rate of the motor. Based on these four sensitive parameters, the relationship between the key parameters of PMSM and the cogging torque fluctuation is established.

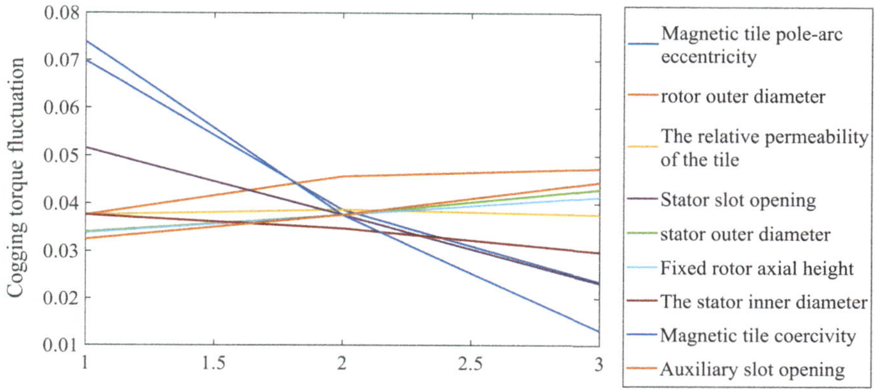

Fig. 8 Sensitivity analysis of cogging torque fluctuation rate

Table 2 Sensitivity range analysis of cogging torque fluctuation rate

Design variable	Parameter	Unit	Reference value	Range
K1	Magnetic tile pole-arc eccentricity	mm	18	0.0059
K2	rotor outer diameter	mm	94	0.00104
K3	The relative permeability of the tile	H/m	1.06396	1.006e−5
K4	Stator slot opening size	mm	2	0.00346
K5	stator outer diameter	mm	135	0.00059
K6	Fixed rotor axial height	mm	55	0.00104
K7	The stator inner diameter	mm	95	0.00173
K8	Magnetic tile coercive force	A/m	−833	0.00779
K9	Width of auxiliary slot opening	mm	2	0.00095

4 Approximate Modeling for Fast Calculation of Gear Torque of PERMANENT Magnet Synchronous Motor

4.1 Approximate Modeling of Traditional Latin Hypercube Sampling

The Latin hypercube sampling method was used to sample within the allowable variation range of four parameters, and 200 groups of samples were obtained. As the initial training set, the initial approximation model was obtained by using the Kriging approximation modeling method. The model is used to approximate 30 sets of Latin hypercube samples with an error of 17.7%. The comparison between simulation and approximate calculation is shown in Fig. 9.

The initial approximation model is used to approximate the single-factor analysis samples of the above four inputs, and the error between the simulation values of the

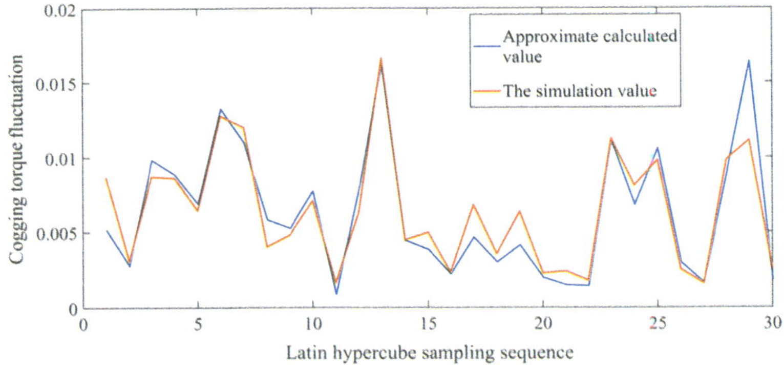

Fig. 9 Comparison of simulation value and initial approximate calculation value

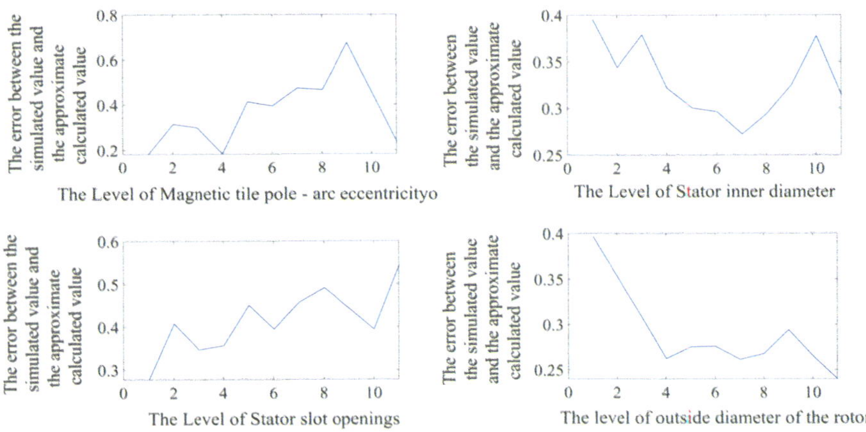

Fig. 10 Error between single factor simulation value and initial approximate calculation of key parameters

above four inputs and the single-factor samples is obtained. The calculated average error is 35%, and the corresponding input value range with a large sample error is obtained when the four inputs vary separately. The approximate calculation errors of the four inputs are shown in Fig. 10.

4.2 *Improve the Approximate Modeling of Latin Hypercube Sampling*

It can be seen from Fig. 9 that the approximate calculation error of the four parameters exceeds 45% is the 7th ~ 10th level of K1 and the 7th ~ 11th level of K4 respectively.

Table 3 Supplementary sample matrix

Magnetic tile pole-arc eccentricity/mm	Rotor outer diameter/mm	Width of stator slot opening/mm	The stator inner diameter/mm
18.36	94	1.95	95
18.468	94	1.96	95
…	94	…	95
19.44	…	2.05	…
17.5	…	1.96	…
17.9	…	1.984	…
…	…	…	…
18.5	94	2.2	95

The maximum error of K2 and K7 is not more than 45%, so the center value is taken. K1 and K4 were respectively 11 groups based on the equidistance intersection method. K2 and K7 were combined to form 22 groups of supplementary sample matrices by taking the central value. The supplementary sample matrix is shown in Table 3.

The sample matrix of supplementary sampling is added to the initial training set to form a new training set. Based on the expanded training set, the improved approximate model is obtained by using Kriging method. The improved approximation model makes an approximate calculation for the single factor analysis sample, and the average error between the approximate calculation result and the simulation value is 6.4%, which is 81.7% lower than before the improvement. The approximate calculation of 30 validation sets of Latin hypercube sampling showed an average error of 10.7%, which was 39.5% lower than that before the improvement, and the number of sample points with an error of more than 30% was 33% lower than that before the improvement. The improved univariate analysis approximate calculation error is shown in Fig. 11, and the comparison figure between the simulation value and approximate calculation value of 30 validation sets is shown in Fig. 12.

5 Conclusion

The cogging torque for permanent magnet synchronous motor, aim at the large approximation modeling error, this paper improved the sampling method of approximate modeling, on the basis of the Latin hypercube sampling, sample points for error exceeds a certain threshold input parameter selection range, an isometric cross complementary sampling method, the approximation model precision is improved. The following conclusions can be drawn:

(1) Through the sensitivity analysis of the cogging torque, the key influencing factors of the cogging torque are obtained. The traditional Latin hypercube

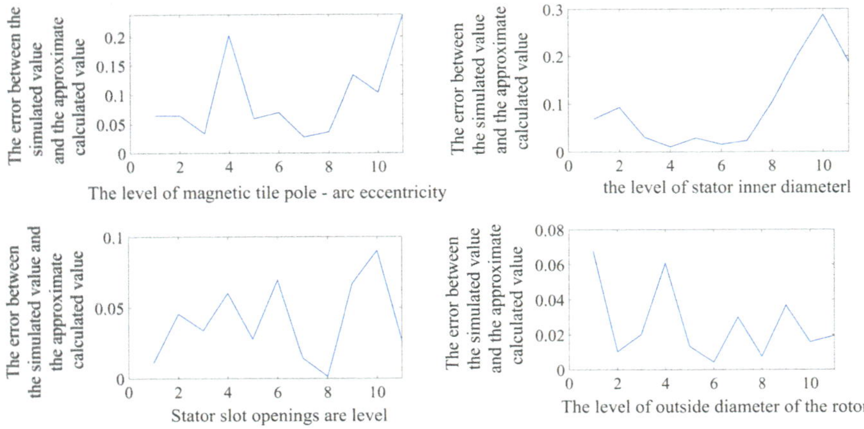

Fig. 11 Error between single factor simulation value of key parameters and improved approximate calculation

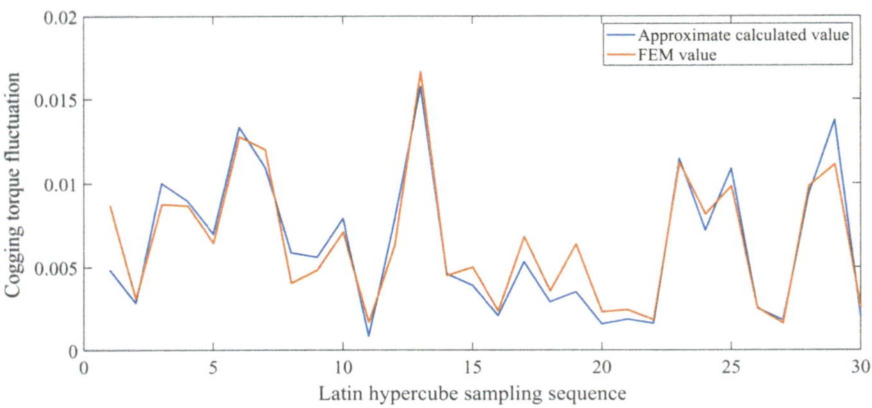

Fig. 12 Comparison of simulation value and improved approximate calculation value

sampling, based on Kriging method to establish the key factors and the cogging torque between the approximate model, sensitivity analysis and the cogging torque of sample points for approximate calculation, get the factors separately under the change of approximation error, accurately lock error exceeds the threshold of sample points of input factors.

(2) An equidistant cross supplementary sampling method is proposed, which carries out supplementary sampling for the value interval of input factors with error exceeding the threshold, expands the training set, and reduces the average error of approximate calculation by 39.5% and the number of sample points with error exceeding from 33 to 30%, compared with before the improvement. Thus, the problem of low order of magnitude, large difference of order of magnitude

and large approximate modeling error of permanent magnet synchronous motor (PMSM) is solved, which lays a foundation for optimization of PMSM's cogging torque.

Acknowledgements National Key R&D Program of China (2017YFB1300800) and National Natural Science Foundation of China (61671172)

References

1. Zhiyong, Wu, Guo Hong, and Qian Hao. 2015. Analysis of torque ripple distribution characteristics of permanent magnet synchronous motor considering noise factor. *Journal of electrical machinery and control.* 19 (1): 8–15. (in Chinese).
2. Wu, Zhiyong. 2015. Robust *Design Theory and Method of High-performance Permanent magnet Synchronous Motor system*, 12–13. Beijing University of Aeronautics and Astronautics (in Chinese).
3. Hong, Guo, and Qian Hao. 2012. Robust design of PMSM with low torque ripple. *Chinese Journal of electrical engineering.* 32 (24): 88–95. (in Chinese).
4. Junguo, Cui. 2014. Study on *Electromagnetic Parameters Optimization of Low-speed and High-torque Submersible Permanent Magnet Synchronous Motor*, 60–63. China University of Petroleum (in Chinese).
5. Shaopeng, Wang, Liu Chengcheng, and Wang Youhua. 2019. 6 σ robust multidisciplinary design optimization method for soft magnetic composite permanent magnet motor. *Transactions of China Electro Technical Society* 34 (4): 637–645. (in Chinese).
6. Chunyan, Li, Yu. Wang, and Meng Tao. 2019. Analysis of flux weakening characteristics of inverse salient permanent magnet synchronous motor. *Transactions of China Electro Technical Society* 34 (S2): 474–483. (in Chinese).
7. Zheng, Li, Chen Qiufa, and Wang Qunjing. 2019. Analysis of rotor dynamic characteristics of multi degree of freedom permanent magnet synchronous motor. *Transactions of China Electro Technical Society* 34 (11): 2269–2276 (in Chinese).
8. Fan, Zhang, Fang Youtong, and Huang Xiaoyan. 2011. Optimization and simulation analysis of a new permanent magnet synchronous traction motor. *Micromotor* 44 (4): 6–10. (in Chinese).
9. Dubas, F., and C. Espanet. 2009. Analytical solution of the magnetic field in permanent-magnet motors taking into account slotting effect: no-load vector potential and flux density calculation. *IEEE Transactions on Magnetics* 45 (5): 2097–2109.
10. Zhu, Z.Q., L.J. Wu, and Z.P. Xia. 2010. An accurate subdomain model for magnetic field computation in slotted surface-mounted permanent-magnet machines. *IEEE Transactions on Magnetics* 46 (4): 1100–1115.
11. Feng-ge, Zhang, Chen Jin-hua, Liu Guang-wei, et al. 2011. Analytical calculation of magnetic field of surface paste type dual rotor permanent magnet Motor. *Journal of Electrotechnics* 26 (12): 28–36 (in Chinese).
12. Jiebao, Li, Jing Libing, Zhou Xiaoyan, et al. 2012. Exact analytical method for surface-mounted permanent-magnet brushless motors. *Transactions of China Electro Technical Society* 27 (11): 83–88. (in Chinese).
13. Zhu, Z.Q., L.J. Wu, D. Staton, et al. 2011. An improved subdomain model for predicting magnetic field of surface-mounted permanent magnet machines accounting for tooth-tips. *IEEE Transactions on Magnetics* 47 (6): 1693–1704.
14. Lijian, Wu, Hao Yin, Dong Wang, et al. 2020. On-load field prediction in SPM machines by a subdomain and magnetic circuit hybrid model. *IEEE Transactions on Industrial Electronics* 47 (4): 7190–7201.

15. Lubin, T., S. Mezani, and A. Rezzoug. 2011. 2-D exact analytical model for surface-mounted permanent-magnet motors with semi-closed slots. *IEEE Transactions on Magnetics* 47 (2): 479–492.
16. Chu, W.Q., and Z.Q. Zhu. 2013. On-load cogging torque calculation in permanent magnet machines. *IEEE Transactions on Magnetics* 49 (6): 2982–2989.
17. Azar, Z., Z.Q. Zhu, and G. Ombach. 2012. Influence of electric loading and magnetic saturation on cogging torque, back-EMF and torque ripple of PM machines. *IEEE Transactions on Magnetics* 48 (10): 2650–2658.
18. Xiaoyong, Ma. 2016. *Lubrication Performance Analysis and Optimization of the Main Bearing of a high-speed Gasoline Engine*. Hunan University (in Chinese).
19. Yuyang, Lai, Jiang Xin, and Fang Liqiao. 2012. *Detailed Explanation of Isight parameter Optimization Theory and Examples*, 1–6. Beijing University of Aeronautics and Astronautics Press (in Chinese).
20. Ding, Han, and Zheng Jianrong. 2012. Approximate model technology in engineering optimization design. *Journal of East China University of science and technology: Natural Science Edition* 38 (6): 762–768. (in Chinese).
21. Zijie, Niu, Sun Zhijun, and Cui Yongjie. 2019. Stator design optimization of 3-DOF Ultrasonic Motor Based on response surface. *Vibration, Test and Diagnosis* 39 (5): 1089–1138 (in Chinese).
22. Zhiyong, Zhang, Zhang Fei, Liu Zhiqiang, and Zhang Xin. 2017. Nonlinear partial least squares regression estimation of output torque of Brushless DC motor. *Transactions of China Electro Technical Society* 32 (5): 62–69. (in Chinese).
23. Zijie, Niu, Sun Zhijun, and Chen Chao. 2014. Structural optimization of flexible rotor of hollow traveling wave ultrasonic motor based on response surface model and adaptive genetic algorithm. *Chinese Journal of Electrical Engineering* 34 (30): 5378–5385 (in Chinese).
24. Bektas, E., K. Broermann, G. Pecanac, et al. 2017. Robust design optimization: on methodology and short review. In *2017 18th International Conference on Thermal, Mechanical and Multi-Physics Simulation and Experiments in Microelectronics and Microsystems*, 1–7.
25. Bangfu, Zhang, Cheng Ming, and Wang Sasa. 2020. Optimal design of flux switched permanent magnet linear motor based on improved surrogate model optimization algorithm. *Transactions of China Electro Technical Society* 35 (5): 1013–1021. (in Chinese).

Degradation Feature Selection Method of AC Conductor Based on Mutual Information

Shuxin Liu, Yanfeng Li, Yang Liu, Yundong Cao, and Jing Li

Abstract The degradation process of AC contactor is analyzed. Aiming at the characteristics of high-dimensionality of degradation parameters, large correlation between parameters and abnormal points, the degradation parameters are regarded as multivariate time series, and a method for selecting degeneration features of AC contactors based on high-dimensional mutual information is proposed. Firstly, through the AC contactor life test platform, the characteristic parameters that characterize the performance degradation process of the AC contactor are obtained. Secondly, the preprocessing of time series analysis was carried out, which verified that the data can be used as input variables of time series model. Finally, the feature selection method of high-dimensional mutual information is used to obtain the correlation ranking of the tested sequence and the degradation process to obtain a strong correlation feature subset, and then use the backward crossover strategy to remove the redundant degradation parameters and obtain the optimal feature set. This method efficiently deletes redundant features and irrelevant features, and has good generalization ability. Moreover, it is not necessary to predict the type of data distribution, which reduces the influence of non-linearity between features on feature selection to a certain extent, and helps to establish a more complete AC contactor degradation state evaluation model.

Keywords AC conductor · Feature selection · Time series analysis · Mutual information

1 Introduction

At present, many scholars have done various researches on the life monitoring and state evaluation of electrical appliances with contacts, and established various methods including mathematical models and physical models to understand the

---

S. Liu (✉) · Y. Li · Y. Liu · Y. Cao · J. Li
School of Electrical Engineering, Shenyang University of Technology, No. 111 West Shenliao Road, Baogong Street, Tiexi District, Shenyang, Liaoning Province, China
e-mail: m15040390042@163.com

W. Chen et al. (eds.), *The Proceedings of the 9th Frontier Academic Forum of Electrical Engineering*, Lecture Notes in Electrical Engineering 743,
https://doi.org/10.1007/978-981-33-6609-1_44

degradation laws of electrical appliances. The development of low-voltage electrical appliances such as AC contactors tends to be intelligent, and it is particularly important to obtain the reliability and real-time status of the equipment, which helps to improve the safety of the electrical system. When studying the degradation state of AC contactors, a data-driven method was adopted. When studying the degraded state of AC contactor, it is an effective method to use the data-driven method to mine the hidden information in the degraded state and build a more accurate model in the case of limited data amount [1].

It is well known that the electrical life of AC contactor is far less than the mechanical life. When studying the degradation process of AC contactor, it is found that the failure form of AC contactor is mainly the failure of the contact system. The failure of the contact system is mainly attributed to the failure caused by electrical wear and electric fusion welding [2]. When quantifying the actual wear of the contact caused by the breaking behavior, because the contact is bonded to the contactor's shell at the end of the life test, the method of calculating the wear by measuring the mass of the contact before and after the breaking is not available. The observation method is relatively complicated, and to wear and useful to ascension [3, 4], can be characterized by the degradation parameters, such as bounce time, average arcing power and electrical wear [5].

At present, the data-driven method is more suitable to deal with the evaluation and prediction problems with unclear physical laws and sufficient collected data than physical methods such as measuring the actual wear amount. A contactor life prediction method based on rough set and evidence theory is proposed [6]. According to the two degradation parameters of arcing energy and contact pressure drop, uncertainty reasoning is used to process random and fuzzy degradation information, and qualitative prediction is made when the number of sample periods is small. This method overcomes to some extent the shortcomings of complex calculations in general data methods. Even if two AC contactors from the same model and batch are different.

In this paper, through the life test of the AC contactor, the degradation parameters that characterize the degradation process of the AC contactor are obtained, including seven degradation parameters such as contact resistance, release time, and arcing energy. Then the degradation parameters are regarded as time series, and relevant pre-processing is carried out to lay a theoretical foundation for the subsequent establishment of a multivariate time series model. Finally, the degradation feature selection method of high-dimensional mutual information is used to obtain the correlation ranking of the tested sequence and the degradation parameters. The mutual information (MI) method eliminates useless and redundant degradation parameters, and finally obtains the optimal feature subset. To prepare for the follow-up work of establishing a multivariate time series model of AC contactor degradation parameters.

Table 1 Test conditions for full life test

Name of test condition	Test conditions
Coil voltage/V	220/AC
Type of work system	AC-4
Load voltage/V	400(380)
Load current/A	222
Load type	Inductive-resistive load
Power factor value	0.35
Breaking frequency/(times/h)	300
Sampling frequency/Hz	1 M

2 Extraction of Degradation Parameters

This article focuses on the CJX2-5011 AC contactor. When the AC contactor is running on the test platform, the platform collects the contact voltage, contact current, coil current and voltage data during the on-off process. The sampling rate of the acquisition card can reach 1 M/s per channel, and it can automatically control the AC contactor and save data.

2.1 Test Conditions

In the current actual use, the main use category of AC contactors is AC-4. Under this use category, the period of equipment reaching the end of life is short, and the degradation trend of the contact system can also be clearly observed. The test platform meets the specific on-off test conditions for AC contactors working in the AC-4 category specified in the national standards. The test conditions are shown in Table 1, and the composition of the test platform is shown in Fig. 1.

2.2 Calculation and Extraction of Degradation Parameters

The distribution of the first phase of different contactless product samples is different, and multiple life tests show that, for the phase with high distribution frequency of the first phase, the degree of electrical wear of the contact is more serious than that of the other two phases. Therefore, the following parameters are taken from the phase with high frequency of the first opening phase in the lifetime process. In the process of breaking, three degradation parameters, namely Arcing time, Arcing energy and Release time, are obtained.

Arcing Energy (E_{arc}) represents the arcing energy generated by a single arc behavior:

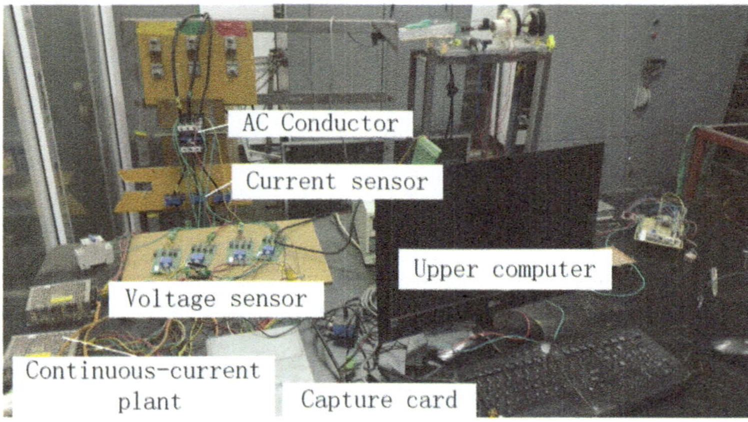

Fig. 1 AC contactor full-life test platform

$$E_{\mathrm{arc}} = \frac{\sum_{k=1}^{N} u_k i_k}{f} \tag{1}$$

Discrete voltage and current signals are collected during the test. Where, u_{k} and i_{k} are the voltage and current values of the contact, N refers to the number of sampling within the Arcing Time in a complete breaking process, and f is the sampling rate.

In addition, in the closing process, the three degradation parameters are obtained: Pick-up time, Bounce time and Closed phase angle.

The ratio of the absorption time to the half-cycle length K of the contact voltage (as can be known from the sampling rate, take 10,000) can be considered as the closing phase angle (α_{c}). In the closing process of AC contactor, the moment when the coil gets electricity is recorded as t_3. The initial time when the current is generated between the contacts is recorded as t_4, which can be expressed as:

$$\alpha_{\mathrm{c}} = \frac{\pi}{2} \cdot \frac{\mathrm{t}_4 - \mathrm{t}_3}{\mathrm{K}} \tag{2}$$

In addition, the Contact Resistance can be calculated as the degradation parameter through the root mean square value of the contact voltage and current under one period. The calculation formula is as follows:

$$R_j = \frac{\sqrt{\frac{1}{\mathrm{N}} \sum_{n=1}^{N} u_n^2}}{\sqrt{\frac{1}{\mathrm{N}} \sum_{n=1}^{N} i_n^2}} = \sqrt{\frac{\sum_{n=1}^{N} u_n^2}{\sum_{n=1}^{N} i_n^2}} \tag{3}$$

where u_n is the contact voltage in a period when the contact is closed steadily, i_n is the contact current corresponding to that period, and N is the number of samples in that period.

To analyze the performance degradation law of AC contactor along with its life process, and to correspond to the performance degradation state of AC contactor, the key is to find out the characteristic parameters that can affect or characterize the electrical life state of AC contactor. The above parameters are related to the degradation law and can be used as the initial set of factors affecting the degradation performance of AC contactor.

3 Time Series Preprocessing

The initial factors that characterize the degradation performance of AC contactor include considerable experimental data for mining and analysis. The variation of these data with the degradation process largely depends on the influence of electrical wear and fusion welding on the contact system. Therefore, compared with the time series consisting of a single parameter, the multiple time series consisting of Arcing Energy, Release Time, etc., contains more degradation information, and can better reflect the law that the degradation state of AC contactor develops and changes with the increase of breaking times.

Time series analysis studies the law of random series, predicts the dynamic path of series, and solves the practical prediction problem [7, 8]. Let the initial factor set of the degradation process be $\boldsymbol{X} = \{\boldsymbol{X}_1, \boldsymbol{X}_2, \ldots, \boldsymbol{X}_7\}^{\mathrm{T}}$, each row vector corresponds to a degenerate parameter, and the number of columns of the matrix is the degree of breaking M. The number of breaks in the life test is $\{\boldsymbol{Y}_{\mathrm{i}}\}$ ($i = 1, 2, \ldots$, M). The degradation parameters in each initial factor set are regarded as the time series of random events, and the stationarity and randomness tests are carried out.

After calculation, when the number of delay periods is 12, the autocorrelation coefficients ρ of all degradation parameters fall within twice the standard deviation range, showing the characteristics of a stationary sequence. After that, the white noise test showed that the return value p of Q statistics at each order of delay was far less than the threshold value, and the hypothesis that the sequence belonged to pure random sequence could be rejected. Therefore, the decision sequences are not white noise sequences, which have the value of carrying out follow-up time series analysis.

4 The Feature Selection Method of High-Dimensional Mutual Information

When analyzing the degradation process of AC contactor, the selected features should have certain physical significance and be closely related to the life state of AC contactor. However, the degradation data of AC contactor has nonlinear and non-periodic chaotic characteristics and strong randomness, so it is impossible to evaluate the degradation with a single variable and a linear model. However, the prediction model based on multivariate time series can better reflect the degradation state characteristics of AC contactor and achieve a higher prediction accuracy [9].

The multivariate time series is used to study the internal change rules of the system and is usually used in econometrics. The high-dimensional characteristics and complex correlation brought by multiple input variables at the same time bring some difficulties to modeling [10], while correlation is an important concept to measure the closeness between two random variables. In view of the friendliness to nonlinear input variables [11], this paper adopts the method based on mutual information for feature selection.

4.1 *Mutual Information and Entropy*

The entropy of a random variable is a function of its distribution, which only depends on the distribution of $\boldsymbol{X}$ and has nothing to do with the value of $\boldsymbol{X}$. The mutual information of two variables $\boldsymbol{X}$ and $\boldsymbol{Y}$ is defined as follows:

$$I(\boldsymbol{X};\boldsymbol{Y})=\sum_{y\in Y}\sum_{x\in X}p(x,y)\log\left(\frac{p(x,y)}{p(x)p(y)}\right) \tag{4}$$

Mutual information I ($\boldsymbol{X}$; $\boldsymbol{Y}$) can measure the information shared by $\boldsymbol{X}$ and $\boldsymbol{Y}$ variables: that is, measure the degree to which one of the two variables reduces the uncertainty of the other. MI can measure any type of relationship between input parameters because mutual information is based on the edge probability density and joint probability density of random variables.

4.2 *High-Dimensional Estimation of Mutual Information*

Z-score standardization is carried out for the data in the initial set, so that the data structure can adapt to subsequent calculations. Mutual information estimation is performed by statistical analysis of all data. The redundant variables $l = 3$ to be eliminated in the multivariate time series, and the remaining variables will participate in the construction of the feature group. The general process is shown in Fig. 2.

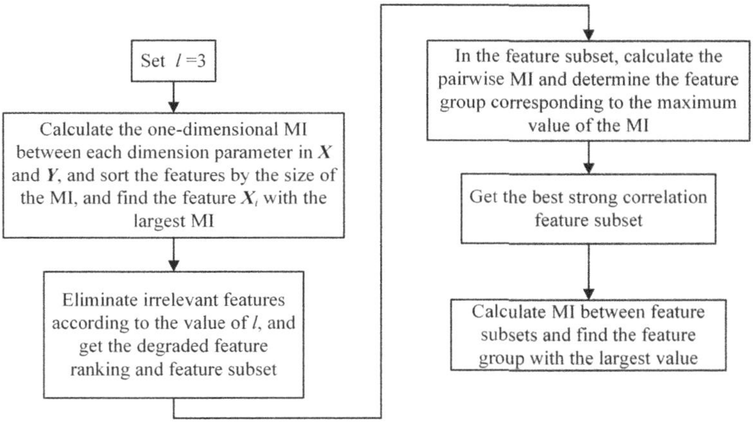

Fig. 2 Mutual information evaluation steps

The mutual information method is friendly to the type of sample distribution and easy to obtain the nonlinear relationship between multiple features. After the first step of calculation, the mutual information values of $\boldsymbol{X}_i$ and the number of breaking times $\boldsymbol{Y}$ in the initial factor set are shown in Table 2.

Due to the excessive number of breaks in the life of the AC contactor, in the calculation of MI, the characteristic datas corresponding to every hundred breaks are averaged. It can be seen from the table that the degradation parameter corresponding to the maximum MI is the Arcing Energy. It is planned to eliminate the three degraded parameters of Closing Phase Angle, Arcing Time and Contact Resistance with low correlation, and the other five degraded parameters as a feature subset. Then calculate the mutual information between the two features, and get the following Table 3.

It is easy to see that the maximum value of the mutual information of the feature subset is the group of features {Arcing Energy, Bounce Time}, that is, the optimal strong correlation feature subset. The Arcing Energy is the degradation parameter corresponding to the maximum mutual information, so the bounce time is regarded as a redundant parameter and eliminated. Finally, the degeneration features composed

Table 2 Mutual information ranking of degradation parameters

Degeneration parameters of the initial factor set	MI (bt)
Arcing energy (AE)	0.9389
Bounce time (BT)	0.9269
Pick-up time (PT)	0.9265
Release time (RT)	0.9255
Closing phase angle (CPA)	0.9223
Arcing time (AT)	0.9176
Contact resistance (CR)	0.8411

Table 3 Mutual information result of feature subset

	RT	PT	BT	AE
RT	–	0.8338	0.8357	0.8514
PT	–	–	0.8375	0.8512
BT	–	–	–	**0.8524**
AE	–	–	–	–

of Arcing Energy, Pick-up Time, and Release Time are obtained, and feature selection is completed.

5 Conclusion

In this paper, the feature selection method based on high-dimensional mutual information is used to select the features of the seven degradation parameters obtained from the life test. The redundant variables in the initial factor concentration are deleted, and the degeneration feature composed of {Arcing energy, Pick-up time, Release time} is obtained as the input feature of the model. And for each degradation parameter, the stationarity and pure randomness are tested, which lays a theoretical foundation for the subsequent establishment of a multivariate time series model. It provides a new feature selection method for the state evaluation of electrical appliances with contacts. However, because the arc and electrical contact theory is still imperfect, the optimal feature subset obtained only by data mining is not necessarily accurate. In the follow-up, it is necessary to combine relevant theories to improve the feature evaluation strategy and obtain a more complete evaluation system.

Acknowledgements This work is supported by the National Natural Science Foundation of China, No 51977132.

References

1. Xiaoting, Gao. 2016. *Research on Life-cycle Condition Monitoring and Degradation State Recognition of AC Contactor*. Shenyang: Shenyang University of Technology. (in Chinese).
2. Kui, Li, Li Xiaobei, Zheng Shumei, et al. 2017. Residual electrical life prediction for AC contactor based on BP neural network. *Transactions of China Electrotechnical Society* 32 (15): 124–131. (in Chinese).
3. Guofu, Zhai, Wang Shujuan, Xu Feng, et al. 2002. Research on double-variable life frecasting based on model-building of super-path time and pick-up time for relays. In *Proceedings of The Chinese Society for Electrical Engineering* 22 (7): 76-80 (in Chinese).
4. Engelbert, Hetzmannseder, F. Rieder Werner. 1996. Make-and-break erosion of Ag/Meo contacts materials. *IEEE Transactions on Components, Hybrids and Manufacturing Technology* 19 (3): 397–403.

5. Chunping, Niu, Chen Degui, Li Xingwen, et al. 2007. Simulation of contact bounce of AC contactor and study of its influence factors. *Transactions of China Electrotechnical Society* 22 (10): 85–90, 108 (in Chinese).
6. Shuguang, Sun, Wang Ruixiong, Du Taihang, et al. 2020. Expected electrical life prediction of AC contactor based on rough set and evidence theory. *Transactions of China Electrotechnical Society* 35 (10): 2158-2169 (in Chinese).
7. Xiaoxin, Liu. 2013. *Correlation Analysis and Variable Selection for Multivariate Time Series based on Mutual Information*. Dalian: Dalian University of Technology. (in Chinese).
8. Mitzenmacher, Michael, and Eli Upfal. 2017. *Probability and Computing: Randomization and Probabilistic Techniques in Algorithms and Data Analysis*, 2nd ed. Cambridge: Cambridge University Press.
9. Enders, Walter. 1996. *Applied Econometric: Time Series*, 4th ed. Hoboken: Wiley.
10. Daniele, Durante, Scarpa Bruno, B. Dunson David. 2014. Locally adaptive factor processes for multivariate time series. *Journal of Machine Learning Research* 15 (1): 1493–1522.
11. Zichun, Liu. 2019. *Research on Life-cycle State Assessment of AC Contactor Based on Data-Driven*. Shenyang: Shenyang University of Technology. (in Chinese).

Comparative Study on the 4-Stage Series-Connected Fast Linear Transformer Driver with Common- and Independent-Induction Cavity

Hao Qiu and Shuhong Wang

Abstract Fast linear transformer driver (FLTD) is a rapidly developing pulsed-power technology with modular and compact structure. The internal structure and media distribution of the FLTD induction cavity is very complicated and the short rise time of the bricks' discharge current will make spatial discretization much denser, leading to a dramatic increase in computational complexity. In this contribution, the transient electromagnetic field distribution of the 4-stage series-connected FLTD with common and independent induction cavity during the bricks' discharge process is investigated based on the time-domain finite integration technique (TD-FIT). FIT is implemented in the computation of electromagnetic transients in FLTD cavity and wave propagations in the water insulated transmission line. Multi-core parallelization and domain decomposition method (DDM) are adopted for reducing computation time, because of the huge amount of hexahedral meshes in the FLTD numerical model. Obtained results demonstrate that TD-FIT can effectively analyze the pulse discharge transient process in the FLTD cavity, output performance and electromagnetic field distribution of the two induction cavities are in good agreement.

Keywords Fast linear transformer driver · Finite integration technique · Domain decomposition method · Electromagnetic transients

1 Introduction

Compared with the traditional Marx generators, fast linear transformer drivers (FLTDs) can directly generate high-power pulses with peak current up to several tens of mega-ampere and rise time less than 100-ns [1]. FLTD is regarded as the most promising technology for the next generation Z-pinch driver. It is widely used

H. Qiu · S. Wang (✉)
State Key Laboratory of Electrical Insulation and Power Equipment, School of Electrical Engineering, Xi'an Jiaotong University, Xi'an 710049, China
e-mail: shwang@mail.xjtu.edu.cn

Shaanxi Key Laboratory of Smart Grid, School of Electrical Engineering, Xi'an Jiaotong University, Xi'an 710049, China

W. Chen et al. (eds.), *The Proceedings of the 9th Frontier Academic Forum of Electrical Engineering*, Lecture Notes in Electrical Engineering 743,
https://doi.org/10.1007/978-981-33-6609-1_45

in Z-pinch inertial confinement fusion (ICF) and inertial fusion energy (IFE), and X-ray radiography, etc. [2]. Currently, the Sandia National Laboratory (SNL) of America, Institute of High Current Electronics (HCEI) of Russia, China Academy of Engineering Physics (CAEP), and the Northwest Institute of Nuclear Technology (NINT) of China and some other scientific research institutions have carried out FLTD technology research and device development. However, the traditional multi-stages FLTD with independent-induction cavity is limited by its complicated triggering and charging systems. To solve this problem, a novel triggering scheme based on an internal brick and azimuthal line in FLTD with common induction cavity is proposed in [3].

Like the finite difference time domain (FDTD) method, the finite integration theory (FIT) utilizes an orthogonally staggered primal-dual grid-pair [4]. However, unlike the FDTD, FIT transforms the integral form, rather than differential form of Maxwell's equations into a set of algebraic equations. Thus it preserves physical properties of Maxwell's equations in the discrete space [5]. FIT is very flexible in dealing with electromagnetic problems range from DC to THz and geometries with complex shape. Given an equivalent number of degrees of freedom (DoFs), FIT is more cost-effective than finite element method (FEM), especially when fast transients and large scale calculations are inevitable [6]. Domain decomposition method naturally enables parallelization in the multicore environment. When dealing with large-scale and spatial multiscale equipment, the DDM transforms the original problem into multiple simpler sub-regions based on geometrical or media properties, and employs appropriate transmission conditions to ensure the continuity of sub-region inter-field and surface current [7]. DDM can be generally divided into three categories: overlapping DDM, non-overlapping DDM, and equivalence theorem based DDM [8]. The first two types are mainly implemented in the calculation of FEM.

Performance of the aforementioned two kinds of 4-stages FLTD topologies will be analyzed numerically based on the time domain FIT. The aim of the research is therefore to investigate the applicability of FIT in electromagnetic analysis for multi-stages series-connected FLTD induction cavity.

2 Numerical Model of the 4-Stage FLTD Cavity

2.1 Time Domain Finite Integration Technique

In FIT, DoF is defined along the edge or through the facet of a grid, where integral of grid edge is called voltage and integral of grid facet is called flux [9]

$$\widehat{e}_i = \int_{L_i} \boldsymbol{E} \cdot d\boldsymbol{s}, \quad \widehat{\widehat{b}}_k = \int_{A_k} \boldsymbol{B} \cdot d\boldsymbol{A} \tag{1}$$

$$\widehat{h}_k = \int\limits_{\tilde{L}_k} \boldsymbol{H} \cdot ds, \quad \widehat{d}_i = \int\limits_{\tilde{A}_i} \boldsymbol{D} \cdot d\boldsymbol{A}, \quad \widehat{i}_i = \int\limits_{\tilde{A}_i} \boldsymbol{J} \cdot d\boldsymbol{A} \tag{2}$$

where $\widehat{e}_i$ and $\widehat{b}_k$ denote electric voltage along the edge and magnetic flux through the facet of primal grid, respectively; $\widehat{h}_k$, $\widehat{d}_i$, and $\widehat{i}_i$ denote magnetic voltage along the edge, electric flux through the facet, and electric current through the facet of dual grid, respectively; L_i is the ith edge of primal grid, $\tilde{A}_i$ is the ith facet of dual grid, A_k is the kth facet of primal grid, and $\tilde{L}_k$ is the kth edge of dual grid.

The Maxwell's equations in integral form are

$$\begin{aligned}
&\oint\limits_{\partial A} \boldsymbol{E} \cdot d\boldsymbol{s} = -\iint\limits_{A} \frac{\partial \boldsymbol{B}}{\partial \boldsymbol{t}} \cdot d\boldsymbol{A} \\
&\oint\limits_{\partial \tilde{A}} \boldsymbol{H} \cdot d\boldsymbol{s} = \iint\limits_{\tilde{A}} \left(\frac{\partial \boldsymbol{D}}{\partial \boldsymbol{t}} + \boldsymbol{J} \right) \cdot d\boldsymbol{A} \\
&\iint\limits_{\partial V} \boldsymbol{B} \cdot d\boldsymbol{A} = 0 \\
&\iint\limits_{\partial \tilde{V}} \boldsymbol{D} \cdot d\boldsymbol{A} = \iiint\limits_{\tilde{V}} \rho dV
\end{aligned} \tag{3}$$

where $\boldsymbol{E}, \boldsymbol{B}, \boldsymbol{H}, \boldsymbol{D}, \boldsymbol{J}$, and ρ denote electric field intensity vector, magnetic flux density vector, magnetic field intensity vector, electric displacement vector, electric current density vector, and electric charge density, respectively.

Based on the integral form of DoF defined in Eqs. (1) and (2), we can obtain the so-called Maxwell's grid equations (MGE)

$$\begin{aligned}
&\mathbf{C}\widehat{\mathbf{e}} = -\frac{d}{dt}\widehat{\mathbf{b}} \\
&\widetilde{\mathbf{C}}\widehat{\mathbf{h}} = \widehat{\mathbf{i}}_C + \widehat{\mathbf{i}}_S + \frac{d}{dt}\widehat{\mathbf{d}} \\
&\mathbf{S}\widehat{\mathbf{b}} = \mathbf{0} \\
&\widetilde{\mathbf{S}}\widehat{\mathbf{d}} = \widehat{\mathbf{q}}
\end{aligned} \tag{4}$$

where matrices $\mathbf{C}$ and $\mathbf{S}$ are discrete curl and divergence operator of primal grid, $\tilde{\mathbf{C}}$ and $\tilde{\mathbf{S}}$ are discrete curl and divergence operator of dual grid, $\widehat{\mathbf{i}}_C$ and $\widehat{\mathbf{i}}_S$ are the loss term and source term of electric current, respectively. $\widehat{\mathbf{q}}$ denotes the charge content.

The material constitutive relations and their discrete counterpart in FIT are

$$\boldsymbol{D} = \varepsilon \boldsymbol{E}, \quad \boldsymbol{B} = \mu \boldsymbol{H}, \quad \boldsymbol{J} = \sigma \boldsymbol{E} \tag{5}$$

$$\widehat{\mathbf{d}} = \mathbf{D}_\varepsilon \widehat{\mathbf{e}}, \ \widehat{\mathbf{b}} = \mathbf{D}_\mu \widehat{\mathbf{h}}, \ \widehat{\mathbf{i}}_C = \mathbf{D}_\sigma \widehat{\mathbf{e}} \tag{6}$$

where ε, μ, σ are permittivity, permeability, and conductivity, respectively; $\mathbf{D}_\varepsilon$, $\mathbf{D}_\mu$, $\mathbf{D}_\sigma$ denote diagonal matrices of permittivity, permeability, and conductivity, respectively.

To achieve second order accuracy for the two curl equations in MGE, temporal discretization is carried out with central difference. Therefore, explicit time-updating is employed by the staggered leap-frog (SLF) algorithm

$$\widehat{\mathbf{e}}^{n+1} = \underline{CAE} \cdot \widehat{\mathbf{e}}^{n} + \underline{CAH} \cdot \left(\tilde{\mathbf{C}} \widehat{\mathbf{h}}^{n+1/2} - \widehat{\mathbf{i}}_S^{n+1/2} \right) \tag{7}$$

$$\widehat{\mathbf{h}}^{n+1/2} = \widehat{\mathbf{h}}^{n-1/2} - \Delta t \mathbf{D}_\mu^{-1} \mathbf{C} \widehat{\mathbf{e}}^{n} \tag{8}$$

where coefficients $\underline{CAE}$ and $\underline{CAH}$ are

$$\underline{CAE} = \left(\frac{\mathbf{D}_\varepsilon}{\Delta t} - \frac{\mathbf{D}_\sigma}{2} \right) \Big/ \left(\frac{\mathbf{D}_\varepsilon}{\Delta t} + \frac{\mathbf{D}_\sigma}{2} \right) \tag{9}$$

$$\underline{CAH} = 1 \Big/ \left(\frac{\mathbf{D}_\varepsilon}{\Delta t} + \frac{\mathbf{D}_\sigma}{2} \right) \tag{10}$$

The transient electromagnetic process can be calculated according to Eqs. (7) and (8). Time step Δt should meet the Courant-Friedrich-Levy (CFL) stability constraint for stable time integration.

2.2 *Simulation Models of the FLTD Cavity*

The simulation model is depicted in Figs. 1 and 2. The basic unit in FLTD is a "brick", which consists of two film capacitors and one gas switch connected in series. In each stage of FLTD cavity, there are 23 main discharging bricks and one triggering brick arranged azimuthally around the magnetic core, and they are parallel-connected to the

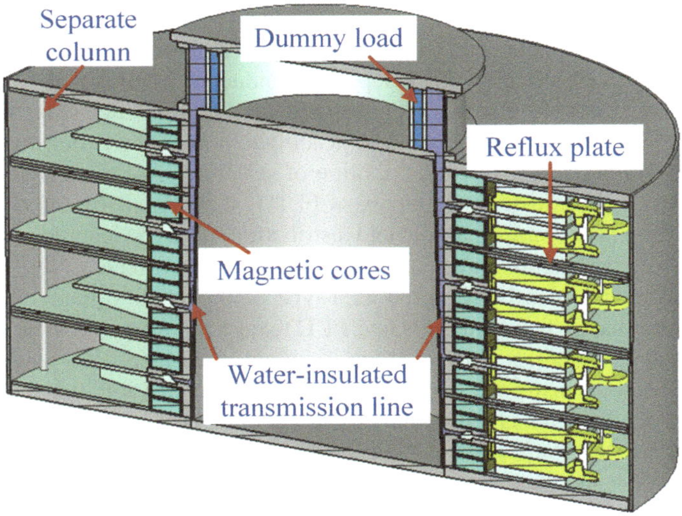

Fig. 1 Four-stage FLTD with sharing common cavity

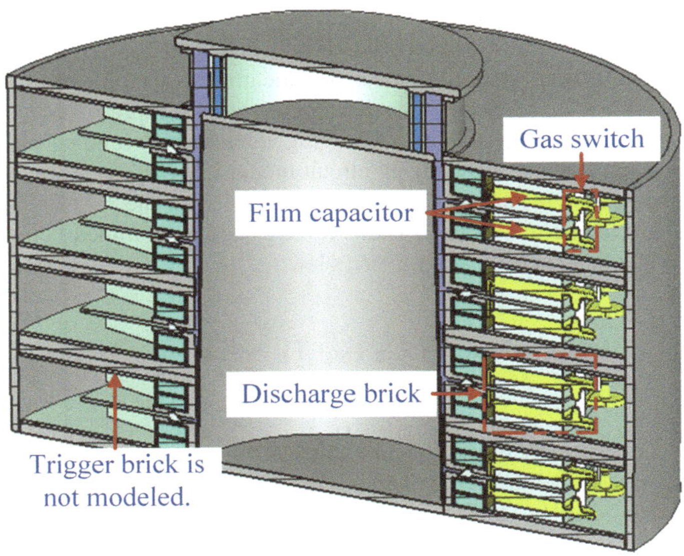

Fig. 2 Four-stage FLTD with independent cavity

secondary side of the internal water-insulated transmission line (WITL) impedance transformer. The outer diameter of the cavity is ~2200 mm. The total height of the 4-stages FLTD with sharing- and independent-induction cavity is ~1324 mm and ~1464 mm, respectively. Hence, compared with the independent cavity, the sharing cavity has a more compact configuration, leading to higher power density.

Considering the complicated distribution of structures and materials and larger differences in spatial-scale in the cavity of FLTD, some details in the engineering model that have little or no impact on field distribution should be ignored or simplified for successful computations, such as film capacitors, gas switches.

A FLTD module can be regarded as a type of induction voltage adder (IVA) [10]. Assume that switches in the same FLTD stage are triggered simultaneously and switches in the adjacent cavities are triggered in an ideal IVA sequence. The WITL impedance transformer, with a varying inner diameter with linear profile and a constant outer diameter, is approximated to be the coaxial TL. Based on the designed structure of TL, only transverse electromagnetic (TEM) mode electromagnetic waves exist. To transfer the maximum electrical power to the load side, the characteristic impedance of WITL impedance transformer at the output of 4-stage series-connected FLTD cavity should be equal to that of the load impedance.

The zero tangential magnetic field symmetry boundary condition is employed for the numerical model with over 20 million primal grids. The outer boundary of computational domain is the convolution perfectly matched layer (CPML). The bottom of the induction cavity is connected to the ground plane.

Since equivalent capacitance of the discharging brick is represented by a lumped element in the circuit model, the initial charging voltage of the lumped capacitor can be set according to requirement. When circuit model is turned on, pulsed discharge current generated by the lumped capacitor is coupled to the 3-D field model as excitations. The field-circuit coupling is achieved by adding a lumped element current term to the right hand side of the Ampère's law

$$\tilde{\mathbf{C}}\widehat{\mathbf{h}} = \widehat{\mathbf{i}}_{C} + \widehat{\mathbf{i}}_{S} + \widehat{\mathbf{i}}_{L} + \frac{d}{dt}\widehat{\mathbf{d}} \tag{11}$$

where $\widehat{\mathbf{i}}_{L}$ denotes the discharge capacitor current.

To improve calculation speed, domain decomposition method is applied to the series-connected 4-stages FLTD. In this research, DDM based on the field equivalence theorem is adopted, whose basic idea is transforming the original model into the problem of equivalent electric- and magnetic-current flowing on the virtual surfaces among different sub-models. In calculating the 4-stage series-connected FLTD, because of the similar structure between the modules of different levels, the module of the same level can be divided into 6 sub-regions, then the whole solution area is divided into 24 sub-regions. Multi-core parallel computing is used on a single workstation, with a separate CPU core for each solved sub-region for accelerated processing of individual computing tasks.

3 Results and Discussion

During the simulation, equivalent capacitance of a single discharging brick is 50 nF and the initial charging voltage is 160 kV (±80 kV for each capacitor). The load resistance is set to 0.4 Ω. In FLTD cavity region, the maximum grid length is set to 1/30 of wavelength, while in background region, the maximum grid length is set to 1/10 of wavelength. The primal grid is chosen to be orthogonal hexahedrons.

The simulated current waveforms are shown in Fig. 3. It can be observed from Fig. 3 that output current of the two configurations is in good agreement. The maximum current is ~826 kA and rise time (10–90%) is ~91 ns. The leakage current flowing through the outer shell of the two FLTD cavities is ~10 kA, which is approximately only 1.2% of the output current amplitude. Thus, the magnetic cores in the primary side of the induction cavity prevent discharging current flowing to the outer shell.

Voltage stress at the end of water-insulated transmission line impedance transformer of each stage of the secondary side FLTD cavity should be validated. The voltage waveforms in Fig. 4 demonstrate that reflections exist in the water-insulated transmission line impedance transformer, especially in the first- and second-stage. Therefore, impedance profile of the transmission line in the secondary side of the FLTD cavity should be further optimized in the future study. Magnitude of electric field intensity at the instant of output voltage peak is illustrated in Figs. 5 and 6, which indicate that electric field intensity are largely distributed in the water-insulated transmission line and water-insulator interface region.

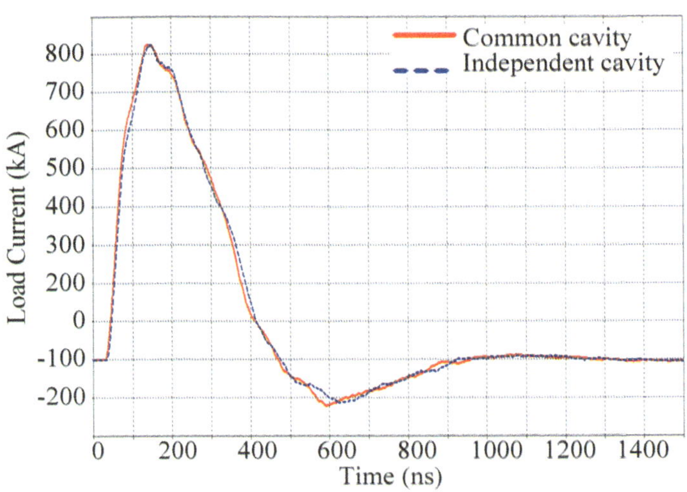

Fig. 3 Load current waveforms of the two FLTD cavity structures

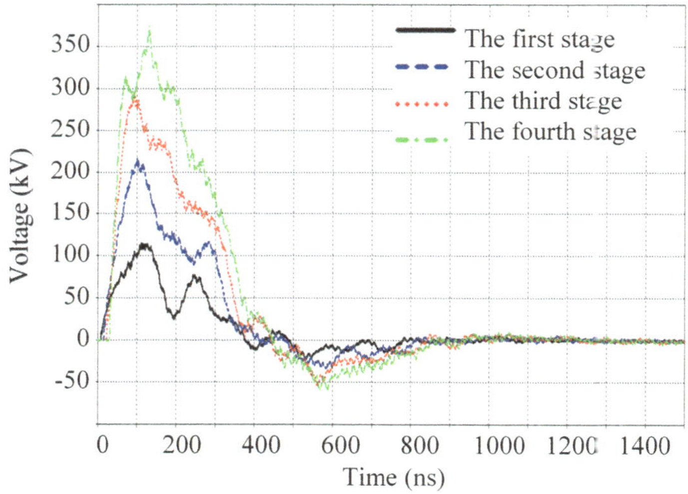

Fig. 4 The voltage waveforms at the end of transmission line of each stage

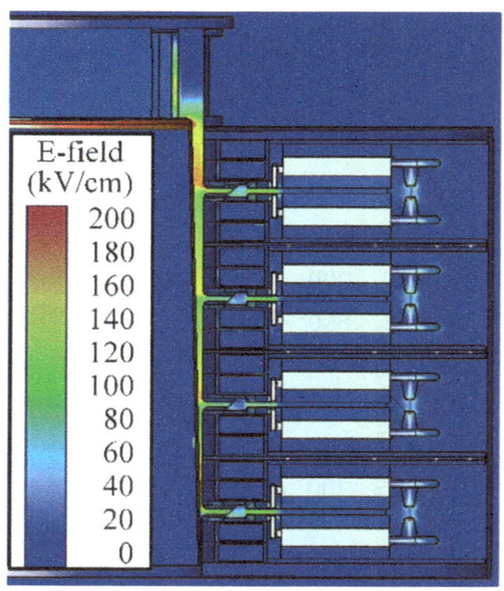

Fig. 5 Cutaway view of electric field intensity magnitude of FLTD with common cavity

4 Conclusion

Numerical model of the 4-stage FLTD with common- and independent-cavity is established by using the circuit-field coupled time-domain finite integration theory. Output performance and electromagnetic field distribution of the two structures are very similar, verifying feasibility of the FLTD with sharing common cavity.

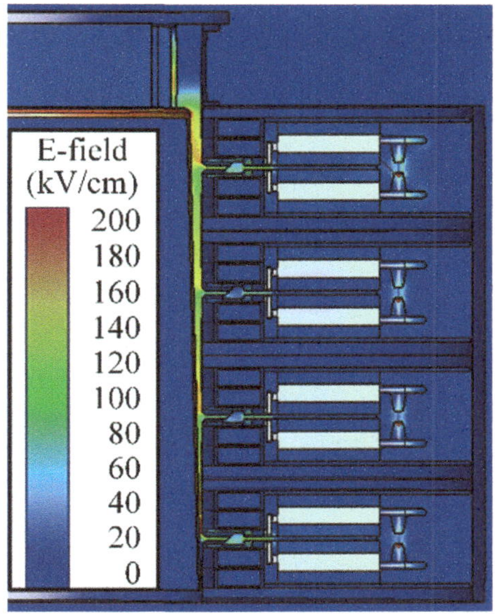

Fig. 6 Cutaway view of electric field intensity magnitude of FLTD with independent cavity

Acknowledgements This work is supported by the National Natural Science Foundation of China under Grant (51790521).

References

1. McBride, R.D., W.A. Stygar, M.E. Cuneo, et al. 2018. A primer on pulsed power and linear transformer drivers for high energy density physics applications. *IEEE Transactions on Plasma Science* 46 (11): 3928–3967.
2. Wang, L., F. Sun, A. Qiu, et al. 2020. Investigating the influence of the wire-arrays' electrical parameters on the load current of the z-pinch drivers. *AIP Advances* 10 (6): 65024.
3. Sun, F., J. Zeng, T. Liang, et al. 2016. A novel triggering technique based on an internal brick and azimuthal line in cavities for linear transformer drivers. *Journal of Applied Physics* 7 (1): 010401. (in Chinese).
4. Weiland, T. 1984. On the numerical solution of Maxwell's equations and applications in the field of accelerator physics. *Particle Accelerator* 15: 245–292.
5. Weiland, T. 2017. Electromagnetic simulators—status and future directions. *IET Science, Measurement and Technology* 11 (6): 681–686.
6. Qiu, H., S. Wang, F. Sun, et al. 2020. Transient electromagnetic field analysis for the single-stage fast linear transformer driver with two different configurations using the finite-element method and finite integration technique. *IEEE Transactions on Magnetics* 56 (4): 7515805.
7. Schulze, S., and U. van Rienen. 2007. Computation of land mine signatures using domain decomposition with Lagrange multipliers. *IEEE Transactions on Magnetics* 43 (4): 1189–1192.
8. Jiang, M., J. Hu, Y. Chen, et al. 2020. The progress of domain decomposition method based on integral equation. *Chinese Journal of Radio Science* 35 (2): 169–177 (in Chinese).

9. Udosen, N.I., and N.J. George. 2018. A finite integration forward solver and a domain search reconstruction solver for electrical resistivity tomography (ERT). *Modeling Earth Systems and Environment* 4 (1): 1–12.
10. Liu, P., F. Sun, H. Wei, et al. 2012. Influences of switching jitter on the operational performances of linear transformer drivers-based drivers. *Plasma Science and Technology* 14 (4): 347–352.

Advanced Frozen Intelligent Control System of PMSM Optimized by Cerebellar Model Articulation Controller

Xiang Zhao, Chengsheng Wang, Wei Duan, Zhiming Lan, and Jun Jiang

Abstract For permanent magnet synchronous motor speed and rotor position easily affected by the mechanical sensor signal transmission precision, to improve the performance of permanent magnet synchronous motor drive, the speed control system satisfies the requirement of fast response, and the complicated condition of frequent load disturbance has stronger resistance, this paper designed a small nerve network optimization of permanent magnet synchronous motor intelligent control system. The intelligent control system will be small neural networks applied to the model reference adaptive control of permanent magnet synchronous motor speed control system. On this basis, the structures, permanent magnet synchronous motor control system simulation, the simulation results show the system can accurately measure the rotational speed, rotor position and compared with the traditional model reference adaptive control, motor start-up speed no overshoot, has the stronger ability to resist load disturbance and load disturbance after speed restore faster (6 ms), in the motor control has strong practical application value.

Keywords PMSM · Cerebellar model articulation controller · Model reference adaptive control · Sensorless control

1 Introduction

In the traditional motor control system, the precise measurement of the rotor position of the motor depends on the signal transmitted by the mechanical encoder, which also leads to the increase of the weight and cost of the motor, which is not suitable for the clever occasions [1, 2]. Sensorless technology solves this problem well, and the model estimation method is used to predict the motor speed and rotor position. Common estimation models include model reference adaptive (MRAS), sliding mode observer (SMO), extended Kalman filter (EKF), fuzzy theory, etc [3]. With the rise of artificial intelligence technology in recent years, more and more researchers

X. Zhao (✉) · C. Wang · W. Duan · Z. Lan · J. Jiang
Automation Research and Design Institute of Metallurgical Industry, Beijing, China
e-mail: 1714101525@qq.com

W. Chen et al. (eds.), *The Proceedings of the 9th Frontier Academic Forum of Electrical Engineering*, Lecture Notes in Electrical Engineering 743,
https://doi.org/10.1007/978-981-33-6609-1_46

will focus their research on the integration of artificial intelligence technology with traditional control methods, so as to improve the control accuracy and robustness of nonlinear control systems such as PMSM. Moreover, cerebellar model articulation controller (CMAC) has prominent advantages in modeling of nonlinear systems. CMAC can suppress the system's nonlinearity, weaken the influence of parameter changes on the system during the motor operation, and ensure the stability and anti-interference of the system [4]. However, there is a over-learning phenomenon, which needs to be suppressed.

This paper designs an intelligent control system which uses small neural network feed-forward control, MRAS for online eliminate the motor speed and tracking error between the reference speed, restrain CMAC over-learning phenomenon at the same time, the PID algorithm to realize the permanent magnet synchronous motor speed control system feedback control, gives the initial response to the all-digital fuzzy, at the same time to ensure the stability of the system. On this basis, this paper set up a permanent magnet synchronous motor control system simulation, simulation results show the system can accurately measure the rotational speed [5], rotor position and rotor position deviation within ± 0.1 rad fluctuations, compared with the conventional model reference adaptive control scheme, speed recovery time shorter (6 ms), no overshoot, has the stronger ability to resist load disturbance, the dynamic performance is better.

2 Mathematical Model of PMSM

In this paper, three-phase PMSM is selected as the research object, ignoring the saturation effect of the iron core [6]. The mathematical model is established as follows. The stator flux linkage equation of the motor is expressed as:

$$\begin{cases} \psi_{\mathrm{d}} = L_d i_d + \psi_f \\ \psi_{\mathrm{q}} = L_q i_q \end{cases} \tag{1}$$

The stator voltage equation of the motor is expressed as:

$$\begin{cases} Ud = Ri_d + Ld\frac{di_d}{dt} - \omega_e L_q i_q \\ U\mathrm{q} = Ri_q + L_q\frac{di_q}{dt} + \omega_e(L_d i_d + \psi_f) \end{cases} \tag{2}$$

The mechanical motion equation of the motor is:

$$J\frac{\mathrm{d}\omega}{dt} = T_e - T_l - B_\omega \tag{3}$$

The electromagnetic torque equation of the motor can be expressed as:

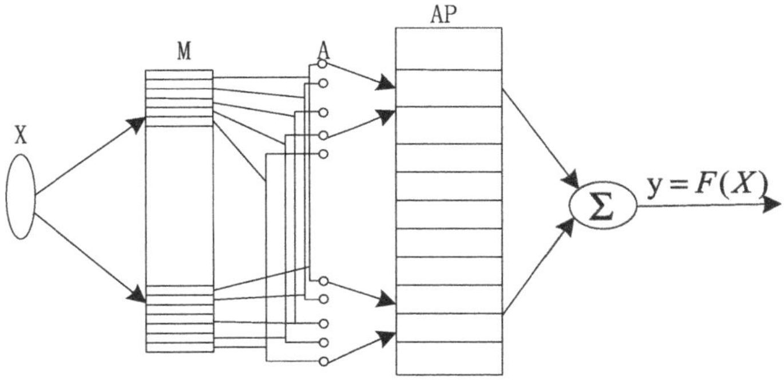

Fig. 1 Structural block diagram of CMAC

$$Te = 1.5P_n i_q[(L_d - L_q)i_d + \psi_{\mathrm{f}}] \tag{4}$$

The speed reference model of PMSM is the curve of expected speed of PMSM In this paper, a first-order inertial system is selected to achieve simple control, and its transfer function is expressed as follows.

$$G(S) = \frac{\mathrm{n}_{ref}(s)}{n*(s)} = \frac{1}{T_{ref}s + 1} \tag{5}$$

3 Intelligent Control System of PMSM Optimized by CMAC

3.1 Cerebellar Model Articulation Controller

PMSM intelligent control system optimized by CMAC makes use of the weight optimization ability of CMAC, so as to classify and store the information. Its structure is shown in Fig. 1.

3.2 Design of Controller for CMAC Optimization

The design of model reference adaptive controller for CMAC optimization is shown in Fig. 2.

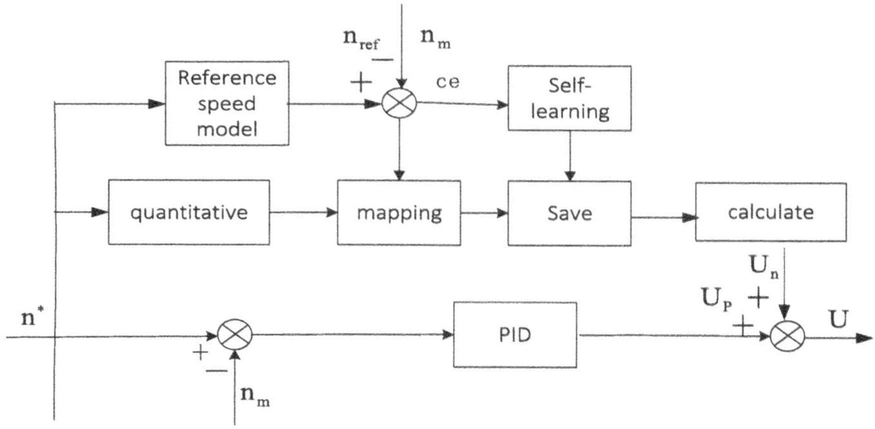

Fig. 2 Model reference adaptive controller structure for CMAC optimization

For control system, each control cycle to calculate the small neural network output Un(k), and the optimized model reference adaptive controller output U(k), according to the difference between doing a weight correction, CMAC self-learning process at the same time, makes the neural network output and the whole the total difference approximation is 0, the output of the control system control algorithm is shown in the following type.

$$U_k = U_n(k) + U_p(k) \tag{6}$$

$$U_{\mathrm{n}}(k) = \sum_{i=1}^{c} \omega_i a_i \tag{7}$$

In the above formula, U(k) represents the total output of the optimized model reference adaptive controller, Un(k) represents the output of the CMAC, Up(k) represents the output of PID control, ai represents the binary selection vector, and C represents the generalization parameter of the network. The principle of mapping and learning algorithm of CMAC is shown as follows.

Principle of CMAC concept mapping: The input space X is divided into N + 2c intervals in the interval $[X_{\mathrm{min}}, X_{\mathrm{max}}]$, namely:

$$\begin{cases} v_1 \cdots v_c = X_{\mathrm{min}} \\ v_j = v_{j-1} + \Delta v_j (j = c+1, \ldots, c+N) \\ v_{N+C+1} + v_{N+2c} = X_{\max} \end{cases} \tag{8}$$

Practical mapping principle of CMAC:

$$\text{aj} = \begin{cases} 1 & Xj \in [vj, vj + c] \\ & j = [c + 1, \ldots, c + N] \\ 0 \text{ others} \end{cases} \quad (9)$$

In order to make the difference between the output of the CMAC and the total output of the optimized model reference adaptive controller approximately 0, after each control cycle, the values of Un(k) and U(k) should be compared, the weights of the CMAC should be modified, and CMAC should enter the self-learning process at the same time, until the control system stops. The gradient descent method was adopted as the weight modification strategy, and the weight optimization was carried out according to Eqs. (10)–(12), with the difference between the reference speed and the actual speed of the PMSM as the variable. Weight adjustment is introduced according to the difference between the reference speed and the actual speed:

$$\text{E(k)} = 12(\text{n}_\text{m}(\text{k}) - \text{n} * (\text{k}))2 \cdot 1\text{c}(13) \quad (10)$$

The adjustment amount of weight in each week is expressed as:

$$\Delta\omega = -\eta\frac{\partial E(K)}{\partial\omega} = \eta\frac{\text{nm}(k) - n * (k)}{c}ai = \eta\frac{ec(k)}{c}ai \quad (11)$$

The final adjustment result of each week's option value is expressed as:

$$\omega(\text{k}) = \omega(\text{k} - 1) + \Delta\omega(\text{k}) + \alpha(\omega(\text{k}) - \omega(\text{k} - 1)) \quad (12)$$

In the formula, the represents the learning speed of CMAC, $\eta \in (0, 1)$; Represents inertia, and $\alpha \in (0, 1)$; Ec(k) represents the difference between the reference speed and the actual speed.

After the control system began to run, first make CMAC weight = 0, at this time Un = 0, U = Up, the speed of the outer ring using PID controller, at the same time to the controlled motor initial response. Through the self-learning process of CMAC, the difference value between the actual speed of the motor and the reference speed gradually becomes 0, thus realizing the output Up of the PID controller finally becomes 0, and the output Un(k) of CMAC is ultimately equal to the total output U(k) of the optimized model reference adaptive controller.

3.3 PMSM Intelligent Control System Design

In this paper, PMSM control system adopts current loop and speed loop double closed loop control structure [7, 8]. Among them, the current inner loop is still regulated by conventional PID controller, which can restrain the disturbance in the current loop, improve the rapid response of the system, and control the upper limit of current to ensure the operation safety of PMSM system. Since the speed of the outer ring can

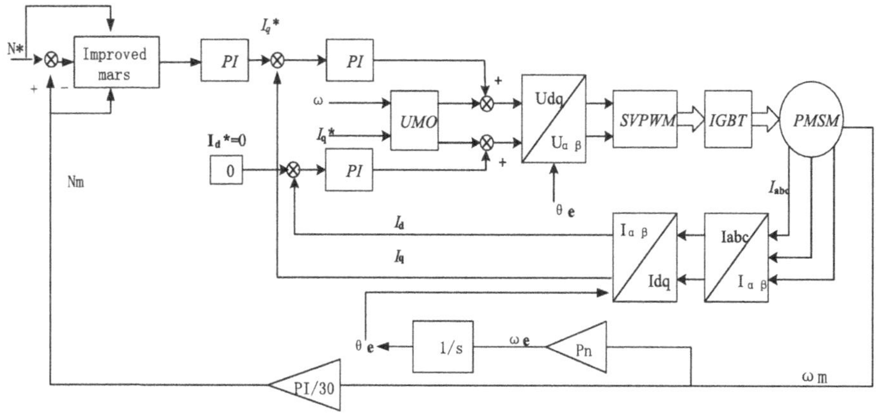

Fig. 3 Model reference adaptive controller structure for CMAC optimization

determine the performance of the system, the interference caused by the inner ring can also be suppressed and compensated by the outer ring. Will be described in this paper, the CMAC optimization model reference adaptive controller applied in the speed loop, used in inverter part id = 0 voltage space vector pulse width modulation (SVPWM) method. The control block diagram of the system is shown in Fig. 3.

4 The Simulation Results

The basic parameters of PMSM used for simulation in this paper are shown in Table 1.

The model reference adaptive controller for CMAC optimization is set for speed ring control, with PID parameters KP = 0.6, Ki = 0, KD = 0.12. Ki zero makes the self-learning of CMAC only dependent on the error value of this time. The generalized value c = 50, the learning step length = 0.6, the quantized series N = 1000, and the inertia = 0.02 to adjust the output weight. Simulation conditions were set as DC side voltage 311 V, SVPWM switching frequency 10 kHz, sampling period 10×10^{-6}s, and variable step size ode23tb algorithm.

Firstly, the PMSM was set to start without load. The reference speed was set as n* = 1000 r/min, and the torque disturbance of 5 N was added to the motor at 0.15 s, that is, TL = 5 N. The simulation results of PMSM based on the model reference adaptive control scheme optimized by cerebellar neural network are shown in Fig. 4.

Table 1 Basic parameters of PMSM

Parameters	U_N	I_N	Rs	Ld	Lq	Pn
Value	380 V	12 A	1.33 Ω	9.16×10^{-3} H	15.03×10^{-3} H	2

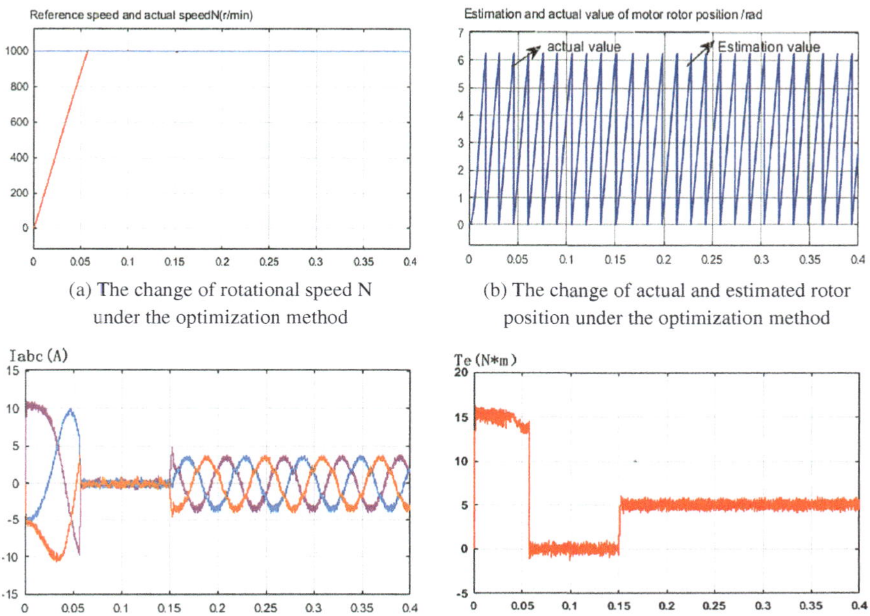

(a) The change of rotational speed N under the optimization method

(b) The change of actual and estimated rotor position under the optimization method

(c) The change of Iabc of three-phase current under the optimization method

(d) The change of electromagnetic torque under the optimization method

Fig. 4 Simulation results of model reference adaptive control scheme for cerebellar neural network optimization

By the simulation results can be seen, when the motor speed rise from 0 to 1000 r/min when small model reference adaptive control scheme of the neural network optimization speed curve almost no overshoot, the rotor position estimation error minimum, make the rotational speed and torque of permanent magnet synchronous motor has a good ability to follow, and resistance to load torque disturbance has a certain effect, good dynamic performance. In order to better analyze the control effect of the model reference adaptive control scheme optimized by cerebrum neural network, the simulation experiment of the traditional model reference adaptive control scheme of permanent magnet synchronous motor was carried out at the same time. The simulation results are shown in Fig. 5.

By comparing the simulation results of the two control schemes, it can be seen that the traditional model reference adaptive control mode makes the motor start torque and speed overshoot larger, and the recovery time is longer when the torque ripple of the reference speed is reached, and the dynamic performance still needs to be optimized. Small nerve network optimization, by contrast, model reference adaptive control scheme has better dynamic performance, the no-load startup phase permanent magnet synchronous motor speed without overshoot, starting torque is reduced greatly, and the load torque change resistance is strong, can more quickly

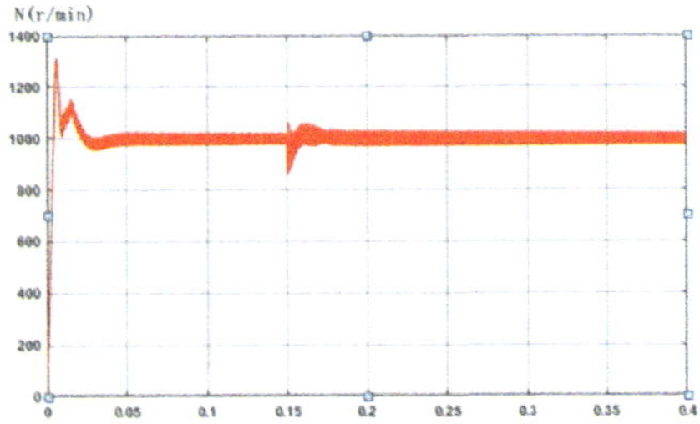

(a)The change of rotational speed N under the traditional method

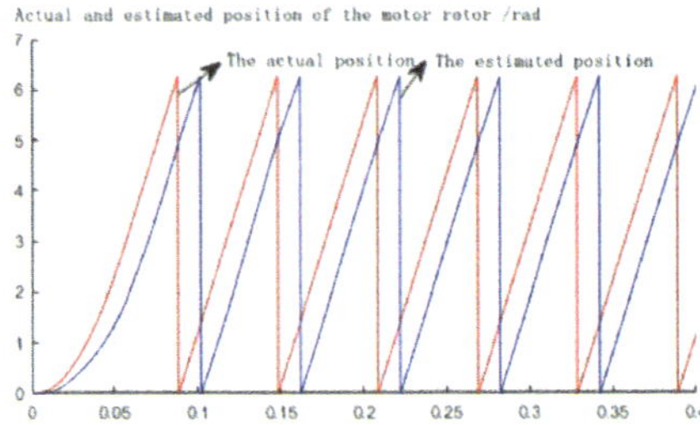

(b)The change of actual and estimated rotor position under the traditional method

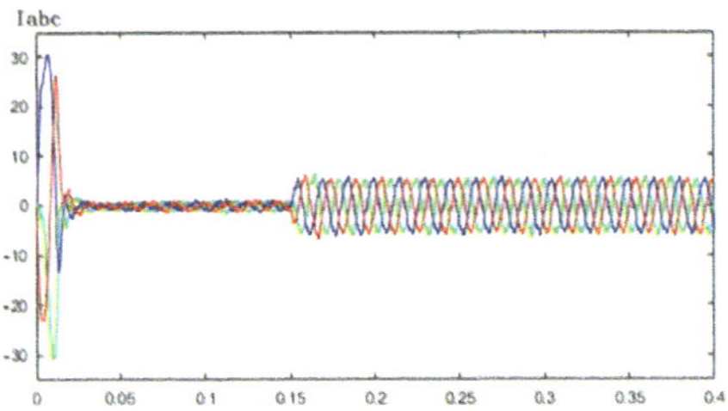

(c)The change of Iabc of three-phase current under the traditional method.

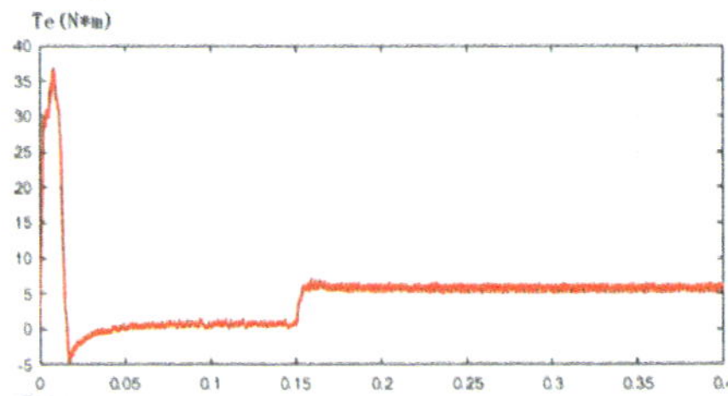

(d)The change of Te under the traditional method

Fig. 5 Simulation results of traditional reference adaptive control scheme for PMSM model are presented

reach a steady state, under load disturbance, produced by the motor speed fluctuation was reduced by 0.22%, only and can recover more quickly to the front of the disturbance state, thus has a better ability to resist load disturbance, but this method is somewhat lacking in terms of speed motor start up time. In general, the PMSM using the optimization method has no speed over-adjustment, which meets the stability requirements of the system and the rapid response requirements of the AC speed control system when the load changes.

5 Conclusion

In order to improve the dynamic performance of PMSM AC speed regulation system, a PMSM model reference adaptive control scheme optimized by PMSM is designed in this paper, which makes full use of the advantages of PMSM network, such as simple structure, fast convergence, good real-time performance and strong robustness. The simulation results show that the optimal control scheme performs well in system stability, performance against load disturbance and robustness, meets the demand of dynamic performance of servo system, and has strong practicability. The future research work will focus more on how to reduce the starting adjustment time during the motor start-up so that the servo system can get better control performance.

References

1. Li, Chong-jian. 2005. Ac synchronous motor speed control system. Science Press (in Chinese).
2. Kivanc, O.C., and S.B. Ozturk. 2018. Sensorless PMSM drive based on stator feed forward voltage estimation improved with MRAS multiparameter estimation. *IEEE/ASME Transactions on Mechatronics* 23 (3): 1326–1337.
3. Yue, Xuelei, Peng Bai, Ruikun Yang, et al. 2016. Application of CMAC-PID in brushless DC motor control system. *Instrument Technology and Sensors* 7: 112–114. (in Chinese).
4. Lin, F.J., K.C. Lu, and B.H. Yang. 2017. Recurrent fuzzy cerebellar model articulation neural network based power control of a single-stage three-phase grid-connected photovoltaic system during grid faults. *IEEE Transactions on Industrial Electronics* 64 (2): 1258–1268.
5. Zhou, Shijiong, Chengsheng Wang, Jun Jiang, Wei Duan, Zhiming Lan, Da Gao. 2019. Improved MRAS control of permanent magnet synchronous motor based on a new voltage model in high power applications. In *2019 21st European Conference on Power Electronics and Applications, EPE 2019 ECCE Europe.*
6. Zhou, Shijiong, Chengsheng Wang, Jun Jiang, Fan Li, Qiongtao Yang, Pan Wang. Improved MRAS control of synchronous motor based on a new reference model in high power inverter. In *2019 22nd International Conference on Electrical Machines and Systems, ICEMS 2019.*
7. Lan, Zhiming, Chongjian Li, Yaohua Yi, Chunyi Zhu, and Chengsheng Wang. 2011. Development of high-power three-level dual-PWM converter based on IGCT. *Journal of Electrotechnics and Technology* 26 (S1): 36–40. (in Chinese).
8. Wang, Chengsheng, Chongjian Li, Yaohua Li, Weihui Sheng, Zhiming Lan. 2007. 7.5MVA high-power three-level IGCT AC-DC converter. *Journal of Electrical Technology* 2007 (08): 24–27 (in Chinese).

Arc Fault Recognition Based on VMD and ELM

Shuxin Liu, Zhenxing Liu, Yang Liu, Yundong Cao, and Jing Li

Abstract The electrical signal of the fault arc in the low-voltage line has its strong characteristics. According to the UL1699 standard, an arc fault generation platform was built to test the common types of loads in life and obtain the electrical signal data of normal and arc faults. This paper proposes a method to obtain the Intrinsic Mode Function (IMF) of the fault arc signal through Variational Mode Decomposition (VMD), and then calculate the sample entropy (SampEn) of the IMF component The analysis found that the first 6 order IMF components can be used as their feature vector input Extreme Learning Machine (ELM) to train a method of identifying fault models to accurately identify and detect arc faults. A new criterion is proposed for arc fault detection and detection. The experimental results show that the method can accurately and quickly detect and identify arc faults.

Keywords Series arc fault · Variational mode decomposition · Extreme learning machine · Fault recognition

1 Introduction

With the development of science and technology, electricity has become an indispensable necessity in people's lives, and electricity safety has also become the top priority in daily life. Data shows that among the identified causes of fire accidents in 2015, 102,000 were caused by electrical failures, accounting for 30.1% of the total, accounting for the highest proportion. In larger fires, the cause of the fire was electrical failure and even reached 56.7%, it can be seen that electrical faults have become the biggest cause of fires [1].

Scholars at home and abroad have done a lot of research on the feature collection and extraction of series fault arc. In terms of experiment, through the simulation experiment of series fault arc, wavelet transform [2–6], Fourier transform [7, 8], EMD decomposition [9] and fractal dimension [10] are used to analyze the signal

S. Liu · Z. Liu (✉) · Y. Liu · Y. Cao · J. Li
School of Electrical Engineering, Shenyang University of Technology, No. 111 West Shenliao Road, Baogong Street, Tiexi District, Shenyang, Liaoning Province, China
e-mail: 353176135@qq.com

W. Chen et al. (eds.), *The Proceedings of the 9th Frontier Academic Forum of Electrical Engineering*, Lecture Notes in Electrical Engineering 743,
https://doi.org/10.1007/978-981-33-6609-1_47

and extract the fault. Characteristics: in terms of simulation, a mathematical model of series arc is proposed for the characteristics of series arc and the simulation verification [11] proves the validity of the model. In terms of the amount of extraction, the current and voltage waveforms of the fault arc are extracted at the same time [12]. After frequency domain analysis of the data, it is found that the high-frequency components of the arc current are mainly concentrated between 10 and 200 kHz, and the sudden change of the power supply voltage can be as a conclusion of the parameters for detecting fault arcs.

2 Key Technology of Fault Arc Recognition

2.1 Variational Mode Decomposition Technique

To solve the variational mode, we must first introduce the inherent mode problem. Starting from the inherent mode problem, by adding another concept of limiting the limited bandwidth, using iterative update to get all the decomposed modes of the signal, and reconstruct the mode of the main signal. Noisy modes are eliminated directly. It not only solves the problem of modal aliasing that may have occurred originally, but also achieves a good denoising effect.

Figure 1 is the IMF component diagram of the current waveform after VMD decomposition when the line fails under resistive load.

Fig. 1 IMF waveform generated by VMD decomposition of current when resistive load fails

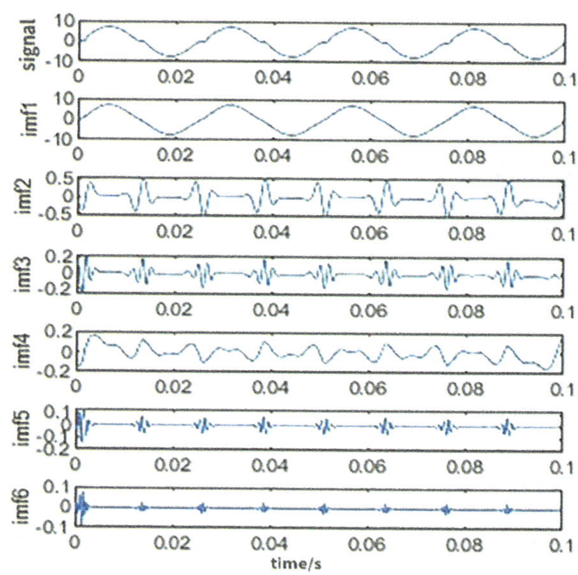

2.2 Sample Entropy Calculation

The number of IMF components obtained by decomposition is huge, and the order of IMF components after EMD decomposition of arc fault current signals with different loads is different, and some IMF components may be noise components or false components (due to insufficient sampling rate and spline The extra components caused by interpolation), these factors are obviously not conducive to the identification of series arc faults. Therefore, the sample entropy is introduced to calculate the IMF component to construct the feature vector. The following briefly introduces the sample entropy calculation method.

Given a set of time series consisting of N points: $\{x(n)\} = x(1), x(1), \ldots, x(N)$, the calculation steps of sample entropy are as follows:

1. N − m + 1 m-dimensional vectors $x_m(i)$ composed of the original sequence:

$$x_m(i) = \{x(i), x(i+1), \ldots, x(i+m-1)\}, i = 1, 2, \ldots, N-m+1 \quad (1)$$

2. Define the distance between the vectors $x_m(i)$ and $x_m(j)$ as $d[x_m(i), x_m(j)]$:

$$d[x_m(i), x_m(j)] = \max_{k=0,\ldots,m-1} [|x(i+k) - x(j+k)|] \quad (2)$$

3. The dimension plus 1 means that when the dimension is m + 1 dimension, repeat the above steps to get $A_i^m(r)$, which is defined as:

$$A_i^m(r) = \frac{1}{N-m+1} A_i \quad (3)$$

4. So when the tolerance r is similar, $B_m(r)$ and $A_m(r)$ are the matching probabilities of the sequence pair m and m + 1 points, respectively. At this time, the sample entropy is defined as:

$$SampleEn(m, r) = \lim_{N \to \infty} \left\{ -\ln\left[\frac{A^m(r)}{B^m(r)}\right] \right\} \quad (4)$$

When N takes a finite value, the estimated value of sample entropy is:

$$SampleEn(m, r, N) = -\ln\left[\frac{A^m(r)}{B^m(r)}\right] \quad (5)$$

The calculation of sample entropy needs to determine the parameters and. Literature pointed out that the calculation of sample entropy does not depend on the length of the data and has very good consistency. Based on experience, this paper takes m = 2, r = 0.1 STD (serial standard deviation).

3 Fault Identification Method Based on VMD Decomposition-Sample Entropy Calculation and ELM

The flowchart of the fault identification method based on VMD decomposition-sample entropy calculation combined with ELM described above is shown in Fig. 2.

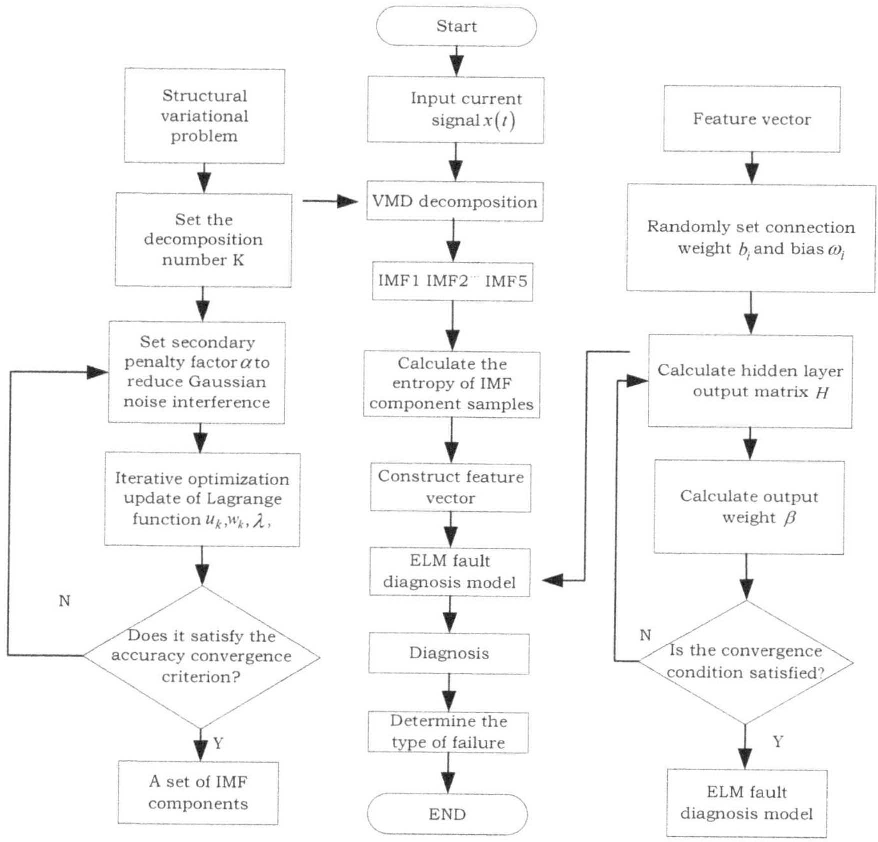

Fig. 2 VMD decomposition-sample entropy calculation combined with ELM fault identification method flow chart

Fig. 3 Principle diagram of the experimental circuit of current measurement

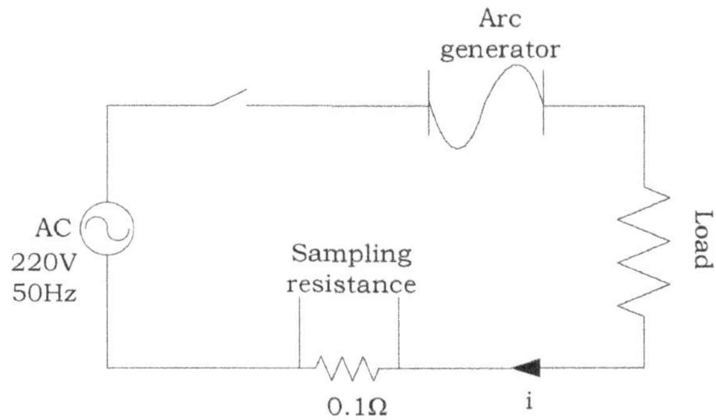

Table 1 Current measurement experiment scheme of load

Load type	R	RL	Hand drill	Computer	Induction
	26.5 Ω	26.5 Ω + 6 mH	200 W	300 W	2000 W
Normal (group)	20	10	10	10	10
Failure (group)	20	10	10	10	10

4 Fault Diagnosis Example Analysis

4.1 Fault Current Detection Experiment

When there are different types of loads in the line, the electrical quantity of the fault arc will be quite different, and the differences between them are more obvious. The experimental principle is shown in Fig. 3. Three types of loads were used in the experiment: resistive load (26.5 Ω), resistive inductive load (26.5 Ω + 6 mH), non-linear load (induction cooker, desktop computer, and electric drill). Because the arc generated in each experiment will interfere with randomness and instability, the experiment uses repeated measurement methods to measure 30 sets of data under normal and fault conditions for the same load. The scheme is shown in Table 1, which can greatly reduce the error caused by the inaccuracy of the data during the training of the diagnostic model.

4.2 Experimental Results and Analysis

Limited space only shows typical waveforms under different working conditions. The hand electric drill experiment uses a gear with a power of 200 W, and the measured current waveform is shown in Fig. 4.

Comparing all the current waveform diagrams above, it can be seen that the current waveform has obvious periodicity under normal conditions. It can be roughly

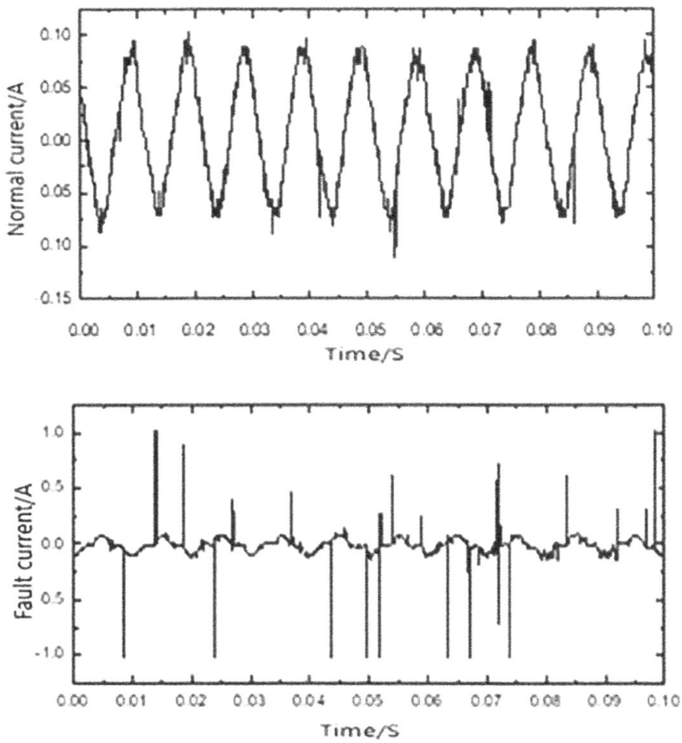

Fig. 4 Hand electric drill normal and fault current waveform

regarded as a sine wave with less distortion. In the event of a fault, there is one thing in common, that is, there are zeros. The phenomenon of suspending, and there will be obvious high-frequency noise when a fault arc occurs.

Table 2 shows the calculation results of the sample entropy of the IMF1–IMF5 components obtained after VMD decomposition of the current data of each load in the normal and fault states.

Table 2 Sample entropy under different loads

	R		RL		Computer		Hand drill		Induction	
	Normal	Fault	Normal	Fault	Normal	Fault	Normal	Fault	Normal	Fault
IMF1	4.6162	5.4913	4.5529	5.4597	4.0552	6.8300	4.5184	6.1640	4.6131	5.3105
IMF2	3.5126	4.9312	3.1720	4.3000	3.3052	5.5144	3.0923	4.1043	3.3310	4.0361
IMF3	1.0776	2.7796	1.1955	1.9038	1.1164	3.2564	1.7520	1.6617	0.2766	0.7496
IMF4	0.2844	0.7448	0.3050	0.7944	0.0922	1.6105	0.7202	0.6284	0.2766	0.7496
IMF5	0.0844	0.0668	0.0416	0.0690	0.0710	0.2959	0.0972	0.0787	0.910	0.849

Table 3 Sample distribution list

Type	Number of samples			
	Training set		Test set	
	Normal	Fault	Normal	Fault
R fault	15	15	15	15
RL fault	15	15	15	15
Hand drill fault	15	15	15	15
Computer fault	15	15	15	15
Induction fault	15	15	15	15

4.3 *Construction of an Arc Fault Diagnosis Model Based on ELM*

After VMD decomposition of the experimentally measured data, the first five-order IMF components are obtained for sample entropy calculation feature vector, and the obtained sample entropy is input as feature vector into ELM for model building. For each load, 30 sets of current signal feature vectors under normal conditions and 30 sets of series arc fault conditions are obtained. The samples are now grouped. 15 sets of feature vectors are proposed as test set samples for each load, and the remaining 15 sets are used as training sets. The sample allocation table is shown in Table 3.

The activation function of the hidden layer of the ELM network is a Sigmoid type function. After testing, when the number of nodes in the hidden layer of the ELM is set to 36, the ELM reaches the maximum accuracy. Table 4 shows the identification test results of this model for various types of loads.

It can be seen from the table that the ELM fault recognition model established by this method can effectively identify the fault, and the overall recognition accuracy rate is over 93%, which is an efficient and feasible method.

Table 4 ELM test results

Working condition	Test	Identify	Correct rate (%)
R fault	15	14	93.3
RL fault	15	13	86.6
Hand drill fault	15	14	93.3
Computer fault	15	14	100
Induction fault	15	14	93.3

5 Conclusion

In this paper, by building an experimental platform for arc fault occurrence, the current signals of different types of loads in the line (resistive load of 26.5 Ω, resistive load of 26.5 Ω + 6 mH, hand drill, computer and induction cooker) are collected during normal operation and faults. Established a fault arc detection database. The VMD decomposition method can not only reduce the noise of the signal, but also extract the characteristics of the signal. Furthermore, in view of the problem that the decomposed IMF components are many and may contain interference, the sample entropy of the IMF components is calculated to extract the respective Features, construct feature vectors and input them into ELM for model training, greatly reducing training time. And from the experimental results, it can be seen that the accuracy of the identification of series fault arc using this method is as high as 93%, and the identification rate of specific loads can reach 100%, and the identification rate of nonlinear loads is higher than that of linear loads.

The results show that this feature extraction method has good feasibility and effectiveness when applied to series fault arc detection.

Acknowledgements This work is supported by the National Natural Science Foundation of China, No. 51977132.

References

1. Lin, Zhang. 2013. *Research on the key technology of fault arc detection and circuit breaker development*. Hangzhou: China Jiliang University. (in Chinese).
2. Liu, Xiaoming, Yang Zhao, Yundong Cao, et al. 2014. Series Arc Fault Diagnosis Based on Wavelet Transform in AC System. *Transactions of China Electrotechnical Society* 29 (1): 10–17. (in Chinese).
3. Oh Y.S., J. Han, G.H. Gwon, et al. 2015. Development of Fault Detector for Series Arc Fault in Low Voltage DC Distribution System Using Wavelet Singular Value Decomposition and State Diagram. *Journal of Electrical Engineering and Technology* 10: 766–776.
4. Tian, Xiaojing, Feng Yu, Yingjie Guo, et al. 2015. Vacuum Arc Morphology Detection Based On Wavelet Transform and Morphological Refinement Algorithm. *Transactions of China Electrotechnical Society* 30 (11): 110–114. (in Chinese).
5. Guo, Fengyi, Kun Li, Changken Chen, et al. 2016. Series Arc Fault Identification Method Based on Wavelet Approximate Entropy. *Transactions of China Electrotechnical Society* 31 (24): 164–172. (in Chinese).
6. Lu, Qiwei, Tao Wang, Zongrui Li, et al. 2017. Detection Method of SERIES Arcing Fault BASED on Wavelet Transform and Singular Value Decomposition. *Transactions of China Electrotechnical Society* 32 (17): 208–217 (in Chinese).
7. Liu, Xiaoming, Yefei Xu, Ting Liu, et al. 2015. The Arc Fault Detection Based on the Current Signal Shorttime Zero Crossing Rate. *Transactions of China Electrotechnical Society* 30(13): 125–133 (in Chinese).
8. Wang, Yao, Yang Li, and Leijiao Ge. 2017. A Series DC Arcing Fault Recognition Method Based on Sliding Discrete Fourier Transform. *Transactions of China Electrotechnical Society* 32 (19): 118–124. (in Chinese).

9. Zhang, Liping, Xiren Miao, and Dunyi Shi. 2016. Research on Low-Voltage Arc Fault Identification Method based on EMD and ELM. *Journal of Electrical Machinery and Control* 20 (9): 58–64. (in Chinese).
10. Yang, Kai, Rencheng Zhang, Jianhong Yang, et al. 2016. Series Arc Fault Diagnostic Method Based on Fractal Dimension and Support Vector Machine. *Transactions of China Electrotechnical Society* 31 (2): 70–77. (in Chinese).
11. Liu, Yanli, Fengyi Guo, Lei Li, et al. 2019. A Mathematical Model of Series Arc Fault. *Transactions of China Electrotechnical Society* 34 (14): 2901–2912. (in Chinese).
12. Liu, Yuan, Shengchang Ji, Lingyu Zhu, Xiu Yao, and Jin Wang. 2015. Research on DC Arc Characteristics and Detection Methods in DC Power Supply System. *High Voltage Apparatus* 51 (02): 24–29. (in Chinese).

Research of Deep Learning Neural Network Based on Regression Analysis in Numerical Simulation Analysis of Motor Stress

Zhang Xin, Wang Wenbin, Song Zihan, Xu Haoyue, and Cai Chenyue

Abstract This paper studies a learning method based on deep learning neural network, and studies the numerical analysis of motor stress simulation, which can obtain the minimum stress value of the required position in the shortest time. A method is proposed to perforate the stator of the motor and fill it with negative magnetostrictive material, change the position and radius of the hole, and find the best position to reduce the noise so as to suppress the vibration and noise of the motor. Through finite element analysis, the stress values on each point of the permanent magnet synchronous motor stator corresponding to different punch positions and radii are obtained as training samples. We established a multiple regression model with 3 fully connected layers, two inputs and one output, and optimized the algorithm to better perform regression analysis on the motor stress value to achieve motor noise optimization.

Keywords Regression analysis · Neural network · Stress analysis of motor · Deep learning · Motor optimization

Z. Xin · W. Wenbin (✉) · S. Zihan · X. Haoyue · C. Chenyue
Tianjin Key Laboratory of New Technology of Electrical and Electric Energy (Tiangong University), Tianjin 300387, China
e-mail: 1831045316@tiangong.edu.cn

Z. Xin
e-mail: zhangxin@tiangong.edu.cn

S. Zihan
e-mail: 1831045291@tiangong.edu.cn

X. Haoyue
e-mail: 1831125470@tiangong.edu.cn

C. Chenyue
e-mail: 1831045324@tiangong.edu.cn

W. Chen et al. (eds.), *The Proceedings of the 9th Frontier Academic Forum of Electrical Engineering*, Lecture Notes in Electrical Engineering 743,
https://doi.org/10.1007/978-981-33-6609-1_48

1 Introduction

In recent years, the rapid development of machine learning and big data has opened up new ways for pattern recognition and curve fitting for complex problems such as dense regression. Most researches are about improving data-driven statistical methods to supplement simulation to solve engineering problems [1, 2]. These studies have been applied in smoke simulation, liquid splash modeling and liquid behavior modeling with obstacles, etc. [3]. Literature [4] studied the feasibility of applying deep learning to solve the two-dimensional Poisson equation, and established a deep convolutional neural network to predict the potential distribution in 2D. The training data generated by the finite difference solver and the powerful approximation ability of the deep convolutional neural network enable it to make correct predictions given the source and distribution information of the permittivity. The prediction error can reach below 1%, and the CPU time is significantly reduced. Literature [5] also studied the feasibility of deep learning model used to predict the Maxwell equation solution of low-frequency electromagnetic equipment.

Noise suppression is one of the important issues that need to be solved in the optimal design and application of motors. More and more researches have taken the magnetostrictive effect as an indispensable factor in the study of motor vibration and noise characteristics. In Literature [6], a dynamic magneto-induced model of silicon steel sheet with DC bias was constructed, and the loss characteristics of silicon steel sheet under DC bias were simulated. Literature [7] measured the permeability and electromagnetic loss of various magnetostrictive materials when the magnetic induction intensity and excitation frequency are not the same. The numerical analysis of motor stress is also particularly important in the analysis of electromagnetic vibration of the motor considering the magnetostrictive effect. At present, there are relatively few studies on applying deep learning neural networks to optimize motor design. The author team proposed to punch holes in the simulation model and add negative magnetostrictive materials to study the influence of the negative magnetostrictive effect on the electromagnetic vibration of the motor under the rotating magnetic field, and find the best hole position to suppress the electromagnetic vibration of the motor [8]. And carried out simulations of different positions and different apertures, and obtained multiple sets of simulation experimental data for training deep learning neural networks. Secondly, simple pre-processing is performed on the exported data so that it can be better loaded into the network. Finally, a BP neural network based on multiple regression is used to process the data. It can fit different punching positions to the corresponding minimum motor stress at that position, and can realize that when the punching position is determined, the minimum stress corresponding to the position can be output within 1 s. The designed neural network model has a prediction accuracy rate of up to 97.3%. It provides a great help to find the best punching position faster and more accurately, and also provides guidance and reference for the work of deep learning in the finite element numerical calculation and analysis of motor stress and motor optimization.

2 Establish a Finite Element Numerical Model

An electromagnetic-mechanical coupling model considering the magnetostrictive effect is established. Through finite element numerical calculation, a calculation program is written for modal analysis to obtain initial transient values and calculate magnetostrictive force and vibration displacement. On the basis of this model, this paper carried out a finite element simulation on a stator core model of a 4-pair pole-mounted permanent magnet synchronous motor. When considering the magnetostrictive effect, it was found that the magnetic circuit. The degree of deflection is large, and the stress on the stator core is relatively large in the magnetic field distortion area of the stator tooth tip and root. Therefore, the finite element simulation of the scanning traversal calculation of the punching position and the punching radius was performed on the stator core, and the area between the stator teeth where the magnetic density was concentrated and the area between the top and the boundary of the stator teeth where the magnetic density was relatively small were scanned respectively. The scanning simulation results are exported for later deep learning neural network model training to achieve the fitting of the minimum stress corresponding to each position, and output the minimum stress at the required position in the shortest time.

2.1 *FEM Model Magnetic Density and Stress Simulation*

Add the motor stator model equations to the finite element simulation software COMSOL Multiphysics, set various parameters of the rotating electromagnetic field and the solid mechanics field, and add a multi-channel sinusoidal AC current source with a frequency of 50 Hz and a controllable amplitude to the motor windings. Produces an alternating rotating magnetic field.

In order to study the reduction of the electromagnetic vibration of the motor, this article first drills holes in the roots of the stator teeth of the motor to fill the negative magnetostrictive effect material, and simulates in COMSOL. The material filled in this article is $Sm_{0.88}Nd_{0.12}Fe_2$.

Set the rotor speed of the motor in COMSOL to 60 r/min, and perform the simulation calculation of the motor stator model considering the magnetostrictive effect. Table 1 lists the simulation [0.01–0.06 s], which is the stress and magnetic density of a half cycle Numerical value. Figure 1 and 2 show the magnetic flux density and stress deformation graphs at different simulation moments of 0, 0.03, and 0.06 s.

Analyzing Fig. 1, and Table 1, it can be seen that when the motor is started and the rotor rotates, the magnetic circuit of the main magnetic field is deflected, resulting in magnetic field distortion, which will cause great changes in the stress and magnetic flux density on the stator teeth. The magnetic flux density is relatively concentrated and obvious at the tip and root of the stator tooth.

Table 1 0.01–0.06 s stress maximum value table

Time (s)	0.01	0.02	0.03	0.04	0.05	0.06
Max (10^6 N/m^2)	2.02	2.02	2	1.97	2.31	2.42
Min (10^3 N/m^2)	6.33	4.33	2.31	4.36	4.32	4.18
Max (T)	1.17	1.36	1.56	1.73	1.84	1.71
Min (10-4T)	0.52	1.40	4.48	4.52	4.52	4.49

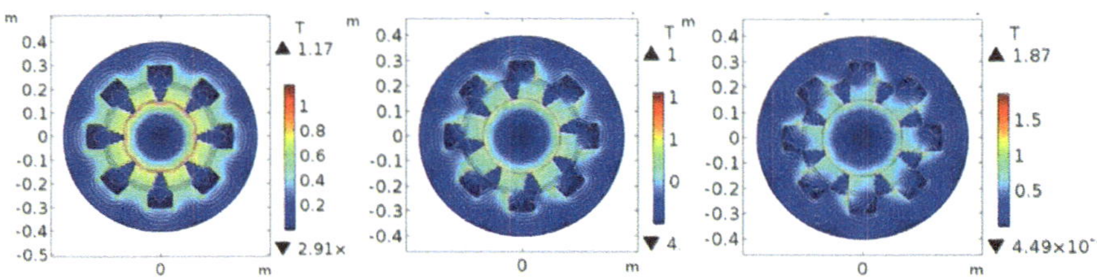

Fig. 1 Stator flux density and Z axis magnetic vector potential at time 0, 0.03, 0.06 s

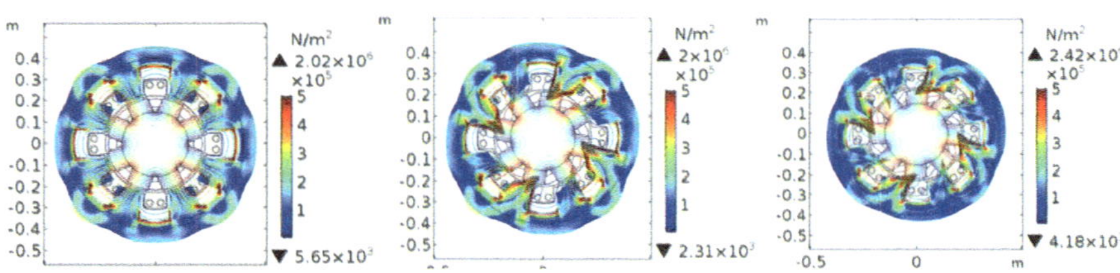

Fig. 2 Stator stress deformation at 0, 0.03, 0.06 s

The simulation shows that the stress on the top and root of the stator of the motor is relatively concentrated, and the degree of deformation is large. In order to reduce the stress in these parts, the stator is punched and filled with negative magnetostrictive material, its characteristics make it can be offset under the action of the alternating magnetic field Part of the deformation caused by the magnetostrictive effect reduces the surface stress of the motor stator teeth and achieves the purpose of suppressing the electromagnetic noise of the permanent magnet synchronous motor. The shape of the hole is set to be circular. The radius of the punched hole is 0.01 m. By changing the angle parameter angle and distance r of the center of the punching circle relative to the center of the stator model, scan the two areas where the magnetic density is concentrated and the magnetic density is relatively small, find the minimum stress, and analyze the stress Whether to reduce and the reduced value to determine the location of the punch. After determining the punch position, change the aperture r1 to scan to determine the size of the best punch hole diameter, and finally determine the best position and radius of the punch hole.

From 0, 0.03, and 0.06 s, the stress deformation at three different moments can be seen from Fig. 2. When the magnetic circuit is deflected, the degree of magnetic field distortion will change, and the stress distribution on the surface and inside of the motor stator teeth will also change. And with the increase of the magnetic field distortion, the deformation degree of the stator ferromagnetic area will also increase. The areas with greater stress are concentrated in the magnetically dense parts of the stator tooth root and top, and the deformation of these areas is also the largest.

2.2 *Experiment of Filling Material*

Design experiment: The scanning area is from the top of the stator teeth to the boundary, and the magnetic density of these areas is relatively small. The shape of the hole is set to a circle, the radius is set to 0.01 m, and the material is filled with negative magnetostrictive effect. The angle parameter angle and distance r of the center of the hole relative to the center of the stator model are changed to determine the position of the hole, where the parameter angle The parameter list of is: range (0, 0.02, 6.28) (starts at 0, ends at 6.28, and the step is 0.02); the parameter list of parameter r is: range (0.315, 0.01, 0.385) (starts at 0.315, It ends at 0.385 and the step size is 0.01).

Analyze the results of the scanning data. After drilling, the stress and deformation of the stator tooth surface will increase in some positions and decrease in some positions. The reduced position can prove that our model is correct, that is, decrease the stress on the surface of the stator teeth caused by the magnetostrictive effect can be reduced by perforating and filling the negative magnetostrictive material, thereby suppressing the electromagnetic noise of the motor. Compared with the point where the stress is reduced, it can be found at the point (0.375, 1.18), where the stress is reduced the most, and it will also produce the best noise suppression effect. The equipotential diagram of the magnetic flux density and the z-axis magnetic vector potential at this point and the stress pattern is shown in Fig. 3. During the sampling period [0.01–0.06 s], the maximum and the minimum stress are shown in Table 2.

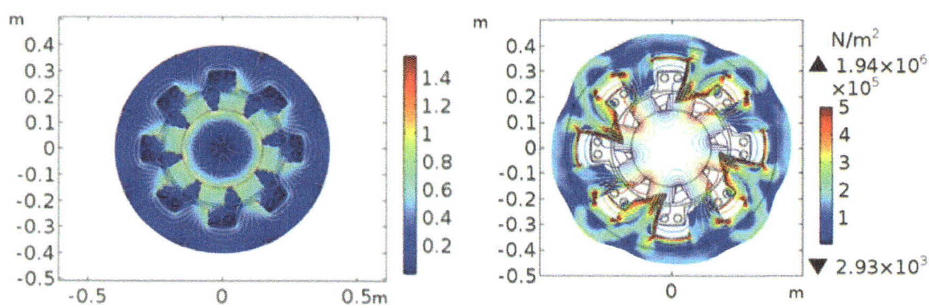

Fig. 3 Stator magnetic density and Stator stress deformation at point (0.375, 1.18)

Table 2 0.01–0.06 s stress maximum value table

Time (s)	0.01	0.02	0.03	0.04	0.05	0.06
Max (10^6 N/m^2)	1.98	1.96	1.96	1.91	2.3	2.42
Min (10^3 N/m^2)	2.14	3.92	2.93	2.02	2.17	2.06

Table 3 Neural network hierarchy

Layer (type)	Output shape	Param #
dense_1 (dense)	(None, 24)	72
dropout_1 (dropout)	(None, 24)	0
dense_2 (dense)	(None, 256)	6400
dense_3 (dense)	(None, 1)	257

As shown in Fig. 3, punch holes at the points (0.375, 1.18) and set the aperture to 0.01 m. After the material is filled, the magnetic field distortion and magnetic circuit offset are reduced, and the stress on the stator tooth surface is the amount of deformation has also been reduced. Comparing Table 2 and Table 3, it can be seen that after perforating the filling material at the appropriate position, the maximum and minimum stress values on the stator surface are reduced, and the maximum stress is reduced by about 5%, which can suppress a part of it to a certain extent Electromagnetic noise caused by magnetostrictive effect.

In the scanning experiment, the center of the punching position is determined according to the angle parameter angle and the distance r relative to the center position of the stator model. Each angle and r will correspond to a motor stator stress deformation diagram (as shown in Fig. 3). And each stress deformation map is actually composed of more than 10,000 coordinate points and the corresponding stress values of these coordinate points.

3 Neural Network Model Construction

In this paper, a multiple regression model with three fully connected layers, two inputs and one output is established to process the motor stator stress simulation data obtained from the finite element calculation. The model is trained and optimized by using the BP neural network algorithm, which can fit the numerical simulation data of the motor stress well, and output the minimum stress value of the required position within 1 s.

3.1 BP Neural Network Analysis

The BP neural network consists of two processes: the forward calculation of the data stream (forward propagation) and the back propagation of the error signal [9, 10]. In the previous process, the propagation direction is the input layer, hidden layer and output layer. The state of neurons in each layer only affects neurons in the next layer. If the expected output cannot be obtained in the output layer, the latter process will begin [9, 11]. Through the alternation of these two processes, the gradient descent strategy of the error function in the weight vector space is realized, and a set of weight vectors are searched through dynamic iteration to minimize the error function of the network, thereby completing the extraction and storage of information. Figure 4 shows the network structure of the BP neural network.

As shown in the figure, x represents the input layer, z represents the hidden layer, and y represents the output layer. Assuming that the input layer of the BP neural network has n nodes, the hidden layer has q nodes, and the output layer has m nodes, the input layer and the hidden layer V_{ki}, and the weight between the hidden layer and the output layer is W_{jk}. The transfer function of the hidden layer is $f_1(g)$, and the transfer function of the output layer is $f_2(g)$. Then, the output of the node in the hidden layer is (put the threshold into the sum term):

$$z_k = f_1(\sum\nolimits_{i=0}^{q} v_{kix_i}), \quad k = 1, 2, 3, \ldots, \mathrm{q} \qquad ((1))$$

The output of the nodes in the output layer is as follows:

$$yi = f_1(\sum\nolimits_{i=0}^{q} w_{jk} z_k), \quad j = 1, 2, 3, \ldots, \mathrm{m} \qquad ((2))$$

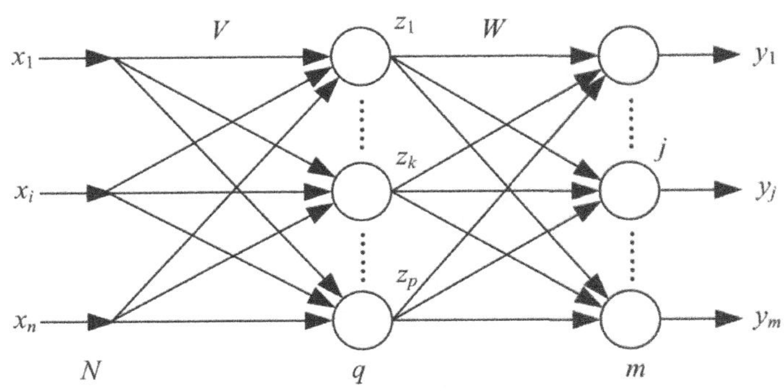

Fig. 4 Network structure of BP neural network

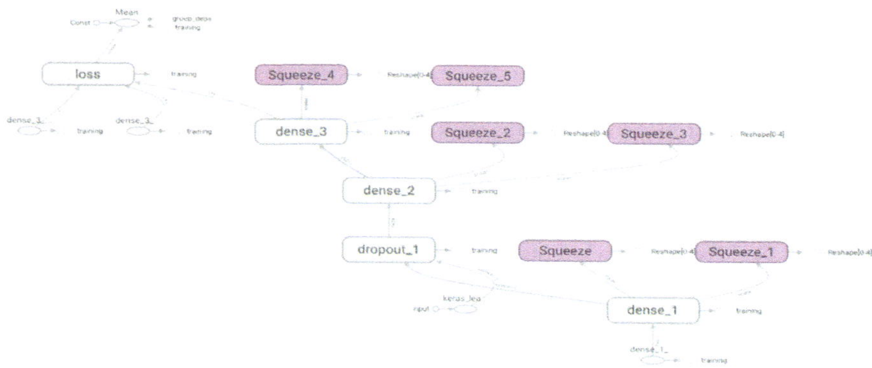

Fig. 5 Data flow graph

3.2 Experimental Steps

This article uses the regression of the fully connected layer neural network, based on the scientific computing environment Anaconda3, and uses the deep learning framework Keras based on the TensorFlow backend for network modeling. It also supports convolutional neural networks and recurrent neural networks, as well as the combination of the two, and can also be used in the CPU Run and switch seamlessly with GPU. This experiment was performed on the Intel(R) Xeon(R) Bronze 3104 CPU

The network consists of 3 fully connected layers, and the number of neurons in each layer is 24, 256, and 1, respectively. The total number of neuron weights is 6729, as shown in Table 3.

Calling Tensorboard in the program, the data flow graph of the network can be generated on the web page, as shown in Fig. 5.

3.3 Analysis of Results

This experiment uses randomly divided 1000 data as the training set, 150 data as the test set, and 800 batches of training. The coordinates x and y of the motor drilling position are used as input variables, and the minimum motor stator stress is used as the output variable. And the error of the test set is output as a file, and 4 sets of test data are selected from the file as shown in Table 4.

Table 4 shows True, Vaild, and MSE represent the true value, predicted value, and error of the stress value, respectively. It can be calculated from the table that the accuracy of the network can reach more than 93%. The program can finally enter the punching coordinate value of any position. The program will give the minimum motor stress corresponding to the position within 1 s, and realize Data fitting and

Table 4 Neural network hierarchy

	x	y	True	Vaild	MSE
1	0.385	0.0	3.971170	4.141764	−0.17059
2	0.325	0.94	2.8920326	3.670611	−0.77857
3	0.315	3.04	4.001280	3.464829	0.53645
4	0.335	0.02	4.811575	4.161636	0.64993

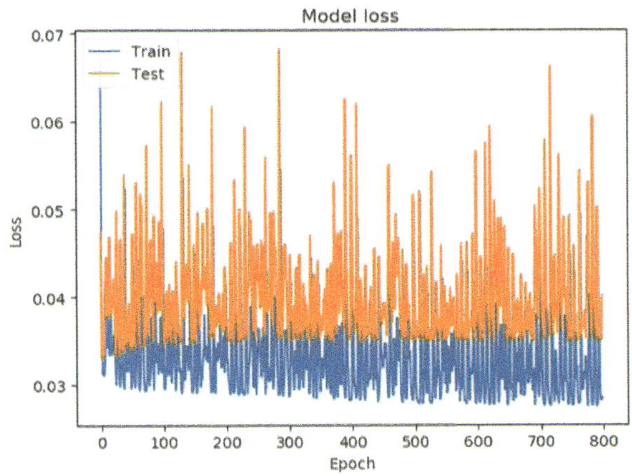

Fig. 6 BP neural network loss rate

accurate prediction in a short time. The loss rate of the specific model is shown in Fig. 6.

4 Conclusion

This paper constructs a numerical multiple regression prediction model of motor stress based on BP neural network. The core algorithm of BP neural network is gradient descent method. After 800 batches of training, the network can finally have an average accuracy of 95%, and can obtain the minimum motor stress at the desired position within 1 s, which has a certain reference for deep learning in finite element calculation and motor optimization direction effect.

References

1. Ghorbanian, V., and D.A. Lowther. 2017. A statistical solution to efficiently optimize the design of an inverter-fed permanent-magnet motor. *IEEE Transactions on Industry Applications* 53 (6): 5315–5326.

2. Ghorbanian, V., A. Salimi, and D.A. Lowther. 2018. A computer-aided design process for optimizing the size of inverter-fed permanent magnet motors. *IEEE Transactions on Industrial Electronics* 65 (2): 1819–1827.
3. Tompson, J., K. Schlachter, P. Sprechmann, K. Perlin. 2017. Accelerating Eulerian fluid simulation with convolutional networks. In *Proceedings of the 34th International Conference on Machine Learning*, 3424–3433.
4. Tang, W., et al. 2017. Study on a Poisson's equation solver based on deep learning technique. In *IEEE Electrical Design of Advanced Packaging and Systems Symposium (EDAPS)*, 1–3.
5. Khan, A., V. Ghorbanian, D.A. Lowther. 2018. Deep learning for magnetic field estimation. In *2018 IEEE Conference on Electromagnetic Field Computation (CEFC)*.
6. Zhao, Xiaojun, Xiaona Liu, Xiao Fan, and Yang Liu. 2020. Simulation of DC bias hysteresis and loss characteristics of oriented silicon steel sheet based on Preisach model. *Acta Electrotechnica Sinica* 35 (09): 1849–1857.
7. Weng, Ling, Zhen Chang, Ying Sun, Bowen Wang, and Wenmei Huang. 2020. Analysis and experimental study on high frequency magnetic energy loss of different magnetostrictive materials. *Acta Electrotechnica Sinica* 35 (10): 2079–2087.
8. Zhang, Xin, Chaoqun Xie, Lihua Zhu, et al. 2017. Numerical simulation and experimental research on stress of motor including magnetostriction effects. *Transactions of China Electrotechnical Society* 32 (S2): 50–55.
9. Qin, A.K., V.L. Huang, and P.N. Suganthan. 2009. Differential evolution algorithm with strategy adaptation for global numerical optimization. *IEEE Transactions on Evolutionary Computation* 13 (2): 398–417.
10. Wang, Y., E. Zio, X. Wei, D. Zhang, and B. Wu. 2019. A resilience perspective on water transport systems: the case of eastern star. *International Journal of Disaster Risk Reduction* 33: 343–354.
11. Wang, H., Z. Wu, and S. Rahnamayan. 2011. Enhanced opposition-based differential evolution for solving high-dimensional continuous optimization problems. *Soft Computing* 15 (11): 2127–2140.

Failure Analysis of Garnet-Type Solid State Electrolyte LLZO by Electrochemical Method

Weichang Guo, Fei Shen, Jie Gao, Chao Yang, and Xiaogang Han

Abstract Garnet-type $Li_7La_3Zr_2O_{12}$ (LLZO) is one of the most attractive candidates for achieving solid state lithium batteries. LLZO pellet is preferred due to its potential to prevent dendritic lithium growth and penetration. However, lithium dendrite can penetrate the LLZO pellets in a low current density, resulting in short circuit of the cell. In this work, we study the failure behavior of LLZO electrolyte through electrochemical method. Meanwhile, we also study the growth of lithium inside the LLZO pellet after failure. The results show that the reason for failure of LLZO is related to the contact between lithium and LLZO, and lithium dendrite may grow from one to another electrode.

Keywords LLZO · Failure · Electrochemical method

1 Introduction

Environment problems and Energy crisis drive our humans to exploit new energy, like solar, wind and tidal energy, to replace the fossil energy, like petroleum. Therefore, we need to develop device to store energy and electric power tool, such as electro mobile, to reduce the usage of fossil energy. Electrochemical power source, especially batteries, can achieve the above two goals, simultaneously. Among all kinds of batteries, lithium ion battery is best choice because of its higher energy density and longer service life than other commercial batteries [1]. However, with the development of society, the energy density of battery is required to be higher. But the energy density of Li-ion battery is just about 200 Wh/kg, which cannot meet

W. Guo · F. Shen · J. Gao · C. Yang · X. Han (✉)
State Key Laboratory of Electrical Insulation and Power Equipment, School of Electrical Engineering, Xi'an Jiaotong University, Xi'an, Shaanxi 710049, China
e-mail: xiaogang.han@xjtu.edu.cn

F. Shen
Xi'an Jiaotong University Suzhou Institute, Suzhou, Jiangsu 215123, China

X. Han
Key Laboratory of Smart Grid of Shaanxi Province, Xi'an, Shaanxi 710049, China

W. Chen et al. (eds.), *The Proceedings of the 9th Frontier Academic Forum of Electrical Engineering*, Lecture Notes in Electrical Engineering 743,
https://doi.org/10.1007/978-981-33-6609-1_49

the need of development. Energy density is related to the capacities of the electrode materials and the voltage of the battery. But capacities of the electrode materials of commercial LIBs are just 372 mAh/g for graphic anode and <200 mAh/g for cathode materials, respectively. To improve the energy density, researchers began to study Li anode (3860 mAh/g) and high voltage cathode materials. However, it is easy to produce Li dendrite in liquid electrolyte which would penetrate separator, resulting in the short circuit of LIBs. And the high voltage cathode materials are not stable under short-circuit. These can cause risk of combustion and explosion, endangering the safety of life and property [2–4].

The root of the above phenomenon is the use of organic electrolyte. Inorganic solid state electrolytes are non-flammable and have wider electrochemical working window and higher mechanical strength. By replacing organic electrolyte with inorganic solid state electrolyte, it will largely improve the safety of batteries. More importantly, it opens the possibility of using Li metal anode (3860 mAh g^{-1}) and high voltage cathode material (up to 5 V), which can provide much higher energy density than commercial LIB [3, 5]. Therefore, it is generally accepted that batteries with organic electrolytes will be replaced by the ones with inorganic electrolytes gradually.

Among different solid-state electrolyte, garnet type $Li_7La_3Zr_2O_{12}$ is one of the most promising one in for application because of its high ion conductivity (10^{-4}–10^{-3} S cm^{-1}), the stability to air and Li metal, and wide electrochemical window, and so on [6, 7]. However, recent published papers reported that lithium dendrite can penetrate the LLZO pellets in a low current density, resulting in short circuit of the cell [8–10]. There are two mainstream views on the causes. Some researcher think that lithium dendrite was formed at the defect of the surface of LLZO and penetrate it gradually [11, 12]. Another think lithium formed inside of LLZO because of the electron conductivity [13, 14]. But there is no unified conclusion. In this work, we study the failure behavior of LLZO electrolyte through the cycling performance and electrochemical impedance spectrum (EIS). Meanwhile, we also study the lithium inside the LLZO pellet with micro morphology analysis. The results show that the failure of LLZO is related to the contact between lithium and LLZO, and Li dendrite may grow from one to another electrode.

2 Experimental

2.1 Synthesis of the $Li_{6.4}La_3Zr_{1.4}Ta_{0.6}O_{12}$ (LLZTO) Pellet

Stoichiometric amount of LiOH (99.99%), La_2O_3 (99.99%), ZrO_2 (99.99%) and Ta_2O_5 (99.99%) were mixed first. 15% wt excess of LiOH was added to compensate for the lithium volatilization loss during sintering. The above mixed materials were ball-milled with isopropanol 300 rpm for 12 h to get primary powder. The primary powder was pressed into pellets before calcined at 800 °C for 10 h and the pre-calcined pellets were then crushed and grilled.

The pre-calcined powder was ball-milled at 300 rpm for another 12 h and dried to obtain secondary powder. Then the powder was pressed into pellets with diameter of 12.5 mm under 500 MPa. The pellets were sintered at 1250 °C for 1 min, then decreased to 1100 °C and kept for 8 h. The LLZTO pellets were stored and polished before use.

2.2 Characterization

X-ray diffraction (XRD, Bruker AXS) was used to analyze the phase composition of the LLZTO. Scanning electron microscope (SEM, Phenom ProX) equipped with energy dispersive spectrometer (EDS) was applied to observe the microstructure of the LLTZO pellet before or after short circuit. AC impedance tester (BioLogic, SP-300) was used to measure the ion conductivity indirectly with the frequency from 6 MHz to 100 Hz. Before testing, a thin layer of gold was sputtered onto the both sides of LLZTO pellet. Li||Li symmetric cells were assembled by first sputtering a thin gold layer on the LLZTO surface and then attaching a 200 μm Li foil and heating at 200 °C for 5 min. The cycling performance of Li||Li symmetric cells was carried out using a LAND battery tester.

3 Results and Discussion

The cross-sectional SEM images of LLZTO pellets are shown in Fig. 1a and b. It can be seen that the micro-structure inside LLZTO was dense, although there are some small pores. XRD pattern of LLZTO was shown in Fig. 1c and the characteristic peaks of the sample match well with the standard PDF card (# 80-0457), indicating that the phase belongs to cubic phase LLZO. The conductivity of LLZTO was 6.2×10^{-4} S/cm calculated indirectly by the EIS (Fig. 1d), according to the formula $\sigma = L/(S*R)$, where L, S and R stand for the thickness, area and resistance, respectively. The intersection of the semicircle in the high frequency and the diagonal in the low frequency in the EIS represents the impedance of the LLZTO electrolyte, including both bulk and grain boundary.

The cycling performance of Li/LLZTO/Li cell at 0.1 mA/cm^2 was exhibited in Fig. 2a, in which Li electrode went through the process of stripping and plating. The potential of the cell increased gradually as the number of cycles increases and the cell was short circuit at 57th cycle after some large polarization. EIS plot after short circuit was shown inside Fig. 2a and the high frequency region is in the fourth quadrant indicating a typical EIS plot of short circuit. The increased potential means the larger impedance which includes the impedance of LLZTO electrolyte and the interface impedance between Li metal and LLZTO. But we cannot identify which

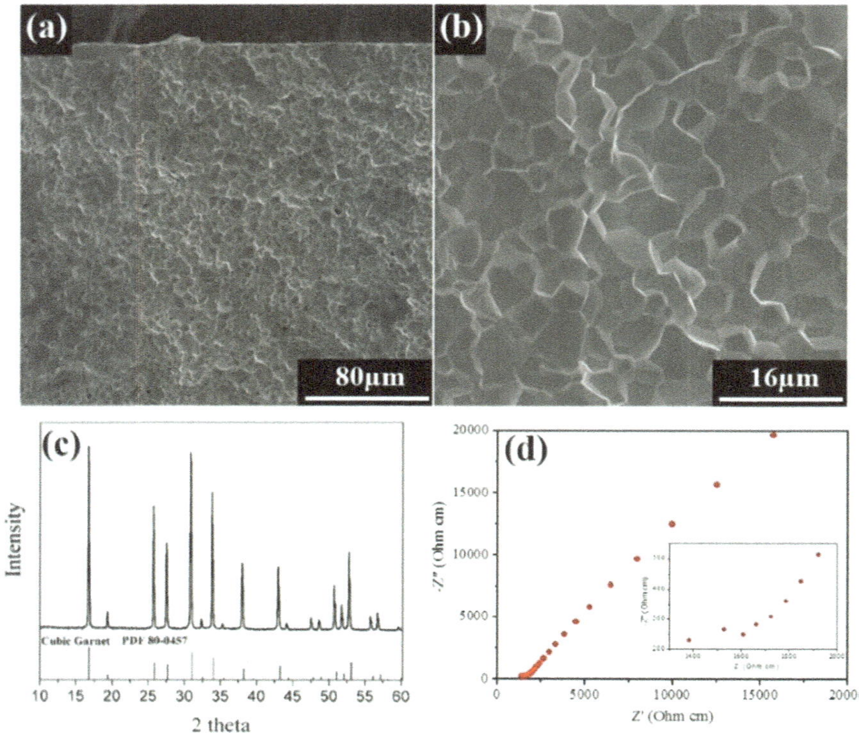

Fig. 1 Cross-sectional SEM images of LLZTO **a** at low magnification and **b** high magnification; **c** XRD patterns and **d** EIS plot of LLZTO

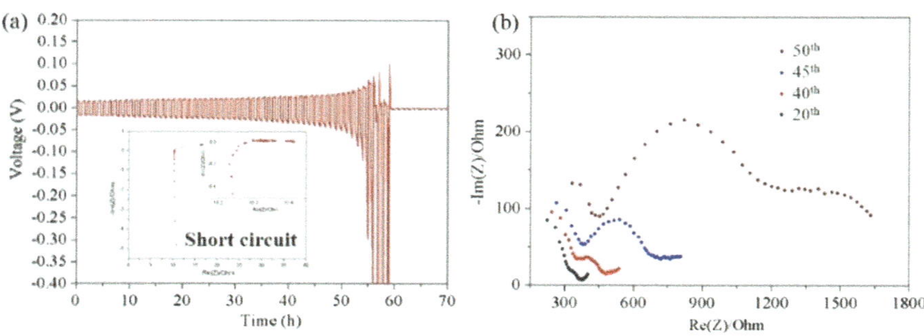

Fig. 2 **a** Cycling performance of Li||Li symmetric cell, inside the EIS after short circuit; **b** EIS at different cycles

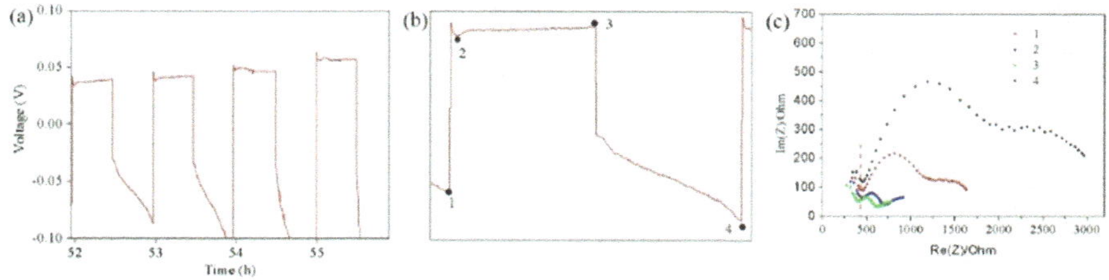

Fig. 3 Cycling performance of **a** last few cycles, and **b** 52nd cycle; **c** EIS plots at different cycles of 52nd

is the decisive factor. Therefore, we tested the EIS of the cell at different cycles and the EIS plots were shown in Fig. 2b. The total impedance increased as the number of cycles increased. However, the amplitude of variation for the impedance of LLZTO electrolyte and the interface impedance is different. The value of the impedance of LLZTO electrolyte increased from 324 to 451 Ω, while the value of interface impedance increased from 52 to 836 Ω. Obviously, interface impedance increased by 784 Ω, which is more than that of the electrolyte impedance (127 Ω), therefore, it played a decisive role in the increasing over potential.

In the last few individual cycles, when the current changes direction after a large polarization, the voltage will increase rapidly, and decrease in the first minutes and then rise (Fig. 3a). After half a cycle, when the current changes direction, there would still be a large polarization. Taking the 52nd cycle as an example, we measured the impedance values at different points shown in Fig. 3b. It can be observed that the interface impedance decreased first (point 2) and basically unchanged in this process (point 3). However, when the whole cycle is over (point 4), the interface impedance increased much more than that of the initial value (point 1). It is worth noting that the electrolyte impedance also decreased from point 1 to point 3, and increased back when it came to point 4. This could be related to the growth of Li dendrite inside LLZTO, and the possible processes will be discussed later.

The value of interface impedance depends on the contact between Li metal and LLZTO which belongs to the contact between solid and solid phases. When they have a good contact, interface impedance is small, otherwise, it becomes large. So the large interface impedance after cycling means a poor contact between LLZTO and Li metal. The possible reason is that the stripped Li cannot return to its original position, thus resulting in the deteriorating interface contact and large interface impedance. The contact area became smaller, as a result, the actual current density through the interface and the polarization voltage increased upon cycling. Particularly, after a large polarization, the contact area is very small which would result in a large current density. And large current density is more likely to cause Li dendrite growth. On this condition, if it started to plate in the smaller areas, the larger current density would trigger the growth of Li dendrite at this end, causing the electrolyte impedance decreased and penetrating the LLZTO electrolyte, which would result in the final short circuit. Therefore, the short circuit is easy to occur after a large polarization.

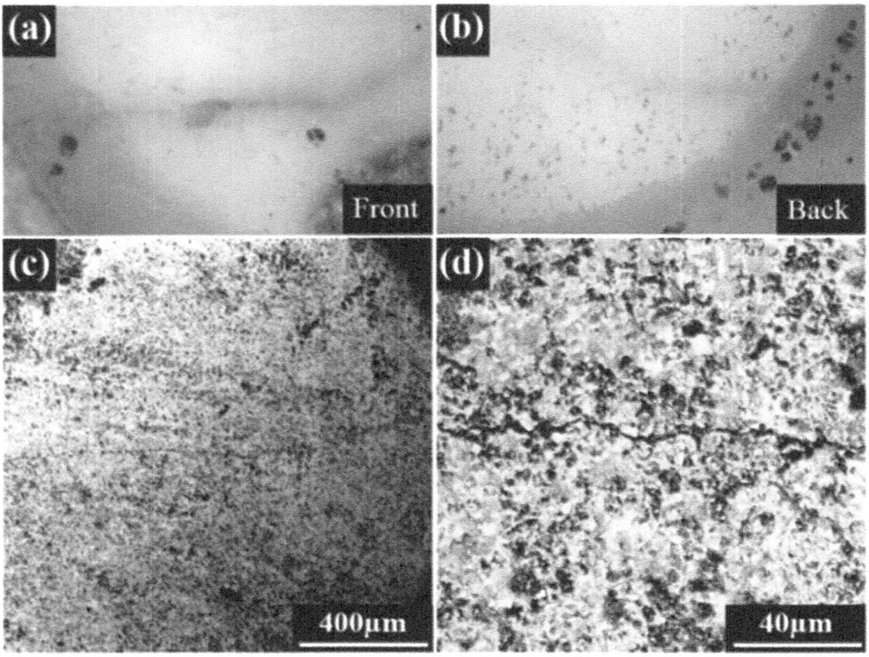

Fig. 4 Photograph of the **a** front and **b** back of LLZTO after short circuit; c, d BSEM images of LLZTO surface after short circuit

Based on the above analysis, it is likely that Li dendrite grew from one end to the other. The LLZTO pellet after failure was studied and it can be seen in Fig. 4a that there was a long crack on the electrolyte surface. The backscatter electron microscopy (BSEM) images in Fig. 4c and d further proved the existence of the crack. While the back in Fig. 4b has only the outline of the crack, which demonstrated that the correctness of the above conjecture about the growth of Li dendrite.

To further testify, we have break off the LLZTO electrolyte along the vertical direction of the crack to observe. It is obvious to see in Fig. 5a and b that the crack originated at one end and spread to the other. And it may generate branches inside the LLZTO (Fig. 5b). When observing the cracks at high magnification, we can find that there is something different inside the cracks. According to contrast, these dark substances in BSEM may be derivatives of lithium dendrite [15]. And it seems that these substances grew along the grain boundary, which is consistent with the literature report [10].

Furthermore, the EDS element analysis of these regions in Fig. 6b shows the dark area contains O element but no La, Zr and Ta. It can be concluded that these area is Li metal originally, and extra O may be introduced by reacting with Li dendrite when exposing to air during sample preparation. Therefore, it is determined that the material in the cracks is lithium metal and the growth pattern that lithium grows from

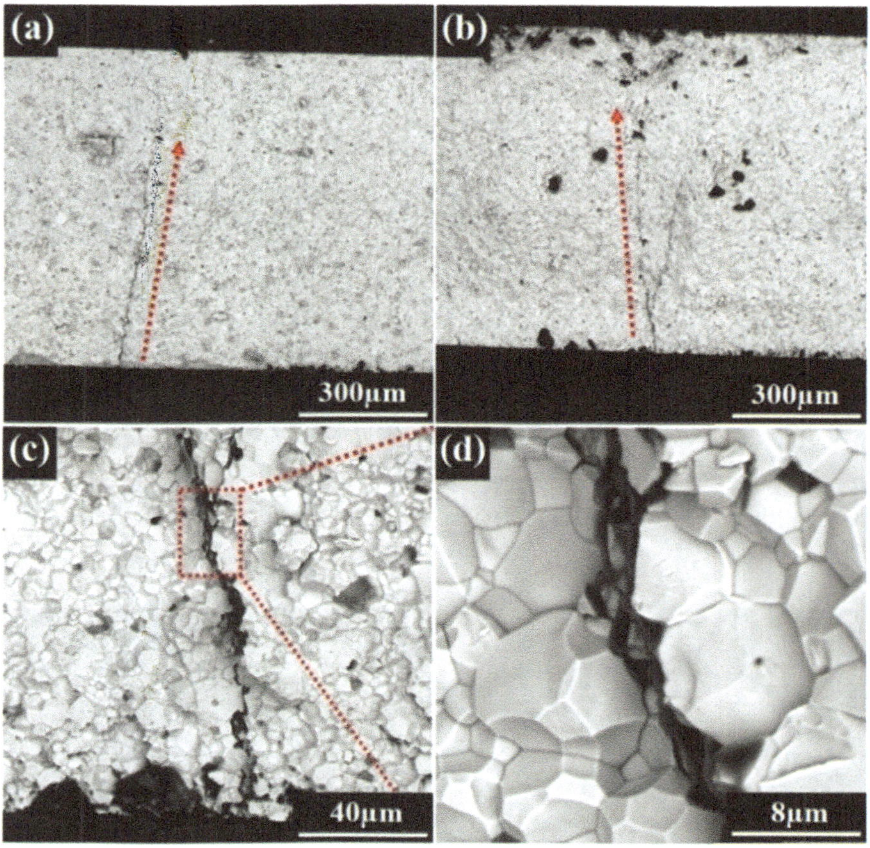

Fig. 5 Cross-sectional BSEM images of LLZTO after short circuit: **a** and **b** at low magnification; **c** and **d** at high magnification

one end to the other, may be the main cause of short circuit. At the same time, lithium was also found in other regions (Fig. 7) that is not within the crack, indicating that lithium may also be generated inside LLZTO, but it is not the decisive reason to lead to the failure of LLZTO.

4 Conclusion

Li/LLZTO/Li cells became short-circuit phenomena after large polarization for some time under an applied direct current with a current density of 0.1 mA/cm^2. The short-circuiting shows correlation to the polarization behavior of Li/LLZTO/Li cells. And the polarization behavior of Li/LLZTO/Li is related to the interface impedance,

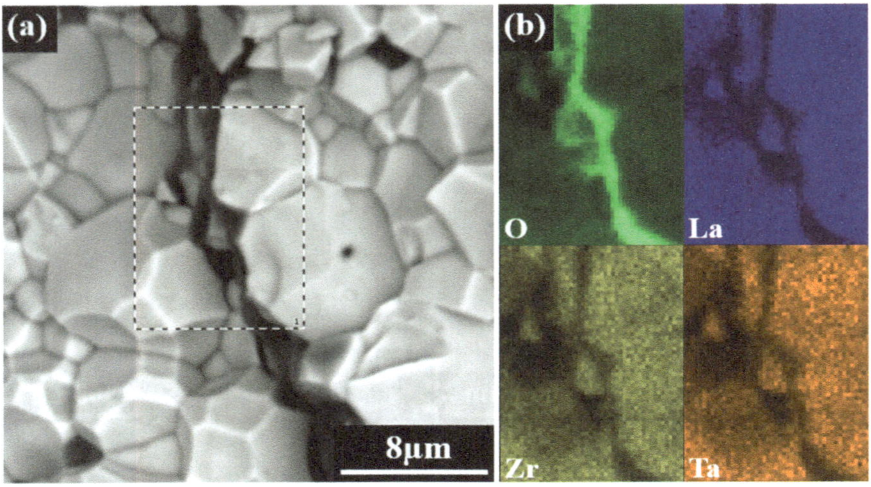

Fig. 6 **a** BSEM image at crack in LLZTO after short circuit and **b** EDS elements mapping of selected area in (**a**)

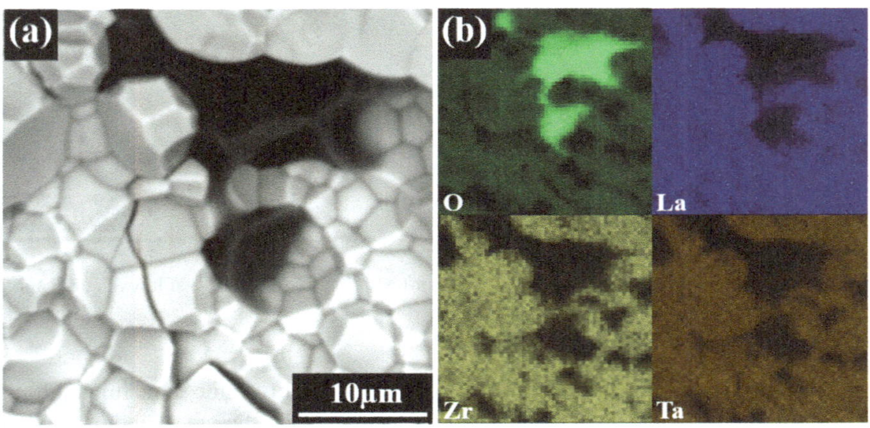

Fig. 7 **a** BSEM image and **b** EDS elements mapping images of LLZTO after short circuit

which is attributed to the contact between Li metal and LLZTO. Direct observation of the short-circuited LLZTO pellet and lithium dendrites inside the LLZTO pellet suggests that the lithium dendrites may grow from one end to the other through grain boundaries and penetrate the LLZTO pellet finally.

Acknowledgements This work was supported by the National Key R&D Program of China (Grant No. 2018YFB0104300), National Natural Science Foundation of China (Grant No. 51772241) and Natural Science Foundation of Jiangsu Province (Grant No. BK20190222). X. Han would like to thank the Independent Research Project of State Key Laboratory of Electrical Insulation and

Power Equipment (Grant No. EIPE19111) for the financial support. F. Shen acknowledges the State Key Laboratory of Electrical Insulation and Power Equipment for financial support. B. Zhao acknowledges the Natural Science Foundation of China (NSFC-51707151) for financial support.

References

1. Winter, M., B. Barnett, and K. Xu. 2018. Before Li Ion Batteries. *Chemical Reviews* 118: 11433–11456.
2. Samson, A.J., K. Hofstetter, S. Bag, and V. Thangadurai. 2019. A Bird's-Eye View of Li-stuffed Garnet-Type $Li_7La_3Zr_2O_{12}$ Ceramic Electrolytes for Advanced All-Solid-State Li Batteries. *Energy & Environmental Science* 12: 2957–2975.
3. Liu, Q., Z. Geng, C. Han, Y. Fu, S. Li, Y. He, F. Kang, and B. Li. 2018. Challenges and Perspectives of Garnet Solid Electrolytes for All Solid-State Lithium Batteries. *Journal of Power Sources* 389: 120–134.
4. Zhang, B., R. Tan, L. Yang, J. Zheng, K. Zhang, S. Mo, Z. Lin, and F. Pan. 2018. Mechanisms and Properties of Ion-Transport in Inorganic Solid Electrolytes. *Energy Storage Materials* 10: 139–159.
5. Takada, K. 2018. Progress in Solid Electrolytes toward Realizing Solid-State Lithium Batteries. *Journal of Power Sources* 394: 74–85.
6. Thompson, T., S. Yu, L. Williams, R.D. Schmidt, R. Garcia-Mendez, J. Wolfenstine, J.L. Allen, E. Kioupakis, D.J. Siegel, and J. Sakamoto. 2017. Electrochemical Window of the Li-Ion Solid Electrolyte $Li_7La_3Zr_2O_{12}$. *ACS Energy Lett* 2: 462–468.
7. Ohta, S., T. Kobayashi, and T. Asaoka. 2011. High Lithium Ionic Conductivity in the Garnet-Type Oxide $Li_{7-x}La_3(Zr_{2-x}, Nb_x)O_{12}$ ($X = 0$–2). *Journal of Power Sources* 196: 3342–3345.
8. Ren, Y., Y. Shen, Y. Lin, and C. Nan. 2015. Direct Observation of Lithium Dendrites inside Garnet-Type Lithium-Ion Solid Electrolyte. *Electrochemistry Communications* 57: 27–30.
9. Sudo, R., Y. Nakata, K. Ishiguro, M. Matsui, A. Hirano, Y. Takeda, O. Yamamoto, and N. Imanishi. 2014. Interface Behavior Between Garnet-Type Lithium-Conducting Solid Electrolyte and Lithium Metal. *Solid State Ionics* 262: 151–154.
10. Cheng, E.J., A. Sharafi, J. Sakamoto. 2017. Intergranular Li Metal Propagation Through Polycrystalline $Li_{6.25}Al_{0.25}La_3Zr_2O_{12}$ Ceramic Electrolyte. Electrochimica Acta 223: 85–91.
11. Kim, S., C. Jung, H. Kim, et al. 2020. The Role of Interlayer Chemistry in Li-Metal Growth through a Garnet-Type Solid Electrolyte. Advanced Energy Materials 10: 1903993.1–1903993.11.
12. Porz, L., T. Swamy, B.W. Sheldon, et al. 2017. Mechanism of Lithium Metal Penetration Through Inorganic Solid Electrolytes. Advanced Energy Materials 7 (20): 1701003.1–1701003.16.
13. Song, Y., L. Yang, W. Zhao, et al. 2019. Revealing the Short-Circuiting Mechanism of Garnet-Based Solid-State Electrolyte. Advanced Energy Materials 9 (21): 1900671.1–1900671.6.
14. Han, F., A.S. Westover, J. Yue, et al. 2019. High electronic conductivity as the origin of lithium dendrite formation within solid electrolytes. *Nature Energy* 4: 187–196.
15. Shen, F., W. Guo, D. Zeng, et al. 2020. A simple and highly efficient method toward high density garnet-type LLZTO solid state electrolyte. *ACS Applied Materials & Interfaces* 12: 30313–30319.

A Flux Linkage Identification Method of PMSM Based on Discounted Least Square Method

Jidong Lai, Mingrui Xie, Jianhui Su, Chenguang Zhou, and Weiwei Zheng

Abstract The control performance of permanent magnet synchronous motor largely depends on the accuracy of motor parameters, where to ensure identification accuracy existing flux identification method normally involves in complex mathematic calculation. In this paper, an online flux identification method based on recursive least square method with discounting factor is proposed for reduction of computation load. This method adopts the fitting recursive least square method to identify flux linkage, which has small amount of calculation and only needs to save the previous estimation value during the operation with small storage space. Aiming at solving the problem of "data saturation" in least square method, forgetting factor is introduced to enhance the proportion of current data in estimation and thus improve the identification dynamic responsiveness. Meanwhile, a discounting method is proposed to reduce the fluctuation of identification results induced by forgetting factor. Simulation results show that the proposed flux linkage identification method achieves a high identification accuracy.

Keywords PMSM · Flux identification · Least square method · Forgetting factor · Discounting factor

1 Introduction

Permanent magnet synchronous motor (PMSM) has been widely used in industry, however, its control performance relies largely on accurate parameters, including winding resistance, d- and q-axis inductance and flux linkage. Particularly, the PMSM flux linkage identification is significant for the control performance, permanent magnet monitoring and demagnetization prevention [1].

The finite element analysis (FEA) has high accuracy [2], but it requires extensive computational analysis, and its real-time identification needs high performance

J. Lai (✉) · M. Xie · J. Su · C. Zhou · W. Zheng
Research Center for Photovoltaic System Engineering of Ministry of Education School of Electrical Engineering and Automation, Hefei University of Technology, No. 193 Tun Xi Street, Bao He District, Hefei 230009, China
e-mail: laijidong@126.com

W. Chen et al. (eds.), *The Proceedings of the 9th Frontier Academic Forum of Electrical Engineering*, Lecture Notes in Electrical Engineering 743,
https://doi.org/10.1007/978-981-33-6609-1_50

processor. A flux linkage adaptive observer is constructed based on the model reference adaptive system (MRAS) and Popov stability theory in literature [3]. As MRAS is sensitive to noise, a common method to reduce it is the threshold denoising method of medium and small waves [4]. Moreover, the particle swarm optimization algorithm (PSO) [5], affine projection algorithm (APA) [6], genetic algorithm [7], differential evolution [8] and other algorithms are also used to track and estimate parameter. However, the intelligent algorithms mentioned above require a rich supply of data for algorithm training and target optimization during rotor flux identification.

In this paper, PMSM's flux linkage is identified online based on the recursive least square method (RLS). To prevent the data saturation caused by RLS, a forgetting factor method called FRLS is introduced to enhance the proportion of real-time data in the flux estimation and reduce the influence of historical data in the estimation. FRLS is easily affected by noise, therefore, a discounted treatment method named DRLS is put forward by introducing the discounting factor in RLS to solve the problem. Based on the proposed identification method, the simulation model are built respectively for verification.

2 The Mathematical Model of PMSM

During stable operation, the current change rate of PMSM can be ignored. Therefore, we can obtain the mathematical model of PMSM under the rotation coordinate system, where u_d, u_q, i_d, i_q, L_d, L_q are the d- and q-axis components of armature voltage, current, and self-inductances, respectively; R_s, ψ_f, p are the armature resistance, permanent magnet flux linkage and the differential operator, respectively; ω_e is the electrical angular velocity.

$$\begin{cases} u_d = R_s i_d - \omega_e L_q i_q u_d = R_s i_d - \omega_e L_q i_q \\ u_q = R_s i_q + \omega_e L_d i_d + \omega_e \psi_f \end{cases} \tag{1}$$

3 Flux Linkage Identification of PMSM Based on DRLS

The system model satisfied by the sample data ($\varphi(k)$, $y(k)$) is as follows (Table 1)

$$y(k) = \varphi^T(k)\theta + \xi(k) \tag{2}$$

Then, we can obtain that

$$\theta = [y(k) - \xi(k)]/\varphi^T(k) \tag{3}$$

Table 1 Parameter and symbol

Parameter	Symbol
$\theta(k), \theta = [\theta_1, \theta_2, \ldots, \theta_n]^T$	Actual parameter value
$\hat{\theta}(k), \hat{\theta} = [\hat{\theta}_1, \hat{\theta}_2, \ldots, \hat{\theta}_n]^T$	Parameter vector to be identified
$\varphi(k)$	Measurable input vector
$y(k)$	Measurable output vector
$e(k)$	Estimation error
$\xi(k)$	White noise vector
$\varsigma(k)$	Relative error
$J(\theta)$	Performance index function
$K(k)$	Adaptive gain matrix
$P(k)$	Covariance matrix
$\Gamma(k, i)$	Discounting factor
$\lambda, \lambda(k)$	Forgetting factor
$\gamma(i)$	Weighted factor
ε	Small positive real number
α	A positive real number, usually 10^4–10^6
I	n-order identity matrix

$$e(k) = y(k) - \varphi^T(k)\hat{\theta} = \varphi^T(k)(\theta - \hat{\theta}) + \xi(k) \tag{4}$$

It can be seen from the output error expression that the system parameter estimation error and noise together cause the system output error.

$$J(\theta) = \sum_{k=1}^{n} e^2(k) = \sum_{k=1}^{n} [y(k) - \varphi^T(k)\hat{\theta}]^2 \tag{5}$$

The expression of recursive least square method (RLS) is shown in Eq. (6).

$$\begin{cases} \hat{\theta}(k) = K(k)(y(k) - \varphi^T(k)\hat{\theta}(k-1)) + \hat{\theta}(k-1) \\ K(k) = P(k-1)\varphi(k)/\left[1 + \varphi^T(k)P(k-1)\varphi(k)\right] \\ P(k) = \left[1 - K(k)\varphi^T(k)\right]P(k-1) \end{cases} \tag{6}$$

As $\hat{\theta}(k)$ of the k_{th} order is modified by the product of $e(k)$ and $K(k)$, $K(k) = P(k)\varphi(k)$, RLS is adopted for parameter estimation, which only needs to save the relevant parameters of the previous time without a large amount of storage space. Through continuous iterative calculation, the identification accuracy meets the requirements.

We can obtain from Eq. (6) that

$$P(k-1) - P(k) = P(k-1)\varphi(k)\varphi^T(k)P(k-1)/\left[1 + \varphi^T(k)P(k-1)\varphi(k)\right] \tag{7}$$

Therefore, $P(k-1) - P(k) > 0$, which means

$$P(k) \leq P(k-1) \tag{8}$$

Formula (8) shows that during the identification process of RLS, $P(k)$ will gradually approach to 0 with the increase of iteration times k. If $P(k) = 0$, then $P(k-1) = P(k) = 0$, $K(k) = 0$, $\hat{\theta}(k) = \hat{\theta}(k-1)$, which means $\hat{\theta}(k)$ is no longer updated, that is, the new data would not update the old data, leading to the loss of the correction ability of the recursive algorithm and the formation of "data saturation".

The problem with data saturation is that parameter estimation is slow when the input changes. If $\varphi(k)$ mutates, due to data saturation, $K(k)$ stabilizes at a small value near zero, that is, the correction speed is slow, and it takes a long time to estimate the estimated value after the input changes. Therefore, there is a problem that RLS has insufficient tracking ability for abrupt and time-varying signals in non-stationary environments. The root cause of this problem is the influence of old data on the estimated value. By introducing the forgetting factor λ, its influence can be eliminated.

The expression of the recursive least square method with forgetting factor (FRLS) is as follows

$$\begin{cases} \hat{\theta}(k) = K(k)(y(k) - \varphi^T(k)\hat{\theta}(k-1)) + \hat{\theta}(k-1) \\ K(k) = P(k-1)\varphi(k)/\left[\lambda + \varphi^T(k)P(k-1)\varphi(k)\right] \\ P(k) = \left[1 - K(k)\varphi^T(k)\right]P(k-1)/\lambda \end{cases} \tag{9}$$

According to expression $P(k)$ in Eq. (9), we can obtain that

$$P(k) > P(k-1) \tag{10}$$

Formula (10) shows that with the increase of iteration times k, the covariance matrix $P(k)$ is always greater than 0 and remains positive, thus it solves the problem of "data saturation" in the traditional RLS.

The smaller the forgetting factor λ is, the better the system is at tracking, but the more sensitive it is to noise, so FRLS is inevitably more susceptible to noise than RLS. Therefore, a method of ψ_f identification using the recursive least square method with discounting factor (DRLS) is proposed, which forms a discounting factor to reduce the influence of noise by introducing the weighted factor into FRLS.

Similarly, the expression of DRLS is as follows

$$\begin{cases} \hat{\theta}(k) = K(k)(y(k) - \varphi^T(k)\hat{\theta}(k-1)) + \hat{\theta}(k-1) \\ K(k) = P(k-1)\varphi(k)/\left[\Gamma(k,i) + \varphi^T(k)P(k-1)\varphi(k)\right] \\ P(k) = \left[1 - K(k)\varphi^T(k)\right]P(k-1)/\lambda(k) \end{cases} \tag{11}$$

where $\Gamma(k,i)=\gamma(i)\prod_{k=i+1}^{n}\lambda(k)$.

If $\Gamma(k, i)$ is not appropriate, it may cause $K(k) < K(k-1)$ at some point, which means gradual "data saturation". Thus, different $\Gamma(k, i)$ may produce different steady-state identification value. If $\varphi(k)$ is to be stable, data saturation phenomenon is formed, which has little influence on the identification. On the contrary, if $\varphi(k)$ is to be changed, data saturation phenomenon is formed, then new stable identification value will be generated, affecting the identification accuracy. To avoid data saturation, improve tracking ability and reduce the influence of noise, the value range should be $0 < \Gamma(k,i) < 1$, where $0.9 < \lambda(k) < 1$.

Define the relative error $\zeta(k)$ to measure the magnitude of the identified value relative to the output fluctuation.

Where $\zeta(k)=\left\|y(k)-\varphi^T(k)\hat{\theta}(k-1)\right\|/\|y(k)\|$.

The value of the discounting factor of DRLS is shown in Eq. (12).

$$\Gamma(k,i)=\begin{cases}\Gamma_1=\gamma_1\lambda_1 & \zeta(k)>\varepsilon\\ \Gamma_2=\gamma_2\lambda_2 & \zeta(k)\le\varepsilon\end{cases} \tag{12}$$

where $0 < \gamma_{1,2}, \lambda_{1,2} < 1$.

It can be seen from Eq. (12) that DRLS can dynamically adjust $\Gamma(k, i)$ according to the magnitude of $\zeta(k)$, so that the identified ψ_f is more accurate than FRLS. Besides, DRLS can ensure the identification accuracy and dynamic tracking performance and reduce the impact of noise.

Particularly, when $\gamma(i) = 1$, DRLS degenerates into FRLS; when $\gamma(i) = 1$ and $\lambda(k) = 1$, DRLS degenerates into RLS.

For the convenience of identification, Eq. (1) is written as follows

$$u_q-\omega_e L_d i_d-R_s i_q=\omega_e\psi_f \tag{13}$$

Consider Eq. (13) as a system to be identified, then flux linkage identification model of PMSM is as follows

$$\begin{aligned}&y(k)=\varphi^T(k)\theta(k)+\xi(k)\\&\varphi^T(k)=[\omega_e]\\&\theta(k)=[\psi_f]\\&y(k)=[u_q-\omega_e L_d i_d-R_s i_q]\end{aligned} \tag{14}$$

4 Simulink Verification

To prove the validity of the proposed algorithm, simulation model is build based on the MATLAB/SIMULINK. Then, simulation verification will be carried out on data saturation of RLS and noise influence. The PMSM flux identification effect under direct calculation method, RLS, FRLS and DRLS will be compared and analyzed respectively as shown in Figs. 3, 4, 5 and 6 according to Fig. 1. The parameters used in simulation are shown in Table 2 but with $\lambda = 0.995$ in FRLS and $\Gamma 1 = 0.995/[0.995 + (1 - 0.995)\zeta^2(k)]$, $\Gamma 2 = 0.999/[0.995 + (1 - 0.999)\zeta^2(k)]$ in DRLS corresponding to $\zeta(k) > 0.1$ and $\zeta(k) < 0.1$ respectively. The motor speed was set as shown in Fig. 2.

It can be seen from Figs. 3, 4, 5 and 6, DRLS has the advantages of both RLS and FRLS, little data saturation and higher identification accuracy.

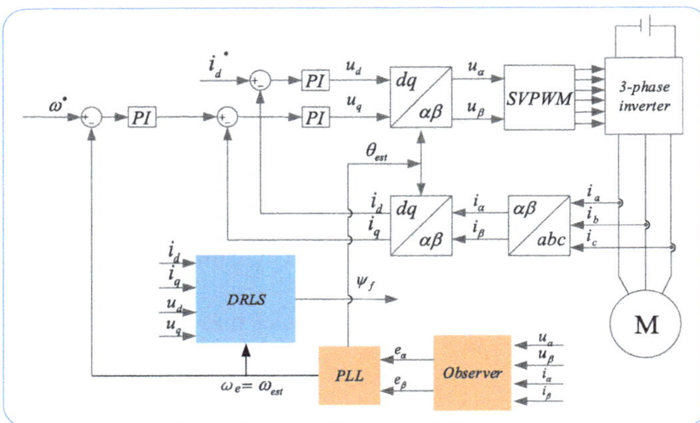

Fig. 1 Control diagram of PMSM flux linkage identification based on DRLS

Table 2 PMSM parameters

Parameter	Value
Rated voltage U_N/N	220
Rated current I_N/A	10
Rated torque T_L/Nm	2
Armature resistance R_s/Ω	3
d-axis inductance L_d/mH	15.67
q-axis inductance L_q/mH	18.27
PM flux linkage $\psi_f/\mathrm{W_b}$	0.0944
No. of polepairs $\mathrm{n_p}$	4
Moment of inertia J/(kg m)	5.33e−4

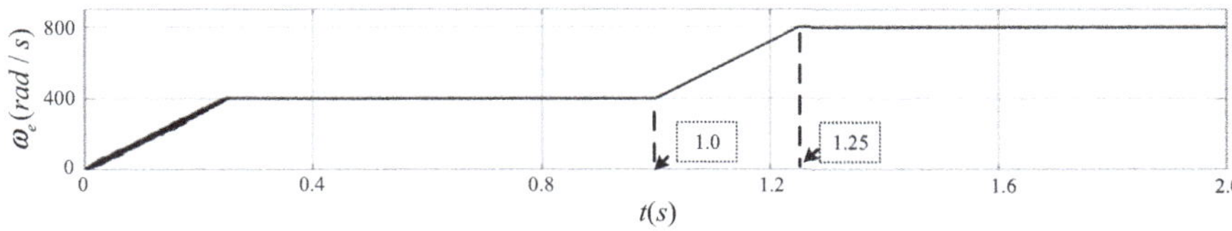

Fig. 2 Waveform of motor speed in flux linkage identification

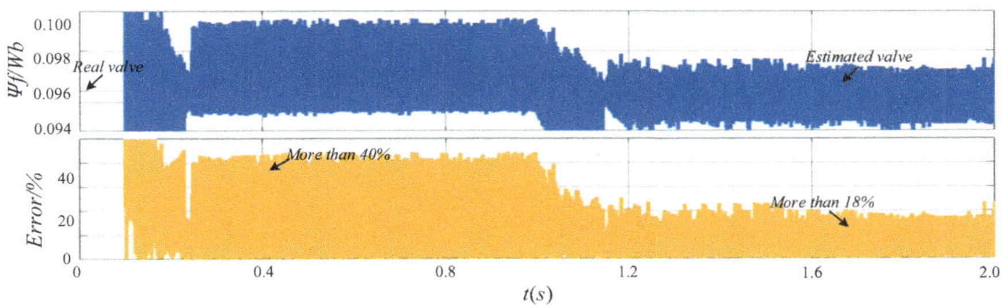

Fig. 3 Identification simulation results of PMSM flux linkage by direct calculation

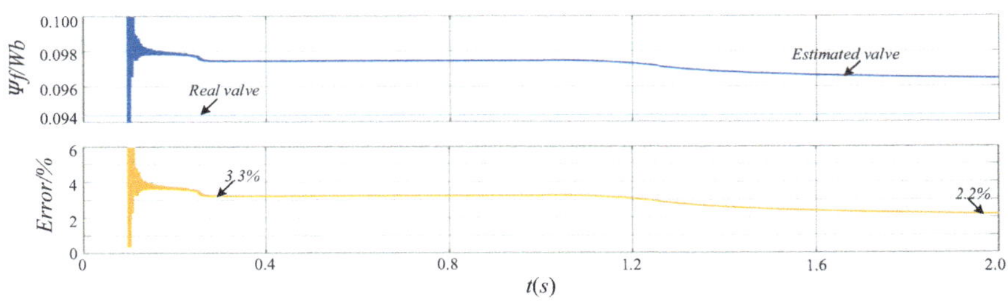

Fig. 4 Identification simulation results of PMSM flux linkage based on RLS

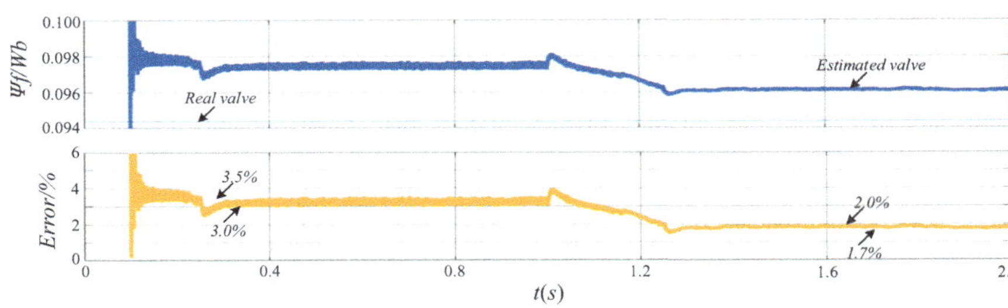

Fig. 5 Identification simulation results of PMSM flux linkage based on FRLS

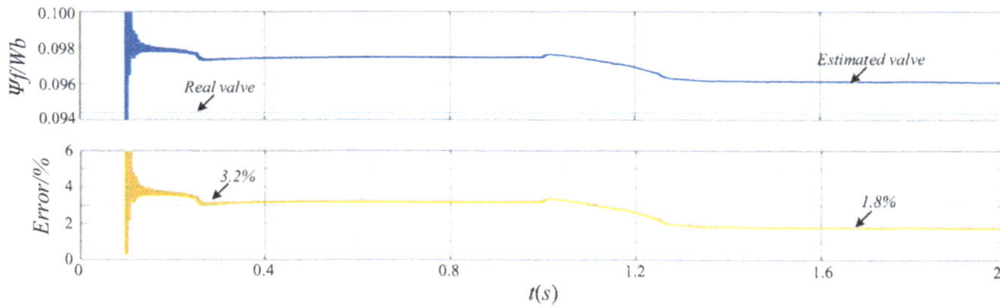

Fig. 6 Identification simulation results of PMSM flux linkage based on DRLS

5 Conclusion

In this paper, an online identification method based on DRLS is proposed for PMSM flux identification. This method adopts the fitting recursive least square method to identify the flux linkage of PMSM. Compared with the relevant intelligent algorithms, less computation is involved and less storage space is required in the identification process, which makes it easy to realize the on-line identification of the flux linkage in the digital controller. To solve the problem of data saturation in RLS, forgetting factor is introduced to improve the identification dynamic tracking ability. Further, DRLS is adopted to increase resistance to noise caused by FRLS. Simulation verifies the effectiveness of the identification method. The method can also be applied to the flux linkage identification of induction motor.

Acknowledgements This work was supported by the basic operating expenses of central scientific research (PA2020GDGP0053), China.

References

1. Liu, K., and Z.Q. Zhu. 2014. Online Estimation of the Rotor Flux Linkage and Voltage-Source Inverter Nonlinearity in Permanent Magnet Synchronous Machine Drives. *IEEE Transactions on Power Electronics* 29 (1): 418–427.
2. Jung, J., K. Jung, B. Lee, et al. 2019. Design and Analysis of Ferrite Magnet Flux Concentrated PMSM with Cross-Laminated Rotor Core Using Equivalent 2-D FEA. IEEE Transactions on Energy Conversion 34: 1623–1631.
3. Qiu, T.F., X.H. Wen, F. Zhao, et al. 2015. Design Strategy of Permanent Magnet Flux Linkage Adaptive Observer for Permanent Magnet Synchronous Motor. Proceedings of the Chinese Society for Electrical Engineering 35 (09): 2287–2294.
4. Sun, S.G., Y. Pang, J.Q. Wang, et al. 2016. EEMD harmonic detection method based on the new wavelet threshold denoising pretreatment. *Power System Protection and Control* 44 (2): 42–48. (in Chinese).
5. Liu, Z., H. Wei, Q. Zhong, et al. 2017. Parameter Estimation for VSI-FED PMSM Based on a Dynamic PSO With Learning Strategies. *IEEE Transactions on Power Electronics* 32 (4): 3154–3165.

6. Rafaq, M.S., F. Mwasilu, J. Kim, et al. 2017. Online Parameter Identification for Model-Based Sensorless Control of Interior Permanent Magnet Synchronous Machine. *IEEE Transactions on Power Electronics* 32 (6): 4631–4643.
7. Liu, K., and Z.Q. Zhu. 2014. Quantum Genetic Algorithm-Based Parameter Estimation of PMSM Under Variable Speed Control Accounting for System Identifiability and VSI Nonlinearity. *IEEE Transactions on Industrial Electronics* 62: 1.
8. Wang, C., Y.C. Liu, X.L. Liang, et al. 2018. Self-adaptive differential evolution algorithm with hybrid mutation operator for parameters identification of PMSM. *Soft Computing* 22 (4): 1263–1285.

Laboratory Preparation and Influence Factor Analysis of $LiFePO_4$ Soft Package Battery

Wenyue Xiao, Zechen Liu, Lei Zhu, and Jinying Zhang

Abstract $LiFePO_4$ soft package battery has been widely used in industry because of its high working voltage, low cost, good thermal stability and high security. However, there are so many steps to prepare a commercial soft package battery and the required equipments are too complex for one to complete all the operations. Also, the proportion of materials in each company is different and not completely public available, which makes it hard to analyze the changes of material structure after various treatments. Therefore, a simplified preparation of the $LiFePO_4$ soft package full battery in laboratory needs to be realized for comparison and analysis in the follow-up study. In this paper, the laboratory simplified preparation process of $LiFePO_4$ soft package battery was introduced in details. The morphology and cycle performance of three batteries with three different ratios of active materials, conductive agents and binders in the cathode material were studied. It was found that the battery with the mass ratio of 92:6.5:7 had the highest discharge platform (3.3 V), the largest specific capacity (153.4 mAh/g) and cycle retention rate. These discussions provide a good reference to repair and improve the industrial soft package full battery.

Keywords $LiFePO_4$ · Soft package battery · Influence of proportion

1 Introduction

The uncontrolled use of fossil fuels in modern industry has destroyed the ecological environment and made human beings faced with a great crisis of energy [1–4]. The shortage of fossil energy and the requirement of environmental protection have

W. Xiao · L. Zhu · J. Zhang (✉)
State Key Laboratory of Electrical Insulation and Power Equipment, School of Electrical Engineering, Center of Nanomaterials for Renewable Energy (CNRE), Xi'an Jiaotong University, No. 28 Xian Ning West Road, Bei Lin District, Xi'an, Shanxi 710049, China
e-mail: jinying.zhang@mail.xjtu.edu.cn

Z. Liu
Center of Nanomaterials for Renewable Energy (CNRE), NARI Technology Co. Ltd, Nanjing, China

W. Chen et al. (eds.), *The Proceedings of the 9th Frontier Academic Forum of Electrical Engineering*, Lecture Notes in Electrical Engineering 743,
https://doi.org/10.1007/978-981-33-6609-1_51

greatly promoted the production and development of modern lithium-ion batteries. Lithium iron phosphate ($LiFePO_4$) was discovered by the research team of John Goodenough of the University of Texas in 1996 as a cathode material for rechargeable lithium battery [4]. It has been widely accepted by the market because of its low raw material price, non-toxic, good thermal stability and high specific capacity (170 mAh/g) [5, 6]. The main obstacle to commercialization is its low electrical conductivity. However, conductive materials, such as carbon, can be effectively covered on $LiFePO_4$ particles by reducing the particle size [7, 8]. The soft package full battery is packed with a layer of aluminum-plastic film instead of metal shell, which can greatly improve the safety [9]. Considering the many advantages of $LiFePO_4$ soft package battery, it has been widely used in industry. However, there are so many steps to prepare a commercial soft package battery, and the required equipments are so complex that it's almost impossible for a single person to complete all the operations. Also, the proportion of materials in each company is different and not completely public available [10]. All these factors make it hard to analyze the changes of material structure after various treatments.

In this paper, the laboratory preparation of $LiFePO_4$ soft package full battery was deeply studied and the factors affecting the battery performance were comprehensively analyzed. The preparation process is introduced in detail in Sect. 1. In Sect. 2, SEM characterization was carried out on three kinds of soft-packed batteries with different ratios of positive materials, and their morphology was compared. The battery cycle performance test results of the three were discussed in Sect. 3. Finally, the best ratio soft pack battery was selected to test its XRD pattern after different cycle times, and the structure was compared and analyzed in Sect. 4.

2 Preparation Process of $LiFePO_4$ Soft Pack Battery

The process of preparing lithium iron phosphate soft pack batteries in the laboratory mainly includes four parts and each will be introduced in turn as following.

2.1 Manufacture of Positive and Negative Plates

The appropriate proportion of the materials is important prerequisite for the manufacture of positive and negative plates, and will directly affect the performance of the battery [11]. The proportion data of cathode and anode materials mostly used in industry are shown in Table 1 and Table 2 [6], The proportion data of three cathode and anode materials tried in laboratory are shown in Table 3 and Table 4. Since this paper mainly studies the influence of cathode materials on soft-packed batteries, and also for the need of controlling variables, the best proportion of anode that we worked out (No.3 Sample in Table 4) was used to combine with different proportion of cathode materials.

Table 1 Proportion data of cathode materials used in industry

Type of battery	Type of cathode	Main material(%)	Conductive agent(%)	Binder PVDF(%)	Solid content(%)
Consumer	$LiCoO_2$	96–98	1.0–2.0	1.5–2.5	55–65
	$LiMn_2O_4$	94–96	2.0–3.0	3.0–4.0	55–65
	$LiNiMnCoO_2$	94–96	2.0–3.0	3.0–4.0	55–65
	$LiFePO_4$	93–94	3.0–4.0	3.0–4.0	40–50
Power	$LiCoO_2$	92–94	3.0–4.0	2.0–3.0	55–65
	$LiMn_2O_4$	92–94	3.0–4.0	4.0–5.0	55–65
	$LiNiMnCoO_2$	92–94	3.0–4.0	4.0–5.0	55–65
	$LiFePO_4$	92–93	3.0–4.0	4.0–5.0	40–50

Table 2 Proportion data of anode materials used in industry

Type of battery	Type of cathode	Main material(%)	Conductive agent(%)	Thickener CMC(%)	Binder SBR(%)	Solid content(%)
Consumer	Graphite	94–96	1–2.5	1.3–2.0	2.0–2.5	40–50
Power	Graphite	91–94	2–4.0	1.3–2.0	2.0–2.8	40–50

Table 3 Proportion data of cathode materials tried in laboratory

Number	$LiFePO_4$ (g)	Conductive graphite (g)	PVDF (g)	NMP (g)	Mixing time (min)	Thickness (μm)	Remark
1	9.2	0.8	0.5	17.5	3 × 10	35	Thin, brittle
2	9.2	0.65	0.5	17.5	3 × 10	40	Brittle
3	9.2	0.65	0.7	18.5	3 × 10	40	√

Table 4 Proportion data of anode materials tried in laboratory

Number	Graphite (g)	Conductive graphite (g)	CMC (g)	SBR (g)	Water (g)	Mixing time (min)	Thickness (μm)	Remark
1	9.3	0.3	0.17	0.3	17.5	3 × 10	25	Thick, brittle
2	9.3	0.3	0.15	0.3	18.5	3 × 10	25	Hard to coat, brittle
3	9.3	0.3	0.17	0.4	18.5	3 × 10	25	√

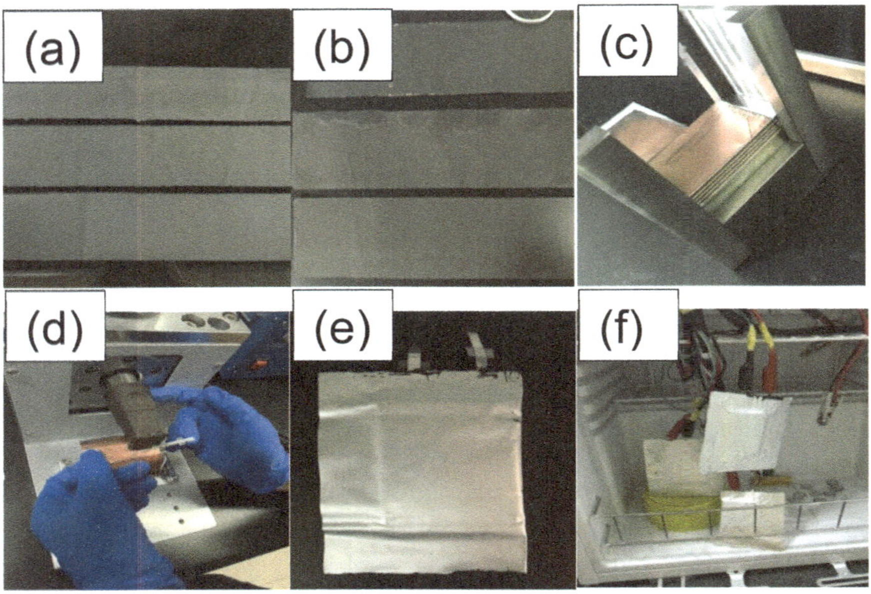

Fig. 1 Preparation process of $LiFePO_4$ soft pack battery. **a** Positive plates. **b** Negative plates. **c** Lamination technology. **d** Ultrasonic welding of a tab. **e** Soft package battery. **f** Pre-forming

To uniformly mix the various materials used in positive and negative plates, a rotation and revolution mixer was used to achieve uniform mixing and removal of bubbles. After homogenization, an adjustable coater was used for coating the plates in the laboratory. Since the clearance of the scraper can be adjusted, the need for various fixed gap coverers can be reduced. The gap can be set between 0 and 3500 microns. The cathode or anode slurry was coated on the aluminum foil and copper foil respectively. After that, the coated positive and negative plates were placed in a vacuum drying oven at 70 °C for drying. 12 hours later, the dried positive and negative plates were shown in Fig. 1a and b. Before the lamination step, the plates were rolled to achieve an appropriate thickness and density.

2.2 Lamination Technology and Ultrasonic Welding of Tabs

The positive and negative plates after rolling were overlapped regularly and neatly with the diaphragm between them, and were folded (equivalent to parallel connection [12]) with laminating auxiliary tools, as shown in Fig. 1c. After that, aluminum and nickel tabs were welded (The aluminum tab was welded to the positive electrode, and the nickel tab was welded to the negative electrode [9]). The specific operation of welding was as follows: (1) The bottom film was placed on the lower side of the tab foil for support; (2) The welding head of the ultrasonic welding instrument was placed on the top of the superimposed tab foil, and then the welding head was used to exert a certain degree of pressure on the tab foil; (3) Then the welding head directly output ultrasonic wave to realize the phase under the condition of high frequency vibration. The resonance of the atoms on the adjacent polar chaff caused the polar chaff to be joined together. The experimental ultrasonic welding was shown in Fig. 1d.

2.3 *Heat Sealing and Liquid Injection of Battery*

Battery heat sealing means that the lithium iron phosphate battery whose electrode tabs have been welded is coated with aluminum plastic film, and the top and side are encapsulated, leaving one side for liquid injection. The aluminum foil composite film used for battery packaging was divided into three layers (nylon layer, aluminum layer and pp layer). After heating to 180°, the aluminum foil of the two layers was completely bonded together by pressure, which achieves the main purpose of the package [13]. The liquid injection was carried out in the glove box in argon atmosphere, because the electrolyte (mainly $LiPF_6$) was easy to be oxidized and hydrolyzed by chemical reaction with water. The electrolyte was dripped directly into the battery from the unsealed side of the battery through a liquid transfer gun, and after standing for 12 h, the last side of the battery was completely tightly sealed with a heat sealing machine. The completed soft package battery is shown in Fig. 1e.

2.4 *Pre-forming of Battery*

Pre-forming was to activate the active substances with positive and negative electrodes inside the battery by charging and discharging the battery for the first time at a small current. At the same time, the positive and negative electrodes inside the battery were promoted to combine the substances with their surfaces to form a good SEI film [14]. After that, the shape of the battery was finally processed, and the battery was heated at a high temperature to rapidly age its outer shell, and a large number of chemical gas impurities were generated during the heating process. The air bag and redundant side edges of the battery case were cut off and all the side edges were folded to complete its final shape. The lithium iron phosphate soft package full battery during the pre-forming procedure is shown in Fig. 1f.

3 Morphology Analysis of $LiFePO_4$ Soft Package Battery

The photos shown in Fig. 2 are SEM images of three mixed cathode materials with different proportions ($LiFePO_4$, conductive graphite and PVDF with mass ratio of

Fig. 2 SEM images of three cathode materials with different mass ratio, 92:8:5 (**a**) and (**d**), 92:6.5:5 (**b**) and (**e**), 92:6.5:7 (**c**) and (**f**). Scale bar is 8 μm (**a–c**), and 2 μm (**d–f**)

92:8:5, 92:6.5:5 and 92:6.5:7, respectively). (a) and (d) are 92:8:5, (b) and (e) are 92:6.5:5, (c) and (f) are 92:6.5:7.

For the 92:8:5 material, the $LiFePO_4$ particles are separated from the conductive agent, and are mixed unevenly. Compared with $LiFePO_4$, conductive graphite's particle size is larger, making it spontaneously agglomerate and form self-adhesion. It may be because the binder ratio is too low, all the solid substances will be tightly bonded together macroscopically, and cannot achieve good bonding effect [15]. For the 92:6.5:5 material, since the PVDF content of the binder is relatively low, there is a physical separation between the $LiFePO_4$ and the surrounding conductive graphite. The grains are highly dispersed and the gap between each other is large. The surface of $LiFePO_4$ is coated with a lot of conductive graphite, and no continuous conductive channel is formed, which is not conducive to electronic conduction [16]. For the 92:6.5:7 material, the volume of $LiFePO_4$ particles and conductive graphite particles are the smallest and uniformly distributed, and the dispersion between the small particles is also the best, The grains of the corresponding materials can be closely contacted during the charging process, and thus making the electronic conductivity improved to a large extent.

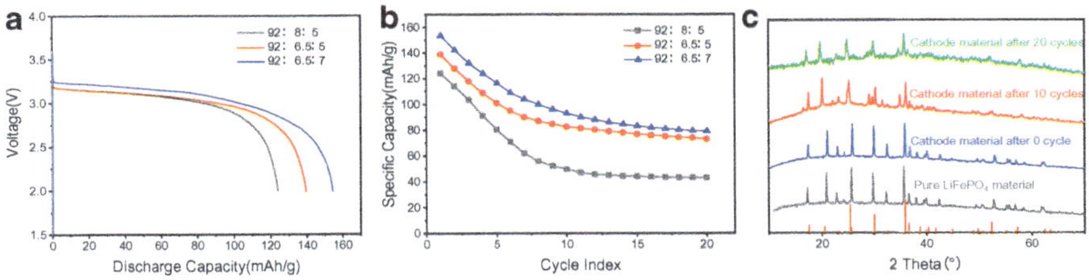

Fig. 3 **a** The first discharge curves and **b** cycle curves of cathode materials with different matching ratios. **c** XRD patterns of $LiFePO_4$ and cathode mixture

4 Electrochemical Performance of $LiFePO_4$ Soft Package Battery

Figure 3a shows the first discharge of three soft package batteries made of cathode materials with three different mass ratios at 0.5 C. The degree of polarization of the battery with the mass ratio of 92:6.5:7 is much lower than that of other similar batteries, and the discharge platform is 3.3 V. It is higher than the lithium iron phosphate battery of the other two ratios (3.2 V). It also shows that the 92:6.5:7 $LiFePO_4$ battery has the highest first discharge capacity, which is 154.3 mAh/g. The first discharge capacity of the 92:6.5:5 battery is in the middle, which is 139.6 mAh/g, and the 92:8:5 battery has the lowest specific capacity, which is 124.1 mAh/g. It can be concluded that the battery can form a better first-time conductive network with the ratio of 92:6.5:7, and its polarization effect of electrochemical material is relatively low and can achieve higher first discharge specific capacity.

Figure 3b shows the cyclic charge-discharge curve of the sample battery made of different kinds of cathode materials. From the figure, we can see that the the specific capacity decreases more slowly after 10 cycles and tends to be stable. The retention rates of battery capacity after 20 cycles were 51.9%, 50.1% and 36.8%, respectively. The sample battery with ratio of 92:6.5:7 has the highest capacity retention after 20 charge-discharge cycles, while the other two batteries have low capacity retention and discharge specific capacity due to the improper amount of conductive graphite and PVDF. The results show that the ratio of positive active materials has great influence on the comprehensive performance of the battery.

5 XRD Analysis of $LiFePO_4$ Soft Package Battery

Figure 3c shows the X-ray diffraction patterns of pure lithium iron phosphate, lithium iron phosphate cathode mixture before recycling and cathode mixture after 10 or 20 battery cycles' charge and discharge. From the figure, we can clearly see that the diffraction peak of pure lithium iron phosphate is consistent with that of $LiFePO_4$ (JCPDS No. 77-0179). However, there is almost no difference between the diffraction peak of lithium iron phosphate positive paste after mixing but not participating in the cycle, indicating that when the positive materials are mixed but no electrochemical reaction has taken place, conductive agent and binder will not affect the crystallinity and the intensity of diffraction peak of lithium iron phosphate. After cyclic testing, the XRD spectra of cathode mixed materials has more messy small peaks compared with those before cycling, and the diffraction peaks become weaker and shift to the left as a whole, which is probably due to the continuous growth of lithium iron phosphate dendrites, the enlargement of crystal layer spacing and the decrease of crystallinity and stability with the reaction of the battery [10]. With the increase of the number of cycles, the X-ray diffraction peaks become weaker and the trend of left shift is strengthened, which once again proves the effect of battery cycle on lithium iron phosphate cathode materials.

6 Conclusion

In this paper, the laboratory simplified preparation process of lithium iron phosphate soft package battery was introduced in detail, and the morphology and cycle performance of the whole battery made of three different cathode materials were studied. Among all the three different ratios, the lithium iron phosphate soft package battery with the mass ratio of 92:6.5:7 had the highest discharge platform (3.3 V), the largest specific capacity (153.4 mAh/g) and cycle retention rate. It can be seen that the ratio of active materials, conductive agents and binders in the cathode material has a key influence on its eletrochemical performance and comprehensive application. These discussions provide a good reference for the laboratory to prepare the soft package full battery with better performance, and even to repair and improve the industrial soft package full battery.

References

1. Sun, Bingxiang, Jia Liu, Zhiqiang Han, et al. 2020. Performance Correlation and Temperature Applicability Analysis of Li-Ion Battery Under Different Interval Decline Paths. *Transactions of China Electrotechnical Society* 035 (009): 2063–2073. (in Chinese).
2. Shengqi Zhang, Bei Yuan, Zhendong Ji, et al. 2019. Secondary frequency modulation control of battery energy storage system based on distributed control principle. *Transactions of China Electrotechnical Society* 034 (0z2): 637–645 (in Chinese).
3. Zhang, Zhenyu, Guangsen Wang, Shixiong Nie, et al. 2019. Estimation of charge state of lithium iron phosphate battery under pulsed high rate discharge. *Transactions of China Electrotechnical Society* 034 (008): 1769–1779. (in Chinese).
4. Lin, X., H.E. Perez, J.B. Siegel, et al. 2013. Online Parameterization of Lumped Thermal Dynamics in Cylindrical Lithium Ion Batteries for Core Temperature Estimation and Health Monitoring. *IEEE Transactions on Control Systems and Technology* 21 (5): 1745–1755.

5. Guo, Z., and Z. Chen. 2015. High-Temperature Capacity Fading Mechanism for $LiFePO_4$/Graphite Soft-Packed Cell Without Fe Dissolution. *Journal of Electroanalytical Chemistry* 754: 148–153.
6. Liao, X., J. Yu, L. Gao, et al. 2012. Electrochemical study on lithium iron phosphate/hard carbon lithium-ion batteries. *Journal of Solid State Electrochemistry* 16 (2): 423–428.
7. Zhong, K., Y. Cui, X. Xia, et al. 2014. Study on the Stability of the $LiFePO_4$ Li-Ion Battery via an Electrochemical Method. *Journal of Power Sources* 250: 296–305.
8. Du, S., M. Jia, Y. Cheng, et al. 2015. Study on the Thermal Behaviors of Power Lithium Iron Phosphate (LFP) Aluminum-Laminated Battery with Different Tab Configurations. *International Journal of Thermal Sciences* 89: 327–336.
9. Yue, Zai. 2008. A Soft Package Lithium Ion Power Battery Module.
10. Zhao-Rong, C., L.U. Hao-Jie, F.U. Xiao-Ning, et al. 2008. High Rate Discharge Performance of Soft Package Li-Ion Battery. *Battery Bimonthly*.
11. Shu-Ting, Y., L. Yu-Xia, Y. Yan-Hong, et al. 2007. Effects of Soft Template on the Performance of $LiFePO_4$/C Cathode Material. *Battery Bimonthly*.
12. Cong, W.Y, Ning, H., W.B. Bei, et al. 2019. *Effect of Separators on the Performance of high-Rate Soft-Pack Lithium-Ion Battery*. Chinese Battery Industry.
13. Wang, K., F. Gao, Y. Zhu, et al. 2018. Internal Resistance and Heat Generation of Soft Package $Li_4Ti_5O_{12}$ Battery During Charge and Discharge. *Energy* 149: 364–374.
14. Guo, Z., and Z. Chen. 2016. Aging Property for $LiFePO_4$/Graphite Cell with Different Temperature and DODs. *Russian Journal of Electrochemistry* 52 (6): 546–554.
15. Yang, Shu-ting, Na-hong Zhao, Yan-hong Yin, et al. 2005. Synthesis of $LiFePO_4$/C Nano-structure Intergrowth by PAM Soft Templating Method. *Battery Bimonthly*.
16. Ke, Lei, Wei Lv, Fang-Yuan Su, et al. 2015. Electrode Thickness Control: Precondition for Quite Different Functions of Graphene Conductive Additives in $LiFePO_4$ Electrode. *Carbon* 92: 311–317.

Permanent Magnet Synchronous Linear Motor Control System Based on Improved Sliding Mode Observer

Zheng Li, Qingshan Zhang, Hexu Sun, Jinfeng An, and Qunjing Wang

Abstract With the improvement of the working requirements of the motor, the closed-loop control of the traditional vector control system of permanent magnet synchronous linear motor (PMSLM) is gradually realized without speed sensor. But the traditional sliding mode observer is based on the switch function of sigmoid, which makes the system produce large chattering. To solve this problem, an improved sliding mode observer is proposed. The sliding mode observer adopts the piecewise tangent switching function based on the super twisting. Finally, the improved sliding mode observer proposed in this paper is simulated and verified by experiments. The simulation results show that the improved sliding mode observer can reduce chattering, and through the thrust force experimental diagram, it can be concluded that this method has better ability to deal with load changes and has better stability.

Keywords Linear motor · Tangent piecewise function · Improved sliding mode observer

1 Introduction

PMSLM has the advantages of low noise, high efficiency, low wear and high precision, and has been widely used in semiconductor, medical, CNC machining and even military industries [1]. However, mechanical sensors are easily damaged.

The current speed sensorless control methods are mainly extended Kalman filter method, model reference adaptive method, high frequency signal injection method and sliding mode observer method [2]. Among other observers, the sliding mode observer method is less affected by parameters and has stronger robustness. This method has greater advantages [3].

Z. Li (✉) · Q. Zhang · H. Sun · J. An
School of Electrical Engineering, Hebei University of Science and Technology, No. 26 Yuxiang Street, Yuhua District, Shijiazhuang, Hebei Province, China
e-mail: Lzhfgd@163.com

Q. Wang
National Engineering Laboratory of Energy-Saving Motor and Control Technique, No. 111 Jiulong Road, Shushan District, Hefei, Anhui Province, China

W. Chen et al. (eds.), *The Proceedings of the 9th Frontier Academic Forum of Electrical Engineering*, Lecture Notes in Electrical Engineering 743,
https://doi.org/10.1007/978-981-33-6609-1_52

Sliding mode control is nonlinear control. It belongs to a special nonlinear control system, that is, a switching characteristic that makes the structure of the system change at any time [4]. The magnetic pole position and speed information required for closed-loop control can be measured in the extended back-EMF.

Because the nonlinearity of the sliding mode observer makes the system chattering, the extra voltage and current noise signals introduced by the linear motor will also aggravate the system chattering. At present, methods to weaken the chattering of sliding mode observers are the mainstream research direction [5].

In this paper, a new type of observer is proposed by combining the piecewise tangent function with the super twisting sliding mode observer. Then through simulation analysis and experimental verification, the desired goal can be achieved.

2 Mathematical Model

The proposed method needs to be modeled under ideal conditions. The mathematical model under the stationary coordinate system is as follows:

$$\begin{cases} \frac{di_\alpha}{dt} = -R\frac{i_\alpha}{L_s} - \frac{e_\alpha}{L_s} + \frac{u_\alpha}{L_s} \\ \frac{di_\beta}{dt} = -R\frac{i_\beta}{L_s} - \frac{e_\beta}{L_s} + \frac{u_\beta}{L_s} \end{cases} \tag{1}$$

The back EMF is as follows:

$$\begin{cases} e_\alpha = -k_e\omega_r \sin(\theta) \\ e_\beta = k_e\omega_r \cos(\theta) \end{cases} \tag{2}$$

where ω_r is the electrical angular velocity, $\omega_r = \pi \upsilon/\tau$, υ is the mover speed; τ is the pole pitch of the linear motor. From Eq. (2), the phase of the back-EMF of the linear motor is closely related to the magnetic pole position. The back EMF of the linear motor can be estimated.

2.1 Traditional Sliding Mode Observer

According to the traditional sliding mode control theory, the current error model of the traditional sliding mode observer can be obtained as follows:

$$\begin{cases} \frac{d\bar{i}_\alpha}{dt} = -R\frac{\bar{i}_\alpha}{L_s} - K\frac{\text{sgn}\left(\hat{i}_\alpha - i_\alpha\right)}{L_s} + \frac{e_\alpha}{L_s} \\ \frac{d\bar{i}_\beta}{dt} = -R\frac{\bar{i}_\beta}{L_s} - K\frac{\text{sgn}\left(\hat{i}_\beta - i_\beta\right)}{L_s} + \frac{e_\beta}{L_s} \end{cases} \tag{3}$$

From the above formula, the current error in the formula can be obtained. When the system reaches the sliding surface, the error is 0. According to the sliding mode dynamic conditions proposed by the sliding mode variable structure control theory:

$$\begin{cases} e_\alpha = K\text{sgn}\left(\hat{i}_\alpha - i_\alpha\right) \\ e_\beta = K\text{sgn}\left(\hat{i}_\beta - i_\beta\right) \end{cases} \tag{4}$$

After the linear motor back-EMF passes through the filter, the estimated magnetic pole position information can be obtained as:

$$\theta_e = -\arctan\frac{e_\alpha}{e_\beta} \tag{5}$$

In the formula, θ_e is the electrical angle; ω_r is the electrical angular velocity.

2.2 Piecewise Tangent Function

In this paper, the tangent function is processed in sections, and the boundary condition in the function is written as a, then the formula is as follows:

$$y(x) = \begin{cases} 1 & x \geq a \\ \tan(wx) & -a < x \leq a \\ -1 & x < -a \end{cases} \tag{6}$$

In the formula, w is the frequency coefficient to be adjusted, and x is the error variable between the actual current value and the observed value [6]. From Eqs. (4) and (6), we can see that the motor back EMF based on the piecewise tangent function is as follows:

$$\begin{cases} e_\alpha = Ky(x) \\ e_\beta = Ky(x) \end{cases} \tag{7}$$

The characteristic curve of this function is shown in Fig. 1. Outside the boundary layer thickness, that is, outside $-a < x \leq a$, the current error is saturated. Within the

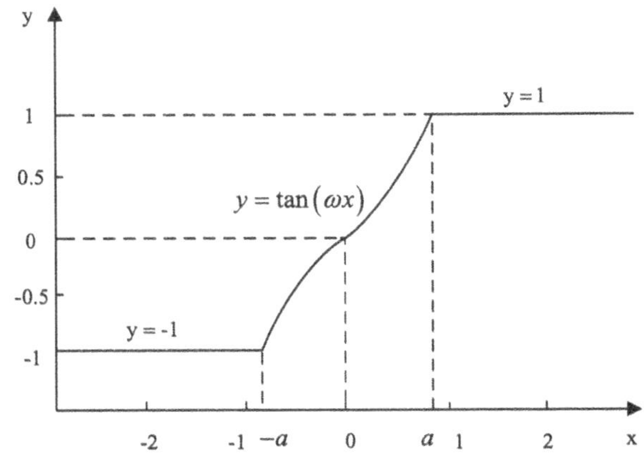

Fig. 1 Piecewise tangent function

boundary layer thickness, the function changes in the form of a tangent function, and the tangent change can make the system quick and stable when changing structure.

2.3 Improved Super Twisting Observer

Reference [7] designed a super-twisting algorithm observer and achieved good control results. Based on the above content, the new super twisting sliding mode observer designed in this paper is as follows:

$$\begin{cases} \dfrac{d\hat{i}_\alpha}{dt} = -\dfrac{R}{L_s}\hat{i}_\alpha + \dfrac{u_\alpha}{L_s} - \dfrac{k\hat{s}_\alpha}{L_s} - \left|\bar{i}_\alpha\right|^{0.5} e_\alpha \\ \dfrac{d\hat{e}_\alpha}{dt} = \lambda e_\alpha \\ \dfrac{d\hat{i}_\beta}{dt} = -\dfrac{R}{L_s}\hat{i}_\beta + \dfrac{u_\beta}{L_s} - \dfrac{k\hat{s}_\beta}{L_s} - \left|\bar{i}_\beta\right|^{0.5} e_\beta \\ \dfrac{d\hat{e}_\beta}{dt} = \lambda e_\beta \end{cases} \tag{8}$$

It can be seen from the above formula that the equivalent feedback signal satisfies the formula:

$$\begin{cases} \hat{s}_\alpha = \displaystyle\int \lambda e_\alpha dt \\ \hat{s}_\beta = \displaystyle\int \lambda e_\beta dt \end{cases} \tag{9}$$

Fig. 2 Structure block diagram (new function)

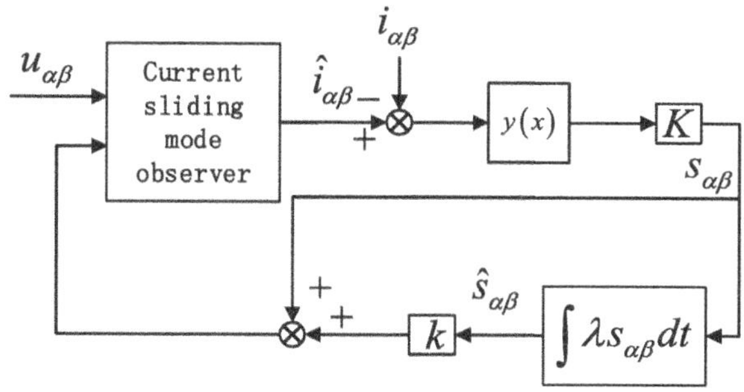

Fig. 3 Equivalent block diagram of PLL

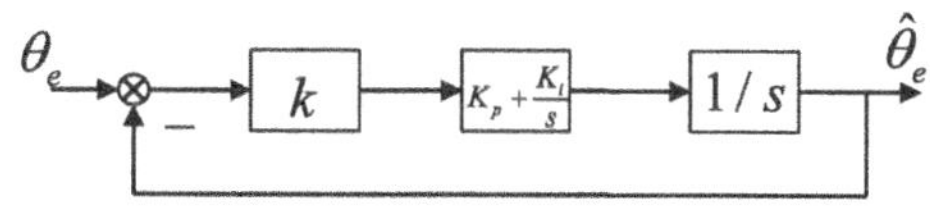

The switching function in the super twisting observer is replaced with a new segmented tangent switching function, and the obtained back-EMF can be filtered to obtain the current magnetic pole position information and current speed information of the motor [8, 9]. Figure 2 is the structural block diagram of the sliding mode observer based on the new function.

3 Phase-Locked Loop and Its DC Bias Suppression

This paper uses a phase-locked loop to obtain magnetic pole position information. The error variable e_r is constructed by the principle of phase-locked loop:

$$e_r = k \sin\left(\theta_e - \hat{\theta}_e\right) \approx k\left(\theta_e - \hat{\theta}_e\right) \quad (10)$$

The equivalent block diagram is shown in Fig. 3.

In order to improve the accuracy of the estimated value, a generalized integrator is used to suppress the DC offset. Figure 4 is a control block diagram.

4 Simulation and Eperimental Results

This paper uses simulation software for simulation experiments. The parameters of the experiment are shown in Table 1. The system simulation time is 0.1 s. The overall control strategy adopts current and speed closed loop. The reference speed is set to 1 m/s. A 5 N load is suddenly added when the simulation time is 0.05 s. The data diagram of the system obtained by the simulation experiment is shown below.

Fig. 4 Block diagram of a new super-twisting sliding mode observer

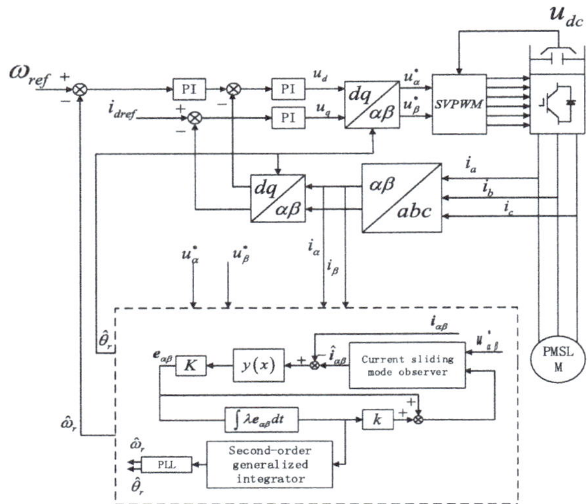

Table 1 Permanent magnet synchronous linear simulation parameters

Parameter	Value
Number of motor pole pairs	4
Stator resistance (Ω)	2.875
Stator inductance (H)	8.5×10^{-2}
Coefficient of viscous friction (Nm s^{-1})	0.001
Stator pole pitch (m)	0.016
Rated speed (m/s)	1
Mover mass (m/kg)	1.425

Figure 5 shows the motor speed waveforms of the new and traditional observers. Observing the waveform, it can be found that the speed waveform of the new sliding mode observer is more stable when the motor is running stably.

Fig. 5 Speed waveform simulation comparison

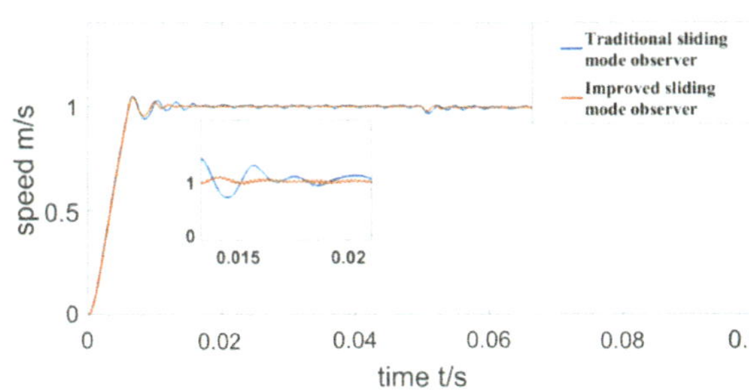

Figures 6 and 7 show the torque ripple waveform. When the motor is running stably, it can be found that the pulsation of the new sliding mode observer is relatively small.

The experiment process and equipment shown in Fig. 8. The STM32F103 series development board used in the experiment, the linear motor model is shown in Fig. 8. This linear motor is ironless. It is used as the controlled object, and the c code of the improved switching function algorithm is generated by a code generation tool, and the algorithm is verified on the STM32 development board.

In Figures 9 and 10, when a large current load is applied, the traditional current will be slightly distorted and the waveform is roughly triangular, while the current of the improved sliding mode observer is closer to a sine wave.

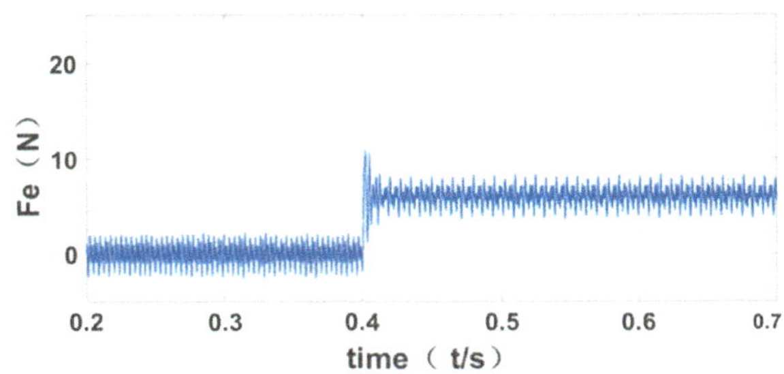

Fig. 6 Torque change (traditional)

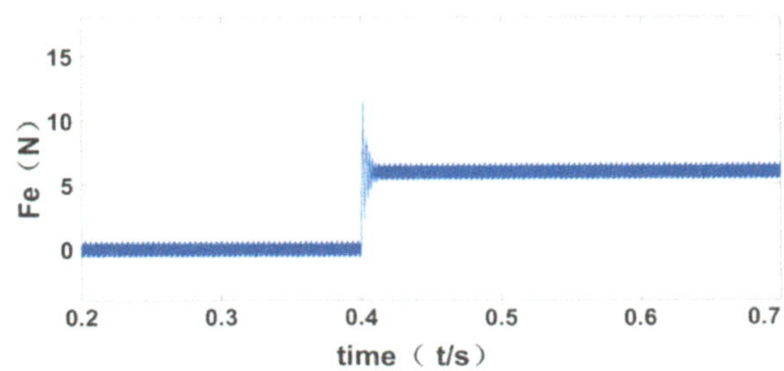

Fig. 7 Torque change (improved)

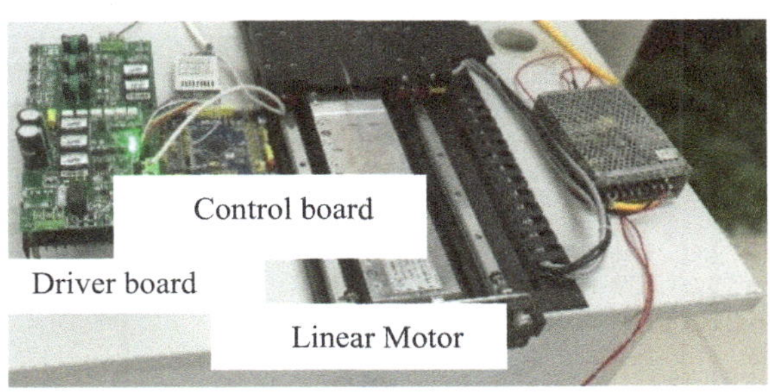

Fig. 8 Experimental platform

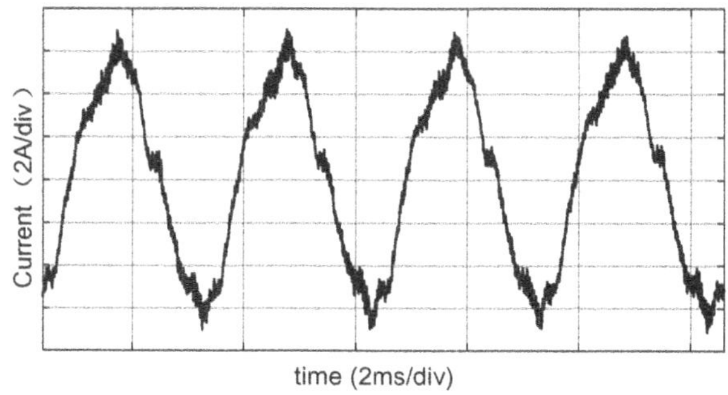

Fig. 9 Single-phase current waveform (traditional)

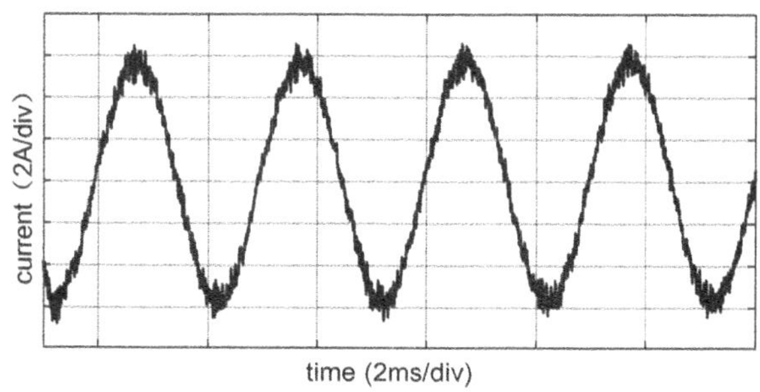

Fig. 10 Single-phase current waveform (improved)

5 Conclusion

In this paper, replace the traditional switch function with an improved switch function. An improved super twisting sliding mode observer is designed to achieve the purpose of weakening chattering. In addition, it also enhances the load capacity of the control system. The robustness of the system is improved. The control system proposed in this paper was simulated and verified by simulation software, and compared with the simulation results of the traditional sliding mode observer.

Acknowledgment This work was supported by the National Natural Science Foundation of China, grant No. 51877070, 51577048, 51637001, the Natural Science Foundation of Hebei Province of China, grant No. E2018208155, the Talent Engineering Training Support Project of Hebei Province, grant No. A201905008, the National Engineering Laboratory of Energy-saving Motor & Control Technique, Anhui University, grant No. KFKT201901, Hebei Province Higher Education Science and Technology Research Key Project, grant No. ZD2018228.

References

1. Hongyan, Jin, Zhao Ximei, and Yuan Hao. 2020. Global boundary sliding mode control for dynamic boundary layer of permanent magnet linear synchronous motor. *Transactions of China*

Electrotechnical Society 35 (9): 1945–1951. (in Chinese).
2. Jun, Hou, Sun Jing, Hofmann Health. 2018. Adaptive model predictive control with propulsion load estimation and prediction for all-electric ship energy management. *Energy* 16 (1): 150–155.
3. Jiang, Jiahao, Xia Yang. 2018. Simulation of permanent magnet synchronous motor control system based on a new sliding mode observer method. In *2018 15th Annual Shenyang Science Conference*, Shenyang, 1–7 (in Chinese).
4. Wang, Haibo, Bo Zhou, and Fang Fang. 2016. Sliding mode control of permanent magnet synchronous motor speed regulation system. *Transactions of China Electrotechnical Socity* 24 (9): 72–77. (in Chinese).
5. Zhang, Qun, Hong Li, and Yong Zheng. 2011. A new type of sliding mode observer for sensorless control of permanent magnet synchronous motor. *Micro Motor* 39 (8): 41–43. (in Chinese).
6. Yu, Wenxin, Dan Jiang, Junnian Wang, Ruiqi Li, Lu Yang. 2020. Rotor-current-based fault detection for doubly-fed induction generator using new sliding mode observer. *Transactions of the Institute of Measurement and Control* (in Chinese).
7. Guo, Leilei, Huaqing Wang, Linwang Dai, et al. 2020. Speed sensorless control method for permanent magnet synchronous motor based on super-helical sliding mode observer. *Power Automation Equipment* 40 (2): 21–31. (in Chinese).
8. Li, Fei. 2019. *Research on Position Sensorless Control of Built-In Permanent Magnet Synchronous Motor in Full Speed Domain*. Northeast Petroleum University (in Chinese).
9. Liu, Huanlong, Shun Li, Chixin Xie. 2020. Motor starting current control method based on hydraulic pump/motor reverse drive. *Journal of Southwest Jiaotong University*, 1–9 (in Chinese).

Research on the Influence of Voltage Polarity on Corona Discharge Characteristics of Insulators

Qiang Li, Chen Liu, Xuehuan Wang, Nana Duan, and Shuhong Wang

Abstract The corona discharge of insulators has an important influence on the electromagnetic environment. In order to study the influence of different voltage polarities on the corona discharge characteristics of insulators, this paper is mainly based on the fluid-chemical reaction equation to establish a two-dimensional simulation model of insulators, which includes a variety of chemical reactions and substances. According to the above mentioned numerical analysis model, the electric field and the particle density of the insulators under different voltage polarities are given through the simulation calculation. Finally, the discharge parameters under different voltage polarities are compared, and the relationship between corona discharge parameters and voltage polarity is obtained.

Keywords Voltage polarity · Insulator · Corona discharge · Particle

1 Introduction

Insulator corona discharge will not only produce radio interference effects and audible noise [1, 2], but also cause certain losses to China's national economy. With the continuous development of high-voltage transmission projects around the world, the design and performance improvement of insulators and conductors required for high-voltage transmission are hot topics of concern to scholars at home and abroad.

Many researchers in China have done some research on corona discharge of insulators. Zhang and Liu [3, 4] combined the theory of gas discharge to study the corona discharge characteristics of AC catenary insulators. Zeng et al. [5] considered the influence of water on corona discharge characteristics, and systematically analyzed the role of water molecules in corona discharge. However, the above

Q. Li · C. Liu
Xi'an High Voltage Apparatus Research Institute Co. Ltd, No. 18 North Section of West Second Ring Road, Xi'an, Shanxi, China

X. Wang · N. Duan (✉) · S. Wang
School of Electrical Engineering, Xi'an Jiaotong University, No. 28 Xianning West Road, Xi'an, Shanxi, China
e-mail: duannana@xjtu.edu.cn

W. Chen et al. (eds.), *The Proceedings of the 9th Frontier Academic Forum of Electrical Engineering*, Lecture Notes in Electrical Engineering 743,
https://doi.org/10.1007/978-981-33-6609-1_53

mentioned studies all equate AC corona discharge to positive corona discharge, without considering the influence of negative corona discharge.

Based on the existing research, this paper establishes a numerical simulation model of insulator corona discharge from a microscopic point of view, and deeply analyzes the influence of different voltage polarities on the corona discharge characteristics of insulators. On this basis, the relationship between different discharge parameters and voltage polarity is given through a simulation model.

2 Fluid-Chemical Reaction Hybrid Numerical Model

2.1 Governing Equation

This paper is mainly based on the theory of gas discharge, establishing a two-dimensional simulation model of fluid-chemical reaction.

First, the electric field and potential during the discharge process are calculated by Poisson's equation [6]:

$$\begin{cases} \nabla^2\phi = -\frac{\rho_0}{\varepsilon_0} \\ \vec{E} = -\nabla\phi \end{cases} \tag{1}$$

Among them, ϕ is the electric potential, ρ_0 is the amount of charge, ε_0 is the dielectric constant, and E is the electric field strength.

The fluid equation is an equation describing the number density of electrons, the average momentum of electrons, and the average energy of electrons. As a function of configuration space and time, the continuity equation of electrons can be described as [7]:

$$\frac{\partial n_e}{\partial t} + \nabla \cdot \left(-\left(\mu_e \cdot \vec{E}\right)n_e - \nabla(D_e n_e)\right) = S_\alpha - S_\eta - S_{ep} + S_{\text{det}} + S_{ph} + S_0 \tag{2}$$

where, n_e is the electron number density, D_e is the diffusion coefficient of electrons, S_α is the electron impact ionization source term, S_η is the electron adsorption reaction source term, S_{ep} is the recombination reaction term of electrons and positive ions, S_{det} is the electron dissociation term, S_{ph} is the photoionization term (not considered in this article) and S_0 means the initial electronic item.

In the insulator discharge process, not only electrons are involved, but also some heavy particles such as ions and gas molecules. In the calculation process, assuming that the heavy particles in the air meet the condition of the average of the mixture, then the diffusion model of the transport mixture can be expressed as:

$$\rho\frac{\partial w_k}{\partial t} + \rho(\vec{u} \cdot \nabla)w_k = \nabla \cdot \vec{j}_k + R_k \tag{3}$$

$$\vec{j}_k = \rho w_k \vec{v}_k \tag{4}$$

Among them, j_k is the vector of the diffusion flux, R_k is the source term that generates the particles, μ is the average velocity vector of the fluid, ρ is the density of the mixture, and w_k is the mass fraction of the particles.

2.2 Plasma Chemical Reaction

The corona discharge process of insulators is complicated. The reaction process includes not only the reactions between electrons and gas molecules such as electron collision reactions and adhesion reactions, but also reactions between heavy particles. In addition, due to the complex composition of the atmosphere, there are more reactive substances in the corona discharge process. The chemical reactions in the article are detailed in the literature [5], including 19 kinds of particles including N_2, O_2 and H_2O, and 41 kinds of chemical reactions of corresponding substances.

2.3 Plasma Boundary Conditions

Using partial differential equations to solve the objective function requires the determination of appropriate boundary conditions. For gas discharge, a secondary electron emission process occurs at the cathode boundary. In the calculations in this paper, only the secondary electron emission generated by positive ions impacting the cathode surface is considered process.

Among them, the secondary electron emission process on the cathode surface usually includes secondary electron reflection and secondary electron emission caused by positive ions colliding with the cathode electrode surface. The electron flux boundary condition formed by the two can be expressed as [8]:

$$\Gamma_e = \frac{1-\gamma_e}{1+\gamma_e}\left(\frac{1}{2}v_{th}n_e\right) - \sum_i \gamma_i \Gamma_i \tag{5}$$

In the formula, Γ_e is the flux of electrons at the boundary; γ_e is the reflection coefficient of electrons on the electrode surface; v_{th} is the thermodynamic velocity; Γ_i is the flux of positive ions at the boundary; γ_i is the secondary electron emission coefficient.

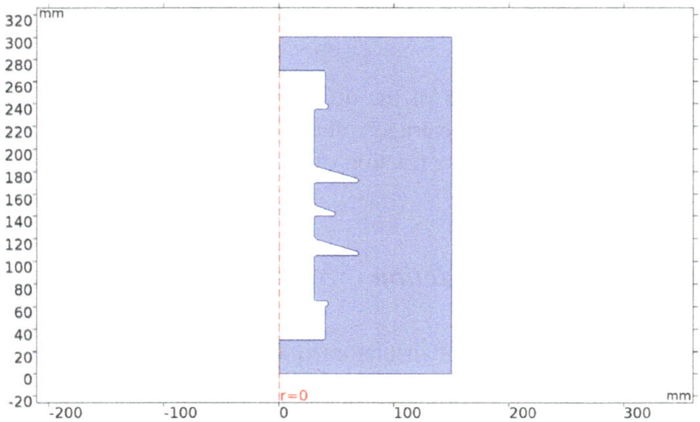

Fig. 1 Insulator calculation model diagram

3 Simulation Calculation Model

According to the literature [5], the calculation model is FQB-25/8 composite insulator in electrified railway. The insulator includes 9 large umbrella skirts and 8 small umbrella skirts. In order to facilitate the calculation, the insulator model is simplified accordingly. The specific calculation model is shown in Fig. 1.

The temperature of the calculated model is 20 °C, a standard atmosphere. Since the time of corona discharge is very short, AC voltage can be calculated as DC voltage. The upper part of the insulator is respectively applied with a voltage of ±39 kV, and the lower part of the insulator is grounded. At the same time, set a blocking capacitance of 1 pF and a protection resistance of 5000 Ω at the position of the metal attachment of the insulator [1].

4 Results Comparison and Analysis

4.1 Graph of Electric Potential and Field Strength

Figure 2 is a graph showing the change of the insulator surface potential and electric field at different times. It can be seen from Fig. 2 that with the continuous development of the corona discharge process of the insulator, the absolute value of the component voltage on the insulator continues to increase, and eventually tends to be consistent.

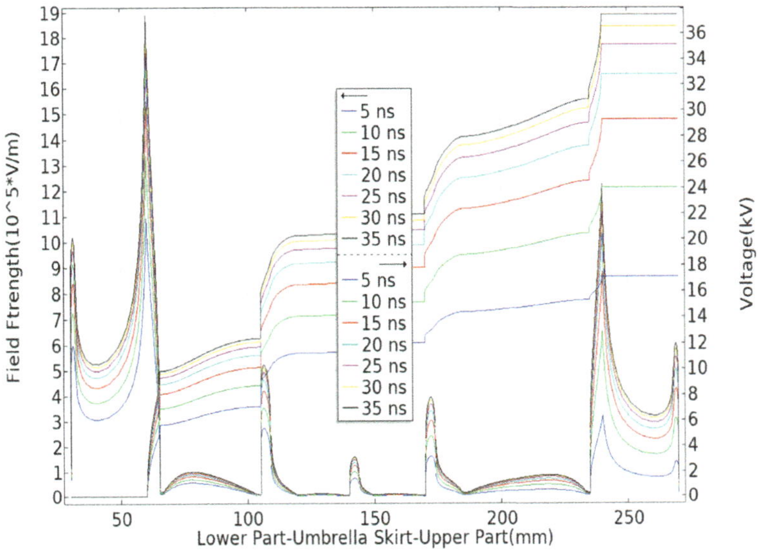

(a) Positive voltage applied to the insulator

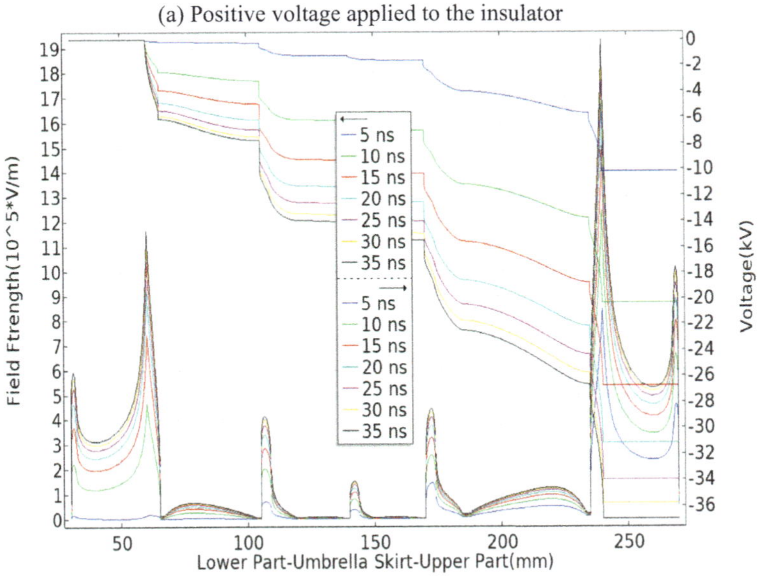

(b) Negative voltage applied to the insulator

Fig. 2 Power battery Thevenin equivalent circuit model

4.2 Electron Number Density Along the Surface of Insulator

Figure 3 shows the electron number density distribution on the surface of the insulator at different times. It can be seen from the figure that the electron density is the largest near the cathode of the insulator. Since positive ions hit the cathode surface can cause secondary electron emission, the electron density near the cathode is the largest. With

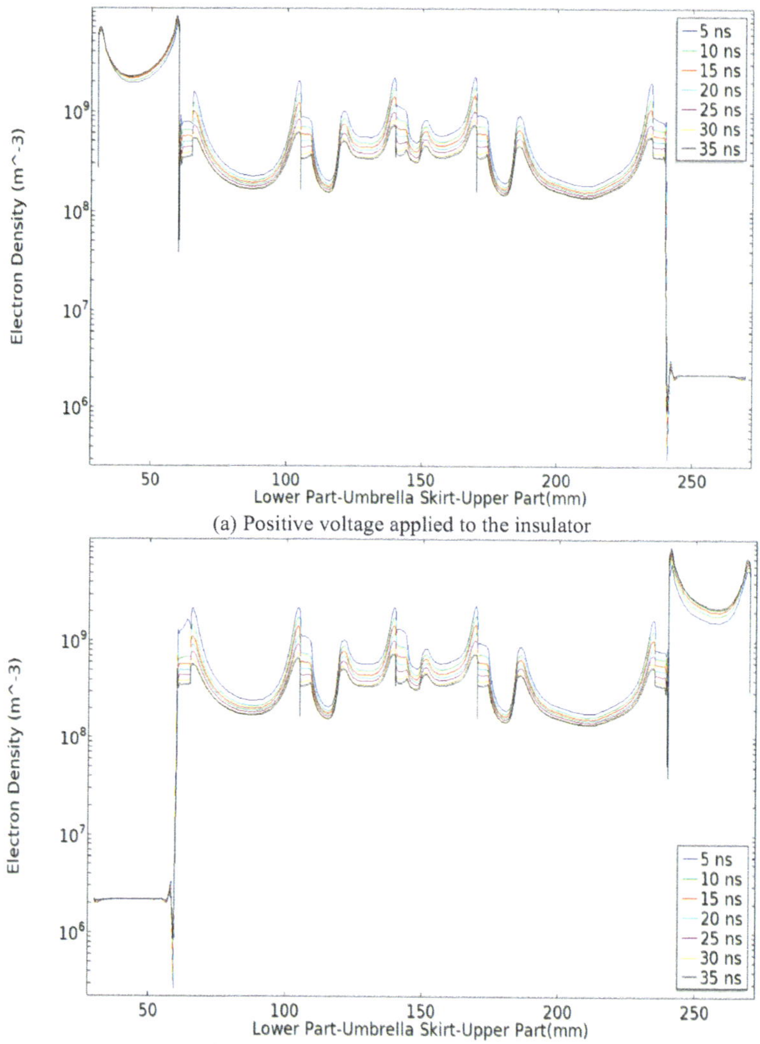

(a) Positive voltage applied to the insulator

(b) Negative voltage applied to the insulator

Fig. 3 Distribution of electron density along the surface of insulator the continuous development of the discharge process, electrons continue to migrate to and around the sun, and the number density of electrons gradually decreases from the cathode to the anode.

At the same time, from the electron density trend graph along the surface of the insulator, it can be seen that the corona discharge process of the insulator is along the surface of the insulator. This conclusion is consistent with the insulator flow path.

5 Conclusion

In this paper, through the establishment of a two-dimensional simulation model of the insulator, the electric field, potential and the distribution of electron density along the surface of the insulator under different voltage polarities are calculated, and the following conclusions are obtained:

(1) Apply voltages of different polarities to the surface of the insulator, and the absolute value of the potential at the metal attachment on the insulator is the same. Nevertheless, the value of the electric field strength of the insulator will be reversed.
(2) From the point of view of the electron density distribution along the surface of the insulator, the influence of different voltage polarities on the electron density on the surface of the insulator is consistent.

Therefore, when voltage values of different polarities are applied to the insulator, the electron distribution law during the corona discharge process is similar, and the voltage change law tends to be consistent. It can be seen that the voltage values of different polarities have little effect on the overall change law of the insulator during corona discharge.

Acknowledgements This work was supported in part by the China National Key Research and Development Projects under Grant (2018YFB0905801), the National Natural Science Foundation of China under Grant (51707142) and the National Natural Science Foundation of China under Grant (52077161).

References

1. Liu, Jiazhi, Youpeng Zhang, Shanpeng Zhao, and Haiyan Dong. 2018. Effect of pressure and initial electron density on corona discharge characteristics of the insulators in overhead contact nets. *Low Temperature Physical Letters* 40 (4): 1–7 (in Chinese).
2. Liao, Ruijin, Feifei Wu, Xinghua Liu, et al. 2012. Numerical simulation of transient space charge distribution of DC positive corona discharge under atmospheric pressure air. *Acta Physica Sinica* 61 (24): 370–380 (in Chinese).
3. Zhang, Youpeng, Jiazhi Liu, Shanpeng Zhao, et al. 2018. Simulation study on corona discharge characteristics of insulators on the contact wire. *Journal of Railway Science and Engineering* 15 (7): 1839–1847 (in Chinese).

4. Liu, Jiazhi. 2018. *Simulation study on corona discharge characteristics of insulators on the contact wire*. PhD, Lanzhou Jiaotong University (in Chinese).
5. Zeng, Han, Jing Zhang, Hongbo Cheng, et al. Numerical simulation of insulator corona discharge considering effect of H_2O. *Guangdong Electric Power* 32 (12): 89–97 (in Chinese).
6. Ma, Xikui. 2018. *Theory and application of electromagnetic field*. Xi'an: Xi'an Jiaotong University Press (in Chinese).
7. Chen, Xiaoyue. 2016. Study on microcosmic physical properties and propagation law of corona discharge in corona cage. PhD, Wuhan University (in Chinese).
8. Xia, Z., Z. Zhu, D. Howe. 2001. Analytical magnetic field analysis of Halbach magnetized PM machines. *IEEE Transactions on Magnetics* 37 (4): 2827–2830.

Simulation Calculation of Loss of Induction Traction Motor

Nana Duan, Xinyu Ma, Shaocong Lu, Shuoyu Wang, and Shuhong Wang

Abstract The finite element method has the characteristics of accurate calculation and convenient analysis of magnetic field and current distribution, so it has a good application prospect in the calculation of motor magnetic field distribution and loss. Taking a squirrel cage induction motor as an example, a two-dimensional finite element model considering skin effect, harmonic effect and saturation characteristics of the motor core is established. By using the finite element method, the transient magnetic field simulation of field-path coupling is carried out, and the copper loss of the motor is calculated. The classical Bertotti constant coefficient trinomial method was used to calculate the core loss. Considering the effect of rotating magnetic field and harmonic in the motor, the calculation method of core loss under the action of both is discussed. Finally, the total loss of the asynchronous traction motor is obtained and compared with the actual measurement results of the prototype. The error is small, which verifies the validity of the model.

Keywords Induction motor · Loss calculation · Finite element method · Rotating magnetic field · Harmonic action

1 Introduction

According to statistics in recent years, about 80% of the industrial power consumption in China is consumed in the operation of motors [1]. Because of its simple structure and low cost, asynchronous motor has been widely used. It not only occupies a major position in the motor market, but also occupies a large proportion of industrial production energy consumption. Therefore, the application of various energy-saving technologies in asynchronous motor to reduce the loss of motor is of great significance to reduce power waste, ease the shortage of power supply, and save national resources.

The biggest problem in the calculation of motor loss is the calculation of core loss.

N. Duan (✉) · X. Ma · S. Lu · S. Wang · S. Wang
School of Electrical Engineering, Xi'an Jiaotong University, No. 28, Xianning West Road, Xi'an, Shaanxi, China
e-mail: duannana@xjtu.edu.cn

W. Chen et al. (eds.), *The Proceedings of the 9th Frontier Academic Forum of Electrical Engineering*, Lecture Notes in Electrical Engineering 743,
https://doi.org/10.1007/978-981-33-6609-1_54

In 1892, Steinmetz first proposed a mathematical model for loss calculation, which included hysteresis loss and eddy current loss; Bertotti found that the calculation results of the model are biased, and he proposed a new constant coefficient trinomial formula for core loss calculation [2, 3]; in Ref. [4], the core loss was regarded as a function of the effective value and the average value of voltage, and the core loss calculation model under the condition of non sinusoidal power supply was proposed. In Ref. [5], another calculation model of non sinusoidal power supply is obtained by calculating non sinusoidal quantities after Fourier decomposition. In Refs. [3, 6], the magnetic flux density waveforms of different measuring points at the same frequency are fitted by polynomial fitting method, and the loss coefficient at any frequency is obtained. The core loss is calculated by using the coefficient, and the variable coefficient core loss model is established.

2 Calculation of Copper Consumption in End Ring

The copper consumption can be divided into stator copper consumption and rotor copper consumption. Among them, the harmonic of stator copper consumption is low, while the harmonic of rotor copper consumption is relatively large. The influence of harmonics on the loss can be calculated by using finite element method.

The stator copper loss and rotor guide bar loss can be calculated directly by using finite element method in finite element software. Another important component of the rotor copper loss is the end ring loss, which needs to be calculated by other way.

The current flow in the rotor guide bar is calculated by finite element theory. Due to the symmetric relationship, it can be considered that the effective value of the current flowing through each guide bar is the same, except for the phase of the current [7]. Due to the difference in the positions of the guide bar when the guide bar is made, the phase difference of the current should be equal to the difference in the electric Angle between the two adjacent rotor bar α.

$$\alpha = \frac{2\pi p}{Z_2} \tag{1}$$

Z_2 means number of rotor slots; p means number of pole logs. The effective value of current in each section of the end ring is the same, and the phase difference between two adjacent sections is also. The current flowing through the end ring is actually obtained by vector operation of the current in the two guide bars connected with the end ring. There is a relationship between the current of the guide bar and the current of the end ring as shown in formula (2).

$$\frac{I_B}{I_R} = 2\sin\left(\frac{\alpha}{2}\right) \tag{2}$$

I_B is represents the effective value of conductor bar current; I_R is represents the effective value of end ring current. According to the effective value of current in the end ring and the resistance of the end ring, the copper loss of the end ring can be calculated. The copper loss of the whole rotor can be obtained by summing the losses obtained by numerical calculation.

3 Calculation of Core Loss

The B-P and B-H characteristic curves of the motor core material can be obtained by actual measurement, and the three loss coefficients of the material can be obtained by fitting the characteristic curve with the least square method, which can be used to calculate the core loss later [8, 9].

The motor model is established in the simulation software, and the magnetic field distribution of stator core is obtained by finite element calculation. The representative characteristic regions are selected in the core model. In these characteristic regions, the solution points are set to replace the magnetic field strength of the whole region with the magnetic field strength of the solution points.

The magnetic field in a motor is a magnetic field that changes in one direction. In order to better analyze the change of magnetic field, it is necessary to decompose it into radial and tangential flux density components according to the direction of magnetic field in the motor. The frequency and flux density of each harmonic are obtained by Fourier decomposition of the two magnetic flux densities. Combined with the loss trinomial coefficient obtained by regression calculation, the calculation formula (3) of Bertotti constant coefficient trinomial formula can be used to calculate the high-frequency loss on the stator core and obtain more accurate core loss of the motor.

$$P_{Fe} = k_h f B^2 + k_e f^2 B^2 + k_c f^{1.5} B^{1.5} \tag{3}$$

P_{Fe} is the core loss of the motor; k_h, k_e, k_c represent the loss coefficient of the three losses; f is the frequency of magnetic field change; B is the amplitude of magnetic flux density.

4 Finite Element Simulation Calculation

4.1 Establishment of Motor Model

In Maxwell software, RMxprt solver is used to simplify the motor modeling process. The software parameters are set according to the prototype parameters: the motor is a 4-pole motor, the rated voltage is 1100 V, the rated power is 200 kW, the working

Table 1 Motor performance report in RMxprt

Parameter	Calculation results	Relative error
Stator tooth flux density	1.43	\
Rotor tooth flux density	1.72	\
Air gap flux density	0.93	\
Working phase current	128.69	127.6
Steady speed	2023.72	2020

frequency is 68 Hz, the winding connection is Y type, the stator slot number is 60 slots, and the rotor slot number is 48 slots. According to the size of the prototype, the RMxprt motor model can be generated by filling in the parameters such as the length of the iron core and the inner and outer diameters of the stator and rotor.

Set the motor working state as constant torque load, rated power of 200 kW, voltage of 1100 V, initial test speed of 2000 rpm, working temperature of 100 °C. After setting, the preliminary magnetic circuit simulation is carried out, and the RMxprt solver is run to calculate. The performance data report is shown in Table 1.

According to the calculation results of the above magnetic circuit method, the 2D finite element solution model is generated. The 1/4 model is used for simplification, and the result is shown in Fig. 1 (Fig. 2).

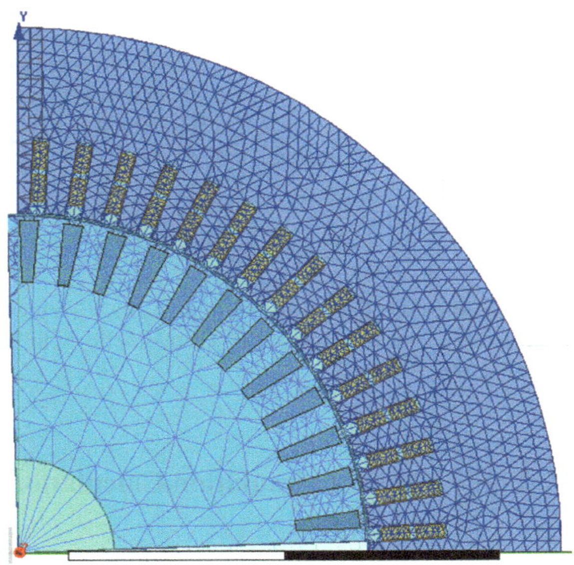

Fig. 1 2D model subdivision diagram of asynchronous moto

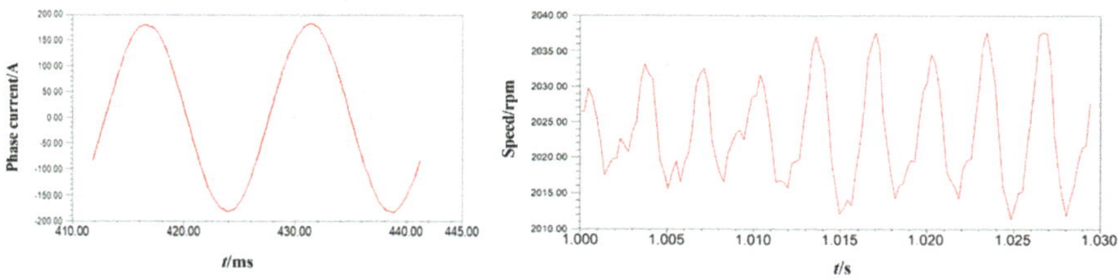

Fig. 2 Waveform of phase current and speed when the motor is running steady

Table 2 The finite element calculation results of copper consumption

Parameter	Calculation results
Stator copper loss	4844.5 W
Guide strip copper loss	854.3 W
Bar current	904.8 A
End ring current	3457.83 A
End ring resistance	6.7E−5 Ω
End ring loss	799.3 W

4.2 Calculation of Copper Loss by Finite Element Method

Through the finite element calculation, the copper loss of stator and rotor bar is obtained in the software.

According to the calculated current of the guide bar, formula (2) is used to calculate the end ring current. According to the resistance of the end ring, the loss of the end ring can be calculated. The results are shown in Table 2.

The core loss is 1269.1w in the initial fundamental frequency state calculated by finite element software. However, there are still many harmonics in the core.

In order to accurately calculate the core loss considering the harmonic effect, the *B-H* and *B-P* characteristic curves of the material are fitted to obtain the loss coefficient of the material.

After that, the iron core is divided and the characteristic points are selected in a certain area. The flux density of the point represents the flux density of the whole area. The specific iron core division is shown in Fig. 3.

The rotating magnetic field in the motor is decomposed into two orthogonal vectors, which are respectively along the radial direction of the motor and the direction perpendicular to it, called tangential direction.

The two variables are defined in the field calculator tool and solved by finite element method. Radial flux density B_j and tangential flux density B_q are shown in formula 4 and 5 (Fig. 4).

$$B_j = B_x \cos\theta + B_y \sin\theta \tag{4}$$

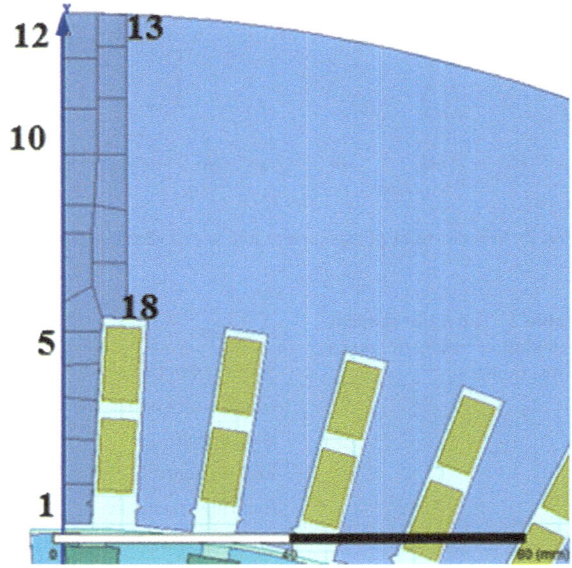

Fig. 3 Model of characteristic region division of iron core

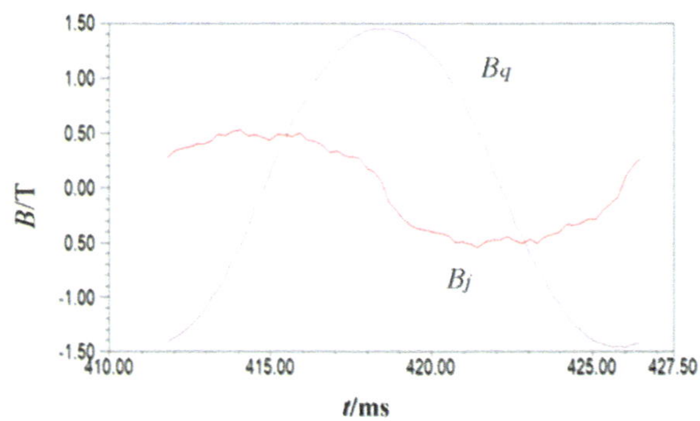

Fig. 4 Magnetic flux density waveform of No. 8 characteristic point

$$B_q = B_x \sin\theta + B_y \cos\theta \tag{5}$$

The calculated results are the radial and tangential flux density waveforms of the characteristic points in each core characteristic region. Fourier decomposition is used to obtain the magnitude of each harmonic. According to the harmonic frequency, the high frequency loss of the iron core under unit mass can be calculated by formula (3). By multiplying the area of each area and the density of iron core material, the high frequency loss of iron core in the region can be obtained (Table 3).

The high frequency loss in the whole core can be obtained by multiplying the calculation result by 120. The total is 535.8 W. The overall motor simulation results are shown in Table 4. High frequency stray loss is added to the total loss measurement.

Table 3 The finite element calculation results of copper consumption

Area number	Hysteresis loss (W)	Eddy current loss (W)	Total loss(W)
1	0.105	1.087	1.192
2	0.025	0.273	0.298
3	0.022	0.222	0.243
4	0.017	0.173	0.190
5	0.017	0.152	0.169
18	0.018	0.119	0.137
Sum of all area	0.460	4.005	4.465

Table 4 Motor simulation results report

	2D FEM	Measurement result	Relative error(%)
Phase current	128.31 A	127.6 A	0.55
Speed	2023.8 rpm	2020 rpm	0.18
Stator loss	4844.5 W	4797 W	0.99
Rotor loss	1656.61 W	1733 W	−4.58
Core loss	1804.92 W	1796 W	0.49
Mechanical loss		874 W	
Fundamental frequency stray loss		282 W	
Total loss	9459.03 W	10,344 W	−8.56

It is difficult to extract by simulation. In the process of the experiment, there are harmonic effects in the copper loss of stator and rotor and core loss, which can't really separate the high-frequency loss, so it is not reflected in the table.

The total loss of the motor also includes mechanical loss and stray loss. Because the software simulation can't directly or indirectly calculate the mechanical loss and fundamental frequency spurious, the actual measurement results are used as the simulation results. The simulation results are summarized, the final simulation results are calculated, and the simulation results are compared with the actual measurement results. The accuracy of magnetic circuit method and finite element method is compared. The relative error between the calculated results of finite element method and the measured results is calculated. According to the results in the table, it can be concluded that the error between the simulation calculation result and the actual measurement value is small, and the maximum error is −8.56%. This proves the validity of the model.

5 Conclusion

The error between the simulation results and the actual measurement value is small, the maximum value is −8.56%. This proves the validity of the model.

Because part of the resistance in the motor model is calculated, there may be a certain deviation between the actual value and the current simulation, which will further affect the simulation of motor loss and lead to errors in the simulation calculation of stator copper loss. Because the discretization of the motor in the calculation process is only the theoretical calculation of partition area. On the other hand, the characteristic curve of iron core material is only obtained by measuring the silicon steel sheet, rather than the whole after processing. The operation in the processing process may have a certain impact on its characteristics, resulting in a certain error in the curve; calculation of core loss. There are some errors in the model formula itself, which can't truly reflect the loss. Therefore, if the finite element method is used, there will be errors, but the errors are acceptable.

Acknowledgements This work was supported in part by the China National Key Research and Development Projects under Grant (2018YFB0905801) and the National Natural Science Foundation of China under Grant (51707142 and 52077161).

References

1. Qin, Hongbo. 2010. Energy saving progress and countermeasures of motor system in China. *Electric Machines & Control Application* 37 (06): 1–4. (in Chinese).
2. Ionel, Dan M., Mocea Popescu, Stephen J. Dellinger, et al. 2010. On the variation with flux and frequency of the core loss coefficients in electrical machines. *IEEE Transactions on Industry Applications* 46 (2): 806–811.
3. Dan M. Ionel, Mocea Popescu, Malcolm I. McGilp, et al. 2007. Computation of core losses in electrical machines using improved models for laminated steel. *IEEE Transactions on Industry Applications* 43 (6): 1554–1564.
4. Boglietti, A., A. Cavagnino, M. Lazzari, et al. 2003. Predicting iron losses in soft magnetic materials with arbitrary voltage supply: an engineering approach. *IEEE Transactions on Magnetics* 39 (2): 981–989.
5. Boglietti, A., P. Ferraris, M. Lazzari, et al. 1991. Iron losses in magnetic materials with six-step and PWM inverter supply. *IEEE Transactions on Magnetics* 27 (6): 5334–5336.
6. Dan M. Ionel, Mocea Popescu, Stephen J. Dellinger, et al. 2010. On the variation with flux and frequency of the core loss coefficients in electrical machines. *IEEE Transactions on Industry Applications* 46 (2): 806–811.
7. Shikun, Chen. 2000. *Motor Desig*. Beijing: China Machine Press. (in Chinese).
8. Gong, Wenjun, Shaogang Huang, Yuwen Yang, Liqiang Rao, yu Zeng. Realization of extracting iron loss coefficient of silicon steel sheet by ANSYS Maxwell. *Electric Machines & Control Application* 43 (04): 82–85 (in Chinese).
9. Huicheng, Mo, Yu. Min Lin, and Zhigang, Yu. Hanfeng. 2008. Research on iron loss of silicon steel sheet for motor. *Micromotors* 41 (11): 5–7. (in Chinese).

Study on Temperature Field of MVA Capacity High Temperature Superconducting Transformer

Nana Duan, Jinqi Zhang, and Shuhong Wang

Abstract It is very important to study the temperature rise of High Temperature Superconducting (HTS) transformer winding in short circuit. In this paper, the distribution of magnetic field and temperature field of MVA three-phase HTS transformer with high current carrying capacity is studied. Based on the finite element meth-od, the magnetic field distribution and AC loss of two-dimensional single-phase pan-cake winding model and cylinder winding model of HTS transformer are calculated respectively. Combined with the thermal parameters of liquid nitrogen and superconducting tape, the temperature field of two models is simulated and analyzed by finite element method. Based on the error analysis results caused by two-dimensional modeling, the electromagnetic field and temperature field distribution of high and low-voltage windings of HTS transformer under the condition of primary side short-circuit are calculated by using appropriate three-dimensional model of HTS transformer. The simulation results show that the heat dissipation performance of superconducting transformer is better than that of conventional transformer, which provides a theoretical basis for the design of HTS transformer.

Keywords High temperature superconducting transformer · Temperature field · Magnetic field

1 Introduction

Power transformer is the key equipment in power system. High temperature superconducting transformer is different from conventional power transformer. High temperature superconducting tape is used as winding and liquid nitrogen insulation is adopted. It not only reduces the loss of transformer, but also improves the safety of transformer. At the same time, it has certain short-circuit current limiting function. Therefore, the important role of superconducting transformer in power transmission and distribution has been widely studied. If the HTS transformer exceeds a certain operating

N. Duan (✉) · J. Zhang · S. Wang
School of Electrical Engineering, Xi'an Jiaotong University, No. 28 Xianning West Road, Xi'an, Shaanxi, China
e-mail: duannana@xjtu.edu.cn

W. Chen et al. (eds.), *The Proceedings of the 9th Frontier Academic Forum of Electrical Engineering*, Lecture Notes in Electrical Engineering 743,
https://doi.org/10.1007/978-981-33-6609-1_55

temperature or is placed in a large background magnetic field, the winding of the HTS transformer will no longer work in the superconducting state, which will greatly reduce the working efficiency of the superconducting transformer, and even affect the safe operation of the transformer. Therefore, it is of great significance to study the electromagnetic field and temperature field distribution of the HTS transformer.

In this paper, the parameters of 25 MVA, 110 kV/10.5 kV designed by China Electric Power Research Institute are used to simulate the magnetic field and temperature field of HTS transformer winding. Based on the finite element method, after the error analysis of magnetic field and temperature field of two-dimensional single-phase pancake winding model and cylinder winding model, the distribution of magnetic field and temperature field of three-phase high-temperature superconducting transformer under the condition of primary side short-circuit is studied by using appropriate three-dimensional model, which provides reference for the design of HTS transformer.

2 Winding Structure of HTS Transformer

HTS transformer is generally composed of iron core, high temperature superconducting winding and low temperature cooling system. The core adopts three column outer iron type, and the winding distribution adopts the mode of low-voltage (LV)-high-voltage (HV)-low-voltage (LV). When cooling, the winding is only placed in the cryogenic vessel where the liquid nitrogen is located, and the iron core is placed in the normal temperature air. The structure of the 25 MVA superconducting transformer is shown in Fig. 1.

The high-voltage winding of HTS transformer adopts the second-generation YBCO coated conductor of Fujikura high-temperature superconducting tape, and the low-voltage winding side needs high current carrying characteristics. It is made of 28 YBCO superconducting tapes with 4-layer copper core, which is made of

Fig. 1 Structure diagram of HTS transformer [1]

cored cable [1]. The structure of the cored cable is shown in Fig. 2. This structure can ensure that the superconductor carries most of the current when it works normally. When the circuit is short circuited and the superconductor loses its superconductivity, the copper core can bear the function of circulating current to avoid the problem of excessive temperature rise caused by excessive current. The parameters of HTS transformer and the design parameters of LV windings are shown in Table 1.

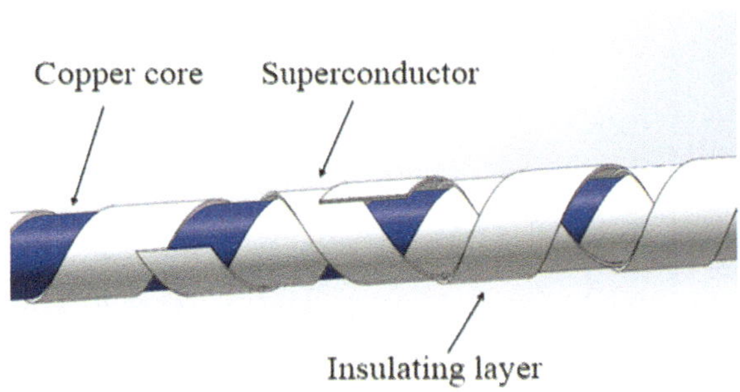

Fig. 2 Winding structure of HTS transformer

Table 1 Parameters of 25 MVA/110 kV HTS transformer and low voltage winding

Parameters	Numerical value
Rated capacity	25 MVA
Rated voltage	110 kV/10.5 kV
Phase number	3
Frequency/Hz	50
Association group	Yd11
Winding material	LV-conductor on round core HV-REBCO
Winding type	Spiral winding
Total turns	120
Layers	2
Turns per layer	60
Diameter of conductor on round core (mm)	10.8
Axial height (mm)	772
Radial height (mm)	113

3 Numerical Simulation of Magnetic Field in HTS Transformer

In order to calculate the magnetic field distribution of HTS transformer, Maxwell's equations are used, and the differential form of Maxwell's equations is selected here.

$$\nabla \times H = J + \frac{\partial D}{\partial t} \tag{1}$$

$$\nabla \times E = -\frac{\partial B}{\partial t} \tag{2}$$

$$\nabla \cdot B = 0 \tag{3}$$

$$\nabla \cdot D = \rho \tag{4}$$

3.1 Comparison of Two-Dimensional Model Magnetic Field Simulation

Take one phase of three-phase HTS transformer and establish two-dimensional finite element pan-cake model and cylinder model. The rated current is applied to the high and low voltage windings respectively according to the rated parameters, and the magnetic field distribution at 0.26 s (after 13 cycles) is obtained as shown in Fig. 3.

Through modeling and comparison, it is found that after the parameters are equivalent, the magnetic field error caused by modeling is about 11.8%, and the error is large. Therefore, it is more accurate to use the pie model to calculate the magnetic field in 3D modeling.

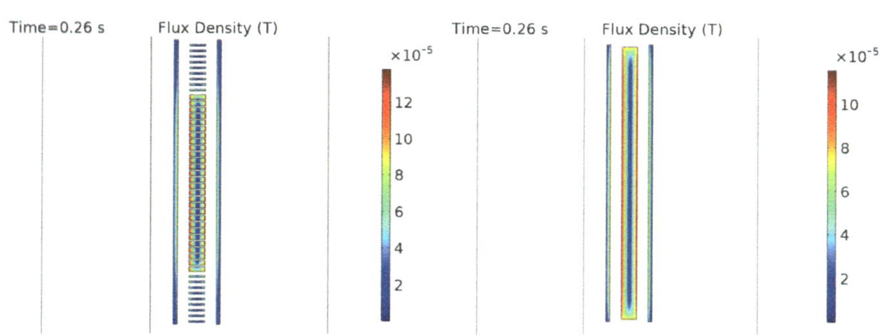

Fig. 3 Magnetic field distribution of pan-cake winding and cylindrical winding at 0.26 s

3.2 Magnetic Field Calculation Results of 3D Model

The three-dimensional pie model of three-phase HTS transformer is established, as shown in Fig. 4.

In the calculation of three-dimensional model magnetic field, the grid divided by the complete pie type 3D model is very large and difficult to calculate. Therefore, it is approximately considered that the transformer is symmetrical in front and back position and the magnetic field distribution is the same. The model of superconducting transformer is cut, only half of the three-dimensional model is calculated, which can improve the calculation efficiency on the premise of ensuring the accuracy of calculation. As the mesh section is shown in Fig. 4, the model is divided into tetrahedron. The number of tetrahedrons is about 170,000. The magnetic field distribution and the magnetic field distribution of winding of three-phase superconducting transformer under primary side short circuit is obtained as shown in Fig. 5.

It can be seen that the magnetic flux density of the high-voltage winding is far greater than that of the low-voltage winding in case of a short-circuit at the primary side. For the second-generation high-temperature superconducting tape, when the

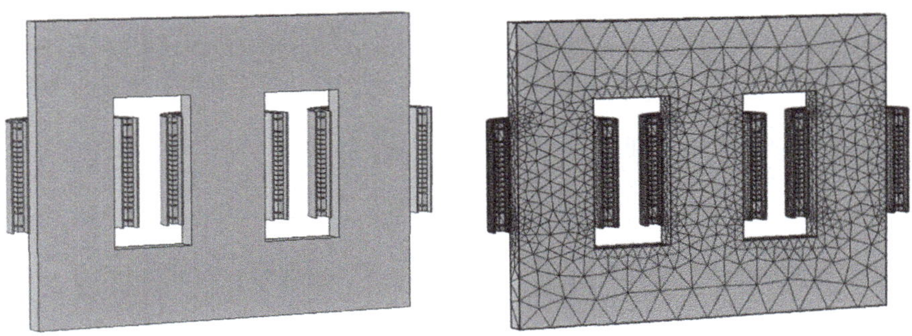

Fig. 4 Three phase HTS transformer model and finite element analysis model

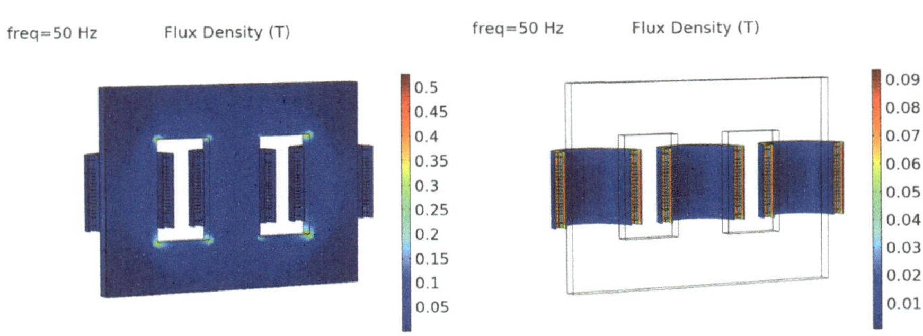

Fig. 5 Magnetic field distribution (left) and distribution of winding (right) in primary side short circuit of three phase HTS transformer

background magnetic field is greater than or equal to 0.1 T, it will be lost When the primary side short circuit occurs, the maximum flux density of the high voltage winding is close to 0.1 T, which has caused the degradation of the critical current of the superconducting tape, but it has not lost its superconductivity.

4 Numerical Calculation of Temperature Field in HTS Transformer

4.1 Magnetic Field Calculation Results of 3D Model

In AC operation of HTS transformer, both high and low voltage windings will produce AC loss. For superconducting tapes, AC loss mainly includes the following three parts [2, 3]: hysteresis loss caused by flux pinning force, eddy current loss caused by eddy current generated by changing magnetic field in metal base material, and coupling loss between metal substrate and superconducting material. Because the coupling loss is very small, only the hysteresis loss and eddy current loss of superconducting tape are calculated in this paper.

The winding of HTS transformer is immersed in a cryogenic vessel containing liquid nitrogen, and its heat transfer modes are mainly heat conduction and forced convection [4].

The heat conduction equation is

$$q = \lambda \cdot \nabla \times T \tag{5}$$

Here q is the heat flow rate passing through the unit area per unit time; λ is the thermal conductivity of the material; T is the temperature.

The heat convection equation is

$$q = h(T_h - T_c) \tag{6}$$

Here q is the heat dissipated by convection per unit area per unit time; h is the convective heat transfer coefficient; T_h and T_c are the temperatures of solids and fluids.

$$Nu = hl/\lambda \tag{7}$$

$$Nu = 0.023 R_e^{0.8} P_r^{0.4} \tag{8}$$

λ is the thermal conductivity of liquid nitrogen, l is the width of Dewar, R_e is Reynolds number, P_r is Prandtl number.

Reynolds number is used to judge the flow mode of liquid nitrogen.

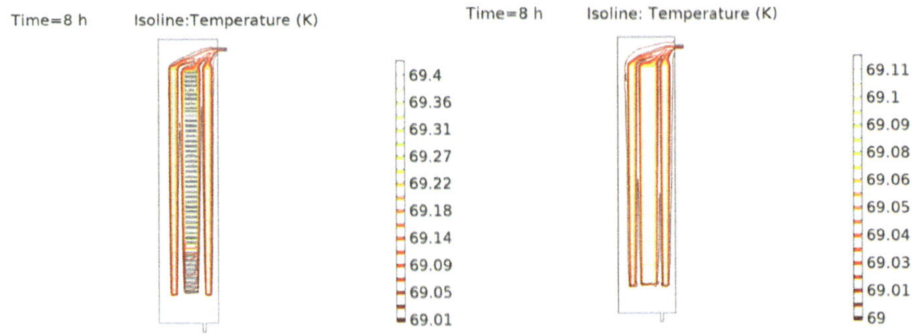

Fig. 6 Temperature distribution of pan-cake winding and cylinder winding in HTS transformer

$$R_e = \rho vl/\mu \tag{9}$$

4.2 Comparison of Two-Dimensional Model Temperature Field Simulation

The AC loss of high and low voltage windings is calculated by using the calculation method of hysteresis loss of superconducting tape and eddy current loss calculation method of superconducting tape [5, 6], and the temperature field is simulated by using the AC loss as the heat source. Two dimensional cake and cylinder axisymmetric models are established, as shown in Fig. 6 [7–9].

The simulation results show that the temperature rise of the cylinder winding is slightly less than that of the pan-cake winding, and the temperature rise error of the cylindrical winding is 0.43%. The reason may be that the contact surface of the pie winding is more, so the heat dissipation is faster, so the temperature of liquid nitrogen is higher. But the error is very small, so in the three-dimensional temperature field model, under the premise of ensuring the calculation accuracy, the cylindrical model can be used for simulation analysis. Simulation Analysis of Temperature Field of 3D model.

According to the calculation method mentioned above [10, 11], the ABC three-phase temperature field distribution results as shown in Fig. 7 are obtained.

According to Fig. 7, the reason why the temperature of the low-voltage winding is higher than that of the high-voltage winding is that the low-voltage winding is the high-voltage current carrying side, which carries a large amount of current, resulting in greater AC loss. The temperature of liquid nitrogen at the lowest inlet is lower, the flow rate is fast, and the heat dissipation performance is good, so the temperature at the outlet of liquid nitrogen is high. The critical temperature of superconducting tape is almost reached in about 5 min in ABC phase of HTS transformer, which indicates that the primary side short circuit will cause great harm to HTS transformer.

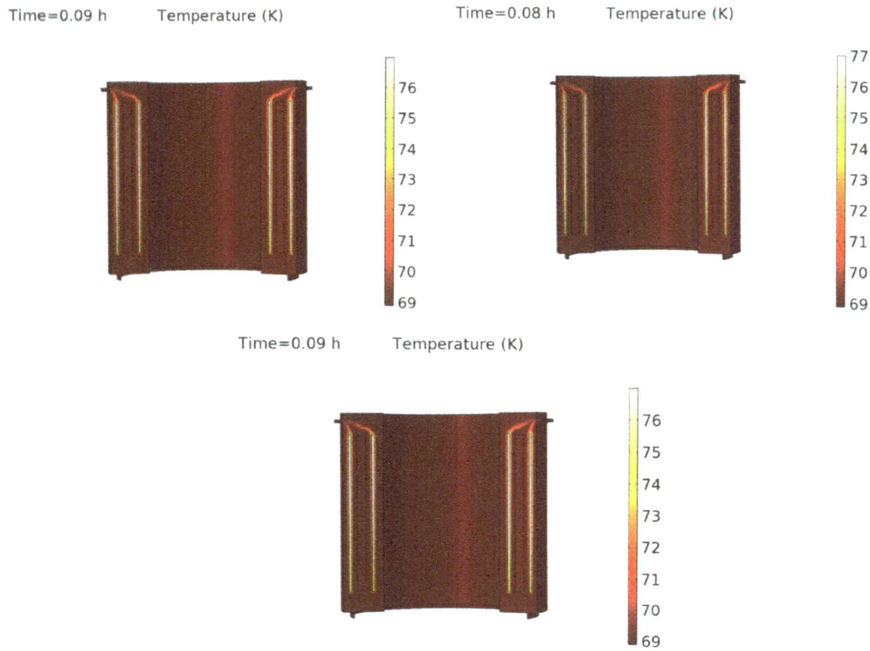

Fig. 7 Three phase temperature rise distribution of HTS transformer

5 Conclusion

Based on the finite element method, a two-dimensional cylindrical model and a cake model are established to simulate the temperature field of the high-temperature superconducting transformer. In the three-dimensional finite element simulation analysis, the magnetic field distribution of HTS transformer under primary side short circuit is calculated by pan-cake model, and the temperature field is calculated by cylinder model. Through the simulation of magnetic field and temperature field, it is found that the heat dissipation performance of superconducting transformer is better than that of conventional transformer, which provides reference for the design of HTS transformer.

Acknowledgements This work was supported in part by the China National Key Research and Development Projects under Grant (2018YFB0905801) and the National Natural Science Foundation of China under Grant (51707142 and 52077161).

References

1. Zhu, Jiahui, Hongjie Zhang, Yanfang Yang, et al. 2019. Characterization modelling and current limiting performance analysis of high temperature superconducting transformer under the fault current impact in power grid. In: *Proceedings of the CSEE* (in Chinses).
2. Takayasu, M., L. Chiesa, L. Bromberg, et al. 2012. HTS twisted stacked-tape cable conductor. *Superconductor Science Technology* 25 (1): 14011–14031.
3. Liu, Erwei, Jin Fang, Yingjie Huo, et al. 2015. Study on AC loss measuring system of YBCO tapes. *Superconductivity* 43 (11): 35–38 (in Chinese).
4. Amemiya, N., S. Murasawa, N. Banno, et al. 1998. Numerical modellings of YBCO tapes. *Physica C: Superconductivity* 310 (1): 16–29. (in Chinese).
5. Liang, Junguo, Xiaochun, Ma, Jin Fang, et al. 2014. 3D Simulation analysis of temperature field of superconducting transformer. *Transformer* 51 (4): 16–19 (in Chinese).
6. Sullican, C.R., and L.W. Losses. 2014. Analytical model for effects of testing on litz-wire losses. IEEE, 1–10.
7. Chen, Lu. 2016. *Design of High Temperature Superconducting Energy Storage Magnet Based on Multi-fields Coupled*. Wuhan: Huazhong University of Science and Technology. (in Chinese).
8. Chen, Min, Yunjia Yu, and Liye Xiao. 2001. Charateristics and Development Prospect of HTS Transformer. *Cryogenics & Superconductivity* 29 (4): 50–55. (in Chinese).
9. Li, Xiaosong. 2005. Electromagnetic design and characteristic of 300kVA/25000V/860V HTS Transformer. PhD, Huazhong University of Science and Technology, Wuhan (in Chinese).
10. Fu, Shanshan. *Electromagnetic Design and Preliminary Stability Study of MVA Superconducting Coated Conductor Transforming Winding*. Beijing: China Electric Power Research Institute (in Chinese).
11. Li, Hailin, Naming Zhang, Shuhong Wang, et al. 2019. An analytical loss model of litz-wire windings of transformers excited by converters with winding configurations considered. *IEEE Transactions on Magnetics* 55 (9): 25–29.

Magnetic Field Analysis and Structural Optimization of Deflection Double Stator Switched Reluctance Generator

Zheng Li, Xuze Yu, Xin Wang, Zhe Qian, and Qunjing Wang

Abstract Switched reluctance generator has been widely used in industry. With the consumption of energy, the requirement of generator is higher and higher. The energy conversion rate of the traditional generator structure is very low, and it can no longer meet the existing technical indicators. In order to solve the problem of low efficiency of existing generators, this paper proposes a deflectable double stator switched reluctance generator (DDSRG), and performed a finite element analysis on this model. The magnetic density of the new generator is analyzed. A comparison between the proposed mathematical model and the finite element method verifies the accuracy of the model. finally, the core loss was used as the target to optimize the structure of the generator using the response surface method. The experimental results were compared with the simulation. The error was within the allowable range, which verified the correctness of the calculation method and the optimization scheme.

Keywords Switched reluctance generator · Finite element · Core loss · Response surface method

1 Introduction

Switched reluctance generators have very reliable performance and have been widely used in the industrial field. Compared with other types of generators, switched reluctance generators have a smaller size. At the same time, the two stators of the switched reluctance generator with dual stator structure can simultaneously induce output voltage, which is more suitable for industrial production [1]. To further improve the working efficiency of the generator, the concept of multiple degrees of freedom can be combined with the generator structure. The deflectable generator structure can

Z. Li (✉) · X. Yu · X. Wang
School of Electrical Engineering, Hebei University of Science and Technology, No. 26 Yuxiang street, Yuhua District, Shijiazhuang, Hebei Province, China
e-mail: Lzhfgd@163.com

Z. Qian · Q. Wang
National Engineering Laboratory of Energy-Saving Motor and Control Technique, No. 111 Jiulong Road, Shushan District, Hefei, Anhui Province, China

W. Chen et al. (eds.), *The Proceedings of the 9th Frontier Academic Forum of Electrical Engineering*, Lecture Notes in Electrical Engineering 743,
https://doi.org/10.1007/978-981-33-6609-1_56

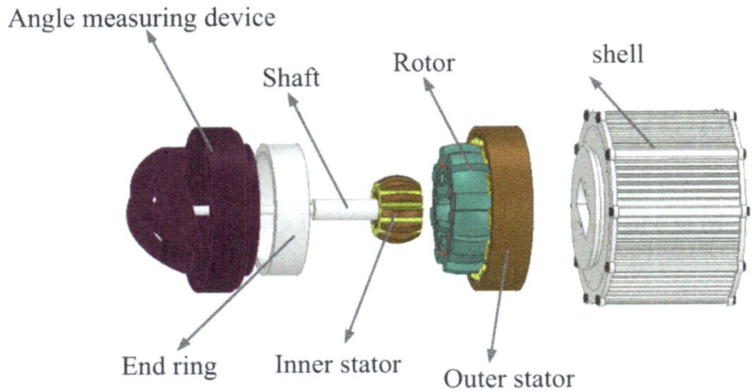

Fig. 1 Structure of deflected double stator generator

improve the power generation efficiency of the generator. Although the dual-stator structure generator has better power generation characteristics, at the same time its iron loss will also increase to a greater extent. This is not negligible for a switched reluctance motor with a double stator structure [2].

This paper proposes a mathematical model for the proposed new generator structure, and compares the test results with the finite element results to verify the accuracy of the model. Finally, the analysis and optimization results are verified through experiments.

2 Structure and Principle of DDSRG

The structure model of DDSRG is shown in Fig. 1. Both stator and rotor are spherical in shape. The parameters of the generator are shown in Table 1.

The tooth pole axis of the inner stator is consistent with that of the outer stator. DDSRG can be regarded as a combination of an inner generator and an outer generator, with a rotor between the inner and outer stators.The rotor part has a magnetic isolation part, which can isolate the magnetic fields of the inner and outer stators from each other.

3 Analytical Calculation of Magnetic Density of DDSR

The magnetic field strength of the main air gap magnetic field, the magnetic density of the main air gap magnetic field, and the magnetic density of the edge air gap magnetic field can be expressed by Eqs. (1)–(3), respectively.

Table 1 Structural parameters of generator

Parameter	Value
Outer stator radius/mm	120
Inner stator radius/mm	19.45
Rotor inner diameter/mm	40.31
Rotor outer diameter/mm	85.5
Inner and outer air gap/mm	1.5
Number of stator teeth	12
Number of rotor teeth	8
Stator pole arc/°	14.9
Rotor pole arc/°	16.1
Core length/mm	60
Unidirectional deflection range/°	0–17

$$H_m = \frac{B_s\left(2\mu_r l_g + 2l_g + l + 2ll_g\right) + N_m i_m \mu_0 \mu_r \left(l + l_g\right)}{2\mu_0 \mu_r l l_g} - \sqrt{\left(\frac{B_s\left(2\mu_r l_g + 2l_g + l + 2ll_g\right) + N_m i_m \mu_0 \mu_r \left(l + l_g\right)}{2\mu_0 \mu_r l l_g}\right)^2 - \frac{N_m i_m\left[B_s\left(\mu_r l - \mu_r l_g + l - l_g\right) + \mu_0 \mu_r N_m i_m\right]}{\mu_0 \mu_r l}} \quad (1)$$

$$B_m = \frac{B_s\left(2\mu_r l_g + 2l_g + l + 2ll_g\right) + N_m i_m \mu_0 \mu_r \left(l + l_g\right)}{2\mu_r l l_g} - \sqrt{\left(\frac{B_s\left(2\mu_r l_g + 2l_g + l + 2ll_g\right) + N_m i_m \mu_0 \mu_r \left(l + l_g\right)}{2\mu_r l l_g}\right)^2 - \frac{\mu_0 N_m i_m\left[B_s\left(\mu_r l - \mu_r l_g + l - l_g\right) + \mu_0 \mu_r N_m i_m\right]}{\mu_r l}} \quad (2)$$

$$B_{\mathrm{f}} = \frac{2\mu_0 \mu_r N_{\mathrm{m}} i_{\mathrm{m}}\left(\frac{A_{\mathrm{s}}}{l - l_{\mathrm{f}}} + A_{\mathrm{f}}\right) + B_{\mathrm{s}}(A_{\mathrm{s}} l_{\mathrm{f}} \mu_r + A_{\mathrm{s}} l_{\mathrm{f}} + A_{\mathrm{f}} l - A_{\mathrm{f}} l_{\mathrm{f}}) - \Phi_{\mathrm{m}} \mu_0 \mu_r l_{\mathrm{f}}}{2\mu_r l_{\mathrm{f}} N_{\mathrm{m}} i_{\mathrm{m}}\left(\frac{A_{\mathrm{s}}}{l - l_{\mathrm{f}}} + 2A_{\mathrm{f}}\right)} \sqrt{\left(\frac{2\mu_0 \mu_r N_{\mathrm{m}} i_{\mathrm{m}}\left(\frac{A_{\mathrm{s}}}{l - l_{\mathrm{f}}} + A_{\mathrm{f}}\right) + B_{\mathrm{s}} l_{\mathrm{f}}(A_{\mathrm{s}} \mu_r + A_{\mathrm{s}}) - \Phi_{\mathrm{m}} \mu_0 \mu_r l_{\mathrm{f}}}{2\mu_r l_{\mathrm{f}} N_{\mathrm{m}} i_{\mathrm{m}}\left(\frac{A_{\mathrm{s}}}{l - l_{\mathrm{f}}} + 2A_{\mathrm{f}}\right)}\right)^2 - \frac{A_{\mathrm{s}}\left(\frac{\mu_0 \mu_r A_{\mathrm{s}} N_{\mathrm{m}} i_{\mathrm{m}}}{l - l_{\mathrm{f}}} + B_{\mathrm{s}} N_{\mathrm{m}} i_{\mathrm{m}} \mu_r\right) - \Phi_{\mathrm{m}}(\mu_0 \mu_r N_{\mathrm{m}} i_{\mathrm{m}} + B_{\mathrm{s}} l - B_{\mathrm{s}} l_{\mathrm{f}})}{\mu_r l_{\mathrm{f}} N_{\mathrm{m}} i_{\mathrm{m}}\left(\frac{A_{\mathrm{s}}}{l - l_{\mathrm{f}}} + 2A_{\mathrm{f}}\right)}} \quad (3)$$

where A_{f1} and A_{f2} are the areas where the lines of force pass through the edges on both sides. $A_f = A_{f1} + A_{f2}$, lf is the average length of the edge magnetic circuit on both sides, l_g is the average air gap length, N_m is the number of turns of the outer stator winding, i_m is the phase current of the winding. B_s is the saturation magnetic density of the material.

4 Finite Element Analysis Results of Generator Magnetic Field

The deflection type double stator switched reluctance generator can be composed of the radial component Br and the tangential component B_θ [3–6].

In the transient field, the composite flux density and magnetic flux density components of each element are simulated. The analysis radius of the outer rotor yoke is 55–70 mm. The analysis radius of the outer rotor teeth is from 70 to 75 mm. The analysis radius of outer stator teeth is 75.5–80.5 mm. The analysis radius of the outer stator yoke is 80.5–110 mm. Figure 2 shows the changes in the magnetic density of different types of generators.

The error between the analytical method and the finite element method is relatively small enough.

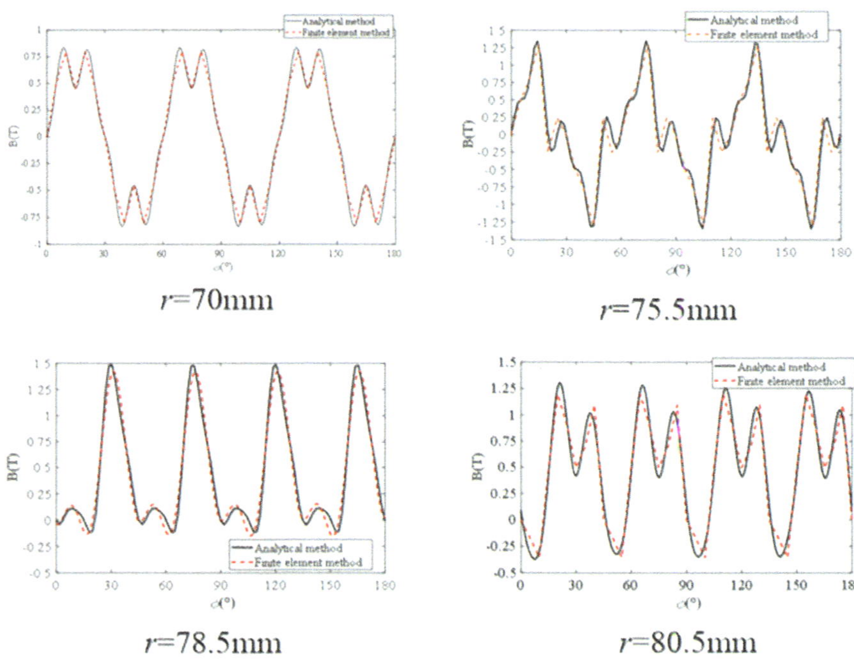

Fig. 2 Comparison of magnetic flux density by analytical method and FEM

5 Structural Optimization of DDSRG

Bertotti et al. believe that the core loss is mainly composed of three types of loss, namely eddy current loss, eddy current loss P_e, hysteresis loss P_h, and residual loss. P_c [7, 8].

$$P_{\mathrm{Fe}} = P_{\mathrm{e}} + P_{\mathrm{h}} + P_{\mathrm{c}} \tag{4}$$

The change of stator and rotor structure has a great influence on hysteresis loss.

Four design variables are selected: core length L_1, stator outer pole width B_1, stator outer yoke height $H1$ and outer air gap G_1.

Figure 3 is the schematic diagram of generator optimization parameters. Each optimized parameter has 3 value variables, respectively, the values of the A, B, and C variables determine the changes of the four parameters. The changes of the four parameters are shown in Table 2.

Taguchi algorithm helps to greatly reduce the iterative method in the experiment and reduce the cost of the experiment [9]. The core loss of the generator model is analyzed using finite element, and the core loss P_{Fe} of each combination in the orthogonal table is obtained, as shown in Table 3.

Table 3 is the orthogonal experiment table, and the finite element method can be used to calculate the iron loss under different conditions.

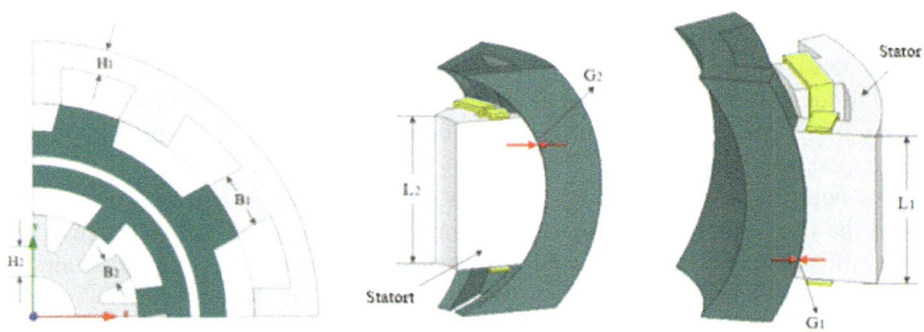

Fig. 3 Generator optimization parameters

Table 2 The optimized parameters and evaluated parameters

Optimization parameters	L_1/mm	B_1/mm	$H1$/mm	G_1/mm
A	60	26	12	0.1
B	75	33	15	0.5
C	90	40	18	0.9

Table 3 Analysis results

	L_1/mm	B_1/mm	$H1$/mm	G_1/mm	P_{Fe}/W
1	A	A	A	A	276.0
2	A	B	B	B	234.6
3	A	C	C	C	172.2
4	B	A	C	C	237.0
5	B	B	B	A	150.0
6	B	C	A	B	212.4
7	C	A	C	B	280.2
8	C	B	A	C	250.8
9	C	C	B	A	156.0

$$YY = 3\sum_{i=1}^{3}[m_{xi}(P_{\mathrm{Fe}i}) - m(P_{\mathrm{Fe}})]^2 \tag{5}$$

where *YY* is the proportion of each parameter variable affecting the performance, x is the optimized parameter variable, namely the core length L_l, the outer stator pole width B_l, the outer stator yoke height *H1*, the external air gap G_l, m_{xi} (P_{Fei}) as the parameters The average value of the core loss of the variable x under the i-th level variable, $m(P_{Fe})$ is the average value of P_{Fe} in 9 experiments [10].

As shown in Table 4, the iron loss has a great correlation with the proposed parameters. Among them, the iron core length L_1 accounts for 7.04%, which has a small influence on the iron loss, and the influence rate of other structural parameters is very high. The largest part of the outer stator pole width B_1 accounts for 46.93% of the total proportion.

The outer stator pole width B_1, the outer stator yoke height $H1$, the outer air gap G_1 as the three parameter variables, the core loss of the generator as the response value, the application of the central composite design (CCD) method to obtain the coding con-version of the parameter variables is shown in Table 5.

The data is fitted by the least square method, and based on the results of the response surface, the regression equations with the generator core loss as the response value are obtained as

Table 4 The Influence rate of optimization parameters on generator core loss

Optimization parameters	YY	Impact rate (%)
L1/mm	1627.44	7.04
G1/mm	3520.32	15.23
H1/mm	7118.16	30.80
B1/mm	10,845.96	46.93
Total	23,111.88	100

Table 5 Comparison of measured and calculated values of generator iron loss

Area	Experimental measurement	Simulation calculated value
Internal iron loss/W	54.34	52.80
External iron loss/W	86.87	79.75
Total iron loss/W	141.21	131.55

$$\begin{aligned} Y = {} & 272.52 + 73.68X_1 - 22.70X_2 - 29.08X_3 \\ & + 12.93X_1X_2 - 17.62X_1X_3 + 0.10X_2X_3 \\ & - 19.53X_1^2 - 11.57X_2^2 - 28.93X_3^2 \end{aligned} \quad (6)$$

where X_1 is the outer stator pole width B_1, X_2 is the outer stator yoke height $H1$, and X_3 are the external air gap G_1, where $X_1, X_2, X_3, X_1X_2, X_1X_3, X_{12}, X_{22}$, and X_{32} have a significant effect on the core loss of the generator Significant influence, other items are not significant.

It can be seen from Fig. 4 that the outer stator pole width B_1, the outer stator yoke height $H1$, and the outer air gap G_1 have a significant effect on the response value of the generator core loss.

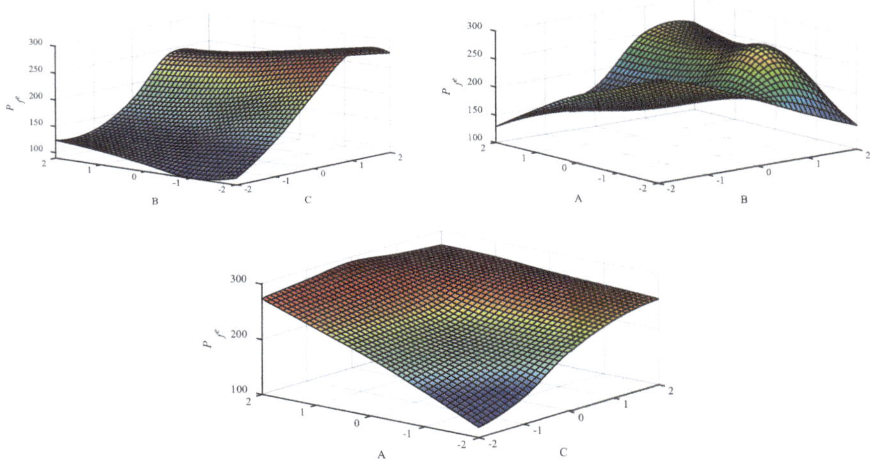

Fig. 4 The relationship between response surface and parameters

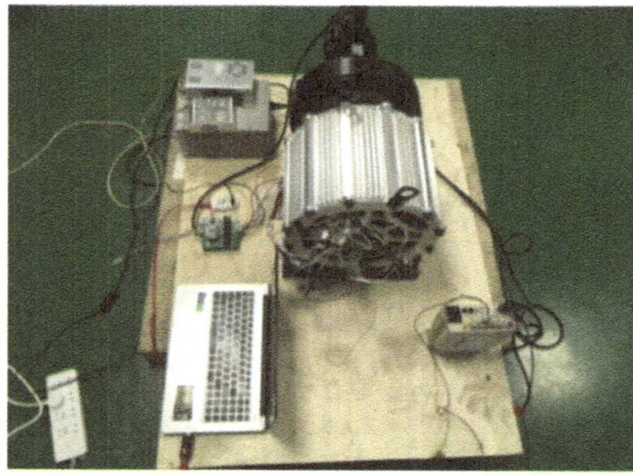

Fig. 5 Experimental device

6 Experimental Verification

When the DDSRG is running in a steady state of rotation, The loss can be calculated by formula (7):

$$P_{\mathrm{Fe}} = P_1 - P_2 - P_{\mathrm{Cu}} - P_{\mathrm{fw}} - P_{\mathrm{s}} \quad (7)$$

where P_1 is input power, P_2 is output power, P_{Cu} is copper loss, P_{fw} is mechanical loss, and P_s represents stray loss. As shown in Fig. 5, the experiment plat-form is composed of generator, power converter and oscilloscope.

7 Conclusion

The deflection type DDSRG has important application value. This paper introduces the basic structure and control principle of the generator, focusing on the magnetic density distribution of the DDSRG under the rotation state. The overall qualitative analysis of the core magnetic density is carried out by using FEM. According to the calculation result of the magnetic density, the Fourier transform is performed to calculate the Iron loss of the DDSRG, and the response surface method is used to optimize the structural parameters of the DDSRG. The indirect measured experimental data verifies the accuracy of the optimization scheme. It provides theoretical support for the subsequent research of the generator.

Acknowledgements This work was supported by the National Natural Science Foundation of China, grant No. 51877070, 51577048, 51637001, the Natural Science Foundation of Hebei Province of China, grant No. E2018208155, the Talent Engineering Training Support Project of Hebei Province, grant No. A201905008, the National Engineering Laboratory of Energy-saving

Motor and Control Technique, Anhui University, grant No. KFKT201901, Hebei Province Higher Education Science and Technology Research Key Project, grant No. ZD2018228.

References

1. Zhou, Yunhong, and Yukun Sun. 2015. A double-stator type bearingless switched reluctance dual-channel full-period generator. *Proceedings of the CSEE* 35 (9): 2295–2303 (in Chinese).
2. Liu, Yongzhi, Zheng Zhou, and Zengjin Sheng. 2015. Research on switching reluctance motor start/power generation switching control strategy. *Electric Machines and Control* 19 (10): 57–63 (in Chinese).
3. Dong, Chuanyou, Yong Li, and Shuye Ding. 2015. Core loss analysis of switched reluctance motor. *Electric Machines and Control* 19 (07): 58–65 (in Chinese).
4. Li, Zheng, Yue Zhang, Qunjing Wang, et al. 2013. Analytical modeling and analysis of magnetic field for a novel 3-DOF deflection type PM motor. *Proceedings of the CSEE* 33 (S1): 219–225 (in Chinese).
5. Tian, Jing, Xuezhong Zhu, and Xiang Zhou. 2015. Analysis and calculation of iron loss of switched reluctance motor. *Mechanical & Electrical Engineering* 32 (2): 256–260 (in Chinese).
6. Narita, K., T. Asanuma, K. Semba, et al. 2015. An accurate iron loss evaluation method based on finite element analysis for switched reluctance motors. In *Energy conversion congress & exposition*. IEEE
7. Heidarian, M., and B. Ganji. 2016. A dynamic simulation model based on finite element method for switched reluctance generator. In *2016 International symposium on power electronics, electrical drives, automation and motion (SPEEDAM)*. IEEE
8. Xu, T., J. Yuan, Q. Wang, et al. 2016. Inductance estimation method for linear switched reluctance machines considering iron losses. *IET Electric Power Applications* 10 (3): 181–188.
9. Cheng, C., F.T. Bao, and H. Xu. 2014. Optimization design for contour of pintle nozzle in solid rocket motor based on response surface method. *Advanced Materials Research* 1016 (1016): 6.
10. Li, Zheng, Zhang Lu, Qunjing Wang, et al. 2015. Optimal design of structure parameters of three-DOF deflection type PM motor based on response surface methodology. *Transactions of China Electrotechnical Society* 30 (13): 134–142 (in Chinese).

Simulation of Pulsed Electro-acoustic Method for Space Charge Measurement Considering Wave Distortion

Penglong He and Bo Zhang

Abstract Pulsed electro-acoustic is a popular nondestructive method to measure the space charge distribution in dielectrics. Experimental device configuration and calibration method both affect the spatial accuracy of the measurement results. And simulation of PEA plays an important role in assessing the PEA device and calibration method. In this paper, a simulation model of the whole PEA system is developed using software COMSOL and MATLAB. The model consists of two electrodes, a specimen, PVDF sensor and absorber material. And acoustic wave propagation in domains with different acoustic impendence, wave distortions of attenuation and overshoot effect are within the consideration of the model. Additionally, arbitrary space charge density distribution inside the specimen can be set and simulated. Using the simulation model, authors have evaluated the traditional calibration method by comparing the simulation results after calibration and the given space charge distribution. The results reveal a poor spatial accuracy of the calibrated data and thus the calibration method need to be improved in the future work.

Keywords Pulsed electro-acoustic method · Space charge measurement · Wave distortion · Simulation model

1 Introduction

Space charge accumulation in insulating materials is an extensive phenomenon, especially in the direct current (DC) circumstance. The accumulated space charge will cause distortion of local electric field distribution, accelerated degradation, and thus life reduction of the insulating material [1]. Consequently, measurement of the space charge in insulating materials is essential, which provides a tool to assess the insulation performance.

Pulsed electroacoustic method (PEA) is a non-destructive method developed by Takada in [2], and nowadays it has been widely used by researchers to evaluate the

P. He · B. Zhang (✉)
Department of Electrical Engineering, Tsinghua University, Beijing, China
e-mail: shizbcn@mail.tsinghua.edu.cn

W. Chen et al. (eds.), *The Proceedings of the 9th Frontier Academic Forum of Electrical Engineering*, Lecture Notes in Electrical Engineering 743,
https://doi org/10.1007/978-981-33-6609-1_57

space charge distribution inside insulating materials [3–5]. A typical PEA device is mainly composed of two electrodes providing a pulsed electric field, a piezoelectric sensor (PZT or PVDF) and an amplifier. The specimen containing internal charge is sandwiched between the electrodes. When a pulsed voltage is applied between the electrodes, a pulsed electric field is generated in the specimen accordingly. Then the charge is slightly shaken due to electric force, and acoustic waves generate. The acoustic waves propagate through the specimen and the electrode. Eventually the piezoelectric sensor attaching to the ground electrode converts the acoustic signal to electric signal that contains space charge information. After applying some calibration process [6, 7], the real space charge distribution in the thickness direction can be derived.

The positional resolution of PEA is limited by the PEA device and calibration method. As for PEA device, the thickness of piezoelectric sensor and the width of the applied pulsed voltage matters. A thinner piezoelectric sensor and narrower pulsed voltage enable the space charge measurement to obtain a better positional resolution [8]. For an established PEA system, the positional resolution is determined by the calibration method, which mainly consists of overshoot distortion calibration, attenuation distortion calibration and charge density calibration [7]. The real space charge distribution in the insulating material is difficult to control precisely. Therefore, we have no accurate information about the space charge inside the specimen during measurement, which causes the problem that we cannot compare the measured space charge with the real one. The comparison will allow us to evaluate the PEA device and calibration process quantitatively and intuitively. For the purpose of realizing these, accurate model and simulation of the whole PEA system will play an important role.

There are already some simulations about PEA. Employing analogy with electrical transmission lines, Chahal [8] established the simulation using PSpice, and Ren [7] also utilized this model to analyze the impact of different setting of acoustic impedances. The limitation of this model is that space charge in the specimen cannot be simulated. Arnaout [9] proposed a PEA model based on commercial software COMSOL, and the model is capable of simulating the signal of piezoelectric sensor for any space charge distribution. However, attenuation and overshoot distortion are not considered. Mulla [10] established a model containing attenuation and overshoot distortion which could precisely predict the output signal. However, charge inside the specimen was not simulated.

In this paper, a simulation model of PEA system was established with software COMSOL and MATLAB, considering attenuation and overshoot distortion. Simulation results showed the effects of acoustic impendence, attenuation distortion and overshoot distortion on the acoustic wave propagation. A simulation with a given space charge distribution inside the specimen was conducted, and the results were calibrated with traditional calibration method deriving a measured space charge distribution by PEA. Finally, the two space charge distributions were compared and a simple assessment of the calibration method was given.

2 Basic Theory of PEA

The fundamental idea of measuring space charge inside an insulating material is to make the charge move, and corresponding signal is detected outside specimen containing the information about the charge. In the PEA system, charge moves slightly due to the pulsed electric field applied on specimen. The force density induced inside dielectrics is demonstrated by equation [11] in one-dimensional form,

$$f = \rho_c E - \frac{1}{2}E^2\frac{\partial \varepsilon}{\partial z} - \frac{1}{2}\frac{\partial (aE^2)}{\partial z} \tag{1}$$

where f is the force density, ρ_c is space charge density within dielectrics, E is the pulsed electric field strength, ε represents permittivity of the dielectric, a is electrostriction coefficient, and z denotes thickness direction of the dielectric. The second term of the right side of Eq. (1) is caused by the inhomogeneous of dielectrics, and the third term represents the electrostriction caused by the spatial variance of electric field. In the real situation, the specimen is very thin resulting a relatively uniform electric field, and is normally homogeneous. As a result, the second and third term of Eq. (1) vanish. Therefore, the force density is equal to the first term, which is Coulomb force proportional to the charge density.

The force density induced by pulsed electric field causes vibrations of the dielectric where space charge is situated. Correspondingly, the vibrations generate acoustic waves which transmit towards both directions. The equation governing the transmission of acoustic waves is demonstrated as bellow,

$$\frac{1}{\rho c^2}\frac{\partial^2 p_t}{\partial t^2} - \frac{1}{\rho}\frac{\partial^2 p_t}{\partial z^2} + \frac{1}{\rho}\frac{\partial q_d}{\partial z} = Q \tag{2}$$

where ρ is density of the medium, c is transmission velocity of acoustic wave, p_t is the sound pressure, q_d represents dipole acoustic source and Q represents monopole acoustic source. The source caused by a force density is dipole acoustic source, which causes acoustic waves transmit in opposite direction with opposite amplitude. It means $q_d = f = \rho_c E$ and $Q = 0$ in PEA model.

3 Modeling of PEA

The model of PEA system was established with the commercial software COMSOL Multiphysics. Similar approach has been used in Ref. [12]. The model configuration is illustrated as Fig. 1. The one-dimensional simulation model mainly consists of five domains: Electrode1, dielectric specimen, Electrode2, PVDF sensor and absorber material. Actually, absorbing boundary is applied at two terminals of the model in

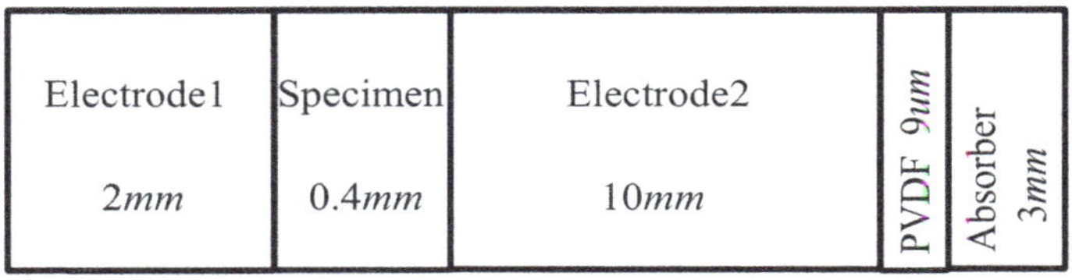

Fig. 1 PEA model configuration

order to cancel the reflections of acoustic waves. In this model, any space charge distribution inside the specimen can be simulated.

Due to the different acoustic impedance of different domains, there are reflection coefficients and transmission coefficients for acoustic waves during propagation and generation coefficients for wave generated at two surfaces between electrode and specimen. These would affect the amplitude of waves received by the PVDF. As the coefficients are determined by acoustic impedance of the nearby domains, it can be implemented in the model by setting different material properties for different domains.

There are three types of acoustic waves generated in the PEA, that is acoustic wave generated at the Electrode1-Specimen interface, Specimen- Electrode2 interface and inside the specimen, respectively. When propagating through the specimen, attenuation occurs and makes the wave a wider width and a lower amplitude. Because the wave generated at the interface of Electrode1-Specimen propagates through the entire specimen, there is more attenuation distortion than the others. A typical raw experimental output data is illustrated as Fig. 2. Peak1 is generated at the Specimen-Electrode2 interface, and peak2 is generated at the Electrode1-Specimen interface. It can be seen that peak2 is wider in time scale and lower in amplitude than peak1.

Considering attenuation effect during acoustic wave propagation, the wave equation is adjusted to,

Fig. 2 Typical raw experimental data

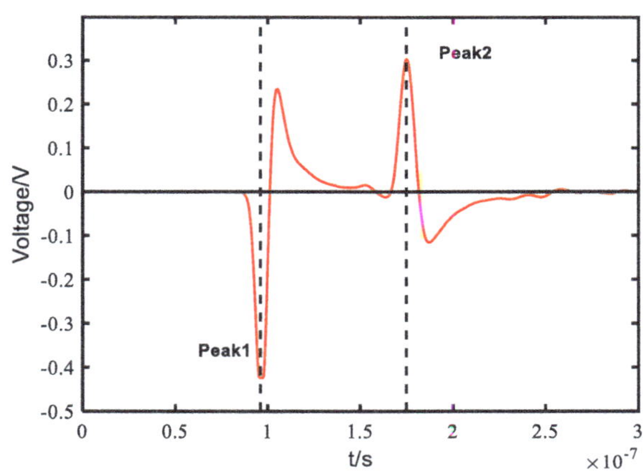

$$\frac{1}{\rho c^2}\frac{\partial^2 p_t}{\partial t^2}-\frac{1}{\rho}\frac{\partial^2 p_t}{\partial z^2}+\frac{1}{\rho}\frac{\partial q_d}{\partial z}+\frac{\delta}{\rho c^2}\frac{\partial}{\partial t}\left(\frac{\partial^2 p_t}{\partial z^2}\right)=0 \quad (3)$$

where δ is a coefficient representing the wave attenuation. Therefore, the attenuation effect of the dielectric is involved in the model.

Another distortion occurring in the PEA system is overshoot distortion. A equivalent high pass filter is formed by capacitance of PVDF sensor and input impedance of the amplifier [13]. And the overshoot distortion is caused by the high pass filter. It is hard to simulate the overshoot distortion effect directly in COMSOL, since a field-circuit coupling is complex. In this model, data from COMSOL calculation is taken for high-pass filtering in MATLAB. The filter parameters are determined by PVDF and amplifier parameters of the experiment. This method has been employed in Ref. [10], detailed information is not given here.

4 Simulation Results and Discussion

In absence of space charges inside a specimen, the output signal is constituted by two converse peaks due to the surface charge caused by DC voltage, like the waveform illustrated in Fig. 2. It is an essential waveform in the calibration procedure, which transfers the voltage signal to the space charge distribution. In traditional calibration procedure, firstly, deconvolution process is conducted to eliminate the overshoot distortion, and secondly, charge calibration process transfers voltage to charge density. The waveform mentioned above is the foundation of these two processes. Additionally, this waveform is the simplest output signal of the PEA system. Therefore, it is convenient to use this signal to check our numerical model.

Equal but opposite charge density is set at the interface of Electrode1-Specimen and Specimen-Electrode2. The simulation result is illustrated in Fig. 3. The wave

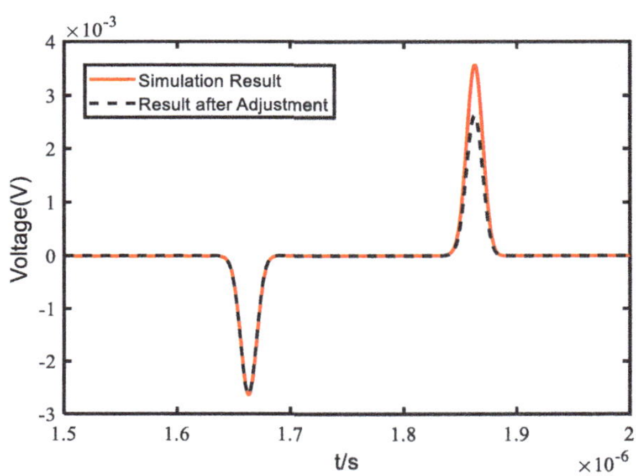

Fig. 3 Simulation results without overshoot and attenuation distortion

distortions including overshoot and attenuation are not involved here. And it is obvious that the amplitudes of the two peaks are not equal. This is the effect of generation coefficients and transmission coefficients which are given as the follow equations,

$$G_{1-2} = \frac{Z_2}{Z_1 + Z_2} \tag{4}$$

$$T_{1-2} = \frac{2Z_2}{Z_1 + Z_2} \tag{5}$$

where G_{1-2} is the generation coefficient at the interface of domain 1 and domain 2 towards the direction of domain 2, T_{1-2} is the transmission coefficient at the interface of domain 1 and domain 2 towards the direction of domain 2, and Z represents the acoustic impedance [14]. And the acoustic impedance can be expressed by Eq. (6),

$$Z = \rho c \tag{6}$$

where ρ is material density, and c is sound velocity in the material. And the material properties of various domains are demonstrated in Table 1.

Supposing the acoustic pressure generated at the two interfaces is $p_0(t)$ and $-p_0(t)$, respectively. Then the acoustic pressures reaching PVDF sensor are expressed as,

$$p_1(t) = -G_{s-E2} T_{E2-PVDF}\, p_0(t + \triangle t_1) \tag{7}$$

$$p_2(t) = G_{E1-s} T_{s-E2} T_{E2-PVDF}\, p_0(t + \triangle t_2) \tag{8}$$

Therefore, the coefficients make the two peak amplitudes differ by a multiple S given in Eq. (9),

$$S = \frac{G_{E1-s} T_{s-E2}}{G_{s-E2}} \tag{9}$$

Table 1 Material properties of domains

Domain	Density/(kg/m^3)	Sound velocity/(m/s)
Electrode1	940	2200
Specimen	2200	2000
Electrode2	2700	6400
PVDF	1780	787
Absorber	1780	787

And the result after adjustment by eliminating the difference of S is also illustrated in Fig. 3. After adjustment, the amplitudes of the two peaks become equal, which means the effect of acoustic impedance of different domains is correctly simulated.

Acoustic wave propagating through the specimen suffers from attenuation, and this process can be expressed by formula [15] (10),

$$P(f, z) = P(f, 0)e^{j\omega t}e^{-\alpha(f)z}e^{-j\beta(f)z} \tag{10}$$

where $\alpha(f)$ is the attenuation coefficient, and $\beta(f)$ is the dispersion coefficient. They are functions of frequency. In this model, the attenuation coefficient is controlled by parameter δ. And a larger value of means a larger value of $\alpha(f)$ at arbitrary frequency.

Here are simulation results considering attenuation distortion in absence of space charge inside the specimen as shown in Fig. 4. There are four curves with different values of δ, which means with different attenuation coefficient, and $\delta = 0$ means no attenuation distortion. Because the first peak don't propagate through the specimen, there is no attenuation distortion with any value of δ. It can be obviously seen that greater value of δ results in a lower amplitude and wider width of the second peak, and it is consistent with theoretical analysis.

Actually, if the two peak waveforms and the distance is known, attenuation coefficient can be calculated with formula [15] (11):

$$\alpha(f) = -\ln(|P(f, d)|/|P(f, 0)|) \times \frac{1}{d} \tag{11}$$

where d is distance between two peaks. Figure 5 shows the results of calculated attenuation coefficients for different parameter settings. Parameter δ is set to 0.001, 0.003, and 0.005, respectively, and three attenuation coefficient curves are derived.

Fig. 4 Simulation results considering attenuation

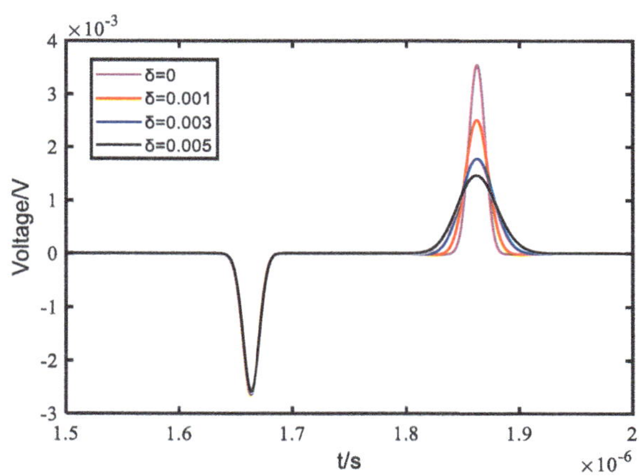

Fig. 5 Attenuation coefficients for different parameters

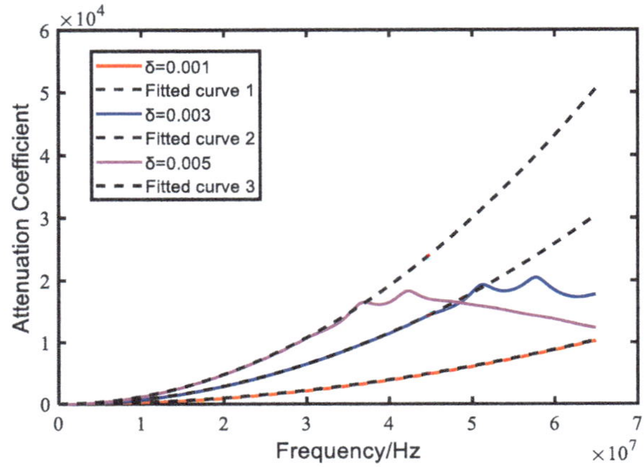

The three black dotted curves are quadratic fitting curves for each attenuation coefficient. From the results in frequency domain, we can derive the same conclusion that greater value of δ gives a greater value of $\alpha(f)$. And the curves are approximately quadratic, which is similar with the experiment result in Ref. [15]. An attractive phenomenon in Fig. 5 is that attenuation coefficient curve deviates from the fitting curve when the frequency exceeds a certain value, acting like a small tail. The reason is that high frequency component in the peak decays faster. By drawing the amplitude-frequency curve, it is shown that the frequency component higher than the frequency where the curve deviates decays close to zero. So the small tail has no effect on the attenuated waveform.

According to Eq. (10), the attenuated waveform is also determined by the propagation distance. A simulation with the same value of δ, but different propagation distance is conducted and illustrated in Fig. 6. As the propagation distance increases, amplitude of the waveform decreases and width increases. However, the attenuation

Fig. 6 Attenuation effect in different propagation distance

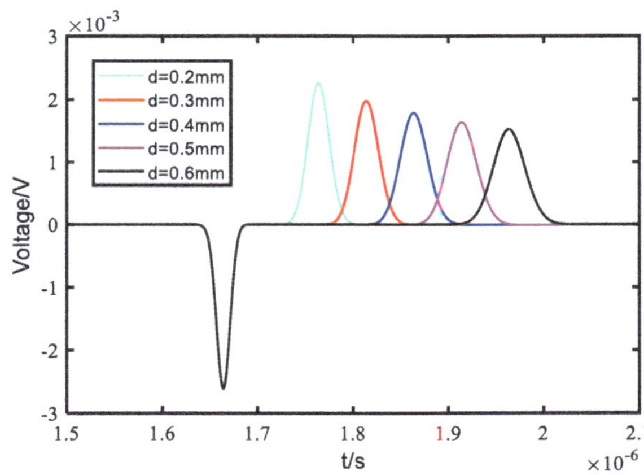

of amplitude is not proportional to the propagation distance. As shown in Fig. 5, higher frequency component bears a greater attenuation coefficient. When the relatively high frequency components are attenuated to zero, the attenuation rate of the peak amplitude will reduce.

The overshoot distortion is not directly simulated in COMSOL, but realizes through post-processing of the signal in MATLAB by adding a filter. And the results are illustrated in Fig. 7. The overshoot is caused by high pass filter effect of PVDF and amplifier, and amplitude reduction is caused by low pass filter because of oscilloscope frequency response.

In the simulation model, arbitrary charge distribution inside the specimen can be set and calculated, but for simplicity a charge density distribution containing two Gaussian shape peaks is illustrated in Fig. 8a. And the simulation result of this charge density configuration considering overshoot and attenuation distortion is shown in Fig. 8b.

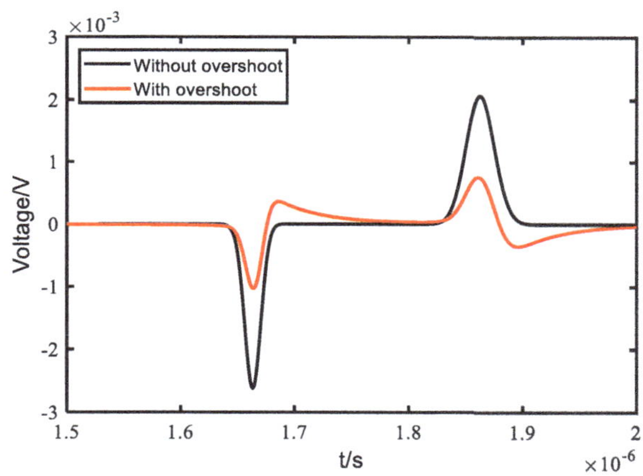

Fig. 7 Simulation results considering overshoot distortion

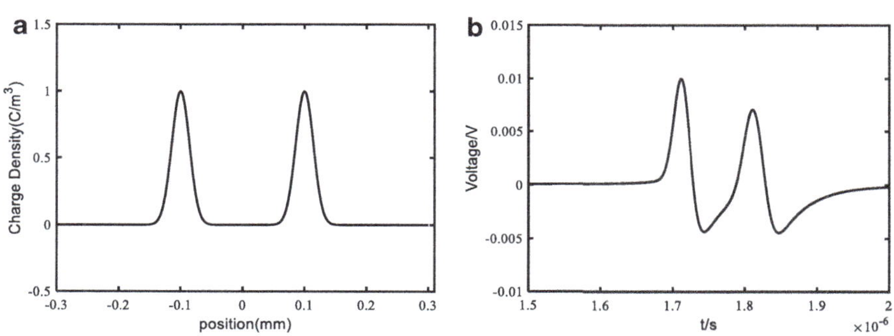

Fig. 8 **a** Charge density distribution, **b** simulation result of internal charge

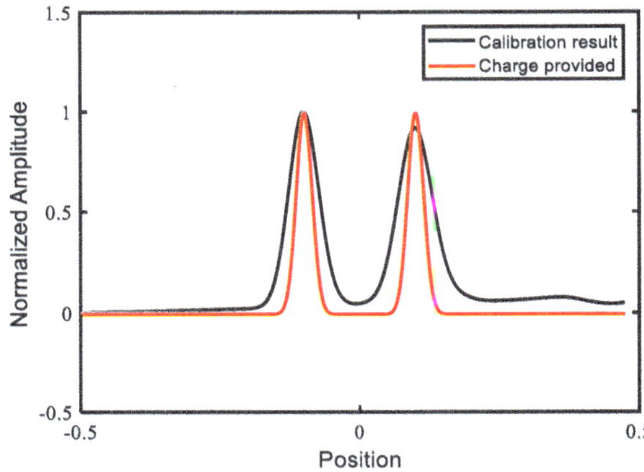

Fig. 9 Calibration result and comparison with the charge provided

5 Calibration of Simulation Results

The first traditional calibration process is to eliminate the overshoot distortion of signals from two electrode-specimen surfaces by means of deconvolution process. The deconvolution process assumes that the charge density distribution on the interface is an ideal pulse, and uses it to calculate a system function from charge density to output signal of the PEA system. After that, another calibration process should be conducted to transfer unit of y-axis from voltage to charge density.

The simulation signal illustrated in Fig. 8b is calibrated by traditional process, and the result is shown in Fig. 9. Because this paper mainly focuses on the spatial distribution of the charge density, the second calibration process is omitted, and a normalized amplitude is adopted instead. From the comparison between calibrated result and the charge density distribution provided to the model, it can be seen that the calibration result become wider. The reason is that the result of deconvolution process is affected by the pulsed voltage. And because the traditional calibration method doesn't consider the effect of attenuation distortion, the second charge density peak has a lower amplitude and wider width. Therefore, in order to obtain a better spatial accuracy of PEA system, calibration method should be improved, such as considering the effect of the pulsed voltage and attenuation distortion. And the simulation model will be an essential tool to achieve the improvement.

6 Conclusion

This paper presented a simulation model of PEA system, which contains acoustic wave propagation, wave distortion of attenuation and overshoot, and arbitrary charge density distribution. This model can quantitatively analyze the influence of various parameters in the system, and make a clear assessment for calibration method by directly comparing the calibration result with the given space charge profiles. A simulation result was obtained by setting a certain space charge density distribution inside the specimen, and after that calibration process was conducted. The comparison between the calibrated result and given space charge profile gives a clear assessment of the spatial resolution. If the parameters of the model can be matched with the experimental device, then the simulation model can be combined with the experiment and become an important tool to improve the device and calibration method to achieve a better space charge measurement method, and this is the direction for our future work.

References

1. Fabiani, D., G. Montanari, C. Laurent, G. Teyssedre, P.H.F. Morshuis, R. Bodega, L.A. Dissado, A. Campus, and U.H. Nilsson. 2007. Polymeric HVDC cable design and space charge accumulation. Part 1: insulation/semicon interface. *IEEE Electrical Insulation Magazine* 23(6): 11–19.
2. Takada, T., T. Maeno, and H. Kushibe. 1987. An electric stress-pulse technique for the measurement of charges in a plastic plate irradiated by an electron beam. *IEEE Transactions on Electrical Insulation* 4: 497–501.
3. Fukunaga, K. 2004. Innovative PEA space charge measurement systems for industrial applications. *IEEE Electrical Insulation Magazine* 20 (2): 18–26.
4. Fukunaga, K. 2008. Progress and prospects in PEA space charge measurement techniques. *IEEE Electrical Insulation Magazine* 24 (3): 26–37.
5. Thomas, C., G. Teyssedre, and C. Laurent. 2010. Space-charge dynamic in polyethylene: From dc to ac stress. *Journal of Physics D: Applied Physics* 44 (1): 015401.
6. Li, Y., M. Yasuda, and T. Takada. 1994. Pulsed electroacoustic method for measurement of charge accumulation in solid dielectrics. *IEEE Transactions on Dielectrics and Electrical Insulation* 1 (2): 188–195.
7. Ren, H., Q. Li, and Z. Wang. 2019. An improved calibration method for the measurement of space charge inside insulating materials. *IEEE Transactions on Instrumentation and Measurement* 69 (4): 1652–1663.
8. Chahal, J.S., and C.C. Reddy. 2013. Modeling and simulation of pulsed electroacoustic measurement method. *IEEE Systems Journal* 8 (4): 1283–1292.
9. Arnaout, M., F. Baudoin, L. Berquez, and D. Payan. 2011. Study of signal treatment for a pulsed electro-acoustic measurement cell: A way of improving the transfer matrix condition number. *Journal of Physics D: Applied Physics* 44 (40): 405403.
10. Mulla, A.A., S.J. Dodd, N.M. Chalashkanov, and L.A. Dissado. 2020. A new numerical approach to the calibration and interpretation of PEA measurements. *IEEE Transactions on Dielectrics and Electrical Insulation* 27 (2): 666–674.
11. Hole, S., T. Ditchi, and A. Lewiner. 2003. Non-destructive methods for space charge distribution measurements: What are the differences? *IEEE Transactions on Dielectrics and Electrical Insulation* 10 (4): 670–677.
12. Arnaout, M., K. Chahine, F. Baudoin, L. Berquez, D. Payan. 2016. Iterative deconvolution of PEA measurements for enhancing the spatial resolution of charge profile in space polymers. *Journal of Polymers*.
13. Bernstein, J.B. 1991. Analysis of the electrically stimulated acoustic-wave method for observing space charge in semi-insulating films. *Physical Review B* 44(19): 10804.

14. Liu, R., T. Takada, and N. Takasu. 1993. Pulsed electro-acoustic method for measurement of space charge distribution in power cables under both DC and AC electric fields. *Journal of Physics D: Applied Physics* 26 (6): 986.
15. Li, Y., M. Aihara, K. Murata, Y. Tanaka, and T. Takada. 1995. Space charge measurement in thick dielectric materials by pulsed electroacoustic method. *Review of Scientific Instruments* 66 (7): 3909–3916.

Appropriate Proportion of Holey Graphene Used as Partial Conductive Agent in Electric Double-Layer Supercapacitors

Chao Yang, Hengrui Yang, Yuge Bai, Xiaodong Wu, and Xiaogang Han

Abstract Graphene is widely used in supercapacitors as active materials and conductive agents because of its large specific surface area, high electrical conductivity and electron mobility. However, there are few studies on microwave treatment of graphene for perforating and using holey graphene mixed with conductive carbon black as conductive agent. Here, we report the best mixing ratio of holey graphene and conductive carbon black treated by microwave, that is, 5%SP + 5%HG has good electrochemical comprehensive properties. We attribute the improved performance to the fact that the holey graphene can reduce the migration distance between the electrode and the electrolyte because of pores. At the same time, the larger specific surface area of graphene can be connected with conductive carbon black, thus improving the comprehensive performance of electric double layer capacitors. Compared with only using graphene as conductive agent, this method reduces the overall cost and is more suitable for industrial application.

Keywords Holey graphene · Supercapacitor · Conductive agent · Electrochemical performance

1 Introduction

Supercapacitor is a new type of electric energy storage device between traditional capacitor and secondary battery [1–4]. It is a special capacitor based on the electrode–electrolyte interface of high specific surface materials such as porous carbon and some metal oxides. Following the same basic principle as the traditional capacitor, it can provide fast pulse power that can not be provided by the main power supply.

According to their different energy storage mechanisms, supercapacitors can be divided into two types [5]: one is pseudocapacitor, which realize the process of charge and discharge based on the redox reaction of active materials, and the commonly used active materials are conductive polymers, transition metal oxides and so on. The other

C. Yang · H. Yang · Y. Bai · X. Wu · X. Han (✉)
State Key Laboratory of Electrical Insulation and Power Equipment, School of Electrical Engineering, Xi'an Jiaotong University, Xi'an 710049, Shaanxi, China
e-mail: xiaogang.han@xjtu.edu.cn

W. Chen et al. (eds.), *The Proceedings of the 9th Frontier Academic Forum of Electrical Engineering*, Lecture Notes in Electrical Engineering 743,
https://doi.org/10.1007/978-981-33-6609-1_58

is electric double layer supercapacitor, which is based on the double layer formed at the electrode material / electrolyte interface [6–8]. The charge and discharge is realized through the absorption and desorption of electrolyte ions on the electrode surface, and the carbon material with high specific surface area is used as the electrode active material.

As electrode materials, activated carbon [9] and transition metal oxides have general electrical conductivity and poor charge adsorption ability, resulting in large electrochemical impedance (EIS) of capacitors [10], which will reduce the power density and rate performance of devices. In order to reduce the EIS of system, it is very important to add conductive agents to the active materials. At present, the conductive agents used as double-layer supercapacitors are graphite, carbon black, metal powder, carbon fiber and carbon nanotubes [11, 12]. Among them, graphene, as a highly conductive and lightweight carbon material, has been used in supercapacitors [13–18] for its unique two-dimensional flake structure and interlaminar conductivity [19]. However, the price of graphene is expensive, and the density is low, which will reduce the overall volume capacitance. Simple use of graphene as a conductive agent in industrial production has some problems, such as difficult preparation, high cost and so on.

We found that microwave can easily make holes in graphene, so we propose a strategy of mixing porous graphene with conductive carbon black after microwave treatment. Graphene after making hole can reduce the migration distance of ions between the electrode and the electrolyte because of the existence of pores. At the same time, the larger specific surface area of graphene can be connected with conductive carbon black, thus improving the comprehensive performance of electric double layer capacitors. Compared with only using graphene, this method of microwave reduces the overall cost. At the same time, it takes advantage of the excellent electrical conductivity and low cost of carbon black. In this paper, we study the effects of different ratios of multilayer microwave radiation graphene and Super P as conductive agent [20] on AC impedance, CV cycle curve, rate discharge characteristics, specific capacitance and cycle performance of electric double layer supercapacitors. As a result, we propose the best ratio to improve the comprehensive performance.

2 Experimental Section

2.1 Experimental Material

See Table 1.

Table 1 Materials needed for experiment

Material name	Model	Manufacturer
Carbon coated aluminum foil	h: 0.014 mm	Showa Denko
Activated carbon	YP-50	KURARAY Co., Ltd.
Graphene	Chemical method of few-layer graphene	Suzhou Hengqiu Technology Co., Ltd.
Polyvinylidene fluoride	Akema 900	Shenzhen Tianchenghe Technology Co., Ltd.
Diaphragm	MAC500LC	NipponPaperIndustries Co., Ltd.
N-Methyl Pyrrolidone	AR,99.0 wt.%	Chemical reagent of Sinopharm group
Button battery case	CR2025	Shenzhen Ming Ruixiang Automation equipment Co., Ltd.
Gasket, shrapnel	h: 0.5 mm d: 16 mm	Shenzhen Ming Ruixiang Automation equipment Co., Ltd.

2.2 *Experimental Step*

2.2.1 Preparation of Electrode

(1) about 0.1 g graphene is put into the quartz grinding bottle without lid, the microwave time is set at 30 s, the interval is 5 s, and the same operation is treated for 4 times.
(2) 0.05 g of graphene treated by microwave was put into a plastic bottle with zirconium beads and dispersed by ultrasonic wave with 2 mL NMP, for 10 min.
(3) Add Super P and PVDF,300r/min to shake 60 s, add activated carbon (YP-50) 0.4 g, add NMP, again, first add 0.5 ml, then add 0.1 ml each time, until there is a clear sound of the slurry in the shaking bottle. 300r/min oscillates for 120 s.
(4) scrape out the slurry, pour it on the carbon-coated aluminum foil, scrape it off at a uniform speed with a 100 μm scraper, dry it in a blast oven for 2 h, take it out and put it in a vacuum oven for 12 h, and cut the pieces with a 12 mm diameter slicer.

2.2.2 Assembly of Symmetrical Supercapacitors

The button battery is assembled in the glove box in the following order: negative electrode shell-electrode sheet-electrolyte-diaphragm-electrolyte–electrode sheet-positive shell-gasket-shrapnel-positive shell. Press the package with a button battery sealing machine with a pressure of 80 MPa.

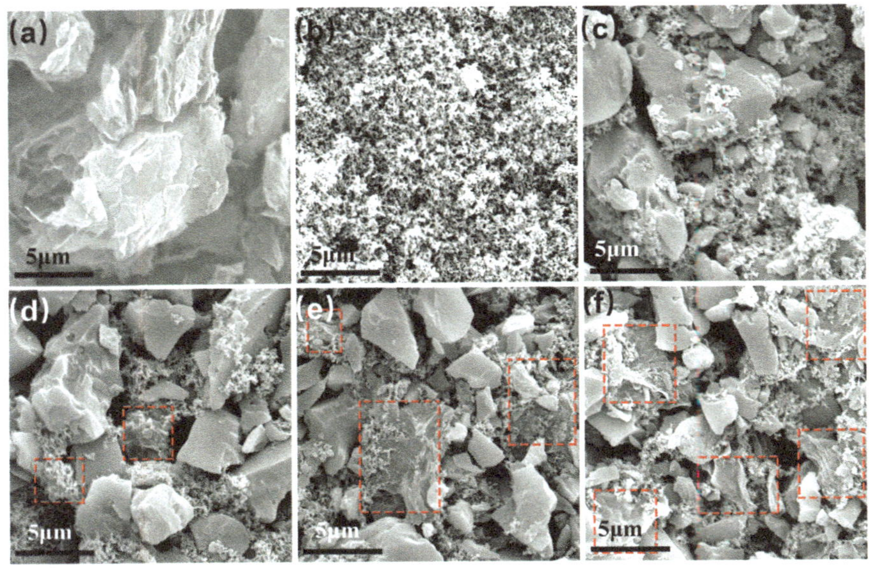

Fig. 1 SEM diagram of electrode material surface

3 Results and Discussion

3.1 Micro-morphology Analysis

The morphologies of graphene and Super P can be observed from Fig. 1a, b. Figure 1c–f of 10% SP, 9% + 1%, 7% + 3%, 5% + 5% SP + HG, is recorded as 10SP, 1HG, 3HG and 5HG in turn. It can be observed that the proportion of graphene is increasing under the unified scale, indicating that the experimental ultrasonic dispersion and oscillation treatment distribute graphene on the pole. Super P particles are uniformly distributed among the bulk activated carbon particles, and Super P particles are connected to the flake graphene. We believe that it is the connection of graphene that makes the dispersed Super P particles form a conductive network, so the electrical conductivity of the dispersed Graphene-Super P is better than individual Super P particles.

3.2 AC Impedance Test

It is observed from Fig. 2 that the semicircle diameter of 10SP is 13, 1HG is 16, 3HG is 15, and the semicircle diameter of HG is 9. The charge transfer impedance of 5HG is the lowest, which is 9 Ω. Because the resistivity of graphene (0.35 Ω · cm)

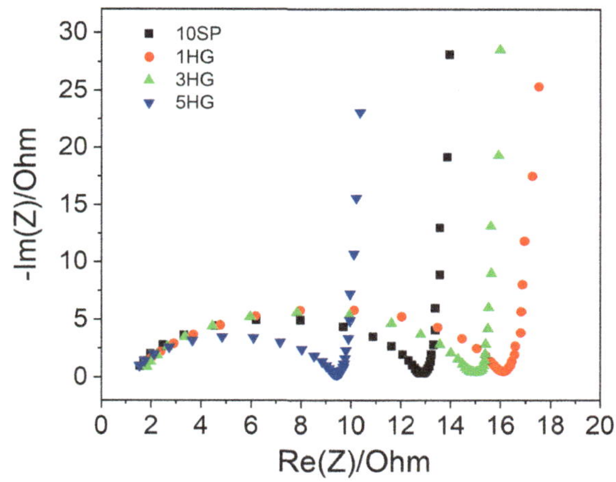

Fig. 2 AC impedance diagram of button battery

powder measured by four-probe powder resistivity meter is higher than that of Super P (0.06 Ω · cm), when a small amount of graphene is added, compared with 10SP, the transfer impedance of the sample with lower proportion of graphene will become larger, because the content of graphene at this time is not enough to form a conductive network, and when porous graphene is added to 3%, it plays a certain connecting role, resulting in a lower transfer impedance than 1HG. However, graphene in 5HG samples has the same mass as Super P, graphene has a larger volume, and more Super P particles are connected to the lamellae, which improves the overall conductivity and enhances the effect of charge transfer.

3.3 Cyclic Voltammetry Test

The four curves are approximately rectangular, which accords with the characteristics of supercapacitors. Because the carbon load is similar, the specific capacitance of the electrode can be approximately calculated by integrating the cyclic voltammetry curve, that is, the area of the curve reflects the specific capacitance. In Fig. 3a, the area increases with the increase of graphene content, and the area of graphene content in 5HG is the largest, so its electrode specific capacitance is the largest. Figure 3b shows the cyclic voltammetry curves of 5HG at different scanning rates, all of which are rectangular waves, indicating that the rate performance of the sample is good.

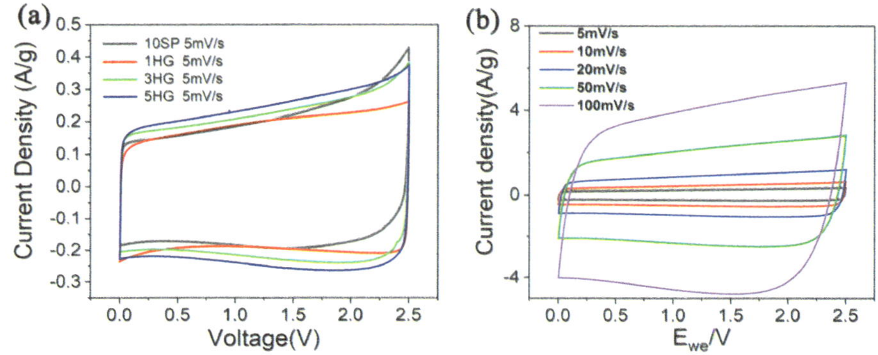

Fig. 3 Cyclic voltammetry of button battery

3.4 Rate Tests and Galvanostatic Charging/Discharging (GCD) Tests

Rate tests of four samples under 1/2/4/8 A/g show that the specific capacitance of the electrode has decreased at high current with the increase of graphene content, and the specific capacitance of 5HG is almost unchanged at high current, because the existence of graphene pores shortens the distance of ions between the electrode and electrolyte, and ions can be fast absorption and desorption at high current, so the retention rate of specific capacitance is higher.

GCD of button battery under the current of 1A/g is shown in Fig. 4b. The four samples are similar to equilateral triangle, indicating that the electrode has high Coulomb charge–discharge efficiency, the longitudinal coordinate range of triangular wave increases gradually with the increase of graphene content, in other words, the specific capacitance increases gradually, and the voltage drop of 5HG in the

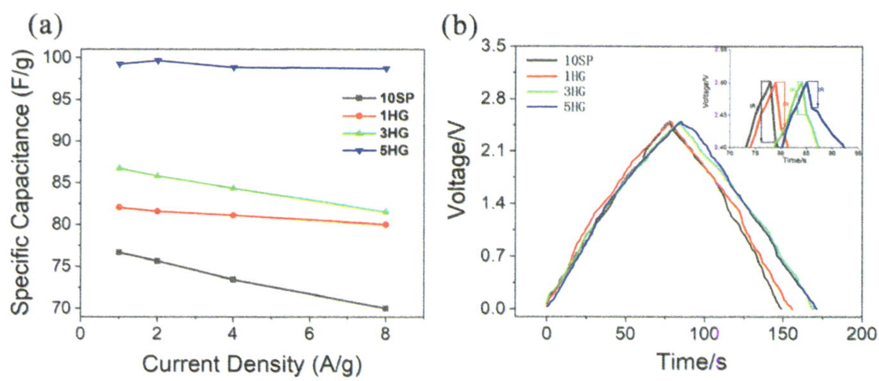

Fig. 4 Specific capacitance under different current and charge–discharge curve under 1A/g

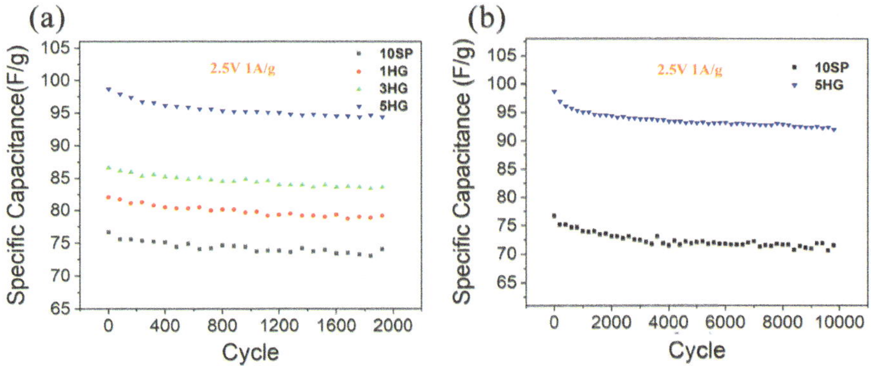

Fig. 5 **a** 1000 cycle tests of 4 kinds of samples and **b** 10SP and 5HG long cycle tests

discharge curve is the lowest, indicating that the internal resistance of the sample is very small and the conductivity has been enhanced.

3.5 Cycle Test

The first 2000 cycles of the four samples under 1A/g current are shown in Fig. 5a, it can be found that the capacitance retention of the four samples is relatively good, but the specific capacitance of the 5HG sample is the largest, so only the long cycles of 10SP and 5HG are compared, and the performance of the 10,000 cycles is shown in Fig. 5b. It can be found that after 10,000 constant current charge–discharge cycles, the capacity retention of 5HG is still more than 90%, with a good capacity retention.

4 Conclusion

In this paper, we explore the different ratios of holey graphene and conductive carbon black. As a result, we get the conclusion that the best comprehensive properties is 5HG. Holey graphene, a highly conductive, lightweight, flaky conductive agent combined with traditional carbon black, can connect granular conductive agents through the unique sheet structure of graphene to form a continuous conductive network, which can improve the internal resistance and rate performance of the electrode. We expect that the minimum use of graphene can not only form a conductive network, but also minimize the cost. It is hopeful to apply this method in industrial production to realize the goal from laboratory to industrialization.

References

1. Jiang, L., J.W. Yan, Y. Zhou, et al. 2013. Activated carbon/graphene composites with high-rate performance as electrode materials for electrochemical capacitors. *Journal of Solid State Electrochemistry* 17 (11): 2949–2958.
2. Bose, S., T. Kuila, A.K. Mishra, et al. 2012. Carbon-based nanostructured materials and their composites as supercapacitor electrodes. *Journal of Materials Chemistry* 22 (3): 767–784.
3. Wang, J.G., B. Wei. 2017. Special issue: materials for electrochemical capacitors and batteries. *Materials (Basel)* 10(4).
4. Xia, H., Z. Yang, Z. Yang, et al. 2017. Control strategy of supercapacitor energy storage system for urban rail transit based on operating status of trains. *Transactions of China Electrotechnical Society* 32 (21): 16–23.
5. Zhao, Y., H. Liang, and Y. Zhang. 2012. Review and expectation of modeling research on electrochemical supercapacitor. *Transactions of China Electrotechnical Society* 27 (3): 188–195.
6. Liu, M.X., M.C. Shi, W.J. Lu, et al. 2017. Core-shell reduced graphene oxide/MnOx@carbon hollow nanospheres for high performance supercapacitor electrodes. *Chemical Engineering Journal* 313: 518–526.
7. Gao, W., N. Singh, L. Song, et al. 2011. Direct laser writing of micro-supercapacitors on hydrated graphite oxide films. *Nature Nanotechnology* 6 (8): 496–500.
8. Pachfule, P., D. Shinde, M. Majumder, et al. 2016. Fabrication of carbon nanorods and graphene nanoribbons from a metal-organic framework. *Nature Chemistry* 8 (7): 718–724.
9. Wei, L., M. Sevilla, A.B. Fuertes, et al. 2011. Hydrothermal carbonization of abundant renewable natural organic chemicals for high-performance supercapacitor electrodes. *Advanced Energy Materials* 1 (3): 356–361.
10. Cao, X.H., Y.M. Shi, W.H. Shi, et al. 2011. Preparation of novel 3D graphene networks for supercapacitor applications. *Small (Weinheim an der Bergstrasse, Germany)* 7 (22): 3163–3168.
11. Toupin, M., D. Belanger, I.R. Hill, et al. 2005. Performance of experimental carbon blacks in aqueous supercapacitors. *Journal of Power Sources* 140 (1): 203–210.
12. Zheng, C., X.F. Zhou, H.L. Cao, et al. 2014. Synthesis of porous graphene/activated carbon composite with high packing density and large specific surface area for supercapacitor electrode material. *Journal of Power Sources* 258: 290–296.
13. Xu, Y., Z. Lin, X. Zhong, et al. 2014. Holey graphene frameworks for highly efficient capacitive energy storage. *Nature Communications* 5: 4554.
14. Xu, Y.X., G.Q. Shi, and X.F. Duan. 2015. Self-assembled three-dimensional graphene macrostructures: synthesis and applications in supercapacitors. *Accounts of Chemical Research* 48 (6): 1666–1675.
15. Chang, L., D.J. Stacchiola, and Y.H. Hu. 2017. An ideal electrode material, 3D surface-microporous graphene for supercapacitors with ultrahigh areal capacitance. *ACS Applied Materials & Interfaces* 9 (29): 24655–24661.
16. Li, C., X. Zhang, K. Wang, et al. 2017. Scalable self-propagating high-temperature synthesis of graphene for supercapacitors with superior power density and cyclic stability. *Advanced Materials* 29(7).
17. Zhang, L., F. Zhang, X. Yang, et al. 2013. Porous 3D graphene-based bulk materials with exceptional high surface area and excellent conductivity for supercapacitors. *Science Report* 3: 1408.
18. Raccichini, R., A. Varzi, S. Passerini, et al. 2015. The role of graphene for electrochemical energy storage. *Nature Materials* 14 (3): 271–279.
19. Pang, S., and X. Liu. 2018. Review on research and application of graphene in electrical field. *Transactions of China Electrotechnical Society* 33 (8): 1705–1722.
20. Wang, R., Y. Qian, W. Li, et al. 2018. Performance-enhanced activated carbon electrodes for supercapacitors combining both graphene-modified current collectors and graphene conductive additive. *Materials (Basel)* 11(5).

Analysis of Influence of Different Propagation Media on Motor Vibration and Noise

Ningning Yang, Jinhua Chen, Yunpeng Gao, Chi Zhang, and Guofu Li

Abstract With the development of marine engineering, the noise of underwater motors has received more and more attention. Aiming at the problem of underwater acoustic radiation caused by electromagnetic force waves of underwater motors through the vibration of the casing, this paper first analyzes the different causes of the vibration caused by the excited shell in water and air; secondly, the same sound pressure and the same sound power are derived respectively In the three cases with the same excitation force, the difference of the sound pressure level of the shell in air and water, focusing on the effect of the underwater acoustic impedance on the difference of the sound pressure level un-der the same excitation force. Finally, taking a 24-pole 36-slot permanent magnet synchronous motor as an ex-ample, the analytical method and finite element simulation are used to analyze the sound pressure level difference of the motor under different media to verify the correctness of the analytical method. The underwater motor is directly analyzed by the noise in the air Provide a reference for predicting underwater noise.

Keywords Excitation force · Vibration · Sound pressure level · Underwater acoustic impedance · Permanent magnet synchronous motor

1 Introduction

The vibration and noise of the motor are one of the main indicators for evaluating the quality of the motor. It reflects the motor design quality, technological level and installation technology. In this regard, motor manufacturers, colleges and universities and research institutes have all carried out experiments and theoretical analysis, and

---

N. Yang (✉) · G. Li
Ningbo University, Ningbo 315211, China
e-mail: yangningning@nimte.ac.cn

J. Chen · Y. Gao · C. Zhang
Zhejiang Provincial Key Laboratory of Robotics and Intelligent Manufacturing Equipment Technology, Ningbo Institute of Materials Technology and Engineering, Chinese Academy of Sciences, Ningbo 315201, China
e-mail: chenjinhua@nimte.ac.cn

W. Chen et al. (eds.), *The Proceedings of the 9th Frontier Academic Forum of Electrical Engineering*, Lecture Notes in Electrical Engineering 743,
https://doi.org/10.1007/978-981-33-6609-1_59

have achieved fruitful results, but most of them have concentrated on the research of vibration and noise of motors in the air.

Electromagnetic noise is the main source of motor noise, and the stator of the motor is the main source of electromagnetic noise, so comprehensive consideration of these two factors is the research focus of most literature [1–3]. The above documents are all analyzing and optimizing the vibration and noise of the motor in the air, which is very different from the vibration and noise of the underwater motor.

In recent years, due to the development of marine engineering, the vibration and noise of underwater motors have been focused on due to their important impact on sound and shadow, underwater communication and navigation. However, due to the complexity of the propagation medium and the underwater environment, the vibration and noise of underwater motors are very different from those in the air [4]. The current technical solutions are not enough to meet the needs of underwater equipment development. Literature [5] gives the displacement vibration formula of a cylindrical shell under the action of electromagnetic excitation force, but this formula does not include frequency parameters, and it is impossible to evaluate the impact of underwater acoustic loads on vibration. Literature [6–9] discussed the influence of the mechanical impedance of the casing and the acoustic impedance of the medium on the vibration and noise by deriving the formula for the vibration and noise radiation of the cylinder under water. To sum up, for the motor, if it is possible to find out the difference in vibration and noise under the same sound source through analysis and simulation on the basis of air noise research, in this way, when analyzing under-water vibration and noise, a more mature The air noise analysis method can indirectly predict the vibration and noise level of underwater motors, which has certain reference significance for engineering experiments.

The main difference between air noise and underwater noise is the difference in the characteristic impedance of the acoustic medium and the coupling problem caused by vibration [7]. The difference between the vibration and noise of the motor in water and air, as well as the new influence brought by quantification, is currently unclear. In this paper, aiming at the noise radiation problem caused by the same excitation force of the motor under water and air, firstly, the difference between underwater and air vibration under the same excitation force is studied; secondly, the difference between water and air-borne media is analyzed. Then the vibration difference and the sound medium difference are comprehensively considered to obtain the analysis method of the noise difference between the water and the air of the motor housing; finally, the finite element simulation and analytical calculation of the noise difference of the 24-pole and 36-slot permanent magnet synchronous motor in the air and water are carried out.

2 Vibration Under Different Propagation Media

Using the vibration of the shell structure as a noise source will generate sound pressure on the surface of the shell structure, and the sound pressure on the surface of the shell structure will have a reaction to the vibrating shell structure. This reaction can be called the sound of the acoustic medium. load. This load is generally related to the density of the sound propagation medium. Air is called light fluid due to its low density, and the acoustic load generated by air can be ignored. Water is called heavy fluid due to its high density, and the load caused by the sound pressure in the sound propagation medium to the vibrating structure is called underwater acoustic load and cannot be ignored [6].

In the analysis, the motor shell or stator core is equivalent to a finite-length cylindrical shell for research, as shown in Fig. 1. The first is the vibration in the air. According to the derivation of the shell theory in [5, 6], When the inner surface of the cylindrical shell is subjected to the n-th order harmonic electromagnetic force density F_0, the radial vibration displacement w in the air Available:

$$w = -\frac{a^2(1-\sigma^2)F_0 \cdot (\Omega^2 - n^2)}{Eh\left[\Omega^2 - \left(\Omega_n^{(1)}\right)^2\right] \cdot \left[\Omega^2 - \left(\Omega_n^{(2)}\right)^2\right]} \tag{1}$$

The maximum vibration displacement in water is:

$$w_s = -\frac{F_0}{1/\xi + Z_s} = -\frac{F_0}{Z_m + Z_s} \tag{2}$$

In the formula: $\Omega = \omega a / c_p$ is a dimensionless frequency parameter, $\left(\Omega_n^{(1)}\right)^2 = \frac{1}{2}\left[1 + n^2 + \beta^2 n^4 + \sqrt{\left(1 + n^2 + \beta^2 n^4\right)^2 - 4\beta^2 n^6}\right]$, $\omega = 2\pi f$ is the circular frequency, $\left(\Omega_n^{(2)}\right)^2 = \frac{1}{2}\left[1 + n^2 + \beta^2 n^4 - \sqrt{\left(1 + n^2 + \beta^2 n^4\right)^2 - 4\beta^2 n^6}\right]$, $\beta^2 = h^2/12a^2$ is the thickness to diameter ratio of the shell, F_0 is the electromagnetic force density amplitude, a is the shell radius, h is the shell thickness, E is the Young's modulus of the shell material, t is the time, σ is the Poisson's ratio of the

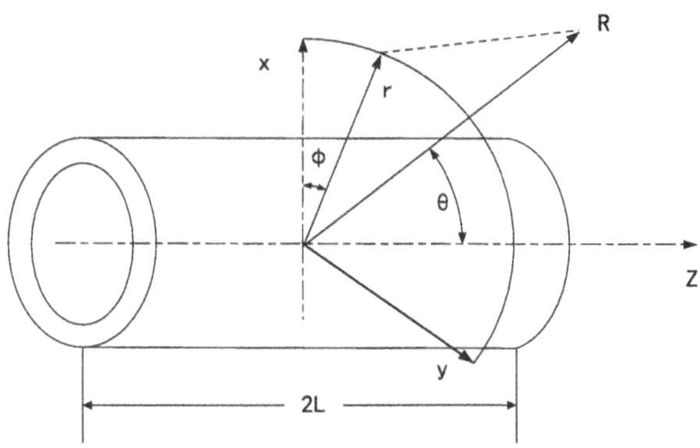

Fig. 1 The coordinate diagram of the far-field sound pressure points of a cylindrical shell

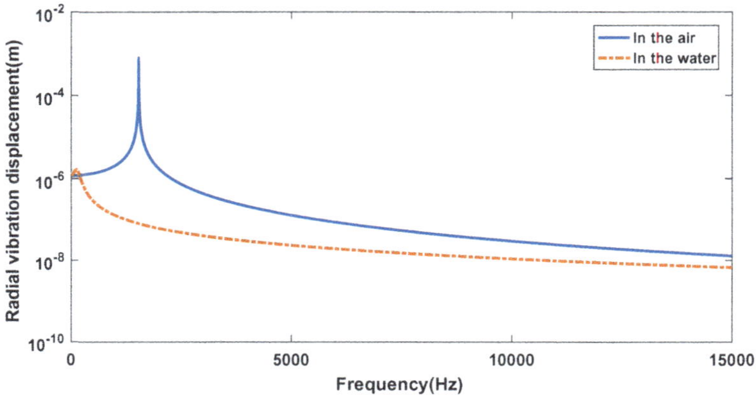

Fig. 2 Comparison of radial vibration displacement in water and air at 0th order pressure

material, n is the order of the circumferential mode, and c_p is the sound velocity of the longitudinal wave in the shell, $Z_s = \rho c\omega \frac{H_n(ka)}{H_n'(ka)}$ is the underwater acoustic impedance, $H_n(ka)$ is the Hank function of the first kind of order n, $H_n'(ka)$ is the first derivative of the Hank function of the first kind, $k = \omega/c$ is the wave number, c is the speed of sound, and ρ is the density, $Z_m = -\frac{Eh\left[\Omega^2-\left(\Omega_n^{(1)}\right)^2\right]\cdot\left[\Omega^2-\left(\Omega_n^{(2)}\right)^2\right]}{a^2(1-\sigma^2)(\Omega^2-n^2)}$ is the mechanical impedance.

Figure 2 shows the radial vibration displacement in the air and the maximum underwater radial vibration under the action of 0-order excitation force of different frequencies when $a = 0.55$ m, $h = 0.055$ m, $F_0 = 50$ kPa, and $f = 0 - 15$ kHz Displacement. It can be seen from Fig. 2 that the vibration of the shell in the water is subject to the sound pressure feedback of the heavy fluid water, and the sound pressure feedback in the air is negligible due to the small air density. Therefore, the shell under the same shell and pressure is affected by the sound pressure feedback. The vibration displacement is much smaller than in the air, and as the frequency gradually increases, the vibration displacement difference will gradually decrease.

3 Noise Under Different Propagation Media

3.1 Noise Difference at the Same Sound Pressure

Table 1 shows the respective reference quantity symbols and reference values of sound pressure level, sound intensity level and sound power level when propagating in air and water. Among them, the selected sound pressure level calculation reference quantity in the air is $p_{01} = 20$ μpa, and the water sound pressure level calculation reference quantity is $p_{02} = 1$ μpa. Therefore, when the sound pressure p is constant,

Table 1 Acoustic standards for water and air

Medium	Sound pressure reference value (μ pa)		Sound intensity reference value (w/m^2)		Sound power reference value (w)	
Air	p_{01}	20	I_{01}	10^{-12}	W_{01}	10^{-12}
Water	p_{02}	1	I_{02}	0.67×10^{-18}	W_{02}	0.67×10^{-18}

the difference between the sound pressure level in the water and the air. The difference in sound pressure level is [10]:

$$L_{\Delta} = 20\log_{10}\frac{p}{p_{02}} - 20\log_{10}\frac{p}{p_{01}} = 26\,\text{dB} \tag{3}$$

Therefore, when the sound pressure is constant, the difference in sound pressure level between underwater and air is about 26 dB.

3.2 *Noise Difference at the Same Sound Power*

When the sound power W is constant, the acoustic impedance ratio ($\rho_c = \rho * c$) will be introduced, which is a different amount in water and air. S is the area of the wave front. Table 2 shows the reference values of density and sound velocity of air and water under common conditions. By directly calculating the difference in sound power level:

The sound pressure level is used to calculate the difference in sound pressure level between the two propagation medias, which can be expressed as follows:

$$W = p^2 S / \rho c \tag{4}$$

Converting sound power level into sound pressure level calculation, you can get the formula as follows:

$$\begin{aligned} L_{W\Delta} = L_{\Delta 1} &= L_{\text{water}} - L_{air} \\ &= 10\log_{10}(\rho_1 c_1 / \rho_0 c_0) + L_{\Delta} = 61.5\,\text{dB} \end{aligned} \tag{5}$$

In summary, when the sound power is constant, the difference in sound pressure level between underwater and air is about 61.5 dB.

Table 2 Basic parameters of air and water

Air density ρ_0 (kg/m^3)	Sound velocity in air c_0 (kg/m^3)	Density of water ρ_1 (kg/m^3)	Sound speed in water c_1 (kg/m^3)
1.25	340	1000	1500

3.3 Noise Difference When the Same Exciting Force

A spherical coordinate system is established for the simplified model of the motor shell with a finite-length cylindrical shell, as shown in Fig. 1. The point with the largest sound pressure value on the cylindrical shell is selected as the research point. The analysis formula found that $\theta = 90°$, $\Phi = 0°$, $R = $ a point sound pressure, that is, $p(a, 90°, 0°)$ is the amplitude point of the sound pressure value [6, 7, 11]. Available formula:

$$|p(a, 90°, 0°)| = \frac{\omega u_0 \rho}{\pi ka} \frac{2L}{H_n'(ka)} = \frac{u_0 \rho c}{\pi a} \frac{2L}{H_n'(ka)} \tag{6}$$

In the formula: ω is the circular frequency (rad/s), R is the radius in spherical coordinates, and L is half the length of the finite-length cylindrical shell, The vibration speed is $u_0 = -j\omega w$.

3.3.1 Regardless of Underwater Acoustic Load

In this section, the difference between sound pressure level and sound power level in water and air is calculated without considering the underwater acoustic load. As follows (Table 3):

$$\begin{aligned} L_{\Delta 2} &= 10\log_{10}\left[\frac{\rho_1 c_1}{\rho_0 c_0}\left(\frac{H_n'(k_0 a)}{H_n'(k_1 a)}\right)^2\right] + L_{W\Delta} \\ &= 20\log_{10}\left[\frac{H_n'(k_0 a)}{H_n'(k_1 a)}\right] + 97\text{ dB} \end{aligned} \tag{7}$$

It can be seen from Fig. 3 that the noise difference between water and air generated by all electromagnetic force waves, when the frequency is small, the sound radiation coefficient has a greater impact on the sound pressure level difference $\gamma\Delta$ (about 97 dB), and the force The higher the wave order, the more obvious the effect. When the point at 1000 Hz is taken, the difference in sound pressure level produced by the 0th and 2nd order electromagnetic force waves is significantly greater than that of the higher order. Therefore, when suppressing noise, attention should be paid to reducing the 0th and 2nd order harmonic currents and the electromagnetic waves produced by them. When the frequency exceeds 5000 Hz, it gradually reaches about

Table 3 Commonly used calculation parameters

Wave number in the air k_0	Wave number in water k_1	Frequency range f (Hz)	Force wave order n
ω/c_0	ω/c_1	0–15,000	0, 2, 4, 6, 8, 12

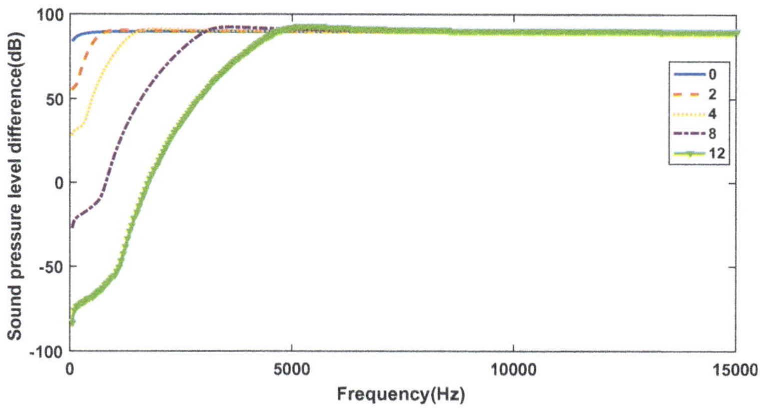

Fig. 3 Difference in sound pressure levels in water and air at different force wave orders

90 dB, so when the sound load is not considered, the difference in sound pressure level between water and air produced by different orders of force waves is about 90 dB.

3.3.2 Regardless of Underwater Acoustic Load

This section mainly considers the influence of the underwater acoustic load in the noise propagation process, that is, the influence of the vibration displacement and speed change of the cylindrical shell in air and water under the same excitation force.

From formula (6), it can be found that after adding the underwater acoustic load, there is a difference between the vibration displacement w in the water and in the air, so the following formula can be obtained:

$$L_{\Delta 3} = 20\log_{10}\left(\frac{|p_{water}(a, 90°, 0°)|}{|p_{air}(a, 90°, 0°)|}\right)+L_{\Delta} = 20\log_{10}\left(\frac{Z_m}{Z_m+Z_s}\right)+L_{\Delta 2} \quad (8)$$

From this, the difference in sound pressure level between underwater and air produced by the 0, 2, and 4 order excitation force can be calculated, as shown in Fig. 4. It can be seen from Fig. 4 that when the frequency is within 7000 Hz, the underwater acoustic load (that is, the influence of the underwater acoustic impedance Z_s on the underwater vibration displacement) has a greater impact on $L_{\Delta 2}$. When the frequency is less than 100 Hz, the sound pressure level difference produced by the 0-order force wave is much greater than that of other orders. Therefore, it is important to suppress the 0-order force wave when reducing vibration and noise. When the frequency gradually increases, the influence of the underwater acoustic

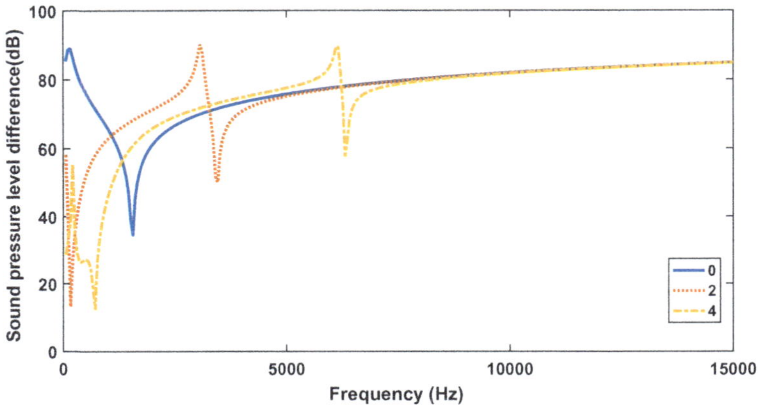

Fig. 4 Sound pressure difference between water and air under sound load

load gradually decreases, and the difference in sound pressure level $L_{\Delta 3}$ between water and air finally stabilizes at about 85 dB.

4 Noise Difference Based on Motor Model

The radial electromagnetic force generated by the 24-pole and 36-slot three-phase permanent magnet synchronous motor is used as the excitation force of the electromagnetic noise of the motor, and the force wave orders are mainly 0 and 12 orders. Table 4 gives the main size parameters of the 24-pole 36-slot permanent magnet synchronous motor. When the underwater acoustic load is not considered, the motor parameters can be put into formula (7) to obtain the noise difference between the water and the air of the 24-pole and 36-slot permanent magnet synchronous motor, as shown in Fig. 5. Through the finite element simulation, the noise difference between the 24-pole and 36-slot permanent magnet synchronous motor in the underwater and in the air is obtained, as shown in Fig. 6.

It can be seen from Figs. 5 and 6 that regardless of the analytical method or the finite element method, when the underwater acoustic load is not considered, the average value of the analytical results in the figure and the simulation results are basically the same (the sound pressure level in water is higher than that in air under

Table 4 Motor parameters

Middle longitude of stator a (m)	Stator thickness h (m)	Half the length of the stator L (m)	Motor's radial electromagnetic force wave order n
0.17875	0.0225	0.0175	0, 12

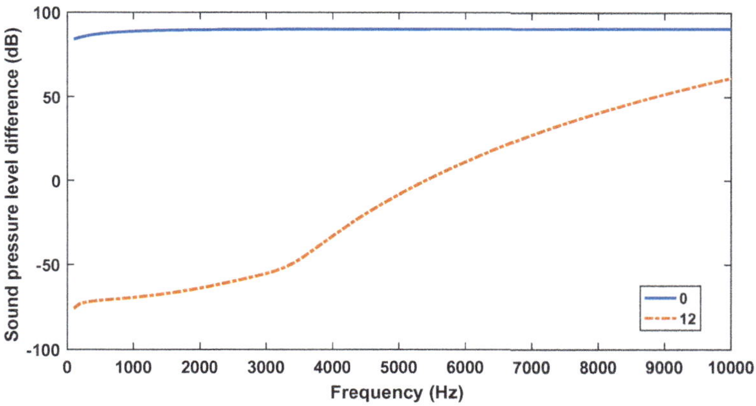

Fig. 5 Difference of noise between water and air in permanent magnet motor

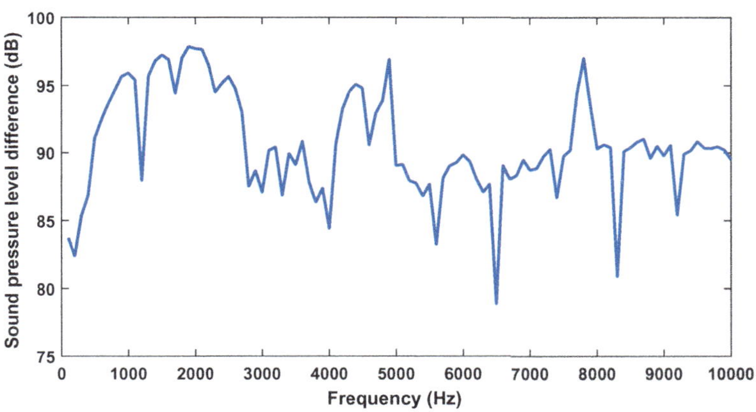

Fig. 6 Difference of noise in water and air of permanent magnet motor based on finite element simulation

the same excitation force About 90 dB.), and the analysis results are consistent with the initial changes of the simulation results. In the simulation results, crests and troughs appear frequently, mainly due to other orders of radial electromagnetic force waves generated by the non-ideal state of the motor.

Furthermore, after considering the displacement change caused by the underwater acoustic load, the noise difference between a 24-pole and 36-slot three-phase permanent magnet motor in water and air is shown in Fig. 7. By comparing with Fig. 5, it can be found that the sound pressure level difference $L_{\Delta 2}$ between air and water produced by the 0-order electromagnetic force wave is greatly affected by the underwater acoustic load in the frequency range of 400–10000 Hz. Water and air the medium sound pressure level difference will drop to 12 dB at the lowest level, and

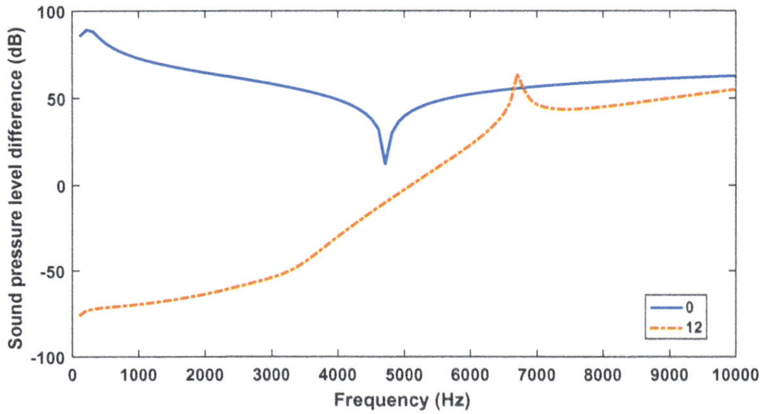

Fig. 7 The difference in noise between water and air between different force wave orders of permanent magnet motors under the influence of acoustic load

it will be stable around 63 dB. The difference in sound pressure level $L_{\Delta 2}$ generated by the 12-order electromagnetic force wave is about 56 dB at the final 10,000 Hz under the influence of underwater acoustic load.

5 Conclusion

Aiming at the difference between the noise of permanent magnet motors in the water and in the air, this paper makes a quantitative analysis of various influencing factors.

1. Under the same sound field energy (sound pressure p), the difference in sound pressure level between underwater and air is 26 dB. 2. Under the same sound source energy (sound power), the difference in sound pressure level between underwater and air is 61.5 dB. 3. When the excitation force is the same, and the underwater acoustic load is not considered, the difference $L_{\Delta 2}$ between the underwater and air sound pressure levels generated by each order of electromagnetic force waves is about 90 dB. After adding the underwater acoustic load, the difference between the underwater and air sound pressure levels produced by the electromagnetic force waves of each order is reduced. 4. The above calculation is verified with a 24-pole and 36-slot permanent magnet synchronous motor, and the maximum sound pressure level under the influence of underwater acoustic load is 78 dB.

Acknowledgements This article was funded by the Zhejiang Basic Public Welfare Research Program (LGG18E070007), the Strategic Pilot Project (XDA22040302), the Ningbo Major Project (2019B10070) and the Ningbo Innovation Team (2016B10016).

References

1. Colby, R.S., F.M. Mottier, and T.J.E. Miller. 1996. Vibration modes and acoustic noise in a four-phase switched reluctance motor. *IEEE Transactions on Industry Applications* 32 (6): 1357–1364.
2. Delaere, K., W. Heylen, and R. Belmans. 2002. Comparison of induction machine stator vibration spectra induced by reluctance forces and magnetostriction. *IEEE Transactions on Magnetics* 38 (2): 969–972.
3. Shenbo, Y., and T. Renyuan. 2006. Electromagnetic and mechanical characterizations of noise and vibration in permanent magnet synchronous machines. *IEEE Transactions on Magnetics* 42 (4): 1335–1338.
4. Zhao, Xu. 2019. *Design of underwater vehicle propulsion motor and research on vibration and noise*. Shenyang University of Technology. (in Chinese)
5. Yang, S.J. 1985. *Low-noise motor*. Science Press.
6. Zhao, Wenfeng, Zhenguo Bai, Yingbo Xu. 2017. Vibration and noise characteristics of underwater cylindrical shells with finite length under electromagnetic excitation. *China Shipbuilding* 58(1): 101–109 (in Chinese).
7. Zhao, Wenfeng. 2018. *Research on vibration and noise of submerged permanent magnet synchronous motor*. China Ship Research Institute (in Chinese).
8. Ruotolo, R. 2002. Influence of some thin shell theories on the evaluation of the noise level in stiffened cylinders. *Journal of Sound and Vibration* 255 (4): 777–788.
9. Bai, Zhenguo. 2014. *Sound-vibration coupling characteristics and control technology of double-layer cylindrical hull shipboard*. China Ship Research Institute. (in Chinese)
10. Yang, Guiheng, Xue Yang, Junqiang He, etc. 2018. *Noise and vibration control technology and its application*, 9–13. Chemical Industry Press. (in Chinese)
11. Junger, M.C. 1986. *Sound, structures, and their interaction*. 2nd ed, 175–177. MIT Press.

Research on Vibration Suppression of Permanent Magnet Synchronous Motor Based on Extended Random Depth PWM Modulation Technology

Xueming Guo, Zhe Jiang, Shuheng Qiu, Chi Zhang, and Jianye Liu

Abstract Based on traditional PWM modulation, there are current harmonics related to the carrier frequency in the drive system of the permanent magnet synchronous motor. These current harmonics affect the control accuracy of the system and cause the high-frequency vibration noise. Random PWM modulation technology can disperse the current harmonics concentrated at the switching frequency and its integer multiple frequencies into a wider frequency spectrum. But traditional random PWM modulation technology cannot disperse the current harmonics evenly, the harmonics near the integer multiple switching frequencies are still obvious. Aiming for solving this problem, a novel method of extending random depth is proposed in this paper. The relationship between high-frequency current harmonics and motor vibration, and the mechanism of random PWM modulation technology are analyzed. On the basis of random switching frequency PWM modulation technology, a variable frequency is implemented to replace the original fixed center frequency. Markov chain algorithm instead of pseudo-random number algorithm is employed to achieve random depth expansion of carrier frequency. Finally, the effectiveness of the proposed method is validated through simulation results.

Keywords Permanent Magnet Synchronous Motor · Switching noise · Current harmonics · Markov chain · RPWM

1 Introduction

Space Vector Pulse Width Modulation (SVPWM) is currently the main method for wide-range speed control of permanent magnet synchronous motors (PMSM). The traditional PWM modulation technology consists of a fixed frequency carrier signal.

X. Guo · J. Liu
Hebei University of Science and Technology, No. 26, Yu xiang Street, Shijiazhuang City, Hebei Province, China

Z. Jiang · S. Qiu · C. Zhang (✉)
Ningbo Institute of Materials Technology and Engineering, Chinese Academy of Sciences, No. 1219, Zhong guan West Road, Zhen Hai District, Ningbo City, Zhejiang Province, China
e-mail: zhangchi@nimte.ac.cn

W. Chen et al. (eds.), *The Proceedings of the 9th Frontier Academic Forum of Electrical Engineering*, Lecture Notes in Electrical Engineering 743,
https://doi.org/10.1007/978-981-33-6609-1_60

The output current therefore contains a large number of harmonic components related to the carrier frequency, which are mainly distributed at the switching frequency and its integral multiple frequencies, thereby causing high-frequency vibration and noise in the motor [1].

In order to reduce the high-frequency switching noise of the motor, American scholar Trzynadlowski et al. proposed a random PWM method. On the basis of maintaining the duty cycle of the PWM signal, the method randomly changes the switching frequency and conduction position, thereby changes the frequency spectrum of the output voltage [2]. Although this method cannot reduce the total harmonic quantity of the current, it can disperse the harmonics to reduce the harmonic amplitudes near the switching frequency and its integer multiple frequencies. Biji Jacob et al. proposed a pulse width modulation technique for randomizing the pulse position, which weakened the maximum amplitude of switching noise, but increased the switching loss [3]; V. Jayamala et al. used a pseudo-random number generator to obtain a random switching frequency PWM signal, but the obtained switching frequency is twice that of the traditional method, which is not suitable for low switching frequency occasions [4]. L. Accardo et al. combined random switching frequency and random pulse position to obtain a hybrid random PWM signal. Compared with a single random PWM technology, it can effectively disperse the current harmonics in the switching frequency and its multiples, and the noise reduction effect has been improved [5]. The carrier frequency of the existing random PWM modulation methods only changes around a center frequency, and the random value generated by the pseudo-random number algorithm is not uniform. It is difficult to achieve uniform harmonic distribution when the concentrated harmonics are dispersed, and there are still obvious harmonics in the frequency band [6]. At the same time, the hybrid random PWM modulation technology has difficulties in engineering realization due to the complicated random quantity and large amount of calculations.

This paper proposes a PWM modulation method with extended random depth. It uses a changing signal instead of the fixed center frequency and Markov chain algorithm instead of pseudo-random number algorithm to generate uniformly varying random numbers. The harmonics are evenly dispersed into a wider frequency band through random depth expansion of the carrier frequency. In this paper, an extended random depth PWM modulation model is established and compared with the traditional random PWM modulation.

2 High Frequency Vibration Noise and Random PWM Modulation Technology

There are sharp noises caused by high-frequency current harmonics in PWM-modulated motor systems. The reason is that the air gap magnetic field induced by the high-frequency current harmonics interacts with the iron core, resulting in high-frequency radial force fluctuations. This causes high-frequency vibration of the stator iron core. Thus, by optimizing the PWM modulation strategy, the sharp noise generated by high-frequency current harmonics can be reduced.

2.1 Vibration Analysis of High Frequency Current Harmonics

When the PMSM is driven by a PWM-based inverter, the distribution of current harmonic in the frequency domain has obvious spikes. The spikes in the spectrum appear at the carrier frequency and its integral multiples, which is shown in Eq. (1) [7]:

$$f_h = k_1 f_c + k_2 f_e \tag{1}$$

where, f_c is the carrier frequency, f_e is the operating frequency, k_1 and k_2 are positive integers with different parities. Among them, $f_e = (n \times p_n)/60$, n is the motor speed, and p_n is the number of pole pairs of the motor.

The amplitude of the current harmonics is related to the carrier frequency f_c and the operating frequency f_e of the motor. Therefore, reducing the amplitude of the current harmonics can be achieved through a control strategy. In summary, the harmonic distribution and amplitude of high-frequency currents are directly related to PWM modulation.

The high-frequency current harmonics excite the high-frequency magnetic field in the air gap, which acts on the stator iron core to generate high-frequency radial force fluctuation. The vibration amplitude caused by radial force can be obtained by Eqs. (2)–(4) [8].

$$b_{dk} = i_{dk}\frac{L_{ad}\cos(p\theta)}{\partial\tau N K_{dp} L_{ef}}, b_{qk} = i_{qk}\frac{L_{aq}\cos(p\theta)}{\partial\tau N K_{qp} L_{ef}} \tag{2}$$

where, i_{dk} and i_{qk} are k-order d/q-axis current harmonics respectively, L_{ad} and L_{aq} are d/q-axis inductance respectively, τ is the pole angle of the harmonic component, and L_{ef} is the length of the motor.

$$P_r = -\frac{(b_r(\theta,t) + b_s(\theta,t)) \cdot b_k(\theta,t)}{2\mu_0} \tag{3}$$

where, P_r is the radial electromagnetic force density, b_r, b_s are the magnetic density generated by the excitation and inducted by the armature, respectively, and μ_0 is the vacuum permeability.

$$A_{n\omega_k} = \frac{\pi D_i L_{ef} P_{n\omega_k}}{n_c\sqrt{\left(\omega_n^2 - \omega_k^2\right)^2 + 4\varepsilon^2\omega_k^2\omega_n^2}} \tag{4}$$

where, ω_n is the n-th order natural frequency of motor stator, $P_{n\omega k}$ is the radial electromagnetic force density at ω_k, and ω_k is an integer multiple of the carrier frequency.

Therefore, the vibration and noise of the PMSMs can be weakened by reducing the radial electromagnetic force fluctuation. And, it can be found that one method to reduce the whistling noise of the PMSMs is to reduce the current harmonic peak at the switching frequency and its integer multiples.

2.2 Random PWM Modulation Technology

The switching frequency and conduction position of PWM modulation can affect the harmonic energy distribution in the voltage and current of the motor. The Fourier expansion of the output voltage is shown in Eq. (5).

$$U(t) = A - \sum_{n=1}^{\infty}\left(\frac{4}{n\pi}\right)\cos\left(\frac{An\pi}{2} + \frac{n-1}{2}\pi\right)\cdot\cos(n\omega_c t) \tag{5}$$

In the above formula, $\cos(n\omega_c t)$ is a unit pulse function after Fourier transform. The unit pulse function causes the harmonics to be concentrated at the switching frequency and its multiples, resulting in obvious harmonic peaks [9]. Therefore, randomly changing the carrier frequency or conduction position can make the originally concentrated harmonic energy evenly dispersed. According to the different implementation methods, modulation techniques can be divided into: Random Switching PWM (RSPWM), Random Pulse Position PWM (RPPPWM), Random Switching Frequency PWM (RSFPWM) [10], as shown in Fig. 1.

In RSPWM modulation, the carrier signal is replaced by a random value, by which the output PWM wave is modulated as Fig. 1a [11]. However when the modulation coefficient is relatively low, RSPWM modulation will cause severe signal distortion and delayed response. In RPPPWM modulation, while keeping the switching period fixed, the on and off positions are randomly selected or the position of the zero vector is changed to realize the random change of the pulse position [12]. But in a system with low modulation and high switching frequency, RPPPWM modulation is relatively weak in dispersing harmonic energy. In RSFPWM modulation, the carrier frequency is randomly changed [13]. Moreover, RSFPWM modulation has advantages over the above two methods in terms of implementation form and dispersion of harmonic energy. Therefore, this paper presents a method of random depth expansion on this basis.

The carrier frequency expression of RSFPWM modulation is shown in Eq. (6).

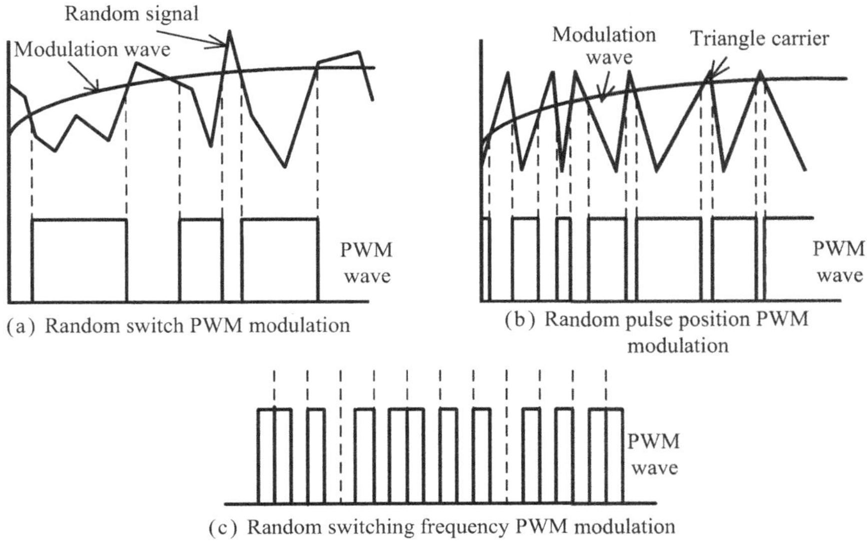

Fig. 1 Random PWM modulation

$$f_s = f_c + R_i \Delta f \tag{6}$$

where, f_c is the carrier center frequency, R_i is a random number varying in [−1,1], and Δf is a fixed frequency band.

Therefore, by randomizing the carrier frequency in the PWM modulation, The harmonics amplitude at the switching frequency and its integer multiples can be reduced, which achieves the purpose of reducing the voltage harmonic peaks. The Fourier expansion of the output voltage is shown in Eq. (7):

$$U(t) = A - \sum_{n=1}^{\infty} \left(\frac{4}{n\pi}\right) \cos\left(\frac{An\pi}{2} + \frac{n-1}{2}\pi\right) \cdot \cos[2\pi n f_a t + \phi(t)] \tag{7}$$

where, f_a is the mean frequency of the carrier, and $\Phi(t)$ is the difference.

It can be seen that the spectrum of the output voltage is divided into a combination of harmonic energy distribution and expanded bandwidth. Therefore the random performance of the carrier frequency determines the uniformity of the harmonic distribution and the degree of peak reduction.

3 Extended Random Depth PWM Modulation Technology

The novel ERDPWM method proposed by this paper is optimized on the basis of RSFPWM modulation. First, the center frequency is replaced a sinusoidal signal to realize the period change of carrier frequency. Secondly, the Markov chain algorithm is used to replace the pseudo-random number algorithm to make the random value more uniform. Finally, the sinusoidal signal and the Markov random value are combined to generate a depth extended random signal.

The carrier frequency in ERDPWM modulation has both randomness and periodicity. Therefore, the periodic changes of the random carrier frequency is realized, which forms a small range randomness in a large range. The ERDPWM modulation reduces the current harmonic amplitude comparing with the fixed carrier frequency modulation. Also, the novel method improves the problem of range limitation in RSFPWM modulation by employing Markov chain, which makes the random value distribution more uniform, which can make the mathematical expected average value of the random switching frequency is approximated to the rated switching frequency. The change range of the random switching frequency can be divided into $[f_s - S, f_s + S]$. The two parts correspond to the two states of less than the rated switching frequency and greater than the rated switching frequency. The sum of the probability that the system's switching frequency is in the two states is 1, that is, $P_1 + P_2 = 1$. The Markov chain algorithm obtains uniform parameters by randomly sampling in the probability space. Random values are continuously generated during the algorithm process, which obeys certain rules that can generate better parameter values. Then the generated random value is combined with the sine signal to achieve an optimized sine random signal.

During the action of the sine signal, the random value of the Markov chain follows the change of the sine signal. In the positive half cycle of the sine signal, the carrier frequency first increases and then decreases, while the action in the negative half cycle is reversed. The random value changes four times in a sine signal cycle. The sum of the change frequency of the sinusoidal signal and the wide-range change frequency of the random value is expressed as the center frequency. So the frequency selection of the sine signal is related to the center frequency. Reflected in the proportional relationship, the frequency of the sinusoidal signal accounts for one-fifth of the center frequency. And the frequency of the sinusoidal signal is positively correlated with the center frequency.

Define the expansion ratio of ERDPWM modulation as:

$$m = \frac{f}{f_c} \tag{8}$$

where f is the frequency of the sinusoidal signal, and f_c is the center frequency of the PWM modulation.

In ERDPWM modulation, the spread ratio is modulated in the linear region to satisfy $0 < m \leq 1$. When $m = 1$, the system oscillates and the total energy of

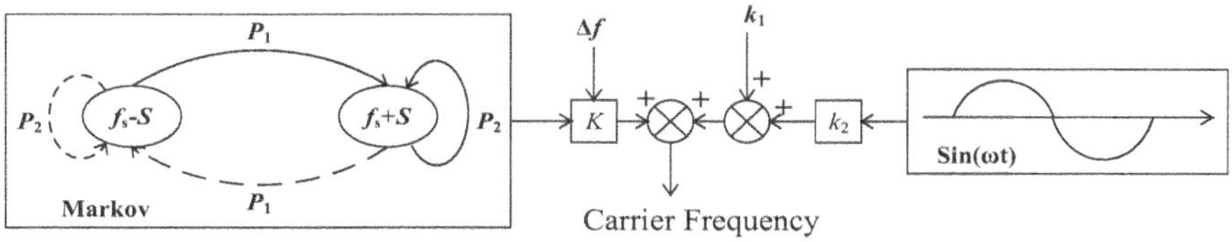

Fig. 2 Block diagram of random carrier frequency implementation

current harmonics increases; when m is close to zero, ERDPWM modulation is approximately RSFPWM modulation. Too large or small m affects the distribution of harmonics in the current spectrum, where the spreading ratio $m = 0.2$.

The carrier frequency realization block diagram is shown in Fig. 2.

The carrier frequency expression of the expanded random switching frequency PWM modulation is shown in Eq. (9).

$$f_s' = [k_1 + k_2 \sin(\omega t)] + S_i \Delta f_s \tag{9}$$

where, $\sin(\omega t)$ is a periodically changing function, where $\omega = 2\pi f$, k_1 and k_2 are natural numbers that are not zero, S_i is a random number generated by the Markov chain algorithm, and Δf_s is a fixed frequency band.

The carrier frequency in the ERDPWM modulation mode is a combination of a sinusoidal signal and a random frequency band. In practical applications, it is necessary to consider the heat dissipation and loss of the switching device. Therefore, the frequency band Δf_s in ERDPWM modulation is selected as 2 kHz, and the expected center frequency is 5 kHz. Thus, the range of the random carrier frequency f_s' in ERDPWM modulation is set to [3, 7 kHz]. The ERDPWM modulation not only inherits the advantages of RSFPWM modulation, but also improves the randomness of the carrier frequency, by which the ERDPWM modulation has a greater harmonic peak reduction degree.

4 Modeling and Analysis

The Matlab/Simulink tool is used to simulate a FOC driven PMSM. And the PWM switching frequency f_c is 5 kHz.

The carrier generation module of ERDPWM modulation is shown in Fig. 3. The triangular carrier generator with random frequency is shown in Fig. 4. The frequency of the random carrier signal in ERDPWM modulation shows a sinusoidal cycle change trend, which makes the carrier frequency more random.

In order to verify the effectiveness of the proposed ERDPWM modulation method, a comparsion between three different modulation is present. The distribution of

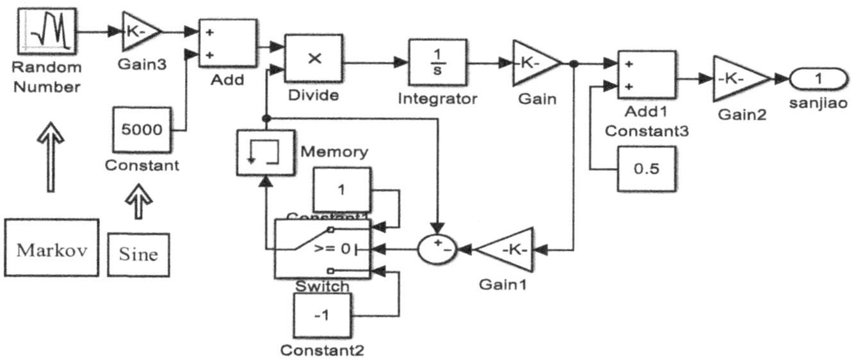

Fig. 3 Random carrier model

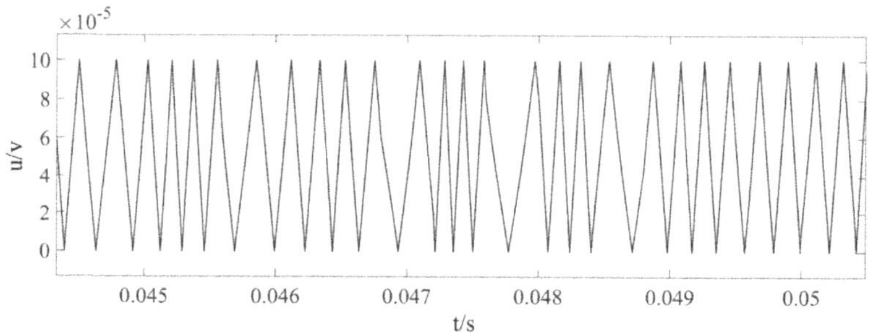

Fig. 4 Random carrier

harmonic energy and the change of harmonic amplitude in the three-phase current under different PWM modulation are shown in Fig. 5.

The FFT analysis of motor current generated by PWM modulation with fixed carrier is shown in Fig. 5a. It can be seen that there are obvious harmonic peaks and harmonic bands at the switching frequency and its integer multiples. The harmonic at the switching frequency and its integer multiple frequencies cause high-frequency noise of the motor, and affect the performance and control accuracy. The RSFPWM modulation effectively reduces the harmonic peaks in the current spectrum, and disperses the energy of the harmonic peaks into a wider frequency band as shown in Fig. 5b. However, there are still some obvious harmonic peaks and harmonic bands still in the frequency spectrum. Thus, the suppression of high-frequency noise is limited. The FFT analysis of the current under ERDPWM modulation is shown in Fig. 5c. It can be found that there is no obvious harmonic peaks in the entire frequency spectrum basically, and the harmonic band is relatively smooth. The ERDPWM modulation has obvious advantages in reducing the harmonic amplitude at the switching frequency and its integer multiples by dispersing the concentrated harmonic energy into a wider frequency spectrum. Moreover, the ERDPWM

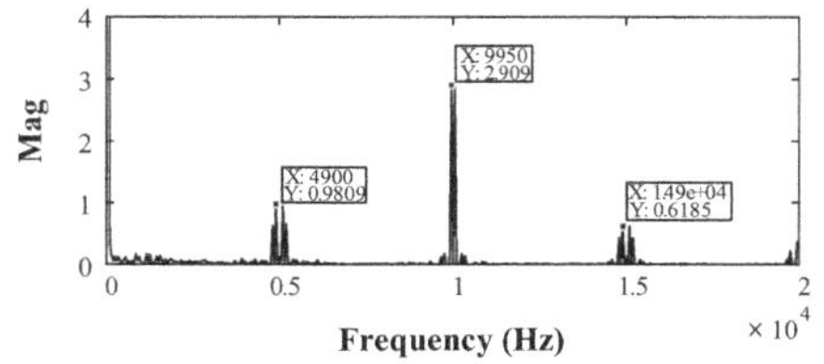

(a) PWM modulation

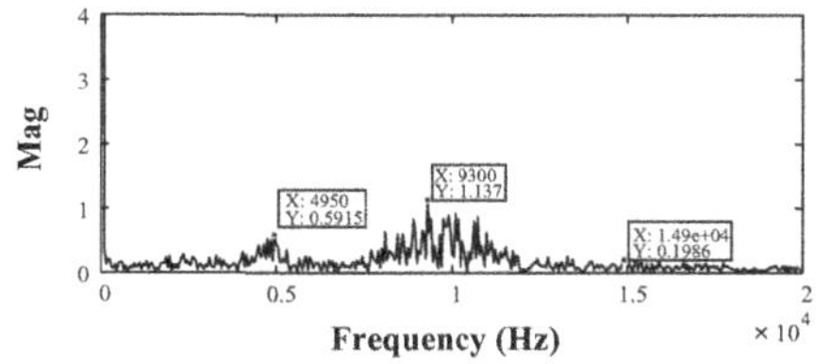

(b) RSFPWM modulation

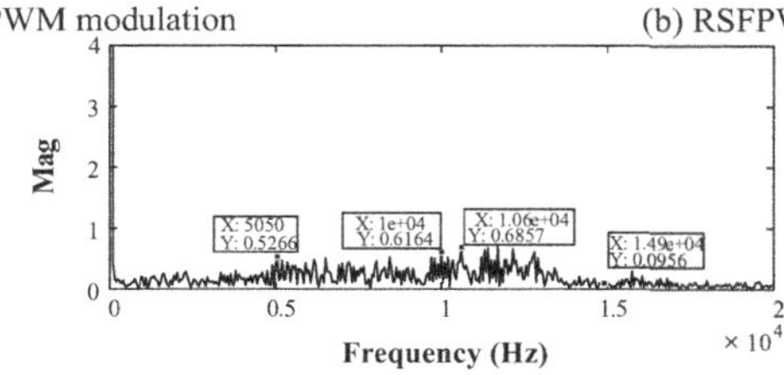

(c) ERDPWM modulation

Fig. 5 FFT analysis of random PWM modulation current

Table 1 Comparison of the Harmonic amplitude

	PWM	RSFPWM	Reduction rate %	ERDPWM	Reduction rate %
f_c	0.981	0.592	39.7	0.526	46.4
$2f_c$	2.909	1.137	60.9	0.616	78.8
$3f_c$	0.619	0.199	68.0	0.096	84.5

modulation does not affect the low-order harmonics, and will not introduce scattered harmonics or increase low-order harmonics.

Comparing and analyzing the harmonic amplitudes at the switching frequency and its double and triple frequency in the current harmonics under PWM modulation, RSFPWM modulation and ERDPWM modulation, the result is shown in Table 1.

The harmonic amplitudes of the three modulation methods gradually decrease at the same rated frequency. The ERDPWM modulation method has the largest reduction rate of the harmonic amplitude. And the harmonic peak dispersing effect of EREPWM is also the best. It is verified that ERDPWM modulation has good advantages in dispersing current harmonic peaks, and has a good suppression effect on the high frequency noise of the motor caused by PWM carriers.

5 Conclusion

This paper analyzes the relationship between current harmonics, vibration, and noise under fixed PWM carrier frequency and regular random carrier frequency. And it is concluded that the motor noise is positively correlated with the randomness of the carrier frequency. Based on the above analysis, this paper proposes an ERDPWM modulation method. This novel method replaces the center frequency with a sinusoidal signal and then combines the random numbers generated by the Markov chain algorithm to obtain an optimized sinusoidal random signal. This method can generate carriers with better frequency randomness. Therefore, the harmonics of the current obtained by ERDPWM modulation can be more evenly dispersed into a wider frequency spectrum, thereby suppressing the amplitude of a single spectral peak. In turn, the motor noise caused by high-frequency switching is well suppressed. Finally, this paper verifies the effectiveness of the method through simulation.

Acknowledgements This work was supported in part by Ningbo Science and Technology Planning Project (2018B10027), Ningbo Science and Technology Planning Project (2019B10071), Ningbo Innovation Team (2016B10016) and Chinese Academy of Sciences STS Project (KFJ-STS-ZDTP-066).

References

1. Wu, Y.X., M.A. Shafi, A.M. Knight, et al. 2011. Comparison of the effects of continuous and discontinuous PWM schemes on power losses of voltage-sourced inverters for induction motor drives. 26(1): 182–191.
2. Trzynadlowski, A.M., S. Legowski, L. Kirlin L. 1990. Random pulse-width modulation technique for voltage-controlled power inverters. 68(6): 1027–1037.
3. Biji, Jacob, M.R. Baiju. 2011. Space vector based pulse density modulation scheme fpr two level voltage source inverter. 6: 1230–1231.
4. Jayamala, V., S. Ramasamy, and S. Jeevananthan. 2011. Investigation of pseudorandom carrier I width modulation technique for induction motor drives. 12(1): 08–10.
5. Accardo, L., M. Fioretto, G. Giannini, and P. Marino. 2006. Techniques of PWM space vector modulation for the reduction of magnetical acoustic noise in traction motros. 25(1): 1084–1089.
6. Zhang, Zhigang, Shoudao Huang, Cungang Hu, et al. 2015. Hybrid method of three-level SHEPWM and SVPWM and its vector smooth-switching. 030(014): 342–349 (in Chinese).
7. Stankovi'c Aleksandar, M. 1997. Randomized modulation of power converters via markov chains. 5(1): 61–73.
8. Blaabjierg, F., J.K. Pedersen, E. Ritchie. 1995. Determination of mechanical resonances in induction motors by random modulation and acoustic measurement. 31(4): 823–829.
9. Kaboli, S., J. Mahdavi, A. Agah. 2007. Application of random PWM technique for reducing the conducted electromagnetic emissions in active filters. 54(4): 2333–2343.
10. Jiang, D., R. Lai, F. Wang, et al. 2011. Study of conducted EMI reduction for three-phase active frontend rectifier. 26(12): 3823–3831.
11. Mohan, V., N. Stalin, and S. Jeevananthan. 2015. A tactical chaos based PWM technique for distortion restraint and power spectrum shaping in induction motor drives. 5(3): 383–392.
12. Gao, Zhan, Qiongxuan Ge, Yaohua Li, et al. 2020. A carrier-based SVPWM begins with the zero voltage vector for three-level neutral point clamped converter. 35(10): 2194–2205 (in Chinese).
13. Jayamala, S., S. Ramasamy, and S. Jeevananthan. 2010. Investigation of pseudorandom carrier pulse width modulation technique for induction motor drives. 25(4): 753–761.

Analysis of Electric Field Focalization Induced by Transcranial Magneto-Acoustical-Electrical Stimulation Parameters

Zhang Shuai, Li Mengdi, Wang Yixiao, and Wang Junjie

Abstract Transcranial magnetic-acoustical-electrical stimulation is a new noninvasive brain modulation method with high spatial resolution and high penetration depth. However, the effect of electric field focalization induced by transcranial magneto-acoustical-electrical stimulation parameters is still unclear. In this paper, the electric field generated by transcranial magnetic-acoustical-electrical stimulation is simulated. The influence of the phased array transducer parameters on the induced electric field distribution is analyzed. The induced electric field distribution caused by different phased array parameters was calculated. The distribution characteristics of the induced electric field are analyzed from the radial and axial directions of the preset focusing point. The simulation results show that induced electric field intensity in the focusing region increases with the length and width of array elements and decreases with the array gap widening. There is a significant negative correlation between the size of focal region and the gap of the array elements. The correlation between phased array parameters and focusing performance is analyzed in this paper, which provides reference and theoretical guidance for research on improving the focusing property of induced electric field.

Keywords Transcranial magnetic-acoustical electrical stimulation · Induced electric field distribution · Electric field simulation

1 Introduction

At present, neurological diseases have become one of the main diseases that endanger human health. Relative lack of domestic mental health service resources, mental illness treatment with high recurrence rate and low treatment problems [1]. Traditional drugs are generally difficult to cure neuropsychiatric diseases and long-term

Z. Shuai (✉) · L. Mengdi · W. Yixiao · W. Junjie
Key Laboratory of Bioelectromagnetic Technology and Neural Engineering of Hebei, Province
Hebei University of Technology, Tianjin 300130, China
e-mail: zs@hebut.edu.cn

Tianjin Key Laboratory of Bioelectrical and Intelligent Health, Tianjin 300130, China

W. Chen et al. (eds.), *The Proceedings of the 9th Frontier Academic Forum of Electrical Engineering*, Lecture Notes in Electrical Engineering 743,
https://doi.org/10.1007/978-981-33-6609-1_61

taking accompanies with side effects. With the increasing demand of mental health, China has made a Brain Initiative to focus on the development of mental discipline. As one of the important means of brain neuromodulation technique is widely used in the treatment of Alzheimer's disease, Parkinson's disease, epilepsy, depression and other mental diseases, with significant efficacy and a wide range of application prospects [2–4].

Noninvasive brain modulation techniques are commonly used in the diagnosis and treatment of mental disorders in the field of brain science. It directly regulates abnormal neural activity based on the principle of electromagnetic induction. By means of magnetic field or induction electric field, it stimulates the specific target area that causes the disease and then changes the discharge activity of nerve tissue, so as to achieve the purpose of regulating brain function. Transcranial magnetic stimulation (TMS), the earliest NBM technology, plays an important role in exploring brain function, remodeling brain network, and treating brain diseases, but the depth of penetration limits the therapeutic effect [5–7]. Transcranial direct current stimulation (tDCS) regulates the activity of nerve tissue by regulating the excitability of neurons through the current transferred from electrodes to brain regions. Transcranial ultrasound stimulation (TUS) uses low-frequency ultrasound to penetrate the skull to stimulate the local brain region, which has the characteristics of high resolution. However, the neuroregulatory mechanism and implementation means of TUS are still in the primary research stage [8–10].

Transcranial magnetic-acoustical-electrical stimulation (TMAES) is a new neuromodulation technique that can stimulate nervous tissue safely and noninvasively. As a new, safe and noninvasive neuromodulation technique, TMAES technology combines the high stimulation depth of TUS with the high spatial resolution of TMS. It is expected to overcome the limitations of TUS and TMS, showing a broad development prospect and research value in brain science research and clinical application [11, 12]. The ultrasonic phase-controlled array probe is composed of several small piezoelectric ceramic pieces embedded together and mainly adopts the electronic phase-controlled focusing method. The focus of the beam can be realized by changing the delay time of the excitation pulse and adjusting the phase difference between the array [13]. The phased array focusing technology can improve the flexibility of ultrasonic focusing to a certain extent and obtain a higher resolution of sound field distribution. The effective stimulus depth and focusing characteristics are affected by the structural parameters of the ultrasonic probe. In this paper, the magneto-acoustic electric stimulation technology based on ultrasonic phased array is explored by changing the structural parameters of the array. It is helpful to understand the effects of different stimulation parameters on the regulation of neuroelectrophysiological activities. It provides some reference for improving the precision of neuromodulation in clinical practice.

2 Model and Methods

2.1 Principle of Transcranial Magnetic Acoustic and Electrical Stimulation

TMAES is a means to apply ultrasound to the nerve tissue in the static magnetic field to make the charged particles vibrate to generate induced current, and then regulate the nerve. The focused ultrasonic wave acts on the charged particles and makes them move along the direction of the vertical magnetic field. Under the action of Lorentz force, positive and negative charges move in the opposite direction, thus generating the induced electric field. The schematic diagram shows in Fig. 1.

Suppose the particle with charge q in the nerve tissue vibrates at velocity v in the ultrasonic focusing region, and the strength of the static magnetic field is B. When the Lorentz force on the ultrasonic wave, the static magnetic field and the particle are orthogonal, the Lorentz force conforms to the following relation:

$$F = qv \times B \tag{1}$$

The amplitude of particle vibration velocity v and the ultrasonic sound pressure p in the ultrasonic focusing region meet the following requirements:

$$v = \frac{p}{\rho c_0} \tag{2}$$

where ρ and c_0 represent the tissue density and the propagation velocity of ultrasonic wave in neural tissue, and can be obtained by combining the first two equations:

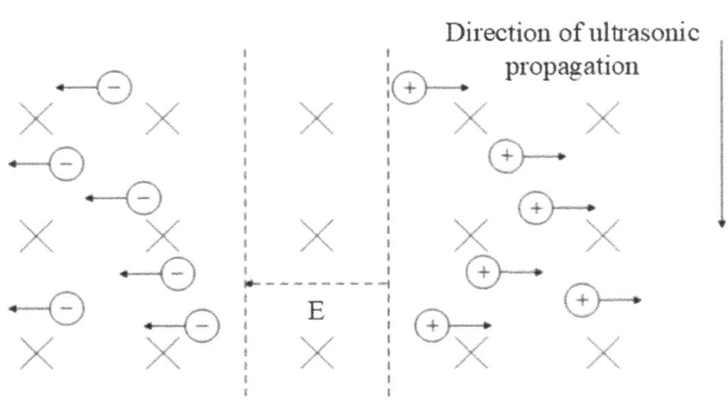

Fig. 1 Schematic diagram of transcranial magnetic acoustic electrical stimulation

$$F = q\frac{Bp}{\rho c_0} \tag{3}$$

According to Eq. (3), the induced electric field E generated in the nerve tissue can be expressed as:

$$E = \frac{Bp}{\rho c_0} \tag{4}$$

According to Eq. (4), the induced electric field distribution can be calculated according to the relationship between the distribution of focused sound field.

2.2 Array Transducer and Parameters

As shown in Fig. 2, there are four main parameters of array transducer, namely, array number N, the length of a single array element: L, the width of a single array element: a and the gap of a single array element: d. The spatial distribution of ultrasound generated by array transducers is related to the natural frequency f, sound velocity c_0, array number N, array length L, array width a and array interval d of ultrasonic transducers, which is called the spatial distribution characteristics of array transducers [14]. The ultrasonic focusing effect generated by array transducer can be obtained by calculating the sound pressure superposition of a single array element [15]. In this paper, the directional coefficient function and the directivity function $D_b(\theta)$ are introduced to represent the beam propagation property. According to the definition of directivity function and sound field distribution, the normalized ultrasonic focusing directivity function $D(\theta,\varphi)$ can be deduced.

Where the direction coefficient function $D_b(\theta)$ is:

$$D_b(\theta) = \frac{\sin(\frac{2\pi b}{\lambda}\sin\theta)}{\frac{2\pi b}{\lambda}\sin\theta} \tag{5}$$

where b, λ and θ are half of the transducer length, the ultrasonic wavelength, and the angle between the sound wave and the central axis.

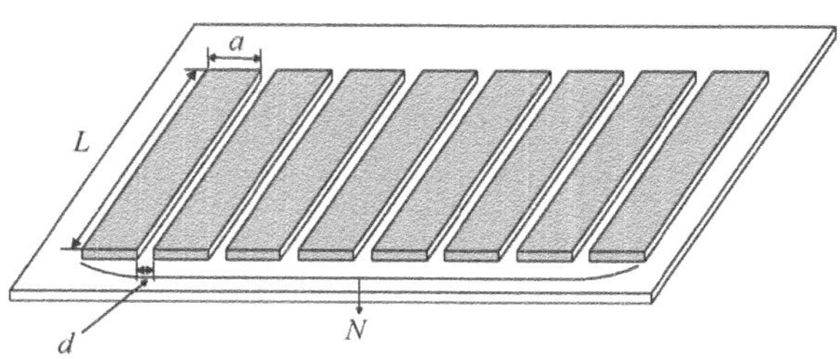

Fig. 2 Array transducer model

Table 1 Organizational parameters

Medium	Density (kg/m^3)	Propagation velocity (m/s)
Scalp	1060	1585
Harnpan	1912	2300
Brain tissue	1030	1550

The normalized ultrasonic focusing directivity function $D_B(\theta,\varphi)$ is:

$$D_B(\theta,\varphi) = \frac{D_{B1}(\theta,\varphi)\cdot D_{B2}(\theta,\varphi)}{D_{\frac{a}{2}}(\theta)} \quad (6)$$

where $D_{B1}(\theta,\varphi)$ and $D_{B2}(\theta,\varphi)$ respectively represent the directivity function of a single rectangular matrix when focusing and the directivity function of a simple point source linear array when focusing.

In order to make the simulation closer to the real situation, the attenuation of ultrasound caused by Skull, scalp and other factors should be considered [16], In this paper, the media consists of scalp, skull and brain tissue. The skull is 15 mm away from the array transducer, and the skull thickness is 5 mm. The tissue parameters are shown in Table 1.

The linear propagation of low-intensity focused ultrasound can be seen in the tissue. Since the amplitude of S-wave is much smaller than that of P-wave, the nonlinear effect and S-wave are ignored in the simulation process [17]. In order to minimize the attenuation of signals in the skull and brain tissue, the ultrasonic frequency should be less than 1 MHz.

3 Influence of Array Transducer Parameters on Electric Field Distribution

3.1 Influence of the Length of a Single Array Element on Electric Field

In this study, the preset focal length F is 3.6 cm, and the array parameters are selected as $N = 16$, $a = 0.5\lambda$, $d = 0.25\lambda$, $c_0 = 1500$ m/s and $f = 0.5$ MHz. The L is taken as 5 times, 7.5 times and 10 times a.

Figure 3 shows the variation of induced electric field distribution with L. It can be seen from the figure that the increase of L can not only improve the induced electric field intensity in the focal area, but also effectively suppress the induced electric field distributed in the near-field area.

Figure 4 shows the radial and axial distribution of the preset focus. It can be seen from the figure that the amplitude of the radial main side lobe is positively

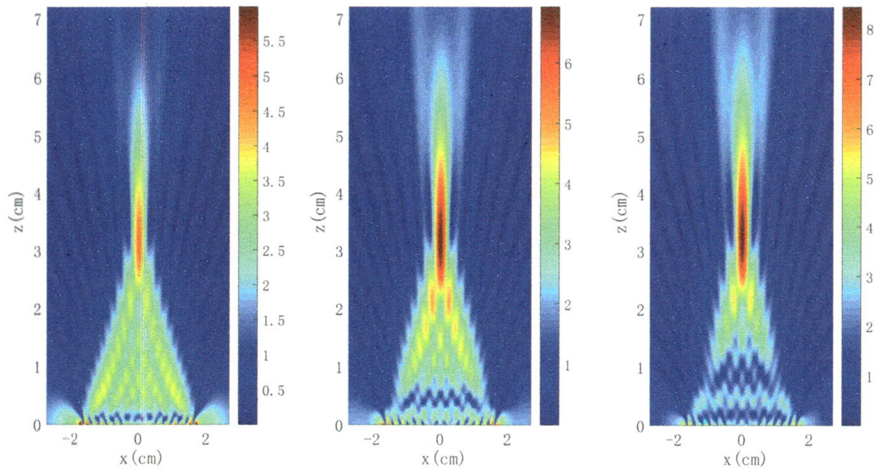

Fig. 3 Distribution of induced electric field when *L* is 5, 7.5 and 10 times *a*

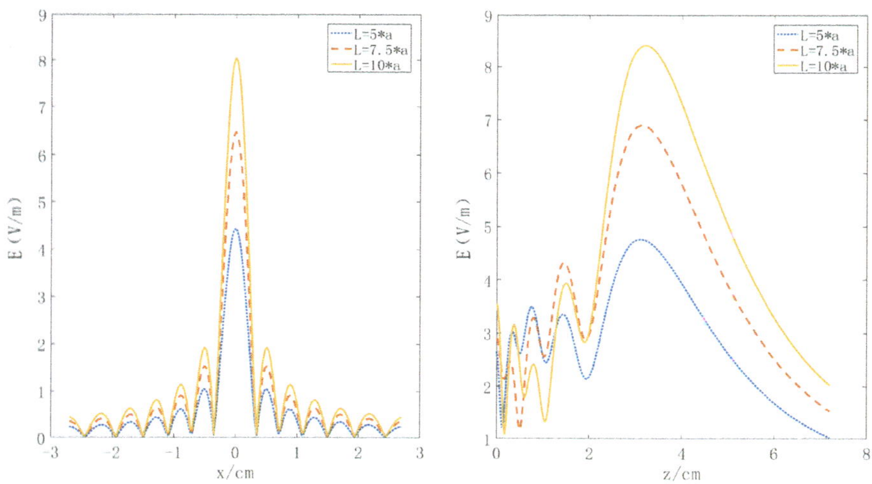

Fig. 4 Radial and axial distribution of induced electric-field

correlated with L. The increase of L can effectively suppress the induced electric field distribution within the axial focal length.

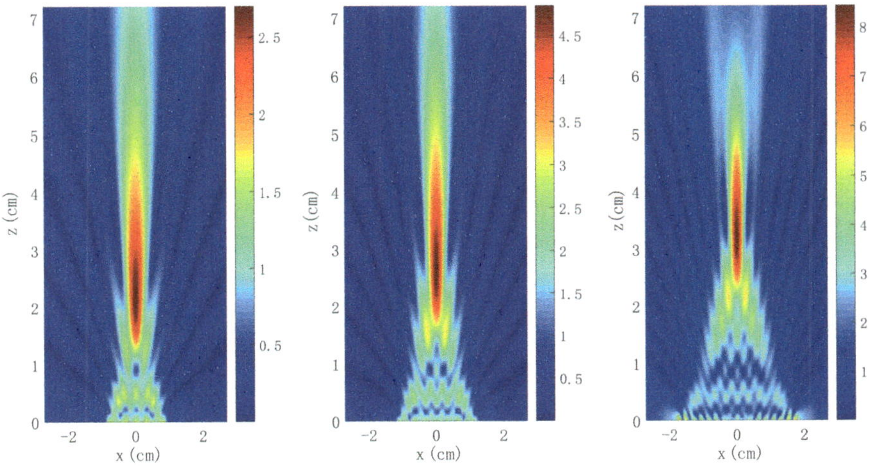

Fig. 5 Distribution of induced electric field when a is 0.125, 0.25 and 0.5 times λ

3.2 Influence of Array Width on Electric Field

In this study, the preset focal length F is 3.6 cm. Select array parameters as $N = 16$, $L = 5\lambda$, $d = 0.25\lambda$, $c_0 = 1500$ m/s and $f = 0.5$ MHz. The array element width a is taken as 0.125, 0.25 and 0.5 times λ.

Figure 5 shows the variation of induced electric field distribution with A. It can be seen from the figure that with the increase of A, the focal area gradually shrinks and the amplitude of the induced electric field doubles.

As the left of Fig. 6 shows, as the width of array element increases, the amplitude of main lobe increases and the distribution range Narrows, the width of the main lobe gradually decreases, and the energy becomes more concentrated.

From the right of Fig. 6, with the increase of the array width, the focusing depth gradually increases while the induced electric field increases, the length of the focusing area gradually decreases, and the attenuation in the skull and scalp decreases gradually. Therefore, in order to improve the intensity and depth of the induced electric field in the beam focusing region, the width of each array element should be appropriately enlarged.

3.3 Influence of Array Gap on Electric Field

In this study, the preset focal length F is 3.6 cm. The array parameters are $N = 16$, $L = 10a$, $a = 0.5\lambda$, $c_0 = 1500$ m/s and $f = 0.5$ MHz. The array element spacing d is taken as 0.67λ, 0.5λ and 0.25λ.

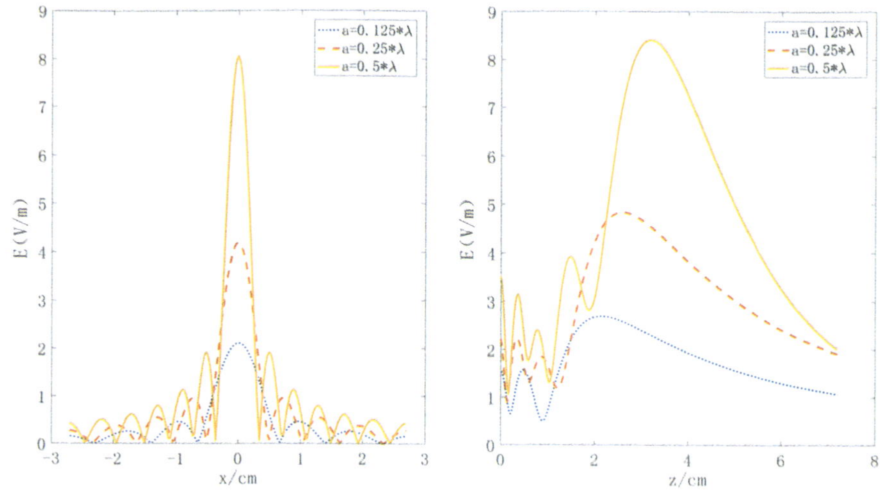

Fig. 6 Radial and axial distribution of induced electric-field

As shown in Fig. 7, the reduction of array clearance can effectively inhibit the distribution of electric field side lobes. But it increases the area of focus.

As the left of Fig. 8 shows, the width of main lobe and side lobe is mainly affected by the reduction of matrix gap. With the increase of gap, the width of the main lobe decreases, the side lobe steps increase, and the energy proportion of the main lobe decreases. From the right of Fig. 8, with the increase of the gap, the focusing depth gradually increases, but the energy attenuation is also obvious, the length of the

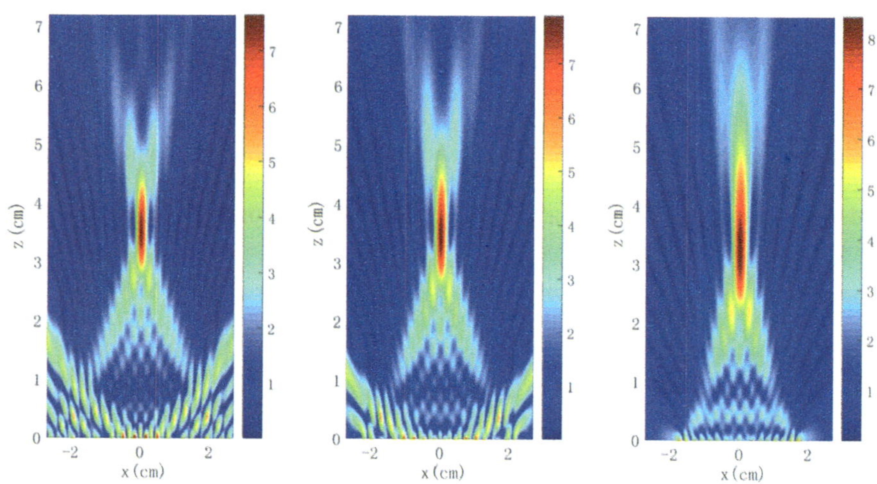

Fig. 7 Distribution of induced electric field when d is 0.67,0,5 and 0.25 times λ

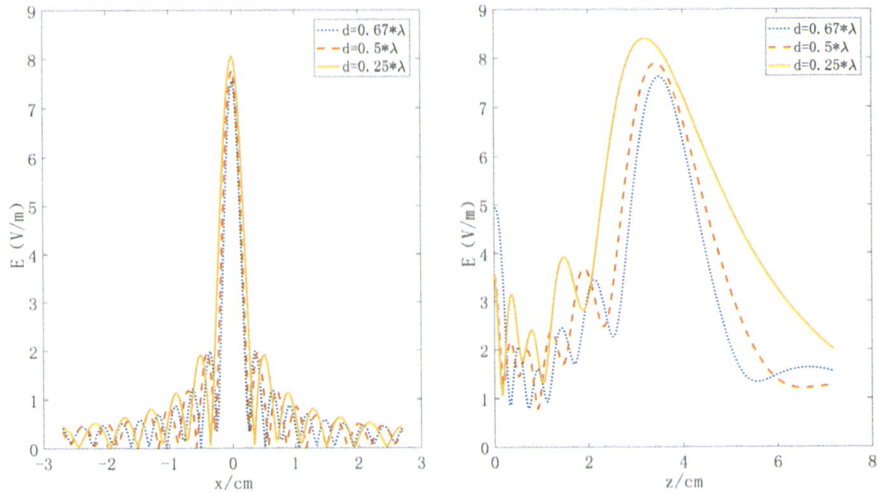

Fig. 8 Radial and axial distribution of induced electric-field

focusing region decreases, and the induced electric field intensity in the focusing region decreases. Therefore, in order to improve the beam focusing, the spacing of each array should be reduced appropriately.

4 Conclusion

In this paper, the induced electric field generated by the ultrasonic phased array transducer and the magnetostatic field acting together in the neural tissue is simulated and analyzed. The induced electric field distribution generated by the different phased array structure parameters in the biological tissue is calculated. The simulation results show that the change of array elements length only impacts the induced electric field intensity about the focusing region. The increase of array elements width can improve the depth of the focal region and increase the induced electric field intensity in the focusing region. It can also reduce the width of the main lobe and improve the focusing effect. Compared with the width of array elements, the increase of the elements gap has less influence on the main lobe and side lobes. As the interval widens, the focal length increases and the intensity of the induced electric field about the focusing region decreases. The simulation results provide reference for further exploring the distribution characteristics of the induced electric field about the ultrasonic focusing region of TMAES. The results also provide guidance and help for the research on improving the focusing property of the induced electric field.

Acknowledgements This work was supported in part by the National Natural Science Foundation of China under Grant (51877069).

References

1. Lin, Youyi, Shanjia Chen, and Haiyan Song. 2020. Advances of non-invasive brain stimulation therapies for disorders of consciousness aftertraumatic brain injury. *Chinese Journal of Practical Nervous Diseases* 23 (10): 917–920 (in Chinese).
2. Weng, Chun-xiao Fan, and Xiao-dong Hou Bing. 2015. Modulation of brain functions by transcranial stimulations. *BME & Clinical Medicine* 19(02): 196–200 (in Chinese).
3. Lin, Yingxuan, Peng Yan, and Lin Hua. 2019. The present situation and future of non-invasing neuro-regulation technology. *Journal of Brain and Nervous Diseases* 27 (02): 114–116 (in Chinese).
4. Li, Guofeng. 2018. *Non-invasive ultrasound brain neuromodulation methods for small animals [Ph.D]*. Chinese Academy of Sciences (Shenzhen Institutes of Advanced Technology) (in Chinese).
5. Zhang, Shuai, Gao, Xinyu, Zhou, Zhenyu, Liu, Yaze, Xu, Guizhi. 2019. Effect of transcranial magnetic-acoustic electrical stimulation on neuronal discharge activity based on GrC model. *Transactions of China Electrotechnical Society* 34(17): 3572–3580 (in Chinese).
6. Zhang, Shuai, Li, Zixiu, Zhang, Xueying, Zhao, Mingkang, Xu, Guizhi. 2019. The simulation and experiment of magneto-motive ultrasound imaging based on time reversal method. *Transactions of China Electrotechnical Society* 34(16): 3303–3310 (in Chinese).
7. Shuai, Zhang, Cui Kun, Shi Xun, Xu. Wang Zhuo, and Guizhi. . 2019. Effect analysis of transcranial magneto-acousto-electrical stimulation parameters on neural firing patterns. *Transactions of China Electrotechnical Society* 34 (18): 3741–3749 (in Chinese).
8. Eloise, Anguluan, Young-Jin, Jung, Axel Yen, Garcia, et al. 2019. Brain functional connectivity changes induced by transcranial ultrasound stimulation. 6(Supl.): S296–S296.
9. Hongchae, Baek, Ki Joo, Pahk, Hyungmin, Kim. 2017. A review of low-intensity focused ultrasound for neuromodulation. *The Korean Society of Medical and Biological Engineering* 7(2).
10. Rezayat Ehsan, Toostani Iman, Ghodrati. 2016. A review on brain stimulation using low intensity focused ultrasound. *Basic and Clinical Neuroscience* 7(3).
11. Wang, Hui-qin, Xiao-qing Zhou, Shi-kun Liu, Zhi-peng Liu, and Tao Yin. 2019. Comparative study of transcranial magnetic stimulation and transcranial ultrasound stimulation on induced EMG motion threshold. *Chinese Medical Equipment Journal* 40 (01): 14–19 (in Chinese).
12. Yuan, Yi, Yudong Chen, and Xiaoli Li. 2016. A new brain stimulation method: Noninvasive transcranial magneto-acoustical stimulation. *Chinese Physics B* 25 (8): 84301–084301 (in Chinese).
13. Bobola, M.S., L. Chen, C.K. Ezeokeke, T.A. Olmstead, C. Nguyen, A. Sahota, R.G. Williams, and P.D. Mourad. 2020. Transcranial focused ultrasound, pulsed at 40 Hz, activates microglia acutely and reduces Aβ load chronically, as demonstrated in vivo. *Brain Stimulation* 13(4) (in Chinese).
14. Wei, Wen-qing, Hong-wei Jing, and Fan Wu. 2018. Analysis of the effect of linear phased transducer array parameters on focusing property. *Technical Acoustics* 37(02): 192–199 (in Chinese).
15. Su, Chang, Zhefan Peng, and Weijun Lin. 2016. Accurate transcranial ultrasound focusing with time reversal based on virtual sources. *Journal of Applied Acoustics* 35 (05): 377–383 (in Chinese).
16. Tayel, M., N. Ismail, and A.A. Talaat. 2006. New ultrasound phased array technique for treat liver cancer using hyperthermia, 1–8. In *Twenty third national radio science conference*. IEEE.
17. Hakan Dogan, Viktor Popov. 2016. Numerical simulation of the nonlinear ultrasonic pressure wave propagation in a cavitating bubbly liquid inside a sonochemical reactor. *Ultrasonics—Sonochemistry* 30.

Torque Ripple Suppression of Low Voltage Permanent Magnet Synchronous Motor Based on Harmonic Voltage Injection

Wang Dongwen, Wang Lianghui, Zhou Wei, and Wang Weishen

Abstract Low voltage AC permanent magnet synchronous motor is widely used in military materials, medical devices, intelligent robots and other occasions because of its excellent performance and the characteristics of battery power supply. But the larger torque ripple also limits its application in some high-precision occasions. In order to solve this problem, based on the mathematical model of PMSM fundamental and harmonic, the physical and mathematical sources of torque ripple are analyzed. Based on the way of compensating harmonic voltage to the control system, the fifth and seventh harmonic components in the three-phase current of the motor are eliminated, so as to suppress the torque ripple. In the Simulink simulation and actual experiment, the output torque ripple amplitude of the motor is effectively reduced, which verifies the reliability of this method. This method is based on the software level algorithm, without adding any hardware equipment. The application of this algorithm effectively improves the sensitivity, torque accuracy and application range of low-voltage permanent magnet synchronous motor.

Keywords Torque ripple · Harmonic voltage · Low voltage · PMSM

1 Introduction

AC PMSM has an important position in the servo field due to its high power density, high efficiency and high reliability [1].The low-voltage high-current AC PMSM is widely used in some special occasions such as medical equipment, robots, AGV logistics vehicles, and military equipment [2]. However, in some occasions with high torque accuracy requirements, due to the design of the low-voltage PMSM body and the dead zone of the controller, the smoothness of the motor output torque is not

W. Dongwen · W. Lianghui (✉)
Automation Research and Design Institution of Metallurgical Industry, Beijing 100071, China
e-mail: 270671559@qq.com

Z. Wei · W. Weishen
A Military Bureau in Beijing, Beijing 100042, China

W. Chen et al. (eds.), *The Proceedings of the 9th Frontier Academic Forum of Electrical Engineering*, Lecture Notes in Electrical Engineering 743,
https://doi.org/10.1007/978-981-33-6609-1_62

ideal [3]. In the conventional control mode, excessive torque ripple cannot meet the application requirements of special occasions.

Experts and scholars at home and abroad have done a lot of research work on improving the current waveform and reducing the torque ripple. The torque ripple of PMSM mainly comes from two aspects. The first one is the cogging torque: asymmetrical stator cogging, imperfect winding distribution and core saturation will cause the air gap magnetic field waveform distortion [4]. The improvement method of cogging torque is mainly by optimizing the design of the motor body to optimize the sine of the air gap magnetic field [5]. The second is harmonic torque: The dead zone effect of power switching devices and the conduction voltage drop [6] will cause the 5th and 7th harmonics in the motor current, which will cause the back EMF and stator current to be distorted during the operation of the motor, which contains a large number of higher harmonics. High-order harmonics will produce torque ripple [7, 8]. The improvement method of harmonic torque is mainly through the use of no dead zone switch control mode, injecting harmonic current into the control loop, parallel resonance controller on the current control loop, and adding LC filter circuit in the three-phase main loop. However, these methods usually have complex algorithms, numerous devices, and poor practicability.

In order to better solve the problem of harmonic torque caused by the dead zone effect of power devices and the conduction voltage drop, this paper proposes a control strategy to compensate the 5th and 7th harmonic voltages in the control loop. The control unit superimposed on the harmonic compensation voltage can effectively eliminate the 5th and 7th harmonic currents in the main circuit, thereby suppressing the 6th pulsation component in the output torque. On the basis of the established PMSM mathematical model, in order to suppress the harmonic components in the motor current, the 5th and 7th harmonic currents are extracted in real time through the coordinate transformation method, and the extracted harmonic currents are respectively closed-loop control to obtain the corresponding harmonic voltage, and feedback this harmonic voltage in the control loop to offset the nonlinear problem of the switching device. This paper verifies the effectiveness of the algorithm through simulation and experiment. It can optimize the output torque ripple of the motor while also improving the sensitivity of the system. Moreover, the algorithm is simple, flexible, and has a wide range of applications, and does not require additional components, which has strong engineering practical significance.

2 PMSM Fundamental Wave and Harmonic Mathematical Model

In general motor theory, the electromagnetic torque of the motor is determined by the motor back EMF and the main loop current. The closer the back EMF and current are to the ideal sine wave, the smoother the electromagnetic torque. As mentioned in the introduction, the non-linear characteristics of the inverter itself will distort the output voltage waveform of the inverter, resulting in the 5th and 7th harmonic currents in the main circuit, which in turn causes the motor to produce 6th torque ripple. Therefore, this article starts from the PMSM fundamental wave mathematical model, and further establishes its 5th and 7th harmonic mathematical models, and proposes a control strategy to suppress torque ripple based on this mathematical model. Since this article mainly focuses on the optimization strategy of harmonic torque, the following assumptions are made in the analysis process: ① The motor body is an ideal model, the three-phase windings are distributed symmetrically, and the star connection; ② The motor core saturation effect can be Ignore and do not consider the various losses of the motor; ③ It is considered that the resistance and inductance parameters are constant during the operation of the motor; ④ The stator three-phase current does not contain even harmonic components.

2.1 Fundamental Voltage Equation of PMSM

The voltage equation in the PMSM three-phase stationary coordinate system is transformed to the d-q synchronous rotating coordinate system by constant amplitude as shown in formula (1):

$$\begin{cases} u_d = R_s i_d + L_d \frac{di_d}{dt} - \omega L_q i_q \\ u_q = R_s i_q + L_q \frac{di_q}{dt} + \omega L_d i_d + \omega \psi_f \end{cases} \tag{1}$$

In the above formula, u_d, u_q are the stator d and q axis voltages; i_d, i_q are the stator d and q axis currents; L_d, L_q are d and q axis inductances; ψ_f is the rotor permanent magnet flux linkage; R_s is tator winding resistance; ω is the rotational angular velocity of the fundamental wave.

Motor electromagnetic torque equation:

$$T_e = 1.5 p i_q \psi_f \tag{2}$$

p is the number of motor pole pairs.

In the case that only the fundamental wave is considered, the above voltage equations are all DC constants. Therefore, the electromagnetic torque is also constant. When the three-phase voltage contains high-order harmonics, the voltage and current are no longer pure direct currents after transformation by formula (1). It will inevitably cause the electromagnetic torque to fluctuate, which is also the mathematical source of torque ripple. It is known that the content of the three-phase voltage is all odd high-order harmonics, and because the winding adopts the star connection method, the motor will not contain the harmonics of multiples of 3 and 3. And the higher the harmonic order, the smaller its amplitude. Therefore, the 5th and 7th harmonics are the main research objects below.

2.2 The 5th and 7th Harmonic Voltage Equation of PMSM

In the stator A-B-C three-phase coordinate system, the rotation direction of the 5th harmonic voltage of the motor is opposite to the fundamental wave, and the rotation speed is 5ω. The same as 7ω. In the three-phase static coordinate system, only the fundamental, 5th and 7th harmonics are considered, the three-phase voltage of the permanent magnet synchronous motor can be expressed as:

$$\begin{cases} u_a = u_1 \sin(\omega t + \theta_1) + u_5 \sin(-5\omega t + \theta_5) + u_7 \sin(7\omega t + \theta_7) \\ u_b = u_1 \sin\left(\omega t + \theta_1 - \frac{2\pi}{3}\right) + u_5 \sin\left(-5\omega t + \theta_5 - \frac{2\pi}{3}\right) + u_7 \sin\left(5\omega t + \theta_7 - \frac{2\pi}{3}\right) \\ u_b = u_1 \sin\left(\omega t + \theta_1 + \frac{2\pi}{3}\right) + u_5 \sin\left(-5\omega t + \theta_5 + \frac{2\pi}{3}\right) + u_7 \sin\left(7\omega t + \theta_7 + \frac{2\pi}{3}\right) \end{cases} \tag{3}$$

In the above formula, u_1, u_5, u_7 are the amplitude of the corresponding harmonic voltage, $\theta_1, \theta_5, \theta_7$ is the initial phase angle corresponding to the harmonic voltage.

The Eq. (3) is transformed to the d-q synchronous rotating coordinate system by equal amplitude value, and then:

$$\begin{cases} u_d = u_{d1} + u_5 \cos(-6\omega t + \theta_5) + u_7 \cos(6\omega t + \theta_7) + \cdots \\ u_q = u_{q1} + u_5 \sin(-6\omega t + \theta_5) + u_7 \sin(6\omega t + \theta_7) + \cdots \end{cases} \tag{4}$$

In the above formula, u_{d1}, u_{q1} are the d and q axis components of the fundamental voltage.

In the same way, the current equation in the d-q synchronous rotating coordinate system is obtained:

$$\begin{cases} i_d = i_{d1} + i_5 \cos(-6\omega t + \theta_{i5}) + i_7 \cos(6\omega t + \theta_{i7}) + \cdots \\ i_q = i_{q1} + i_5 \sin(-6\omega t + \theta_{i5}) + i_7 \sin(6\omega t + \theta_{i7}) + \cdots \end{cases} \tag{5}$$

In the above formula, i_{d1}, i_{q1} are the d and q axis components of the fundamental current. i_5, i_7 are the amplitude of the 5th and 7th harmonic currents. θ_{i5}, θ_{i7} are the initial phase angle of the 5th and 7th harmonic currents.

It can be seen from Eqs. (4) and (5) that the distortion of the stator voltage and current waveform is due to the high-order harmonic components, and the most important component is the 5th and 7th harmonics. And in the fundamental synchronous rotating coordinate system, u_d, u_q, i_d, i_q all contain the 6th harmonic component, and then from the formula (2), the electromagnetic torque mainly contains the 6th harmonic component, which is the source of the 6th torque ripple in the previous section.

The current Eq. (5) is transformed into d-q coordinate and then substituted into Eq. (1) to obtain the voltage equation with harmonic components:

$$\begin{cases} u_d^* = -\omega L_q i_{q1} + R_s i_{d1} + 5\omega L_q i_5 \sin(-6\omega t + \theta_5) + R_s i_5 \cos(-6\omega t + \theta_5) \\ \quad -7\omega L_q i_7 \sin(6\omega t + \theta_7) + R_s i_7 \cos(6\omega t + \theta_7) \\ u_q^* = \omega L_d i_{d1} + R_s i_{q1} - 5\omega L_d i_5 \cos(-6\omega t + \theta_5) + R_s i_5 \sin(-6\omega t + \theta_5) \\ \quad +7\omega L_d i_7 \cos(6\omega t + \theta_7) + R_s i_7 \sin(6\omega t + \theta_7) \end{cases} \quad (6)$$

As the above formula, in the fundamental wave d-q synchronous rotating coordinate system, the fundamental wave component is the direct current, and the 5th and 7th harmonics are the alternating current.

2.3 Principle of 5th and 7th Coordinate Transformation

The voltage Eq. (6) contains the 5th and 7th harmonic currents. According to the rotating coordinate theory, the 5th harmonic current is AC in the fundamental wave and the 7th rotating coordinate system, and DC in the 5th rotating coordinate system; The 7th harmonic current is AC under the fundamental wave and the 5th rotating coordinate system, and DC under the 7th rotating coordinate system. As a result, the 5th and 7th harmonic currents can be extracted by rotating coordinate transformation, so as to perform corresponding harmonic voltage compensation respectively.

The transformation matrix from the fundamental synchronous rotating coordinate system to the 5th and 7th harmonic rotating coordinate system is shown in formulas (7) and (8):

$$T_{d^1q^1-d^5q^5} = \begin{bmatrix} \cos(-6\theta) & \sin(-6\theta) \\ -\sin(-6\theta) & \cos(-6\theta) \end{bmatrix} \quad (7)$$

$$T_{d^1q^1-d^7q^7} = \begin{bmatrix} \cos(6\theta) & \sin(6\theta) \\ -\sin(6\theta) & \cos(6\theta) \end{bmatrix} \quad (8)$$

Substituting the harmonic mathematical model of Eq. (6) into Eq. (7) and removing the AC component, the equation of the 5th harmonic current in the 5th rotating coordinate system can be obtained:

$$\begin{cases} u_{d5} = 5\omega L_q i_{q5} + R i_{d5} \\ u_{q5} = -5\omega L_d i_{d5} + R i_{q5} - 5\omega\psi_{f5} \end{cases} \quad (9)$$

In the same way, the equation of the 7th harmonic current in the 7th rotating coordinate system can be obtained:

$$\begin{cases} u_{d7} = -7\omega L_q i_{q7} + R i_{d7} \\ u_{q7} = 7\omega L_d i_{d7} + R i_{q7} + 7\omega\psi_{f7} \end{cases} \quad (10)$$

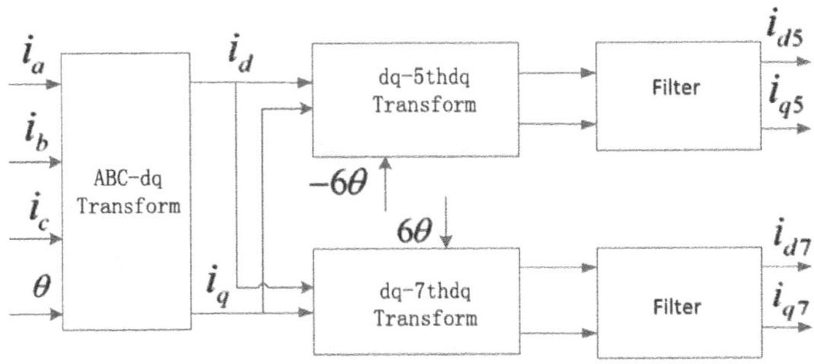

Fig. 1 Harmonic current extraction

3 Harmonic Voltage Compensation Strategy

3.1 Harmonic Current Extraction

To obtain the harmonic compensation voltage, the 5th and 7th harmonic currents must be extracted first. From Eqs. (7) and (8), the d and q axis components of the 5th and 7th currents can be obtained through coordinate transformation and filtering links. The harmonic current extraction module is shown in Fig. 1.

3.2 Harmonic Voltage Compensation

The compensation strategy is to superimpose the harmonic voltage feed forward compensation on the basis of the 5th and 7th harmonic current closed loop. The 5th harmonic voltage compensation module is shown in Fig. 2.

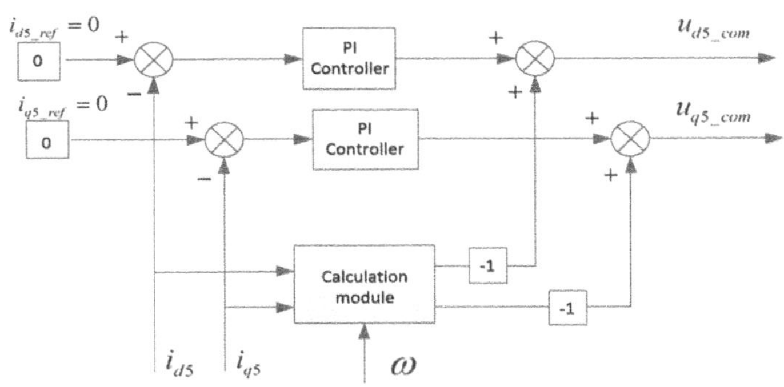

Fig. 2 Compensation voltage generation

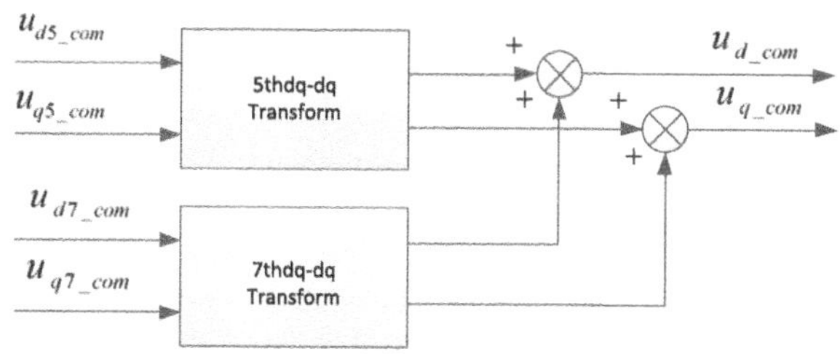

Fig. 3 Compensation voltage injection

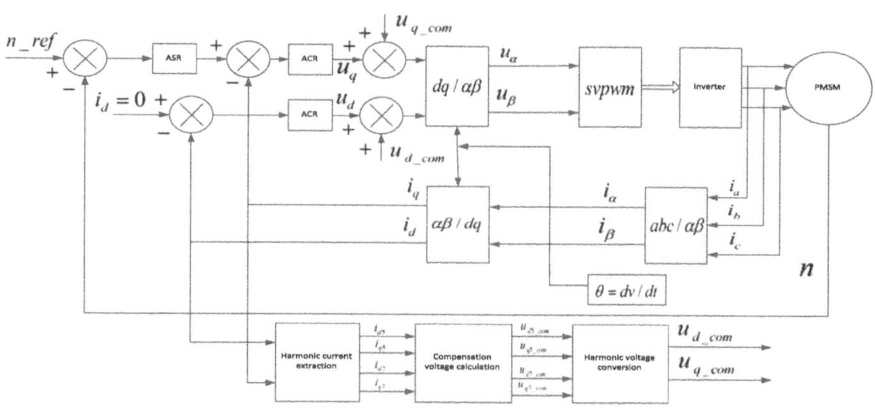

Fig. 4 Control system with voltage compensation

The same method can get the 7th harmonic voltage u_{d7_com}, u_{q7_com}. The 5th and 7th feedforward compensation voltage calculation modules are Eqs. (9) and (10).

Inverse transformation of 5th and 7th compensation voltage can get the voltage component in the fundamental wave coordinate system, Then it can be added to get the final compensation voltage u_{d_com}, u_{q_com}. Module as shown in Fig. 3.

In summary, the control strategy in this article is to add a harmonic current loop composed of a harmonic voltage compensation module on the basis of the double closed loop PI control.

Then inject the compensation voltage into the corresponding d-axis and q-axis voltage to form the entire voltage compensation system (Fig. 4).

4 Simulation

In order to verify the effectiveness of the above control strategy, this paper establishes a low-voltage PMSM model in Matlab/Simulink environment, and compares the compensation algorithm system described in the text with the original system, and analyzes the torque ripple amplitude and 5, THD value of 7th harmonic current.

The saliency PMSM is selected for simulation, and high-order harmonic current is generated by setting the inverter dead time and power device conduction. The relevant parameters are shown in Table 1.

The given speed is 1000 r/min and the load torque is 3 Nm. The output torque before and after the harmonic voltage compensation algorithm is added is shown in Figs. 5 and 6.

The algorithm improves the torque smoothness. Before optimization, the torque ripple amplitude is close to 1.2 Nm, and after optimization, the torque ripple amplitude becomes 0.3 Nm, which effectively reduces the torque ripple amplitude.

It can be seen from the waveform comparison before and after U-phase current optimization in Fig. 7 that the sine of the waveform is significantly improved, and the harmonic glitch suppression effect is obvious. Performing fast Fourier transform to obtain the current spectrum shows that the total harmonic distortion (THD) is reduced from 9.68 to 2.21% after adding the algorithm. The 5th and 7th harmonic components are reduced from 3.38 and 3.49% before the algorithm is added to 0.09 and 0.16%. It can also be found from the current comparison in Fig. 8 that the q-axis components of the 5th and 7th currents have dropped from the previous 1A and −1.1A to about 0A respectively.

Table 1 Motor and inverter parameter table

Parameter	Value
Stator resistance R/Ω	0.00177
Q-inductance $L_q/\mu H$	19.5
D-inductance $L_d/\mu H$	17.97
Number of pole pairs P	3
Load torque T_L/Nm	3.0
IGBT turn-on voltage drop/V	0.7
Diode conduction voltage drop/V	1.1
Dead time (μs)	2

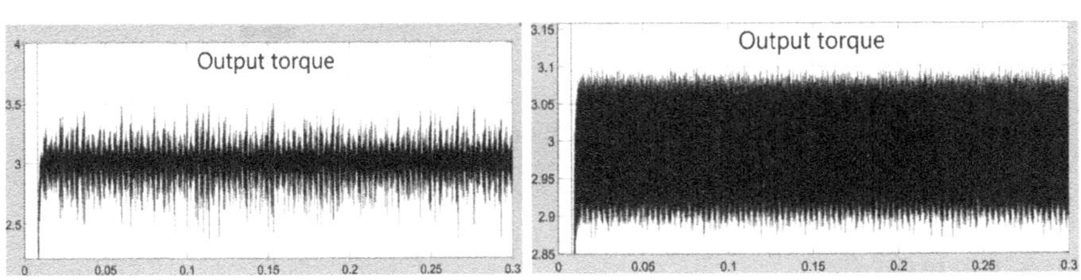

Fig. 5 Optimized torque waveform

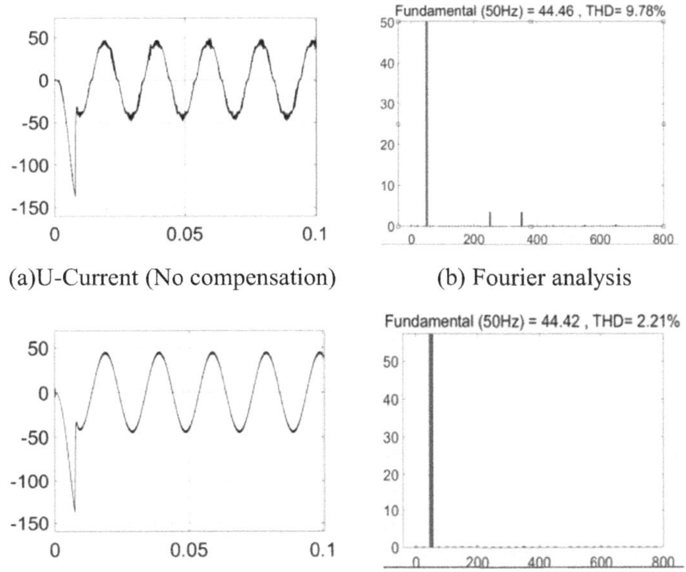

(a)U-Current (No compensation)
(b) Fourier analysis
(c) U-Current (Compensated)
(d) Fourier analysis

Fig. 6 U-phase current waveform and spectrum analysis results **a** 5th current (no—com.), **b** 7th current (no—com.), c 5th current (compensated), **d** 7th current (compensated)

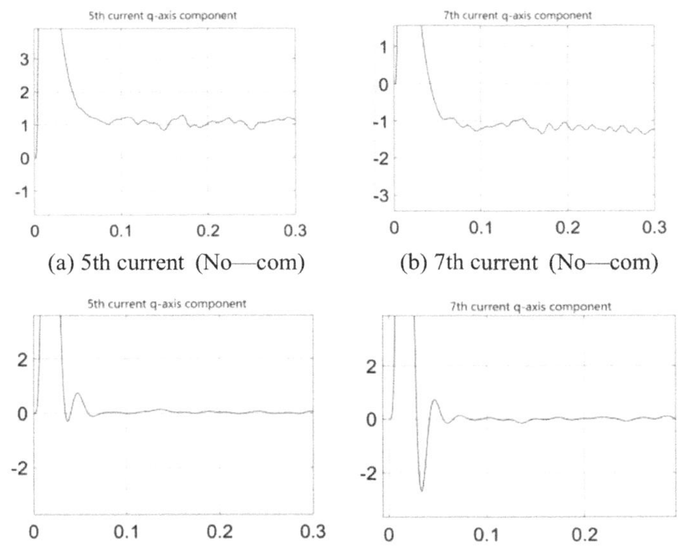

(a) 5th current (No—com)
(b) 7th current (No—com)
(c) 5th current (Compensated)
(d) 7th current (Compensated)

Fig. 7 5, 7 times q-axis current before and after optimized

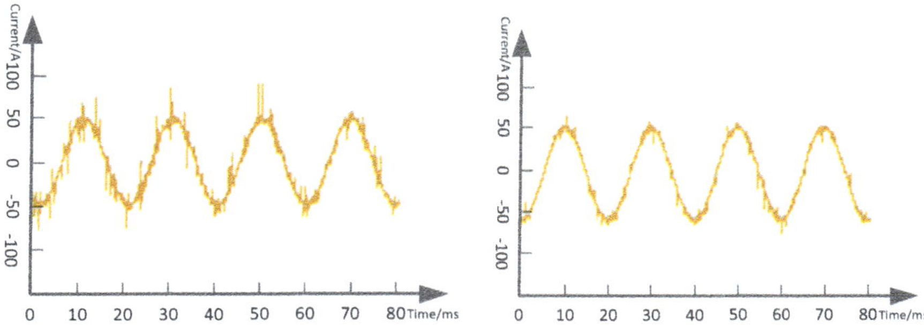

Fig. 8 Before and after U phase current is added to the algorithm

5 Experiment

In order to further verify the effectiveness and practicability of the above-mentioned harmonic suppression method, this paper uses a 1.6 kW PMSM vector control platform to experimentally verify the algorithm, and compares and analyzes the current waveform when the algorithm is not added.

The main control chip of the experiment adopts DSP-TMS320F283335 of TI Company, and the PWM frequency is set to 10 kHz. In the above experimental platform and parameter settings, the given speed is 1000r/min, the load torque is 3 Nm, and the U-phase current waveform before and after adding the algorithm is as follows.

It can be seen from the figure that the current harmonic content is significantly reduced after adding the harmonic suppression algorithm, and the sine of the waveform increases. The Fourier analysis of the waveform shows that the harmonic content is reduced from 15.76% before optimization to 7.97%. The optimization effect is more obvious, which reflects the practicality of the algorithm.

6 Conclusion

In order to reduce the amplitude of torque ripple and increase the scope of application of low-voltage permanent magnet synchronous motors, this paper analyzes the mathematical sources of torque ripple and proposes to inject harmonic compensation voltage to suppress the 5th and 7th harmonics in the phase current. Then cancel the 6th pulsation component of torque. This paper establishes a motor voltage model in a high-order rotating coordinate system, and superimpose it with the 5th and 7th harmonic voltage under id = 0 closed-loop control mode to obtain the d and q axis compensation voltage. After the compensation voltage is superimposed on the original control system, the simulation and experimental results show that the addition of this algorithm can well improve the current waveform distortion caused by the nonlinearity of the inverter, and also effectively suppress the torque ripple of the PMSM. So as to verify the feasibility of the algorithm.

References

1. Wu, Haokun, Keyuan Huang, Wei Lü, Xiaoling Mo, Shoudao H. Z-source Inverter DC-link voltage control strategy for high speed permanent magnet motor [J/OL]. *Transactions of China Electrotechnical Society*, 1–10 (in Chinese).
2. Han Jiahui. 2019. *Research and implementation of low voltage AC permanent magnet synchronous motor controller* (in Chinese).
3. Bao, Xiaohua, Jiwei Liu, Yue Sun, Changjiang Wu. Review and prospect of low-speed high-torque permanent magnet machines. *Transactions of China Electrotechnical Society* 34(6): 1148–1160 (in Chinese).
4. Luo, Honghao, Jun Wu, Wen-sen Chang. 2007. Minimization of cogging force in moving magnet type PMBLDCLM. *Proceedings of the CSEE* 6: 12–16 (in Chinese).
5. Bao, Xiaohua, Changjiang Wu, Jinlong Fang. 2018. Cogging torque reduction in surface-mounted permanent magnet synchronous motor by combining different permanent magnets in axial direction. *Transactions of China Electrotechnical Society* 33(18): 4231–4238 (in Chinese).
6. Pinzhi, Zhao, Yang Guijie, and Li. Yong. 2011. Third harmonic current restraining of five-phase permanent magnet synchronous motor based on dual synchronous rotating coordinates. *Proceedings of the CSEE* 31 (12): 71–76 (in Chinese).
7. Wu, Maogang, Rongxiang Zhao, Xinzhou Tang. 2005. Study of vector-controlled permanent magnet synchronous motor at low speed and light load. *Transactions of China Electrotechnical Society* 7: 87–92 (in Chinese).
8. Islam, M.S., S. Mir, T. Sebastian, et al. 2005. Design considerations of sinusoidally excited permanent-magnet machines for low-torque-ripple applicatons. *IEEE Transactions on Industry Applications* 41 (4): 955–962.

Low-speed Sensorless Control Method of SPMSM for Oil Pump Based on Improved Pulsating High-Frequency Voltage Injection

Xudong Liu, Pengjie Li, Peipei Dong, Guoqiang Xu, and Minghui Wang

Abstract Aiming at the problem that the sensorless surface-mounted permanent magnet synchronous motor has a certain failure rate and poor low-speed control performance, an improved sensorless rotor initial position detection and low-speed position tracking method is proposed. First of all, the pulsating high-frequency voltage is injected into the estimated coordinate system, and the relationship between current response and position error is obtained through the principle of flux linkage cross-coupling. The composite phase-locked loop rotor position observer is established by optimizing the filter and performing phase compensation of the modulation signal. Then, combined with pulse voltage vector injection, the conditions for determining the rotor polarity are given. This observer can accurately detect and track the rotor position from start to low speed in sensorle ss control. Finally, build a test platform for oil pump experiments, the experimental results verify that the method can realize the closed-loop start of the surface-mounted permanent magnet synchronous motor with large torque at rest, which improves the efficiency and robustness of low-speed operation.

Keywords Surface-mounted permanent magnet synchronous motor · Sensorless · Compound phase-locked loop · Pulse high-frequency voltage injection · Position detection

---

X. Liu (✉) · P. Li · G. Xu
No. 208 Research Institute of China Ordnance Industries, Beijing, China
e-mail: XudongLiu@No.208.com

P. Li
e-mail: PengjieLi@No.208.com

G. Xu
e-mail: GuoqiangXu@No.208.com

P. Dong
Huayu Sanden Automotive Air Conditioning Co., Ltd., Shanghai, China

M. Wang
School of Electrical Engineering and Automation, Harbin Institute of Technology, Harbin, China

W. Chen et al. (eds.), *The Proceedings of the 9th Frontier Academic Forum of Electrical Engineering*, Lecture Notes in Electrical Engineering 743,
https://doi.org/10.1007/978-981-33-6609-1_63

1 Intruduction

Surface-mounted permanent magnet synchronous motors (SPMSM) have the advantages of simple structure, wide speed range, high power density, etc. Therefore, they are widely used in the electric industry [1–3]. In order to overcome the shortcomings caused by the position sensor to the system, many scholars and engineers have devoted themselves to the research of position sensorless control technology. Commonly used position observers based on fundamental excitation models, such as sliding-mode observer [4, 5], luenberger observer [6], model reference adaptive control [7], etc. However it is only applicable in the middle and high speed section, so static startup and low speed operation are the hot and difficult points of this kind of technology research.

The high-frequency injection method based on non-ideal characteristics can demodulate the rotor position information from the feedback current/voltage according to the salient pole effect [8–10]. It does not rely on information such as the motor parameters and back electromotive force in the fundamental excitation model, and get better results in start-up and low-speed operation [11].

Reference [12] studies a SPMSM's rotor initial position detection method of pulse voltage injection, and proposes an improved five-voltage vector injection method. Select the optimized voltage vector for secondary injection and calculate, but there are higher requirements on the bandwidth and accuracy of the current sampling circuit. The traditional pulsating high-frequency voltage injection method extracts rotor angle information by injecting high-frequency voltage excitation into the stator. [13, 14], which can realize the start of SPMSM under heavy, but there are problems of phase delay and failure to judge the polarity of special positions.

In order to further improve the accuracy of rotor position estimation, this paper proposes an improved pulsating high-frequency voltage injection method. Starting from the analysis of the cross-coupling of the flux linkage of the motor, the high-frequency equivalent mathematical model of the motor is established. By optimizing the filter and performing phase compensation of the modulation signal, the rotor position estimation error is indirectly compensated. On this basis, combined with pulse voltage vector injection, the conditions for determining the rotor polarity are given. Emphasize the sequence of signal injection to ensure program coherence and improve the success rate of judgment. Experiments on the SPMSM for oil pump verify the effectiveness of this method and the advantages compared to traditional high frequency pulsation injection.

2 Saturation Salient Pole Effect and High Frequency Mathematical Model of SPMSM

2.1 Analysis of Saturation Salient Pole Effect of SPMSM

The working point of the d-axis magnetic circuit is usually set at the inflection point of the curve $\psi - i$ [15], and the working point of the q-axis magnetic circuit is set at the origin, which makes the d-axis magnetic link very sensitive [16, 17], but q-axis is not.

When the magnetic circuit of SPMSM is not saturated, the d-axis and q-axis flux linkages are not coupled [18, 19]; When the magnetic circuit of SPMSM is saturated, the d-axis and q-axis flux linkages are coupled, and there is mutual inductance, so their flux linkage formula is as follows:

$$\begin{cases} \psi_d = L_d i_d + M_{dq} i_q + \psi_{dr} \\ \psi_q = L_q i_q + M_{qd} i_q + \psi_{qr} \end{cases} \quad (1)$$

where ψ_d, ψ_q, L_d, L_q, i_d, i_q are the flux linkage, inductance and current of the d-axis and q-axis, respectively;M_{dq}, M_{qd} are the mutual inductance between d-axis and q-axis, $M_{dq} = M_{qd}$;ψ_{dr} and ψ_{qr} are the flux linkage components produced by the permanent magnet on the d-axis and q-axis. As shown in Fig. 1, it can be seen that the saturated saliency effect will cross-couple the magnetic chains, the d-axis self-inductance changes approximately sinusoidally with the rotor position.

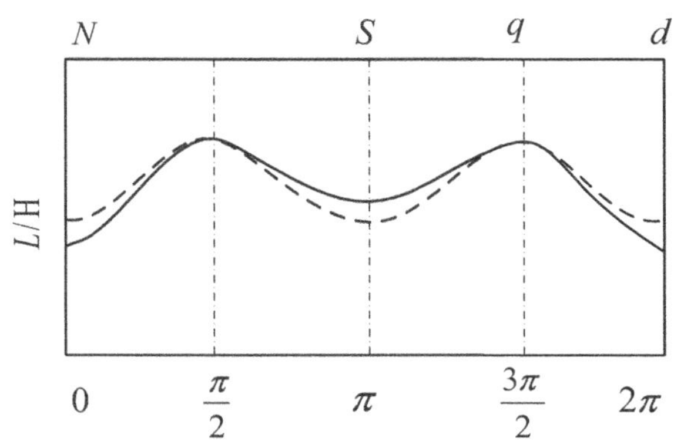

Fig. 1 Inductance change curve of SPMSM

2.2 High-Frequency Mathematical Model of SPMSM

When the injected voltage angular frequency ω_h is much higher than the motor rotation frequency ω_r, and is much lower than the chopping frequency of the inverter, The cross-coupling term of the SPMSM in the coordinate system $d-q$ is very small, The simplified voltage equation in the steady state in the form of impedance is:

$$\begin{bmatrix} u_{dh} \\ u_{qh} \end{bmatrix} = \begin{bmatrix} Z_{dh} i_{dh} \\ Z_{qh} i_{qh} \end{bmatrix} = \begin{bmatrix} R_{dh} + j\omega_h L_{dh} & 0 \\ 0 & R_{qh} + j\omega_h L_{qh} \end{bmatrix} \begin{bmatrix} i_{dh} \\ i_{qh} \end{bmatrix} \tag{2}$$

In the above expression, u_{dh}, u_{qh}; i_{dh},i_{qh}; R_{dh}, R_{qh}; L_{dh}, L_{qh} are the high-frequency components of the voltage, current, stator resistance and stator inductance under the axis system of $d-q$, respectively, p is the differential operator, Z_{dh} and Z_{qh} represent the high-frequency impedance of the d-axis and q-axis, respectively.

3 Position Detection Method Based on Pulsating High-Frequency Voltage Signal Injection

3.1 Principle of Pulsed High-frequency Voltage Injection

Define the real rotor position as θ and the estimated rotor position as $\hat{\theta}$, then the angle between the two coordinate systems is $\Delta\theta = \theta - \hat{\theta}$. The high-frequency current response is transformed into the estimated coordinate system $\hat{d} - \hat{q}$ as:

$$\begin{bmatrix} \hat{\mathrm{i}}_{\mathrm{dh}} \\ \hat{\mathrm{i}}_{\mathrm{qh}} \end{bmatrix} = \begin{bmatrix} \cos\Delta\theta & \sin\Delta\theta \\ -\sin\Delta\theta & \cos\Delta\theta \end{bmatrix} \begin{bmatrix} 1/Z_{dh} & 0 \\ 0 & 1/Z_{qh} \end{bmatrix} \begin{bmatrix} \cos\Delta\theta & -\sin\Delta\theta \\ \sin\Delta\theta & \cos\Delta\theta \end{bmatrix} \begin{bmatrix} \hat{u}_{dh} \\ \hat{u}_{qh} \end{bmatrix} \tag{3}$$

In the above formula, $\hat{u}_{dh}$, $\hat{u}_{qh}$ and $\hat{\mathrm{i}}_{\mathrm{dh}}$, $\hat{\mathrm{i}}_{\mathrm{qh}}$ respectively represent the injected voltage component and response current component in the coordinate system $\hat{d} - \hat{q}$, and then define:

$$\begin{bmatrix} Z_{avg} \\ Z_{dif} \end{bmatrix} = \frac{1}{2} \begin{bmatrix} Z_{dh} + Z_{qh} \\ Z_{dh} - Z_{qh} \end{bmatrix} \tag{4}$$

In Eq. (4), Z_{avg} and Z_{dif} represent the average and half-difference of high-frequency impedance between the d-axis and q-axis. Substituting Eq. (4) into Eq. (3), the following equation can be obtained as:

$$\begin{bmatrix} \hat{\mathrm{i}}_{\mathrm{dh}} \\ \hat{\mathrm{i}}_{\mathrm{qh}} \end{bmatrix} = \frac{1}{Z_{dh}Z_{qh}} \begin{bmatrix} (Z_{avg} - Z_{dif}\cos(2\Delta\theta))u_{dh} + (Z_{dif}\sin(2\Delta\theta))u_{qh} \\ (Z_{dif}\sin(2\Delta\theta))u_{dh} + (Z_{avg} + Z_{dif}\cos(2\Delta\theta))u_{qh} \end{bmatrix} \tag{5}$$

In order to reduce the torque ripple, the high-frequency voltage $U_{inj}\cos(\omega_h t)$ is injected into the $\hat{d}$-axis, U_{inj} and ω_h are the amplitude and angular frequency of the injected voltage signal, put them into Eq. (5) to get:

$$\begin{bmatrix} \hat{\mathrm{i}}_{\mathrm{dh}} \\ \hat{\mathrm{i}}_{\mathrm{qh}} \end{bmatrix} = \frac{U_{inj}\cos(\omega_h t)}{Z_{dh}Z_{qh}} \begin{bmatrix} Z_{avg} - Z_{dif}\cos(2\Delta\theta) \\ Z_{dif}\sin(2\Delta\theta) \end{bmatrix} \tag{6}$$

Obviously, $\hat{\mathrm{i}}_{\mathrm{dh}}$ and $\hat{\mathrm{i}}_{\mathrm{qh}}$, both of them contain rotor position error information. In addition to the injection signal parameters, $\hat{\mathrm{i}}_{\mathrm{dh}}$ is related to Z_{avg} and Z_{dif}, $\hat{\mathrm{i}}_{\mathrm{qh}}$ is only related to Z_{dif}. There is a difference between the real d-axis and q-axis high-frequency impedance, so when $\Delta\theta$ approaches 0, $\hat{\mathrm{i}}_{\mathrm{qh}}$ also approaches 0, and $\hat{\mathrm{i}}_{\mathrm{dh}}$ cannot approach 0. Therefore, select $\hat{\mathrm{i}}_{\mathrm{qh}}$ as the input signal of the rotor position observer to obtain the real rotor position error and speed information.

3.2 Rotor Position Observer Based on Compound Phase-Locked Loop

The current signal sampled by the circuit contains torque current $\hat{\mathrm{i}}_{\mathrm{qr}}$, high-frequency response current $\hat{\mathrm{i}}_{\mathrm{qh}}$, and high-frequency harmonic components $\hat{\mathrm{i}}_{\mathrm{qi}}$. The traditional second-order BPF will cause the signal amplitude to be attenuated by Λ_{b} times and the phase shifted by φ_r while extracting $\hat{\mathrm{i}}_{\mathrm{qh}}$, and because the motor is an inductive load, there is a phase delay angle $\varphi_{\hat{d}} + \varphi_{\hat{q}}$ between the injected $\hat{u}_{dh}$ and the response $\hat{\mathrm{i}}_{\mathrm{qh}}$. In the steady state, the high-frequency resistance is much smaller than the high-frequency inductive reactance. Ignoring the difference in high-frequency resistance, the simplified $\hat{q}$-axis high-frequency current response component can be expressed as:

$$\hat{\mathrm{i}}_{\mathrm{qh}} = \frac{\Lambda_{\mathrm{b}} U_{inj} L_{dif}\sin(\omega_h t - \varphi_b - \varphi_{\hat{d}} - \varphi_{\hat{q}})}{\omega_h L_{dh} L_{qh}}\sin(2\Delta\theta) \tag{7}$$

In the formula, $L_{dif} = (L_{dh} - L_{qh})/2$, which means the half-difference high-frequency inductance. When the rotor position error is very small, $\sin(2\Delta\theta) \approx 2\Delta\theta$, $L_{dif} < 0$,the other items are all positive values, but the polarity of $\sin(\omega_h t - \varphi_b - \varphi_{\hat{d}} - \varphi_{\hat{q}})$ is uncertain. This problem directly determines the pole distribution of the observation system and affects the stability of the system. Therefore, phase compensation is required for $\hat{\mathrm{i}}_{\mathrm{qh}}$.

In order to solve the phase shift problem caused by BPF, we no longer extract $\hat{i}_{qh}$, but directly multiply the sampled current signal with the sinusoidal modulation signal and passed through the LPF, $\hat{i}_{qr}$ becomes the AC component with the same frequency as the modulated signal and is filtered out, and $\hat{i}_{qi}$ is still the AC component with a frequency much higher than ω_h, which will also be filtered out. The rotor position error function $f(\Delta\theta)$ obtained is expressed as:

$$\begin{aligned} f(\Delta\theta) &= \mathrm{LPF}\left(\frac{U_{inj}L_{dif}}{2\omega_h L_{dh} L_{qh}}\left(1-\cos(2\omega_h t - 2\varphi^*)\right)\sin(2\Delta\theta)\right) \\ &= \frac{U_{inj}L_{dif}}{2\omega_h L_{dh} L_{qh}}\sin(2\Delta\theta) \end{aligned} \tag{8}$$

In the formula, $\varphi^* = \varphi_{\hat{d}} + \varphi_{\hat{q}})$. The determination of this value will be discussed later. When $f(\Delta\theta)$ is stabilized through PLL, the frequency difference between the input value and the feedback value of the speed tends to 0, the phase error converges to 0, and $f(\Delta\theta)$ linearization is: $f(\Delta\theta) \approx \frac{U_{inj}L_{dif}}{\omega_h L_{dh} L_{qh}}\Delta\theta = K_\varepsilon \Delta\theta$.

Where K_ε is the position error coefficient, which is only related to the SPMSM's own parameters and the injected high-frequency voltage parameters. At this time, the estimated rotor position $\hat{\theta}$ is the true rotor position.

The modulation signal needs to be strictly synchronized with $\hat{i}_{qh}$ to compensate the phase delay angle φ^*. From Eq. (6), when $\Delta\theta = 0$, $\hat{i}_{qh} = 0$, $\hat{i}_{dh} \neq 0$, and they have the same phase, then $\hat{i}_{dh}$ can be used as a PLL input signal to determine the phase delay angle φ^*, also ignoring the high-frequency resistance, $\hat{i}_{dh}$ can be expressed as:

$$\begin{aligned} \hat{i}_{dh} &= \frac{j \cdot U_{inj}\cos(\omega_h t - \varphi^*)}{-\omega_h L_{dh} L_{qh}}\left(L_{avg} - L_{dif}\cos(2\Delta\theta)\right) \\ &= \frac{U_{inj}\sin(\omega_h t - \varphi^*)}{\omega_h L_{dh}} \end{aligned} \tag{9}$$

Among them, $L_{dif} = \left(L_{dh} + L_{qh}\right)/2$ represents the average high frequency inductance. The $\hat{i}_{dh}$ is processed similarly to $\hat{i}_{qh}$, and the phase delay angle error function is obtained:

$$\begin{aligned} g(\Delta\varphi^*) &= \mathrm{LPF}\left(\hat{i}_{dh}\cdot\cos\left(\omega_h t - \hat{\varphi}^*\right)\right) \\ &= \mathrm{LPF}\left(\frac{U_{inj}}{2\omega_h L_{dh}}\left(\sin(2\omega_h t - \varphi^* - \hat{\varphi}^* + \sin\left(-\varphi^* + \hat{\varphi}^*\right)\right)\right) \\ &= \frac{U_{inj}}{2\omega_h L_{dh}}\sin(\hat{\varphi}^* - \varphi^*) = K_\xi \sin(\Delta\varphi^*) \end{aligned} \tag{10}$$

where K_ξ is the phase error coefficient, $\hat{\varphi}^*$ is the estimated phase delay angle. It can be seen that $g(\Delta\varphi^*) = 0 \Rightarrow \hat{\varphi}^* = \varphi^* + k\pi$, $(k = 0, 1, 2\ldots)$. The analysis shows that the system response is stable. $\hat{\varphi}^*$ can be used in the sinusoidal modulation signal

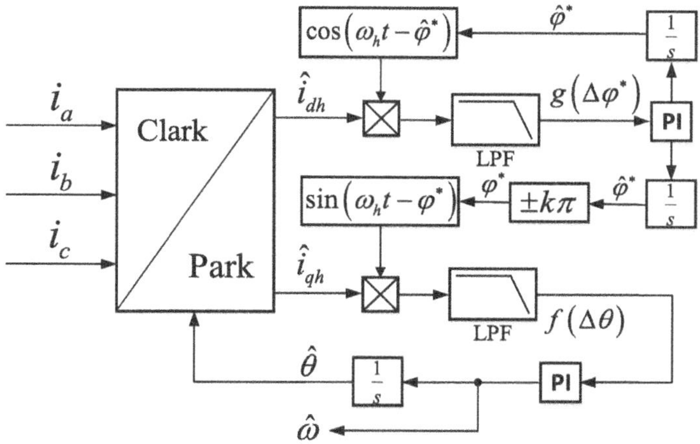

Fig. 2 Rotor position observer of compound phase-locked loop

of $\hat{i}_{qh}$ to reduce the current deviation caused by the current loop bandwidth. After omitting the BPF and improving the modulation signal, the composite phase-locked loop rotor position observer is shown in Fig. 2.

3.3 Judgment of Rotor Polarity

The saturated salient pole effect is symmetrical, and the condition that can make the Eq. (8) converge to 0 is $\Delta\theta = n\pi\,(n = 0, 1, 2, \ldots)$, so the actual N pole angle of the rotor may be $\hat{\theta}$ or $\hat{\theta}+\pi$. For this reason, a judgment method that avoids interruption of high-frequency signal injection and can perform multiple polarity checks is proposed to improve the success rate of polarity judgment. It should be noted that the pulse voltage vector injection must be performed before the high frequency injection to obtain the interval where the rotor N pole is located.

According to Fig. 1, the closer to the d-axis, the smaller the inductance, in the d-axis direction, when the applied voltage is in the same direction as the rotor N pole, the saturation increases and the inductance decreases; when the applied voltage is opposite to the rotor N pole, the main magnetic circuit desaturates and the inductance increases. Therefore $L_S > L_N$.

Under the condition of not causing the motor to rotate, use voltage vectors u_x with the same amplitude and the same duration at different angles to inject in the order shown in Fig. 3a. Calculate from the current components under the two-phase stationary shaft system to obtain the true current response, that is, $i_s^2 = i_\alpha^2 + i_\beta^2$. Compare the current response amplitude $i_{\max}$, the voltage vector corresponding to the maximum value is marked as $A_{\max}$; in order to improve the success rate of polarity judgment, take $A_{\max}$ as the center and inject the voltage vector according to the order

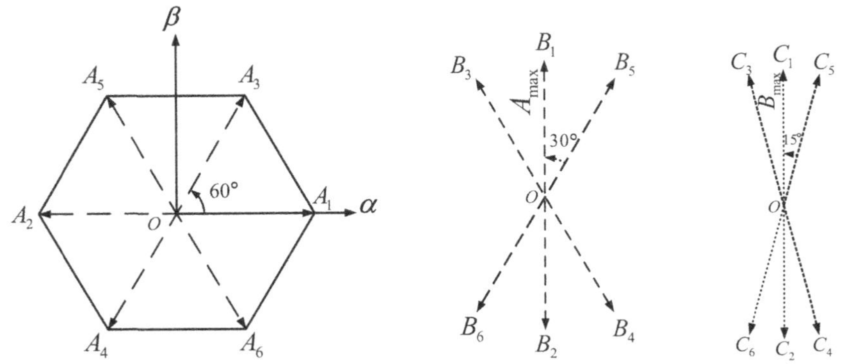

(a) 60°interval (b) 30°interval (c) 15°interval

Fig. 3 Injection sequence of pulse voltage vector

and angle shown in Fig. 3b, c. Similarly. Record the last position of the maximum current response as θ_e. Make the difference between θ_e and the angle of the maximum value of the current response injected in the previous rounds. If the angle-difference >90° more than <90°, then it will be injected again; If the angle-difference >90° less than <90°, save θ_e as the final rotor N pole position.

Combine $\hat{\theta}$ and θ_e, if $\left|\hat{\theta} - \theta_e\right| < 90°$, then $\hat{\theta}$ is the true N pole angle of the rotor; if $\left|\hat{\theta} - \theta_e\right| \geq 90°$, then $\hat{\theta} + \pi$ is the true N pole angle of the rotor. During this period, there is no need to interrupt the high-frequency voltage injection, the rotor polarity is determined at the moment when the rotor position $\hat{\theta}$ is located, and the starting step is entered.

4 Experimental Results and Analysis

In order to verify the correctness and feasibility of the above analysis, method verification was carried out on a 300 W oil pump load, and a motor sensorless low-speed control experimental test platform based on improved pulsating high-frequency voltage injection was built. The relevant parameters are as follows (Table 1).

The vector control strategy is $i_{\mathrm{d}} = 0$, Fig. 4 is the block diagram of the control system.

Figure 5a shows the injected voltage waveform U_{inj} and the current fundamental wave response i_{d} and i_{q} when the motor is at no-load 200 RPM. Figure 5b shows the actual rotor position and the estimated rotor position waveform of the motor, with an average error of 1.8°, which meets the needs of oil pump motor control.

The experimental waveform of the rotor initial position detection is shown in Fig. 6. Due to the influence of the motor cogging, the rotor usually stops at a certain

Table 1 Motor and control system parameters table

Motor parameters		Control system parameters	
Winding voltage	8.49 V	Injected signal frequency	625 Hz
Rated current	45 A	Injected signal amplitude	2.5 V
Torque constant	0.02 Nm/A	LPF cut-off frequency	100 Hz
Poles	8	Speed loop frequency	1 kHz
Rated speed	2250 RPM	Carrier frequency	10 kHz
Phase resistance (±12%)	0.03424 Ω	Microcontroller	Freescale-MPC5604
d-axis inductance (±20%)	0.000042 H	Upper computer	Freemaster
q-axis inductance (±20%)	0.000042 H	Absolute encoder	CJW-HE-1501A
Permanent magnet flux	0.00335 Wb		

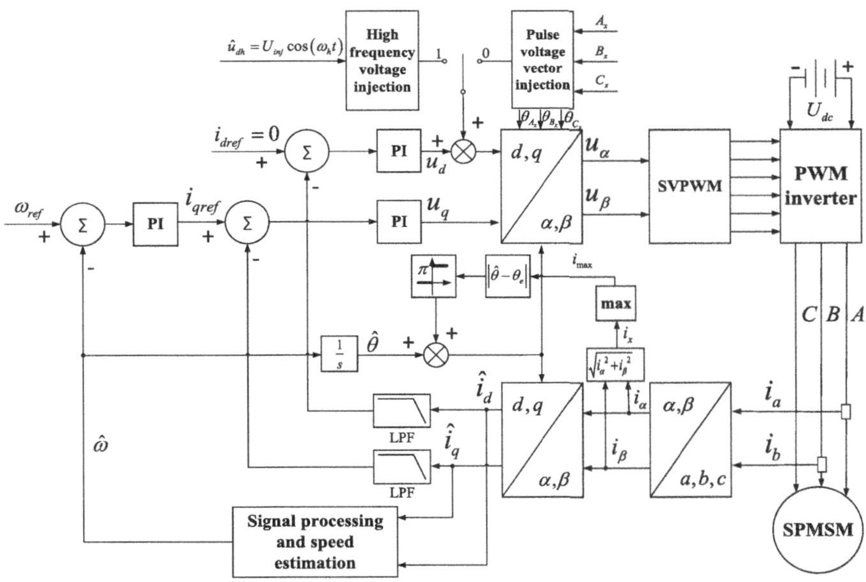

Fig. 4 Block diagram of low-speed sensorless control system for SPMSM

cogging. To verify the versatility of the detection method, the actual initial rotor position θ is selected as: 30°, 90°, −90°, −30°. In the figure, the pulse voltage vector injection time lasts about 0.1 s, the estimated rotor initial position $\hat{\theta}$ quickly converges to the actual position θ, and the polarity judgment is completed instantly. Among them, Fig. 6a, d do not need compensation, and the actual rotor initial position $\theta = \hat{\theta}$; Fig. 6b, c need to be compensated by 180°, and the actual rotor initial position $\theta = \hat{\theta} + \pi$. The maximum error of initial position identification is about 2.5°, which meets the requirements of initial position identification.

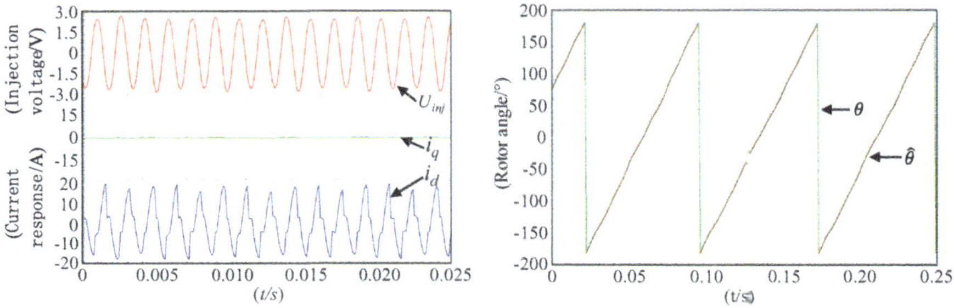

(a) Experimental results of voltage and current waveforms

(b) Experimental results of the positional relationship between θ and $\hat{\theta}$

Fig. 5 Experimental results of no-load running under 200 RPM

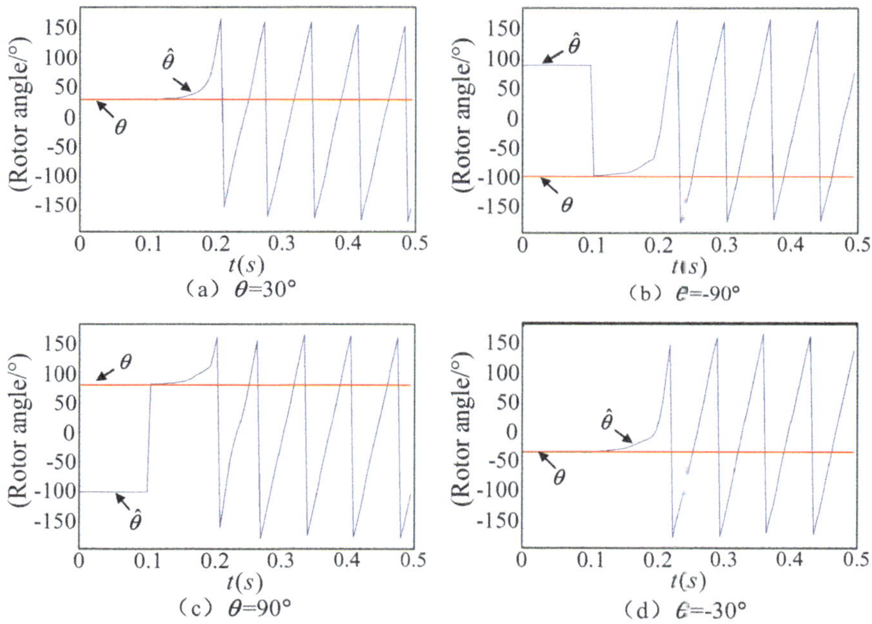

Fig. 6 Experimental results of rotor initial position detection

After the motor is started, it can run stably in the range of 50–1500 RPM. Figure 7 shows the motor speed and electrical angle waveforms at ±50 RPM and 30% load, speed fluctuates within the range of ±10 RPM, the angle is smooth and stable.

In order to verify the speed regulation performance, a motor forward and reverse switching experiment was carried out. The experimental results are shown in Fig. 8a. The motor can switch from 300 to −300 RPM in 1.3 s, which can meet the needs of a large range of speed changes in a short time; In order to verify the anti-disturbance

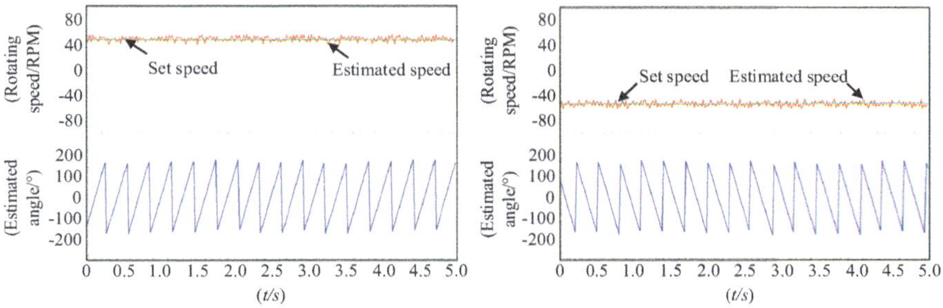

(a) Experimental results of forward 50 RPM (b) Experimental results of reverse 50 RPM

Fig. 7 Experimental results of low-speed running under 30% load

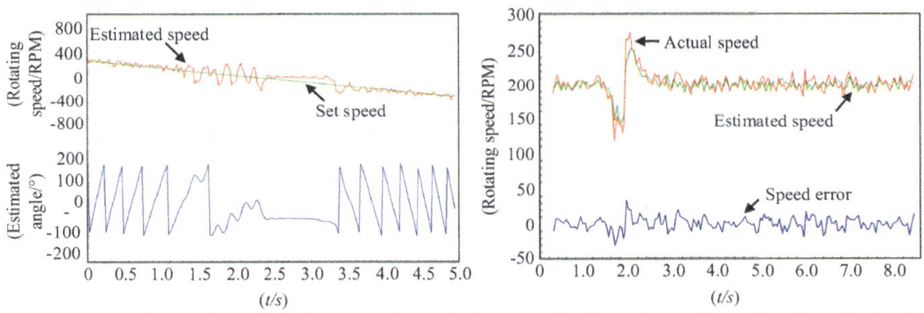

(a) Experimental results when steering switch (b) Experimental results when the load changes suddenly

Fig. 8 Experimental results during steering switching and sudden load changes

ability of the system, a sudden load experiment was carried out, and the no-load speed of the motor was set to 200 RPM, and 80% load was suddenly added at 1.5 s. The experimental results are shown in Fig. 8b. The average speed estimation error before loading is 10 RPM, and the speed change during the loading phase is about 120 RPM, and it recovers to around 200 RPM within 1 s. After that, the average speed estimation error is 18 RPM. When the method is used, the load adaptability is strong at low speed, the load effect is good, and it shows strong robustness.

5 Conclusion

In this paper, the saturated salient pole effect is analyzed and the principle of pulsating high-frequency voltage injection is studied, and proposes an improved rotor initial position detection and tracking technology for low-speed sensorless control of SPMSM for oil pump. The pulsating high-frequency voltage signal is injected into the synchronous rotating coordinate system, and the characteristic of permanent magnet synchronous motor cross saturation under high-frequency excitation is utilized. The frequency and phase characteristics of current response are analyzed, and the filter is optimized to compensate for the phase delay of the modulation signal. Then, a rotor position observer with a composite PLL structure is established, which improves the demodulation SNR and the stability of the control system. Finally, the polarity judgment condition is established by pulse voltage vector injection, which effectively improves the success rate of rotor polarity judgment. It is verified by oil pump platform that this method can identify and track rotor position well, with stable and reliable operation at the low speed.

References

1. Yuan, Qian, Zhongping Yang, Fei Lin, etc. 2013. Sensorless control of permanent magnet synchronous motor with stator flux estimation. *Journal of Computers* 8(1).
2. Aymen, F., H. Kraiem, S. Lassaad. 2012. Robust high speed control algorithm for PMSM sensorless drives. In *2012 9th International Multi-Conference on Systems, Signals and Devices (SSD)*.
3. Belkacem, S., B. Zegueb, F. Naceri. 2010. Robust non-linear direct torque and flux control of adjustable speed sensorless PMSM drive based on SVM using a PI predictive controller. *Journal of Engineering & Technology Review* 3(1).
4. Zhang, X., L. Sun, K. Zhao, et al. 2013. Nonlinear speed control for PMSM system using sliding-mode control and disturbance compensation techniques. *IEEE Transactions on Power Electronics* 28 (3): 1358–1365.
5. Bossoufi, B., M. Karim, S. Ionita, et al. 2012. Nonlinear non adaptive backstepping with sliding-mode torque control approach for PMSM motor. *Journal of Electrical Systems* 8 (2): 236–248.
6. Gong, D., Y. Huang. 2016. Exactly decoupled robust passivity-based control of PMSM based on Luenberger observer. In *2016 35th Chinese Control Conference (CCC)*. IEEE.
7. Ling-Xia, Z., H. He, Z. Kai. 2017. Simulation of hardware-in-loop for PMSM based on speed sensorless. *Journal of Mechanical & Electrical Engineering*.
8. Liu, J.M., and Z.Q. Zhu. 2014. Novel sensorless control strategy with injection of high frequency pulsating carrier signal into stationary reference frame. *I-EEE Transactions on Industry Application* 50 (4): 2574–2583.
9. Tang Q.P., A. Shen, X. Luo, et al. 2018. IPMSM sensorless control by injecting bi-directional rotateng HF carrier signals. *IEEE Transactions on Power Electronics* 99: 1.
10. Seilmeier, M., and B. Piepenbreier. 2015. Sensorless control of PMSM for the whole speed range using two-degree of-freedom current control and HF test current injection for low-speed range. *IEEE Transactions on Power Electronics* 30 (8): 4394–4403.
11. Accetta A., M. Cirrincione, M. Pucci, et al. 2009. PMSM drives sensorless position control with signal injection and neural filtering. *Electric Machines and Drives Conference, 2009. IEMDC'09*. IEEE International. IEEE.
12. Xie, G., K. Lu, S.K. Dwivedi, J.R. Rosholm, and F. Blaabjerg. 2016. Minimum-voltage vector injection method for sensorless control of PMSM for low-speed operations. *IEEE Transactions on Power Electronics* 31 (2): 1785–1794.
13. Xiang, Xiaodong, Yikang He. 2007. Sensorless vector control operation of a PMSM by rotating high-frequency voltage injection approach. *International Conference on Electrical Machines & Systems*. IEEE.

14. Shinnaka, S. 2007. A new generalized high-frequency voltage injection method and mirror-phase estimation method for sensorless drive of salient-pole PMSMs. *IEEJ Transactions on Industry Applications* 127 (3): 973–986.
15. Wu, X., Y. Feng, X. Liu, et al. 2017. Initial rotor position detection for sensorless interior PMSM with square-wave voltage injection. *IEEE Transactions on Magnetics* 11: 1–1.
16. Liu, Y., B. Zhou, S. Li, et al. 2011. Initial rotor position detection of surface mounted permanent magnet synchronous motor. *Proceedings of the CSEE* (in Chinese).
17. Wei, J., T. Guo-Jun, Y.E. Zong-Bin. 2010. Modeling of salient-pole synchronous motor considering saturation effect. *Electric Machines & Control*.
18. Zhang, L., C. Gao, J. Zhang, et al. 2012. Electromagnetic vibration characteristics of PMSMs with salient effect. *Diangong Jishu Xuebao/Transactions of China Electrotechnical Society* 27 (11): 89–96 (in Chinese).
19. Lu, Jiadong, Jinglin Liu, Lichao Wei. 2015. Estimation of the initial rotor position for permanent magnet synchronous motors. *Transactions of China Electrotechnical Society* 30(7): 105–111 (in Chinese).

Research on Short Time Scale Failure Mechanism of MMC Sub-module Busbar

Nana Duan, Shaocong Lu, Xinyu Ma, and Shuhong Wang

Abstract Laminated busbar have the advantages of low impedance, good anti-interference, high reliability, space saving, easy installation, fast and so on. As a high-power modular connection component, it is widely used in wind power, photovoltaic, electric power and hybrid traction, power generation Systems, power switch systems, power conversion modules of electric equipment and other places. And it is indispensable part for the development of the power industry. Therefore, in the connection part of MMC sub-modules, laminated busbar are also used. This article first uses FEM Software to extract the stray inductance of the busbar of the module, highlighting the advantage of low stray inductance of the laminated busbar. Taking into account the DC short-circuit fault between the sub-modules, the temperature rise and deformation of the busbar caused by voltage changes are analyzed by using the electric field-heat transfer-structure three-field coupling method. And by changing the voltage level of the busbar and the operating time under a certain voltage level, the safe operating domain of the laminated busbar was judged, and the safe operating time and voltage were obtained.

Keywords Large capacity · MMC sub-module · Laminated busbar · Short time scale failure · Safe operation domain

1 Introduction

With the emergence of new power electronic devices, emerging power electronic technologies represented by flexible DC transmission products have been widely used, and are showing the development trend of high voltage and large capacity [1]. China has built a number of flexible DC transmission demonstration projects [2]. Commercial insulated gate bipolar transistors (IGBTs) can have a voltage level of 6500 V, a current level of 3600A, and a switching time of about 0.2 μs. Therefore, the larger di/dt during the IGBT switching action will induce a higher voltage spike

N. Duan (✉) · S. Lu · X. Ma · S. Wang
School of Electrical Engineering, Xi'an Jiaotong University, No. 28 Xianning West Road, Xi'an, Shanxi, China
e-mail: duannana@xjtu.edu.cn

W. Chen et al. (eds.), *The Proceedings of the 9th Frontier Academic Forum of Electrical Engineering*, Lecture Notes in Electrical Engineering 743,
https://doi.org/10.1007/978-981-33-6609-1_64

under the action of the stray inductance in the loop. The voltage spike will increases the voltage stress of the switching device [3]. Laminated busbar have become an indispensable and important component for the development of the power industry due to their advantages of low impedance, good anti-interference, high reliability, space saving, easy installation, and fast speed. Therefore, in the connection part of MMC sub-modules, laminated busbar are also used.

Most of the research on laminated busbar has focused on the study of its stray inductance. The Literature [4–11] mainly studied the influence of the physical structure of the laminated busbar on its stray inductance, including the size of the opening, the position of the opening, the shape of the boundary and other characteristics, and proposed their own optimization plan. The literature [12–15] mainly studied the method of extracting the stray inductance is optimized. However, there is little discussion about the safe operation domain of the laminated bus.

This article will use FEM software to find the safe operation domain of busbar.

2 Laminated Busbar Structure and Its Connection in MMC

The laminated busbar is a kind of electrical connector with a multi-layer composite structure. The use of a composite busbar can provide a power distribution system that is easy to design, easy to install, fast, and has a clear structure. It is suitable for use in large-capacity MMC circuits.

2.1 Basic Connection of Laminated Busbar

Laminated busbar is to press multiple layers of wide and thin copper plates together, and use insulating materials such as polyester film and epoxy resin to separate them. As shown in Fig. 1. IGBTs and other devices are connected to the busbar with bolts as shown in Fig. 2.

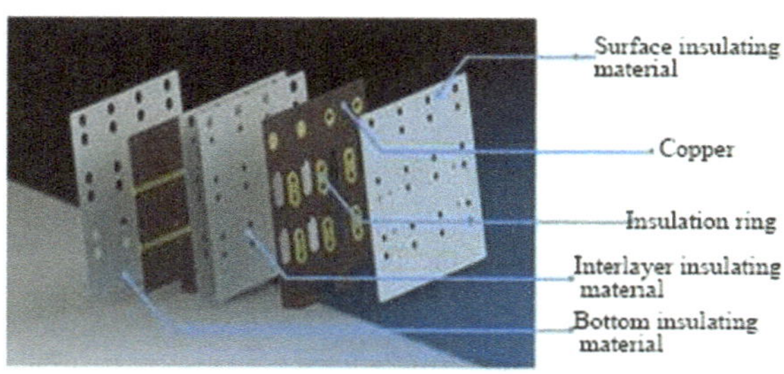

Fig. 1 Exploded view of laminated busbar

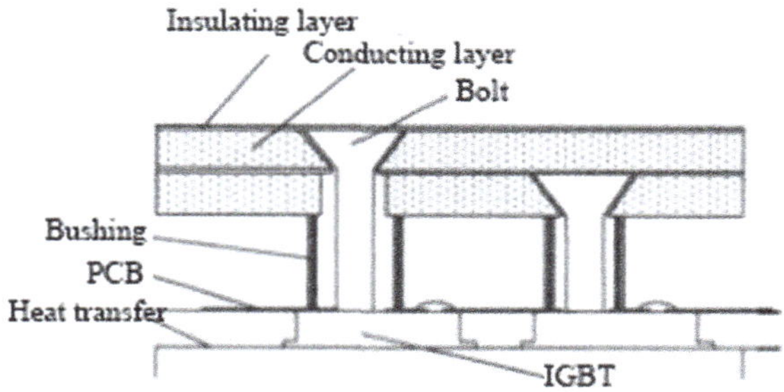

Fig. 2 Laminated busbar connection method

2.2 *MMC Submodule*

The three-phase topology of MMC is shown in Fig. 3.

3 Stray Inductance Extraction of Laminated Busbar

The busbar model used in the MMC sub-module is shown in the Fig. 4. Use FEM Software to extract the stray inductance of the busbar, apply Source and Sink to the position shown in the Fig. 5. The current frequency is increased from a low frequency

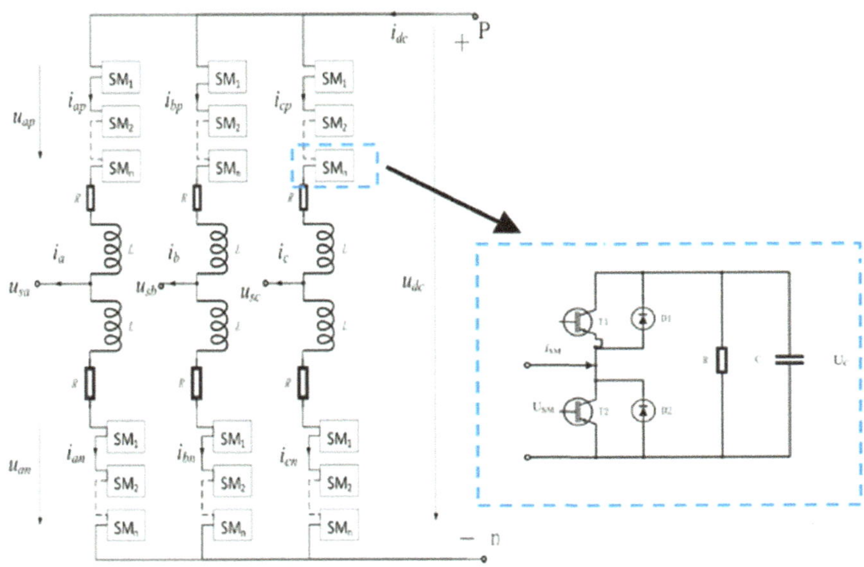

Fig. 3 The topology of MMC

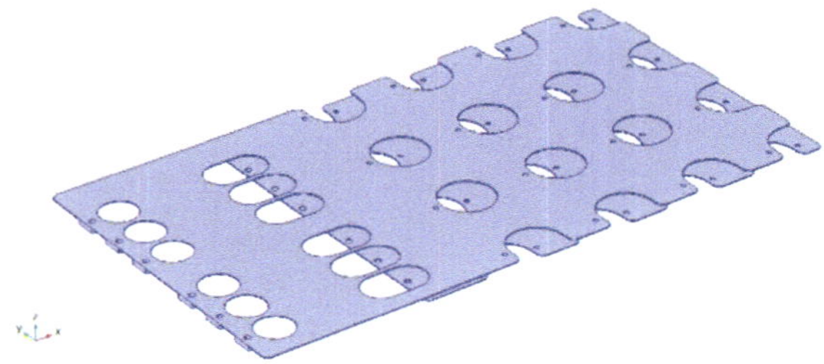

Fig. 4 Laminated busbar model

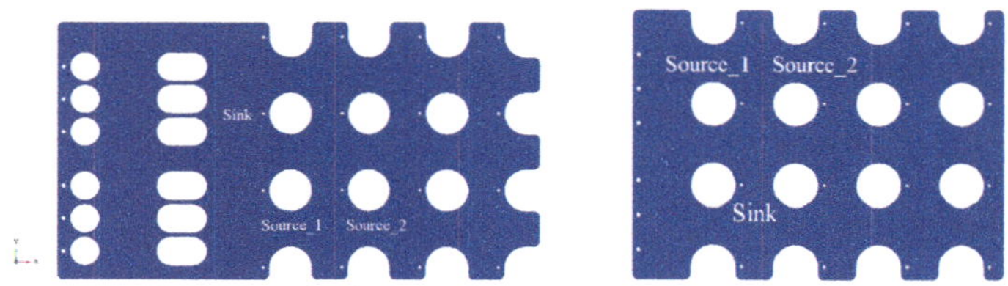

(a) Upper busbar **(b)**.Lower busbar

Fig. 5 Laminated busbar current application position

of 50 Hz to a high frequency of 10 MHz, and the change of stray inductance is observed.

In a laminated busbar, the higher the frequency of the current, the more obvious the skin effect and proximity effect. Table 1 shows the inductances extracted from all paths on the busbar at partial frequencies. Since the area of the upper busbar is slightly larger than the lower busbar, the stray inductance will also be larger than the lower busbar, as shown in Fig. 6.

Table 1 Inductance when frequency changes (nH)

Freq: KHz	bottom_1	bottom_2	top_1	top_2
0.30	10.75	26.51	16.94	34.56
10.00	8.15	22.60	13.50	29.24
370.00	7.61	21.90	12.81	28.31
820.00	7.57	21.86	12.77	28.25
1000.00	7.56	21.85	12.76	28.23
10,000.00	7.52	21.79	12.70	28.16

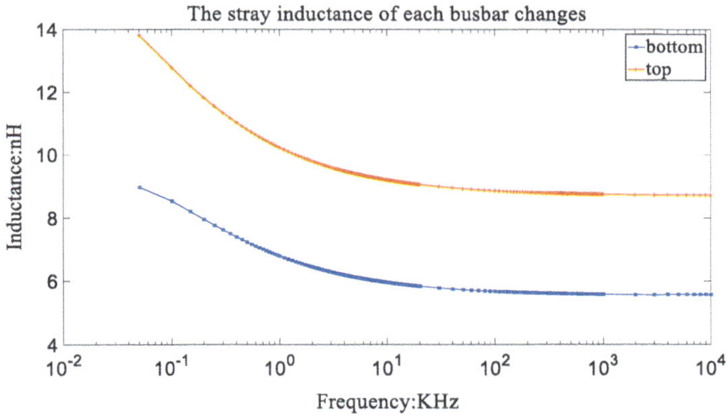

Fig. 6 The stray inductance of each busbar changes

4 Research on Safe Operation Region of Laminated Busbar

The FEM software is used to simulate and analyze the safe operation domain of the laminated busbar, which is divided into two parts: safe operation voltage and safe operation time. When a short-circuit fault occurs in the module, a huge current is generated on the high-power power electronic device and its various parts of the connector, and the heat generated by the current will cause the connector to deform. Once the laminated busbar is deformed, the positions of the connection holes on the upper and lower layers may be misaligned, which may result in failure to use, or short-circuits during use. Therefore, it is necessary to study the safe operation domain of laminated busbar.

4.1 Connection Status in MMC Submodule Electric-Heat-Solid Three-Field Coupling

Three fields of electric field, solid heat transfer and solid mechanics are used in the simulation to observe the temperature rise and deformation of the laminated busbar. Voltage is applied to the electric field, and the energy consumed is used as the heat source in formula (2), and thermal expansion causes the busbar to deform.

$$\begin{aligned} \boldsymbol{J} &= \sigma \boldsymbol{E} + \frac{\partial \boldsymbol{D}}{\partial t} + \boldsymbol{J}_e \\ \boldsymbol{E} &= -\nabla V \end{aligned} \tag{1}$$

$$\rho C_p \frac{\partial T}{\partial t} + \rho C_p \boldsymbol{u} \cdot \nabla T + \nabla \cdot \boldsymbol{q} = Q$$

$$\boldsymbol{q} = -k\nabla T \tag{2}$$

$$0 = \nabla \cdot S + \boldsymbol{F}v \tag{3}$$

where ρ is the density of the medium, C_p is the heat capacity of the medium, k is the thermal conductivity, $\boldsymbol{u}$ is the velocity, here is the air flow velocity, and Q is the heat source.

4.2 Safe Operating Voltage of Laminated Busbar

The temperature range of the busbar product specification is 0–40 °C. In an environment with a room temperature of 20 °C, apply a voltage of 50 Hz to the upper layer of the laminated busbar, and the simulation time is 0–1 s to observe the temperature rise of the laminated busbar.

The voltage level is increased from 1200 V until the maximum temperature of the busbar exceeds 40 °C. Within this range, look for the critical voltage of temperature rise (Fig. 7).

As can be seen from the above figure, when the voltage exceeds 4 kV, there is a very obvious change, and the temperature has a step increase. It can be inferred that 4 kV is the maximum safe operating voltage of the laminated busbar. On this basis, we can continue to explore the safe operating time of laminated busbar.

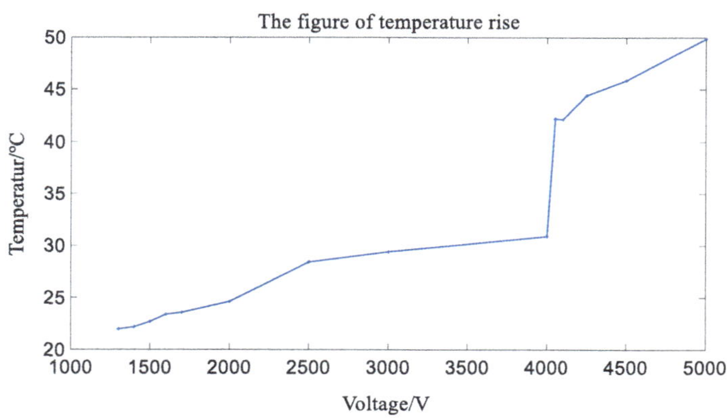

Fig. 7 Busbar temperature rise graph

4.3 Safe Running Time of Laminated Busbar

As can be seen from the above figure, when the voltage exceeds 4 kV, the temperature will exceed the maximum temperature range of the busbar. Set the voltage to 4.05 kV, and its level setting is slightly higher than the safe operating voltage, which can better study the safe operating time of the laminated busbar. Take the maximum temperature point used in Fig. 8 to simulate the safe running time. Figure 9 shows the temperature rise at the highest temperature of the busbar within 0–1 s. The temperature of the busbar only reaches 40°C when it is close to 1 s. After more careful judgment, the time to reach 40 °C is 0.974 s. The distribution of busbar temperature at 1 s is shown in Fig. 10.

At this temperature, the deformation of the busbar is not obvious. The maximum allowable temperature is just reached at 1 s, so the effect is not very obvious as shown in Fig. 11. Extend the running time to 5 s, the temperature will continue to rise, and the busbar will produce more obvious temperature changes and deformation (Figs. 12 and 13).

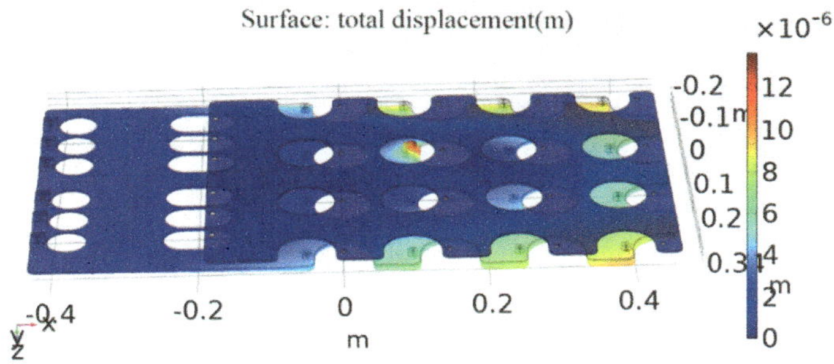

Fig. 8 Forced deformation of the busbar at 5 kV

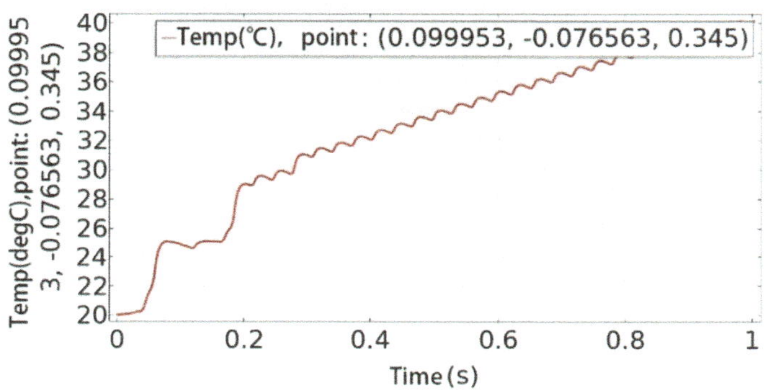

Fig. 9 Safe operating time when the voltage is 4.05 kV

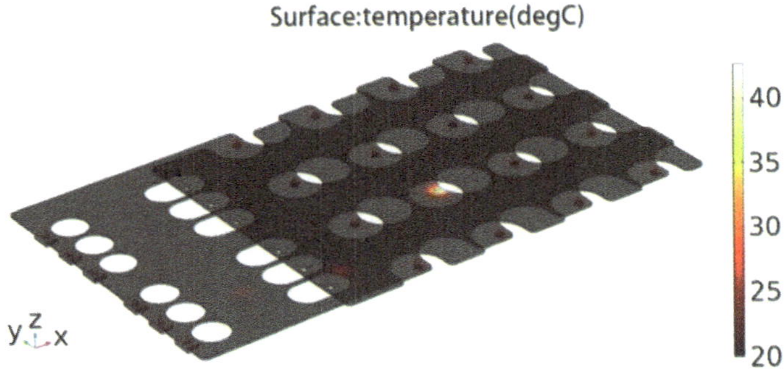

Fig. 10 Busbar surface temperature when running at 4.05 kV for 1 s

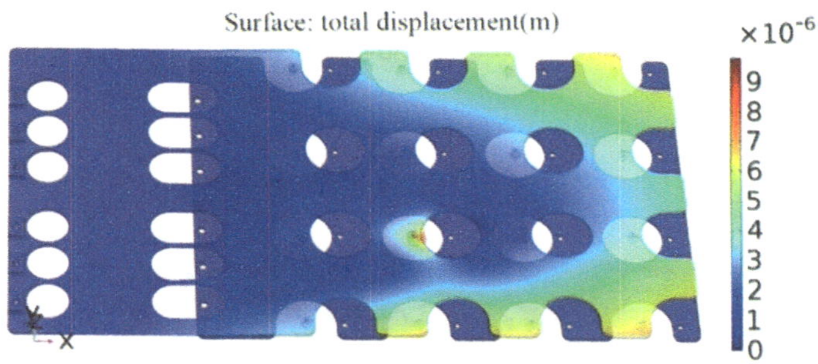

Fig. 11 The force and deformation of the busbar when running at 4.05 kV for 1 s

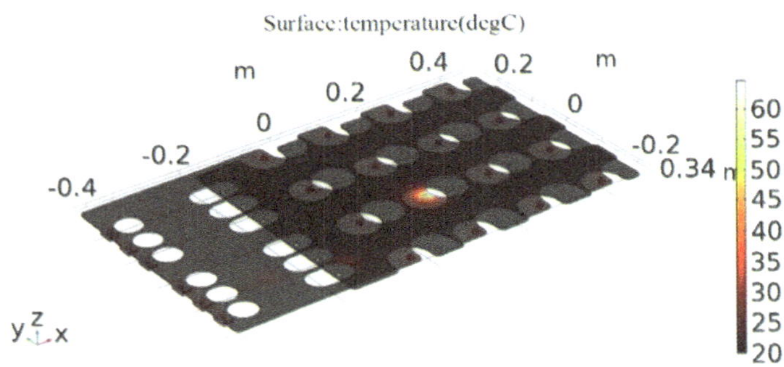

Fig. 12 Busbar surface temperature when running for 5 s

5 Conclusion

This paper has carried out a more comprehensive simulation analysis on the laminated busbar. The stray inductance of the laminated busbar is extracted. Compared with

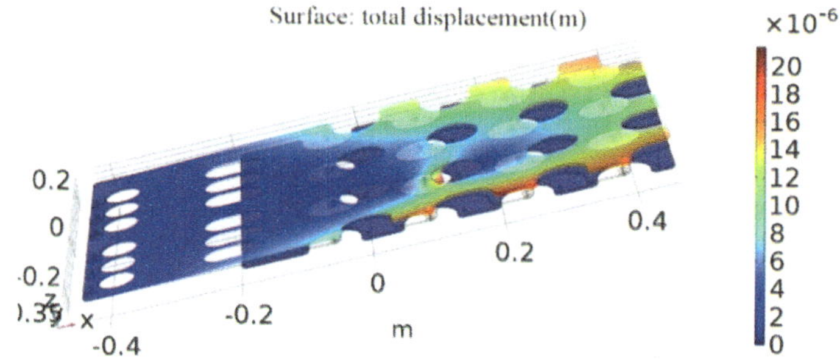

Fig. 13 The force and deformation of the busbar when running at 4.05 kV for 5 s

other types of connection bars, the inductance stray inductance is smaller and the stray inductance decreases as the switching frequency increases.

Afterwards, the safe operating range of the laminated busbar was studied at a room temperature of 20 °C. In 1 s of operation time, if the voltage exceeds 4 kV, the temperature of the busbar will exceed 40 °C, exceeding the safe operation standard of the busbar. When the voltage level is 4.05 kV, the operating time of the busbar exceeds 0.974 s, the temperature will reach 40 °C, and it will not have a big impact in a short time, and the busbar will not produce obvious deformation. If the time is too long or the voltage level is too high, it will cause the busbar to deform and affect normal use.

Acknowledgements This work was supported in part by the China National Key Research and Development Projects under Grant (2018YFB0905801) and the National Natural Science Foundation of China under Grant (51707142 and 52077161).

References

1. Bo, Le, Nian Mei, Siyuan Liu, et al. 2014. Overview of HVDC flexible. *China Electric Power (Technology Edition)* 5(14): 50–54 (in Chinese).
2. Taiyuan, Wang Yue, Duan Guochao, etc. 2019. Zero DC voltage control based DC fault ride-through strategy for hybrid modular multilevel converter in HVDC. *Transactions of China Electrotechnical Society*, 34(S1): 343–351 (in Chinese).
3. Zhao, Zhengming, Haitao Zhang, Liqiang Yuan, etc. 2006. Failure mechanism and protection strategy of high voltage three-level inverter based on IGCT. *Transactions of China Electrotechnical Society* 21(5): 1–6 (in Chinese).
4. Ruan Jie, Chang Liu, Guangzhuo Li, et al. 2019. Design method of low-inductance laminated busbar for device-parallel ANPC circuit. *High Voltage Engineering* 045 (007): 2093–2100 (in Chinese).
5. Du, Litian, Junrong Peng, Binchuan Yin. 2017. Design and optimization of laminated busbar of power module for high-power pulse power converter. *Marine Electric* 11: 74–77 (in Chinese).
6. Yu, Shaolin, Xing Zhang, and Jianing Wang. 2018. Effect of key physical structures on laminated bus bar inductance of inverter. *Acta Energiae Solaris Sinica* 39 (11): 116–122 (in Chinese).

7. Zhu, Junjie, Jingxin Yuan, Ziling Nie, et al. 2019. Optimum design of planer busbar based on all-silicon carbide power module. *Proceedings of the CSEE* 39 (21): 6383–6393 (in Chinese).
8. Wang, Qing, Linhui Lu. 2013. Design of laminated busbar for power convertor in switched reluctance motor. *Power & Energy* 01: 34–38 (in Chinese).
9. Jing, Chen. 2016. Design of low_inductance laminated busbar based on ansys. *Electrical Engineering* 6: 78–80 (in Chinese).
10. Zhu, Yifeng, Jingle Zheng. 2017. Research on optimization method of stray inductance converter busbar based on structure. *Journal of Mechanical Electrical Engineering* 34(003): 293–297 (in Chinese).
11. He, F.Y., S.Z. Xu, and C.F. Geng. 2016. Improvement on the laminated busbar of NPC three-level inverters based on a supersymmetric mirror circulation 3D cubical thermal model. *Journal of Power Electronics* 16 (6): 2085–2098.
12. Feng, Gaohui, Liqiang Yuan, Zhengming Zhao, et al. 2014. A novel stray inductance extraction method for bus bars based on turn-on/off transient process. *Proceedings of the CSEE* 34 (36): 6442–6449 (in Chinese).
13. Jin, Zhufeng, Weichen Li, Sideng Hu, et al. 2017. Optimized stray inductance extraction method of bus bar in large-capacity power electronic equipment. *Transactions of the China Electrotechnical Society* 32(014): 1–7 (in Chinese).
14. Yang, Wenbin, Luhua Zhang, Wei Li. 2015. Testing method for stray inductance of laminated bus-bar used in three-level power converter based on LC resonant principle. *Low Voltage Apparatus* 012: 22–26, 31 (in Chinese).
15. Zhao, Dong, Shangbin Ye, Jiajia Zhang, et al. 2014. Extraction method of bus parameters for the high power electronic devices. *Electrical & Energy Management Technology* 000 (021): 12–16 (in Chinese).

Research on Fault Diagnosis Technology of AC Medium Voltage Vacuum Circuit Breaker Based on Wavelet Packet

Aijun Yang, Jiajun Guo, Sheng Xiong, Jiaming Tan, Yijun Ye, Huan Yuan, Aijun Yang, Xiaohua Wang, and Mingzhe Rong

Abstract Since mechanical failure is one of the main failures of the circuit breaker, on-line monitoring of mechanical condition has important practical significance. Mechanical vibration can reflect the mechanical state of circuit breaker in the whole time series and spectrum range. Besides, the vibration signals produced by the same type of circuit breaker are similar, and the operation status of the equipment can be diagnosed and studied by the vibration signals. Therefore, the vibration signal can be used to extract features to diagnose the running state of the equipment. In this paper, for a 12 kV AC medium volt-age vacuum circuit breaker, a variety of fault conditions are designed and the vibration signal of the breaker under normal and fault conditions is collected. The original signals are decomposed by wavelet packet and processed by method of band energy, energy entropy and Euclidean distance respectively, and then the eigenvectors including vibration characteristics are obtained. Finally Support Vector Machine is used to distinguish the normal and fault states, which provides a basis for the realization of on-line fault diagnosis technology for circuit breakers.

Keywords Fault diagnosis · Vibration signal · Wavelet packet · Band energy · Energy entropy · Euclidean distance

1 Introduction

The circuit breaker is one of the most important power equipment in the power system. It has the dual functions of control and protection. It is also one of the equipment that requires the most work in the primary equipment maintenance of the power

A. Yang (✉) · J. Guo · Y. Ye · H. Yuan · A. Yang · X. Wang · M. Rong
State Key Laboratory of Electrical Insulation and Power Equipment, Xi'an Jiaotong University, Xi'an 710049, China
e-mail: yangaijun@mail.xjtu.edu.cn

S. Xiong
School of Electrical Engineering Naval, University of Engineering, Wuhan 430033, China

J. Tan
State Grid Jiangsu Electric Power Co. Ltd. Maintenance Branch Company, Jiangsu 211106, China

W. Chen et al. (eds.), *The Proceedings of the 9th Frontier Academic Forum of Electrical Engineering*, Lecture Notes in Electrical Engineering 743,
https://doi.org/10.1007/978-981-33-6609-1_65

system. Overhaul and maintenance of circuit breakers are expensive and improper overhaul can easily lead to major accidents [1]. Therefore, online monitoring and fault diagnosis of the state of the circuit breaker has very important practical significance.

The failure statistics of circuit breakers at home and abroad show that the mechanical failure of the operating mechanism is the main cause of circuit breaker failure [2, 3]. The opening and closing process of the circuit breaker will produce strong vibration, and the vibration signal contains a large amount of mechanical state information. By comparing and processing the vibration waveform and data, extracting appropriate fault characteristics for analysis, the circuit breaker fault diagnosis can be realized. There are many mature vibration signal analysis methods, such as Fourier transform method, empirical mode decomposition, wavelet analysis, etc. Fourier transform is a traditional mechanical vibration signal processing method, which is suitable for processing stationary signals. But circuit breaker vibration is a transient non-stationary signal [4]. This method cannot take into account the overall and localized characteristics of the signal in the time or frequency domain at the same time [5]. The empirical mode decomposition (EMD) can adaptively decompose the signal into a series of intrinsic mode functions (IMF), and the frequency is arranged from high to low [6]. The key information of the original signal can be extracted by analyzing the IMF containing the local characteristics of the signal, but EMD may have modal aliasing and end effects [7]. The IMF obtained when modal aliasing occurs is meaningless, and it is difficult to determine the dimensionality of the intrinsic modal function for vibration signals in different states [8, 9]. Wavelet analysis is a time–frequency localized analysis method with a fixed window area but a variable shape, that is, time and frequency windows can be changed. It is adaptive to signal analysis and it has been widely used in power equipment fault diagnosis and noise elimination [10–12].

The specific process of diagnosis in this paper is as follows. First we use wavelet denoising to preprocess the vibration signal. Then we perform 3-layer wavelet packet decomposition on the preprocessed vibration signal and we use the frequency band energy algorithm, energy spectrum entropy, and Euclidean distance algorithm to extract signal features. Then we normalize the feature quantity and use the support vector machine to diagnose the normalized feature quantity. Finally, we compare the pros and cons of the two signal feature extraction methods to achieve the purpose of identifying the type of fault and optimizing the feature parameters.

2 Basic Principles of Wavelet Packet

Wavelet transform is a time–frequency localization analysis method in which both the time window and the frequency window can be changed. The window size is unchanged but the shape is variable. On the basis of wavelet analysis, wavelet packet analysis further decomposes the high-frequency parts that are not subdivided. It can adaptively selects the corresponding frequency band according to the characteristics

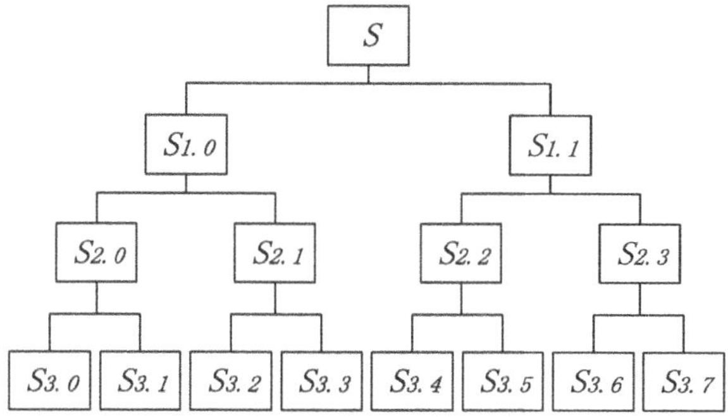

Fig. 1 The structure of wavelet packet decomposition

of the analyzed signal to match it with the signal spectrum to achieve improved frequency resolution purpose [13].

For vibration signals, wavelet packet decomposition can be carried out according to formula (1) and formula (2).

$$u_{2n}(t) = \sqrt{2}\sum_{k\in Z} h(k)u_n(2t-k) \tag{1}$$

$$u_{2n+1}(t) = \sqrt{2}\sum_{k\in Z} g(k)u_n(2t-k) \tag{2}$$

In the formula, h(k) is the high-pass filter coefficient of wavelet packet decomposition, and g(k) is the low-pass filter coefficient. Assuming that the signal undergoes three-layer decomposition, the three-layer decomposition structure of the wavelet packet is shown in Fig. 1.

Wavelet packet decomposition decomposes the original signal into high and low frequency bands. The two frequency bands decomposed by each node do not overlap each other, and the bandwidth is halved. Repeat the above steps for the decomposed signal to decompose the original signal into a combination of signals of different frequency bands. When the circuit breaker fails, the frequency and energy of the vibration signal will change accordingly in a certain frequency band or several frequency bands.

3 Extraction and Analysis of Energy Features Based on Wavelet Packet

3.1 Data Preprocessing

The output signal of the vibration sensor contains a certain DC component. When calculating the frequency band energy, the accumulation of the DC component will cause the result to be distorted, which is not conducive to analysis. In addition, the collected vibration signal contains various noises, in addition to the high-frequency noise in the vibration process, there are also various disturbances in the environment. When processing the signal, these noises will pollute the real signal and cause interference to the fault diagnosis. Therefore, the original signal needs to be preprocessed. In this paper, the average method and wavelet denoising method are used for data preprocessing.

3.1.1 Average Method.

The average method is to subtract the average value of the signal from the original signal to eliminate the DC component in the original signal. The specific process is as follows. Suppose the collected signal is x(i), i = 1,2,3, ..., n, n is the number of collected signal points, find the signal average.

$$ave = \sum_{i=1}^{n} x(i)/n \tag{3}$$

Subtract the average value from the original signal.

$$y(i) = x(i) - ave \tag{4}$$

3.1.2 Wavelet Denoising

The principle of wavelet denoising is similar to low-pass filter, but it can retain the characteristics of the signal after denoising, so it is better than low-pass filter.Wavelet denoising decomposes the signal into low frequency band and high frequency band through wavelet decomposition. Noise is usually expressed as a high-frequency signal, so only the wavelet coefficients in the high-frequency band need to be processed accordingly, and then the signal can be reconstructed to achieve the purpose of eliminating noise [14]. In this paper, for the same vibration signal, different wavelet basis functions and wavelet decomposition layers are selected for denoising. The signal-to-noise ratio SNR and the root mean square difference RMSE are calculated and compared to select appropriate parameters. The results are shown in Table 1.

3.2 Band Energy Extraction

The wavelet packet transform is a linear transform that satisfies the law of conservation of energy.

Table 1 Signal-to-noise ratio and RMSE with different wavelet basis functions and wavelet decomposition layers

Wavelet decomposition		db10	db9	db8	db7	db6
3 layers	SNR	64.51	78.21	63.56	67.89	63.69
	RMSE	0.0129	0.0135	0.0129	0.0129	0.013
4 layers	SNR	63.41	65.98	74.67	67.74	67.07
	RMSE	0.0138	0.0145	0.0138	0.0142	0.014

$$\int_{-\infty}^{+\infty} |f(t)|^2 dt = \sum_j \sum_k |c_{j,k}|^2 \tag{5}$$

Therefore, the wavelet packet coefficient has the dimension of energy, and the energy of each frequency band can be calculated by the wavelet packet coefficient.

Assuming that the signal has a total of N sampling points, a 3-layer wavelet packet decomposition is performed on the signal, and the decomposition structure is shown in Fig. 1.

In Fig. 1, $S_{i,j}$ represents the j-th node of the i-th layer, where $i = 0, 1, 2, 3, j = 0,1,2, \ldots, 7$. Each node coefficient corresponds to a section of the waveform of the corresponding frequency band, and this section of the waveform contains sampling points, so its energy can be obtained by formula (6).

$$E_{3,j} = \int |S_{3,j}(t)| dt = \sum_{k=1}^{n} |d_{j,k}|^2 \tag{6}$$

3.3 *Extraction of Energy Spectrum Entropy and Euclidean Distance*

Information entropy is a measure of the degree of system disorder in a certain state. When it is applied to signal analysis, the uniformity and complexity of the signal can be measured. After decomposing the signal with 3-layer wavelet packet, the nodes $S_{3,j}$ are obtained, where $j = 0, 1, \ldots 7$. According to the time characteristics of the signal, the signal is divided into N segments, and the energy of each segment can be calculated separately by formula (7).

$$Q_{i(3,j)} = \int_{t_{i-1}}^{t_i} |A_i(t)|^2 dt \tag{7}$$

$Q_{i(3,j)}$ is the energy of the i section of the signal and $A_i(t)$ is the amplitude of the i-th segment of the signal. The calculated energy is normalized, and the wavelet packet energy entropy is calculated, and the wavelet packet energy spectrum entropy of the j-th node in the third layer is defined as $H_{3,j}$.

$$H_{3,j} = -\sum_{i=1}^{N} \varepsilon_{3,j}(i) \lg \varepsilon_{3,j}(i) \tag{8}$$

$\varepsilon_{3,j}(i)$ is the normalized value of the energy of each segment of the signal.

$$\varepsilon_{3,j}(i) = \frac{Q_{i(3,j)}}{\sum_{i=1}^{N} Q_{i(3,j)}} \tag{9}$$

Euclidean distance is used to describe the distance between two vectors in mathematics. Euclidean distance is used to measure the state information of the vibration signal as follows. Take two sets of vibration signals, one set is the reference signal *NB*(i), where i is the sampling point number, the other is the reference signal *NR*(i). Assuming that the signal to be analyzed is *TS*(i), then the Euclidean distance between the reference signal *NB*(i) and *NR*(i) and that between the reference signal and the signal *TS*(i) are $D1$ and $D2$.

$$D_1 = \left\{\sum_{k=1}^{N} [NB(k) - NR(k)]^2\right\}^{\frac{1}{2}} \tag{10}$$

$$D2 = \left\{\sum_{k=1}^{N} [NB(k) - TS(k)]^2\right\}^{\frac{1}{2}} \tag{11}$$

$$FF = D_2 / D_1 \tag{12}$$

FF represents the ratio of the Euclidean distance between the state under test and the normal state and the Euclidean distance between the normal state category. The larger the *FF*, the greater the difference between the state under test and the normal state.

4 Circuit Breaker Fault Simulation Verification

4.1 Circuit Breaker Fault Simulation Experiment Scheme

In this paper we simulate the common faults of opening and closing on a 12 kV AC medium voltage vacuum circuit breaker. The fault simulation scene is shown in Fig. 2. There are two closing springs, and one of them is removed to simulate

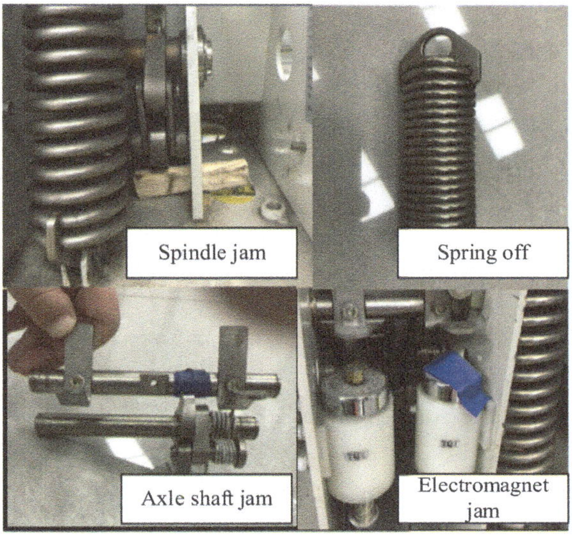

Fig. 2 Fault simulation field

the spring falling off. The small spring is replaced to simulate the reduction of the closing spring force. Spindle jamming is simulated by adding foreign objects under the spindle and half-shaft jamming is simulated by wrapping insulating tape on the half-shaft. Blocking electromagnet jamming is simulated by wrapping insulating tape on the electromagnet to hinder its core movement. There are three opening springs, one of which is removed to simulate the reduction of opening spring force.

4.2 Result Analysis

The vibration signal of the opening and closing process under the normal state and the vibration signal under each fault state are measured respectively. Perform a 3-layer wavelet packet frequency band energy decomposition on the original vibration signal. In the analysis, the db10 wavelet is selected as the mother wavelet function, and the optimal wavelet packet decomposition tree structure is selected using Shannon entropy as the selection criterion. The third layer is calculated according to the formula. Since the sampling rate of the vibration signal is 50 kHz, according to the sampling theorem, the highest frequency that can be resolved in the original signal is 25 kHz, so the bandwidth of each frequency band after wavelet packet decomposition is 3.125 kHz. Take the eight frequency band characteristic waveforms of the closing process under normal conditions as an example, as shown in Fig. 3. The wavelet packet decomposition results of the other 6 types of fault vibration signals can be obtained in the same way. Figure 4 is the characteristic waveforms of 8 frequency bands in the reduced state of the closing spring force. When the circuit breaker is closing, if the closing spring force is reduced, the closing speed will slow down,

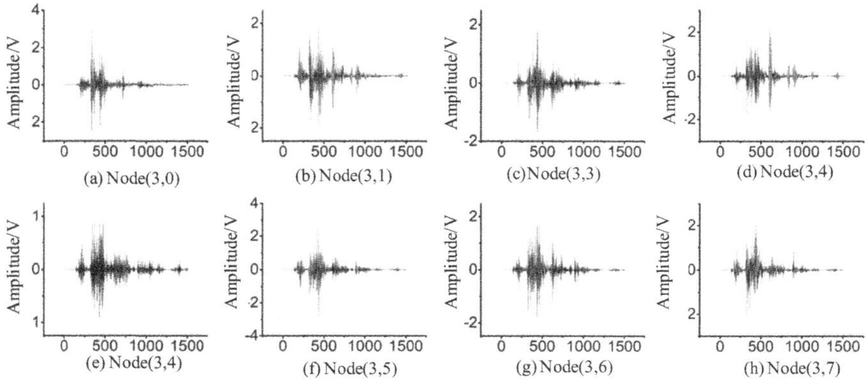

Fig. 3 The characteristic waveforms of 8 frequency bands in the closing process under normal conditions

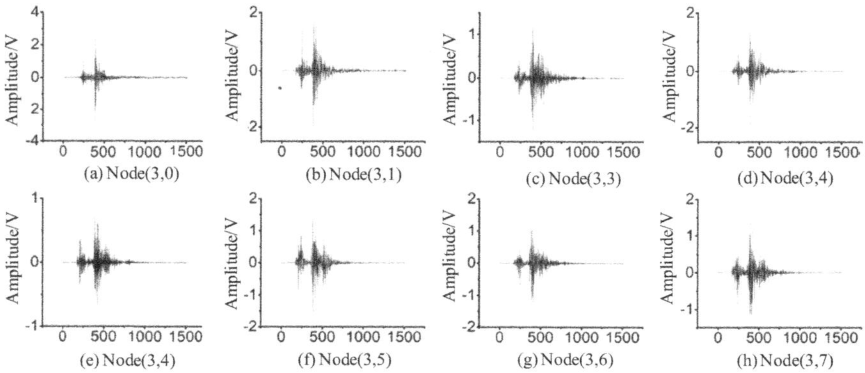

Fig. 4 The characteristic waveforms of 8 frequency bands under the condition of reducing the closing spring force

and in severe cases, the closing will fail. Comparing the above two waveforms, it can be seen that when the closing spring force is reduced, the vibration waveform has changed significantly compared to the normal state. The number of peaks of the characteristic waveform of each node is significantly reduced and the vibration has basically ended near the 1000 point, but there is still a slight vibration at the 1000 point under normal circumstances, which is consistent with the reduction of the closing force.

4.3 Band Energy

In order to identify faults, quantitative analysis of each characteristic waveform is required. The vibration signal is decomposed by wavelet packet, and the energy of each frequency band under the normal state and the five fault conditions is calculated. The results are shown in Table 2.

From Table 2 we can draw the following conclusions. In the case of closing, the frequency band energy change of each fault state is very obvious. Compared with the frequency band energy in the normal state, the energy of each frequency band in the state of the closing spring force reduction and the closing spring missing state gradually decreases. Taking the first node in the low frequency band as an example, the energy of the first node during normal closing is 119.97. When the small spring is replaced and the spring force is reduced, the energy is reduced to 55.69, and the circuit breaker closing time is longer at this time. When the closing spring is missing, the stored energy of the spring is further reduced and the energy is reduced to 14.28. At this time, the circuit breaker is prone to closing difficulties and closing failures. The jamming fault of the axle shaft also has good discrimination. Due to the half-axis jamming fault, the lock of the circuit breaker cannot be locked after closing, and the phenomenon of just closing and opening occurs.

In the case of opening, because the electromagnet jams, it will be difficult for the iron core to open and trip, and the circuit breaker cannot perform normal opening actions. Only small vibrations are generated, so the frequency band energy of the electromagnet jamming fault is only 0.14. It is necessary to use the support vector machine to further train the distinction.

Table 2 Band energy in normal and five fault states

Category	(3, 0)	(3, 1)	(3, 2)	(3, 3)	(3, 4)	(3, 5)	(3, 6)	(3, 7)
Normal closing	119.97	97.23	72.64	134.6	33.54	114.21	84.26	92.64
Closing spring force decreases	55.69	45.62	22.74	34.26	12.29	36.24	28.69	32.15
One less closing spring	14.28	12.23	6.11	11.75	3.17	8.37	5.59	7.55
Axle shaft jam	319.54	218.03	159.06	222.46	208.6	259.74	171.38	234.88
Normal opening	130.47	122.62	74.59	118.14	88.85	133.46	101.28	117.45
One less opening spring	146.39	122.22	92.32	94.51	73.51	174.07	64.58	145.18
Electromagnet jam	0.14	0.14	0.13	0.14	0.12	0.13	0.14	0.13

Table 3 Energy entropy and euclidean distance in normal and five fault states

Category	H3, 0	H3, 1	H3, 2	H3, 3	H3, 4	H3, 5	H3, 6	H3, 7	Euclidean distance
Normal closing	0.54	0.77	0.73	0.72	0.77	0.71	0.72	0.66	0.11
Closing spring force decreases	0.43	0.6	0.64	0.55	0.6	0.65	0.65	0.57	0.35
One less closing spring	0.63	0.63	0.61	0.61	0.65	0.61	0.52	0.64	0.35
Axle shaft jam	0.62	0.72	0.73	0.76	0.85	0.76	0.76	0.72	0.16
Normal opening	0.33	0.28	0.3	0.31	0.26	0.26	0.28	0.26	0.07
One less opening spring	0.37	0.4	0.35	0.37	0.34	0.34	0.37	0.35	0.17
Electromagnet jam	0.99	1	1	0.99	0.99	1	1	1	1.95

4.4 *Energy Entropy and Euclidean Distance*

According to the principle of energy entropy, each energy entropy of the normal state and the five types of faults is calculated. And according to the calculation formula of Euclidean distance, the corresponding Euclidean distance is calculated. The results are shown in Table 3.

It can be seen from Table 3 that there is still a good distinction between the fault state and the normal state. Compared with the normal situation when the closing spring force is reduced, the energy entropy of the $H_{3,0}$ frequency band and the $H_{3,1}$ frequency band are both smaller, and the Euclidean distance is larger. When one of the closing springs is jammed with the half shaft, the energy entropy of the $H_{3,0}$ frequency band is larger and the energy entropy of the $H_{3,1}$ frequency band is lower, but the Euclidean distance of the half shaft jamming is significantly smaller than the spring missing, which can pass the Euclidean distance Distinguish these two faults. The opening fault can also be identified by the energy entropy and Euclidean distance of $H_{3,0}$ and $H_{3,1}$ frequency bands. When there is one less opening spring, the entropy of the $H_{3,1}$ frequency band and the Euclidean distance are larger, and when the electromagnet is stuck, the energy entropy and Euclidean distance of each frequency band is much larger than normal. Therefore, the energy entropy and Euclidean distance of the H3, 0 and H3, 1 frequency bands can be selected as the feature vector, and the fault judgment can be performed through the support vector machine.

4.5 *SVM-Based Fault Diagnosis*

Perform trend item removal and wavelet denoising preprocessing on the collected 100 sets of data, and extract the corresponding frequency band energy, energy spectrum

Table 4 Fault identification results of band energy, energy entropy and Euclidean distance

Fault identification	Close		Open	
	E (%)	S (%)	E (%)	S (%)
First	100	100	95	97.5
Second	100	95	100	100
Third	97.5	92.5	97.5	100

entropy and Euclidean distance as input feature quantities. The test results are shown in Table 4. E represents the accuracy of fault classification using frequency band energy as the feature vector, and S represents the accuracy of fault classification using energy entropy and Euclidean distance as the feature vector.

Combining the advantages of the two methods, frequency band energy is used as the characteristic vector of closing fault, and energy entropy and Euclidean distance are used as the characteristic vector of opening fault. This will effectively improve the recognition accuracy.

5 Conclusion

In this paper, the energy feature extraction method based on wavelet packet is adopted to analyze and process the vibration signal to realize the diagnosis and identification of the specific fault of the circuit breaker. After experiment and comparison, the appropriate wavelet denoising parameters are selected, which effectively improves the signal-to-noise ratio of the signal, reduces the root mean square error, and obtains a better denoising effect. Based on the three-layer wavelet packet decomposition of the vibration signal, the band energy, energy entropy and Euclidean distance of the signal are extracted as features, and the support vector machine is used to intelligently identify the fault. For the identification of closing faults, the frequency band energy is more intuitive and the accuracy rate is higher. For opening faults, the recognition effect of energy entropy and Euclidean distance is better. The use of characteristic vectors with strong applicability for opening and closing will effectively improve the accuracy of circuit breaker fault recognition and provide new ideas for the diagnosis of mechanical faults of circuit breakers.

References

1. Rong, M.Z., S.L. Jia, and X.H. Wang. 2007. *Condition monitoring of switchgear*. Beijing, China: China Machine Press (in Chinese).
2. Sun, Y., J.W. Wu, S.J. Lian, et al. 2014. Extraction of vibration signal feature vector of circuit breaker based on empirical mode decomposition amount of energy. *Transactions of China Rlectro-Technical Society*. 29 (03): 228–336 (in Chinese).

3. Chen, W.G., B.F. Deng, and B. Yang. 2009. Fault recognition for high voltage circuit breaker based on EMD of vibration signal and energy entropy characteristic. *High Voltage Apparatus.* 45(02): 90–93+96 (in Chinese).
4. Jiang, W.T., J.Y. Zhen, and J. Mei. 2010. Simulation research on vibration signal fault diagnosis in high-voltage circuit-breakers based on wavelet transform. *Electrotechnics Electric* 01: 5–6+24 (in Chinese).
5. Xu, J.Y., B. Zhang, X. Lin, et al. 2012. Application of energy spectrum entropy vector method and RBF neural networks optimized by the particle swarm in high-voltage circuit breaker mechanical fault diagnosis. *High Voltage Engineering.* 38 (06): 1299–1306 (in Chinese).
6. Yang, W., and M.Z. Rong. 2002. A new method of high-voltage circuit breaker's vibration signal processing. *Advanced Technology of Electrical Engineering and Energy.* 21 (3): 57–61 (in Chinese).
7. Huang, Y.L., and J.L. Qian. 1998. Mechanical condition monitoring for high voltage circuit breakers. *Journal of Tsinghua University (Science and Technology)*: 79–81 (in Chinese).
8. Meng, Y.P., B. Zhong, and S.L. Jia. 2005. Application and development of the vibration analysis in the condition monitoring of electrical equipment. *High Voltage Apparatus.* 06: 461–546 (in Chinese).
9. Zhao, K., F. Wang, Y.W. Yang, et al. 2018. Mechanical condition evaluation of high voltage circuit breaker based on the features fusion and optimization of different signals. *High Voltage Ap-Paratus* 54 (04): 14–19 (in Chinese).
10. Zhong, J.Y., Y. Liu, X. Lin, et al. 2018. Mechanical fault diagnosis technology research of high-voltage circuit breaker based on the vibration signal characteristic. *High Voltage Apparatus.* 49 (09): 49–54 (in Chinese).
11. Sun, L.J., X.G. Hu, Y.C. Ji, et al. 2016. Application of wavelet packet-eigen entropy in fault diagnosis of high voltage circuit breakers. *Automation of Electric Power Systems*: 62–65 (in Chinese).
12. Pan, H., Z.X. Song, and B. Niu. 2014. Design and development of condition monitoring system for circuit breaker based on vibration signa. *High Voltage Apparatus*: 83–88 (in Chinese).
13. Hu, X.G., J.M. Dai, Y.C. Ji, et al. 2001. The fault diagnosis of high voltage breakers based on wavelet singularity detection. *Proceedings of the CSEE*: 68–71 (in Chinese).
14. Yang, M., J. Wang, X.F. Zhou, et al. 2018. De-noising method based on CEEMD and wavelet packet. *Journal of Nanjing University of Posts and Telecommunications (Natural Science Edition).* 38 (02): 41–47 (in Chinese).

Relationship Between Digital Twin and Software Eco-chain

Shuhong Wang, Ruting Tang, and Naming Zhang

Abstract To adapt to the 4.0 era of industry, all kinds of enterprises must use digital means to accelerate product development and improve the efficiency of development and production. All of these require enterprises to have the complete digitizing ability, which is based on Digital Twin. To build a scientific, rational, professional, and accurate digital twin structure, the most critical technology is the industrial simulation software. By analyzing the key technologies of the popular digital twin concept, this paper points out the shortcomings of the domestic software market compared with foreign countries, and on this basis, analyses the relationship between building a perfect software eco-chain and the application of Digital Twin technology, and concludes that it is very important for China's industrial development to build a technology-based software industry ecological chain.

Keywords Digital twin · Software eco-chain

1 Introduction

After 2014, with the rapid development of the Internet of things, artificial intelligence, 5G, and virtual reality technology [1], Digital Twin has rapidly become one of the hot topics in the industry 4.0 era. In 2019 alone, more than 500 papers related to Digital Twin have been published. Digital Twin, together with big data and artificial analysis, artificial intelligence, cloud computing, information physical systems, and augmented reality, is listed as the six major technological trends in the industry 4.0 era [2]. Based on the advantages of model, data, and service, Digital twin is expected to become the development trend of Intelligent Manufacturing in the future. For the construction of a scientific and reasonable, professional, and accurate digital twin structure, the most critical technology is the industrial simulation software [3]. However, China's software industry has not yet formed a complete ecological chain of the software industry, which is very disadvantageous for mastering the cutting-edge

---

S. Wang · R. Tang · N. Zhang (✉)
School of Electrical Engineering, Xi'an Jiaotong University, No.28, Xianning West Road, Xi'an, Shaanxi, China
e-mail: namingzhang@xjtu.edu.cn

W. Chen et al. (eds.), *The Proceedings of the 9th Frontier Academic Forum of Electrical Engineering*, Lecture Notes in Electrical Engineering 743,
https://doi.org/10.1007/978-981-33-6609-1_66

technology of industrial production. Starting from the key technologies of Digital Twin, this paper discusses the shortcomings of mastering these technologies in China and makes a preliminary exploration of the influence of building a complete software industry ecological chain on Mastering Digital twin technology, to provide a reference for relevant researchers.

2 Concepts of Digital Twin and Its Core Technology

2.1 Concepts of Digital Twin Technology Structure of the System

In October 2019, a set of domestic digital twin standard system was set up in an article [4] published jointly by Beijing University of Aeronautics and Astronautics, China Institute of Standardization of Electronic Technology, and other relevant research institutions and enterprises. In April this year, Professor Lin [5] proposed for the first time that Digital Twin is a digital model of physical objects by comprehensively comparing the explanations on Digital Twin in various related fields. The model can evolve in real-time by adding data from physical objects, to keep consistent with the whole life cycle of physical objects. Digital Twin is a digital model corresponding to physical objects in real-time.

2.2 Application of Digital Twin Technology

The technology of Digital Twin has been used to rebuild the physical world in many fields, including smart manufacturing, smart city management, healthcare, and so on [6].

In the field of smart manufacturing, the military industry is still the main application field of Digital Twin. Represented by the U.S. Navy's Digital Twin Technology, the program encourages small and medium-sized contractors to actively create digital copies of components, equipment, and systems for testing new components, operational concepts, etc. That will be installed on ships or aircraft in the laboratory or other conditions, to identify potential problems ahead of time and improve performance; followed by high-end equipment manufacturing, such as aircraft and aircraft large engine, represented by Boeing 777, is the world's first airliner without drawings, models and digital twins. Digital Twin technology shortens the research and development cycle of the project by 40%; there are also heavy machinery industries and manufacturing enterprises.

In the field of smart city management, the "Earth Cloning Plan 3" conference held in February this year announced that the application of Digital Twin technology-driven microcosmic simulation, industrial simulation, and other applications are increasingly penetrating the research and development of technologies such as traffic management, smart transportation, and automobile driving. Digital Twin technology will play an important role in future urban transportation. The conference also launched the platform of automatic construction of full feature scenes and the "Shanghai Digital Twin System" [7].

In the healthcare industry, Japan's Takeda Pharmaceutical Company expects end-to-end production automation through digital twins. Pistek and his development team built complex virtual presentations of manufacturing processes in laboratories, created digital twins for each step, connected all parts through the overall digital twins, and automated control of processes between steps to complete end-to-end simulation of manufacturing processes. At present, the digital twin architecture of Wutian Pharmaceutical has been set up and applied to biological preparations [8].

2.3 The Core Technology of Digital Twin

Gartner believes that Digital Twin technology consists of at least four elements: a digital model, associated data, identity recognition, and real-time monitoring [9]. It can be seen that Digital Twin technology is a comprehensive technology. To build a digital twin for physical entity objects and put them into actual production, we need to first set up a digital model with computer-aided technology (CAX). By using sensor technology, we can collect all data of physical objects in real-time and synchronize them into the digital model. The parameters of this model correspond to the displayed physical objects one by one, which needs to meet the accuracy requirements. Such a model is called an object. The digital twin of the physical object and will run through the entire life cycle of the physical entity. Then through the analysis and observation of twins, the problems that products may encounter during operation are predicted and adjusted to help the physical objects to optimize and make decisions. In this way, physical objects, digital twins, and simulation and feedback based on Digital Twin constitute an information physical system.

In this process, it needs to involve many fields such as the Internet of Things technology [10], virtual reality technology, large data/artificial intelligence, 5G technology [11], and so on. The relationship between them is shown in Fig. 1.

Among these key technologies, the most important link is to connect the real world with the virtual digital world through the Internet of things technology, which

Fig. 1 Schematic diagram of key technologies of digital twin

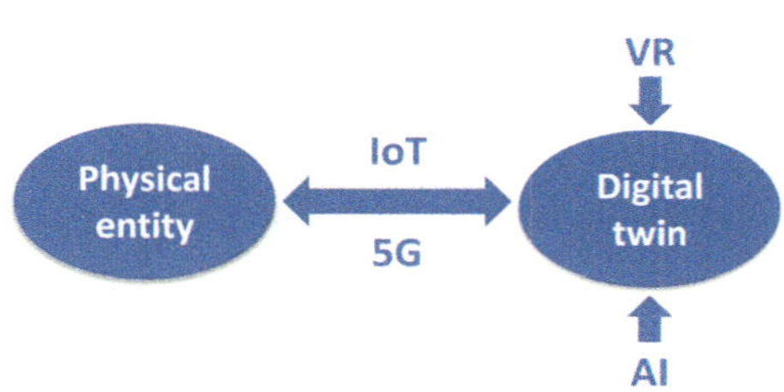

inevitably requires engineers to model the real physical objects, to observe and evaluate the digital model in the virtual world [12]. It can be said that the Internet of things technology is the core foundation of Digital Twin technology.

3 Viewing China's Software Industry Ecology from the Digital Twin Perspective

3.1 *The Importance of Software for Digital Twin*

As the definition says, behind Digital Twin is modeling and simulation technology, and the biggest flashpoint of Digital Twin is to combine digital models with the Internet of Things to polish them closer to reality. If we say that the Internet of Things technology is the core of the Digital Twin technology, then the software is the core and soul of the Internet of Things.

In recent years, although China's software industry has made considerable progress, there are still many problems. People in the industry generally believe that China's software industry lacks a harmonious ecological chain built by the government, enterprises, research institutions, institutions of higher learning, and users.

3.2 *Comparison of Domestic and Foreign Status of Software Industry Eco-chain*

At present, the global software market has formed an international software industry division system dominated by the United States, Europe, India, Japan, China, and other countries. The United States holds the core technology, standard system, and product market of the global software industry in the upstream. Most of the basic platform software enterprises such as the operating system and database are located in the United States. The middleware environment is concentrated in Ireland, India, Japan, and other countries and regions, and the application software set is in Germany, China, the Philippines, and other countries and regions. The core of the software industry—operating system, middleware, and database are occupied by American enterprises (Figs. 2 and 3).

China's software industry eco-chain construction is not yet perfect. There are many problems in coordination and coordination among government, software enterprises, research and development institutions, colleges and universities, customers, and the functions of each link have not been fully played.

Fig. 2 Global market share of mainstream operating systems in 2018

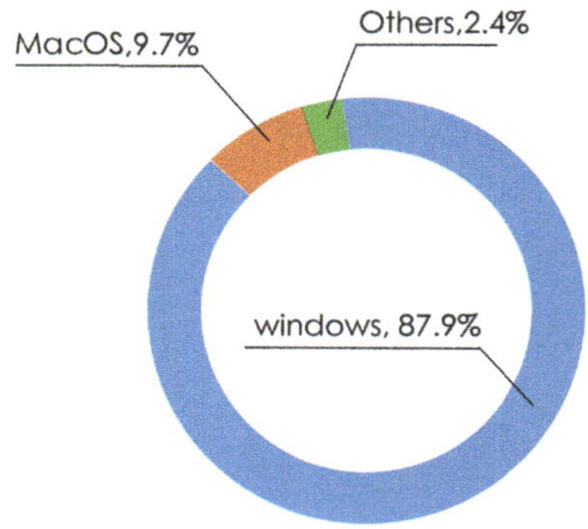

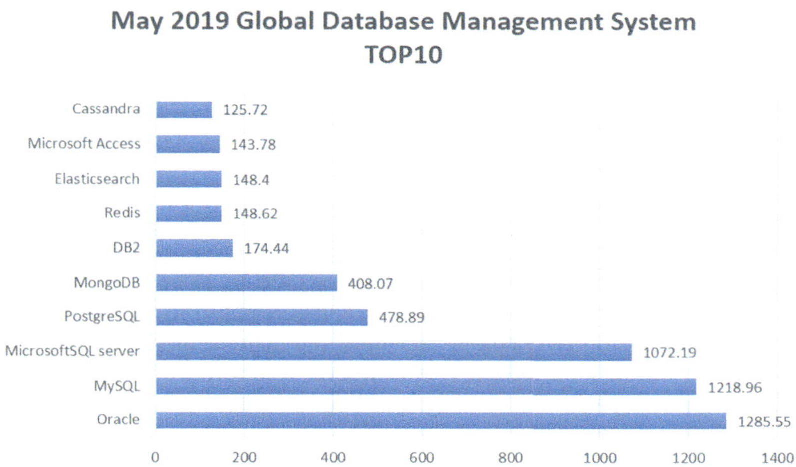

Fig. 3 Database management system score ranking published by DB engines in May 2019

3.3 *Prospect of Software Industry Eco-chain*

Competition among software enterprises has evolved into competition in the software industry chain. Foreign software giants have increased industrial concentration through mergers and acquisitions, optimized the software industry structure, and established a new pattern characterized by competition in the industrial ecological system. Vertical integration of the industry chain will become an important trend in future development [13]. Acquiring new business and superior resources quickly through merger and reorganization will be the inevitable trend of software industry development. The development of the software industry is an industry chain consisting of five main links: government, enterprise, research and development unit, colleges and universities, and users. The five links must play their respective functions

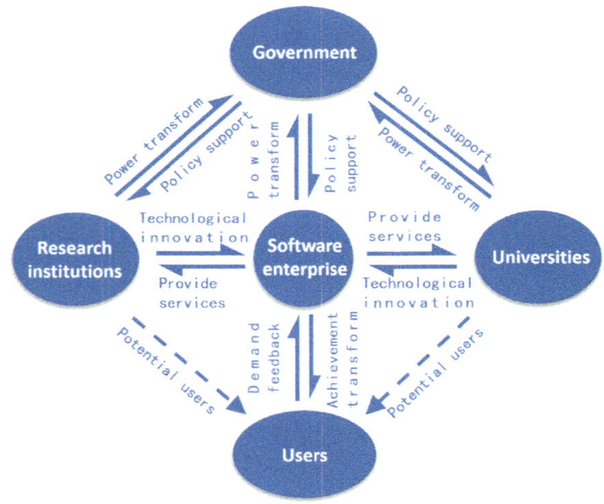

Fig. 4 Prospect of China's software industry eco-chain structure

and coordinate development so that the whole software industry can truly develop [14].

At present, China is accelerating the implementation of "Made in China 2025", which is the Chinese version of "Industry 4.0", and making every effort to promote the transformation and upgrading of the manufacturing industry to smart manufacturing. Industrial software should develop to the basic software platform for building a deep integration of information technology and advanced manufacturing technology and controlling and managing the entire production mode. Previous industrial software emphasized knowledge and technology, while future industrial software should be platform-based and support the entire process of the industry, which is completely consistent with the concept of Digital Twin (Fig. 4).

4 Conclusion

As one of the key technologies to determine the future, Digital Twin is bound to cause profound changes in industrial technology worldwide, and it is one of the most strategic cutting-edge technologies in the world. The combination of industrial 4.0 and Digital Twin will prompt industries to quickly launch new products, improve operations, and introduce new business models to boost revenue growth. Digital twin related core technologies, such as sensor technology, CAD technology, can not be separated from multi-domain and multi-scale fusion modeling. Most of these core technologies are in the hands of foreign enterprises. With the change of the international situation, Sino-US trade frictions bring great challenges to the domestic industrial software industry. Therefore, it is the only way for our country to achieve high-speed development to build a perfect software eco-chain, solve the current imbalance of the software industry and excessive dependence on foreign software, and form its core competitiveness.

Acknowledgements This work was supported by the State Grid Corporation of China through the Science and Technology Project under Grant 5500-201999543A-0-0-00.

References

1. Chen, Yong, Chen Yi, Pei Zhi, etc. 2020. Research progress of digital twin based on bibliometry. *China Mechanical Engineering* 31(7): 797–807. https://doi.org/10.3969/j.issn.1004-132X.2020.07.005 (in Chinese).
2. Technology Trends in the Industrial 4.0 Era. *China Elevator* 31(03): 1 (2020) (in Chinese).
3. Xie, Keqiang. 2020. Breaking the eco-bottleneck and standing the industrial ecology-thinking on developing design and simulation industrial software. *China Industry & Information Technology* 03: 44–49 (in Chinese).
4. Lin, Zhang. 2020. Cold thoughts on digital twin and the modeling and simulation technology behind it. *Journal of System Simulation* 32 (04): 1–10 (in Chinese).
5. Taofei, Ma Xin, Tianliang Hu, Zuguang Huang, Jiangfeng Cheng, Qinglin Qi, Meng Zhang, Weiran Liu, He Zhang, Shanggang Wang, Ruijuan Xue, Xiaodong Li, Weisha, Liu Mo, Difei Liu, Jian Zhou, Jian Zhang, Qin Li, Jinsong Ouyan, Wen Hu, Jun Ge, Lijuan Yan, Dongdong He, Luchen Chen, Hui Xu, Kui Liu, Wangmin Yi, and Hu Chen. 2019. Digital twin standard system. *Computer Integrated Manufacturing Systems* 25(10): 2405–2418 (in Chinese).
6. Liao, Xiaohong. 2019. Digital twin and its application tracking. *Guangdong Communication Technology* 39(07): 13–16+38 (in Chinese).
7. Digital twin technology simulates the optimal solution of urban transportation. *Automobile Parts* (02): 36 (2020) (in Chinese).
8. Digital twin: Connecting reality and digital world. *Software and Integrated Circuit* (05): 78–85 (2020) (in Chinese).
9. Huang, Changqing, Wuhan Wise, and Wuhan. 2012. Yangtze River Publishing House, Nov 2012: 158 (in Chinese).
10. Zhu, Hongbo, Longxiang Yang, and Qi Zhu. 2011. Progress and application of the internet of things technology. *Journal of Nanjing University of Posts and Telecommunications (Natural Science)* 31 (01): 1–9 (in Chinese).
11. Yang, Ling, and Gao Nan. 2017. 5G mobile communication key technologies and application trends. *Telecommunications Technology* (5) (in Chinese).
12. Gong, Xiaodong, Peng Bian, and Wenjing Wei. 2016. *Interactive design*, 34. Hefei University of Technology Press (in Chinese).
13. Tan, Zhanglu, and Xiao Chen. 2016. Study on the localization development strategy of china's software industry. *Journal of Technical Economics & Management* 08: 104–108 (in Chinese).
14. Tao, Zhang. 2006. Study on the construction of china software industry chain. *Journal of Liaoning Administration College* 07: 186–187 (in Chinese).

A Fast Calculation Method of Electromagnetic Parameters of Superconducting Magnet Based on Generating Convolution Network

Lingfeng Zhu, Yinshun Wang, Guangyi Zhang, and Yueyin Wang

Abstract In the process of designing magnets stacked by the second-generation superconducting ReBCO annular plates, the magnetic field distribution under the assumed current distribution can often determine the relevant performance of the magnet. However, when the number of annular superconducting plates stacked along the axis of magnet becomes large, the time cost of electromagnetic analysis of magnet by finite element method (FEM) increased at a geometric level, even can up to many days. In order to improve the calculation speed in this process, this paper proposes a fast calculation method of electromagnetic parameters of superconducting magnet based on generating convolution network. According to the size of the magnet, the generalized graph of the current distribution is established and input to the convolution neural network to obtain the magnetic field distribution of the magnet. When the number of conductor pieces is inconsistent with the number of training pieces, the method of input zeroing and dislocation superposition are used to realize the compatibility of the model. Then, the calculation time of magnetic field distribution of superconducting magnet with different number of rings is measured by this method, and the calculation accuracy of this method is compared with that of finite element method. The results show that the method can complete the calculation faster than the FEM, and the calculation accuracy can meet the engineering requirements.

Keywords Superconducting magnet · High speed magnetic field calculation · Convolution neural network · Generalized graph · Input compatibility

1 Introduction

The superconducting magnet staked by the second generation HTS REBCO annular plates has the advantages of high energy storage density, compact structure and simple processing, which has attracted many people's attention [1]. When designing superconducting magnet, the magnetic field distribution under the assumed current

L. Zhu · Y. Wang (✉) · G. Zhang · Y. Wang
State Key Laboratory of New Energy (North China Electric Power University), Beijing, China
e-mail: yswang@ncepu.edu.cn

W. Chen et al. (eds.), *The Proceedings of the 9th Frontier Academic Forum of Electrical Engineering*, Lecture Notes in Electrical Engineering 743,
https://doi.org/10.1007/978-981-33-6609-1_67

distribution usually needs to be calculated first before calculating the stress energy storage density to optimize the structure of the magnet. At present, the methods to calculate the magnetic field distribution under the assumed current distribution are mainly numerical algorithm, such as finite element method [2], boundary element method and so on. However it is not rational, especially in the early stage of magnet design, to take so much time to calculate the magnetic field distribution of a magnet via numerical algorithm and may affect the progress of the design [3, 4].

Fortunately, with the rapid development of artificial intelligence, artificial neural network algorithm is widely used in computer, communication, power, aerospace, medicine, chemistry, and other fields which also provides a new idea for the calculation of magnetic field distribution. However, the existing neural network algorithms have strict requirements on the dimension and length of input data, which means that the structure of a trained network can't be flexibly changed with the actual engineering requirements [5, 6]. In other words, the dimension of the data to be calculated should be highly consistent with the data used for training, which limits the application of neural network algorithm in the engineering design stage, leading to the neural network algorithm can not complete some tasks that numerical calculation method can.

Facing this situation, Xu Chao, Wang Changlong et al. proposed a finite element neural network model for 3D magnetic field calculation, which used conjugate gradient method as the training algorithm of neural network to analyze the simulation model [7]. However, although this algorithm can simplify the computational complexity of the finite element method, the computation load of the algorithm is still very large, and it can not flexibly change the input dimension of the network according to the actual engineering requirements [8].

In order to improve the speed of magnetic field calculation of superconducting magnet, at the same time, make the network adapt to the input data of different dimensions, this paper proposed a fast calculation method of electromagnetic parameters of superconducting magnet based on generating convolution network. Firstly, according to the geometric size of the superconducting magnet stacked by superconducting plates, the current distribution in the plates is transformed into a generalized graph matrix of current distribution. The generalized graph matrix is input into the convolution layer neural network, and then the magnetic field output is obtained by the network. Secondly, considering that the number of plates required in practical engineering may not be the same as that specified in network training, this paper adopts the method of zeroing the input matrix and shifting the summation to realize the magnetic field calculation when the dimension of input data is different from that required by network. Thirdly, the time and accuracy of the proposed algorithm and the finite element method are compared with each other when the number of superconducting plates is different. In addition, the accuracy of the proposed algorithm is also verified when the dimension of input data changes. The results show that the proposed algorithm can meet the requirements of magnetic field calculation accuracy in the early stage of design, and the calculation time can be reduced from a few hours or even days of finite element method to several seconds. In addition, the

Fig. 1 Structure of magnets

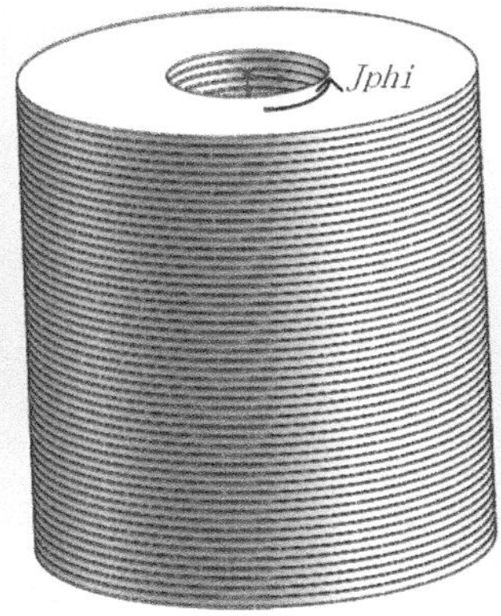

algorithm can overcome the dimension limitation of the network to the input data, which makes the algorithm more flexible in engineering calculation.

2 Generating Convolution Network

2.1 Magnets Stacked by ReBCO Annular Plates

The magnet calculated in this paper is composed of ReBCO annular plates stacked along the axial direction. There is a circular hole in the center of the superconducting plates which can be used for excitation. Its structure is shown in Fig. 1.

As is shown in Fig. 1, The current of the magnet flows along the phi direction around the central hole in the superconducting plates. Considering the small size of each superconducting plates, we can believe that the material properties of each plates are uniform along the phi direction if they have the same distance from the center of the circle. On this basis, the current distribution of the superconducting magnet can be simplified into a two-dimensional axisymmetric model as is shown in Fig. 2.

In the two-dimensional axisymmetric magnet model, we can use the current distribution as input to construct neural network.

2.2 Structure of Generating Convolution Network

The network structure is shown in Fig. 3.

As we can see in the picture above, J is the graph matrix derived from the current distribution among the plates which is input of the convolution network. And then

Fig. 2 two-dimensional axisymmetric model

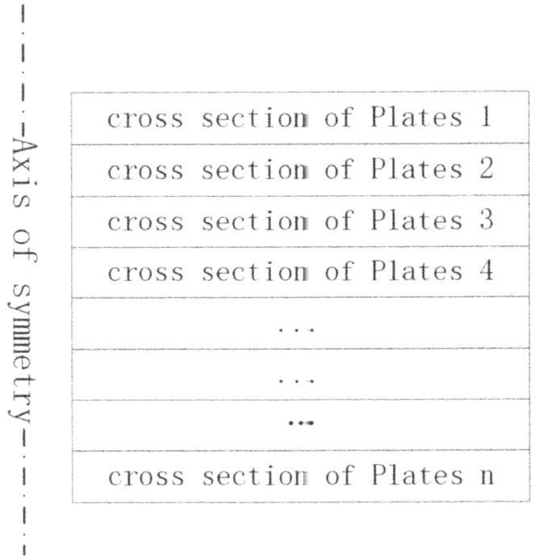

Fig. 3 structure of generating convolution network

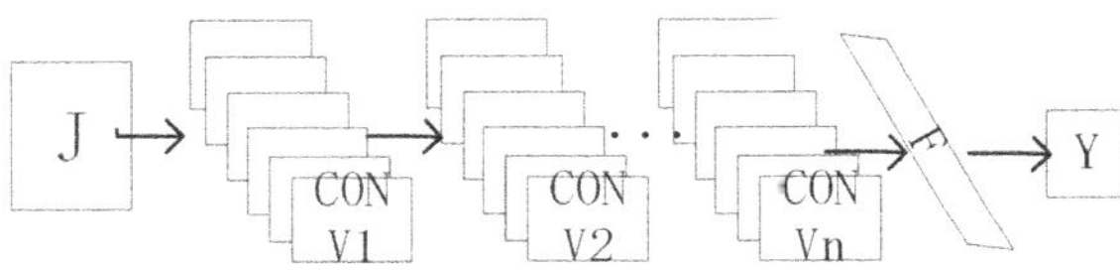

the output of the network is obtained through four convolution layers (Conv) and one flatten layer. Among them, convolution layer adopts depth separable structure to reduce network parameters. It is similar to Mobile_ Net that each convolution kernel convolutes with a row of the input matrix. Assuming that the input current distribution matrix is j, and each row of the matrix corresponds to the current in each superconducting plates in Fig. 2, the activation value h_i of each convolution kernel i is given as Eq. 1:

$$h_i = J(i, :) * \text{kernel}_i \tag{1}$$

The specific convolution correspondence is shown in Fig. 4.

The output of each convolution kernel is only connected in the last layer (flatten layer), and the flatten layer is directly connected to the output layer to obtain the magnetic field distribution.

3 Network Adaptation

Neural network requires that the dimension of input data must be the same as the standard dimension specified in network training. When the number of plates to be calculated is not equal to the number of superconducting magnets used in network

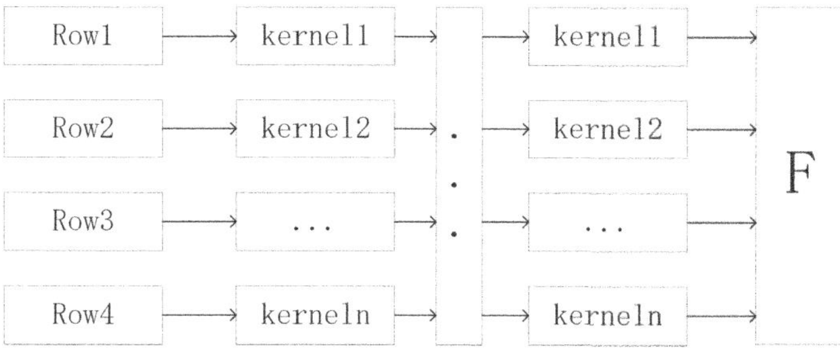

Fig. 4 The structure of convolution layer

training, the network will not be able to calculate normally. In view of the situation that the number of superconducting plates is less than or more than the number of standard input, this section adapts the proposed network, so that the network can be flexibly applied in engineering.

3.1 The Number of Plates is Less Than the Standard Number

According to Maxwell's equations, we know that the magnetic field can be calculated via this equation:

$$\nabla \times H = J \tag{2}$$

This equation is a linear partial differential equation. Therefore, the magnetic field constrained by the equation can be superposed.

When the number of superconducting plates is less than the standard input specified by the network, for example, the standard input required by the network is the current matrix corresponding to superconducting plates which number is m, while the actual input is the current distribution matrix generated by the superconducting plates which number is m–n. If we input matrix J directly into the network, the network will report an error. However, we can think that the number of superconducting sheets that make up the magnet is still m, but the current in n superconducting plates is zero. Therefore, we add n lines of 0 vector after the current matrix J to form a new current matrix J_1:

$$J_1 = \begin{bmatrix} J \\ J_0 \end{bmatrix} \tag{3}$$

J_0 is an all zero matrix with n rows.

$$\nabla \times H_1 = \nabla \times \begin{bmatrix} H \\ H_0 \end{bmatrix} = \begin{bmatrix} J \\ J_0 \end{bmatrix} = J_1 \tag{4}$$

Based on Eq. 2, we can get the results that:

$$H_1 = H + H_0 \tag{5}$$

Among them, H and H_0 can be derived from J and J_0 according to Eq. 2.

3.2 The Number of Plates is More Than the Standard Number

When the number of superconducting plates is more than the standard input specified by the network. We can't solve this problem by simply zeroing the input matrix. Assuming that the standard input required by the network is the current matrix corresponding to superconducting plates which number is m, while the actual input is the current distribution matrix generated by the superconducting plates which number is m + n. To solve this problem, we have to divide the magnetic field region to be solved and superconducting magnet into three regions, as is shown in Fig. 5.

As is shown in the figure, the number of superconducting plates to be calculated is significantly larger than that of standard input size. Therefore, the region of magnetic field distribution to be solved is also larger than that of standard input. Therefore, we divide the magnet into three regions and calculate them separately. According to the linear superposition principle, the calculated magnetic field distribution is superimposed, and then the magnetic field distribution of superconducting magnet is obtained.

When dividing the region of superconducting magnet, we make the height of region 1 plus region 2 the same as that of standard input, and the height of region 2 plus region 3 is the same as that of standard input as well. Therefore, we can solve the magnetic field of the three regions according to the following addition method.

For regions, we can solve the magnetic field distribution exactly as shown in the following Fig. 6.

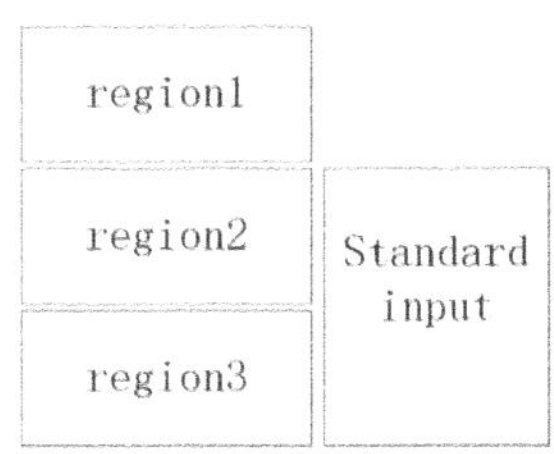

Fig. 5 Schematic diagram of area division

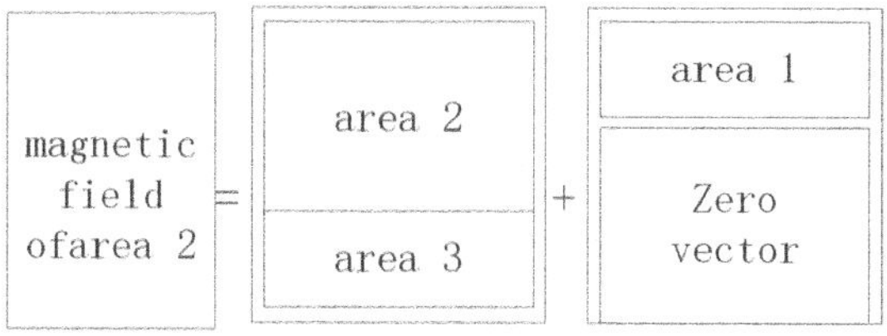

Fig. 6 Method of calculating the magnetic field at region 2

As is shown in Fig. 6, the magnetic field at region 2 can be thought as a superimposed value of the magnetic field generated by the currents in the superconducting annular plates at region 1, region 2 and region 3. First we input the matrix generated by the current in the superconducting plates at regions 2 and 3 into the neural network, and then we can obtain the magnetic fields at regions 2 and 3 generated by the superconducting plates. Second, the current distribution matrix for region 1 is fed into the network to calculate the magnetic field generated by this current at region 2. Since the height of the current distribution matrix for region 1 is less than the height of the standard input, it is necessary to set the current value corresponding to the superconducting plates of region 2 to zero and input the current distribution of region 1 and zeroed region 2 together into the neural network to obtain the magnetic field out of region 2. Finally, by summing the magnetic field values derived from the two calculations for region 2, the actual magnetic field values at region 2 generated by the superconducting annular plates of the three regions are obtained.

Since the geometry of regions 1 and 3 of a superconducting magnet is symmetrical with respect to region 2, magnetic field values at regions 1 and 3 are calculated in the same way, and we use region 1 as an example to show how to calculate the magnetic field in this region.

The magnetic field calculation for region one is schematically shown in Fig. 7.

The magnetic field at region 1 is generated by the current from all three regions together, but when the network is actually calculated, the magnetic field generated by region 3 can only be calculated up to region 2, and the magnetic field at region 1 is not in the output range of the network. However, given the longer distance between region 1 and region 3, the magnetic field generated by the currents in region 3 has

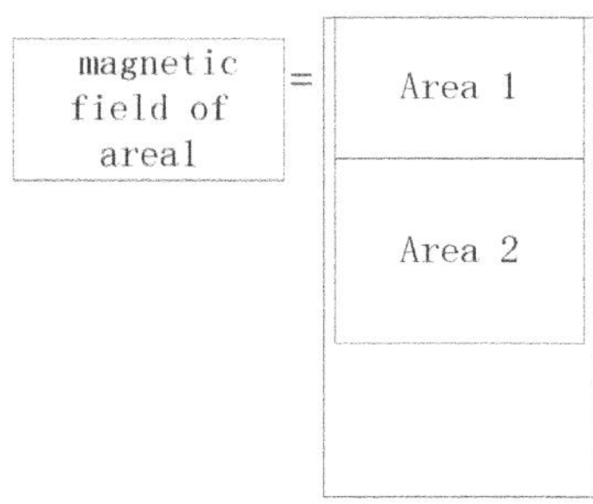

Fig. 7 Method of calculating the magnetic field at region 1

a large attenuation at one of the regions. Therefore, it can be approximated that the magnetic field at area 1 is generated by the currents in areas 1 and 2.

4 Simulation

The magnetic field distributions of magnets with different numbers of superconducting plates were calculated separately using the proposed algorithm, and the results were compared with the finite element algorithm as follows.

The results of the magnetic field distribution of a superconducting magnet composed of 100, 200, 300, and 400 superconducting loops under the proposed algorithm and finite element algorithm are given in Fig. 8a–d respectively.

In the figure, we can see that the calculation results of the proposed algorithm in this paper are almost identical to the FEM algorithm for both large and small magnetic field gradients, and its accuracy can fully meet the engineering needs.

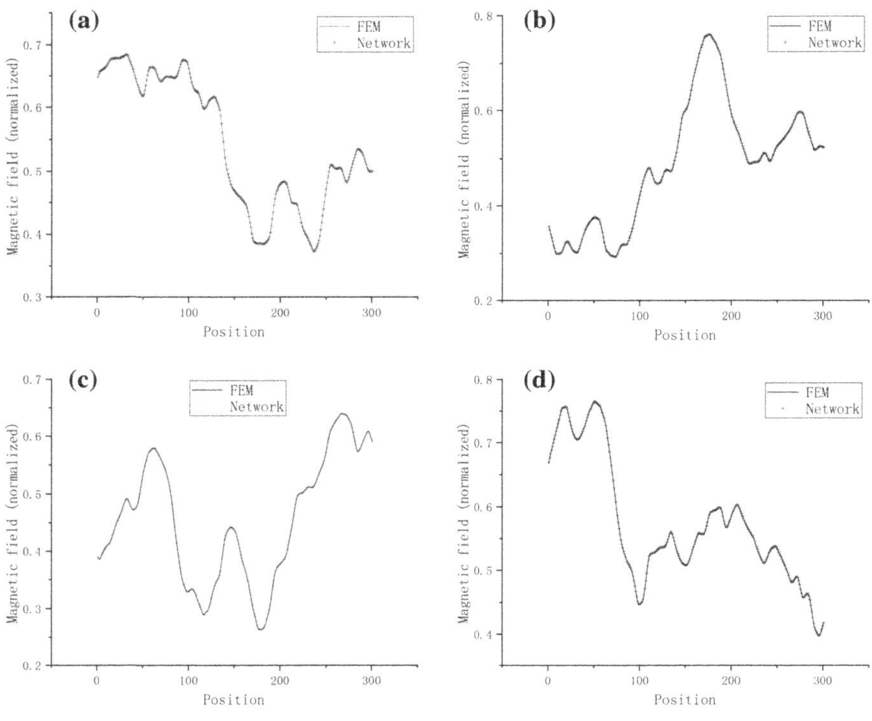

Fig. 8 Magnetic field distributions of magnets, **a** 100 loops, **b** 200 loops, **c** 300 loops, **d** 400 loops

Adaptation the network whose standard input is a current matrix corresponding to 250 superconducting plates to a magnet composed of 200 and 300 superconducting plates. The results are shown in Fig. 9.

It can be seen that the accuracy of the calculation is not affected when the magnet to be calculated is smaller than the magnet specified by the network, but when the magnet to be calculated is larger than the magnet specified by the network, the magnetic field is distorted at positions close to the top and bottom of the magnet, and the closer the position is to the ends, the lower the distortion of the calculation.

The computational speed of the proposed algorithm in this paper and the FEM algorithm are compared in Fig. 10.

As can be seen in the figure, although the computation time required for the proposed algorithm in this paper rises as the number of superconducting plates increases, the proposed algorithm in this paper is very much faster when compared to the time required for the finite element algorithm computation.

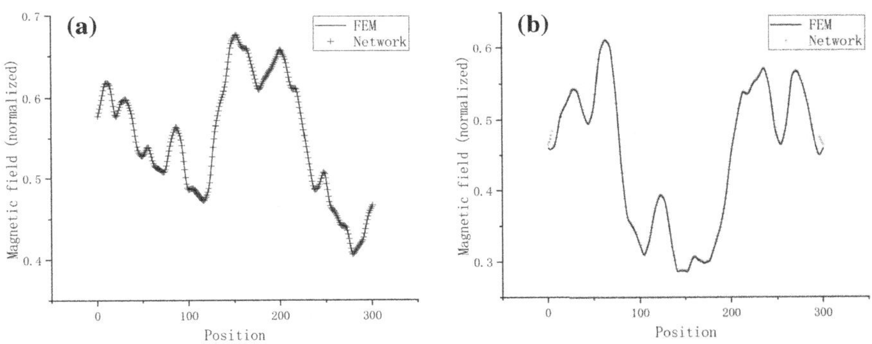

Fig. 9 Magnetic field calculation when the network is adapted, **a** 200 plates, **b** 300 plates

Fig. 10 Calculation speed

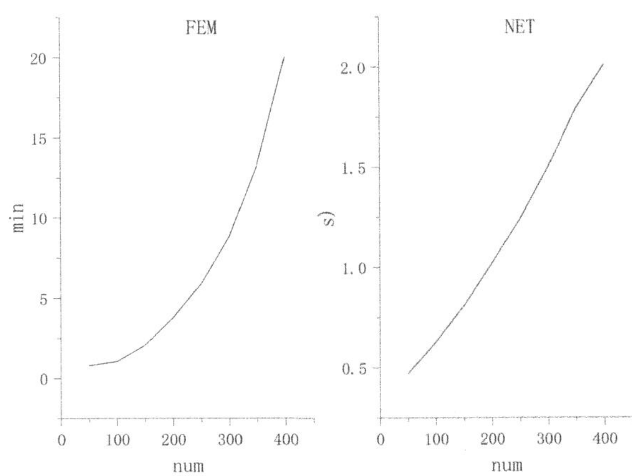

5 Summary

In this paper, we propose a fast calculation method for the magnetic field distribution of superconducting magnets, which can realistically calculate the magnetic field distribution of superconducting magnets under a hypothetical current distribution within a few seconds, greatly saving computation time. In addition, when the input data dimension changes, this paper adapts the algorithm to improve its practicality for engineering applications. The results show that the proposed algorithm can calculate the results with almost the same accuracy in a very short time compared to the finite element algorithm.

Acknowledgement This work is supported in part by the National Natural Science Foundation of China Under Grant No.51977078.

References

1. Yuan, X., Y. Wang, Y. Hou, C. Kan, C. Cai and M. Sun. 2018. Conceptual design of a bitter-like superconducting magnet stacked by REBCO annular plates and magnetized by flux pump. *IEEE Transactions on Applied Superconductivity* 28(3): 1–5.
2. Satiramatekul, T., and F. Bouillault. 2010. Finite element modeling of superconducting magnet. *IEEE Transactions on Applied Superconductivity* 20(3): 1634–1637.
3. Chen, Y., and D. Wang. 2011. Finite element method analysis of surface-inserted permanent magnets synchronous generator, 665–668. In 2011 *international conference on electrical and control engineering*. Yichang.
4. Yu, M., D. Shu, and X. Tao. 2018. Finite element calculation of normal electromagnetic force at feature key-points of permanent-magnet synchronous motor, 2059–2063. In *2018 Chinese automation congress (CAC)*. Xi'an, China.
5. Zhang, K.Y., M.K. Hu, C.P. Du, and M.Y. Xia. 2018. Detection of magnetic dipole target signals by using convolution neural network, 1–3. In 2018 *cross strait quad-regional radio science and wireless technology conference (CSQRWC)*. Xuzhou.
6. Ma, H., and E. Li. 2018. Prediction of IC equivalent magnetic dipoles using deep convolutional neural network, 1–3. In *2018 IEEE electrical design of advanced packaging and systems symposium (EDAPS)*. Chandigarh, India.
7. Wiegert, R., and J. Oeschger. 2006. Portable magnetic gradiometer for realtime localization and classification of unexploded ordnance. *Oceans*: 1–6.
8. Ginzburg, B., L. Frumkis, and B.Z. Kaplan. 2002. Processing of magnetic scalar gradiometer signals using orthonormal functions. *Sensors Actuators A: Physical* 102(1): 67–75.

Surface Modification of Low Temperature Plasma Electrode in Air Atmosphere

Yan Li, Fei Zhang, Xi Liu, Ning Luo, Yang Liu, Xiaodong Wu, and Xiaogang Han

Abstract Research on plasma surface modification of electric double layer supercapacitors has been widely used. In this paper, the surface of the electrode is modified by low temperature plasma in air atmosphere. It is found that the functional groups grafted on electrode by the surface treatment are mainly oxygen-containing functional groups, which can effectively reduce the interface transfer impedance of the capacitor and increase the specific capacitance (9.34%). The gas used in the treatment method is cheap and easy to obtain, environmentally friendly and non-polluting. It can rapidly process the electrode on a large scale and has practical value.

Keywords Electric double layer supercapacitors · Low temperature plasma · Oxygen-containing functional groups · Specific capacitance

1 Introduction

Climate change and the limited supply of fossil fuels have greatly affected the world economy and ecology. With the rapid growth of the portable electronic device market and the development of hybrid electric vehicles [1–6], the demand for environmentally friendly high-power energy is growing and pressing. In recent years, with the shortage of fossil energy and the increasingly serious climate problems, the use of new-type power cars and portable electronic devices has been rapidly increasing. At the same time, high-power power supplies are an important source of energy

Y. Li · F. Zhang · X. Liu · N. Luo
State Grid Xianyang Electric Power Supply Company, Xi'an, China

Y. Liu
International Electrotechnical Commission, System Committee Smart Manufacturing, System Committee Smart Cities, Xi'an, China

X. Wu · X. Han (✉)
State Key Laboratory of Electrical Insulation and Power Equipment, Xi'an Jiaotong University, No. 28 Xianning West Road, Xi'an, China
e-mail: xiaogang.han@xjtu.edu.cn

W. Chen et al. (eds.), *The Proceedings of the 9th Frontier Academic Forum of Electrical Engineering*, Lecture Notes in Electrical Engineering 743,
https://doi.org/10.1007/978-981-33-6609-1_68

for these devices, and market demand has also grown tremendously. Supercapacitors, also known as electrochemical capacitors, are a kind of energy density and power density between traditional capacitors and secondary batteries [7]. They are based on high specific surface materials such as porous carbon and metal oxides and are charged on the electrode–electrolyte interface. Discharged new type electric energy storage device. From its energy storage principle, it can be divided into electric double layer capacitors, pseudocapacitors and hybrid capacitors. Electric double layer capacitors mainly store charge through the electric double layer at the interface [8], and are currently the only commercially available types of supercapacitors. The electrode materials are graphene, carbon nanotubes, activated carbon and other carbon materials. The currently used commercial supercapacitors mainly use activated carbon [9], which has the advantages of wide range of raw materials, low price, large specific surface area, and abundant pores. However, poor conductivity and poor structural stability limit the energy density of products.

Due to the low energy density of electric double layer capacitors, researchers have designed from the aspects of electrode materials and electrolytes to improve their energy density. The research work on electrode materials mainly focuses on the pore design of carbon materials [10] and the surface treatment of materials. Since its discovery, plasma has been widely used in various aspects. Surface modification of carbon materials using low temperature plasma [11] has been widely concerned.

Plasma is an ionized gas-like substance composed of positive and negative ions produced by atoms and atomic groups after partial electron deprivation. The scale is a macroscopically neutral ionized gas larger than Debye length [12–15]. In the plasma chemical reaction, the energy states of electrons and ions are important parameters, which are expressed by the electron temperature Te and the ion temperature Ti. When the plasma is formed by discharge, the electrons are accelerated in the electric field and become high-energy electrons. The collision caused by the high-energy electrons is the ionization of the gas, which is generally divided into two cases: Te $\approx$ Ti and Te $\gg$ Ti. The former is called equilibrium plasma or high temperature plasma, and the latter is called non-equilibrium plasma or low temperature plasma. Low temperature plasma effectively avoids the thermal decomposition of reactants and products caused by the high temperature in the balanced plasma system, and has been widely used in the surface treatment of materials in recent years [11, 16, 17]. Low-temperature plasma reaction devices can be basically divided into two types: flow tube type and bell jar type. According to the position of the electrode, it can be divided into internal electrode type and external electrode type. Usually the device uses a high-frequency power supply, which can remove the segregation of charged particles in the plasma and obtain a uniform plasma space. Because the 13.56 MHz frequency band is easy to couple with the discharge gas, the stability of the tuning circuit is also good, and the operation is convenient. It is the most commonly used high-frequency power supply (Fig. 1).

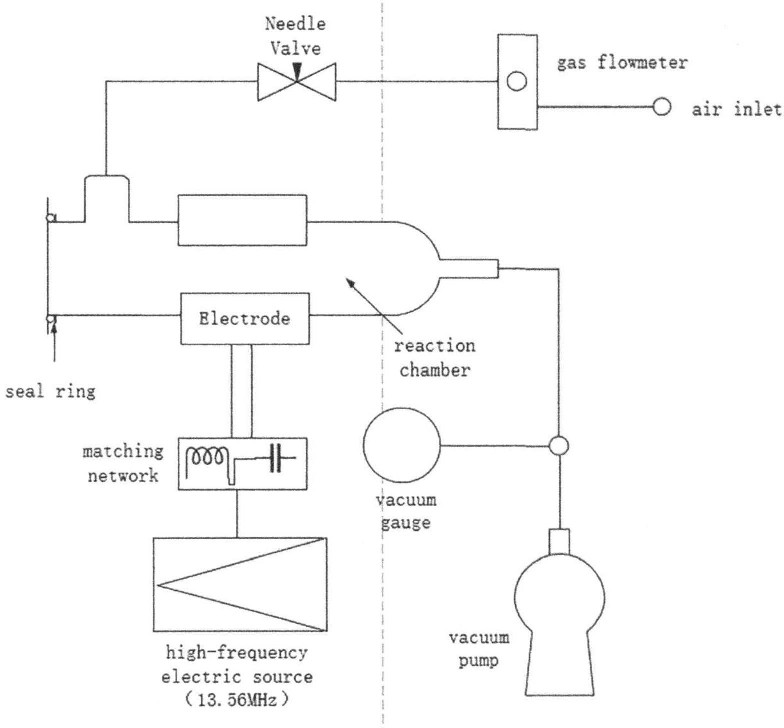

Fig. 1 Flow tube type internal electrode type plasma reaction device

2 Experimental

2.1 Materials

The capacitor electrode used in the experiment is made of a mixture of activated carbon (AC), SuperP and polyvinylidene fluoride (PVDF). The activated carbon uses YP50 produced by Japan Coca-Cola Company, and it is used without further treatment. The conductive agent is SuperP produced by TIMCAL. PVDF was purchased from Shenzhen Tianchenghe Technology Co., Ltd. (Arkema 900). AC:SuperP:PVDF was 80:10:10% wt. The slurry was coated on carbon-coated aluminum foil and dried in the vacuum oven at 70 °C. The electrode was then pressed and punched into a sheet with a diameter of 12 mm and a thickness of 50 μm (including an 18 μm aluminum foil current collector). The electrode liquid used for assembling the capacitor is DLC306 of Shenzhen Xinzhoubang Company.

2.2 *Measurement*

The surface treatment was performed on the electrode using a plasma reaction apparatus (CIF International Group Co, Ltd. CPC-B-13.56) under an air atmosphere. The reaction power is 80 W, the air pressure is 80 Pa, and the treatment time is 1 min and 5 min, respectively. The untreated and treated two sets of electrodes were made into capacitors respectively, and the electrochemical impedance spectroscopy (EIS) and cyclic voltammetry (CV) of the three sets of capacitors were tested using an electrochemical workstation (Bio-logic SP200, France), where CV Measured at 5 mv/s, 10 mv/s, 20 mv/s, 50 mv/s and 100 mv/s scan rates respectively. The optical contact angle measuring instrument (KRUSS DSA100, German) was used to measure the contact angle of the electrolyte of the pole piece, and the change of the first 1 s contact angle of the electrolyte on the pole piece was observed. Scanning electron microscope (NEC Electronics Corporation JEM2100) was used to perform energy spectrum analysis on the pole piece, and X-ray photoelectron spectrometer (AXIS ULtrabld) was used to analyze the carbon, oxygen and nitrogen elements on the surface of the eletrode.

3 Results and Discussion

Electrochemical impedance spectroscopy tests were carried out on the untreated and treated capacitors for 1 min and 5 min respectively, and it was found that the interface transfer impedance of the capacitor was reduced to a certain extent after plasma treatment in an air atmosphere (Fig. 2).

The Rct value of the interface transfer resistance of the control group was 23.56 Ω. The Rct of air plasma surface treatment for 1 min (hereinafter referred to as 1

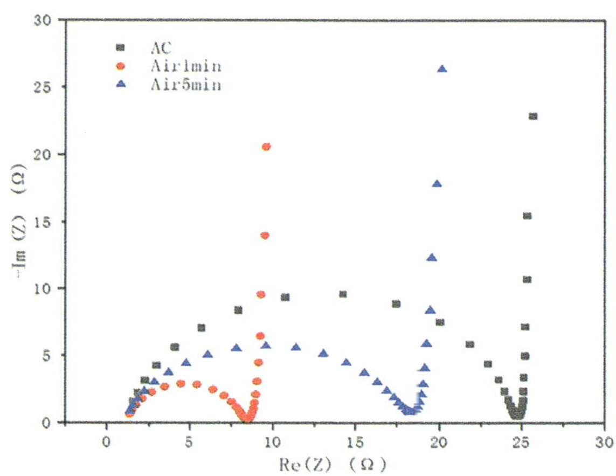

Fig. 2 Comparison of EIS of three groups of capacitors before and after air atmosphere plasma treatment

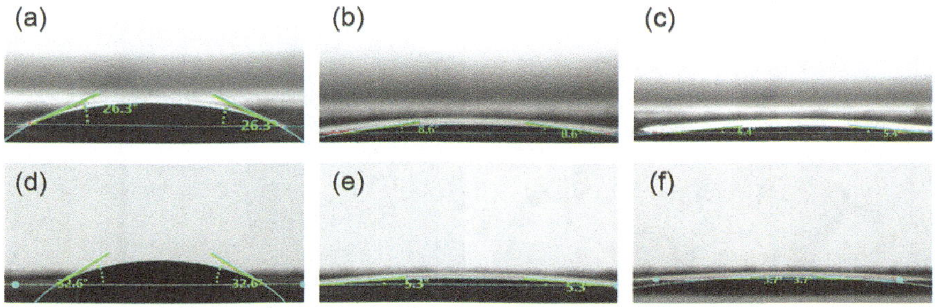

Fig. 3 The dynamic contact angle of the untreated and treated 1 min pole piece **a** AC 0 s, **b** AC 0.5 s, **c** AC 1 s, **d** 1 min 0 s, **e** 1 min 0.5 s, **f** 1 min 1 s

Min) and surface treatment for 5 min (referred to as 5 Min) were 7.195 Ω and 17.77 Ω, respectively (Fig. 3).

The contact angle of the electrolyte in the control group changed from 26.3° to 5.4° in 1 s, and the contact angle of the 1 min electrolyte changed from 32.6° to 3.7°. After the surface treatment of the pole piece, the wettability of the electrolyte is improved.

The interface transfer impedance is mainly related to the surface morphology and materials of the electrode. Due to the plasma surface treatment, the functional groups containing N and O will be grafted on the surface of the pole piece. These functional groups can improve the contact between the electrolyte and the electrode surface to a certain extent. Large electrolyte wettability reduces interface transfer resistance.

Elemental analysis by EDS and XPS further proves this point. The schematic diagram of EDS scanning corresponding to the pole piece with surface treatment for 1 min is shown in Fig. 4.

Through the comparison of the content of surface elements (Tables 1 and 2), the surface treatment effectively grafted nitrogen-containing and oxygen-containing functional groups on the electrode, mainly oxygen-containing functional groups. The content of nitrogen and oxygen on the surface of the 1 min group is more than that of the 5 min group. After 5 min of surface treatment, the number of surface functional groups decreased compared with 1 min of treatment. Compared with the EIS test results of electrochemical impedance spectroscopy above: the transfer resistance of the electrode interface at 5 min. The pole piece greater than 1 min is consistent. This is mainly because when the plasma treats the surface of the pole piece, the original structure on the surface of the pole piece is destroyed first, and the nitrogen and oxygen functional groups are grafted on the electrode surface. These two processes occur at the same time, and the appropriate power and air pressure need to be adjusted. And the processing time to achieve the best performance of the electrode surface. Comparing the EDS and XPS element content data, since the EDS test depth is about several hundred nanometers and the XPS test depth is about 10 nm, the grafted nitrogen and oxygen functional groups are mainly distributed

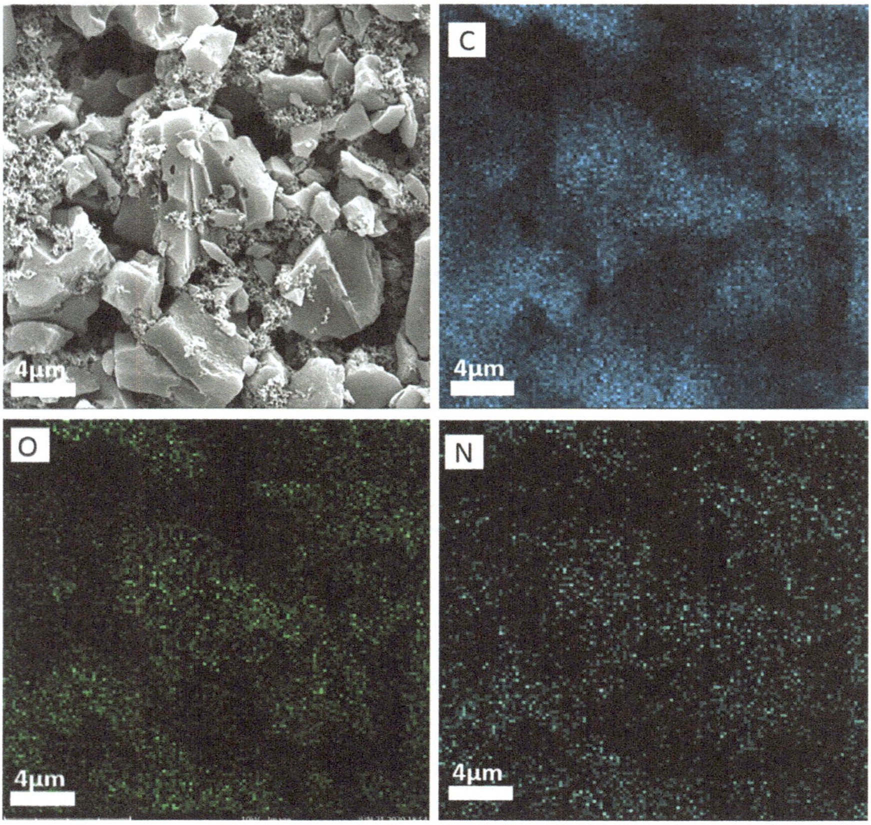

Fig. 4 The EDS corresponding SEM image and C, O, N element analysis of electrode with a processing time of 1 min

Table 1 EDS control group, 1, 5 min electrode percentage content

Name	C	O	N
AC	100	0	0
1 MIN	91.80	5.77	2.43
5 MIN	92.87	5.38	1.75

Table 2 XPS control group, 1, 5 min electrode surface element percentage content

Name	C	O	N
AC	93.69	5.59	0.73
1 MIN	80.29	16.22	3.49
5 MIN	81.14	16.21	2.66

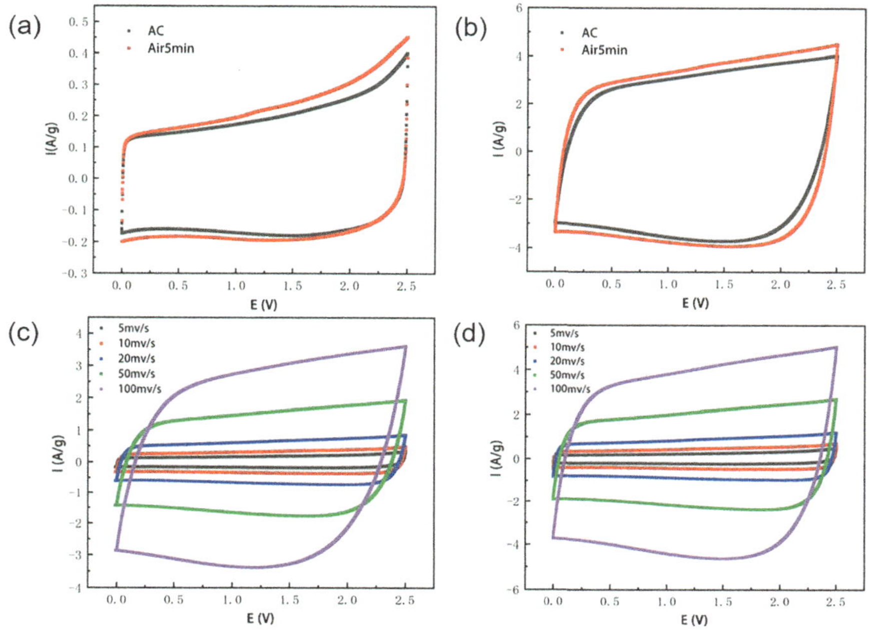

Fig. 5 Cyclic voltammetry curves of untreated group, 1 min group and 5 min group **a** Scan rate 5 mv/s, **b** scan rate 100 mv/s different scan rates of CV curve, **c** untreated group, **d** Process CV curve of 5 min group

on the electrode surface. As the depth increases, the content of O and N element is reduced.

Three groups of capacitors were tested by cyclic voltammetry (CV) (Fig. 5). The comparison of CV curves revealed that the surface treatment was carried out for 5 min at low sweep speed (5 mv/s) and high sweep speed (100 mv/s). The curve of the capacitor is closer to rectangular, indicating that its electric double layer performance is better. The treated pole piece was subjected to constant current charge and discharge test, and it was found that the capacitance value of the capacitor increased to a certain extent. The initial capacity of the untreated group was 34.47 F/g; the one-minute treatment group was 37.66 F/g, the capacity increased by 9.25%; the five-minute treatment group was 37.69 F/g, and the capacity increased by 9.34%. In the CV curve, the envelope area treated for 5 min is smaller than the untreated group because of the difference in the quality of the electrode active material. On the area of 4.5 mm^2, the mass of activated carbon is 1.936 mg in the former and 1.552 mg in the latter.

4 Conclusion

This work proves that the use of low-temperature plasma to treat activated carbon-based electrode materials in an air atmosphere can increase the energy density of the capacitor (9.34%) and reduce the interface transfer resistance of the capacitor, which is conducive to improving its stability during cycling. Sex. Because air is cheap and readily available, it is relatively easy to realize large-scale and rapid processing of electrode materials in production, and has strong practical value.

References

1. Zhang, L.L., and X.S. Zhao. 2009. Carbon-based materials as supercapacitor electrodes. *Chemical Society Reviews* 38 (9): 2520–2531.
2. Zhijun, J., W. Jun, and W. Yi. 2014. Research progress of the electrode materials for electrochemical capacitors. *Energy Storage science & Technology*.
3. Kang, W., Y.U. Ai-Mei, and Z. Hua-Jun. 2010. Electrode materials for super-capacitor. *Zhejiang Chemical Industry* (in Chinese).
4. Chao, X., T. Bin, W.U. Meng-Qiang, et al. 2007. Research progress of supercapacitor electrode materials. *Insulating Materials*.
5. Lei, Y., W. Chaoyang, F.U. Zhibing, et al. 2010. Research progress in electrode materials for supercapacitor. *Materials Review*.
6. Xiu-Feng, Z., J. Xiao-Yan, Z. Mi-Lin. 2002. Progress on metal oxides -based supercapacitors and their applications. *Journal of Functional Materials & Devices*.
7. Lili, Y.U., Z. Junjie, and Z. Jingtai. 2015. The present situation and development trend of supercapacitors. *Chinese Journal of Nature* (in Chinese).
8. Hong-Sheng, Y., Z. Xiao, F. Tian-Fu, et al. 2003. Recent advances in the study on electrochemical capacitors I. Electric double-layer capacitors. *Electronic Components and Materials*.
9. Gao, Y., L. Li, Y. Jin, et al. 2015. Porous carbon made from rice husk as electrode material for electrochemical double layer capacitor. *Applied Energy* 153: 41–47.
10. Chmiola, J., G. Yushin, Y. Gogotsi, et al. 2013. Anomalous increase in carbon capacitance at pore sizes less than 1 nanometer. *Science* 313.
11. Shan, Q.J. 2001. Application of low temperature plasma in surface treatment of carbon materials. *New Carbon Materials*.
12. Guijian, Z., L.I. Kai, L. Qiang, et al. 2015. Research progress of removing atmospheric pollutants by non-thermal plasma technology. *Materials Review*.
13. Xi, Y., Y.I. Honghong, T. Xiaolong, et al. 2017. Research progress in treatment of toluene in industrial waste gas by non-thermal plasma technology. *Safety and Environmental Engineering*.
14. Zhang, K., R. Wang, W. Han, et al. 2016. Progress of heavy oil processing by plasma technology. *Transactions of China Electrotechnical Society* 31 (24): 1–13 (in Chinese).
15. Fan, L., et al. 2016. Response regularity between glow discharge plasma and static pressure change. *Transactions of China Electrotechnical Society* 031 (024): 54–61 (in Chinese).
16. Zhang, X., H. Xiao, and Y. Huang. 2016. A review of degradation of SF_6 waste by low temperature plasma. *Diangong Jishu Xuebao/Transactions of China Electrotechnical Society* 31 (24): 16–24 (in Chinese).
17. Liu, Y., L. Jiang, H. Wang, et al. 2019. A brief review for fluorinated carbon: Synthesis, properties and applications.

Numerical Analysis and Research on Rectangular Crack Detection of Basin Insulator Based on Laser Ultrasonic

Haifei Wu, Hui Xia, Guoqiang Liu, Cong Chen, and Xin Huang

Abstract In order to improve the surface defect detection efficiency of GIS basin insulator materials and ensure the reliability and safety of the final product, a non-contact non-destructive detection method for rectangular crack defects based on laser ultrasonic technology was studied. The finite element numerical analysis method (FEM) was used to theoretically study the laser guided ultrasonic guided waves in the GIS basin insulator material. The interaction between the laser ultrasonic and rectangular crack defects with different widths and depths on the surface of the material was simulated. The numerical results showed that when the crack was located at the excitation point and the reception point, the width and depth information of the rectangular crack can be qualitatively characterized by the received signal amplitude, and the defect depth can be quantitatively characterized by whether the longitudinal wave signal can be received.

Keywords Basin insulator · Laser ultrasonic · The finite element method · Rectangular crack

1 Introduction

Gas Insulated Metal-enclosed Switchgear (GIS) equipment is widely used in my country's power system due to its integrated structure, high reliability, convenient maintenance, and small footprint [1–5]. Basin insulator is an indispensable and important part of GIS. It has the functions of insulation, isolation of air chamber,

H. Wu (✉) · H. Xia · G. Liu · C. Chen
Institute of Electrical Engineering, Chinese Academy of Sciences, NO. 6 Beiertiao, Hai Dian District 100190, Beijing, China
e-mail: xiahui@mail.iee.ac.cn

H. Wu · C. Chen
University of Chinese Academy of Sciences, NO. 80 Zhongguancundonglu, Hai Dian District 100190, Beijing, China

X. Huang
Shandong Jiaotong University, 5001 Haitang Road, Changqing District, Jinan 250357, China

W. Chen et al. (eds.), *The Proceedings of the 9th Frontier Academic Forum of Electrical Engineering*, Lecture Notes in Electrical Engineering 743,
https://doi.org/10.1007/978-981-33-6609-1_69

support and fixation. It is mainly composed of metal inserts, outer flanges and epoxy resin parts [6, 7]. The safety performance of the insulator determines whether the entire GIS equipment can operate stably and safely. When its structure is damaged, it will cause partial discharge and even insulation breakdown, endangering the safety of the power system. In severe cases, it will cause a large-scale power outage. The loss will be immeasurable [8–10]. Therefore, the defect detection of basin insulators is of great significance to ensure the safe operation of the power system and extend the service life of power equipment.

Surface cracks are one of the main defects of basin-type insulators, which may be caused by the temperature change during curing and the mechanical extrusion of the centre conductor and the outer shell that cause the insulator to withstand huge stress, which leads to cracks on the surface; or transportation and installation Unbalanced forces and environmental temperature differences in the process may cause surface cracks [11]. According to the current operating experience analysis of GIS equipment, the main impact on the safe operation of the equipment is the cracking failure caused by the surface cracks of the insulators [12]. Therefore, for the safe operation of power grid equipment, it is necessary to effectively detect whether the basin insulators have surface cracks.

At present, the defect detection methods for basin insulators are mainly divided into electrical detection methods and non-electric detection methods [13–16]. Electrical detection mainly includes pulse current method and ultra-high frequency method. Non-electric detection methods mainly include local ultrasonic detection method. X-ray is commonly used for factory inspection [17–19]. Imaging technology detection. The pulse current method can quantitatively detect partial discharge, but it is susceptible to external electromagnetic interference, so this method is mainly used in the laboratory; the ultra-high frequency method is mainly used for on-site detection, which can effectively avoid the influence of external electromagnetic interference and can be detected 5% discharge phenomenon. However, the defect location ability of this method is not accurate enough, and the hardware and software are not mature enough. Further research is needed. The local ultrasonic detection method excites the basin insulator through an external sensor, which is a contact detection method, And the characteristics of single acoustic wave mode and narrow frequency band make the detection sensitivity not high, which has certain limitations for the detection of crack defects of basin insulators. Therefore, the current state monitoring and defect fault detection technologies for basin insulators are still not mature enough, and it is necessary to improve the existing methods or explore better methods for detecting insulators [20].

This article introduces the use of laser ultrasonic technology to detect surface cracks in insulators. Laser ultrasonic technology has the advantages of non-contact, diverse modes, bandwidth, and high detection resolution. It is widely used in non-destructive testing of composite materials and sheet metal materials [21–23].

2 Principle of Laser Ultrasonic Testing

2.1 Heat Transfer Process

532 nm Pulsed laser irradiated surface of the insulator, as shown in Fig. 1, due to the thermoelastic effect material produces a strong directivity Rayleigh waves on the surface, characterized by surface cracks to analyse signal characteristics of waves, the two-dimensional Cartesian coordinate system employed in the analysis, the heat conduction equation is expressed as:

$$\rho C_p \frac{\partial T(x, y, t)}{\partial t} = \kappa \left(\frac{\partial^2 T(x, y, t)}{\partial x^2} + \frac{\partial^2 T(x, y, t)}{\partial y^2} \right) \tag{1}$$

where $T(x, y, t)$ represents the temperature distribution field at time t, ρ represents the insulator density, C_p is the heat capacity at constant pressure, and κ is the thermal conductivity coefficient.

The pulsed laser energy can be expressed by a heat source function Q, so the boundary condition at $y = 0$ can be obtained as:

$$-\kappa \frac{\partial T(x, y, t)}{\partial y} \bigg|_{y=0} = Q \tag{2}$$

The remaining boundaries are "Thermal Insulation", the boundary condition heat flux is 0, and the initial test temperature is room temperature $T = 293.15$ K. For the heat source function Q can be written as the product of the spatial distribution and the time distribution, as:

$$Q = I_0(1 - R) f(x) g(t) \tag{3}$$

where is the spatial distribution function of $f(x)$ pulsed laser, $g(t)$ is the time distribution function of laser, I_0 is the power density of laser, R is the reflectivity of the

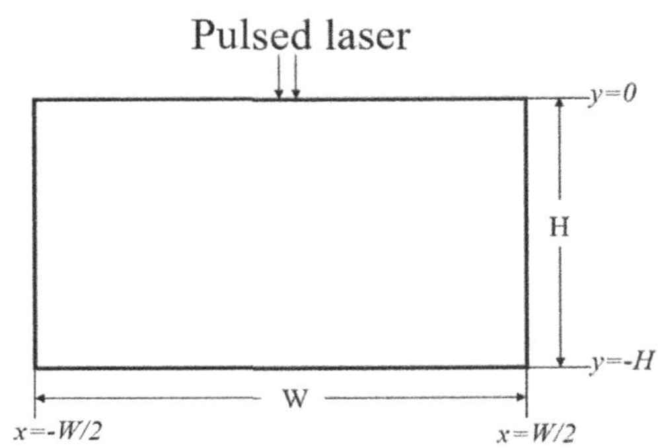

Fig. 1 The geometry of the material illuminated by the laser

basin insulator composite material surface to light, $f(x)$ and $g(t)$ can be expressed by an exponential function, as:

$$f(x) = e^{-\frac{x^2}{w^2}} \tag{4}$$

$$g(t) = \frac{t}{t_{ex}} e^{-\frac{t}{t_{ex}}} \tag{5}$$

Among them, w is the half-width of the laser beam, and t_{ex} is the rise time of the laser energy. Generally, the width of the pulsed laser used for non-destructive testing is nanoseconds, and the pulse width of 10 ns is adopted in this article.

2.2 *Thermoelastic Process*

When a material is irradiated by a pulsed laser, an elastic wave is generated due to the expansion mechanism. The equation of motion of the elastic body (Navier equation) is:

$$\nabla \cdot \mathbf{S} + \mathbf{f} = \rho \frac{\partial^2 \mathbf{u}}{\partial t^2} \tag{6}$$

where $\boldsymbol{S}$ is the stress tensor, its divergence $\nabla \cdot \mathbf{S}$ is the stress vector, $\boldsymbol{f}$ is the external force of the volume, and $\boldsymbol{u}$ is the displacement field. Considering the thermal stress and ignoring the external force of the volume, there are:

$$(\lambda + 2\mu)\nabla^2 \mathbf{u} + (\lambda + \mu)\nabla \times \nabla \times \mathbf{u} = \rho \frac{\partial^2 \mathbf{u}}{\partial t^2} + (3\lambda + 2\mu)\alpha \nabla T \tag{7}$$

where λ and μ are the Lame coefficients of the material, and α is the thermal expansion coefficient of the material. The displacement field is coupled with the heat conduction equation. Pulsed laser irradiation produces a temperature gradient, resulting in a transient displacement field and generating elastic waves to propagate outward. The upper and lower surfaces are "free" boundary conditions, and the left and right boundaries are "low reflection boundary" conditions. The initial conditions are:

$$\mathbf{u}|_{t=0} = \left.\frac{\partial \mathbf{u}}{\partial t}\right|_{t=0} = 0 \tag{8}$$

2.3 Thermoelastic Process

According to the finite element principle, the discrete form governing equations of Eqs. (1) and (6) can be established by neglecting the effect of damping:

$$\mathbf{C_p}\{\dot{T}\} + \mathbf{K_1}\{T\} = \mathbf{Q} \tag{9}$$

$$\mathbf{M}\{\ddot{\mathbf{U}}\} + \mathbf{K_2}\{\mathbf{U}\} = \mathbf{F} \tag{10}$$

where $\boldsymbol{C_p}$ is the heat capacity coefficient matrix, $\boldsymbol{K}_1$ is the heat conduction matrix, $\boldsymbol{Q}$ is the heat source vector, $\{T\}$ is the nodal temperature, $\{\dot{T}\}$ is the temperature change rate with time, $\boldsymbol{M}$ is the mass matrix, $\boldsymbol{K}_2$ is the stiffness matrix, and the $\boldsymbol{F}$ is the node load vector, $\{\boldsymbol{U}\}$ is the displacement vector, $\{\ddot{\mathbf{U}}\}$ is the acceleration vector.

3 Numerical Simulation Analysis of Rectangular Crack

3.1 Laser Parameters and Material

Based on the above-mentioned theory, calculate the ultrasonic guided wave of the laser pulse emitted by the Nd:YAG solid-state laser in the basin-type insulator material, and the receiving position and the excitation position are located at y = 0 on the upper surface. Pulsed laser is irradiated on the surface of the insulator material, part of the energy is reflected, and part of the energy is absorbed by the insulator material. Assuming that the heat generated by the laser energy is concentrated on the irradiated surface, the "heat flux" boundary condition is adopted, considering that the damage threshold energy of the insulator should not be Too high, so the magnitude of the heat amplitude is $I_0 = 1 \times 10^5$ MW/m^2, the rise time t_{ex} of the pulsed laser is 10 ns, and the beam radius w of the laser is 100 μm. Set up a two-dimensional model, the geometric width W of the solution area is 6 mm, and the height H is 3 mm. In order to accurately solve the distribution of the sound field, the maximum grid length of the excitation boundary is 20 μm, the domain grid far away from the laser incident center is maximum 100 μm, and the time step is 2 ns. In addition, the material parameters of the basin insulator are shown in Table 1.

3.2 Numerical Simulation Results Without Defects

Figure 2 shows the displacement curve of the receiving position 3 mm away from the excitation point. The receiving position is on the same side as the source. The received signal features include grazing longitudinal waves (L), grazing transverse waves (S)

Table 1 Material parameters of basin insulator

Parameter attributes	Values
Density $\rho/(\mathrm{kg\ m^{-3}})$	2200
Young's modulus E/(GPa)	20
Poisson's ratio υ	0.3
Thermal conductivity $\kappa/(\mathrm{W\ m^{-1}\ K^{-1}})$	0.25
Heat capacity $C_p/(\mathrm{J\ kg^{-1}\ K^{-1}})$	1500
Thermal expansion coefficient $\alpha/(1/\mathrm{K})$	18×10^{-6}

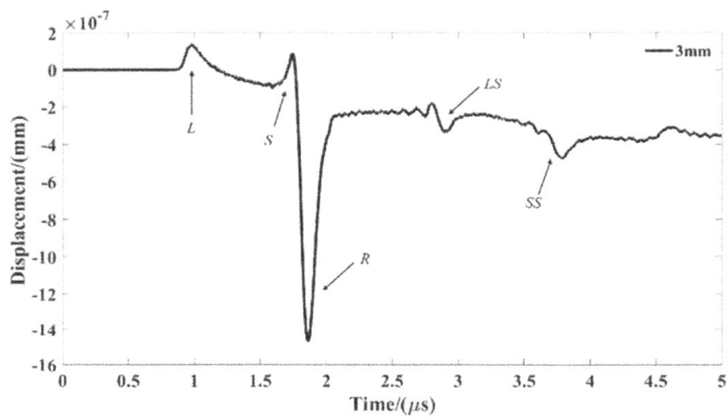

Fig. 2 Displacement curve at 3 mm

and Rayleigh waves (R). In addition, the longitudinal and transverse waves generated by the laser excitation will be reflected when they meet the bottom surface. After the longitudinal wave is reflected, a part of the mode will be converted into a transverse wave (LS), causing the displacement at the receiving position to fluctuate, and the transverse waves will also be reflected into transverse waves (SS) and propagate back to cause changes in the signal. Therefore, it can be judged whether there is a defect based on the change of the received signal characteristics.

According to the duration of longitudinal waves and surface waves, the wavelengths of longitudinal waves and surface waves can be estimated. From Fig. 2 we can see that the duration of longitudinal waves is 0.13 μs, the dominant frequency is 7.7 MHz, the duration of surface waves is 0.20 μs, and the dominant frequency is 5 MHz. The velocity of longitudinal wave and surface wave can be calculated by the length and time of the propagation path, and the velocity of longitudinal wave c_L = 3498.3 m/s and the velocity of surface wave c_R = 1734.7 m/s. Therefore, the longitudinal wave wavelength is about λ_L = 0.445 mm, and the surface wave wavelength is about λ_R = 0.350 mm.

3.3 Simulation Results of Rectangular Cracks with Different Widths

As shown in Fig. 3, a rectangular crack defect is set at a distance of 1.5 mm from the center of the laser excitation position. The depth *d* of the crack is the length λ_L of a longitudinal wave wavelength. Different crack widths are changed, and the crack widths are respectively set to $\lambda_L/4$, $\lambda_L/2$ and λ_L.

Place the signal receiving point on the right side 1 and 3 mm away from the excitation point. At 1 mm, both the excitation point and the receiving point are located on the left side of the crack, while the receiving point is at 3 mm. The excitation point is on the left side of the crack and the receiving point is on the right side of the crack. Side, the observed displacement signal is shown in Fig. 4.

It can be seen from Fig. 4 that when the receiving position and the excitation point are on the same side of the crack, the signal characteristics received for rectangular cracks of different widths are almost coincident, and when the receiving position is moved to a distance of 3 mm from the excitation point, that is, it is separate from the excitation point. When located on different sides of the crack. It can be found that as the width increases, the amplitude of the Rayleigh wave received for the first-time decreases. Because the width increases, the ability of the Rayleigh wave to bypass the rectangular crack to the receiving point decreases. Therefore, the energy received decreases accordingly. When the depth d remains the same, continue to increase the crack width, the received signal amplitude will continue to decrease, but will not be zero.

Fig. 3 Location of crack

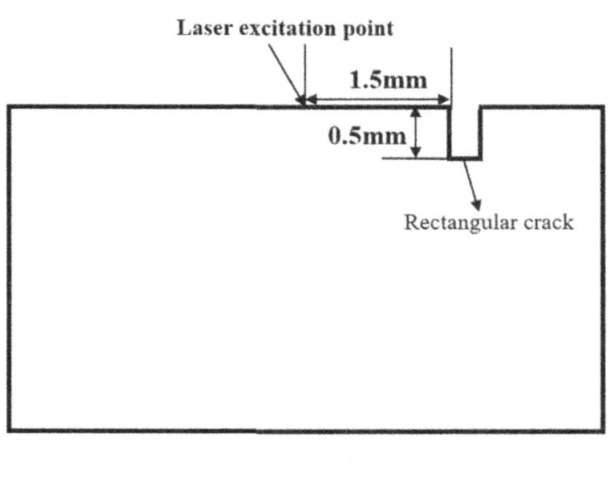

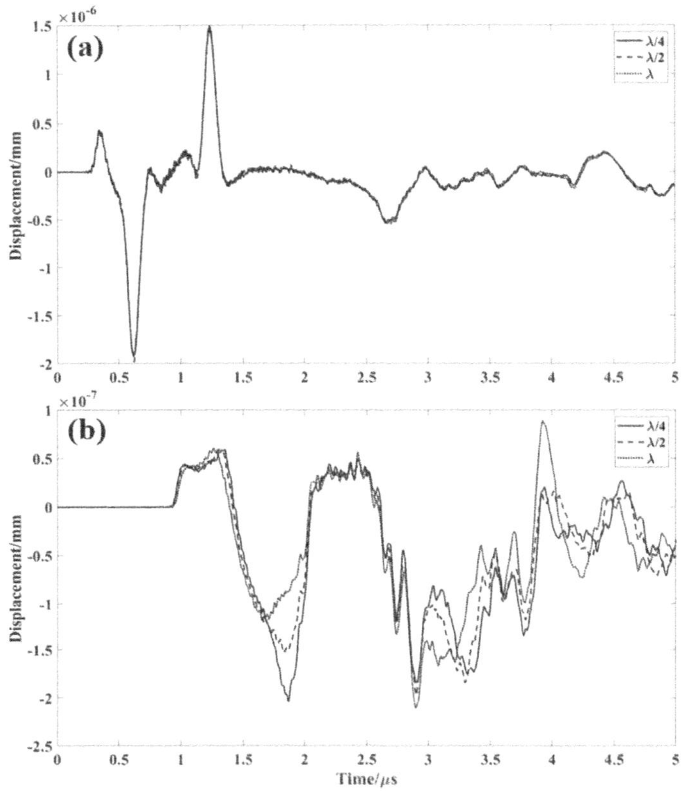

Fig. 4 Rectangular defects of different widths, the receiving point is located at **a** 1 mm, **b** 3 mm

3.4 Simulation Results of Rectangular Cracks with Different Depths

Similarly, a rectangular crack defect is set at a distance of 1.5 mm from the excitation point. The width l of the defect is half the longitudinal wave wavelength $\lambda_L/2$, and the depth is set to $\lambda_L/2$, λ_L, $2\lambda_L$ and $4\lambda_L$ respectively. The received signal is shown in Fig. 5. (a) The receiving position is located at 1 mm, and there is no obvious difference between the received signals of different depths; (b) The displacement signal changes of cracks at different depths at 3 mm are shown in the figure. It can be seen that the deeper the depth, the smaller the amplitude of the received Rayleigh wave signal, and when the crack depth d is $4\lambda_L$, the amplitude is reduced by 91.4% compared with the depth of $\lambda_L/2$. When the depth is $4\lambda_L$, the amplitude of the received glancing longitudinal wave (L) signal is almost zero. Therefore, it can be determined that when the depth is further increased, the amplitude of the received Rayleigh wave will continue to decrease or even be 0, and the received glancing longitudinal wave

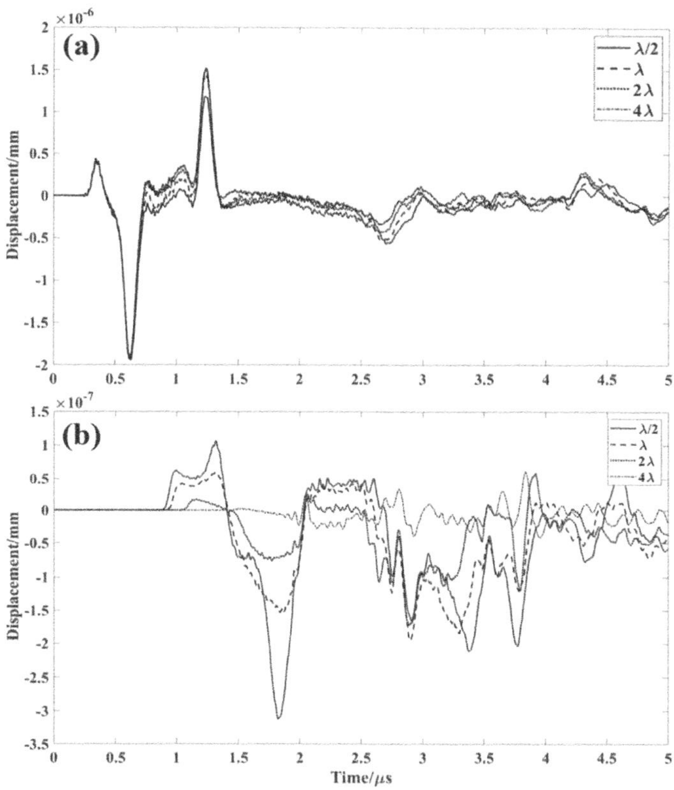

Fig. 5 Rectangular defects of different depths, the receiving point is located at **a** 1 mm, **b** 3 mm

amplitude will continue to remain at 0. Therefore, when the receiving point and the excitation point are on different sides of the crack, it is judged whether the crack depth exceeds 1.5 mm by judging that the glancing longitudinal wave signal can be received.

4 Conclusion

This paper uses the finite element method to numerically simulate the laser thermoelastic mechanism to excite the ultrasonic guided waves in the basin insulator material, and verify that the pulsed laser irradiation on the insulator material can produce multiple modes of waves such as longitudinal waves, transverse waves, and Rayleigh waves, and the frequency of the excited sound waves up to a few MHz, it is judged whether there is a defect based on the characteristics of the received wave signal, which provides a theoretical and simulation basis for laser ultrasonic technology to detect basin insulators. By setting rectangular crack defects with different widths and depths, when the excitation point and the receiving point are on the same side of the crack defect, the received signal characteristics are almost the same, and when they are located on different sides, the crack can be qualitatively analyzed by different amplitudes. The size of the defect can be quantitatively detected by judging whether the longitudinal wave signal can be received to determine whether the depth exceeds 1.5 mm.

Acknowledgements This work was supported in part by the National Grid Technology Project (520530190008), China, and National Natural Science Foundation of China, (No. 61771448 & No. 51937010).

References

1. Liu, Zhenya. 2005. *Extra high voltage power grid*. Beijing: China Economic Publishing Press. (in Chinese).
2. Liu, Ronghai, Chunyan Zang, Yingchun Yang, et al. 2019. Non-destructive testing technology of the cracking failure of the basin-type insulator. *High Voltage Apparatus* 55 (03): 139–143. (in Chinese).
3. Jiang, Xingliang, Yang Pan, Quanlin Wang, and Yuyao Hu. 2017. Research on icing characteristics of composite insulator and structural parameter analysis based on equivalent diameter. *Transactions of China Electrotechnical Society* 32 (07): 190–196. (in Chinese).
4. Zhang, Qian, Jianping Wang, and Weitao Li. 2019. Insulator state detection of convolutional neural networks based on feedback mechanism. *Transactions of China Electrotechnical Society* 34 (16): 3311–3321. (in Chinese).
5. Lv, Yukun, Weiping Zhao, Guanglu Pang, et al. 2018. Simulation of contamination deposition on typical shed porcelain and composite insulators. *Transactions of China Electrotechnical Society* 33 (01): 209–216. (in Chinese).
6. Qiu, Zhibin, Wenjun Yao, Huihui Li, et al. 2013. Mechanical properties study of porcelain post insulator for high voltage dis connector. *Insulating Materials* 46 (5): 37–42. (in Chinese).
7. Guo, Rui, Qijun Guo, Yanfei Guo, et al. 2015. The electric field simulation of the metal flange pouring hole of disc-type insulator in the 126 kV GIS. *Advances of Power System and Hydroelectric Engineering* 31 (15): 27–31. (in Chinese).
8. Zhao, Dongbo, Dengwei Ding, and Senjing Yao. 2015. Investigation of frequency characteristics of typical PD and the propagation properties in GIS. *IEEE Transactions on Dielectrics and Electrical Insulation* 22 (3): 1654–1662.
9. Qin, Kai, and Jian Zhao. 2015. Research status of flaw detection technology for post porcelain insulator. *Insulating Materials* 48 (2): 6–11. (in Chinese).
10. Wang, Haoran, Zihao Guo, Siyu Zhang, et al. 2018. Influence of UHV AC GIS spacer defects on electrical field distribution. *High Voltage Engineering* 44 (3): 982–992. (in Chinese).
11. Tian, Fangyuan, Yanpeng Hao, Zhouyiao Zou, et al. 2019. Ultrasonic LCR method for detecting micro-cracks on the surface of epoxy composite insulation. *Guangdong Electric Power* 32 (12): 106–112. (in Chinese).
12. Yuan, Weiqi. 2014. *The study of crack discharge in ultra-high voltage basin-type insulator*. Beijing: North China Electric Power University. (in Chinese).
13. Xiao, Yan, and Weiyong Yu. 2005. Present status and prospect of research of on-line partial discharge monitoring system in GIS. *High Voltage Engineering* 01: 47–49. (in Chinese).

14. Guo, Cuijuan, Mingzhu Yang, Feng Rong, et al. 2016. Basin insulator damage detection system based on vibration response analysis. *Chinese Journal of Sensors and Actuators* 29 (10): 1606–1612. (in Chinese).
15. Wang, Feng, Pingjun Liao, Jun Huang, et al. 2016. Novel method for electric field calculation considering the thickness of contaminant layer on the surface of composite insulator. *Transactions of China Electrotechnical Society* 31 (10): 77–84. (in Chinese).
16. Lv, Fangcheng, Hongyu Liu, Fochi Wang, et al. 2017. Deposit criterion of pollution particles on composite insulators surface under high speed aerosol. *Transactions of China Electrotechnical Society* 32 (01): 206–213. (in Chinese).
17. Xu, Huanqing, Junpeng Ma, Chengliang Wang, et al. 2017. Study on digital X-ray imaging technology for detecting typical defects in GIS equipment. *Power System Technology* 41 (5): 1967–1972. (in Chinese).
18. Tehlar, D., U. Riechert, and G. Behrmann. 2013. Pulsed x ray induced partial discharge diagnostics for routine testing of solid GIS insulators. *IEEE Transactions on Dielectrics and Electrical Insulation* 20 (6): 2173–2178.
19. Yoshida, M., H. Kojima, and N. Hayakawa. 2011. Evaluation of UHF method for partial discharge measurement by simultaneous observation of UHF signal and current pulse waveforms. *IEEE Transactions on Dielectrics and Electrical Insulation* 18 (2): 425–431.
20. Tian, F., Y. Hao, Z. Zou, et al. 2019. An ultrasonic pulse-echo method to detect internal defects in epoxy composite insulation. *Energies* 12 (24): 4804.
21. Li, C.H., S.N. Li, C. Wei, et al. 2012. A comparison of laser ultrasound measurements and finite element simulations for evaluating the elastic properties of tissue mimicking phantoms. *Optics and Laser Technology* 44 (4): 866–871.
22. Ma, Baoquan, and Zhenggan Zhou. 2014. Progress and development trends of composite structure evaluation using noncontact nondestructive testing techniques in aviation and aerospace industries. *Acta Aeronautica et Astronautica Sinica* 35 (7): 1787–1803. (in Chinese).
23. Shen, Zhonghua, Ling Yuan, Hongchao Zhang, et al. 2015. *Laser ultrasound in solids*. Beijing: Posts & Telecom Press. (in Chinese).

Magnetostrictive Simulation of Amorphous Alloy Based on Dynamic Jiles-Atherton Model

Xiaoyu Zhou and Lihua Zhu

Abstract Magnetostriction is the main cause of vibration and noise of transformer core. In order to obtain a more accurate magnetostrictive model suitable for amorphous alloys, a new magnetostrictive model is established based on the im-proved Jiles-Atherton model and the quadratic domain rotation model. Among them, the eddy current loss and additional loss are considered, and the parameter v is introduced to reflect the change of k with the magnetization state in the actual magnetization process, all of which make the model have stronger physical significance. Particle swarm optimization algorithm is used to extract the parameters of the model. The data of amorphous alloy strip under different excitation sizes were measured and compared with the results of model calculation. The results show that the calculated results of the magnetostrictive model based on the hysteresis model which is more consistent with the physical basis are close to the measured data. The model established in this paper can be applied to the simulation of the magnetostrictive characteristics of amorphous alloys.

Keywords Magnetostrictive model · Amorphous alloy · Particle swarm optimization · J-A model

1 Introduction

Magnetostriction is the main cause of transformer vibration and noise [1]. The magnetic field caused by the material of magnetic susceptibility changes will change the material of magnetostrictive strain until it reaches its saturation value. At present, the magnetostrictive measurement is difficult, there is no unified national or international standards. In order to more accurately study the magnetostrictive characteristics of power transformers and simulate their vibration and noise, it is necessary to establish a practical magnetostrictive model for amorphous alloys.

X. Zhou (✉) · L. Zhu
Tianjin Key Laboratory of Advanced Technology of Electrical Engineering and Energy, Tiangong University, Tianjin, China
e-mail: 1831045300@tiangong.edu.cn

W. Chen et al. (eds.), *The Proceedings of the 9th Frontier Academic Forum of Electrical Engineering*, Lecture Notes in Electrical Engineering 743,
https://doi.org/10.1007/978-981-33-6609-1_70

The key to the modeling of magnetostrictive properties of magnetic materials lies in the physical description and mathematical expression of magnetic hysteresis, which can be divided into four categories based on magnetic domain theory, phenomenology, thermodynamics and elastic mechanics. Jiles derived a linear combination of squared and quadrature magnetostrictions for magnetization based on the principle of energy minimization [2]. J. Zhu realizes the simulation of magnetic properties of core materials by Preisach hysteresis model, and simulates the magnetic properties of annular core by combining dynamic circuit model with Preisach model [3]. T. Hilgert introduced the dynamic description of magnetostriction under sinusoidal excitation, but this model has high requirements for the quantity and accuracy of experimental data [4], so it is difficult to achieve. At present, some new magnetostrictive models have been developed, but amorphous alloys have not yet been involved [5–7].

In this paper, an improved magnetostrictive model is proposed based on the quadratic domain rotation model and Jiles-Atherton (J-A) model. In order to conform to the fact that k changes with the magnetization state during the actual magnetization process, the parameter v is introduced. Eddy current loss and residual loss are added to lay a foundation for the study of high frequency dynamic magnetostrictive model. In order to determine the relationship between model parameters and magnetostrictive energy, a preliminary experimental and theoretical study on amorphous alloys was carried out. Particle swarm optimization (PSO) is used to solve the magnetostrictive model parameters. In order to evaluate the accuracy of the results, the calculated magnetostriction results are compared with the experimental results of a monolithic tester.

2 Derivation of the Basic Model Modeling

2.1 Traditional J-A Hysteresis Model

In this model, the relationship between the applied magnetic field H and the magnetization M is established through five steps:

1. Determine the effective magnetic field of magnetic materials, H_e
2. Determine the magnetization without hysteresis, M_{an}
3. Determine the irreversible magnetization generated by the domain wall displacement, M_{irr}
4. Determine the reversible magnetization intensity generated by domain wall bending, M_{rev}
5. The total magnetization M is the sum of M_{irr} and M_{rev}.

The specific expression is as follows:

$$H_e = H + \alpha M \tag{1}$$

$$M_{an} = M_s[\coth(H_e/a) - (a/H_e)] \tag{2}$$

$$\frac{\mathrm{dM_{irr}}}{\mathrm{d}H} = \frac{M_{an} - M_{irr}}{\delta k - \alpha(M_{an} - M_{irr})} \tag{3}$$

$$M_{\mathrm{rev}}=c(M_{an} - M) \tag{4}$$

$$M = M_{rev} + M_{irr} \tag{5}$$

where c is the bending coefficient related to the domain wall, a is the shape parameter of anhysteretic curve, α is coupling coefficient, k is the pinning coefficient and δ is defined as $sign(\mathrm{d}H/\mathrm{d}t)$.

The energy balance equation obtained is as follows:

$$\mu_0 \int M\mathrm{d}H_e = \mu_0 \int M_{an}\mathrm{d}H_e - \mu_0 k\delta \int \frac{\mathrm{d}M_{irr}}{\mathrm{d}H_e}\mathrm{d}H_e \tag{6}$$

2.2 *Derivation of Magnetostrictive Mode*

In earlier studies, dynamic hysteresis loop considers the dynamic loss with frequency variation, and has a good effect on simulation. In the actual magnetization process, k changes with the change of magnetization state. In order to get a better result, a parameter v is introduced to explain the change of k with the magnetization state in the actual magnetization process, and eddy current loss and residual loss are considered. This will complicate the model and calculation, but it is more in line with the physical process of magnetization.

The modified energy balance expression in the magnetization process can be expressed as:

$$\int M\mathrm{d}H_e = \int M_{an}\mathrm{d}H_e - \int k\left[1 - v\left(\frac{M_{irr}}{M_S}\right)^2\right]\delta\frac{\mathrm{d}M_{irr}}{\mathrm{d}H_e}\mathrm{d}H_e$$
$$- \int \frac{e^2\sigma}{2\beta}\left(\frac{\mathrm{d}B}{\mathrm{d}t}\right)^2\mathrm{d}t - \int (GSV_0\sigma)^{1/2}\left|\frac{\mathrm{d}B}{\mathrm{d}t}\right|^{3/2}\mathrm{d}t \tag{7}$$

where H_e is the effective magnetic field intensity, M_an is the anhysteresis magnetization, M_irr is the irreversible magnetization, and v is the parameter that can correct k to some extent.

Similar to the original J-A model, the term with k illustrates the domain wall pinning effect, while the term with v modifies k to some extent, making the physical meaning of k changing with the magnetization state more perfect.

$$\lambda = 1.5\lambda_s M^2 / M_s^2 \tag{8}$$

where M is the magnetization, M_s is the saturation magnetization and λ_s is the saturation magnetostriction of the material.

Therefore, by combining (7) and (8) with the modified J-A model, the following model is established:

$$\frac{\mathrm{d}\lambda}{\mathrm{d}B} = \frac{3\lambda_S M}{M_S^2} \frac{M - M_{an} - \frac{c\delta}{1-c}k\left[1 - \upsilon\left(\frac{M_{irr}}{M_S}\right)^2\right]\frac{\mathrm{d}M_{an}}{\mathrm{d}H_e} + k_c\frac{\mathrm{d}B}{\mathrm{d}t} + k_c\delta\left|\frac{\mathrm{d}B}{\mathrm{d}t}\right|^{\frac{1}{2}}}{\mu_0(\alpha - 1)\left(M_{an} - M + \frac{c\delta}{1-c}k\left[1 - \upsilon\left(\frac{M_{irr}}{M_S}\right)^2\right]\frac{\mathrm{d}M_{an}}{\mathrm{d}H_e}\right) - \frac{\mu_0\delta}{1-c}k\left[1 - \upsilon\left(\frac{M_{irr}}{M_S}\right)^2\right]} \tag{9}$$

3 The Experimental Measurements

In order to verify the accuracy of the model, the magnetic property and magnetostrictive property test system produced by a German company was used to measure the magnetization property and magnetostrictive property of the sample. The magnetostrictive measurement system is shown in the Fig. 1a. The sample is made of

Fig. 1 Experimental chart. **a** One-dimensional single sheet tester system. **b** Cured processed strip

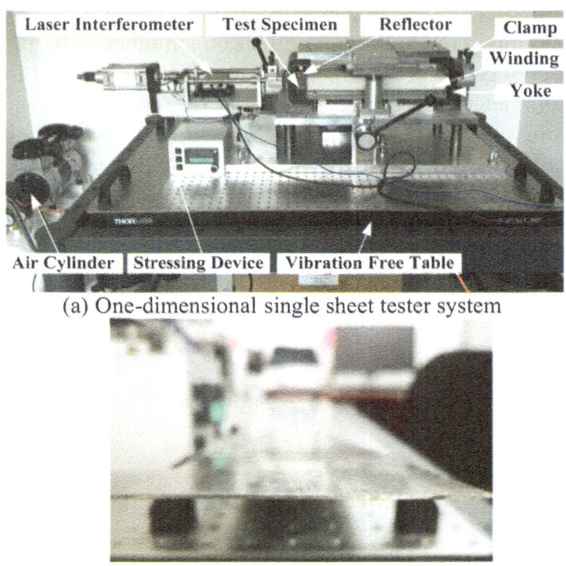

(a) One-dimensional single sheet tester system

(b) Cured processed strip

35 pieces of 1K101 amorphous alloy strip cured and superposition by lacquer dip process, as shown in Fig. 1b, whose length, width, thickness is 600 mm, 100 mm, 1 mm, respectively.

The hysteresis loop and magnetostrictive curve of amorphous alloy samples were measured and calculated. The measurement results are shown as follows in Fig. 2:

Fig. 2 Hysteresis loop and magnetostrictive curve under different flux densities.
a Hysteresis loop at different flux densities.
b Magnetostrictive curve at different flux densities

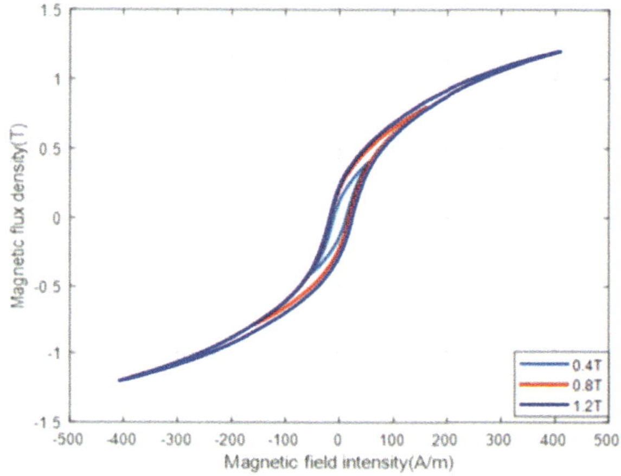

(a) Hysteresis loop at different flux densities

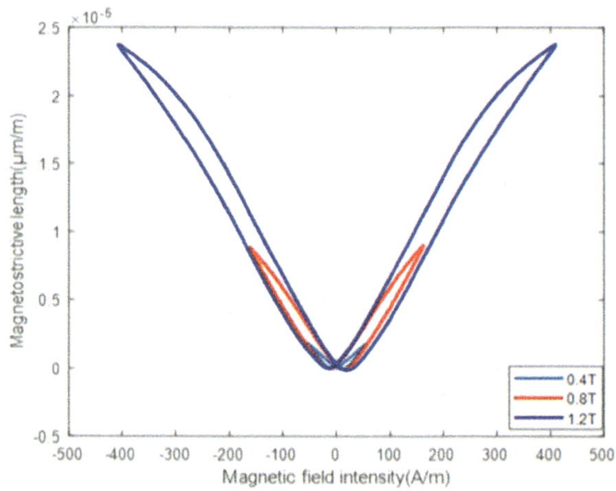

(b) Magnetostrictive curve at different flux densities

4 Simulation and Validation of the Model

4.1 Particle Swarm Optimization Algorithm

Particle swarm optimization algorithm is a kind of evolutionary optimization algorithm [8], where each particle in the group is defined to identify a potential solution to the problem in n-dimensional space. The parameters of the magnetostrictive model can be derived directly from the measured data of the reference point. The hysteresis model and magnetostrictive model parameters of amorphous alloy are determined by particle swarm optimization.

The error function can be expressed as:

$$Fitness = \left[\sum_{sample} \frac{(\lambda_0 - \lambda_m)^2}{N} \right]^{1/2} \tag{10}$$

where N is the number of points included in the hysteresis loop calculated by the J-A model, λ_0 is the data measured by the experiment, and λ_m is the fitness function, which is the magnetostriction calculated by the dynamic magnetostrictive model.

4.2 Parameter Analysis

The parameters of the traditional J-A hysteresis model can be calculated by the following formula:

$$c = \frac{3a\chi_{in}}{M_s} \tag{11}$$

$$a = \frac{M_s}{3}\left(\frac{1}{\chi_{an}} + \alpha\right) \tag{12}$$

$$k = \frac{M_{an}(H_c)}{1-c}\left[\alpha + \frac{1}{\left(\frac{1}{1+c}\right)\chi_{Hc} + \left(\frac{c}{1+c}\right)\frac{dM_{an}(H_c)}{dH}}\right] \tag{13}$$

$$\alpha = \frac{k(1-c)}{M_r - M_{an}(M_r)} - \frac{1-c}{\chi_r - c\frac{dM_{an}(M_r)}{dH}} \tag{14}$$

$$M_m = M_{an}(H_m) - \frac{(1-c)k\chi_m}{\alpha\chi_m + 1} \tag{15}$$

In the calculation, the traditional parameter is done by the traditional parameter determination method, and the dynamic parameter has different values under different excitations, and the change rules of the two parameters are consistent. Therefore, by introducing the modified parameter s in the calculation, according to the dynamic correction parameter k_c, k_e, the modified parameter determination formula can be expressed as:

$$k_c = e^2 \sigma s / 2\beta \tag{16}$$

$$k_e = (G S V_0 \sigma s)^{1/2} \tag{17}$$

In this way, the two dynamic solving problems are transformed into solving the correction parameter s, which reduces the computational dimension of the optimization algorithm and increases the computational efficiency.

The whole solving process is actually the optimal combination of solving parameters to minimize the value of the objective function. The specific optimization flow chart of its parameters is shown in Fig. 3.

The simulation parameters were applied to the model, and the calculated results were compared with the experimental measurement results to obtain a comparison diagram, as shown in the Fig. 4.

Figure 4 shows the simulation comparison of hysteresis loop measurement under different magnetic flux conditions. Figure 5 shows the simulation comparison of magnetostrictive measurement under different magnetic flux conditions. It can be

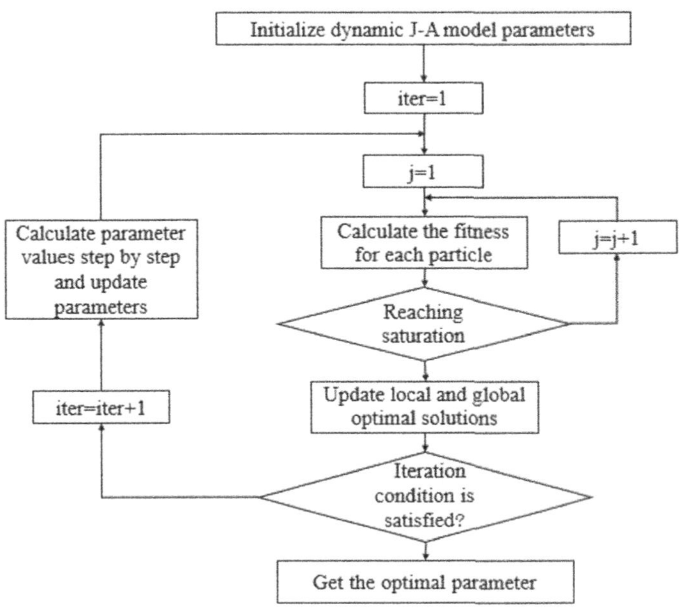

Fig. 3 Flow chart of particle swarm optimization algorithm

Fig. 4 Simulation comparison diagram of hysteresis loop measurement under different magnetic flux conditions. **a** Comparison diagram of 0.8 T hysteresis loop experiment and simulation. **b** Comparison diagram of 1.2 T hysteresis loop experiment and simulation

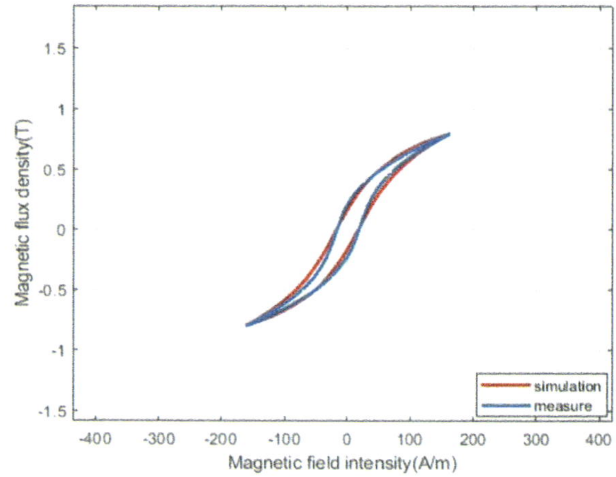

(a) Comparison diagram of 0.8T hysteresis loop experiment and simulation

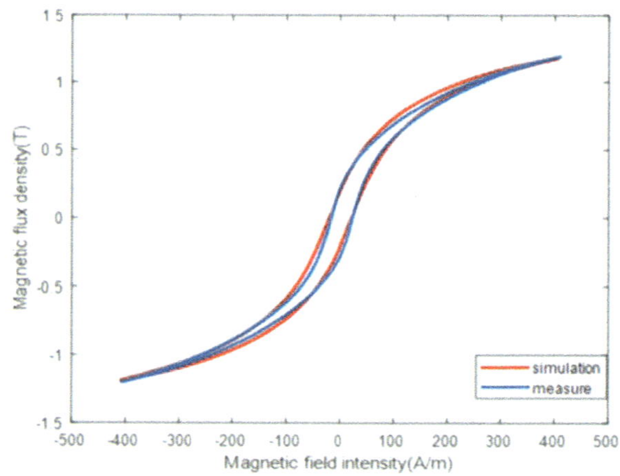

(b) Comparison diagram of 1.2T hysteresis loop experiment and simulation

seen that the calculated results are very close to the experimental results and have a good consistency. The results show that the model can simulate the magnetostriction of amorphous alloy with different magnetic flux.

Fig. 5 Experimental and simulated comparison of magnetostrictive curves with different flux densities. **a** Comparison diagram of 0.8 T magnetostrictive curve experiment and simulation. **b** Comparison diagram of 1.2 T magnetostrictive curve experiment and simulation

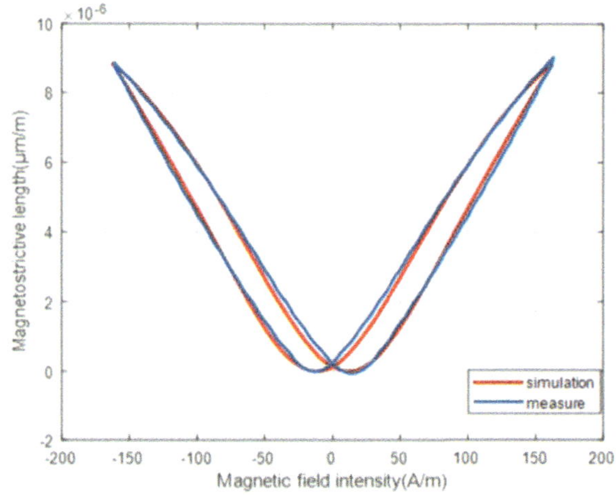

(a) Comparison diagram of 0.8T magnetostrictive curve experiment and simulation

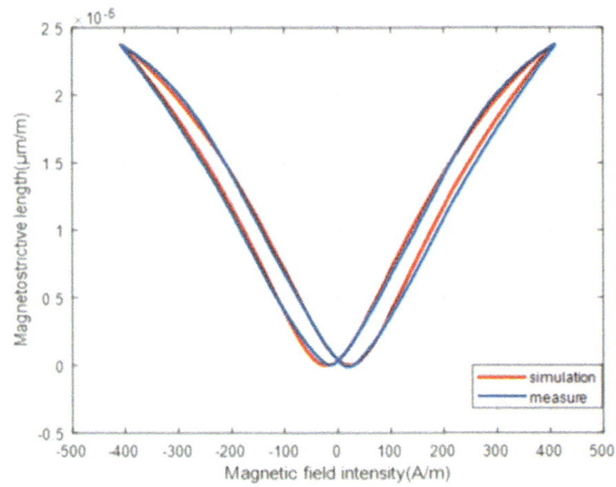

(b) Comparison diagram of 1.2T magnetostrictive curve experiment and simulation

5 Conclusion

In this paper, the traditional hysteresis parameter k is modified by introducing the parameter v, and the eddy current loss and residual loss are introduced to simulate the dynamic magnetostriction of amorphous alloy strip more accurately. Particle swarm optimization (PSO) algorithm is used to extract the parameters of the model. The data of amorphous alloy strip under different excitation sizes were measured and compared with the results of model calculation.

The results show that the calculated results of the magnetostrictive model based on the hysteresis model which is more consistent with the physical basis are close to the measured data. The model established in this paper can be used to simulate the magnetostrictive characteristics of amorphous alloy materials, which provides some references for magnetostrictive and noise calculation under high frequency conditions.

References

1. Liu, D., J. Li, R.K. Noubissi, S. Wang, X. Xu, and Q. Liu. 2019. Magnetic properties and vibration characteristics of amorphous alloy strip and its combination. *IET Electric Power Applications* 13 (10): 1589–1597.
2. Jiles, D.C. 1999. Theory of the magnetomechanical effect. *Journal of Physics. D. Applied Physics* 28 (8): 1537.
3. Zhu, J.G. 1994. *Numerical modelling of magnetic materials for computer aided design of electromagnetic devices*. Sydney: University of Technology Sydney.
4. Hilgert, T., L. Vandevelde, and J. Melkebeek. 2007. Neural-network-based model for dynamic hysteresis in the magnetostriction of electrical steel under sinusoidal induction. *IEEE Transactions on Magnetics* 43 (8): 3462–3466.
5. Li, Y., J. Zhu, L. Zhu, Y. Li, and G. Lei. 2020. A dynamic magnetostriction model of grain-oriented sheet steels based on Becker-Döring crystal magnetization model and Jiles-Atherton theory of magnetic hysteresis. *IEEE Transactions on Magnetics* 56 (3): 1–5.
6. Lihua, Z., L. Jingjing, Y. Qingxin, Z. Jianguo, and C. Koh. 2020. An improved magnetostriction model for electrical steel sheet based on Jiles-Atherton model. *IEEE Transactions on Magnetics* 56 (3): 1–4.
7. Li, Y., L. Zhu, and J. Zhu. 2018. Core loss calculation based on finite-element method with Jiles-Atherton dynamic hysteresis model. *IEEE Transactions on Magnetics* 54 (3): 1–5.
8. Kennedy, J., R. Eberhart. 1995. Particle swarm optimization. In *Proceedings of the IEEE International Neural Network*, 1942–1948.

Design of High Sensitivity Foreign Object Detection System in Wireless Charging Based on the Variation of Detection Coils Impedance

Ying Sun, Jixing Liu, Ce Liang, Guo Wei, Chunbo Zhu, and Kai Song

Abstract Metal foreign objects could enter the wireless power system easily due to the contactless characteristics of primary side and secondary side. Because of eddy current loss, metal foreign objects would not only reduce the system output power and efficiency, but also cause the potential safety hazard. To solve the problem that the conventional approach has low detection sensitivity and blind zone, a design method of high sensitivity foreign object detection system is proposed. Firstly, according to the magnetic induction intensity distribution of the primary coil, the distribution of detection area and the symmetrical structure of detection coils are designed. Secondly, the relationship between the parameters of detection coils and detection sensitivity is analyzed in detailed through the finite element simulation. Then, the excitation frequency of detection circuit is optimized to realize the highest detection sensitivity. Finally, the proposed design is verified by experiments on a LCC-LCC 6.6 kW prototype. As for 1 RMB yuan coin in the position of alignment, corner and 5 mm high, the variation of amplitude of detection signal are 94.1%, 31.8% and 23.18%, respectively.

Keywords Wireless charging · Foreign object detection · Detection coils design · Finite element analysis

1 Introduction

Wireless charging technology has gradually become a hotspot of research in recent years by virtue of its ability to wirelessly deliver high power in a short to medium distance with high efficiency. The technology is becoming increasingly perfect in theory [1–7]. But when it comes to the practical application, a variety of working

Y. Sun · J. Liu · G. Wei · C. Zhu · K. Song (✉)
School of Electrical Engineering and Automation, Harbin Institute of Technology, Harbin, China
e-mail: kaisong@hit.edu.cn

C. Liang
School of Instrument Science and Engineering, Harbin Institute of Technology, No. 92 Xi Da Zhi Street, Nan Gang District Harbin, China

W. Chen et al. (eds.), *The Proceedings of the 9th Frontier Academic Forum of Electrical Engineering*, Lecture Notes in Electrical Engineering 743,
https://doi.org/10.1007/978-981-33-6609-1_71

conditions are required to be considered, especially the metal foreign object in the charging area would produce serious security risks [8–12]. Therefore, FOD (Foreign Object Detection) is one of the indispensable technologies to enhance the safety and stability of wireless charging system [13, 14]. At present, magnetic flux detection [15–17] and coil impedance detection [18–20] are the commonly used methods for metal foreign object detection in electric vehicle wireless charging systems. These two methods have the advantages of high integration, low cost, fast response, and good detection effect on metallic foreign objects. Both methods make use of detection coils, but there has been little research on the optimal design of the detection coils.

In this paper, a design method of high sensitivity foreign object detection system has been proposed. According to the finite element simulation, the effect of metal foreign object on the impedance variation of different size and structure detection coils was analyzed in detailed. And the excitation frequency of detection circuit was optimized by theoretical analysis and experimental test to realize high detection sensitivity. The proposed method could significantly improve the sensitivity of foreign object detection and also have good detection effect for some small size foreign objects and the certain height foreign objects.

2 Detection Coil Sensitivity Analysis

According to Faraday's law of electromagnetic induction, when a bulk conductor is placed in an alternating magnetic field or moves in a fixed magnetic field, an induced current is generated and closed in the conductor. Therefore, the eddy current effect can be equated to a short-circuit loop current model, where the metallic foreign object can be equated to a coil and the effect of the metallic foreign object on the impedance of the detection coil can be analyzed using the mutual inductance model.

As shown in Figs. 1 and 2, L_D and R_D are the self-inductance and internal resistance of the detection coil. L_m and R_m as the equivalent inductance and equivalent internal resistance of the metal foreign object. Mm is the mutual inductance between the metal

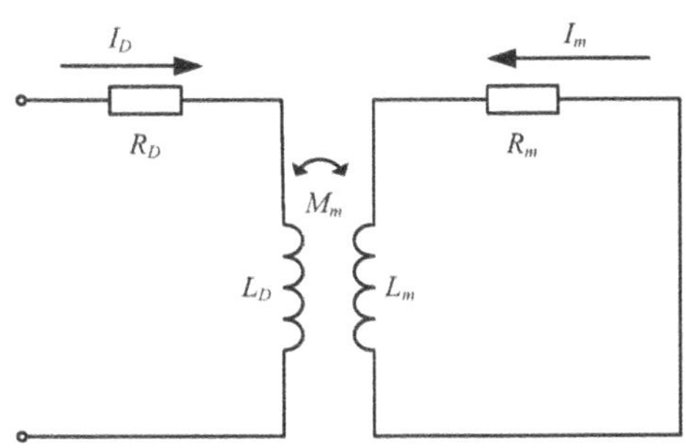

Fig. 1 Model of metal foreign object coupling to detection coil

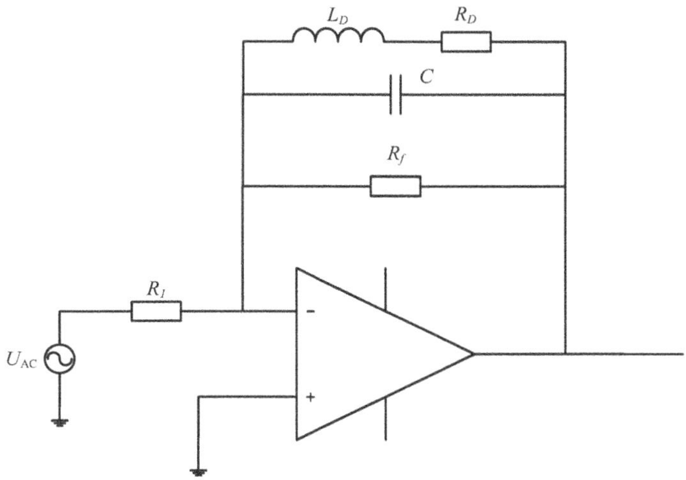

Fig. 2 Detection of coil impedance variation based on parallel resonance

foreign object and the detection coil. Z_{NONE} and Z_{FOD} are the equivalent impedance of the detection coil when there is no metallic foreign object and when there is metallic foreign object, respectively. Their expressions could be derived as follows:

$$Z_{NONE} = R_D + j\omega L_D \tag{1}$$

$$Z_{FOD} = \left[1 + \frac{R_m}{R_D} \cdot \frac{\omega^2 M_m^2}{R_m^2 + \omega^2 L_m^2}\right] R_D + j\omega \left[1 - \frac{L_m}{L_D} \cdot \frac{\omega^2 M_m^2}{R_m^2 + \omega^2 L_m^2}\right] L_D \tag{2}$$

$$\beta = 1 + \frac{R_m}{R_D} \cdot \frac{\omega^2 M_m^2}{R_m^2 + \omega^2 L_m^2} \tag{3}$$

$$\alpha = 1 - \frac{L_m}{L_D} \cdot \frac{\omega^2 M_m^2}{R_m^2 + \omega^2 L_m^2} \tag{4}$$

where α and β are the rate of change in inductance and the rate of change in internal resistance of the detection coil, respectively. So that Eq. (2) can be simplified to:

$$Z_{FOD} = \beta R_D + j\omega\alpha L_D \tag{5}$$

Define δ as the percentage variation of the impedance of the detection coil after introduction of the foreign object.

$$\delta = \frac{|Z_{FOD}| - |Z_{NONE}|}{|Z_{NONE}|} = \frac{\sqrt{\beta^2 R_D + \omega^2 \alpha^2 L_D^2} - \sqrt{R_D^2 + \omega^2 L_D^2}}{\sqrt{R_D^2 + \omega^2 L_D^2}} \tag{6}$$

The quality factor of the detection coil Q_D is brought into (6) for further analysis. Thus, δ could be expressed as:

$$\delta = \sqrt{\frac{\beta^2 + \alpha^2 Q_D^2}{1 + Q_D^2}} - 1 = \sqrt{\alpha^2 + \frac{\beta^2 - \alpha^2}{1 + Q_D^2}} - 1 \tag{7}$$

From (7), the percentage variation of impedance of the detection coil δ is related to the mutual inductance M_m of the metal foreign object so that M_m, α, β can be improved by the reasonable design and optimization of the structure and size of the detection coil.

If the quality factor Q_D of the detection coil is high enough even $Q_D \gg 1$, then (7) could be simplified to:

$$\delta \approx \left|\sqrt{\alpha^2} - 1\right| = |\alpha - 1| \tag{8}$$

From (8), even if the quality factor Q_D of the detection coil is high enough, the percentage variation in impedance δ of the detection coil is still limited. As a result, it is difficult to determine whether there is a foreign object only based on the impedance variation of the detection coil. Therefore, this paper adopts a parallel resonant amplifier circuit with an excitation source to amplify the percentage variation in impedance δ of the detection coil, enhance the noise suppression ability, and significantly improve the detection sensitivity. The structure is shown in Fig. 2. The principle is to convert the effect of foreign object on the impedance of the detection coil into a change in the amplification ratio of the output signal, and then the foreign object detection is carried out by detecting the change in the amplification circuit amplitude. The detection sensitivity S can be defined as:

$$S = \left|\frac{U_{FOD} - U_{NONE}}{U_{NONE}}\right| = \left|\frac{U_{in}}{U_{NONE}}\frac{U_{FOD}}{U_{in}} - 1\right| \tag{9}$$

$$S = \left|\frac{1}{\sqrt{1 + \omega^2 C^2 R_f^2 \frac{\delta^2}{(\delta+1)^2}}} - 1\right| \tag{10}$$

where ω, C, R_f is the excitation source angular frequency of the detection circuit, the resonant capacitor of the detection circuit, and the feedback resistance, respectively.

From (10), a reasonable optimization of excitation source angular frequency ω, detection circuit resonant capacitance C and feedback resistance R_f could make sensitivity S much larger than the percentage variation in impedance of the coil δ. As a result, a high detection sensitivity can be achieved.

3 Detection Coil Design

In this paper, the WPT3Z2 magnetic coupler in SAEJ2954 standard [13] has been selected as the experimental prototype and the output power is 6.6 kW. The design process of the detection coil is shown in Fig. 3.

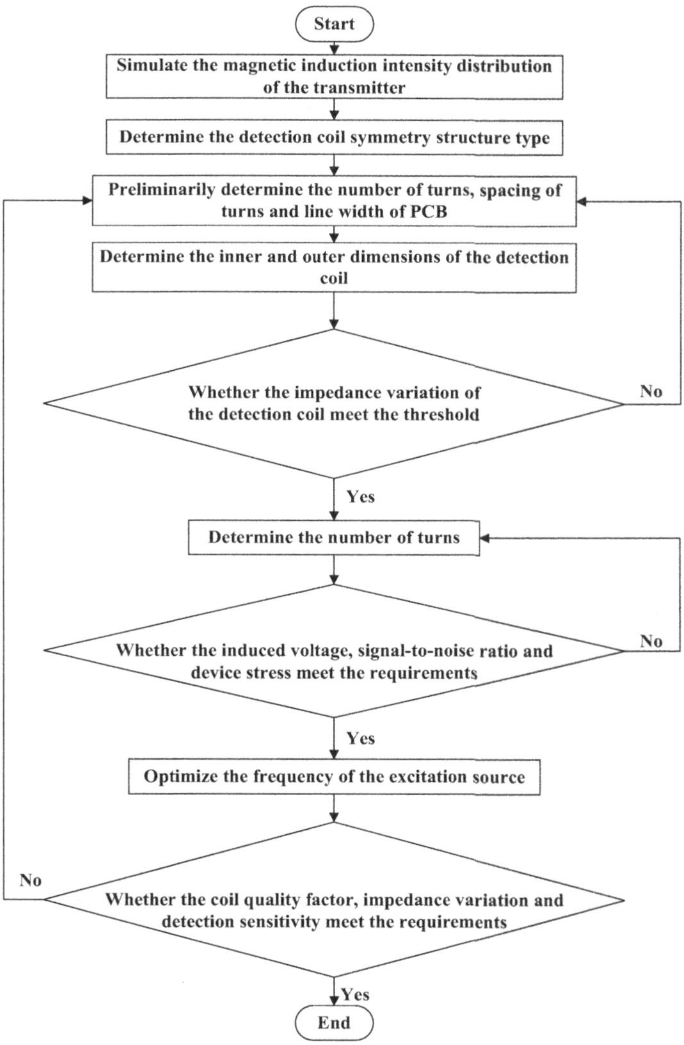

Fig. 3 The process of the detection coil design

Firstly, in order to reduce the influence of the induced voltage generated by the power magnetic field on the detection sensitivity, the detection coil utilize the series-opposition structure to reduce the net magnetic flux. The series-opposition structure of the detection coil is shown in Fig. 4. The two-substructure coils of each detection coil should be placed in the area where the magnetic flux of two-substructure coils are similar. As a result, the induced voltage is similar to 0 V and the effect of the induced voltage would have minimum effect on the detection sensitivity. According to the simulation results of the magnetic induction intensity distribution of primary coil, the arrangement of detection coils could be designed and shown in Fig. 5. The detection coils in area where coordinate axis X meet the condition: -125 mm < X <+ 125 mm utilize the upper and lower axial symmetry. And the detection coils in other area utilize the left and right axial symmetric.

Secondly, considering the detection sensitivity of the edge of detection coil, the coupling degree between metal object and detection coil, the internal resistance of

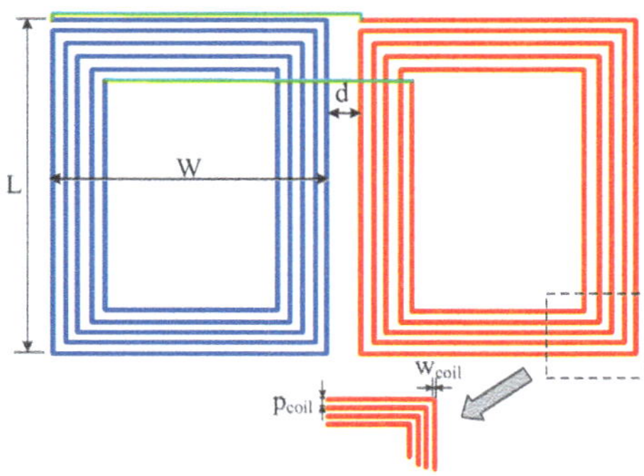

Fig. 4 The series-opposition structure of detection coil

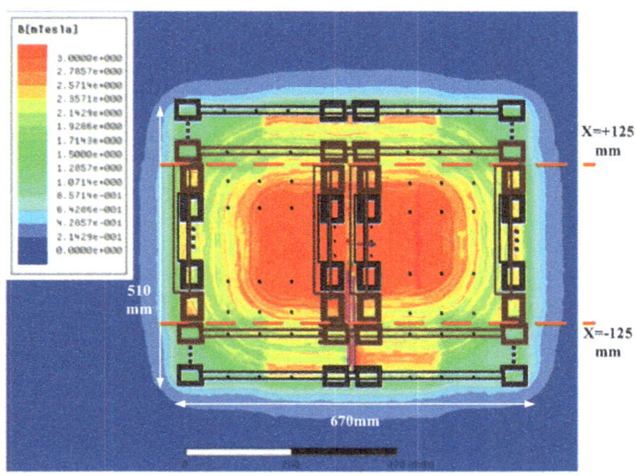

Fig. 5 The arrangement of detection coils

detection coil and system EMC, the coil gap d, spacing of turns p_{coil} and the width of PCB coil line w_{coil} are preliminary selected as 1.5 mm, 0.25 mm and 0.5 mm, respectively.

Thirdly, the finite element simulation model of detection coil is established by using a copper cylinder of which the diameter is 30 mm as metal foreign object. With adjusting the outer length L, outer width W, number of turns N, simulate the impedance variation percentage of the detection coil δ under the two conditions where foreign object is aligned to the detection coil center and placed on the corner edge respectively. The simulation results are shown in Figs. 6 and 7. In order to

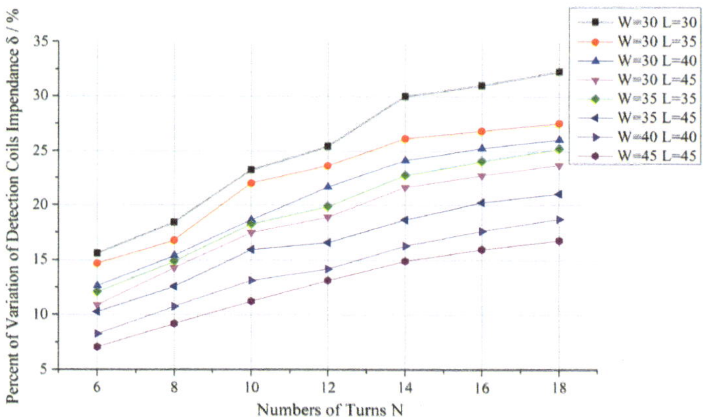

Fig. 6 δ with foreign object aligned to the detection coil center

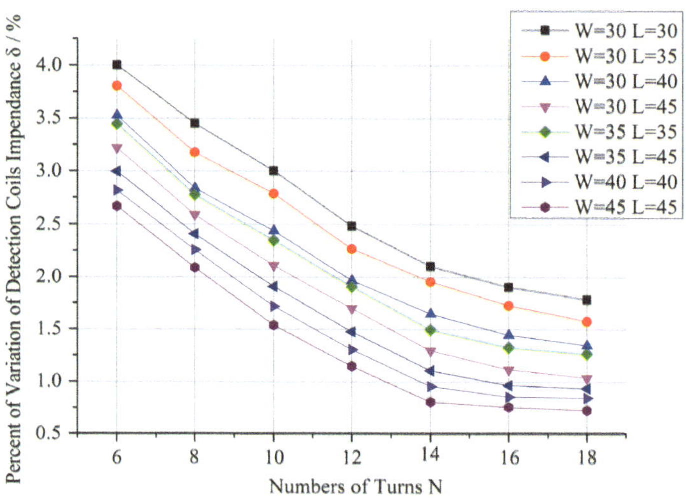

Fig. 7 δ with foreign object placed on the corner edge

extract signal conveniently and achieve high detection sensitivity, the threshold of impedance variation percentage δ of two conditions where foreign object is aligned the detection coil center and placed on the corner edge are selected as 15 and 2.5%.

Finally, comprehensively considering the signal to noise ratio, total number of detection coils, device stress and the parameter consistency of detection coils, the parameters of detection coil has been selected. The outer length L, outer width W, number of turns N are 40 mm, 30 mm and 10, respectively.

4 Excitation Source Frequency Selection

As shown in Fig. 2, the root cause of the amplification ratio of the detected signal is the metallic foreign object affecting the resonant cavity impedance value of the feedback loop. According to (1), the resonant cavity in the absence and presence of the foreign object is given by the following equation.

$$|Z_f| = \frac{1}{\sqrt{\left(\frac{1}{R_f} + \frac{R_D}{\omega^2 L_D^2 + R_D^2}\right)^2 + \left(\omega C - \frac{\omega L_D}{\omega^2 L_D^2 + R_D^2}\right)^2}} \quad (11)$$

$$\frac{\Delta|Z_f|}{|Z_f|} = \frac{||Z_{f\_FOD}| - |Z_f||}{|Z_f|} \times 100\% \quad (12)$$

where Z_f, $Z_{f_FOD}$is the feedback loop RLC parallel resonant cavity impedance without foreign object and its impedance with foreign object.

Combined with simulation results and actual measurement data of detection coil, provided that the feedback loop is resonant when there is no foreign object, respectively, to plot the feedback loop RLC parallel resonant cavity impedance change $\Delta Z_f/Z_f$ with frequency change curve cluster in the case of the foreign object is located in the alignment position and the corners. The analysis results are shown in Figs. 8 and 9 respectively.

After the frequency reaches 10 MHz, ΔZ_f no longer increases with frequency regardless of whether the foreign object is in the alignment or corner position. The ΔZ_f is 92 and 63% for the alignment and corner positions respectively, but the excitation frequency is too high to process the signal. Considering the parasitic parameters, device selection, electromagnetic compatibility and other factors, the frequency is not convenient for subsequent signal processing. When the frequency is about 2.5 MHz, the ΔZ_f of the positive pair and the corner position are more than 77 and 32%, which can accurately identify the presence of foreign objects. Therefore 2.5 MHz is selected as the best frequency for the system.

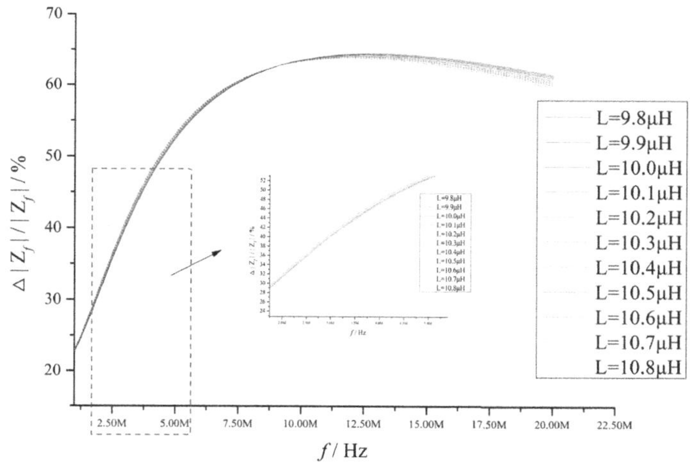

Fig. 8 Percentage change in resonant cavity impedance of the feedback loop for the foreign object corner detection coil

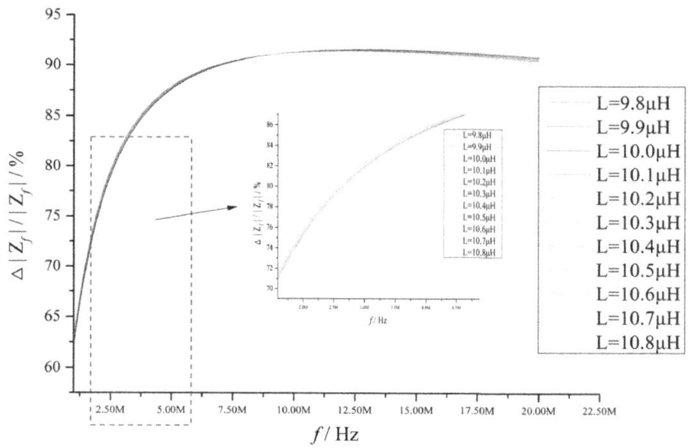

Fig. 9 Percentage change in resonant cavity impedance of the feedback loop when the foreign object is in the detection coil alignment

5 Experimental Verification

A 6.6 kW LCC-LCC wireless charging system is shown in Fig. 10, which is used as the experiment prototype. The wireless power transmission system in this paper is designed and built according to the parameter requirements of WPT3Z2 [13].

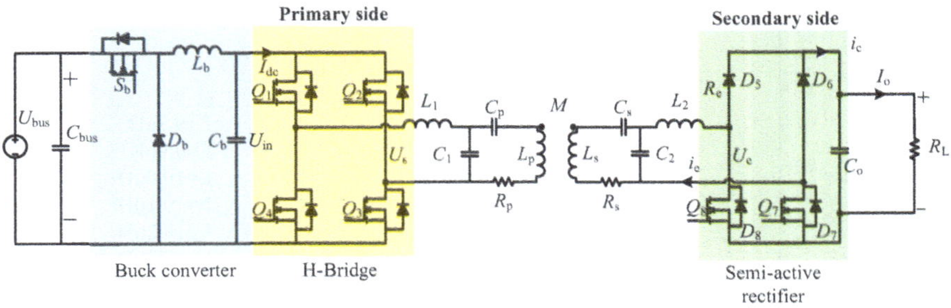

Fig. 10 Double-sided LCC 6.6 kW wireless power transfer system

Table 1 System parameters

Parameters	Value
L_1	19.5 μH
C_p	181.1 nF
C_1	177.3 nF
L_2	7.1 μH
C_s	110.6 nF
C_2	493.8 nF
L_p	40.1 μH
L_s	38.5 μH
M	7.75 μH
$I_{p(RMS)}$	34.02 A
$I_{s(RMS)}$	51.71 A
R_L	6.9 Ω
Gap	12 cm

Detailed parameters are shown in Table 1. Where, *Gap* is the size of the air gap. $I_{p(RMS)}$ and $I_{s(RMS)}$ are the RMS values of the primary coil current and the secondary coil current, respectively.

In order to extract high-frequency signals for foreign object detection, the detection circuit needs to include a bandpass filter to remove the induced voltage and high-frequency burr noise generated by the magnetic field on the detection coil. The AD8066 was selected as the operational amplifier for the detection circuit, considering the induced voltage amplitude, gain, pass band, and device stress. The pass band of the bandpass filter is 1.5–3 MHz, and the attenuation at 85 kHz is 65 dB, so that the high-frequency signal can be extracted for foreign object detection. The structure and parameters of the detection circuit are shown in Fig. 11.

The center region of the transmitter is the most difficult region to detect because it has the highest magnetic inductance, the detection coil has the highest induction

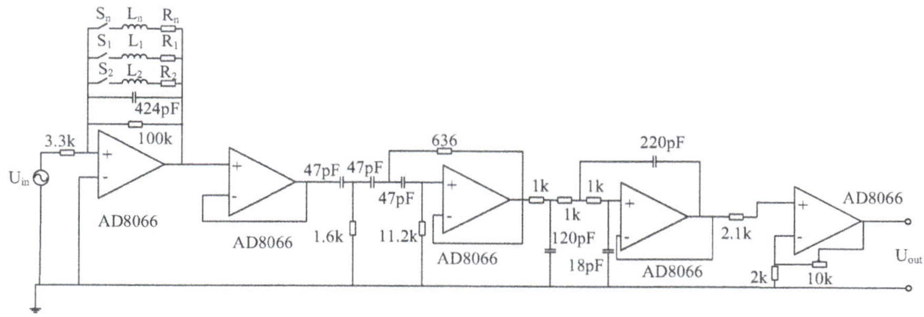

Fig. 11 Foreign object detection circuit

voltage generated by the power magnetic field, and the device has the highest withstand voltage. The experimental system is shown in Fig. 12. The center of the emitting end region is taken as an example.

In order to verify the detection performance of the designed system, a variety of high-purity materials, metal objects of different shapes, as well as common coins and clips are selected as the metal foreign objects to be detected (where D is the diameter of the cylindrical foreign object and *a* is the length of the cubic foreign object edge, in mm). The geometric centers of the metallic foreign objects are placed at the center of the detection coils and at the junction of the four detection coils (detection coil corners) to investigate the optimal and worst-case detection performance of the foreign object detection system.

The detection results are shown in Figs. 13 and 14. Figure 13a, b show the output signal waveform of the detection circuit at the center and edge of the detection coil of

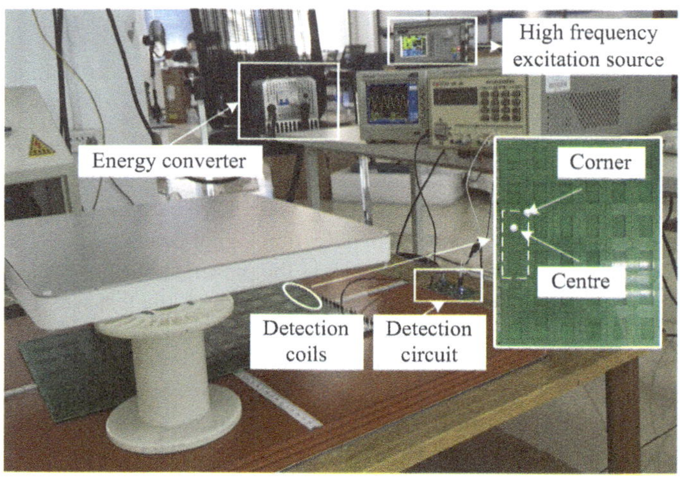

Fig. 12 Foreign object detection circuit

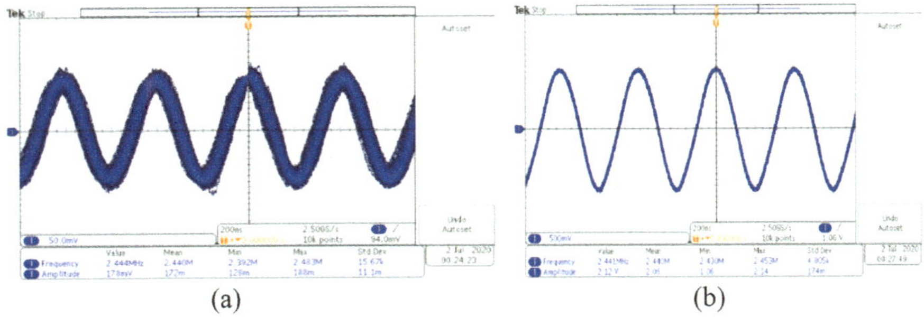

Fig. 13 Peak output voltage of the detection circuit when foreign object is 1-yuan coin **a** 1-yuan coin at the center of the detection coil. **b** 1-yuan coin at the edge of the detection coil

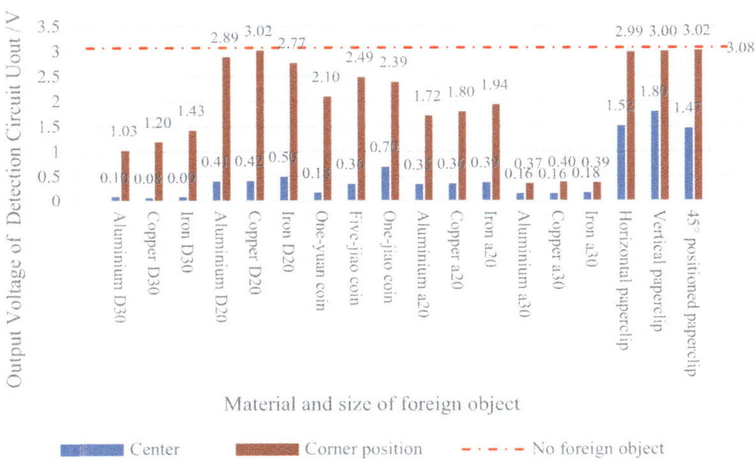

Fig. 14 The peak-to-peak output voltage of the detection circuit at the center of the detection coil and at the edge of the detection circuit

a 1 yuan coin respectively. Correspondingly, the detection circuit output signal peaks decreased from 3.08 V to 178 mV and 2.12 V when there was no foreign object. And the corresponding percentage change in detection signal peaks was 94.1 and 31.8%. In addition, taking into account factors such as the presence of a cover plate on top of the foreign object detection system in the actual product, the test was conducted by placing a 1 yuan coin in the center of the detection coil at a height of 5 mm, and the peak-to-peak value of the detection circuit was 0.714 V, with a percentage change of 23.18%.

In Fig. 14, the dashed red line shows the peak value (3.08 V) when there is no metallic foreign object. The blue and deep red curves show the peak value when the foreign object is located at the center and corner of the detection coil. The greater the difference between the red dashed line and the histogram, the greater the influence of the metallic foreign object on the peak value of the output signal, i.e., the easier to detect. The experiment results show that:

(a) When different types of foreign objects are located in the center of the detection coil, the three types of placed paperclips cause the peak-to-peak variation of the output signal of the detection circuit to exceed 41.5%, and the other foreign objects cause the peak-to-peak variation of the output signal of the detection circuit to exceed 77.3%, so that the presence of foreign objects can be accurately detected.
(b) When different types of foreign objects are located in the corner position of the detection coil, except the cylinders of three different materials with a diameter of 20 mm (D = 20 mm) and paperclips are difficult to detect, other foreign objects make the detection circuit output signal peak amplitude change of more than 19.2%, can achieve accurate detection.

6 Conclusion

This paper proposed a design method for high-sensitivity detection coil based on finite element simulation (FEM), which solved the problems of low detection sensitivity and the existence of blind spots in traditional detection methods.

Firstly, the detection area is divided and the symmetric structure of the detection coil is designed according to the magnetic induction distribution of the transmitter coil. Secondly, the influence of the size and structure of the detection coil on the impedance variation percentage is simulated by FEM and the parameters of detection coil was selected according to the simulation results. Thirdly, the effect of excitation source frequency on the detection sensitivity was analyzed and optimized in detailed. Finally, the proposed design is verified experimentally on a LCC-LCC 6.6 kW system prototype.

The experiment results illustrate that the proposed foreign object detection system has good sensitivity to detect coins and other foreign objects, and the detection system can accurately detect small-sized foreign objects such as paperclips when they are aligned to the detection coil. The experiments prove that the proposed design method can realize the high-sensitivity and accurate detection of many types of foreign objects in wireless charging systems, and at the same time, the proposed design method with universal applicability can be applied to different types and power levels of wireless charging systems.

Acknowledgements This work was supported by General Program of National Natural Science Foundation of China No. 51977043.

References

1. Zhu, C.B., J.H. Jiang, K. Song, and Q.F. Zhang. 2017. Research progress of key technologies for dynamic wireless charging of electric vehicle. *Automation of Electric Power Systems* 41 (2): 60–65 (+72). (in Chinese).
2. Wu, L.H., and B. Zhang. 2020. Overview of static wireless charging technology for electric vehicles: Part I. *Transactions of China Electrotechnical Society* 35 (6): 1153–1165. (in Chinese).
3. Wu, L.H., and B. Zhang. 2020. Overview of static wireless charging technology for electric vehicles: Part II. *Transactions of China Electrotechnical Society* 35 (8): 1662–1678. (in Chinese).
4. Zeng, Y.F., D.Y. Qiu, and B. Zhang. 2019. Review of coil designs for magnetic resonant wireless power transfer system. *Journal of Power Supply* 17 (4): 94–104. (in Chinese).
5. Zhao, Z.M., F. Liu, and K.N. Chen. 2016. New progress of wireless charging technology for electric vehicles. *Transactions of China Electrotechnical Society* 31 (20): 30–40. (in Chinese).
6. Zhao, Z.M., Y.M. Zhang, and K.N. Chen. 2013. New progress of magnetically-coupled resonant wireless power transfer technology. *Proceedings of the CSEE* 33 (3): 1–13 (+21). (in Chinese).
7. Huang, X.L., L.L. Tan, Z. Chen, H. Qiang, Y.L. Zhou, W. Wang, and W.J. Cao. 2013. Review and research progress on wireless power transfer technology. *Transactions of China Electrotechnical Society* 28 (10): 1–11. (in Chinese).
8. Gao, Y., X. Zhang, Q.X. Yang, B. Wei, and L. Wang. 2019. Bio-electromagnetic safety assessment of wireless charging environment for electric vehicles. *Transactions of China Electrotechnical Society* 34 (17): 3581–3589. (in Chinese).
9. Zhang, X., Z.H. Wang, B. Wei, S.C. Wang, and Q.X. Yang. 2019. Analysis of the influence of electric shield on space magnetic field in electric vehicle wireless charging system. *Transactions of China Electrotechnical Society* 34 (8): 1580–1588. (in Chinese).
10. Chen, C., X.L. Huang, W.H. Sun, L.L. Tan, and H. Qiang. 2014. Impact of metal obstacles on wireless power transmission system based coupled resonance. *Transactions of China Electrotechnical Society* 29 (9): 22–26. (in Chinese).
11. Chen, C., X.L. Huang, L.L. Tan, F. Wen, and W. Wang. 2015. Electromagnetic environment and security evaluation for wireless charging of electric vehicles. *Transactions of China Electrotechnical Society* 30 (19): 61–67. (in Chinese).
12. Ma, Z.Y., C.L. Liao, and L.F. Wang. 2017. Analysis of metal foreign object setting on electric vehicle wireless power transfer system. *Transactions of China Electrotechnical Society* 36 (2): 14–20. (in Chinese).
13. http://standards.sae.org/J2954_201904. Accessed 10 Sept 2020.
14. IEC TC 69: Electric road vehicles and electric industrial trucks IEC 61980-3 ED1.
15. Sonapreetha, M.R., S.Y. Jeong, S.Y. Choi, and C.T. Rim. 2015. Dual-purpose non-overlapped coil sets as foreign object and vehicle location detections for wireless stationary EV chargers. In *2015 IEEE PELS Workshop on Emerging Technologies: Wireless Power (2015 WoW)*, Daejeon, 1–7. https://doi.org/10.1109/wow.2015.7132803.
16. Jang, G.C., S.Y. Jeong, H.G. Kwak, and C.T. Rim. 2016. Metal object detection circuit with non-overlapped coils for wireless EV chargers. In *2016 IEEE 2nd Annual Southern Power Electronics Conference (SPEC)*, Auckland, 1–6. https://doi.org/10.1109/spec.2016.7846132.
17. Meichle, David Paul. 2017. Foreign object detection in wireless energy transfer systems. WO2017/0700227A1 April 27.
18. Jeong, S.Y., V.X. Thai, J.H. Park, and C.T. Rim. 2019. Self-inductance-based metal object detection with mistuned resonant circuits and nullifying induced voltage for wireless EV chargers. *IEEE Transactions on Power Electronics* 34 (1): 748–758. https://doi.org/10.1109/TPEL.2018.2813437.

19. Jeong, S.Y., V.X. Thai, J.H. Park, and C.T. Rim. 2018. Metal object detection system with parallel-mistuned resonant circuits and nullifying induced voltage for wireless EV chargers. In *2018 International Power Electronics Conference (IPEC-Niigata 2018 -ECCE Asia)*, Niigata, 2564–2568. https://doi.org/10.23919/ipec.2018.8507535.
20. Widmer, H., L. Sieber, A. Daetwyler, M. Bittner. 2017. Systems, methods, and apparatus for radar-based detection of objects in a predetermined space. U.S. Patent 9,772,401 B2, Sep 26.

Harmonic Suppression Method of High Speed PMSM Based on LC Filter and Adaptive Notch Filter

Xiaodong Zhao and Jinhua Du

Abstract The stator inductance of high-speed PMSM is usually very small. When voltage source inverter (VSI) is used to drive, a large number of high-frequency current harmonics will be generated. A high speed PMSM harmonic suppression method based on LC filter and adaptive notch filter is proposed in this paper. Firstly, the mathematical model of high speed PMSM is established. Secondly, the LC filter circuit is introduced between the three-phase winding and the inverter, and the transfer function of the system is deduced, and the amplitude frequency characteristics of the system are analyzed. It is found that the introduction of LC filter can suppress the high frequency harmonics generated by the system, but the resonance peak appears in the system. After that, an adaptive notch filter is set in the current loop of the motor, and the LMS algorithm is used to suppress the resonance peak generated by the system. Finally, the simulation results show the feasibility and effectiveness of this method, which makes the current harmonics of high-speed PMSM control system smaller, dynamic response performance and anti-interference ability enhance.

Keywords High speed PMSM · LC filter · Resonance peak · Adaptive notch filter

1 Introduction

High speed permanent magnet synchronous motor (HPMSM) has the advantages of high power density, high reliability, fast operation speed. And the high-speed permanent magnet motor can be directly connected with the load. So it is widely used in compressor machine tool, vacuum cleaner, generator and other fields [1]. However, due to the structure characteristics and operation state of the motor, the stator inductance of high-speed permanent magnet synchronous motor is usually very small, generally less than 0.5 mH. When the voltage source inverter is used for driving, because of the high switching frequency of the inverter, a large number of current harmonics will be generated near the switching frequency and its integer

X. Zhao (✉) · J. Du
School of Electrical Engineering, Xi'an Jiaotong University, No.28 Xianning West Road, Beilin District Xi'an 710049, China
e-mail: zxd152130@stu.xjtu.edu.cn

W. Chen et al. (eds.), *The Proceedings of the 9th Frontier Academic Forum of Electrical Engineering*, Lecture Notes in Electrical Engineering 743,
https://doi.org/10.1007/978-981-33-6609-1_72

multiple frequency, which will cause the current distortion in the motor winding. A large amount of harmonic current will produce large noise, core loss and torque ripple, which will reduce the motor performance and system efficiency [2, 3]. In order to suppress the current harmonics of the motor, the most direct method is to connect three identical inductors in series in the three-phase stator winding of the motor. Using this method, the equivalent inductance of the motor can be increased, and the harmonic component in the electronic current can be reduced in Ref. [4]. However, with this method, the volume and weight of the system will be increased due to the inductors in series. In Ref. [5], a LC filter is introduced between the three-phase stator winding and the inverter. Compared with the three-phase inductors in series in the stator winding of the motor, the volume and weight of the system can be reduced. But at the same time, the introduction of LC filter makes the system have resonance frequency. In order to suppress the resonance peak generated by LC filter, many papers have proposed damping technology for suppressing resonant peak in Ref. [6–8]. In Ref. [9], a passive damping method with parallel resistors at both ends of LC filter capacitor is proposed. Although the method can suppress the resonance peak, it also has a strong suppression effect on the system harmonics in the high frequency band. However, the parallel resistance will produce excess energy loss in the resistor, which will reduce the efficiency of the system. In Ref. [10], an active damping method is proposed by using transfer function equivalent method to feedback the current or voltage of the filter capacitor to simulate the resistance, which does not need to consume the extra energy of the system, and can achieve the suppression of the system resonance peak. However, this method needs to add additional sensors, which increases the volume of the system and increases the possibility of interference.

In this paper, a high-speed PMSM harmonic suppression method based on LC filter and adaptive notch filter is proposed. The LC filter is used to suppress the high frequency harmonic current component produced by the motor system, and the adaptive notch filter is set to suppress the resonance peak generated by the LC filter, and the anti-interference ability and dynamic response performance of the motor control system are enhanced. In the second part of this paper, the mathematical model of high speed permanent magnet synchronous motor is established. In the third part, the high-speed permanent magnet synchronous motor system with LC filter is analyzed, and the damping technology of resonance peak suppression is analyzed. The fourth part introduces the principle of adaptive notch filter. In the fifth part, the simulation is carried out. The effectiveness and feasibility of the proposed method are verified.

2 Mathematical Model of HPMSM

In order to simplify the analysis of high-speed PMSM and establish its reliable mathematical model, the following assumptions are made. The saturation of magnetic circuit is ignored; eddy current and hysteresis loss are ignored; the rotor of PMSM is not damped; symmetrical three-phase sinusoidal waveform current.

In the three-phase static coordinate system, the voltage equation of three-phase winding of high-speed PMSM can be expressed as:

$$u = Ri + \frac{d\psi}{dt} \tag{1}$$

where u are phase voltage of three phase winding. And i are phase current of three phase winding. ψ are flux linkage of three phase winding. R is stator winding resistance.

Because PMSM is a complex nonlinear system, it is necessary to establish HPMSM mathematical models in different coordinate systems to design the later controller by using vector control method.

In the two-phase rotating coordinate system, according to the transformation of equal amplitude coordinate, the following current expression is obtained as:

$$\begin{bmatrix} i_d \\ i_q \end{bmatrix} = \frac{2}{3}\begin{bmatrix} \cos\theta_e & \cos\left(\theta_e - \frac{2\pi}{3}\right) & \cos\left(\theta_e + \frac{2\pi}{3}\right) \\ -\sin\theta_e & -\sin\left(\theta_e - \frac{2\pi}{3}\right) & -\sin\left(\theta_e + \frac{2\pi}{3}\right) \end{bmatrix}\begin{bmatrix} i_A \\ i_B \\ i_C \end{bmatrix} \tag{2}$$

In the two-phase rotating coordinate system, the flux linkage equation is as follows:

$$\begin{cases} \psi_d = L_d i_d + \psi_f \\ \psi_q = L_q i_q \end{cases} \tag{3}$$

The voltage equation in the two-phase rotating coordinate system is obtained as:

$$\begin{cases} u_d = Ri_d + L_d \frac{di_d}{dt} - \omega_e L_q i_q \\ u_q = Ri_q + L_q \frac{di_q}{dt} + \omega_e (L_d i_d + \psi_f) \end{cases} \tag{4}$$

The equation of electromagnetic torque Te is as follows:

$$T_e = \frac{3}{2} p_n i_q \left[i_d \left(L_d - L_q \right) + \psi_f \right] \tag{5}$$

where u_d, u_q are stator voltage component on d-q axis; so as i_d, i_q for stator current component; and ψ_d, ψ_q for flux linkage component; ψ_f is flux linkage which is generated by permanent magnet; L_d, L_q are winding inductance component.

3 HPMSM System Based on LC Filter

A group of three-phase LC filter circuit is connected between PWM inverter circuit and high-speed PMSM, which can be used to suppress high-frequency harmonics in current. The A-phase structure of motor drive system is shown in Fig. 1.

From Fig. 1, the following equation can be obtained:

$$u_{\mathrm{c}} = L\frac{\mathrm{d}i_{\mathrm{a}}}{\mathrm{d}t} + Ri_{\mathrm{a}} + e_{\mathrm{a}} \tag{6}$$

$$\begin{aligned} i_{\mathrm{a1}} &= i_{\mathrm{a}} + C\frac{\mathrm{d}u_{\mathrm{c}}}{\mathrm{d}t} \\ &= LC\frac{\mathrm{d}^2 i_{\mathrm{a}}}{\mathrm{d}t} + RC\frac{\mathrm{d}i_{\mathrm{a}}}{\mathrm{d}t} + i_{\mathrm{a}} + C\frac{\mathrm{d}e_{\mathrm{a}}}{\mathrm{d}t} \end{aligned} \tag{7}$$

$$\begin{aligned} u_{\mathrm{a}} &= u_{\mathrm{c}} + L_1\frac{\mathrm{d}i_{\mathrm{a1}}}{\mathrm{d}t} \\ &= LL_1C\frac{\mathrm{d}^3 i_{\mathrm{a}}}{\mathrm{d}t^3} + RL_1C\frac{\mathrm{d}^2 i_{\mathrm{a}}}{\mathrm{d}t^2} \\ &+ (L_1 + L)\frac{\mathrm{d}i_{\mathrm{a}}}{\mathrm{d}t} + Ri_{\mathrm{a}} + L_1C\frac{\mathrm{d}^2 e_{\mathrm{a}}}{\mathrm{d}t^2} + e_{\mathrm{a}} \end{aligned} \tag{8}$$

The transfer function of the motor control system can be obtained as:

$$\phi(s) = \frac{I_{\mathrm{a}}(s)}{U_{\mathrm{a}}(s)} = \frac{1}{LL_1Cs^3 + RL_1Cs^2 + (L_1 + L)s + R} \tag{9}$$

The amplitude frequency characteristic curve is shown in Fig. 2.

It can be concluded from Fig. 2 at low frequency, the system can be simplified into a first-order inertial system, the transfer function of the system is equivalent to:

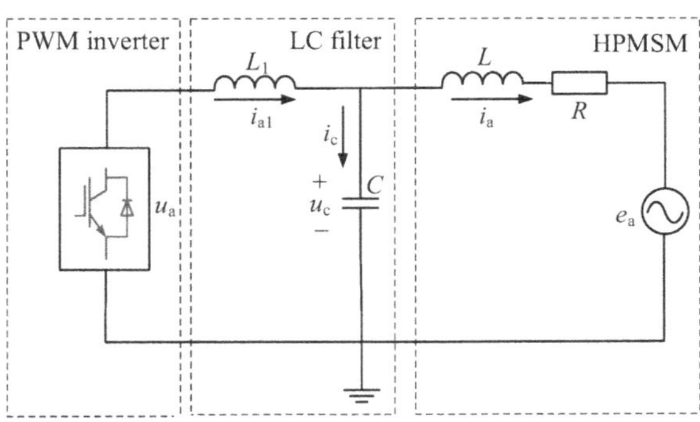

Fig. 1 Structure diagram of A-phase drive system of high speed PMSM based on LC filter

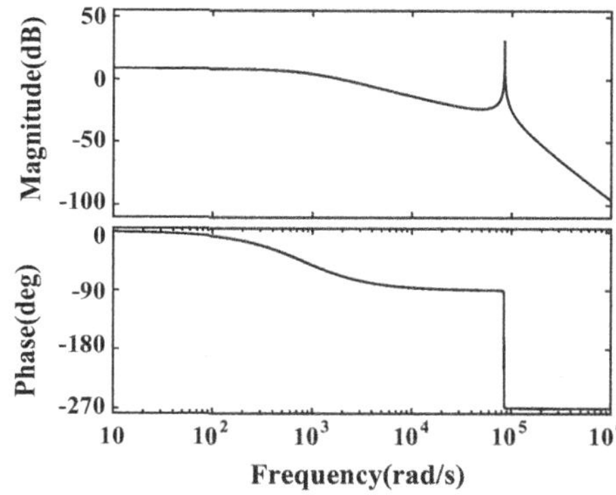

Fig. 2 Amplitude frequency characteristic curve of high speed PMSM control system using LC filter

$$\phi(s) \approx \frac{1}{(L_1 + L)s + R} \tag{10}$$

From Eq. (10), the cut-off frequency of the motor drive system is obtained as follows:

$$\omega_b \approx \frac{L + L_1 - \sqrt{(L + L_1)^2 - 4R^2 L_1 C}}{2RL_1 C} \approx \frac{R}{L + L_1} \tag{11}$$

It can be seen from Fig. 2 that the LC filter causes the motor control system to have resonance frequency at high frequency, and the phase changes by $-180°$ at the resonance frequency. Because the phase jump $-180°$ is a negative crossing, it will produce a pair of closed-loop poles in the right half plane, which makes the whole system unstable. Therefore, it is necessary to analyze the resonant frequency. At high frequency, the system can be simplified as an integral link and a second-order oscillation link in series. The equivalent transfer function is as follows:

$$\phi(s) \approx \frac{1}{s\left(LL_1 Cs^2 + RL_1 Cs + L + L_1\right)} \tag{12}$$

From Eq. (12), it can be concluded that the resonant frequency of the motor drive system is:

$$f = \frac{1}{2\pi}\sqrt{\frac{L + L_1}{LL_1 C} - \frac{R^2}{2L^2}} \tag{13}$$

4 Adaptive Notch Filter

4.1 Adaptive Notch Filter

In order to suppress the resonant peak of high-speed PMSM system under LC filter, the active damping method based on capacitor current proportional feedback needs to add sensors at the filter capacitor, which increases the cost of the system and reduces the reliability of the system. An adaptive notch filter is used to suppress the resonance peak generated by LC filter without adding additional sensors. The adaptive notch filter can also adjust the adaptive coefficient according to the adaptive algorithm and the change of the external environment, so that the characteristics of the notch filter can be changed with the changes of harmonic and noise, so that the system has a certain degree of adaptability and achieves an optimal filtering effect. Figure 3 is a block diagram of the adaptive notch filter.

Among them, $x(t)$ is the input signal, which can be regarded as the original signal superimposed with the interference signal, $s(t)$ is the target signal, $n(t)$ is the interference signal to be filtered by the adaptive notch filter, $r_s(t)$ and $r_c(t)$ are the reference signals with the same frequency as the interference signal. ω_1 and ω_2 are the adaptive coefficient, $y(t)$ is the weighted signal of $r_s(t)$ and $r_c(t)$, and $e(t)$ is the output signal of the adaptive notch filter. The signal is the original signal after filtering out the specific frequency harmonic.

The adaptive algorithm of the adaptive notch filter is LMS algorithm. According to the principle of minimizing the mean square error $e^2(t)$, the LMS algorithm will adjust the adaptive coefficient ω_1 and ω_2 through the error signal $e(t)$ and the reference input signal $r_s(t)$ and $r_c(t)$. The specific algorithm of LMS can be sorted into formula:

$$\begin{cases} y(k) = \omega_1(k)r_s(k) + \omega_2(k)r_c(k) \\ e(k) = x(k) - y(k) \\ \omega_1(k+1) = \omega_1(k) + \mu e(k)r_s(k) \\ \omega_2(k+1) = \omega_2(k) + \mu e(k)r_c(k) \end{cases} \tag{14}$$

where μ is the iteration step size of the algorithm, k is the iterative sequence, and there is $t = k\tau$ and τ is the sampling interval. According to Eq. (14), the equivalent transfer function of the adaptive notch filter can be obtained:

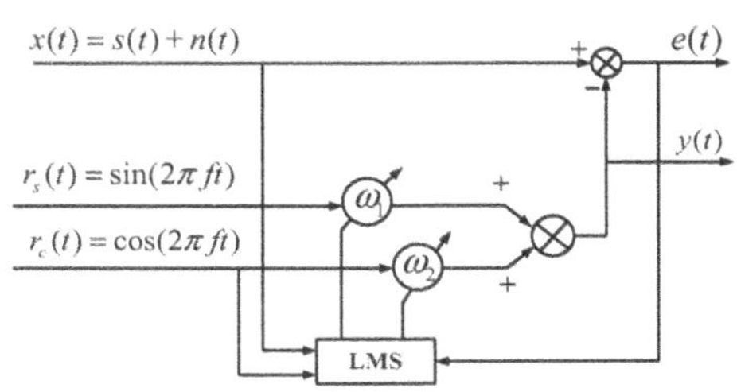

Fig. 3 Structural block diagram of adaptive notch filter

$$\frac{e(s)}{x(s)} = \frac{s^2 + \omega_0^2}{s^2 + \frac{\mu A^2}{\tau}s + \omega_0^2} \quad (15)$$

4.2 Control System Design

The control system structure of the system is shown in Fig. 4. The inner loop of the control system is the current loop and the outer loop is the speed loop. The PI control is used in the inner and outer loops. In order to suppress the high-order harmonics in the system, a LC filter is added between the inverter and the motor. In order to suppress the resonance peak generated by the introduction of LC filter, d-axis and q-axis adaptive notch filters are added in the current loop.

Take the q-axis adaptive notch filter as an example, its structure is shown in Fig. 5.

1. Calculation of resonant frequency

From Eq. (13), it can be seen that the resonant frequency value of motor control system in three-phase static coordinate system, but due to the addition of adaptive

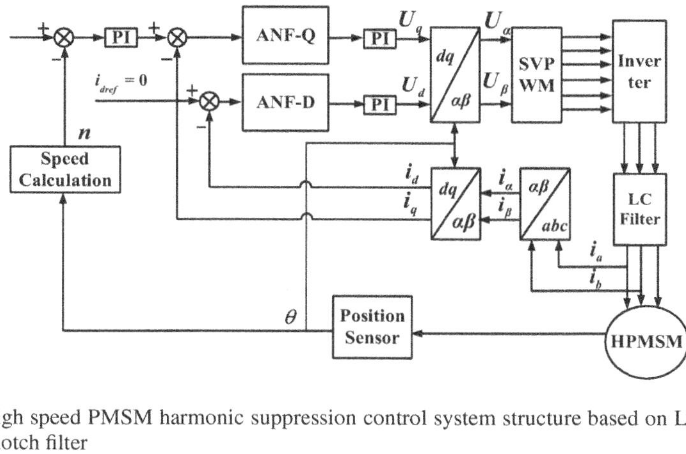

Fig. 4 High speed PMSM harmonic suppression control system structure based on LC filter and adaptive notch filter

Fig. 5 Q-axis adaptive notch filter structure

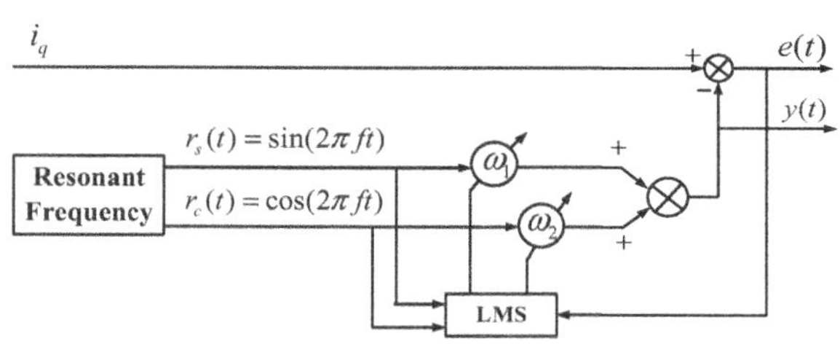

notch filter in d-axis and q-axis, it is necessary to convert the resonant frequency value from three-phase static coordinate system to dq coordinate system. Because the three-phase winding of PMSM generally adopts star connection, three-phase symmetry and half wave symmetry, the three-phase current of PMSM generally does not contain even harmonics and odd harmonics with integral times of 3. Considering the harmonics, the three-phase current of PMSM in three-phase static coordinate system can be expressed as follows:

$$\begin{cases} i_a = i_1\cos(\omega t + \theta_1) + i_5\cos(-5\omega t + \theta_2) \\ \quad + i_7\cos(7\omega t + \theta_3) + i_{11}\cos(-11\omega t + \theta_4) \\ \quad + i_{13}\cos(13\omega t + \theta_5) + \cdots \\ i_b = i_1\cos\left(\omega t + \theta_1 - \frac{2}{3}\pi\right) + i_5\cos\left(-5\omega t + \theta_2 - \frac{2}{3}\pi\right) \\ \quad + i_7\cos\left(7\omega t + \theta_3 - \frac{2}{3}\pi\right) + i_{11}\cos\left(-11\omega t + \theta_4 - \frac{2}{3}\pi\right) \\ \quad + i_{13}\cos\left(13\omega t + \theta_5 - \frac{2}{3}\pi\right) + \cdots \\ i_c = i_1\cos\left(\omega t + \theta_1 + \frac{2}{3}\pi\right) + i_5\cos\left(-5\omega t + \theta_2 + \frac{2}{3}\pi\right) \\ \quad + i_7\cos\left(7\omega t + \theta_3 + \frac{2}{3}\pi\right) + i_{11}\cos\left(-11\omega t + \theta_4 + \frac{2}{3}\pi\right) \\ \quad + i_{13}\cos\left(13\omega t + \theta_5 + \frac{2}{3}\pi\right) + \cdots \end{cases} \tag{16}$$

In Eq. (16), i_1, i_5, i_7, i_{11} and i_{13} represent the components of fundamental wave, 5th harmonic, 7th harmonic, 11th harmonic and 13th harmonic in three-phase static coordinate system. According to the rotating magnetic field theory of AC motor, the fundamental current of PMSM is the same as the rotation direction of the rotor, the 5th harmonic current is opposite to the rotation direction of the fundamental wave, and the rotation speed is -5ω; the 7th harmonic current and the fundamental rotation direction are the same, the rotation speed is 7ω, and so on. According to the coordinate transformation formula of Eq. (2), the current is transformed into dq coordinate system, which can be expressed as follows:

$$\begin{cases} i_{\mathrm{d}} = i_{\mathrm{d}1} + i_5\cos(-6\omega t + \theta_2) + i_7\cos(6\omega t + \theta_3) \\ \quad + i_{11}\cos(-12\omega t + \theta_4) + i_{13}\cos(12\omega t + \theta_5) + \cdots \\ i_{\mathrm{q}} = i_{\mathrm{q}1} + i_5\sin(-6\omega t + \theta_2) + i_7\sin(6\omega t + \theta_3) \\ \quad + i_{11}\sin(-12\omega t + \theta_4) + i_{13}\sin(12\omega t + \theta_5) + \cdots \end{cases} \tag{17}$$

From Eq. (17), it can be obtained that for the 5th and 7th harmonic of PMSM, the rotation angular velocity is -5ω and 5ω respectively in dq rotating coordinate system; in dq rotating coordinate system, the rotation angular velocity of the 11th and 13th harmonic becomes -12ω and 12ω; for (6n − 1) and (6n + 1) harmonics, the rotation angular velocity becomes $-6n\omega$ and $6n\omega$.

Therefore, the values of filter inductance L and filter capacitor C in LC filter can be adjusted appropriately so that the resonant frequency obtained by Eq. (16) is as close as possible to (6n − 1) and (6n + 1) harmonic frequencies of motor current. In dq coordinate system, the resonant frequency of motor current is converted into 6n harmonic frequency.

2. Generate reference input signal

After calculating the resonant frequency f of the motor control system, the reference input signals $r_s(t)$ and $r_c(t)$ with frequency f are generated by the controller.

3. LMS module

According to the formula of LMS algorithm shown in Eq. (14), the adaptive coefficient ω_1 and ω_2 is continuously adjusted is updated iteratively so that the cosine weighted signal $y(t)$ is close enough to the harmonics to be filtered out, and the output $e(t)$ does not contain harmonics.

5 Simulation Experiment

In this paper, a high-speed PMSM vector control model with speed and current double closed-loop in $i_d = 0$ is built. On this basis, LC filter and adaptive notch filter are introduced to compare and analyze the speed, current and THD of the motor during operation. In addition to the initial load of 0.8 N·m, the motor is loaded to 2.3 N·m when t = 0.6 s. The parameters used in the simulation are shown in the Table 1.

1. The traditional high-speed PMSM vector control model with speed and current double closed loop in $i_d = 0$ is built, and LC filter and adaptive notch filter are not introduced. The simulation results of speed, current and THD during motor operation are shown in Fig. 6.
2. The traditional high-speed PMSM vector control model with speed and current double closed loop in $i_d = 0$ is built, and the LC filter is introduced. The simulation results of speed, current and THD during motor operation are shown in Fig. 7. From the analysis and simulation results of current and THD, it is

Table 1 Parameters used in simulation of high speed PMSM control system

Parameters	Value
Pole number	2
Permanent magnetic linkage/Wb	0.0414
Rated speed/(r/min)	15,000
Stator resistance/Ω	0.37
Stator inductance L_d/mH	0.427
Stator inductance L_q/mH	0.427
Moment of inertia/(kg m^2)	0.0005
Adaptive notch filter step size	0.005
Filter inductance/mH	0.7
Filter capacitance/μF	5
Switching frequency/kHz	10

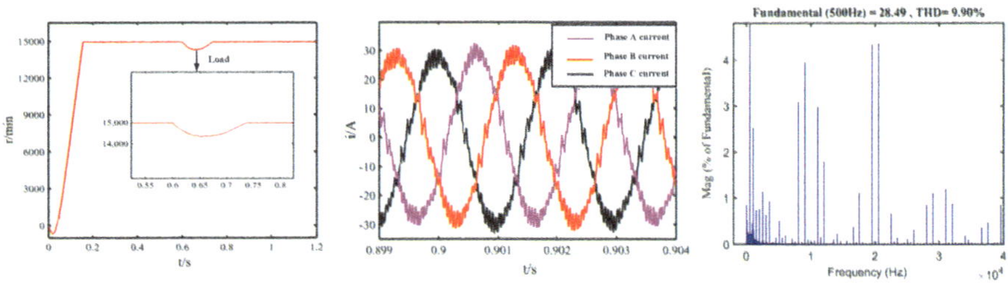

(a) Speed curve (b) Current curve (c) Current THD analysis

Fig. 6 Correlation waveform of high speed PMSM under traditional control. **a** Speed curve. **b** Current curve. **c** Current THD analysis

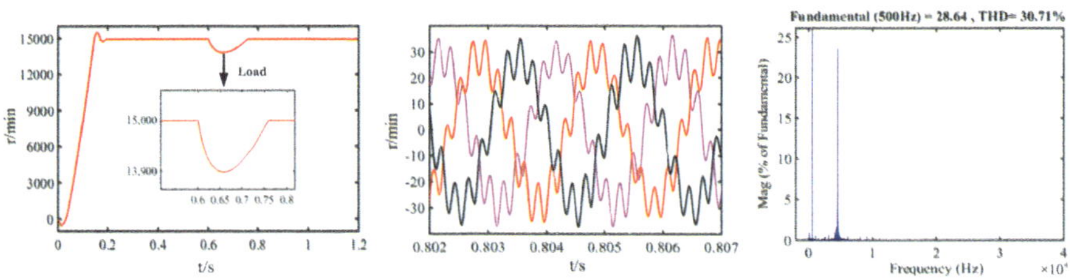

(a) Speed curve (b) Current curve (c) Current THD analysis

Fig. 7 High speed PMSM system correlation waveform based on LC filter. **a** Speed curve. **b** Current curve. **c** Current THD analysis

concluded that the introduction of LC filter can suppress the high frequency harmonics of the system. However, the resonance frequency of the system deteriorates the performance of the system. At the resonance frequency, the resonance peak appears in the current. The harmonic reaches 28% of the fundamental wave and the THD reaches 30.71%, which is far more than the harmonic content before the filter is introduced.

Compared with the speed waveforms in Figs. 6 and 7, there is overshoot in the speed regulation after the introduction of LC filter. When the sudden load is applied, the time for the speed to decline and the system to return to the stable state is increased, and the dynamic response performance and anti-interference ability of the system become worse.

From the simulation results of current and THD in Fig. 8, it is concluded that the introduction of adaptive notch filter can suppress the resonance peak generated by the introduction of LC filter. The THD of the current is only 3.66%, and the harmonic content of the current is greatly reduced. Compared with the speed waveforms in Figs. 6 and 8, when the sudden load is applied, the speed drop is only 120 r/min, and the time for the system to recover to the stable state is 0.05 s, so the dynamic response performance and anti-interference ability of the system are significantly enhanced.

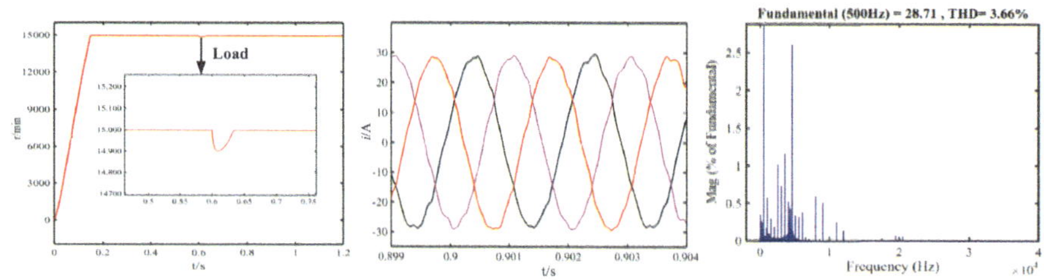

(a) Speed curve (b) Current curve (c) Current THD analysis

Fig. 8 Correlation waveform of high speed PMSM system based on LC filter and adaptive notch filter. **a** Speed curve. **b** Current curve. **c** Current THD analysis

6 Conclusion

This paper presents a high-speed PMSM harmonic suppression method based on LC filter and adaptive notch filter. The introduction of LC filter can suppress the high-frequency harmonics of the system, but the system also introduces the resonant frequency. At the resonance frequency, the system produces a resonance peak, which worsens the performance of the motor system. An adaptive notch filter is set in the d-axis and q-axis of the motor, which can suppress the resonance peak generated by the introduction of LC filter, so as to suppress the harmonic current. The simulation results show that compared with the traditional double closed-loop control method, this method not only suppresses the harmonic current, but also enhances the dynamic response performance and anti-interference ability of the system, which verifies the effectiveness and feasibility of this method.

Acknowledgements This work was supported by the State Key Program of National Natural Science of China Grant (51737010) and the National Natural Science Foundation of China Grant (51877172).

References

1. Zhang, Fengge, Guanghui Du, Tianyu Wang, and Guangwei Liu. 2016. Development and design of high speed motor. *Transactions of China Electrotechnical Society* 31 (07): 1–18. (in Chinese).
2. Yu, Jikun, Liyi Li, Pengcheng Du, et al. 2016. Armature current harmonic analysis of high speed permanent magnet synchronous motor. *Electric Machines and Control* 5: 28–36. (in Chinese).
3. Li, Liyi, Jikun Yu, Jiwei Cao, Jiangpeng Zhang, and Pengcheng Du. 2015. A new universal harmonic algorithm for voltage and current of permanent magnet synchronous motor fed by PWM inverter circuit. *Proceedings of the CSEE* 35 (23): 6203–6213. (in Chinese).
4. Zwyssig, C., M. Duerr, D. Hassler, et al. 2007. An ultra-high-speed, 500000 rpm, 1 kW electrical drive system. In *2007 Power Conversion Conference*, Nagoya, 1577–1583.

5. Li, Liyi, Guangjun Tan, Jiaxi Liu, et al. 2013. Inductance-capacitance filter circuit for reducing current harmonics of high-speed motor. *Micromotors* 46 (7): 38–44. (in Chinese).
6. Xu, Jinming, Shaojun Xie, Liuliu Huang, et al. 2017. Design of LCL-filter considering the control impact for grid-connected inverter with one current feedback only. *IET Power Electronics* 10 (11): 1324–1332.
7. Busada, Claudio A., and Sebastian Gomez Jorge. 2015. Full-state feedback equivalent controller for active damping LCL-filtered grid-connected inverters using a reduced number of sensors. *IEEE Transactions on Industrial Electronics* 62 (10): 5993–6002.
8. Zhang, Xing, Peng Chen, Fei Li, et al. 2017. Study of a current control strategy based on multisampling for high-power grid-connected inverters with an LCL filter. *IEEE Transactions on Power Electronics* 32 (7): 5023–5034.
9. Wang, X., F. Blaabjerg, and P.C. Loh. 2015. Virtual RC damping of LCL-filtered voltage source converters with extended selective harmonic compensation. *IEEE Tractions on Power Electronics* 30 (9): 5023–5034.
10. Zhao Rende, Bin Zhao, Hailiang Xu, Pengxiao Sun, and Xuemei Dong. 2019. Research on control system and strategy of permanent magnet synchronous motor with LC filter. *Transactions of China Electrotechnical Society* 34 (S1): 79–86. (in Chinese).

Research on Key Position of Hot Spot Temperature of Dry Type Transformer in Distribution Network

Xin Liu, Pengfei Song, Youqin Zhang, and Naming Zhang

Abstract The power transformer is one of the important equipment in the distribution network system and its safe and stable operation has an important impact on the reliability of the entire distribution network system [1]. The hot spot temperature of the transformer is one of the important factors that affect the insulation life of the distribution network transformer. In this paper, the magnetic field distribution of the dry-type transformer is obtained by using finite element software through electromagnetic field simulation calculation, so as to obtain its core loss and coil loss, and the temperature distribution of the dry-type transformer can be accurately calculated through the heat flow field, thereby obtaining the hot spot position of the transformer, which provides theoretical basis for online hot spot temperature monitoring of transformers.

Keywords Dry-type transformer · Magnetic field · Temperature field · Hot spot temperature

1 Introduction

The temperature of the dry-type transformer is one of the important factors affecting the insulation life of the transformer, and the temperature of the key point directly affects the operating life of the transformer. Under common conditions, the transformer is most prone to aging and insulation problems at the location with the highest temperature. The research shows that the unstable operation or damage of dry-type transformer is mainly due to the winding temperature of the transformer exceeds the heat-resistant limit of the insulating material [2]. For the transformer with A-class insulation, the aging speed of the insulation will be double [3], For the former distributed optical fiber sensor temperature measurement system, or the use of thermal imager, although the internal temperature or external surface temperature of the dry-type transformer can be quickly obtained, the highest temperature of

X. Liu · P. Song · Y. Zhang · N. Zhang (✉)
School of Electrical Engineering, Xi'an Jiaotong University, Xi'an, China
e-mail: 329880173@qq.com

W. Chen et al. (eds.), *The Proceedings of the 9th Frontier Academic Forum of Electrical Engineering*, Lecture Notes in Electrical Engineering 743,
https://doi.org/10.1007/978-981-33-6609-1_73

transformer has always existed in the transformer, which can't be measured accurately. How to get the maximum temperature of dry-type transformer is an important subject.

At present, the hotspot temperature calculation methods for transformers at home and abroad are mainly divided into four categories: multi-physics modeling calculation methods, artificial intelligence algorithms, empirical formula methods and thermal circuit model methods [4]. The team of Professor Chen Weigen from Chongqing University applied generalized neural networks and support vector machines to the prediction of transformer hotspot temperature, and achieved better results [5]. The team of Professor Ruan Jiangjun from Wuhan University adopted the three-dimensional electromagnetic fluid temperature field coupling calculation method combining finite element method and finite volume method to analyze the internal oil flow distribution of oil immersed transformer under different working conditions According to the characteristics of winding temperature distribution, a hot spot temperature inversion model of oil immersed transformer based on multi physical field calculation and fuzzy neural network algorithm is proposed [6]. Overheating of winding will not only cause accelerated aging of insulation, and then affect its life, but also cause serious safety problems. A series of malignant accidents will occur every year at home and abroad because of winding overheating, this led to huge economic losses [7]. The simulation object is a three-phase dry-type transformer with a rated power of 100 KVA and a rated voltage of 315 V/400 V. The connection mode is triangle star connection. The low-voltage and high-voltage windings are all made of aluminium foil. Each aluminium foil is insulated by ceramics. Ceramics are often made of non-silicate chemical raw materials or synthetic raw materials, and have many advantages such as excellent insulation, corrosion resistance, high temperature resistance, high hardness, low density, and radiation resistance [8]. Insulation paper is used for insulation between winding layers of transformer. The high-voltage winding of dry-type transformer has three layers with 105 turns, and the secondary side winding has three layers of 78 turns. The cross-sectional area of high-voltage winding is 34.04 mm, and the cross-sectional area of secondary winding is 44.47 mm. The three-dimensional structure of dry-type transformer is shown in (Fig. 1).

2 Dry-Type Transformer Structure Modeling

During the operation of the transformer, there is an alternating magnetic flux in the iron core with the same frequency as the power supply, resulting in eddy current loss and hysteresis loss in the iron core [9]. There is a load current in the winding, resulting in load loss on the primary and secondary side coils. The rest are DC loss, eddy current loss on the winding and eddy current loss on metal devices.

Fig. 1 Three-dimensional structure diagram of dry-type transformer

2.1 Magnetic Field Calculation and Loss Analysis of Dry Type Transformer Under Short Circuit Test

In the case of transformer secondary side short circuit, when the primary side voltage is applied to the primary side and the secondary side current is the rated value, this case is the transformer short circuit test. In this case, due to the small resistance of the secondary side winding, when the primary side voltage is low, the secondary side can generate the rated current. From formula (1), it can be seen that because the voltage of the primary and secondary windings is low, the magnetic field in the iron core is relatively small, and the primary voltage of the transformer is 7.6 V through simulation.

$$\begin{cases} E_1 = 4.44 f N_1 \phi_m \\ E_2 = 4.44 f N_2 \phi_m \end{cases} \quad (1)$$

According to the simulation, the magnetic field distribution is shown in Fig. 2. The magnetic field in the iron core is small and the loss is small. The current of primary and secondary winding is rated value, so the loss of transformer is mostly copper loss of winding. The loss on the winding can be obtained as 1570 W by integrating the volume loss density with the finite element software.

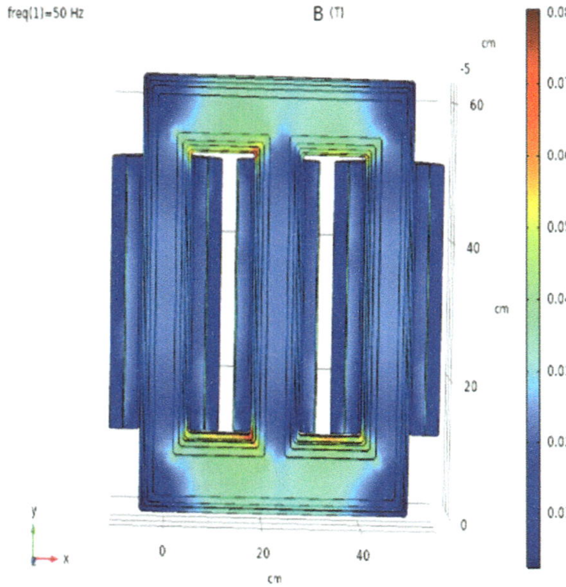

Fig. 2 Distribution of effective value of magnetic field intensity in short circuit experiment

2.2 Magnetic Field Calculation and Loss Analysis of Dry-Type Transformer Under no-Load Test

In the no-load test, the applied voltage of the primary side is 315 V, and the voltage of the secondary side is 400 V. According to formula (1), the magnetic field generated is similar to that of the normal operation, and the loss of the iron core is basically the same as that of the normal operation. In the no-load condition, the current of the primary and secondary sides of the transformer is relatively small, so the loss of the primary and secondary windings is ignored. The simulation results show that the primary side current is 8.9 A. The magnetic field density distribution is shown in the Fig. 3.

Under no-load condition, the maximum magnetic field density of the core column is 2.5 T, the maximum flux density is distributed at the corner of the transformer core, and the main flux range of the rest is about 1.5 T. The tangent plane of the magnetic core is shown in Fig. 4, and the magnetic field in the coil is shown in Fig. 5.

As the no-load loss is mainly the core loss, there are two commonly used loss calculation methods for the iron core made of silicon steel sheet. One is the empirical formula as follows:

$$\begin{cases} P_h = \eta f B_{\max}^{1.6} V * 10^{-3} \\ P_e = \eta d^2 f^2 B_{\max}^2 V * 10^{-3} \\ Pc = P_h + P_e \end{cases} \tag{2}$$

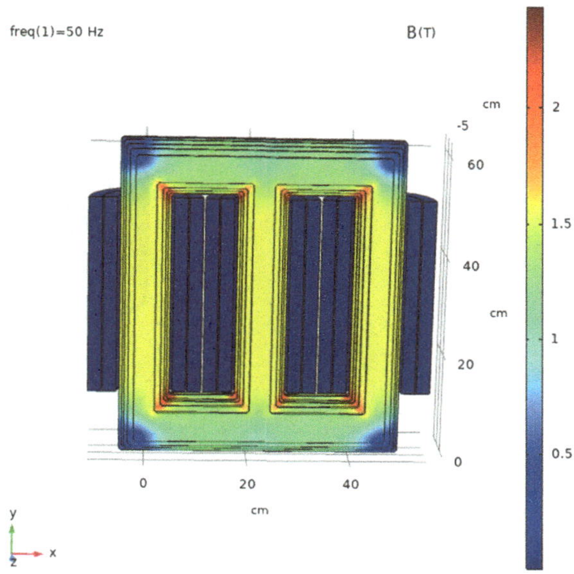

Fig. 3 Distribution of effective value of magnetic field intensity in no-load experiment

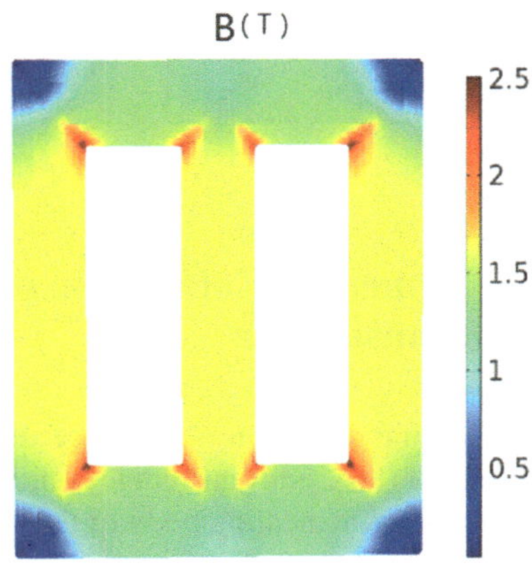

Fig. 4 Effective value of magnetic field strength of iron core in no-load experiment

The other is obtained from the loss curve [10]. The loss curve of silicon steel sheet used in this paper is shown in the Fig. 6. In the finite element software, the core loss is calculated as follows:

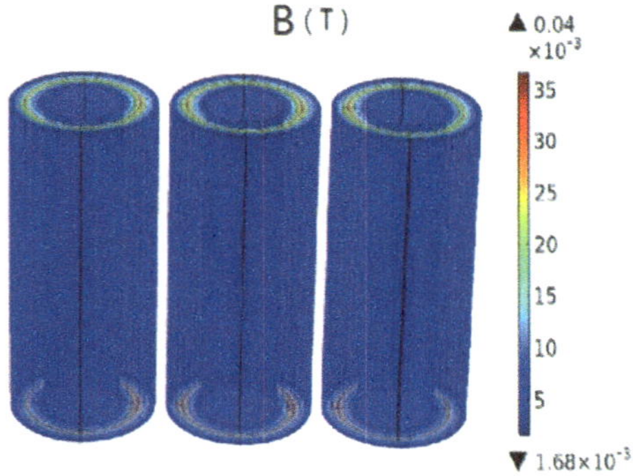

Fig. 5 Effective value of magnetic field strength of windings in no-load experiment

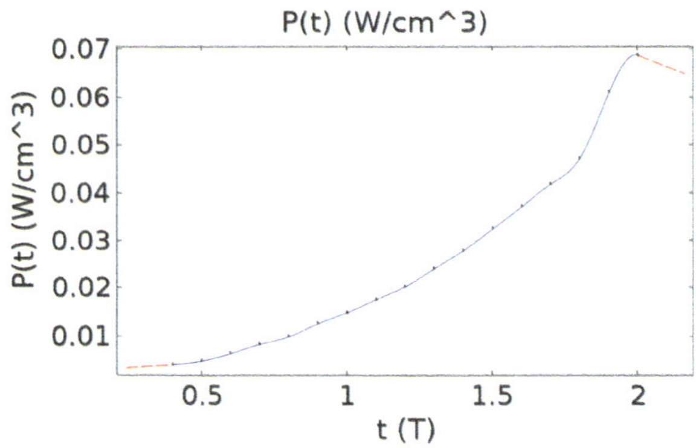

Fig. 6 Ceramic insulation dry-type transformer core loss curve

3 Numerical Calculation of Temperature Field of Dry-Type Transformer

During the operation of dry-type transformer, heat is generated by the core due to hysteresis loss, eddy current loss and winding resistance and lead loss. Part of the heat is stored in the heating body of the dry transformer, which makes the temperature of the heating body rise, and the other part of the heat disperses into the surrounding medium to increase the temperature of the surrounding medium [11]. The solid heat transfer layer flow field of transformer is to calculate the temperature of transformer through the loss of transformer. The winding loss and core loss are calculated by electromagnetic field calculation, and the temperature field of transformer is calculated as heat source. Heat transfer to the surrounding can be divided into three basic forms: heat conduction, thermal convection and thermal radiation [12]. The heat transfer in the dry-type transformer is mainly realized by heat conduction. At the same time, there are heat convection and heat radiation between the equipment surface and the outside air. The temperature rise of dry-type transformer mainly includes the temperature rise of iron core and winding [13].

The low-voltage winding and the high-voltage winding of the transformer transfer heat by means of heat conduction. The heat conduction formula is as follows. The surface of the transformer in the air transfers heat in the form of thermal convection, and the formula is as follows (Figs. 7 and 8):

$$q = -k\nabla T \tag{3}$$

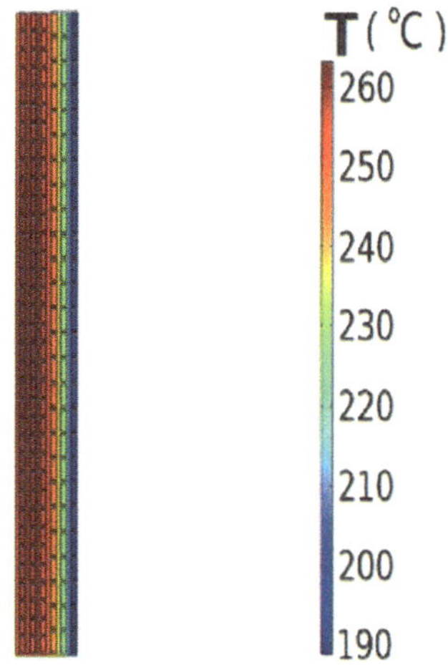

Fig. 7 Winding temperature distribution of ceramic dry-type transformer

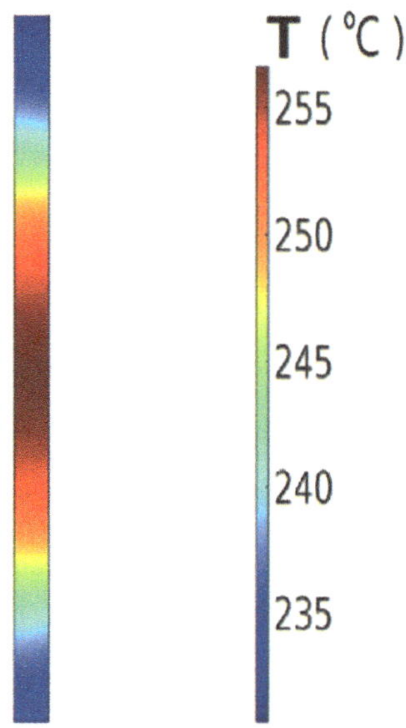

Fig. 8 Temperature distribution of ceramic dry-type transformer core

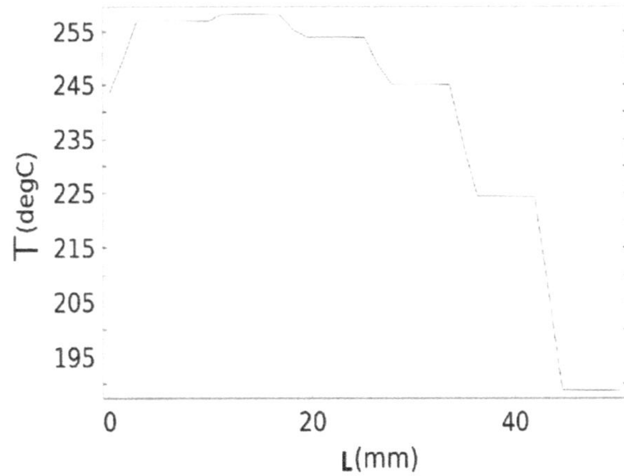

Fig. 9 Transverse temperature distribution of dry-type transformer winding

$$\rho C_p u \cdot \nabla T + \nabla \cdot q = Q \tag{4}$$

The axial temperature distribution of dry-type transformer is shown in Fig. 9. The aluminum strip and ceramic in the dry-type transformer dissipate heat faster than the insulation paper, so there is a certain temperature difference in each layer of winding, and there is a large temperature gradient at the insulation paper.

From the above Fig. 9, it can be seen that the temperature of high-voltage winding, low-voltage winding and iron core is unevenly distributed [14]. Low-voltage winding temperature is higher, high-voltage winding temperature is relatively minimum. The hottest spot temperature of the transformer is 260 °C, which is located in the low voltage winding.

The correctness of the model is verified by measuring the temperature distribution of the transformer as shown in Fig. 10.

4 Conclusion

Based on the finite element simulation, this paper analyzes the magnetic field of ceramic dry-type transformer, studies the magnetic field distribution and loss size of the transformer in short-circuit test and no-load test, and studies the temperature field of ceramic dry-type transformer based on this. The results show that the magnetic field mode value at the inner corner of the transformer is larger than that at the outer corner, and the core loss is larger at the inner corner. The temperature distribution of ceramic dry-type transformer is higher than that of common resin cast dry-type transformer. However, due to the high-temperature resistance of ceramic material, it can operate normally. The hot spot temperature of a dry-type transformer is not on

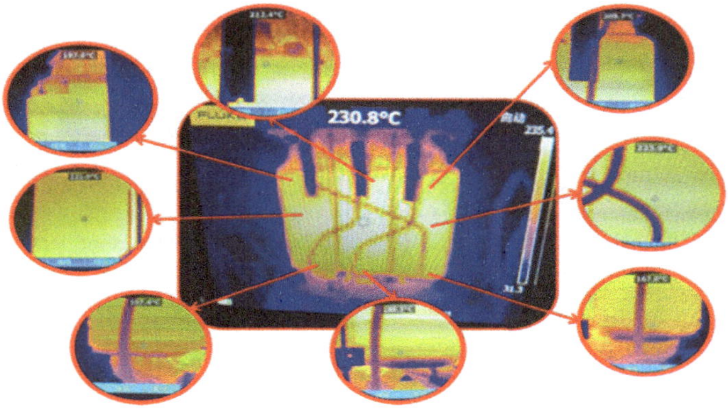

Fig. 10 Temperature distribution of transformer in actual operation

its surface or core [15], it is located on the second layer of the low-voltage winding, and the maximum temperature is 263.21 °C.

Acknowledgement This work was supported by the State Grid Corporation of China through the Science and Technology Project under Grant 5500-201999543A-0-0-00.

References

1. Li, Mengli, Weigen Chen, Shudi Peng, et al. 2010. Simulation model of transformer winding hotspot temperature calculation based on thermoelectric analogy. In *Proceedings of the 2010 Academic Conference of Chongqing Electrical Engineering Society*.
2. Yao, Yucheng, Dingguo Cai, Yebin Xu, et al. 2012. Numerical simulation of temperature field of dry type transformer. *Transformer* 49 (8): 31–33.
3. Chen, Hongbo, and Min Li. 2014. *Condition monitoring technology of power transmission and transformation equipment based on sensors*. Xi'an: Xi'an Jiaotong University Press.
4. Liao, Caibo, Jiangjun Ruan, Chao Wei, et al. 2018. Review of research methods for transformer hot spot temperature. *High Voltage Apparatus* (7).
5. Chen, Weigen, Li Teng, Jun Liu, et al. 2014. Prediction model of transformer winding hot spot temperature based on genetic optimization support vector machine. *Journal of Electrical Technology* 29 (001): 44–51.
6. Luo, Xiaoman, Jiangjun Ruan, Yongqing Deng, et al. 2020. Inversion of transformer hot spot temperature based on multi physical field calculation and fuzzy neural network algorithm. *High Voltage Technology* 46 (3).
7. Eslamian, M, B. Vahidi, and A. Eslamian. 2011. Thermal analysis of cast-resindry-type transformers. *Energy Conversion and Management* 52 (3): 2479–2488.
8. Qian, Jun. 1991. Status and development of high performance ceramics. *Acta Internatioal techno economic research* 02: 37–43.
9. Mei, Suzhen. 1994. Eddy current loss and hysteresis loss of transformer. *Journal of Hubei Normal University: Philosophy and Social Science Edition*.
10. Mu, Zhang, Liye Gao, Juan Wei, et al. 2015. Three dimensional temperature field simulation calculation of resin cast dry-type transformer. *Journal of Tianjin University of technology* 000 (003): 62–66.

11. Li, Yan, Yongzhi Liu, and Yongteng Jing. 2012. Finite element analysis of temperature field of flameproof dry-type transformer. *Journal of power system and automation* 24 (4): 59–63.
12. Chen, Shisheng, and Xiuju Chen. 2002. Analysis of the hottest spot position of the winding in resin cast dry type transformer. *Transformer* 39 (009): 8–12.
13. Zhao, Yanlong. 2013. *Numerical calculation and on-line monitoring system for temperature field of resin cast dry-type transformer*. Chongqing University.
14. Xu, Lingxiao, Baobao Ma, and Jianping Gui. 2019. Temperature field analysis of resin cast dry type transformer. *Electrical Technology* 000 (009): 56–57.
15. Liu, Guojian, and Fenghua Wang. 2016. Calculation and analysis of temperature field distribution of resin cast dry type transformer. *High voltage apparatus* 8: 83–89.

The Effect of Laser Parameters on the Ultrasonic Signal of Laser Ultrasonic Detection of GIS Basin-Type Insulator

Guanliang Li, Xiaojing Li, Haifei Wu, Shuai Li, Xin Huang, and Hui Xia

Abstract Laser ultrasonic technology has the advantages of non-contact, remote detection, wide frequency band and high spatial resolution, and is widely used in the field of material nondestructive testing. At present, GIS basin insulator defect detection method is affected by environmental electromagnetic interference and cannot be on-line. As a non-contact detection method, laser ultrasonic detection technology has the potential advantage of on-line detection. However, the detection technology of basin-type insulators based on laser ultrasonic is still in its infancy. Firstly, the theoretical model of laser ultrasonic excitation GIS basin insulator is established. Then the characteristics of excited ultrasonic signal are studied according to different laser energy, pulse width and laser rise time. Finally, it is concluded that the incident laser parameters will directly affect the amplitude and frequency of the excitation ultrasound through simulation analysis, which provides a basis for selecting the optimal laser excitation parameters for the laser ultrasonic testing of the basin-type insulator.

Keywords Nondestructive testing · Laser ultrasonic detection technology · Multi-physical field coupling modeling · Ultrasonic signal characteristics

1 Introduction

Gas Insulated Switchgear (GIS) is as important as transformer in transmission and distribution system [1, 2]. Basin insulator is the core component of GIS system, which plays the role of supporting conductor, isolating gas chamber and electrical insulation. Statistical analysis shows that basin-type insulators are the weakest insulation link in GIS, and their defects account for a large proportion in GIS faults [3, 4]. Especially,

G. Li · X. Li · S. Li
State Grid Shanxi Electric Power Research Institute of SEPC, Shanxi, Taiyuan 030001, China

X. Huang
Shandong Jiaotong University, 5001 Haitang Road, Changqing District, Jinan 250357, China

H. Wu · H. Xia (✉)
Institute of Electrical Engineering of the Chinese Academy of Sciences, Beijing 100190, China
e-mail: xiahui@mail.iee.ac.cn

W. Chen et al. (eds.), *The Proceedings of the 9th Frontier Academic Forum of Electrical Engineering*, Lecture Notes in Electrical Engineering 743,
https://doi.org/10.1007/978-981-33-6609-1_74

the flashover voltage will decrease rapidly when metal particle pollution and GIS internal surface burr and protrusion occur. Moreover, in case of failure of basin-type insulators, the maintenance time is much longer than that of traditional air-insulated substations [5], so it is very important to detect the defects of GIS basin-type insulators.

At present, there are many researches on nondestructive testing of metal materials [6–10], but there are no reports on laser ultrasonic testing of epoxy resin materials for basin type insulators. In this paper, a multi-physical field coupling model for laser ultrasonic testing is established, taking the basin-type insulator material as the detection object. Then the ultrasonic signal characteristics of different laser energy, pulse width and laser rise time are studied to provide the optimal excitation parameters for further on-line laser ultrasonic testing of basin-type insulators.

2 Laser Ultrasonic Multi-physical Field Coupling Modeling

The process of laser ultrasonic excitation of base-type insulators is described. The insulator is irradiated by high-energy pulsed laser, which causes vibration on the surface and inside of the insulator based on photoelastic effect. Considering that the insulator is a solid material, ultrasonic modes such as longitudinal wave, shear wave and surface wave are generated simultaneously inside the insulator after laser excitation.

The pulsed laser energy is irradiated on the surface of the material, which can be expressed by a heat source function Q. The control equation is composed of the heat transfer equation and the displacement field equation, as follows

$$\rho_0 C_p \frac{\partial T}{\partial t} = \nabla \cdot (\kappa \nabla T) + Q \tag{1}$$

$$\mu \nabla^2 \boldsymbol{u} + (\lambda + \mu)\nabla(\nabla \cdot \boldsymbol{u}) = \rho_0 \frac{\partial^2 \boldsymbol{u}}{\partial t^2} + \beta \nabla T \tag{2}$$

where ρ_0 is the material density of the uniform basin-type insulator, C_p is the isobaric heat capacity constant, κ is the material thermal conductivity, T is the temperature field distribution, λ, μ are the material's Lame coefficient, $\boldsymbol{u}$ is the distribution of the displacement field generated in the laser irradiated solid, β is the thermoelastic coupling coefficient, also known as the volume expansion coefficient, which can be written as $\beta = (3\lambda + 2\mu)\alpha$, where α is the coefficient of thermal expansion, where α is the coefficient of thermal expansion, sometimes also called linear expansivity, which indicates the degree of expansion or contraction of the material.

For the laser heat source function Q:

$$Q = \beta e^{-\beta z} I_0 (1 - R) f(x, y) g(t) \tag{3}$$

$$f(x, y) = e^{-\frac{x^2+y^2}{w^2}}$$

$$g(t) = \frac{t}{t_{ex}} e^{-\frac{t}{t_{ex}}}$$

where $f(x, y)$ is the spatial distribution function of the pulsed laser, $g(t)$ is the time distribution function of the pulsed laser, and β is the attenuation coefficient, its reciprocal $1/\beta$ is the depth of light penetration. I_0 is the laser power density whose unit is W/m^3. R is the reflectivity of light on the surface of the basin-type insulator composite material. w is the half width of the laser beam, t_{ex} is the rise time of the laser energy, also known as the pulse width of the laser.

As shown in Fig. 1, the two-dimensional simulation model is established. The model is a rectangle with a length of 6 mm and a width of 3 mm. The material characteristics are shown in Table 1.

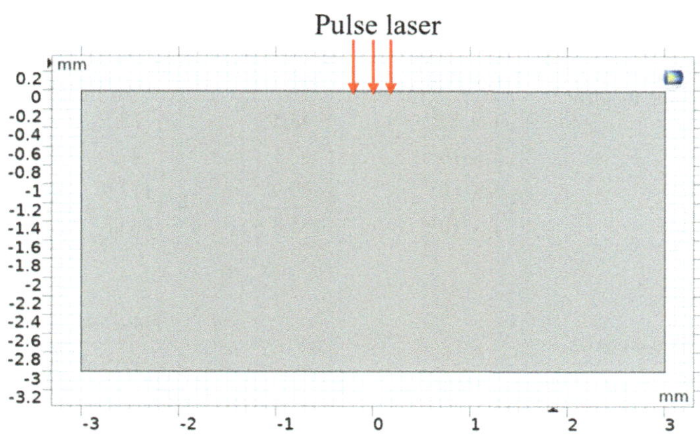

Fig. 1 The geometric region of the simulation model

Table 1 Material properties

Parameter properties	Numerical value
Density ρ/(kg m^{-3})	2200
Young's modulus E/(GPa)	20
Poisson's ratio υ	0.3
Coefficient of thermal conductivity κ/(W m^{-1} K^{-1})	0.25
Constant pressure heat capacity C/(J kg^{-1} K^{-1})	1500
Coefficient of thermal expansion α/(1/K)	18 × 10^{-6}

3 Influence Analysis of Laser Excitation Parameters

3.1 The Effect of Laser Energy

The laser energy density I_0 is set to 1×10^{11} W/m^2, 2×10^{11} W/m^2, 5×10^{11} W/m^2, 10×10^{11} W/m^2, the laser rise time t_{ex} is set to 10 ns, the laser half-width w is 0.2 mm, and the model is parametrically scanned and analyzed. With the increase of laser energy density, the instantaneous temperature rise at the laser incident point is shown in Table 2.

It can be seen from Table 2 that the increase of temperature is approximately linearly related to the laser energy. Moreover, displacement changes at distances of 1, 2 and 3 mm from the laser excitation point are extracted through simulation, as shown in Fig. 2.

It can be seen from Fig. 2 that the longitudinal wave and shear wave can be clearly distinguished by the displacement curve. The displacement diagram clearly shows

Table 2 The maximum temperature at the incident point

I_0 (W/m^2)	T_{max}(°C)	$\Delta T = T_{max} - T_0$
1×10^{11}	43.5	23.5
2×10^{11}	67.1	47.1
5×10^{11}	137.9	117.9
1×10^{12}	255.8	235.8

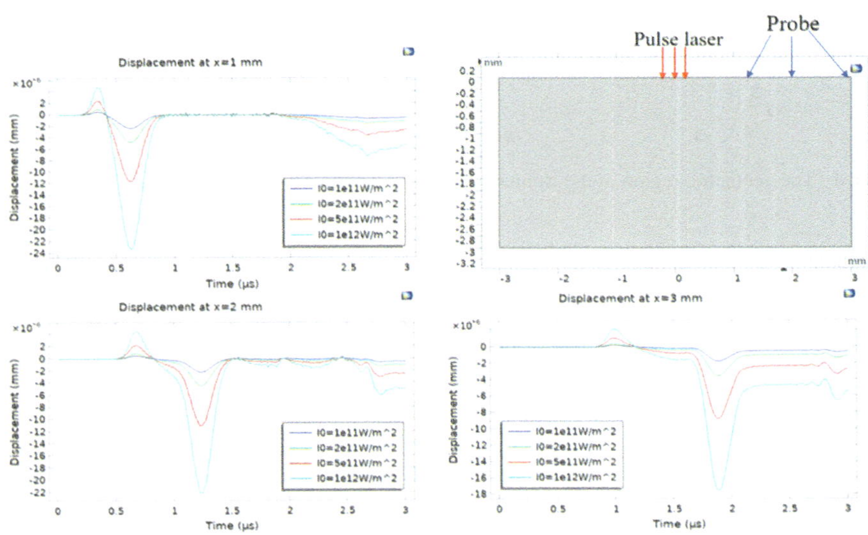

Fig. 2 Vibration displacement signals under different laser energy parameters at different receiving points

the position of the longitudinal and surface waves. Moreover, the laser energy has a linear relationship with the peak vibration displacement generated. When the energy is doubled, the vibration displacement is also doubled.

3.2 The Effect of Laser Rise Time

The laser rise time t_{ex} is set to 7, 10, 15 ns, the laser energy I_0 is set to 1×10^{11} W/m^2, the laser half width w is 0.2 mm, and the model is parametrically scanned and analyzed. And the temperature change at the incident point was observed, as shown in Table 3.

As shown in Table 3, with the increase of rising time, the temperature of laser excitation point also increases. And the laser rise time and the temperature change approximate linear relationship.

As shown in Fig. 3, vibration displacement curves at different receiving points are obtained. As can be seen from the figure, the peak value of shear wave excited

Table 3 Maximum temperature at different rise times at the incident point

t_{ex} (ns)	T_{max}(°C)	$\Delta T = T_{max} - T_0$
7	36.5	16.5
10	43.5	23.5
15	55.3	35.4

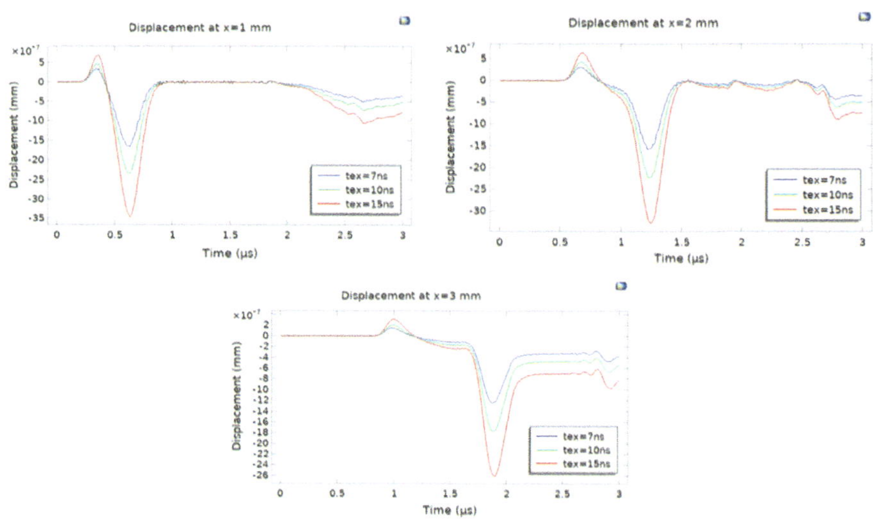

Fig. 3 Displacement signals with different rise times and different receiving points

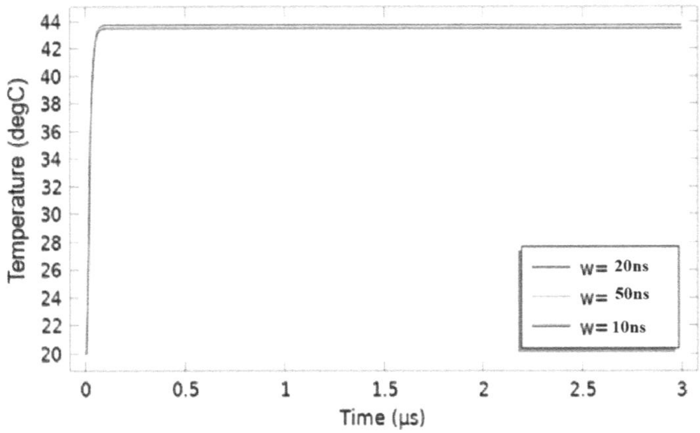

Fig. 4 Temperature change curve at laser excitation point

by laser ultrasound increases with the increase of rising time, and the relationship between them is approximately linear.

3.3 The Effect of Laser Half Width

The laser half width w is set to 10, 20, 50 ns, and the laser energy I_0 is set to 1×10^{11} W/m^2. The rise time t_{ex} is set to 10 ns.

As shown in Fig. 4, the temperature change curve with time at the excitation point corresponding to different laser half-width. As can be seen from the figure, the laser half width has no obvious effect on the temperature change.

As shown in Fig. 5, when the laser half width changes, the displacement signals received at different positions will also change. When the laser half width is increased, the frequency of vibration displacement signal produced by laser ultrasound decreases, which will lead to the resolution reduction of defects. Therefore, laser excitation parameters should be considered for defect types in actual detection.

4 Conclusion

In this paper, a two-dimensional laser ultrasonic multi-physical field coupling model is established for the epoxy resin material parameters of the basin-type insulator. Then, based on this model, the vibration displacement signal characteristics of epoxy resin materials under different laser excitation parameters were studied, and the following three conclusions were drawn. First, the incident laser energy and the laser rise time will directly affect the laser excitation temperature and the amplitude of

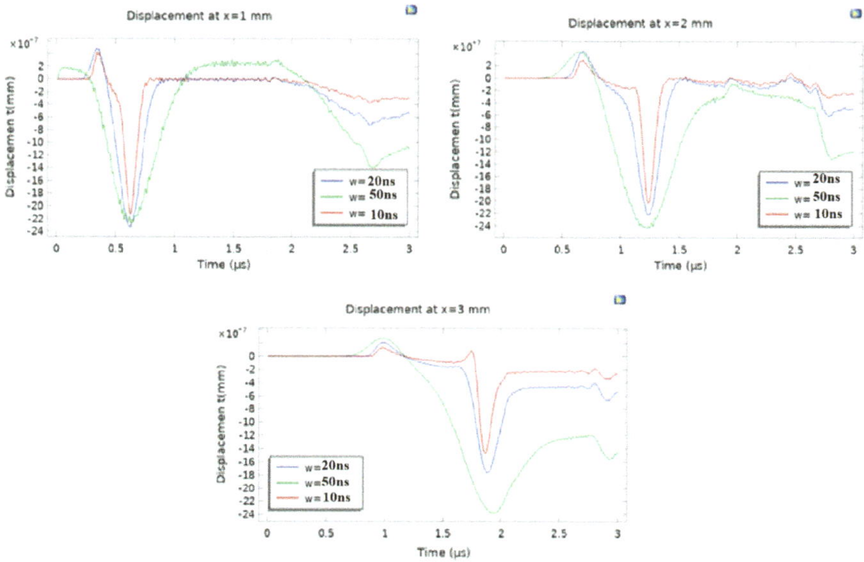

Fig. 5 Vibration displacement signal curves at different receiving points with different laser half-width

the laser ultrasonic signal. Second, the laser half width directly affects the vibration displacement frequency of laser excitation. With the increase of the laser half width, the lower the laser ultrasonic frequency is, the lower the defect detection resolution will be, but the detection depth will increase. Therefore, it is necessary to select an appropriate beam half-width for excitation to obtain effective vibration displacement and velocity images for epoxy resin composite materials and defect characteristics. In addition, the effect of laser energy density should also be taken into account. Otherwise, laser energy rice vinegar will make the temperature of the center too high and cause damage to the material.

Acknowledgements This work was supported by the National Natural Science Foundation of China (Nos. 61771448 & Nos. 519370104) and State Grid Technology Project (Nos. 520530190008).

References

1. Li, Junhao, Xutao Han, Zehui Liu, and Yanming Li. 2015. Review on partial discharge measurement technology of electrical equipment. *High Voltage Engineering* 41 (8): 2583–2601. (in Chinese).

2. Shen, Qiang, Hailong Yang, Hang Liu, Gang Wei, Chao Gao, and Junwei Yang. 2017. Application of ultrasonic partial discharge detection technology on the particle defect diagnosis of a 500 kV GIS Basin-insulator. *High Voltage Apparatus* 53 (9): 228–233. (in Chinese).
3. Liu, Yunpeng, Yansong Li, Shilong Huang, Tao Zhao. 2016. Ultrasonic detection system for gas insulated switchgear partial discharge based on optical fiber transmission. *High Voltage Engineering* 42 (1): 186–191. (in Chinese).
4. Niu, Bo, Dingge Yang, Yiliang Yu, and Xiaotao Wang. 2018. Simulation of surface dirt detection on GIS insulator based on ultrasonic method. *High Voltage Apparatus* 54 (11): 97–102. (in Chinese).
5. Ding, Dengwei, Wensheng Gao, and Weidong Liu. 2013. Insulation defects discrimination in GIS by fisher discriminant analysis of partial discharge. *High Voltage Engineering* (4): 805–812. (in Chinese).
6. Lee, Sang Eon, Peipei Liu, Young Woo Ko, et al. 2019. Study on effect of laser-induced ablation for Lamb waves in a thin plate. *Ultrasonics* 91: 121–128.
7. Tam, A.C. 1986. Applications of photo acoustic sensing techniques. *Reviews of Modern Physics* 58 (2): 381–431.
8. Seruby, C.B., et al. 1982. Laser generation of ultrasound in metals. In *Research Techniques in NDT*, ed. R.S. Shape, vol. V, 281–327. London UK: Academic Press.
9. Rose, L.R.F. 1984. Point-source representation for laser generated ultrasound. *The Journal of the Acoustical Society of America* 75 (3): 723–732.
10. Wu, Lei, Chen Jian Chun, and Zhang Shu Yi. Mechanisms of laser-generated ultrasound in Plates. *Journal of Physics D: Applied Physics* 28: 957–964.

An Analytical Core Loss Admittance Calculating Method and an Equivalent Circuit Modeling Method of Induction Motors Fed by Converter Considering Space-Time Harmonics

Meihui Jiang, Jiawei Yi, Dongdong Zhang, Xinzhi Guo, and Qiang Qin

Abstract Induction motors fed by PWM inverter have great advantages in speed control. However, the output voltage waveform of PWM inverter have a great influence on the loss characteristics of induction motor. For control the overall loss of motor and inverter effectively, a fast-analytical calculation model of iron losses based on the output voltage of inverter is proposed. In the proposed model, the influence of the output harmonic voltage of the inverter and the spatial harmonic components on the iron losses are considered. Based on the proposed analytical calculation method of iron loss, the equivalent circuit model of induction motor considered the iron losses is obtained in this paper. To verify the validity of this model, the iron losses of a 5.5 kW inverter-fed induction motor, under different operation conditions are calculated using the proposed model, classical iron loss model and piecewise variable coefficient model (based on time-step finite element method), respectively. The results show that the accuracy of the proposed model is acceptable. The conclusions of this paper can provide theoretical support for the optimal control of overall losses of the inverter and motor.

Keywords PWM · Induction motors · Core losses · Harmonic fields · Inverter

1 Introduction

With the development of frequency conversion drive technology, the motor is mostly driven by PWM voltage source inverter to meet the need of frequency conversion speed regulation [1–3]. The output voltage of PWM inverter contains a lot of high-order harmonics, which makes it very difficult to calculate the iron loss of induction

M. Jiang · J. Yi · D. Zhang
School of Electrical Engineering, Guangxi University, Nanning 530004, China

X. Guo
Electric Power Economic Research Institute of State Grid (Henan), Zhengzhou 450052, China

Q. Qin (✉)
Capital Construction Department, Guangxi University, Nanning 530004, China
e-mail: 67085499@qq.com

W. Chen et al. (eds.), *The Proceedings of the 9th Frontier Academic Forum of Electrical Engineering*, Lecture Notes in Electrical Engineering 743,
https://doi.org/10.1007/978-981-33-6609-1_75

motor quickly. At present, the most accurate method for iron loss calculation is generally based on the finite element method [4–6]. But the calculation amount of the iron loss model based on the finite element method is too larger. In order to effectively control the overall loss of motor and inverter, it is necessary to put forward an analytical calculation method of iron loss, which takes the output voltage or current of the inverter as variables.

There are a lot of iron loss models. Most of these models are derived from Steinmetz equation [7]; subsequently, Jordan divides the iron loss into hysteresis loss and eddy current loss, which is called the classical iron loss model [8]. However, the classical iron loss model is more accurate in predicting the iron loss of NiFe alloy, while the accuracy of iron loss prediction of SiFe alloy is poor [9]. In addition, the accuracy of the classical iron loss model can only be guaranteed when the magnetic density and frequency of the motor core are low [10, 11]. In order to take into account nonlinear characteristics of ferromagnetic materials, [10] and [11] propose an iron loss model in which an additional magnetic density term is added to the classical eddy current loss term. Ref. [12–14] adopts compensation algorithm to take into account the iron loss of harmonic. In addition, many scholars have studied the core loss caused by local hysteresis loop distortion [15, 16] and loss caused by rotating magnetic field [17], and put forward corresponding improvement measures. In recent years, many scholars have put forward many new ideas for the complex analysis of the motor magnetic field [18–20], which provides a theoretical basis for the motor iron loss model.

In order to effectively control the overall loss of motor and inverter, a fast analytical calculation model of iron loss with inverter output voltage as variable is proposed in this paper. Based on the piecewise variable coefficient model, the flux density and frequency variables in the model are replaced by voltage and speed variables, and the influence of time and space harmonic is considered. Based on this method, an equivalent circuit model for induction motor fed by PWM inverter are presented. Finally, this paper calculates the iron loss of a 5.5 kW induction motor fed by PWM inverter under different supply voltage and switching frequency by using the different models, The results show that the proposed model has higher accuracy.

2 Calculation Method of Iron Loss of Variable Coefficient Induction Motor Considering Time Harmonic

2.1 Piecewise Variable Coefficient Iron Loss Model Considering Harmonics

According to [13], the expressions of eddy current loss P_{EL} (W) and hysteresis loss P_{HL} (W) of the piecewise variable coefficient iron loss model are shown in expressions (1) and (2) respectively.

$$P_{\mathrm{EL}} = k_{\mathrm{e}}\rho l_{\mathrm{m}} \sum_{j} \left[A_j \sum_{n} n^2 \left(1 + k_{2(j,n)} B_{\mathrm{m}(j,n)}^{\beta_{2(j,n)}}\right) B_{\mathrm{m}(j,n)}^2 f_1^2 \right] \tag{1}$$

$$P_{\mathrm{HL}} = k_{\mathrm{h}}\rho l_{\mathrm{m}} \sum_{j} \left[A_j \sum_{n} n \left(k_{1(j,n)} B_{\mathrm{m}(j,n)}^{\beta_{1(j,n)}}\right) B_{\mathrm{m}(j,n)}^{\alpha} f_1 \right] \tag{2}$$

where A_j is the area of the jth grid element in the finite element model, $B_{\mathrm{m}(j,n)}$ is the equivalent axial length of the induction motor, $K_{1(j,n)}$ is the amplitude of the nth harmonic flux density of the jth grid element. The values of $K_{1(j,n)}$, $\beta_{1(j,n)}$ and $K_{2(j,n)}$, $\beta_{2(j,n)}$ are determined by nf_1 and $B_{\mathrm{m}(j,n)}$.

2.2 *Relationship Between Supply Voltage and Core Flux Density*

Under the condition of neglecting the stator winding voltage drop and leakage flux, according to the law of electromagnetic induction,

$$B(t) = \frac{1}{N^*S} \int v(t)\mathbf{d}t \tag{3}$$

where N* is the equivalent turns per phase of the stator in series, S is the equivalent cross-sectional area of the motor core, v(t) is the phase voltage of the stator winding, t is the time, and B(t) is the magnetic flux density. By applying Fourier transform to the output voltage of the inverter and according to the relationship between speed and frequency, the amplitude of the Nth harmonic flux density can be obtained from the following formula:

$$B_m(n) = \frac{30}{N^*S\pi p} \frac{V_n}{n\Omega_1} \tag{4}$$

where Ω_1 is the synchronous speed of the induction motor, f_1 is the fundamental frequency of the supply voltage, p is the number of poles of the motor.

2.3 *Calculation of Stator Iron Loss Considering Time Harmonic*

When calculating the equivalent area of the motor core, the nonuniformity of the magnetic density distribution in the tooth and yoke of the induction motor is ignored. Under the above conditions, the stator eddy current loss and hysteresis loss expressed can be calculated by substituting Eq. (4) into Eqs. (1) and (2) respectively,

$$P_{\mathrm{EL1}} = \frac{k_{\mathrm{e}}\rho l_{\mathrm{m}} S_1 C_\tau^2 p^2}{3600} \sum_n \left(1 + k_{2(n)}\left(C_\tau \frac{V_n}{n\Omega_1}\right)^{\beta_{2(n}}\right) V_n^2 \tag{5}$$

$$P_{\mathrm{HL1}} = \frac{k_{\mathrm{h}}\rho l_{\mathrm{m}} C_\tau^\alpha p S_1}{60} \sum_n \left(k_{1(n)}\left(C_\tau \frac{V_n}{n\Omega_1}\right)^{\beta_{1(n)}}\right)\left(\frac{V_n}{\Omega_1}\right)^{c} (n)^{1-\alpha}\Omega_1 \tag{6}$$

where S_1 is the equivalent cross-sectional area of the stator core.

2.4 *Calculation of Rotor Iron Loss Considering Time Harmonic*

Considering the time harmonic, the rotor eddy current loss P_{EL2} expressed by the supply voltage is shown in the following formula,

$$P_{\mathrm{EL2}} = \frac{k_{\mathrm{e}}\rho l_{\mathrm{m}} S_2 p^2 C_\tau^2}{3600} \sum_n \vartheta_{1(n-1+s)} V_n^2 \left(\frac{n-1+s}{n}\right)^2 \tag{7}$$

where s is the slip and S_2 is the equivalent cross-sectional area of the rotor core.

Similarly, the rotor hysteresis loss P_{HL2} can be calculated, as shown in the following formula,

$$P_{\mathrm{HL2}} = \frac{k_{\mathrm{h}}\rho l_{\mathrm{m}} C_\tau^\alpha p S_2}{60} \sum_n \vartheta_{2(n-1+s)} \left(\frac{V_n}{\Omega_1}\right)^\alpha \left(\frac{n-1-s}{n^\alpha}\right)\Omega_1 \tag{8}$$

3 Calculation Method of Additional Iron Loss of Induction Motor Considering Space Harmonics

3.1 *Surface Additional Loss*

Compared with the iron loss generated by the first harmonic magnetic field, the iron loss caused by the 5th, 7th order space harmonic MMF can be ignored [21]. Because the rotor of induction motor generally adopts semi-closed slot or closed slot, the surface loss caused by the rotor tooth harmonic on stator can be ignored. The surface loss (W) caused by the stator tooth harmonic on rotor can be obtained as follow [22, 23],

$$P_{\mathrm{surfL}} = \left(0.5k_0 K_{\mathrm{L1}} \beta_{01}^2 K_{\sigma 1}^2 \left(\frac{2p}{4.44\pi f N^* \alpha_{\mathrm{p}}^{'}(D_2 + l_\delta) l_{\mathrm{m}}}\right)^2 t_1^2 Z_1^{1.5} D_2 l_{\mathrm{m}} \frac{t_2 - b_{02}}{t_2}\right) V_1^2 \Omega^{1.5} \quad (9)$$

where Ω is the motor speed;K_0 is the coefficient related to the material and processing factors of silicon steel sheet [22]; β_{01} is a function of the slot width of the stator; $K_{\delta 1}$ is the air gap coefficient when the stator is slotted and the rotor surface is smooth, and. the specific values can be obtained from the table [22], t_1 and t_2 are the tooth pitch of stator and rotor respectively; D_2 is the radius of the rotor; l_{m} is the effective shaft length of the motor; b_{02} is the slot width of rotor; $\alpha_p^{'}$ is the calculated pole arc coefficient, which is related to the saturation degree of the stator and rotor teeth magnetic density; l_δ is the air gap width of the motor, K_{L1} is the load factor of surface loss.

3.2 *Calculation of Tooth Pulsation Loss*

According to [22, 23], the tooth pulsation loss P_{psL} of the motor stator and the tooth pulsation loss P_{prL} of the motor rotor are shown in formulas (10) and (11), respectively.

$$P_{\mathrm{psL}} = \left[0.0175 K_{\mathrm{L1}} Z_2^2 \left(\frac{N_\beta \gamma_2 l_\delta}{t_1 - b_{01}}\right)^2 G_{\mathrm{t1}} \times 10^{-6}\right] V_1^2 \Omega^2 \quad (10)$$

$$P_{\mathrm{prL}} = \left[0.0175 K_{\mathrm{L2}} Z_1^2 \left(\frac{N_\beta \gamma_1 l_\delta}{t_2 - b_{02}}\right)^2 G_{\mathrm{t2}} \times 10^{-6}\right] V_1^2 \Omega^2 \quad (11)$$

4 Equivalent Circuit Model Construction of Induction Motor Considering Space-Time Harmonics

The equivalent circuit model of induction motor considering iron loss is shown in Fig. 1. In the figure, R_{s} is stator winding resistance, R_{r} is rotor winding resistance, L_{ls} is stator winding leakage inductance, and L_{lr} is rotor winding leakage inductance. In this model, the iron loss of the motor is equal to the loss caused by the iron loss admittance on the induced potential Em1. Taking stator eddy current loss as an example, the equivalent deformation of formula (7) can be obtained,

$$P_{\mathrm{EL1}} = G_{\mathrm{EL1}} E_{\mathrm{m1}}^2 = \frac{k_{\mathrm{e}} \rho l_{\mathrm{m}} S_1 C_\tau^2 p^2}{3600} \sum_n \left(1 + k_{2(n)} \left(C_\tau \frac{V_n}{n\Omega_1}\right)^{\beta_{2(n)}}\right) V_n^2 \quad (15)$$

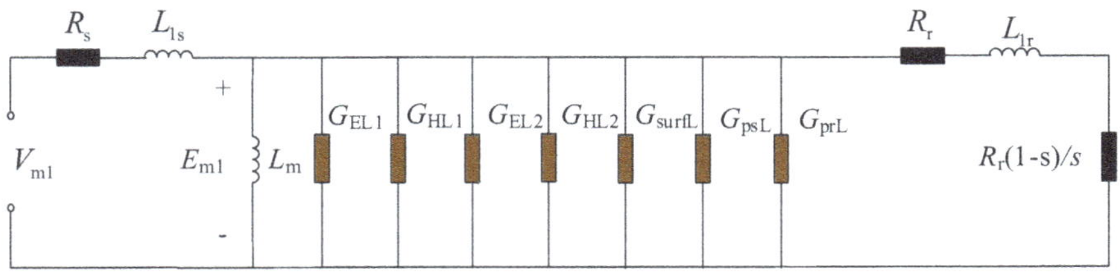

Fig. 1 Equivalent circuit diagram of induction motor considering iron loss

where G_{EL1} is the equivalent admittance of stator eddy current loss considering time harmonics. From Eq. (16), the expression of G_{EL1} can be obtained as follows:

$$G_{EL1} = \frac{\frac{k_e \rho l_m S_1 C_\tau^2 p^2}{3600} \sum_n \left(1 + k_{2(n)} \left(C_\tau \frac{V_n}{n\Omega_1}\right)^{\beta_{2(n)}}\right) V_n^2}{E_{m1}^2} \tag{16}$$

According to the above formula of iron losses of each component, the other equivalent admittances G_{HL1}, G_{HL1}, G_{EL2}, G_{HL2}, G_{surf}, G_{psL}, G_{prL} can be obtained in the same way.

5 Experimental Verification

The experimental platform is shown in Fig. 2. In this paper, the iron loss of a 5.5 kW induction motor fed by PWM inverter with specifications as shown in Table 1 is calculated respectively by using the model presented in this paper, the classical model based on time-stepping finite element method and the piecewise variable coefficient model.

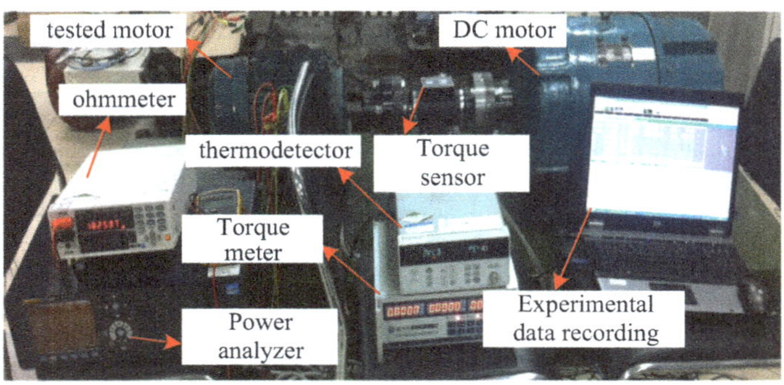

Fig. 2 Test bench of 5.5 kW induction motor

Table 1 Specification of the 5.5 kW induction motor

Poles	4
Phase	3
Slot number of stator	36
Slot number of rotor	32
Outer diameter of the stator	210 mm
Inner diameter of the rotor	136 mm
Effective shaft length of motor	115 mm
The air gap length	0.4 mm
Rated current	11.64 A
Rated voltage	380 V
Stator winding connection mode	Delta
Rated speed	1440 r/min

The comparison between actual measurement and simulation is shown in Figs. 3 and 4. It can be seen that:

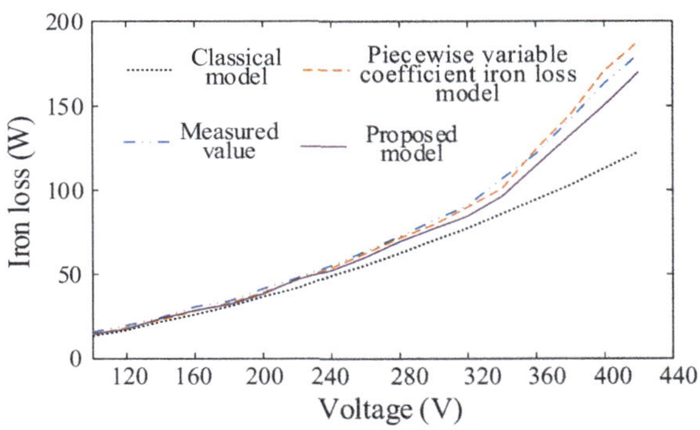

Fig. 3 Comparison of predicted and tested iron losses under sinusoidal voltage

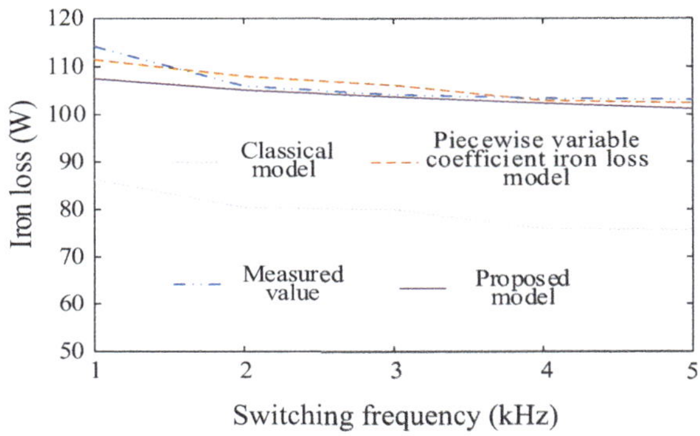

Fig. 4 Comparison of predicted and tested iron losses under PWM voltage

Table 2 The simulation time of the CIL model, the PVCIL model and the proposed model

Model type	Classical model	Piecewise variable coefficient iron loss model	Proposed model
Simulation time	3.70 h	3.71 h	1.52 min

1. When the sinusoidal supply voltage is low, the accuracy of the three models is high; when the voltage is high, the classical model has a large error due to not taking into account the harmonic loss, while the calculated values of the analytical calculation model in this paper and the piecewise variable coefficient model are still close to the measured values, as shown in Fig. 3. When the induction motor is powered by PWM inverter, the error of analytical calculation model and piecewise variable coefficient model is still small, but the accuracy of classical model becomes poor, as shown in Fig. 4.
2. When the sinusoidal supply voltage is high and the PWM inverter supplies power, although the analytical calculation model in this paper has higher accuracy, the error is larger than the piecewise variable coefficient model. This is mainly because the model ignores the mismatch of the core magnetic density distribution of induction motor stator and rotor.

In order to verify the calculation efficiency of the iron loss model based on the analytical method, we use a high-performance computer (two Xeon e5-2690 V3 CPUs, 256 GB RAM) for simulation. The calculation step is 4 μs. The calculation steps are 55,000. The comparison of simulation time is shown in Table 2. It can be seen from Table 2 that the calculation time of iron loss model based on analytical method is more than 90% less than that based on finite element method.

6 Conclusion

In order to effectively control the overall loss of motor and inverter, a fast analytical calculation model of iron loss with inverter output voltage as variable is proposed in this paper. The main conclusions are as follows:

1. In this paper, the iron loss model is based on the piecewise variable coefficient model. The influence of harmonic voltage output by the PWM inverter and the additional surface loss and pulse loss caused by the spatial tooth harmonic component of the stator and rotor teeth are considered.
2. The equivalent admittance equation takes the motor speed, voltage and other centralized parameters as variables, which can provide a model basis for the overall energy saving of the inverter motor drive.
3. A comparison test of 5.5 KW asynchronous motor fed by PWM inverter is carried out The results show that the analytical model has high accuracy on the basis of reducing calculation time by more than 90%.

Acknowledgements This work was supported in part by National Key Research and Development Program of China under Grant (2017YFB0901900).

References

1. Zhu, Yesheng, et al. 2017. PWM-based direct instantaneous torque control of switched reluctance machine. *Transactions of China Electrotechnical Society* 32 (7): 31–39. (in Chinese).
2. Kun, Xia, et al. 2017. Research on hybrid PWM control method with current feedback for torque-ripple reduction in BLDCM. *Transactions of China Electrotechnical Society* 32 (17): 172–179. (in Chinese).
3. Tan, Jixin, Yongteng Jing, Hao Guo, Zhanyang Yu, and Yan Li. 2020. Voltage harmonic optimization control of PMSM variable frequency power supply based on boost transform. *Transactions of China Electrotechnical Society* 1000–6753 (in Chinese).
4. Zhao, H., D. Zhang, et al. 2017. Piecewise variable parameter piecewise variable loss model of laminated steel and its application in fine analysis of analysis of iron loss of inverter-fed induction motors. *IEEE Transactions on Industry Applications* PP (99): 1–8.
5. Zhao, H., Y. Wang, D. Zhang, Y. Zhan, G. Xu, and Y. Luo. 2017. Piecewise variable parameter model for precise analysis of iron losses in induction motors. *IET Electric Application* 11 (3): 361–368.
6. Zhan, Y., B. Li, H. Zhao, G. Xu, and D. Zhang. 2020. Fast solution of rotor losses in inverter-fed cage induction motors with skewed slots. *IEEE Transactions on Industry Applications* 56 (2): 1284–1292.
7. Steinmetz, C. 1984. On the law of hysteresis (originally published in 1892). *Proceedings of the IEEE* 72 (2): 197–221.
8. Jordan, H. 1924. Die ferromagnetischen konstanten fur schwache wechselfelder. Elektr. Nach. Techn.
9. Boll, R. 1990. *Weichmagnetische Werkstoffe*, 4th ed. Publicis Corporate Publishing.
10. Eggers, D., S. Steentjes, and K. Hameyer. 2012. Advanced iron-loss estimation for nonlinear material behavior. *IEEE Transactions on Magnetics* 48 (11): 3021–3024.
11. Simon, Steentjes, and Pfingsten von Georg. 2013. Iron-loss model with consideration of minor loops applied to FE-simulations of electrical machines. *IEEE Transactions on Magnetics* 49 (7): 3945–3948.
12. Tong, Wenming, Xiaofeng Zhu, Jianguo Jia, et al. 2015. Influence law of additional losses induced by time harmonic in permanent magnet synchronous motors. *Transactions of China Electrotechnical Society* 30 (6): 60–99. (in Chinese).
13. Zhang, Dongdong, et al. 2016. A piecewise variable coefficient model for precise analysis on iron losses of electrical machines. *Transactions of China Electrotechnical Society* 31 (15): 16–24. (in Chinese).
14. Zhao, H., D. Zhang, et al. 2016. No-load iron loss distribution characteristics and its fine analysis for inverter-fed induction motors. *Proceedings of the CSEE* 36 (8): 2261–2269. (in Chinese).
15. Luo, Fuqiang, Changliang Xia, et al. 2012. Model of equivalent iron loss resistance of induction motor fed by PWM. *Transactions of China Electrotechnical Society* 27 (7): 101–108. (in Chinese).
16. Zhang, D., H. Dai, H. Zhao, and T. Wu. 2018. A fast identification method for rotor flux density harmonics and resulting rotor iron losses of inverter-fed induction motors. *IEEE Transactions on Industrial Electronics* 65 (7): 5384–5394.
17. Belahcen, Anouar, Paavo Rasilo, and Antero Arkkio. 2014. Segregation of iron losses from rotational field measurements and application to electrical machine. *IEEE Transactions on Magnetics* 50 (2): 893–896.

18. Ye, Pinzhou, Hongwei Li, Wentao Yu, and Hui Zhong. 2020. Modeling of equivalent magnetic circuit of radial electromagnetic bearing considering material nonlinearity and eddy current. *Transactions of China Electrotechnical Society* 35 (09): 1858–1867. (in Chinese).
19. Zhao, Xiaojun, Xiaona Liu, Fan Xiao, and Yang Liu. 2020. Simulation of dc magnetic bias hysteresis and loss characteristics of oriented silicon steel based on Preisach model. *Transactions of China Electrotechnical Society* 35 (09): 1849–1857. (in Chinese).
20. Zhang, Dongdong, Xinzhi Guo, Ruichi An, Lixiao Pu, and Haifeng Li. 2019. High efficiency separation method of harmonic flux density of induction motor rotor based on DFT and rotor iron loss characteristics of variable frequency motor under load condition. *Transactions of China Electrotechnical Society* 34 (01): 75–83. (in Chinese).
21. Heller, B., and V. Hamata. 1977. *Harmonic field effects in induction machines*. Amsterdam: Elsevier Scientific Publ. Corp.
22. Chen, Shikun. 2008. *The motor design*. Beijing: Machinery Industry Press. (in Chinese).
23. Boldea, Ion, and Syed A. Nasar. 2010. *The induction machines design handbook* 2nd ed. Boca Raton: CRC Press, Taylor & Francis Group.

The Application of Genetic Algorithm in the Structural Optimization of Permanent Magnet Synchronous Motor

Song Huang, Tian Sun, Shuhong Wang, Nana Duan, and Bowen Shang

Abstract The objective and constraint conditions for the optimal design of permanent magnet synchronous motors (PMSM) are generally non-linear numerical functions and multi-extreme functions of design variables. It is difficult to obtain optimal design parameters. This paper has completed the compilation of the electromagnetic calculation program for the motor, and introduced the genetic algorithm into the optimization design program, and we specifically use the genetic algorithm to optimize the magnetization length of the permanent magnet, the width of the permanent magnet, the air gap and the pole arc coefficient in the motor, and the goal was set to obtain the maximum efficiency. From the results, the optimization effect is significant, which fully demonstrates that the genetic algorithm can effectively improve the efficiency of the permanent magnet synchronous motor and shorten the design cycle.

Keywords PMSM · Genetic algorithm · Parameter optimization · Efficiency optimization · Electromagnetic calculation program

1 Introduction

The increasing performance of rare earth permanent magnet materials and the advancement of power electronic equipment have created good conditions for the design, manufacture and application of permanent magnet synchronous motors. Compared with traditional electric excitation motors, permanent magnet motors, especially rare earth permanent magnet motors, have the advantages of simple structure and reliable operation; small size and light weight; low loss and high efficiency;

S. Huang · S. Wang (✉) · N. Duan · B. Shang
Faculty of Electrical Engineering, Xi'an Jiaotong University, Xi'an 710049, China
e-mail: shwang@mail.xjtu.edu.cn

T. Sun
Systems Engineering Research Institute, Beijing 100036, China

W. Chen et al. (eds.), *The Proceedings of the 9th Frontier Academic Forum of Electrical Engineering*, Lecture Notes in Electrical Engineering 743,
https://doi.org/10.1007/978-981-33-6609-1_76

the shape and size of the motor can be flexible and diverse [1–4]. However, the design procedure of PMSM is complicated and depends on the accumulation of design experience. Therefore, it is absolutely necessary to optimize the design process of PMSM.

The optimization of the permanent magnet synchronous motor design program is a multi-objective non-linear optimization process. The traditional optimization methods can no longer meet the needs of motor design. In the literature [1], genetic algorithm, simulated annealing algorithm and differential evolution algorithm are compared, and it is concluded that the approximation strategy of artificial intelligence algorithm is sufficient and extraordinary for the design and optimization of PMSM; The Taguchi method is used to optimize the air gap length, pole arc coefficient, permanent magnet thickness, stator tooth width and slot width of the fractional slot concentrated winding permanent magnet synchronous motor [3], but Taguchi algorithm requires a lot of orthogonality experiment; Ref. [5] compares the application of differential evolution algorithm and response surface algorithm in motor design. When there are many optimization variables, the differential evolution algorithm is a better choice. This paper introduced genetic algorithm to optimize the design parameters of permanent magnet synchronous motors. Genetic Algorithms (GA) is a search algorithm based on the principle of natural selection and natural genetic mechanisms. It simulates the evolutionary mechanism of life in nature. Optimizing design parameters to achieve specific goals in artificial systems. The basic idea of GA is to map the problem space to the coding space through coding operations, and then perform three genetic operations of selection, crossover, and mutation in the coding space and its cyclic iterative operations to simulate the genetic evolution mechanism of organisms to obtain the optimal solution [6, 7]. This paper applied genetic algorithm to the optimal design of PMSM, which improves efficiency and shortens the cycle of the motor design.

In this paper, we equate the permanent magnets as the source of magnetomotive force and establish the magnetic circuit model of the permanent magnet synchronous motor. The design program of the permanent magnet synchronous motor was designed through the magnetic circuit calculation model, and the accuracy of the design program was verified by comparison with the finite element calculation results. Then the genetic algorithm is introduced into the motor design program, and the magnetization length, width of the permanent, pole arc coefficient, and air gap size are optimized to obtain maximum efficiency. The optimization results are verified by the finite element calculation, which fully shows that the genetic algorithm is effective for the optimization of the design parameters of PMSM. It can simplify the design process and shorten the motor design cycle.

2 Structure of PMSM

The structure of PMSM consists of three parts: a fixed part, a rotating part and an auxiliary part. The fixed part includes stator frame, stator core, stator winding, end

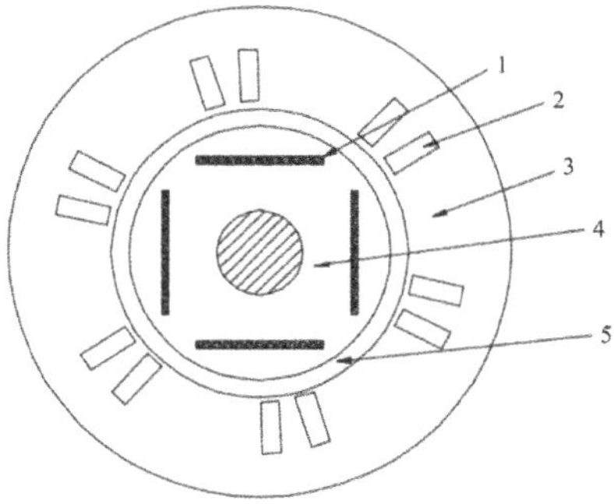

1-permanent magnet, 2-stator winding, 3-stator, 4-rotor core, 5-air gap

Fig. 1 The structure of PMSM

cover and bottom plate. The rotating part includes the rotating shaft, the rotor core, the rotor winding, and the cooling fan. The rotor core adopts a solid structure or a laminated structure. Figure 1 depicts the cross section of a PMSM, which consists of two parts: the rotor and the stator, including permanent magnets, windings, stator core, rotor core, air gap. When the three phase current flows into the three phase symmetric windings of the PMSM stator, the magnetomotive force generated by the current synthesizes a rotating magnetomotive force with a constant amplitude. Because its amplitude is constant, the trajectory of this rotating magnetomotive force forms a circle, which is called a circular rotating magnetomotive force. The magnetic field generated by the permanent magnets on the rotor and the rotating magnetic field generated by the circular rotating magnetomotive force of the stator interact at the air gap. The result of the interaction of the two magnetic fields is an electromagnetic torque that drives the motor to rotate.

The magnetic field generated by the permanent magnet interacts with the rotating magnetic field generated by the armature winding to produce torque. Therefore, the selection of the parameters of the permanent magnet has a significant impact on the performance of PMSM. In the optimization of the motor structure, it is necessary to optimize the size and shape of the permanent magnet. In theory, the smaller the air gap, the higher the efficiency of the motor. However, due to the limitation of the production process, the air gap cannot be infinitely small, so the size of the air gap is also a parameter that needs to be optimized. This paper has completed the optimization of the permanent magnet size, air gap size and pole arc coefficient of the motor.

3 Equivalent Magnetic Circuit of PMSM

We established the electromagnetic calculation model of PMSM by the method of magnetic circuit equivalent, and completed the permanent magnet synchronous motor design program. The equivalent magnetic circuit method equated with the permanent magnet as a source of magnetomotive force. According to different magnetic circuit structures, the magnetic circuit was divided into linear magnetic circuit sections. According to the ampere loop rule of the magnetic circuit, List the magnetomotive force balance equation, and then determined the specific size of the permanent magnets according to the magnetic density requirements of each section of the magnetic circuit, and obtained the original design data of PMSM.

The equivalent process of the permanent magnet equivalent to the magnetomotive force source, the magnetic field provided by the permanent magnet to the external magnetic circuit is represented by the magnetomotive force, that is, is the magnetomotive force from both ends of the permanent magnet. The virtual intrinsic magnetic flux comes from the permanent magnet. The permanent magnet is equivalent to a permanent magnetomotive force source in series with a constant internal permeance [4], as shown in Fig. 2. It can be expressed as

$$\phi_m = \phi_r - \phi_0 = \Lambda_0(F_c - F_m) \tag{1}$$

ϕ_m is the total magnetic flux provided by the permanent magnet to the external magnetic circuit, which can be divided into two parts, one is the main flux (the air gap flux) ϕ_δ, and the other is the leakage flux ϕ_σ; The circuit is divided into a main magnetic circuit and a leakage magnetic circuit, and its permeance is respectively the main permeance Λ_δ and the leakage permeance Λ_σ.

The equivalent magnetic circuit under no load is shown in Fig. 3, and the equivalent magnetic circuit under load is shown in Fig. 4. The main magnetic circuit increases the armature magnetomotive force under load, and the armature magnetomotive force F_a comes from the magnetic field generated by the armature. It is convenient for calculation and analysis. We can simplified Fig. 4A to Fig. 4B according to Thevenin's theorem. Combining Fig. 4B with Fig. 2, that is, the permanent magnet magnetic circuit and the external magnetic circuit are connected to form a whole magnetic circuit, and the total equivalent magnetic circuit of the permanent magnet motor under load is obtained, as shown in Fig. 5. Set $F_a = 0$ to obtain the equivalent

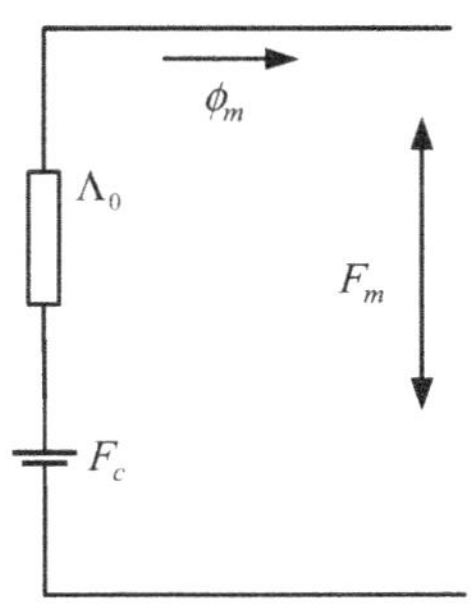

Fig. 2 Equivalent of permanent magnet

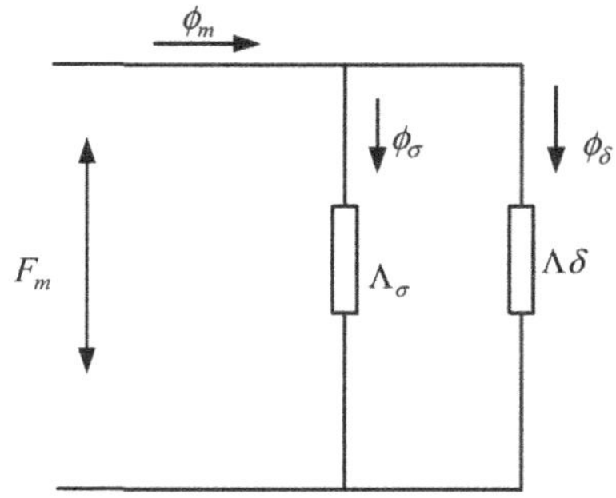

Fig. 3 No-load external magnetic circuit equivalent

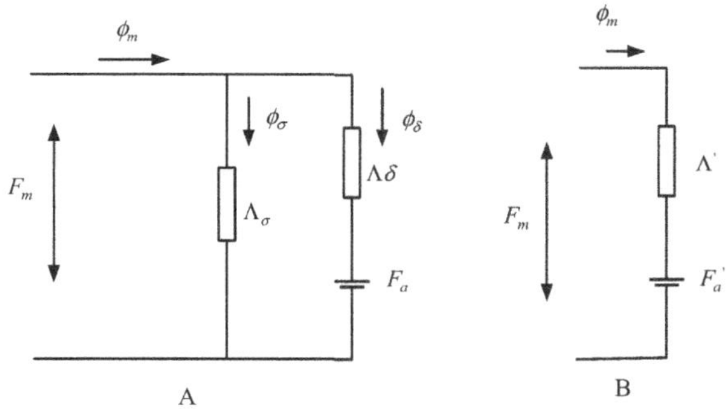

Fig. 4 Equivalent external magnetic circuit under load. (**A**) Equivalent magnetic circuit. (**B**) Simplified magnetic circuit

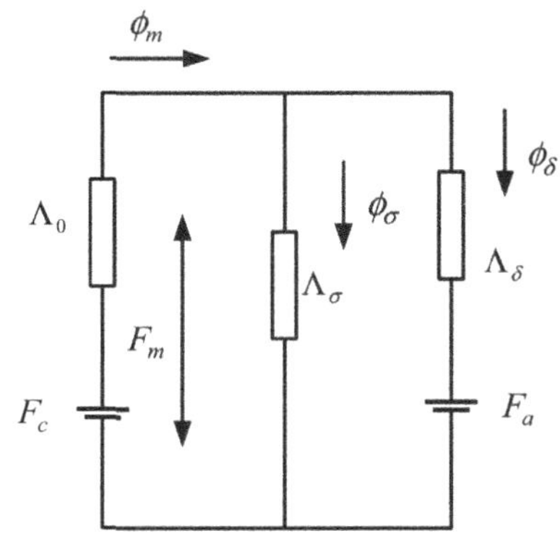

Fig. 5 Equivalent magnetic circuit of PMSM

magnetic circuit at no load. For different permanent magnet motors, the specific composition of the equivalent magnetic circuit will be different.

We use the magnetic circuit method to program the electromagnetic calculation code of PMSM with MATLAB, and carry on the electromagnetic calculation to the calculation example. In order to verify the accuracy of our calculation program, we

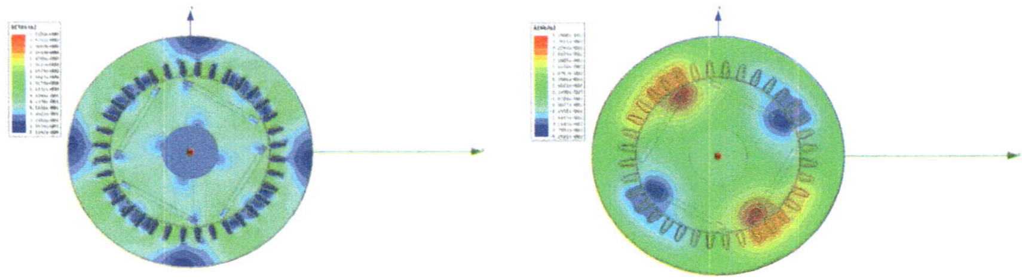

A B

Fig. 6 Finite element calculation results. (**A**) Magnetic flux density mode distribution. (**B**) Magnetic potential distribution

established a finite element calculation model, and performed a finite element analysis on the electromagnetic field of the example motor. The finite element calculation results are shown in Fig. 6, Fig. 6A is the distribution of magnetic flux density, and Fig. 6B is the distribution of magnetic potential.

The comparison between the calculation results obtained by using our own MATLAB program and the results calculated by the finite element method is shown in Table 1. It can be seen in the table that the error between our self-programming calculation result and the finite element calculation result is within 10%, which is a completely acceptable error range. Only the calculation results of loss differ greatly. This is because the calculation of loss itself is related to the nonlinearity of the material, and it is difficult to obtain an accurate model of the magnetic properties of the material. From the comparison results, the program we completed is accurate and reliable for the electromagnetic field analysis results of the permanent magnet synchronous motor.

Table 1 Comparison of self-programming and finite element calculation

	Self-programming	FEM	Difference value $\|\Delta\|$
Stator slot fill factor (%)	34.85	33.66	1.19(3.5%)
Air gap flux density (T)	0.71	0.69	0.02(2.8%)
Xad (ohm)	3.45	3.47	0.02(0.5%)
Xaq (ohm)	12.32	12.41	0.09(0.7%)
R1 (ohm)	0.99	0.98	0.01(1%)
Total loss (W)	2507.85	2280.84	227.01(10%)
Output power (W)	15,115.48	14,997.8	117.68(0.8%)
Input power (W)	17,623.32	17,278.6	344.72(2%)
Efficiency (%)	85.77	86.80	1.03(1.2%)

4 Optimization of Motor Parameters by Genetic Algorithm

Based on the self-programming of the electromagnetic calculation of the motor, we successfully realized the optimization of the design parameters of PMSM using the genetic algorithm, and obtained the best design parameters, which provided a new idea for the design of the motor.

The genetic algorithm maps the search space into a genetic space. Each possible solution is encoded as a vector, called a chromosome, which is expressed as a string of binary or decimal numbers. Each element of the vector is called a gene. All chromosomes form a population, and each chromosome is evaluated according to a predetermined objective function, and a fitness value is given according to the result. The algorithm first randomly generates some chromosomes, calculates their fitness, selects, exchanges, and mutates the chromosomes according to the fitness, and removes the chromosomes with low fitness, leaving the chromosomes with high fitness to obtain a new population. Since the members of the new population are outstanding individuals of the previous generation, they have inherited the fine traits of the previous generation and are thus significantly better than the previous generation. In this way, the genetic algorithm iterates repeatedly, evolving towards a better solution, until it meets a certain predetermined optimization index [8–12].

We use genetic algorithm to optimize the length of the permanent magnet of the motor in the magnetization direction, the width of the permanent magnet, the size of the air gap and the pole arc coefficient. The goal is to maximize the efficiency of PMSM. The mathematical model of efficiency optimization can be expressed as

$$\max \eta = \mathrm{f(hm, bm, gap, ap)} \tag{2}$$

$$\begin{aligned} s.t. \ 6.5 &\le hm \le 7.5 \\ 0.45 &\le bm \le 0.55 \\ 0.04 &\le gap \le 0.08 \\ 0.6 &\le ap \le 0.85 \end{aligned}$$

where η represents the efficiency of the permanent magnet synchronous motor, *hm* is the length of the permanent magnetization direction, *bm* represents the width of the permanent magnet, *gap* represents the size of the air gap, and *ap* denotes the pole arc coefficient.

After 500 iterations of calculations, the optimized results can be obtained as $hm = 7.4985$, $bm = 0.5466$, $gap = 0.0400$, $ap = 0.6666$. At this time, the efficiency of the permanent magnet synchronous motor is 89.76%. The optimization results are depicted in Fig. 7.

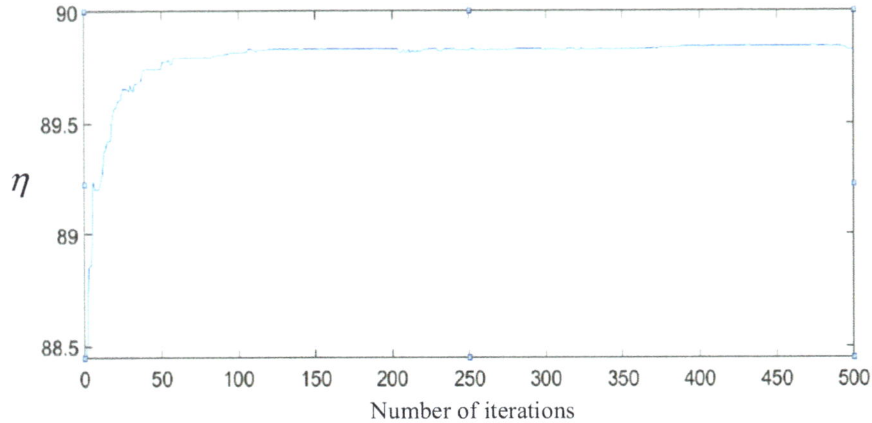

Fig. 7 Relationship between genetic algebra and efficiency (η)

5 Conclusion

This paper has completed the self-programming of the electromagnetic calculation program of PMSM, and compared it with the finite element calculation results, which proved the accuracy of our PMSM electromagnetic calculation program. Based on the electromagnetic calculation program, the genetic algorithm was introduced into the optimization of the design parameters of PMSM, and the efficiency can reach 89.76% after optimization. The process of optimizing the calculation of motor design parameters through the combination of magnetic circuit equivalence and genetic algorithm can simplify the motor design process and shorten the motor design cycle.

Acknowledgements This work was supported in part by the National Natural Science Foundation of China under Grant (51737010).

References

1. Mutluer, M., and O. Bilgin. 2011. Comparison of stochastic optimization methods for design optimization of permanent magnet synchronous motor. *Neural Computing and Applications* 21 (8): 2049–2056.
2. Tanaka, C.N., and I.E. Chabu. 2020. Flux reversal free splittable stator core doubly salient permanent magnet motor. *IEEE Latin America Transactions* 18 (8): 1329–1336.
3. Wen, Jiabin, and Lan Yu. 2019. Application of Taguchi method in permanent magnet synchronous motor structure optimization. *Journal of Harbin University of Science and Technology* 024 (005): 64–69. (in Chinese).
4. Tang, Renyuan. 2016. *Theory and design of modern permanent magnet motors*. China Machine Press. (in Chinese).

5. Duan, Y., and D.M. Ionel. 2013. A review of recent developments in electrical machine design optimization methods with a permanent magnet synchronous motor benchmark study. *IEEE Transactions on Industry Applications* 49 (3): 1268–1275.
6. Fang, Ruiming, Xuping Zhai, and Qiansheng Hu. 2002. Application of multi-turns evolution genetic algorithm and its application in optimization design of electrical machines. *Micro Motors* 035 (003): 6–9, 20. (in Chinese).
7. Yang, Lina, Gang Liu, and Qiusheng Wang. 2005. An improved genetic algorithm and its application. *Journal of Zhengzhou University: Engineering Edition* 03: 101–104. (in Chinese).
8. Mutluer, M., and O. Bilgin. 2012. Design optimization of PMSM by particle swarm optimization and genetic algorithm. In *International Symposium on Innovations in Intelligent Systems and Applications*. IEEE.
9. Li, Renhou. 1999. *Theory and method of intelligent control*. Xidian University Press. (in Chinese).
10. Guo, P., X. Wang, and Y. Han. 2010. The enhanced genetic algorithms for the optimization design. In *2010 3rd International Conference on Biomedical Engineering and Informatics (BMEI)*. IEEE.
11. Kim, J., and B.P. Zeigler. 1996. Hierarchical distributed genetic algorithms: A fuzzy logic controller design application. *IEEE Expert* 11 (3): 76–84.
12. Whitley, D. 1994. A genetic algorithm tutorial. *Statistics & Computing* 4 (2): 65–85.

Design and System-Level Optimization of Switched Reluctance Motors for Electric Vehicles Oriented to Complex Scenarios

Yuhan Yu, Shuhong Wang, Hao Qiu, and Yanlou Song

Abstract The characteristics of the switched reluctance motor (SRM) such as the highly nonlinear magnetic circuit, severe local saturation and so on restrict the accurate calculation and analysis of its performance. Therefore, the construction of a mathematical model based on the circuit and electromagnetic field theory, that is accurate and effective within the acceptable calculation cost is the key and basis for the smoothly progress of the subsequent electromagnetic characteristics and speed regulation characteristics analysis and optimization design of the motor. Taking the three-phase 12/8 SRM as an example, a general calculation program for the initial calculation of the parameters and performance indicators of the motor is developed. With the help of the inductance curve family and the flux curve family in the calculation program, the speed and current double closed loop dynamic simulation model is established. The simulation results are consistent with the law, which proves the entire nonlinear dynamic simulation model effectively realize the unification of accuracy and speed. For SRM for electric vehicles, research and propose indicators describing its complex application scenarios, analyze and obtain the influence of motor structure parameters and control parameters on motor performance to obtain optimization variables. The hierarchical genetic algorithm of multi-objective optimization strategy is designed for system-level optimization, so that SRM has the characteristics of small torque ripple, high efficiency and best robustness in a variety of complex scenarios.

Keywords Switched reluctance motor · Nonlinear model · Optimization

1 Introduction

With the increasing scarcity of natural resources and the increasing demand for protecting the health of the ecological environment, it is a general trend to use green

---

Y. Yu · S. Wang (✉) · H. Qiu · Y. Song
State Key Laboratory of Electrical Insulation and Power Equipment, School of Electrical Engineering, Xi'an Jiaotong University, Xi'an 710049, People's Republic of China
e-mail: shwang@mail.xjtu.edu.cn

W. Chen et al. (eds.), *The Proceedings of the 9th Frontier Academic Forum of Electrical Engineering*, Lecture Notes in Electrical Engineering 743,
https://doi.org/10.1007/978-981-33-6609-1_77

and clean energy to replace traditional petroleum energy. Regardless of the application of clean energy to any electric drive product, it is inseparable from the research on the drive motor as the core component.

Switched reluctance motor (SRM) has the characteristics of simple and firm structure, small starting current, good controllability, low cost and high fault tolerance, which is very suitable for electric vehicles. The drive system composed of it has a wide range of speed regulation, so it has broad application prospects and huge development potential in military and civilian fields such as all-electric aircraft, electric vehicles, and wind power generation [1]. However, it is well known that the reluctance motor with the characteristics of multi-variable strong coupling and has the characteristics of high non-linearity of the magnetic circuit, severe local saturation, large torque ripple, and large electromagnetic noise due to its double salient pole structure and numerous control parameters. It restricts the accurate calculation and analysis of its performance, and also limits its application in the field of electric vehicle drive motors.

For electric vehicles that use SRM as the drive motor, scholars have continuously invested in research and achieved some results in the optimization design of the motor and control, the entire vehicle electric drive system, the construction of driving conditions and parameter matching, etc. In [2], a 350 W low-power switched reluctance motor was designed according to the related theory of motor design. The Taguchi method was used to optimize the design with high motor efficiency as the optimization goal. In [3], the torque ripple of the reluctance motor was optimized. Establish a model in finite element and analyze the sensitivity of structural parameters to it, and use integrated optimization software for optimization analysis. In [4], the author analyzes a special two-phase 4/2-pole SRM and proposes a optimized air gap structure, in [5], first obtained the preliminary design parameters of the SR motor according to the traditional design method, and then used the improved genetic algorithm to optimize the design with high efficiency and small pulsation as the optimization goals and obtain the Global optimal solution; However, most of the above optimization design work for traditional SRM or SRM with new topology and special structure did not design from the actual application background, but only focused on the design optimization of the motor itself, without considering the actual driving conditions and Practical application problems such as matching of motor parameters for electric vehicles.

In [6–8], the main focus is on the collection and analysis of driving condition data, and the characteristic parameters that can characterize the working condition are extracted through principal component analysis and clustering, which verifies the effectiveness of the construction of the working condition. Based on this, the parameters of the motor are matched, and the optimal design is made with the best energy consumption index, the best power distribution method of the hybrid system, or the best transmission speed ratio. In [9], the author proposes some design indicators Characterize the performance of the motor, and optimize the structure and size to achieve the optimal goal, without considering the impact of the actual parameter matching of the motor. In [10–12], the main analysis is the selection of the rated parameters and peak parameters in the motor parameter matching. The impact of performance has guiding significance for the selection, parameter matching and optimization of electric vehicles. However, the above-mentioned research is conducted on a single level such as actual working condition construction and motor parameter matching, and most of them pay attention to the performance indicators of the whole vehicle, and have not optimized the performance of the motor in various scenarios, showing certain limitations.

This paper presents a nonlinear model of SR motor is constructed on the basis of circuit and electromagnetic field theory, and a general calculation program for the initial calculation of the parameters and performance indicators of the reluctance motor is developed in (2). The model is verified by the calculation result of the finite element method. In (3), the field characteristic diagram and dynamic simulation results are summarized, and the influence and sensitivity of the motor parameters and control parameters on the motor performance index and sensitivity are analyzed, which lays the foundation for the optimization designed. In (4), Using random numbers to define the wind resistance coefficient C_D, rolling resistance coefficient f and slope i in the working condition to simulate different road conditions and different working condition types, so as to put forward different requirements for the vehicle dynamics equation, In the parameter matching link, the motor structure parameters are used as the optimization variables of the motor construct layer, and the indicators describing the scene are optimized under the premise of meeting the dynamic requirements, and then the optimized motor construct parameters are transferred to the control layer for optimization, and finally get The global optimal solution of the motor construct and control parameters in this scenario. The summary and conclusion are presented in (5).

2 SRM Design and Nonlinear Modeling

The SRM speed regulation system (SRD) consists of four parts: a switched reluctance motor, a power converter, a control circuit and a position detector. As shown in Fig. 1, the operation of SRM follows the "principle of minimum reluctance", that is, the magnetic flux is always along the path with the least reluctance; after the controller comprehensively processes the speed feedback signal and the feedback information of the current sensor and the position sensor, it will control the working state of the main switching device in the power converter to realize the control of the SRM operation state, asymmetric The two freewheeling diodes in the bridge circuit play the role of freewheeling current, and feed the current back to the power supply during freewheeling. Therefore, SRM has a regenerative effect and high system efficiency [13], Among them, the three-phase 12/8-pole SRM is composed of 4 coils on the stator poles 90° in series. The generated reluctance torque is evenly distributed on the entire circumference, but due to its special breaking control and easy localization Saturated features, therefore, harmful torque ripples are generated in the commutation area where each phase is turned on and off and when the current fluctuations caused by the relative position changes of the stator and rotor teeth are large.

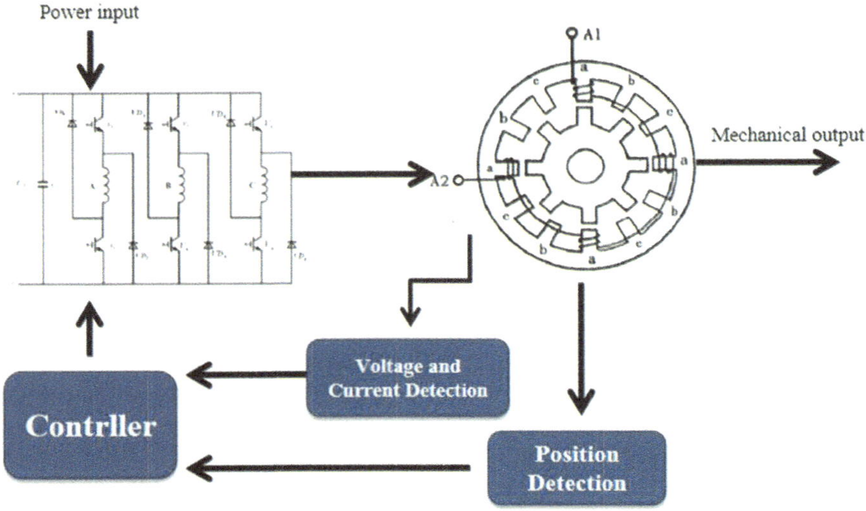

Fig. 1 SRD speed control system

Mathematically, it can be regarded as a non-linear model composed of circuit equations, mechanical equations, and electro-mechanical equations with a mechanical-electrical port.

$$\begin{cases} U_A = R_A i_A + \frac{d\psi_A}{dt} \\ \psi_A = \psi_A(i_A, \theta) \\ J\frac{d\omega}{dt} = T_e - T_L - F\omega \end{cases} \tag{1}$$

$$T_e = \left.\frac{\partial W'}{\partial \theta}\right|_{i=const} \tag{2}$$

U_A is the phase voltage of the A-phase winding, R_A is the resistance of the A-phase winding, i_A is the current of the A-phase winding, and ψ_A is the flux linkage of the A-phase winding. Due to the non-linearity of the SRM magnetic circuit, the flux linkage of each phase winding is a function of the relative position angle of the stator and rotor and the current of the stator winding; the electromagnetic torque T_e can be expressed as a function of magnetic co-energy W' as shown in Fig. 2, where F is the damping coefficient, ω is the angular velocity of the motor.

Although the above mathematical formula theoretically completely and accurately describes the electromagnetic and mechanical relationship of SRM, it is difficult to analyze due to the nonlinearity of the SRM magnetic circuit. However, the accurate description of the electromagnetic characteristics of SRM directly affects the analysis, design and application of the motor. Therefore, to build a nonlinear model that

Fig. 2 Magnetic co-energy of the area enclosed by the flux linkage-current curve during one cycle

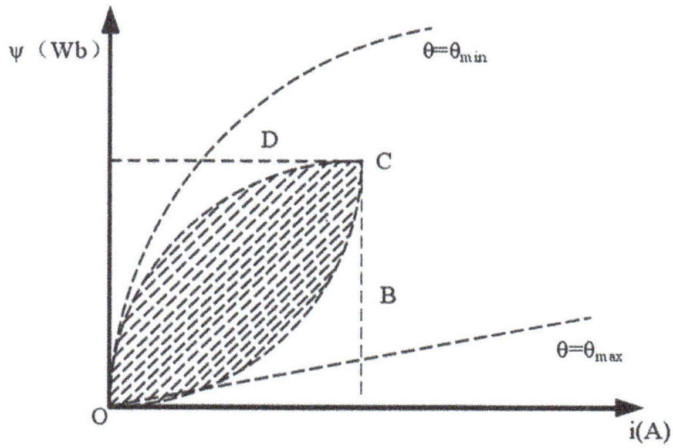

is accurate and effective within the acceptable calculation cost based on the circuit and electromagnetic field theory is the foundation of all future work.

The flux linkage-current magnetization curve family is the basis for the calculation of various performances of the motor. After continuous exploration by predecessors, it is found that the SR motor needs to obtain the magnetization curves of four special positions to quickly and accurately fit the magnetization under other rotor positions' curve.

$$L_{\theta x} = 2\frac{\psi}{i} = \frac{4}{i}(\psi_1 + \psi_2 + \psi_3 + \psi_4 + \psi_5) \tag{3}$$

Take the two special positions in Fig. 3 relative to phase A as an example. The calculation principle of the magnetization curve at this position is to divide the

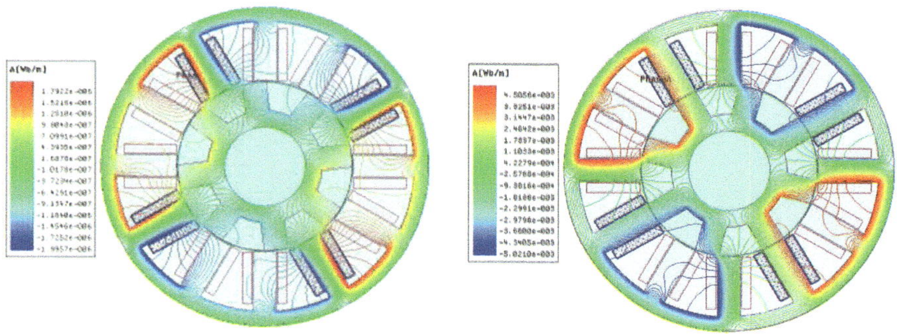

(a) Aligned (b) Unaligned

Fig. 3 Two of the special positions: aligned and unaligned. **a** Aligned. **b** Unaligned magnetic field lines in the model into several different parts and calculate the inductance of the path under each part as magnetic The accumulation of the lead component, as shown in the formula 3, can express the permeance of each part by the parameters related to the motor structure parameters into an analytical formula through the variables defined in the magnetic circuit structure diagram. Figure 4 shows the flow of the flux linkage curve family program construction.

Take the three-phase 12/8 SRM as an example, the structure is shown in Fig. 5. The initial design parameter and other calculation parameter are shown in Table 1.

With the help of the developed calculation program, the other structural size parameters of the motor and the inductance and magnetization curves at four special

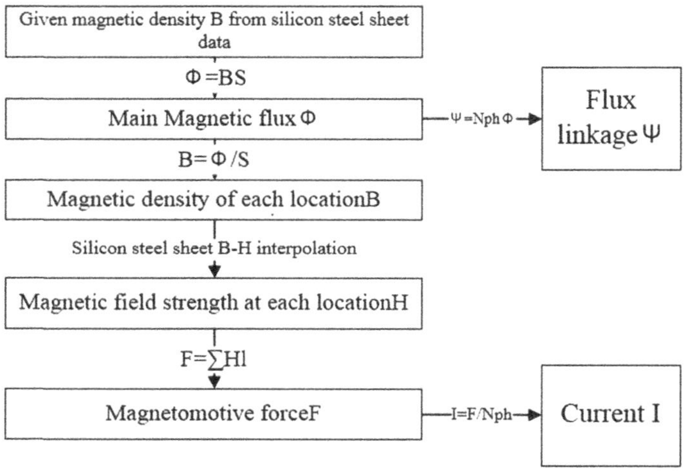

Fig. 4 Programming process of flux linkage-current curve family

Fig. 5 Prototype structure diagram

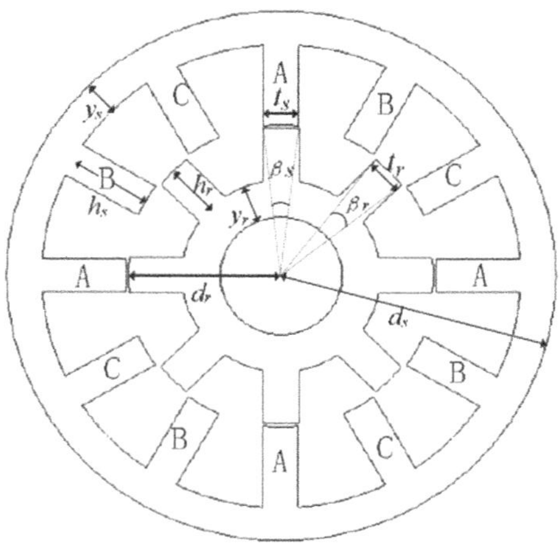

Table 1 Some parameters of the prototype

Parameters	Value
Ns/Nr	12/8 pole
d_s	60.5 mm
β_s/β_r	15/16 deg
d_r	34.5 mm
y_s	6 mm
g_l	0.4 mm
h_s	20 mm
y_r	6.7 mm

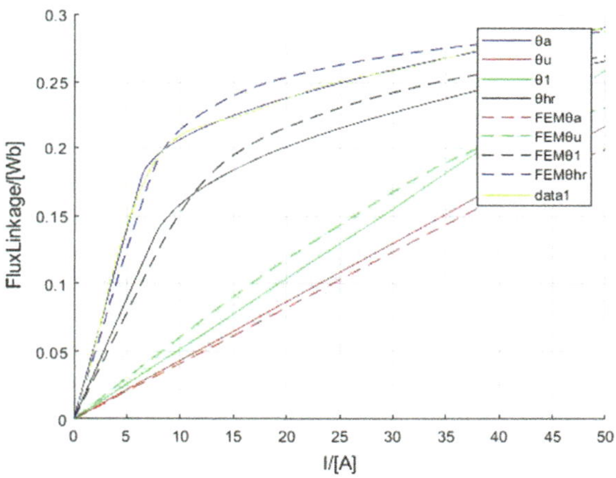

Fig. 6 Four special position magnetization curve families

rotor positions are obtained. Based on this, the traditional magnetization curve family is modeled into the $\psi-\theta-i$ form, and the fourth-order RK is used to solve the winding voltage balance differential equation, The calculated phase current, composite torque, and phase inductance curves are shown in Figs. 6 and 7, and compared with the results of the two-dimensional finite element method to verify the effectiveness and accuracy of the program.

3 Numerical Calculation and Dynamic Simulation

This chapter establishes a two-dimensional switched reluctance motor finite element model, analyzes the characteristic curve and field distribution diagram of the switched reluctance motor obtained under static field conditions, and examines and analyzes the static characteristics of the selected prototype and the influence of the motor

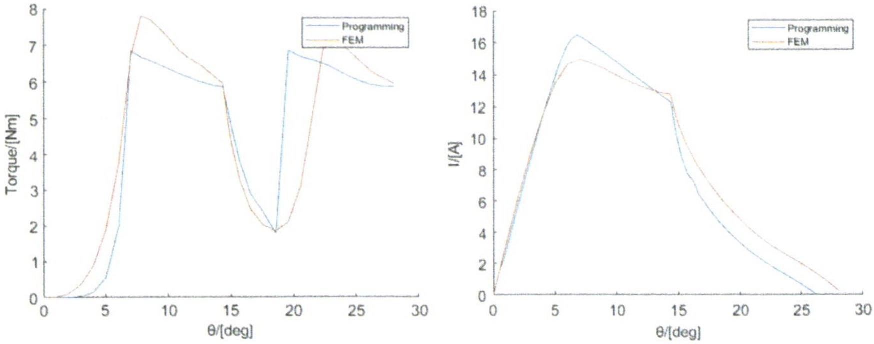

Fig. 7 One-phase current, synthetic torque waveform and comparison with FEM results

structure parameters on the performance It provides a theoretical basis for optimized design and development research and subsequent research work of switched reluctance motors.

As shown in Fig. 8, taking Phase A as an example, when the relative position of the stator and rotor starts from 0° to 1/6 cycle, that is, 7.5°, as the rotor position increases, the overlap range of the stator teeth and the rotor teeth increases, and the magnetic flux leakage decrease, the air gap reluctance is reduced, at this time, the stator and rotor tooth tips will have local saturation.

As shown in Fig. 9, when the alignment position is reached, that is, 22.5°, the air gap reluctance is the smallest and the inductance is the largest, and most of the magnetic lines of force pass and close along the stator teeth and the rotor teeth. At this time, for the A-phase stator rotor teeth, the bending of the magnetic field lines is the lowest.

For the static characteristics of SRM, as shown in Fig. 10, near the unaligned position, because the motor does not enter the saturated working area, the inductance hardly changes with the current change, as shown in the figure as a plane; near the

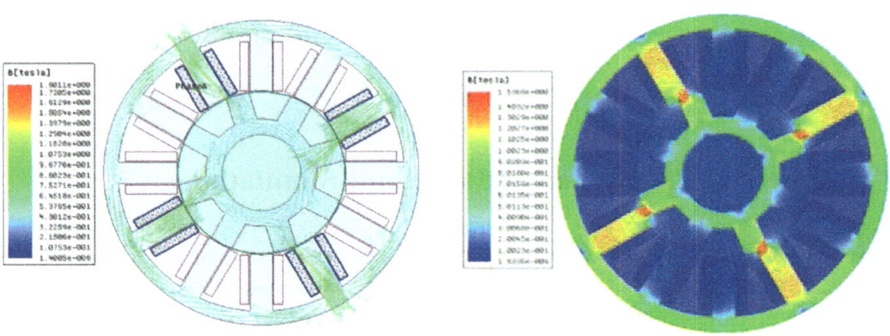

Fig. 8 The magnetic density distribution and cloud map when the position is 7.5°

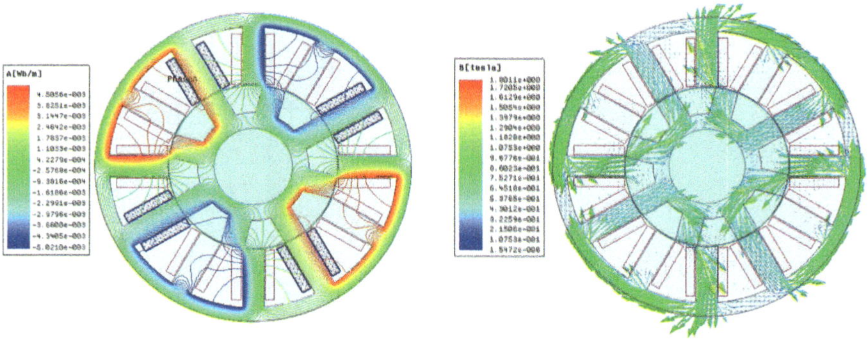

Fig. 9 The flux-line and magnetic density distribution when the position is 22.5°

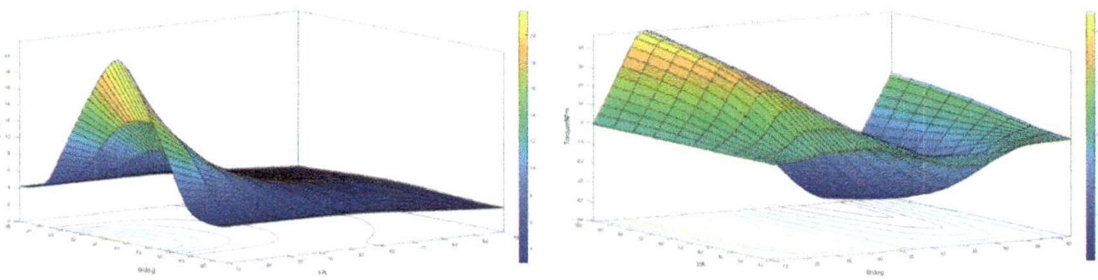

Fig. 10 Inductance and torque static characteristic surface

aligned position, The inductance decreases as the current increases. In the unaligned position and the aligned position, the motor is at the working state exchange point, and the electromagnetic torque is zero; the motor produces the maximum torque in the middle area.

In order to apply the corresponding control algorithm, a dual closed-loop dynamic simulation model of speed and current based on a two-dimensional look-up table of inductance and flux linkage is built in Simulink, as shown in Fig. 11.

In order to avoid the excessive pulse current peak value exceeding the allowable value at low speed, the SR motor adopts current chopping control at low speed to limit the amplitude of the current and provide the largest possible torque, as shown in Fig. 12.

In the high-speed situation after entering the steady state, the angular position control method is adopted, as shown in Fig. 13.

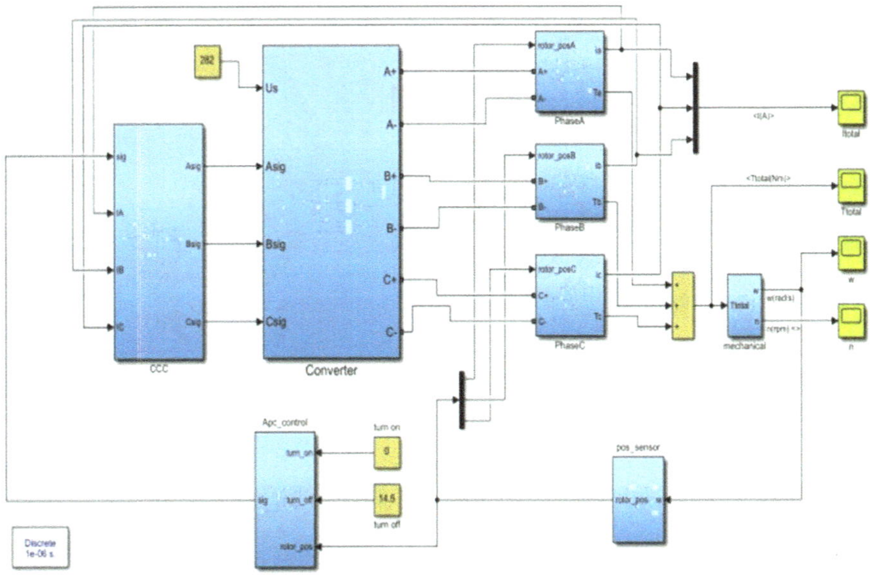

Fig. 11 Double closed loop dynamic simulation model

4 Bi-Level Multi-objective Optimization Based on Complex Scenarios

Figure 14 is a graph of the ideal driving characteristics of a vehicle. The motor maintains a low speed and constant torque below the base speed, and is in a constant power operating mode above the base speed. SRM has excellent speed regulation performance, low speed, high torque, and high speed and constant power mechanical characteristics. The characteristics of maintaining high efficiency in a wide rotational speed and torque working area are also very suitable for electric vehicle drive system motors.

The actual operating conditions of the vehicle are of vital significance to the matching of the vehicle power system. This paper uses random numbers to define the wind resistance coefficient C_D, rolling resistance coefficient f and slope i in the operating conditions to simulate different road conditions and different types of operating conditions. Therefore, different requirements are put forward for the vehicle dynamics equations. For example, formulas 4–7 respectively determine the lower limit of the rated power at the highest vehicle speed, check the peak power with the full load climbing performance and the acceleration performance of one hundred kilometers, and determine the peak motor speed at the highest vehicle speed.

$$P_{\max 1} = \left(m_h g f + \frac{C_D A v_{\max}^2}{21.15} \right) \frac{v_{\max}}{3600 \eta_t} \tag{4}$$

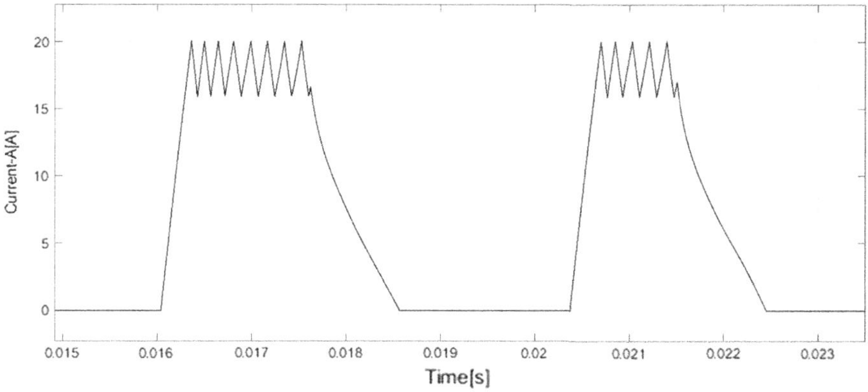

(a) Phase A current under CCC

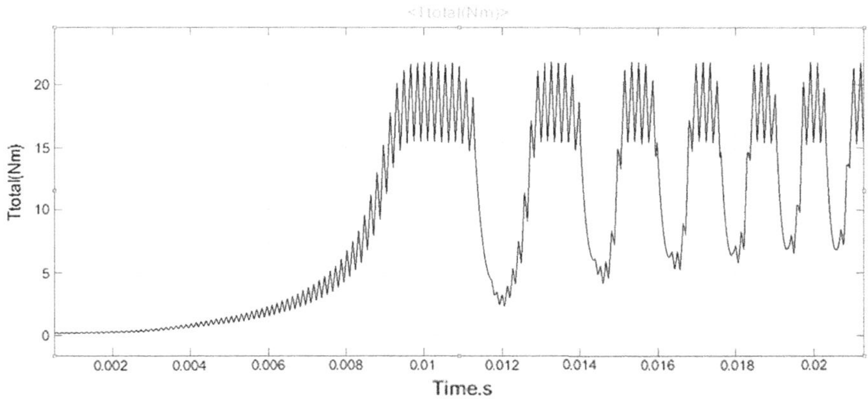

(b) Total torque under CCC

Fig. 12 Low-speed current chopper control. **a** Phase A current under CCC. **b** Total torque under CCC

$$P_{\max 2} = \left(m_l g f + \delta m_l \frac{dv}{dt} + \frac{C_D A v^2}{21.15} \right) \frac{v}{3600\eta_t} \tag{5}$$

$$P_{\max 3} = \left\{ m_f g f \cos\alpha + m_f g \sin\alpha + \frac{C_D A v^2}{21.15} \right\} \frac{v}{3600\eta_t} \tag{6}$$

$$n_{\max} = \frac{v_{\max} i}{0.377 r_w} \tag{7}$$

$$T_{\max} = \left\{ m_f g f \cos[\tan^{-1}(i_{\max}/100)] + m_f g \sin[\tan^{-1}(i_{\max}/100)] + \frac{C_D A v^2}{21.15} \right\} \frac{r}{i\eta_t} \tag{8}$$

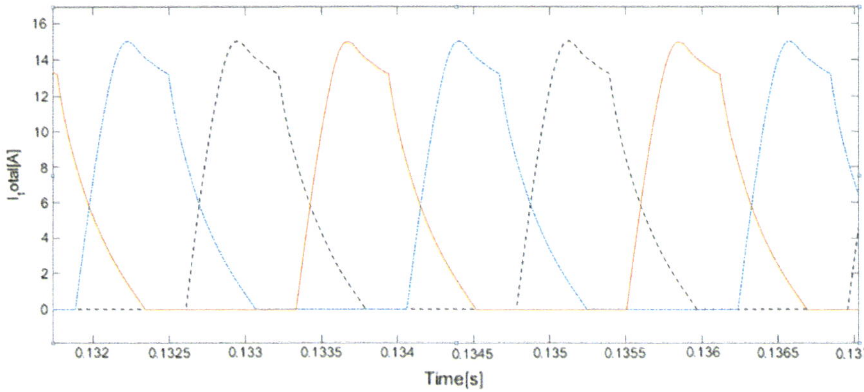

(a) Three-phase current under APC control mode

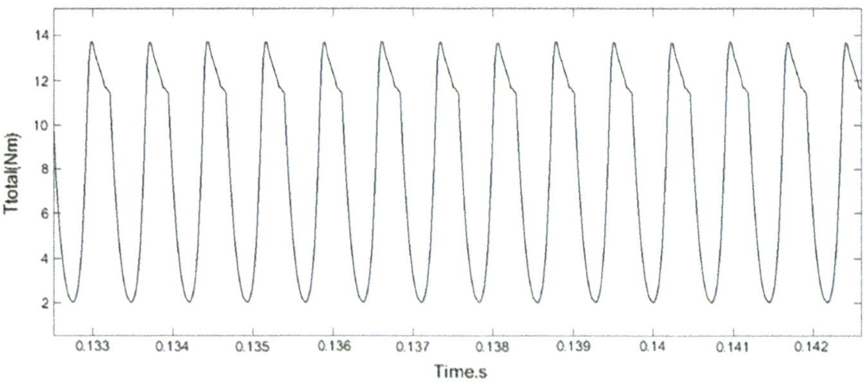

(b) Total torque under APC control mode

Fig. 13 Angle position control in steady state. **a** Three-phase current under APC control mode. **b** Total torque under APC control mode

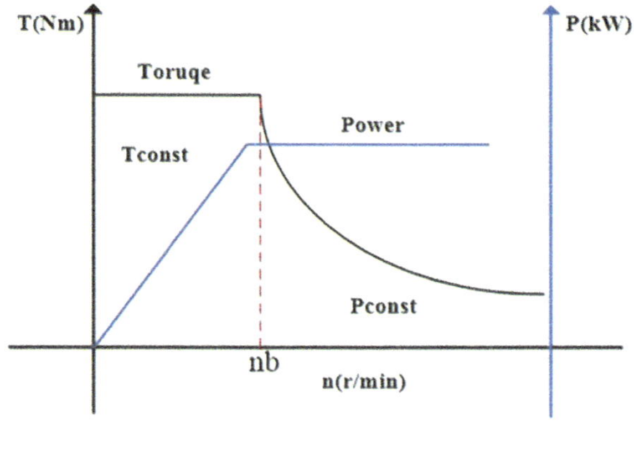

Fig. 14 Vehicle ideal driving characteristic diagram

Formula 8 is used to check the peak torque to meet the climbing performance; through the above formulas, the selection of the rated and peak characteristics is finally transformed into the matching of the performance parameters of the motor.

The peak operating characteristics of the motor are closely related to the expansion of the constant power zone coefficient. The parameter matching process in this article is different from the conventional method. First, after determining the rated power and peak speed of the motor, determine the optimal rated speed, that is, determine the optimal expansion constant power zone coefficient β and the optimal overload coefficient T_{ol} as the target, so as to meet the acceleration performance conditions as much as possible while reducing the peak power demand of the motor, avoid unnecessary waste caused by excessive motor power. It can be seen from the driving characteristics of the motor that in the case of determining the peak speed, appropriately increasing the constant power zone coefficient can make the motor meet the acceleration performance while selecting a smaller peak power parameter, and can improve the motor rotation in the low speed zone. The torque output characteristics enable the vehicle to have faster acceleration response capabilities.

The coefficient β and T_{ol} can be defined as:

$$\beta = \frac{n\max}{n_b} \tag{9}$$

$$T_{ol} = \frac{T\max}{T_{avg}} \tag{10}$$

However, an excessive value of β will increase power consumption and decrease the overall efficiency of the motor; it will also cause excessive starting current, exceeding the threshold of power electronic equipment; and excessive peak torque will cause the motor The mass and volume increase, and the load on the motor rotating bearing is increased, and its service life is shortened. Therefore, in the parameter optimization of the body layer, the goal is to find the optimal expansion constant power zone coefficient and overload coefficient, that is, to match the dynamic demand, and the following constraints need to be met:

1. The motor efficiency is not lower than the design value, $\eta \geq \eta_{rate}$
2. The output torque is not higher than the limit value, $T \leq T_{constraint}$

After determining the optimal solution of the motor body structure, it is passed to the control layer, with the opening angle and the closing angle as the optimization variables, and the motor performance data efficiency and torque pulse action as the parameters to consider whether the economic and comfort indicators are met, using genetic algorithms Implement multi-objective optimization.

The implementation process of the optimization scheme in this paper is shown in Fig. 15. According to the natural sequence, the optimization is divided into the sequence of the structure parameters first and then the control parameters. The structure layer meets the dynamic demand as the optimization goal, and the stator and rotor pole arc coefficients and stator yoke height and the rotor outer diameter are

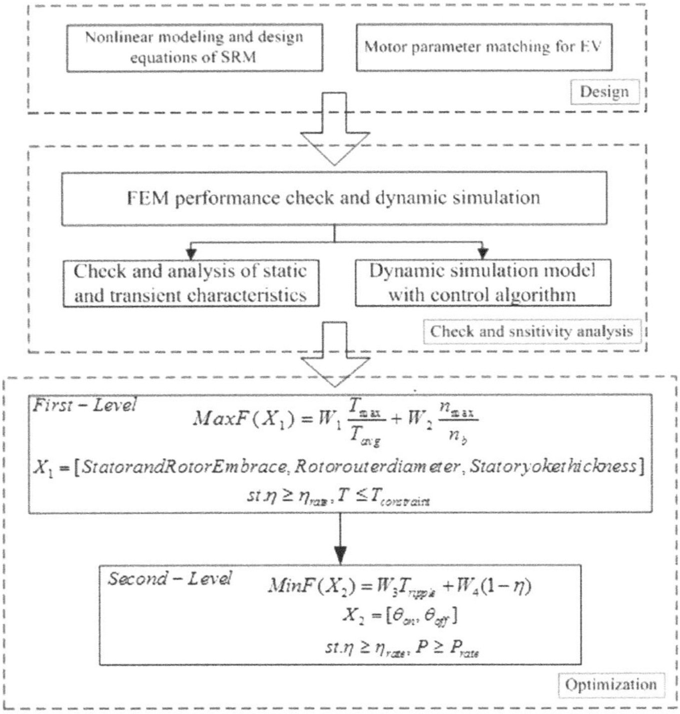

Fig. 15 Design-check-optimize process

optimized variables, trying to determine the optimal expansion constant power zone coefficient within the constrained range, and then transfer the optimized structure parameters to the control layer, taking the highest efficiency and the smallest torque ripple as the optimization goals, using multi-objective Genetic algorithm to finds the global optimal solution.

Taking a certain type of electric vehicle on the market as the basic parameters, the initial matching results of the driving motor parameters are shown in Table 2 under the complicated working conditions of randomly given wind resistance, rolling resistance and slope.

Table 2 Parameter initial matching result

Three-phase 12/8 SRM	Value
Rated power	7.5 kW
Peak power	16 kW
Rated speed	2000 rpm
Peak speed	5500 rpm
Peak torque	80 Nm

Table 3 Optimized parameters

Parameter	Value
T_{max}	91.7332 Nm
n_{max}	5500 rpm
n_b	1421 rpm
T_{avg}	29.18 Nm
R_r	91.27 mm
S_hcs	8.22 mm
S_emb	0.542
R_emb	0.429

Using the developed design program to calculate the motor parameters and establish a layered optimization model. The optimized motor parameters and performance indicators are shown in Table 3. Among them, S_emb and R_emb represent the stator and rotor pole arc coefficients, R_r represents the rotor outer diameter, and S_hcs represents The stator yoke thickness. The results are shown in Figs. 16 and 17. The results show that the design parameter compromise value found in the first level enables the motor to have a wide speed range in this scenario, and the rapid peak torque response capability is the acceleration performance of low speed and large torque.

After the optimization of the first level is completed, the structure size will be fixed and the optimization results will be transmitted to the control layer, and the turn-on and turn-off angles will be used for multi-objective optimization to reduce the motor torque ripple coefficient and obtain higher efficiency as much as possible, as shown in Fig. 18. After optimization, the motor efficiency can reach 89.36%, while the torque ripple coefficient is 1.55.

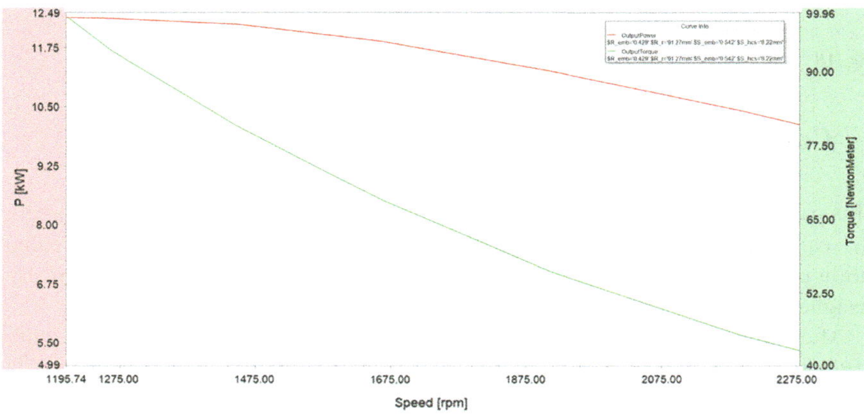

Fig. 16 Optimized motor external-characteristics curve

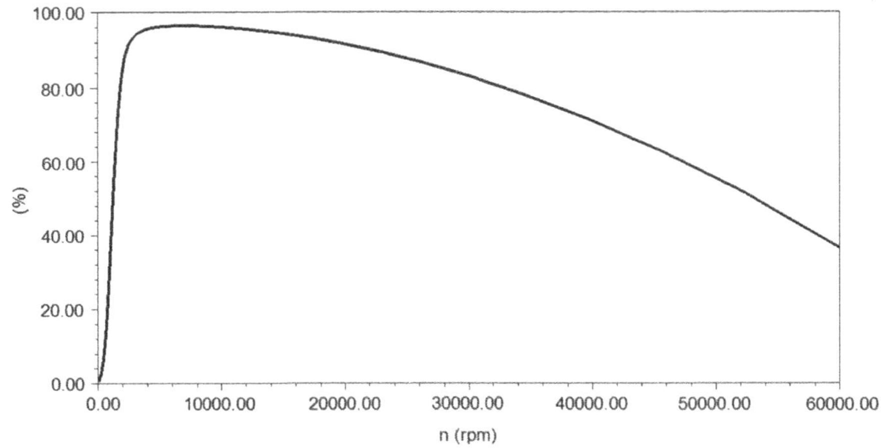

Fig. 17 Wider and more efficient working area after optimization

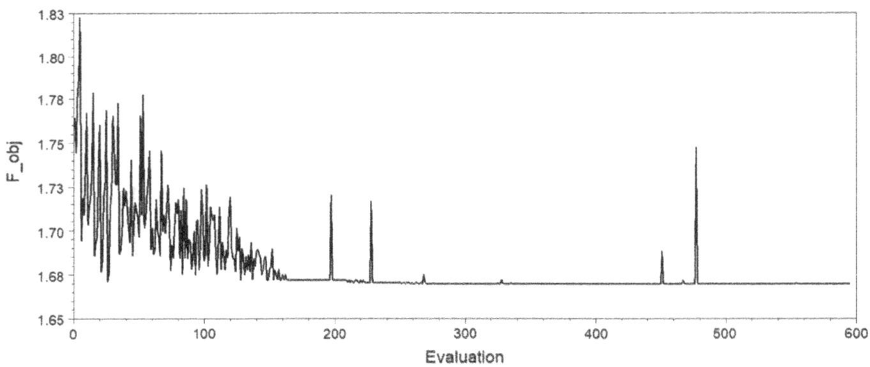

Fig. 18 Multi-objective optimization after passing to the control level

5 Conclusion

This paper establishes a scheme of SR motor nonlinear modeling and initial parameter calculation through the traditional motor design method. Based on this calculation program, the motor performance and structural parameters are preliminarily designed.

Afterwards, the FEM was used to check the influence and sensitivity of the motor structure and control parameters on the motor performance index, and lay the foundation for the optimization design. Using random numbers to define the wind resistance, rolling resistance and slope in the conditions to simulate random and complex scenes, and propose several parameters representing dynamics, economy and comfort. In the bi-level optimization, the first level uses the motor structure parameters as optimization variables, and optimizes the design of the indicators describing the scene on the premise of meeting the dynamic requirements, and then transmits the optimized motor structure parameters to the control layer for optimization. Obtain the global optimal solution of the motor structure and control parameters in this scenario.

Obviously, under actual working conditions, the design and optimization of the motor and the vehicle drive system are much more complicated than the above examples, but the ideas in this article can be used as a basis for a preliminary optimization design plan for reference.

Acknowledgements This work was supported in part by the National Natural Science Foundation of China under Grant (51737010).

References

1. Bilgin, B., et al. Making the case for switched reluctance motors for propulsion application. *IEEE Transactions on Vehicular Technology*. https://doi.org/10.1109/tvt.2020.2993725.
2. Yang, Gong. 2019. *Design and optimization method research of a low-power switched reluctance motor*. South China University of Technology. (in Chinese).
3. Liang, Qiaodan. 2019. *Research on structural torque ripple suppression method of switched reluctance motor*. South China University of Technology. (in Chinese).
4. Zheng, Yicheng. *Design optimization and simulation research for two-phase 4/2-pole high-speed switched reluctance motor*. Zhejiang University China. (in Chinese).
5. Song, Shoujun, Lefei Ge, Hucheng Liu, and Weiguo Liu. 2014. Design and multi-objective optimization method of switched reluctance machines. *Transaction of China Electrotechnical Society* 29 (05): 197–204. (in Chinese).
6. Sen, Zhan, Datong, Qin, and Yuping, Zeng. 2016. Energy management strategy of HEV based on driving cycle recognition using genetic optimized K-means clustering algorithm. *China Journal of Highway and Transport* 29 (04): 130–137+152. (in Chinese).
7. Xue, X.D., K.W.E. Cheng, T.W. Ng, and N.C. Cheung. 2010. Multi-objective optimization design of in-wheel switched reluctance motors in electric vehicles. *IEEE Transactions on Industrial Electronics* 57 (9): 2980–2987. https://doi.org/10.1109/TIE.2010.2051390.
8. Zhou, Baohua, Minghui Hu, Jianjun Hu, and Xi Wang. 2011. Parameters design of powertrain system of electric vehicle with two-speed gearbox. 34 (01): 1–6. (in Chinese).
9. Cheng, H., H. Chen, and Z. Yang. 2015. Design indicators and structure optimisation of switched reluctance machine for electric vehicles. *IET Electric Power Applications* 9 (4): 319–331.
10. He, Hongwen, Xiaojiang Yu, and Fengchun Sun. 2006. Study on power performance of traction motor system for electric vehicle. *Proceedings of the CSEE* 06: 136–140. (in Chinese).
11. Song, Ke, and Tong Zhang. 2013. Parameter matching of electric motor drive system for electric and series hybrid electric vehicles. *Automotive Engineering* 35 (06): 559–564. (in Chinese).

12. Ehsani, M., K.M. Rahman, and H.A. Toliyat. 1997. Propulsion system design of electric and hybrid vehicles. *IEEE Transactions on Industrial Electronics* 44 (1): 19–27. https://doi.org/10.1109/41.557495.
13. Wu, Jianhua. 2000. *Design and application of switched reluctance motor*. Beijing: Machinery Industry Press. (in Chinese).

Yuhan Yu Major research interests in Design, simulation and optimization of energy-efficient electric energy conversion and transmission equipment and special electromagnetic devices.

System-Level Optimization of Permanent Magnet Synchronous Motors for Electric Vehicles

Yanlou Song, Song Huang, Hongjing He, Yuhan Yu, and Shuhong Wang

Abstract In recent years, affected by factors such as air pollution and fossil energy shortages, the automobile industry is facing a transition from traditional fuel vehicles to electric vehicles. Electricity and intelligence have become frontier issues in the automobile industry. Electric drive system is the core part of electric vehicles. The actual driving environment of an electric vehicle is a very complex scene. Therefore, in order to obtain the best performance of the electric drive system, the motor and its control system should be designed and optimized at the system level combined with the driving scene of the vehicle. This paper proposes a system-level optimization design method for permanent magnet synchronous motors of electric vehicles based on scenario optimization. Random piecewise functions are used to simulate the driving scenes of electric vehicles. The system-level optimization is implemented in a layered manner.

Keywords Electric vehicles · Permanent magnet synchronous motor · System-level optimization · Scene optimization · Particle swarm optimization

1 Introduction

Batteries, motors, and electronic control systems are the core parts of electric vehicles. In particular, power endurance is the main factor restricting the performance of electric vehicles [1]. Therefore, improving the efficiency of the motor can effectively improve the endurance of electric vehicles. The motor will not always work near the rated point in the actual driving scene [2].

With the emergence of high-energy permanent magnet materials, and the continuous improvement of the performance price ratio of rare earth permanent magnet materials, such as neodymium iron boron (NdFeB), permanent magnet synchronous motors have become more and more attractive. Coupled with the continuous progress

Y. Song · S. Huang · H. He · Y. Yu · S. Wang (✉)
School of Electrical Engineering, Xi'an Jiaotong University, Xi'an 710049, China
e-mail: shwang@mail.xjtu.edu.cn

W. Chen et al. (eds.), *The Proceedings of the 9th Frontier Academic Forum of Electrical Engineering*, Lecture Notes in Electrical Engineering 743,
https://doi.org/10.1007/978-981-33-6609-1_78

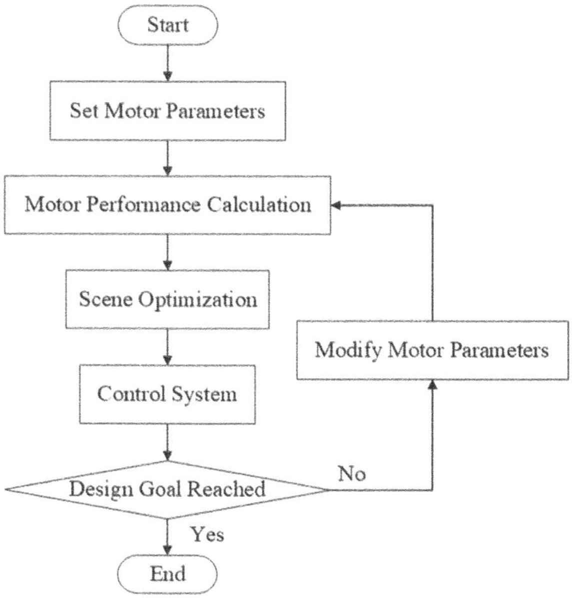

Fig. 1 System-level optimization process of electric vehicle motors

of new motor topologies and control strategies, permanent magnet synchronous motors are generally considered to be the most promising solution for electric vehicle.

Permanent magnet synchronous motor (PMSM) has the advantages of high power density, high efficiency and reliable overall performance. It is an ideal execution unit for efficient drive and is very suitable for electric vehicle drive systems [3].

The use of permanent magnet synchronous motors in the drive system of electric vehicles can reduce the weight, size, and efficiency of the vehicle. Permanent magnet synchronous motors have been widely used in electric vehicles and other demanding speed control drive systems [4].

Therefore, this paper proposes a system-level optimization method for permanent magnet synchronous motors. Figure 1 shows the system-level optimization process of electric vehicle motors. The optimization process is divided into motor performance calculation layer, scene optimization layer and control system layer.

2 Motor Performance Calculation

The motor performance calculation layer is the basis of system-level optimization. The main work of this layer is to do magnetic circuit calculation and finite element simulation of the motor according to the basic parameters. Then draw the efficiency

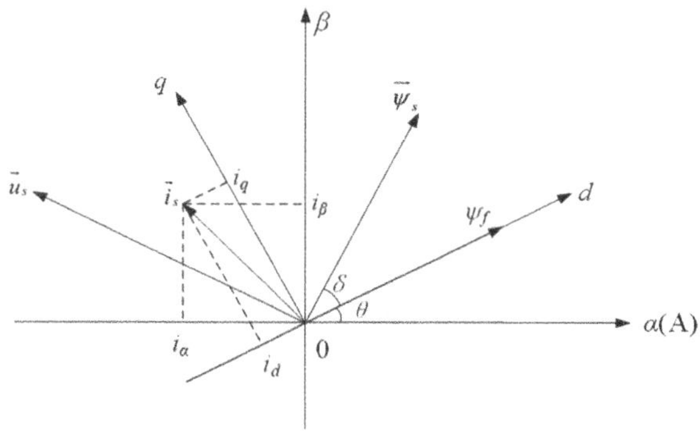

Fig. 2 PMSM vector diagram

MAP and transfer the efficiency data under different speeds and torques to the scene optimization layer, transfer the resistance, inductance, flux and other data to the control layer.

The rotor of a permanent magnet synchronous motor is excited by permanent magnets [5]. The rotor flux vector is consistent with the position of the rotor poles [6]. The *d*-*q* rotor coordinate system is usually used (fix the *d*-axis in the direction of the rotor excitation flux $\overrightarrow{\psi}_f$, and the *q*-axis is the anticlockwise direction leading the *d*-axis by 90 electrical degrees, the *d*-*q* coordinate system rotates synchronously with the rotor) or $\alpha\beta$ stator coordinate system (the α-axis coincides with the a-axis of the three phase *a*-*b*-*c* stator coordinate system, and the β-axis is 90 electrical degrees ahead of the a-axis in the anticlockwise direction, the $\alpha\beta$ coordinate system is stationary) [7].

The space vector of the permanent magnet synchronous motor is shown in Fig. 2 [8]. Establish a permanent magnet synchronous motor *d*-*q* rotor coordinate coefficient model. Voltage, flux linkage, torque and mechanical motion equations are formulas (1)–(6) [9].

$$u_d = \frac{\mathrm{d}\psi_d}{\mathrm{d}t} - \omega\psi_q + R_s i_d \tag{1}$$

$$u_q = \frac{\mathrm{d}\psi_q}{\mathrm{d}t} - \omega\psi_d + R_s i_q \tag{2}$$

$$\psi_d = L_d i_d + \psi_f \tag{3}$$

$$\psi_q = L_q i_q \tag{4}$$

$$T_e = \frac{3}{2}p(\psi_d i_q - \psi_q i_d) = \frac{3}{2}p[\psi_f i_q + (L_d - L_q)i_d i_q] \tag{5}$$

$$J\frac{\mathrm{d}\omega_m}{\mathrm{d}t} = T_e - T_L - B\omega_m \tag{6}$$

In the formula, u_d and u_q are the d-axis and q-axis components of the stator voltage respectively. id and iq are the d-axis and q-axis components of the stator current respectively. ψ_d and ψ_q are the d-axis and q-axis components of the stator flux linkage respectively. L_d and L_q are the d-axis and q-axis components of the stator inductance respectively. R_s is stator resistance. ψ_d is the flux linkage produced by the permanent magnets of the rotor. T_e is the electromagnetic torque of the motor. T_L is the load torque. J is the moment of inertia. B is friction coefficient. ω_m is rotor mechanical angular velocity. p is the number of pole pairs. $\omega = p\omega_m$ is the electrical angular velocity of the rotor.

From formulas (1)–(6), the state equation of the permanent magnet synchronous motor is shown in formula (7).

$$\begin{bmatrix} \frac{\mathrm{d}i_d}{\mathrm{d}t} \\ \frac{\mathrm{d}i_q}{\mathrm{d}t} \\ \frac{\mathrm{d}\omega_m}{\mathrm{d}t} \end{bmatrix} = \begin{bmatrix} -\frac{R_s}{L_d} & p\omega_m & 0 \\ -p\omega_m & -\frac{R_s}{L_q} & -\frac{p\psi_f}{L_q} \\ 0 & \frac{p\psi_f}{J} & \frac{B}{J} \end{bmatrix} \begin{bmatrix} i_d \\ i_q \\ \omega_m \end{bmatrix} + \begin{bmatrix} \frac{u_d}{L_d} \\ \frac{u_q}{L_q} \\ -\frac{T_L}{J} \end{bmatrix} \tag{7}$$

3 Scene Optimization

The core of this method is to simulate the driving scene of the motor. The work of this layer is to use random function to simulate the actual driving scene of the electric motor. First, a random piecewise function is used to simulate the road slope, friction, wind resistance and other parameters. Then, convert these parameters into the torque required to drive the car at different speeds. Next, draw the efficiency MAP based on the data of the motor layer and calculate the total efficiency of the motor passing the road at a certain speed. Finally, with the speeds as variables, use particle swarm algorithm to optimize the total efficiency of the motor. Transmit a set of speed and torque to the control layer and realize system-level optimization of motor. The basic process of scene optimization is shown in Fig. 3.

3.1 Electric Vehicle Driving Scene

Step 1, use random piecewise function to simulate electric vehicle driving scene. Due to the complicated road conditions in the actual driving of electric vehicles, it is difficult to accurately describe the driving scene with fixed parameters [10].

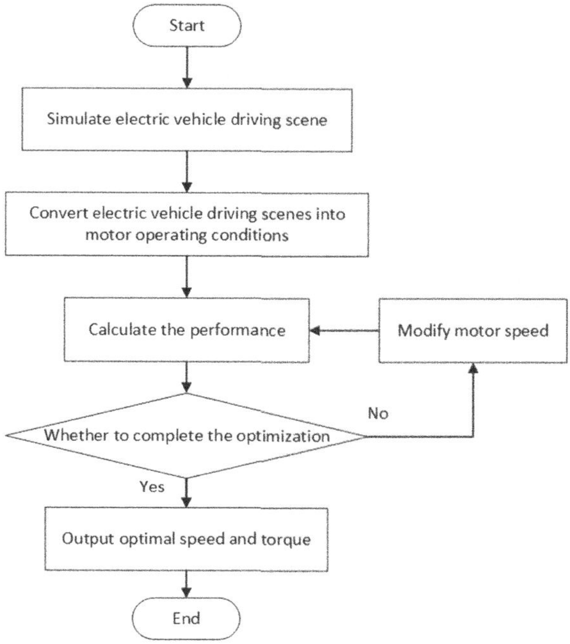

Fig. 3 Scene optimization

Therefore, multiple sections of roads are simulated in a random manner to generate a set of parameters such as slope, ground friction, wind resistance and so on to describe the driving scene of electric vehicles.

3.2 Operating Condition of Motor

Step 2, convert the driving scene of electric vehicle into the operating condition of motor. Use formula to express the torque as a function of speed, that is, calculate the torque required for electric vehicle to pass through the road at this speed.

This paper mainly considers the central motor of electric vehicles. The best working area of traditional vehicle engine is 1000–4000 rpm. The output torque is too low at low speed, and the output torque at high speed attenuates sharply. Therefore, the traditional engine needs multiple reducers with different reduction ratios, namely gearboxes. So that the engine speed can be maintained in the best working area. The speed regulation performance of the electric motor is better than that of the traditional engine. The electric vehicle does not need a complicated transmission structure, and only a fixed reduction ratio reducer is enough. Thus, the speed of the electric vehicle and the speed of the motor are determined by the main reduction ratio and the size of the hub, and the two parameters are proportional.

In order to express the torque as a function of speed, vehicle specific power needs to be introduced, as shown in formula (8). The vehicle specific power (VSP) is defined as the unit mass of vehicle transportation (including its own weight, passengers and objects), the instantaneous power output by the engine, in kW/t [11]. It represents the work done by the vehicle to overcome the frictional resistance of tires, air resistance, kinetic energy and potential energy changes.

$$VSP = v(a + gs + gC_R) + \frac{1}{2}\rho_a \frac{C_D}{m} Av^3 \tag{8}$$

In this formula, v is the driving speed of the vehicle in m/s; a is the acceleration in m/s^2; g is the acceleration of gravity, taking 9.81 m/s^2; s is the road gradient; C_R is the rolling resistance coefficient; ρ_a is the ambient air density; C_D is the drag coefficient; A is the windward area, in m^2; m is the total mass of the vehicle, in kg.

The VSP is defined as the power required by the vehicle to transport a unit of mass, so the power can be expressed as formula (9).

$$P = v(a + gs + gC_R)m + \frac{1}{2}\rho_a C_D Av^3 \tag{9}$$

The relationship between motor power, speed and torque is shown in formula (10).

$$P = \frac{n \times T}{9549} \tag{10}$$

In this formula, P is the power, n is the speed of motor, T is the output torque.

Since the motor speed is proportional to the speed of the electric vehicle, it is only related to the reduction ratio and the radius of the hub. So the relationship between the motor speed and the speed can be expressed as formula (11).

$$n = i \times v \tag{11}$$

Combining formulas (9), (10), and (11), the relationship between motor speed and torque can be expressed as formula (12).

$$T = 9549\left[\frac{1}{i}(a + gs + gC_R)m + \frac{1}{2i^3}\rho_a C_D An^2\right] \tag{12}$$

Formula (12) is divided into two parts. One part represents the torque required for vehicle driving on this road condition, its main parameters are slope and road friction. The other part represents the torque required to overcome wind resistance, its main parameters are air density, windward area and wind resistance coefficient. Simplify the constant in formula (12) to get formula (13).

$$T = k_1 + k_2 n^2 \tag{13}$$

In this formula, the constant term k1 represents road conditions, such as slope, friction, etc. And the quadratic term k2n2 represents wind resistance, that is, the higher the vehicle speed, the greater the torque required to overcome the wind resistance.

3.3 Motor Performance Evaluation

Step 3, motor performance evaluation. After the derivation of the above formula, in a given driving scene, the torque required to pass the road can be calculated by the speed. Therefore, the variable for scene optimization is attributed to the motor speed. For multi-segment roads simulated by a random segment function, motor rotation speed is given for each road segment to calculate the torque, and then the efficiency of the motor passing each segment of the road can be calculated from the efficiency MAP data.

The scene optimization layer uses speed as the variable, and the objective function is divided into three parts: total efficiency, time factor, and energy consumption factor. The total efficiency is the ratio of the sum of output power to the sum of the input power through all road sections [12].

The higher the vehicle speed, the greater the energy required to overcome the wind resistance. In order to reduce the energy loss, define energy consumption factor (*ECF*) as formula (14).

$$ECF = \frac{e}{e_{\max}} \tag{14}$$

In this formula, e is the energy loss of passing the road at the target speed, and $e_{\max}$ is the energy loss of passing the road at the highest speed (120 km/h), at this time, the energy consumed by overcoming wind resistance is the largest.

The efficiency of the motor will decay at high speed, and the energy consumed by the car at high speed to overcome wind resistance is large. Passing the road at a lower speed can reduce energy loss, but the low speed of the vehicle can not meet the actual requirements of transportation, which wastes time, so define time factor (*TF*) as Eq. (15).

$$TF = \frac{t_{\min}}{t} \tag{15}$$

In this formula, t is the time to pass the road at the target speed, and $t_{\min}$ is the time to pass the road at the highest speed (120 km/h).

3.4 Particle Swarm Optimization

Step 4, particle swarm optimization. Particle Swarm Optimization (PSO) is a classic intelligent algorithm [13, 14]. It starts from a random solution and finds the optimal solution through iteration. The algorithm evaluates the quality of the solution through fitness, and finds the global optimum by tracing the optimal value currently found [15]. The speed and position update formulas are shown in formulas (16) and (17) [16].

$$v_{id} = w \times v_{id} + c_1 r_1 (p_{id} - x_{id}) + c_2 r_2 \left(p_{gd} - x_{id}\right) \tag{16}$$

$$x_{id} = x_{id} + v_{id} \tag{17}$$

In the formula, v_{id} is particle velocity, x_{id} is particle position, w is inertia weight, c_1 and c_2 are learning factors, p_{id} is individual optimal value, p_{gd} is group optimal value, r_1 and r_2 are uniform random numbers in the range.

4 Control System

The main work of the control layer is to build a motor module based on the data from the motor performance calculation layer and build its control circuit. Use a set of speed values from the scene optimization layer as the target speed, and use the torque value as the load torque for simulation.

Since the flux linkage produced by the permanent magnets of the rotor is constant, the rotor flux linkage oriented FOC is adopted in this paper [17]. Set the d-axis of the two-phase synchronous rotating internal coordinate system in the direction of the rotor flux linkage $\overrightarrow{\psi}_r$, and the q-axis is anticlockwise leading the d-axis $\pi/2$ electrical angle, as shown in Fig. 4 [18].

The stator flux, voltage and torque equations of the PMSMS in the d-q coordinate system are shown in formulas (1)–(5). The torque of a PMSM basically depends on the q-axis component of the rotor flux linkage and the stator current [19]. Since the rotor structure of the PMSM is a permanent magnet, the amplitude of the generated flux linkage is basically constant, so the rotor flux orientation method can be used to achieve high-performance control of the PMSM.

When the $i_d = 0$ control strategy is adopted, there is only q-axis component in the stator current, and the space vector of the stator magnetomotive force is orthogonal to the space vector of the magnetic field generated by the rotor permanent magnet [20]. At this time, the stator voltage equation can be rewritten as:

$$u_d = -\omega L_q i_q \tag{18}$$

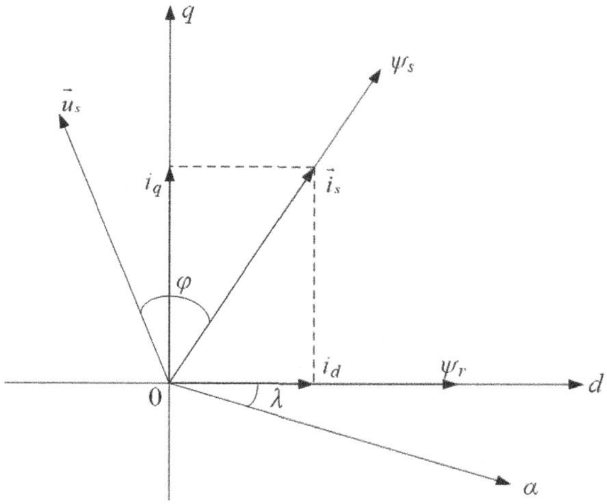

Fig. 4 Rotor flux linkage oriented vector diagram

$$u_q = R_s i_q + L_q \frac{\mathrm{d}i_q}{\mathrm{d}t} + \omega\psi_{PM} \tag{19}$$

The torque equation can be rewritten as:

$$T_e = p\psi_{PM} i_q \tag{20}$$

It can be seen from the above equations that when the permanent magnet synchronous motor adopts the rotor flux linkage oriented control strategy, the torque is proportional to the rotor flux linkage ψ_{PM} and the torque component i_q of the stator current. As long as the stator excitation current i_d can be well controlled, the amplitude of the rotor flux linkage ψ_{PM} is constant, the torque T_e is only controlled by the torque component i_q of the stator current.

5 Example

This paper takes a three-phase 4-pole permanent magnet synchronous motor with rated power of 15 kW and rated speed of 1500 rpm as an example. The key parameters of the motor are shown in Table 1.

The main calculation results of the motor are shown in Table 2.

The magnetic flux density distribution of the motor is shown in Fig. 5.

Change the input current to control the torque, and change the speed, run the two-dimensional finite element simulation multiple times, use the results to plot the efficiency map of the motor as Fig. 6 [21].

According to the efficiency map, run the scene optimization program, and get a set of results as shown in Table 3.

Table 1 Key parameters of the motor

Parameter	Value	Parameter	Value
Rated power	15 kW	Rated power factor	0.95
Number of phases	3	Winding configuration	Y
Rated frequency	50 Hz	Rated phase voltage	220 V
Number of pole pairs	2	Rated phase current	25.66 A
Rated speed	1500 rpm	Rated torque	95.49 Nm

Table 2 Calculation results

Parameter	Value	Parameter	Value
Input power	15,605.5 W	Armature copper loss	283.399 W
Output power	14,998.8 W	Iron-core loss	163.272 W
Efficiency	96.1124%	Air gap flux density	0.667467 T
Total loss	606.671 W	RMS fundamental induced voltage	222.761 V
Friction and windage loss	160 W	Terminal resistance	0.1651 Ω

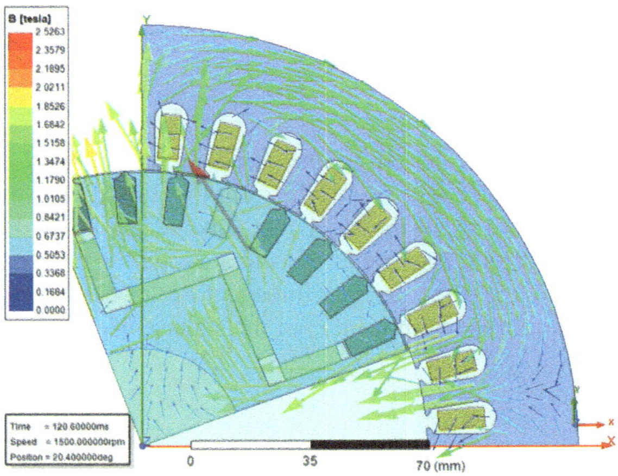

Fig. 5 Magnetic flux density distribution

For this electric vehicle operating scenario, using the optimized speed to pass the road can get a higher total efficiency. Due to the high vehicle speed, the time factor is not too small, but the energy consumption factor is therefore higher. Overall, the electric vehicle scene optimization has achieved its purpose, ensured the total efficiency of the motor passing this road.

The motor control system uses field oriented control strategy, and its control system model is shown in Fig. 7. Use the data of the parameter calculation layer to establish the motor module of the control system. Two PI regulators are used to achieve double closed-loop control of speed and current.

Fig. 6 Efficiency map

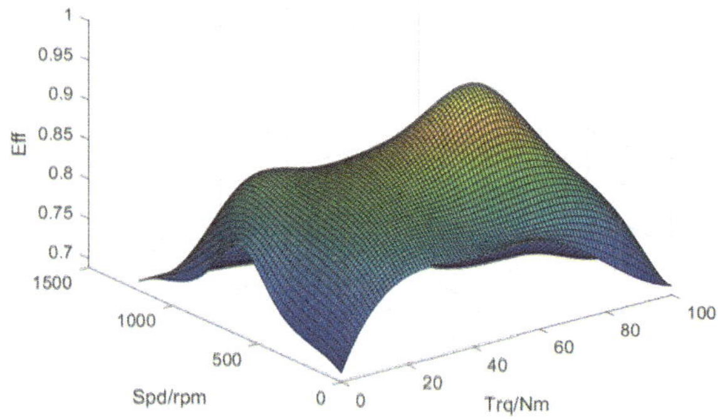

Table 3 Scene optimization results

Parameter	Value				
Number of road sections	5				
Road length (km)	20.8	11.8	16.7	10.5	10.6
Base torque (Nm)	55.4	64.3	62.1	62.5	58.8
Transmission ratio	10				
Wind resistance coefficient	0.002				
Speed (km/h)	94.6	106.4	65.7	99.2	60.1
Output torque (Nm)	70.3	83.2	69.3	78.9	64.8
Total efficiency	87.1%				
Time factor	0.676				
Energy consumption factor	0.865				

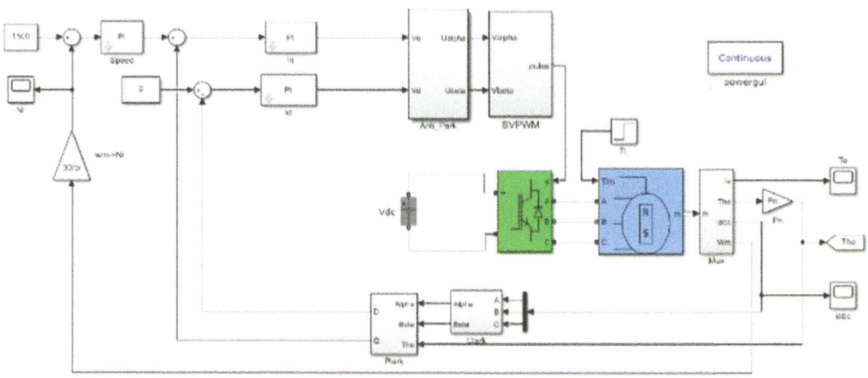

Fig. 7 Control system

In order to verify its control effect, the rated motor speed of 1500 rpm is set as the target speed, and the load torque is 10 Nm when starting, and it is increased to 30 Nm after 0.5 s. The torque and three-phase current waveforms are shown in Figs. 8 and 9.

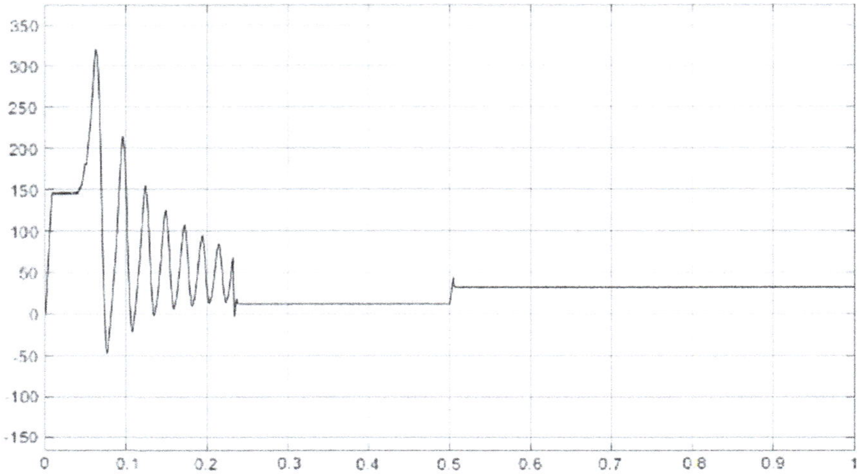

Fig. 8 Output torque

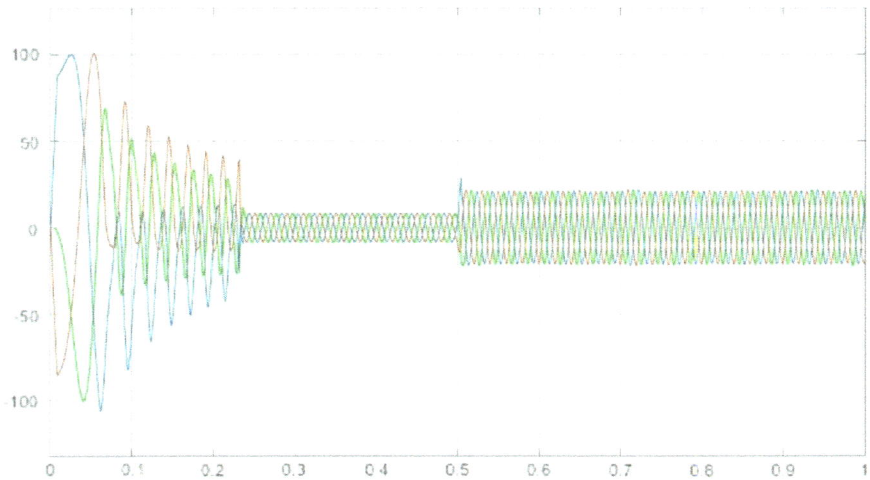

Fig. 9 Current

6 Conclusion

This paper proposes a system-level optimization method for permanent magnet synchronous motors for electric vehicles based on scene optimization, which uses hierarchical optimization to achieve system-level optimization of the motor. The performance of the example motor was calculated by the finite element method and the efficiency map was drawn. As the core part of this method, the scene optimization layer was verified by programming, and a control system was built for the motor. It provides a design idea for the optimal design of electric vehicle motors.

Acknowledgements This work was supported by the National Natural Science Foundation of China under Grant (51737010).

References

1. Zhou, F.K. 2013. *Research on the power system parameter matching and vehicle control strategy of pure electric vehicles*. Jilin University. (in Chinese).
2. Qu, R.H., and C. Qin. 2016. Development status and prospects of electric vehicles and drive motors. *Southern Power System Technology* 10 (3): 82–86. (in Chinese).
3. Zhang, T.P. 2020. *Research on modeling and simulation of pure electric vehicle drive system*. Jilin University. (in Chinese).
4. Wei, D., H.W. He, and J.F. Cao. 2020. Hybrid electric vehicle electric motors for optimum energy efficiency: A computationally efficient design. *Energy* 203.
5. Finken, T., M. Hombitzer, and K. Hameyer. 2010. Study and comparison of several permanent-magnet excited rotor types regarding their applicability in electric vehicles. In *2010 Emobility—Electrical Power Train*, Leipzig, 1–7.
6. Văscan, I., F.P. Piglesan, C. Marţiş, and I. Bârsan. 2019. PMSM rotor topologies for automotive HVAC system. In *2019 Electric Vehicles International Conference (EV)*, Bucharest, Romania, 1–4.
7. Chan, C.C., and K.T. Chau. 1996. An advanced permanent magnet motor drive system for battery-powered electric vehicles. *IEEE Transactions on Vehicular Technology* 45 (1): 180–188.
8. Sant, A.V., V. Khadkikar, W. Xiao, and H.H. Zeineldin. 2015. Four-axis vector-controlled dual-rotor PMSM for plug-in electric vehicles. *IEEE Transactions on Industrial Electronics* 62 (5): 3202–3212.
9. Lin, W.J. 2005. *Research on the control strategy of permanent magnet synchronous motor servo system*. Zhejiang University.
10. Liang, H., Y. You, and L. Yang. 2015. Research on electric vehicle cluster model based on scenes simulation. In *International Conference on Renewable Power Generation (RPG 2015)*, Beijing, 1–5.
11. Zhai, Z.Q. 2019. *Uncertainty analysis of vehicle specific power distribution for road section traffic emission evaluation*. Beijing Jiaotong University. (in Chinese).
12. Chen, Y.Q., W. Zhou, and W.G. Zhang. 2019. Optimal driving decision analysis of electric vehicles based on dynamic programming. *Control and Information Technology* 06: 1–7. (in Chinese).
13. Shi, Y., and R. Eberhart. 1998. A modified particle swarm optimizer. In *1998 IEEE International Conference on Evolutionary Computation Proceedings. IEEE World Congress on Computational Intelligence (Cat. No.98TH8360)*, Anchorage, AK, USA, 69–73.

14. Shi, Y., and R.C. Eberhart. 1999. Empirical study of particle swarm optimization. In *Proceedings of the 1999 Congress on Evolutionary Computation-CEC99 (Cat. No. 99TH8406)*, Washington, DC, USA, Vol. 3, 1945–1950.
15. del Valle, Y., G.K. Venayagamoorthy, S. Mohagheghi, J. Hernandez, and R.G. Harley. 2008. Particle swarm optimization: Basic concepts, variants and applications in power systems. *IEEE Transactions on Evolutionary Computation* 12 (2): 171–195.
16. Zhan, Z., J. Zhang, Y. Li, and H.S. Chung. 2009. Adaptive particle swarm optimization. *IEEE Transactions on Systems, Man, and Cybernetics, Part B (Cybernetics)* 39 (6): 1362–1381.
17. Abassi, M., A. Khlaief, O. Saadaoui, A. Chaari, and M. Boussak. 2015. Performance analysis of FOC and DTC for PMSM drives using SVPWM technique. In *2015 16th International Conference on Sciences and Techniques of Automatic Control and Computer Engineering (STA)*, Monastir, 228–233.
18. Wang, Z., J. Chen, M. Cheng, and K.T. Chau. 2016. Field-oriented control and direct torque control for paralleled VSIs fed PMSM drives with variable switching frequencies. *IEEE Transactions on Power Electronics* 31 (3): 2417–2428.
19. Sun, X., et al. 2019. MPTC for PMSMs of EVs with multi-motor driven system considering optimal energy allocation. *IEEE Transactions on Magnetics* 55 (7): 1–6. (Art no. 8104306).
20. Tang, X.Z. 2005. *Vector control system of permanent magnet synchronous motor*. Zhejiang University.
21. Yang, Z., F. Shang, I.P. Brown, and M. Krishnamurthy. Comparative study of interior permanent magnet, induction, and switched reluctance motor drives for EV and HEV applications.

Modelling and Experimental Verification on Magnetic Hysteresis Properties of Soft Magnetic Composite Material

Weijie Xu, Nana Duan, Song Huang, and Shuhong Wang

Abstract This paper in magnetic troduces some research results about the magnetic characteristics on soft composite (SMC) material. Based on the mechanisms of magnetic material magnetization, a vectorial hysteresis model is proposed to simulate the hysteresis characteristics of SMC material. Meanwhile, by using the three dimensions magnetic properties test platform, the magnetic features of SMC material under different applied magnetic fields, such as alternating field and rotational field, have been measured. To verify this presented vectorial hysteresis model, a given SMC material is employed to be simulated and measured under rotating and alternating excitations, respectively. Compared with the two results, a good agreement can be obtained, which shows the validity and practicability of the proposed hysteresis model.

Keywords Hysteresis model · Soft magnetic composite (SMC) material · Magnetization mechanisms · Hysteresis characteristics

1 Introduction

The SMC materials, which made from highly iron powder with surface coating, are widely adopted in various scientific and commercial applications for their properties like high electrical resistivity, low eddy current loss, 3-D magnetic flux, and great design flexibility [1]. SMC materials can be used for fabrication of various electromagnetic devices, such as claw pole and transverse flux machines [2]. Therefore, the magnetic characteristics of SMC, such as hysteresis, need to be accurately measured and simulated before used.

W. Xu
TeBian Electric Apparatus Stock Co.,LTD, Changji 831100, China

State Grid Shaanxi Electric Power Research Institute, Xi'an 710010, China

N. Duan (✉) · S. Huang · S. Wang
School of Electrical Engineering, Xi'an Jiaotong University, No.28 Xianning West Road, Xi'an, Shanxi 710049, China
e-mail: duannana@xjtu.edu.cn

W. Chen et al. (eds.), *The Proceedings of the 9th Frontier Academic Forum of Electrical Engineering*, Lecture Notes in Electrical Engineering 743,
https://doi.org/10.1007/978-981-33-6609-1_79

Magnetic hysteresis model has been under research and development since the 30's, and many models have been presented [3]. The Preisach model, which based on the superposition of finite set of rectangular operators, is universally accepted for its simplicity and simplicity. The Preisach model is effective in the simulation of scalar hysteresis properties, but invalid to describe the vectorial properties [4]. In addition, the Stoner-Wohlfarth model is other widely recognized hysteresis model. In this model, a single-domain particle with uniaxial crystal anisotropy is assumed as a Stoner-Wohlfarth operator. The Stoner-Wohlfarth operator is vectorial inherently, but inconsistent with the phenomena in cubic textured magnetic materials which have biaxial anisotropy in each crystal plane [5]. In this paper, based on the physical mechanism of magnetic materials, an elemental operator with biaxial anisotropy is introduced to overcome the shortcomings of the previous models.

2 The Measurement System

With the help of the three dimensions magnetic properties test platform, the magnetic properties of SMC material under alternating magnetization and rotational magnetization can be fully measured. As depicted in Fig. 1, this measurement system consists of a three dimensions yoke, and three groups of computer controlled excitation coils around on the yoke can produce different magnetic flux patterns such as alternating, rotating in a plane and rotating in a three dimensions pattern [6, 7]. The SMC specimen in the center of the system. The components of the sample surface field intensity H and the flux density B are measured by the sensing coils and exploring coils around the sample, respectively [8, 9]. Therefore, both the alternating and rotational magnetic

Fig. 1 The real structure of the three dimensions magnetic measurement platform

hysteresis characteristics of SMC material can be obtained by this three dimensions magnetic property measurement platform.

3 Modelling of Soft Magnetic Composite Material

The research of accurate modelling on magnetic properties is crucial to develop high performance electromagnetic devices, accurate modelling of the magnetic properties, such as magnetic hysteresis, is crucial. In this paper, the iron powder in SMC material is regarded as a microscopic crystalline particle with magnetocrystalline anisotropy, as shown in Fig. 2. Thus, based on the physical mechanism, a vectorial elemental operator with biaxial anisotropy can be presented to simulate the iron particle in magnetic material [10]. Therefore, the energy of this proposed vectorial operator can be described as

$$E = K \sin^2 \theta \cos^2 \theta - \mu_0 H m \cos(\theta_H - \theta) \quad (1)$$

where **m** is the magnetization of the operator, K is the biaxial anisotropy coefficient, θ_H and θ are the applied field and the angles of the resultant magnetization, respectively, both with respect to the operator's easy axis.

In the crystal structure, the stable orientation of the magnetization can be calculated by minimizing the total energy of the elemental operator, which similar to the determination method in Stoner-Wohlfarth model. Thus, the critical curve, which separates the regions with different energy extremes can be obtained from

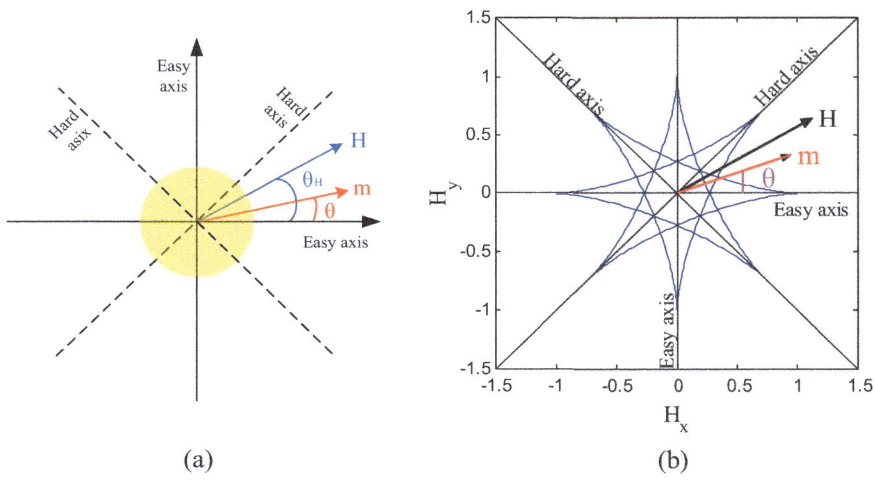

Fig. 2 **a** The elemental operator under different magnetization axis and **b** the corresponding biasteroid curve

$$\frac{\partial E}{\partial \theta} = \frac{1}{2}K \sin 4\theta - \mu_0 H m \sin(\theta_H - \theta) = 0 \tag{2}$$

$$\frac{\partial^2 E}{\partial \theta^2} = 2K \cos 4\theta + \mu_0 H m \cos(\theta_H - \theta) = 0 \tag{3}$$

The solution of this energy formulas related to Hx and Hy, can be regarded as the equation of a biasteroid, as shown in Fig. 2b.

$$H_x = \frac{2K}{\mu_0 m} \cos^3 \theta (5 - 6 \cos^2 \theta) \tag{4}$$

$$H_y = \frac{2K}{\mu_0 m} \sin^3 \theta (5 - 6 \sin^2 \theta) \tag{5}$$

where H_x and H_y are the two constituent parts of the applied magnetic field H along the two orthogonal easy axes.

The biasteroid curve demonstrates biaxial anisotropy, as shown in Fig. 2, with two orthogonal hard axes and two orthogonal easy axes. This characteristic strictly agrees with physical mechanism of biaxial anisotropy in each crystal plane for the cubic textured magnetic.

Different number of the energy minima corresponding to different region. The number of local energy minima decreases from four to one when the applied field gradually increases from the central octagonal region to the outside region, as shown in Fig. 3.

With the new symmetry considerations on the magnetic anisotropy, an analytical method to demonstrate the vectorial operator has been proposed. A modified

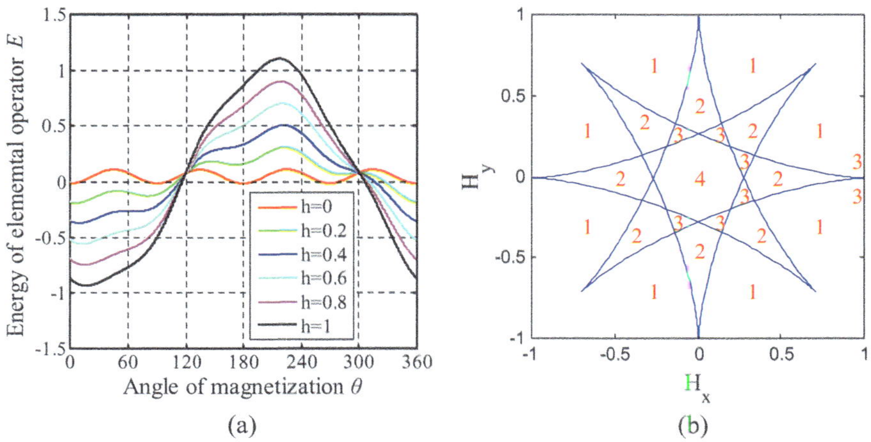

Fig. 3 **a** The energy curves of the elemental operator under different applied magnetic fields and **b** the corresponding number of the energy minima in the critical biasteroid curve

analytical expression of the hysteresis loop can be obtained by partial approximate substitutions.

$$m(H,\theta_H) = \pm\frac{\sin\theta_H}{\sqrt{\frac{(H+h_i)}{h_k}^2 \pm 2\frac{(H+h_i)}{h_k}\cos\theta_H + 1}} \tag{6}$$

where the minus sign is for the lower branch and the plus sign for the upper branch in a given hysteresis loop, h_i is the interaction field on each single elemental operator, and h_k is the normalized anisotropy field varying with the overall anisotropy coefficient.

Then, the total magnetization **M** of a magnetic specimen for a given applied field **H** can be obtained by the following integral.

$$M(H) = \iiint m(H,\theta)\mu(h_i, K)\cos\theta \mathrm{d}h_i \mathrm{d}K \mathrm{d}\theta \tag{7}$$

where $\mu(h_i, K)$ is a Gaussian-Gaussian distribution function of the elemental operator in this model.

4 Results and Conclusion

The magnetic characteristics of SOMALOY™ 500, a kind of SMC material, are measured by this 3-D magnetic property measurement system and simulated, respectively. Good agreements are observed between the measurement and simulation, as shown in Fig. 4.

In this paper, the research of a hysteresis model, including the principle, numerical implementation, which based on a biaxial elemental operator, has been proposed to simulate the magnetic vectorial hysteresis. With the employed of the three dimensions magnetic properties test platform, the simulation results are fully verified. The simulations agree well with the experimental results of a cubic SMC sample measured by the test platform. The comparison demonstrates that the hysteresis model presented in this paper is acceptable and valid for developing high performance electromagnetic devices.

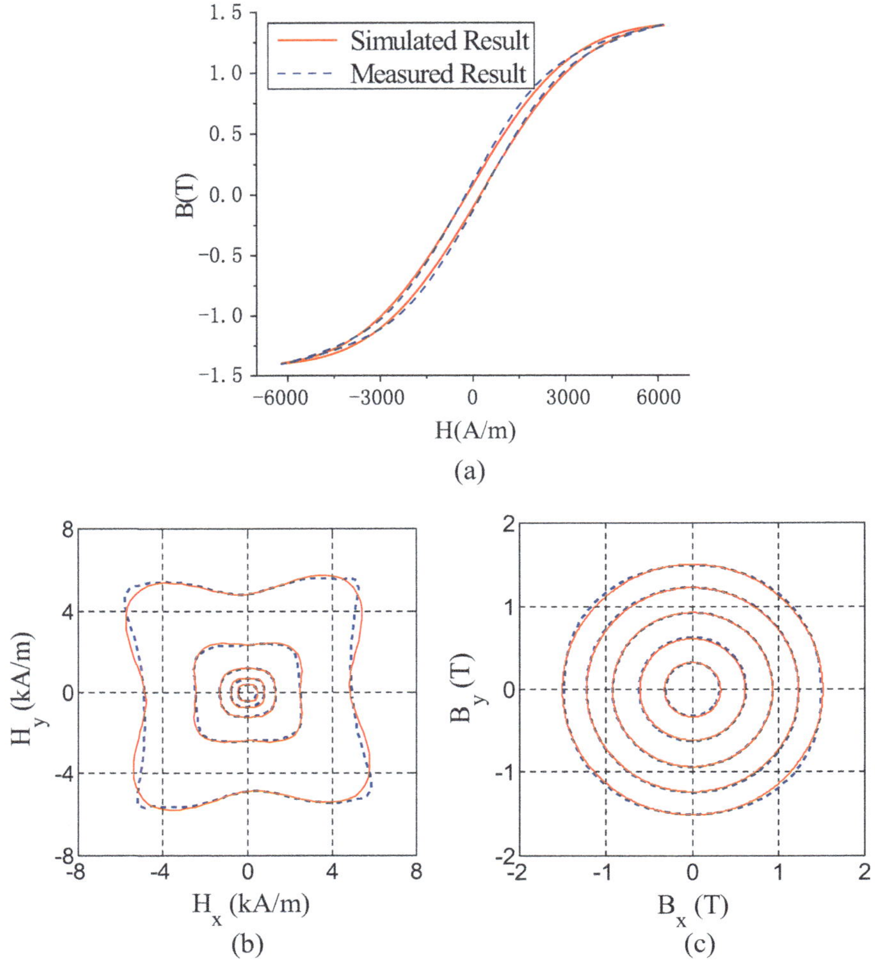

Fig. 4 The comparison of simulation results (red solid line) and measurement results (blue dashed line) about **a** hysteresis loop under alternating field, **b** the loci of $\boldsymbol{H}$ and **c** the loci of $\boldsymbol{B}$ under rotating field

Acknowledgements This work was supported in part by the National Natural Science Foundation of China under Grant (52007141).

References

1. Li, Yongjian, Qingxin Yang, Jianguo Zhu, et al. 2010. Measurement of soft magnetic composite material using an improved 3-D tester with flexible excitation coils and novel sensing coils. *IEEE Transactions on Magnetics* 46 (6): 1971–1974.
2. Li, Yongjian, Qingxin Yang, Jianguo Zhu, et al. 2010. Research of three-dimensional magnetic reluctivity tensor based on measurement of magnetic properties. *IEEE Transactions on Applied Superconductivity* 24 (3): 1932–1935.
3. Bertotti, G. 1998. *Hysteresis in magnetism: For physicists, materials scientists, and engineers.* San Diego: Academic press.
4. Duan, Nanan, Weijie Xu, Shuhong Wang, et al. 2015. A new Preisach type hysteresis model of high temperature superconductors. *Journal of applied physics* 117 (arts. 17A71).
5. Duan, Nanan, Weijie Xu, Shuhong Wang, et al. 2015. Hysteresis modeling of high-temperature superconductor using simplified Preisach model. *IEEE Transactions on magnetics* 51 (3) (arts. 7300904).
6. Zhu, Jianguo, Youguang Guo, ZhiWei Lin, et al. 2011. Development of PM transverse flux motors with soft magnetic composite cores. *IEEE Transactions on Magnetics* 47 (10): 4376–4383.
7. Zhu, Jianguo, Jinjiang Zhong, ZhiWei Lin, et al. 2003. Measurement of magnetic properties under 3D magnetic excitations. *IEEE Transactions on Magnetics* 39 (5): 3429–3431.
8. Li, Yongjian, et al. 2014. Magnetic anisotropic properties measurement and analysis of the soft magnetic composite materials. *IEEE Transactions on Applied Superconductivity* 24 (5): 1–4.
9. Guo, Youguang, Jianguo Zhu, et al. 2009. Design and analysis of a claw pole permanent magnet motor with molded soft magnetic composite core. *IEEE Transactions on Magnetics* 45 (10): 4582–4585.
10. Xu, Weijie, Nanan Duan, Shuhong Wang, et al. 2015. A stress-dependent magnetic hysteresis model for soft magnetic composite materials. *IEEE Transactions on Applied Superconductivity* 26 (7) (arts. 0611305).

The manufacturer's authorised representative in the EU is Springer Nature Customer Service Centre GmbH, Europaplatz 3, 69115 Heidelberg, Germany. If you have any concerns regarding our products, please contact ProductSafety@springernature.com

Printed and bound by CPI Group (UK) Ltd, Croydon, CR0 4YY

30/09/2026

02225320-0020